Microwaves: Industrial, Scientific, and Medical Applications

For a complete listing of the *Artech House Microwave Library*,
turn to the back of this book . . .

Microwaves: Industrial, Scientific, and Medical Applications

Jacques Thuéry

Edited by
Edward H. Grant
King's College London

Artech House
Boston • London

Library of Congress Cataloging-in-Publication Data

Thuéry, Jacques.
   [Micro-ondes et leurs effets sur la matiere. English]
   Microwaves: thermal properties applications / Jacques Thuéry.
    p.   cm.
   Rev. translation of: Les micro-ondes et leurs effets sur la matiere.
   Includes bibliographical references and index.
   ISBN 0-89006-448-2
   1. Microwave devices. 2. Microwaves. 3. Microwave devices–Industrial
applications. I. Title.
TK7876.T5313   1991   91-39971
621.381'3–dc20   CIP

© 1992   ARTECH HOUSE, INC.
685 Canton Street
Norwood, MA 02062

International Standard Book Number: 0-89006-448-2
Library of Congress Catalog Card Number: 91-39971

10 9 8 7 6 5 4 3 2 1

Table of Contents

Preface to the English Edition   **xv**

Foreword   **xvii**

Part I - MICROWAVES

1   Electromagnetism and Radiation   3

   1.1   Electromagnetic spectrum, ISM bands   4

   1.2   Electromagnetism   8

   1.3   Radio broadcasting   10

   1.4   Electromagnetic detection   14

   1.5   Thermal applications   17

   1.6   Microwaves in industry   19

2   The Laws of Radiation   33

   2.1   Basic definitions   34

   2.2   Maxwell's equations   36

   2.3   Propagation equation   37

   2.4   Plane wave   40

   2.5   Spherical and cylindrical waves   42

   2.6   Propagation media   44

   2.7   Boundary conditions   48

   2.8   Reflection and transmission   49

   2.9   Guided propagation   52

   2.10   Stationary wave   57

2.11	Electromagnetic cavities	58
	2.11.1 Resonant modes	58
	2.11.2 Energy balance	61
	2.11.3 Power loss in the walls	64
	2.11.4 Quality factor	66
2.12	Radiation sources	69
	2.12.1 Characteristics	69
	2.12.2 Radiation from a slot	70
	2.12.3 Radiation of an aperture	72
	2.12.4 Radiation from a horn	75
	2.12.5 Radiation zones	76
	References	

3	Microwaves and Matter	83
3.1	Dielectric polarization	84
3.2	Polarization by dipole alignment in a static field	85
	3.2.1 Polar and nonpolar media	85
	3.2.2 Induced dipole moment	85
	3.2.3 Permanent dipole moment	88
3.3	Dipole alignment polarization in an alternating field	90
3.4	Dielectric relaxation	92
	3.4.1 Hysteresis	92
	3.4.2. Debye equation	93
	3.4.3 Intermolecular bonds	93
	3.4.4 Relaxation time	94
	3.4.5 Debye and Cole-Cole diagrams	95
3.5	Different types of dielectrics	97
	3.5.1 Permittivity measurements	97
	3.5.2 Lowloss dielectrics	98
	3.5.3 Aqueous dielectrics	98
	3.5.4 Mixtures	100
	3.5.5 Saline solutions and biological constituents	101
3.6	Heat generation	103
3.7	Thermal runaway	106
	References	

4	Generators and applicators	117
4.1	Introduction	117
4.2	Microwave generators	118

	4.2.1	The magnetron	118
	4.2.2	Klystron and TWT	125
	4.2.3	RF energy transmission	127
4.3		Applicators	128
	4.3.1	Different types	128
	4.3.2	Design constraints	132
4.4		Conclusion	147

Part II - INDUSTRIAL APPLICATIONS

1	Drying		159
	1.1	Humidity and drying	159
	1.2	Drying kinetics	161
	1.3	Microwave drying	162
	1.4	Paper and printing industries	163
	1.4.1	Paper	163
	1.4.2	Printing inks	164
	1.4.3	Glued products	165
	1.5	Leather and textile industries	166
	1.5.1	Leathers	166
	1.5.2	Tufts and yarns	166
	1.5.3	Dyeing and finishing	168
	1.5.4	Tufted carpets	170
	1.6	Construction	172
	1.6.1	Wood and plywood	172
	1.6.2	Plaster, concrete, and ceramics	172
	1.7	Foundries	173
	1.8	Rubbers and plastics	175
	1.8.1	Drying of polymers	175
	1.8.2	Photographic film and magnetic tape	175
	1.9	Pharmaceutical industry	176
	1.10	Drying of tobacco	177
	1.11	Regeneration of zeolites	178
		References	

2   The treatment of elastomers

   2.1   Macromolecules and   188
       2.1.1   Principles of interaction   188
       2.1.2   Relaxation mechanisms   188
       2.1.3   Dielectric properties of elastomers   189

   2.2   Vulcanization   191

   2.3   Microwave vulcanization   194
       2.3.1   Formulation of mixtures   194
       2.3.2   Advantages and disadvantages of microwave vulcanization   196
       2.3.3   Materials available   199

   2.4   Thawing and preheating of rubber   201

   2.5   Microwave devulcanization   203
       References

3   Miscellaneous applications   207

   3.1   Polymerization   207
       3.1.1   Thermosetting and thermoplastic polymers   207
       3.1.2   Microwave reticulation of thermosetting resins   208
       3.1.3   Thermoplastic polymers   212

   3.2   Fusion   213
       3.2.1   Dewaxing of casting moulds   213
       3.2.2   Viscous materials in metal   214
       3.2.3   Oil and shale oil   215
       3.2.4   Road repairs   215
       3.2.5   Defrosting of soil   216

   3.3   Consolidation   216
       3.3.1   Hardening of foundry mouldings   216
       3.3.2   Fast-setting concrete   217
       3.3.3   Sintering of ferrites and ceramics   217

   3.4   Emulsification   220

   3.5   Crushing   220

   3.6   Purification of coal   221

   3.7   Nuclear waste treatment   222

   3.8   Cellulosic waste treatment   224
       References

Part III - APPLICATIONS IN THE FOOD INDUSTRY

1   Cooking   231

   1.1   Mechanisms   232

   1.2   Animal products   235
      1.2.1   Red meat   235
      1.2.2   Poultry   239
      1.2.3   Bacon and fat   243
      1.2.4   Meat patties   245
      1.2.5   Fish   246
      1.2.6   Dairy products   248

   1.3   Vegetable products   249
      1.3.1   Vegetables   249
      1.3.2   Cereals and soya   251
      1.3.3   Roasting   252

   1.4   Catering   253

   1.5   Baking   255
      1.5.1   Bread   255
      1.5.2   Doughnuts   256

   1.6   Digestibility of foods cooked by microwaves   257

   References

2   Thawing and tempering   275

   2.1   Conventional thawing   276

   2.2   Mechanisms of microwave   278
      2.2.1   Dielectric properties of frozen products   278
      2.2.2   Thermal runaway   281
      2.2.3   Energy limitations   283
      2.2.4   Surface cooling   285

   2.3   Available equipment   290
      2.3.1   896 and 915 MHz   290
      2.3.2   2.45 GHz

   2.4   Advantages of microwave processing   296
      2.4.1   Industrial aspects   296
      2.4.2   Qualitative aspects   297

   References

3   Drying   305

    3.1   Vaporization   306

    3.2   Drying at atmospheric pressure   307
        3.2.1   Final drying of potato chips   308
        3.2.2   The drying of pasta   310
        3.2.3   Miscellaneous food products   311

    3.3   Drying at low pressure   313
        3.3.1   Freeze drying   313
        3.3.2   Expansion in vacuum   317
        3.3.3   Various processes   324

    3.4   Determination of dry content   324
        References

4   Preservation   339

    4.1   Enzymatic inactivation   339
        4.1.1   Blanching of fruits and vegetables   341
        4.1.2   Inactivation of α-amylase in wheat   344
        4.1.3   Treatment of grains and soya beans   345

    4.2   Sterilization   346
        4.2.1   Animal products   350
        4.2.2   Vegetable products   351
        4.2.3   Prepared meals   355
        References

5   Miscellaneous applications   365

    5.1   Disinfestation   365

    5.2   Soil treatment   369

    5.3   Germination   377

    5.4   Crop protection   378

    5.5   Wine-making by carbonic fermentation   379

    5.6   Opening of oysters   380
        References

Part IV - BIOLOGICAL EFFECTS AND MEDICAL APPLICATIONS

1   Interactions with the organism   391
     1.1   Dielectric behavior of biological material   392
            1.1.1   Biomolecules   392
            1.1.2   Cells and membranes   395
            1.1.3   Tissues   396
     1.2   Quantum aspects   397
     1.3   Basic interaction with cell membranes   399
            1.3.1   Continuous wave   399
            1.3.2   The modulated wave   403
            1.3.3   Pearl chain formation   404
     1.4   Thermal interaction with the living organism   404
            1.4.1   Reflection and transmission   405
            1.4.2   Absorption and dosimetry   407
            1.4.3   Experimental aspects   408
            1.4.4   Modeling   410
            1.4.5   Near-field interaction   415
            1.4.6   Main results   416
            References

2   Biological effects   443
     2.1   Cells and micro-organisms   446
     2.2   Blood and hematopoiesis   451
     2.3   Immune system   455
            2.3.1   Natural resistance   459
            2.3.2   Lymphopoiesis   460
            2.3.3   Multiplication of lymphocytes
                     FcR^+ and CR^+ 460
            2.3.4   Stimulation of the response of lymphocytes
                     to mitogens   461
            2.3.5   Modulation of the activity of activator
                     T lymphocytes   463
     2.4   Nervous system   465
            2.4.1   Fluxes of calcium ions   465
            2.4.2   Neurons and synapses   469
            2.4.3   Blood-brain barrier   471
            2.4.4   Central nervous system   474
            2.4.5   Peripheral nervous system and sensory
                     perception   476

		2.4.6	Auditory perception	478
		2.4.7	Autonomic nervous system	481
		2.4.8	Psychophysiology	485
	2.5	Endocrine system		488
		2.5.1	Pituitary-thyroid axis	488
		2.5.2	Pituitary-suprarenal axis	490
		2.5.3	Pituitary-ovarian and pituitary-testicular axes	492
		2.5.4	Growth hormones	493
	2.6	Thermal regulation and metabolism		494
	2.7	Effects on growth		496
		2.7.1	Insects	497
		2.7.2	Birds	498
		2.7.3	Mammals	499
	2.8	Lesions and cataracts		502
		References		
3	Safety standards			553
	3.1	Soviet Union		556
	3.2	United States of America		557
	3.3	Eastern Europe		565
	3.4	Canada		567
	3.5	Australia		568
	3.6	Sweden		568
	3.7	European Community		568
	3.8	International organisations		569
		References		
4	Biomedical applications			585
	4.1	Hyperthermia for cancer treatment		586
		4.1.1	Historical development	586
		4.1.2	Mode of action	591
		4.1.3	Applicators	595
		4.1.4	Integrated systems	605
		4.1.5	Clinical results	608
	4.2	Specific effects		611
		4.2.1	Bioelectric vibrations	611
		4.2.2	Antigenicity	612

	4.2.3	Immune response	613
4.3	Miscellaneous applications		616
	4.3.1	Clinical	616
	4.3.2	Biological	617

Addresses   641

Index   665

Preface to the English Edition

The use of microwaves on the industrial and domestic scenes has increased dramatically over the past thirty years. Microwave ovens in the home have become commonplace and major contributions are made by microwaves in the food industry. Throughout the manufacturing industries new processes are all the time being developed using latest microwave technology and in the last decade microwaves have been employed in medicine for treating cancer and other clinical conditions. As with any new technique, particularly one involving radiation, the possibility of undesirable biological effects is always present, and thus the question of microwave hazards needs to be considered.

Mr Thuéry should be congratulated on producing in one volume a detailed and comprehensive treatment of all the above topics. The whole area has been very thoroughly researched and this is witnessed by the inclusion of more than four thousand references.

The text has been translated from the original French by Dr Camelia Gabriel who herself has had considerable experience in research relating to the industrial applications and biological effects of microwaves.

Edward H Grant
King's College London

Foreword

Microwave frequencies occupy the three decades of the electromagnetic spectrum (300 MHz to 300 GHz) that lie between VHF radiowaves and the far infrared. Their applications fall into two categories, depending on whether the wave is used to transmit information or just energy. The first category includes terrestrial and satellite communication links, radar, radioastronomy, microwave thermography, material permittivity measurements, and so on. In all cases, the transmission link incorporates a receiver whose function is to extract the information, that in some way modulates the microwave signal.

The second category of applications is the subject of this book. Here, there is no modulating signal and the electromagnetic wave interacts directly with certain solid or liquid materials known as *lossy dielectrics*, among which water is of particular interest. A few related topics are not discussed. These include microwave plasma (interaction with gases) and energy transmission by microwave links, e.g. P.E. Glaser's *Solar Power Station*, or the supply of energy from ground to an aircraft or spacecraft, e.g., the *sharp* prototype of the Canadian Communication Research Centre.

According to C. Olsen, microwave heating is the first fundamentally new heating technique since the discovery of fire. It has become familiar to the general public because of the ubiquitous domestic microwave oven, but its principles are often poorly understood. Part I of this book is therefore a theoretical introduction to this technique in its historical and regulatory contexts, and covers electromagnetic theory, wave-matter interaction, and industrial applicator design.

Part II and III describe industrial applications of microwaves with special reference to the food industry. Microwaves are currently used for drying, vulcanizing, polymerizing, melting, sintering, hardening , cooking and baking, thawing, blanching, pasteurizing and sterilizing, and in many other processes of lesser importance. The actual installations that are described below provide only part of the story because many manufacturers and users tend to be secretive.

Part IV examines microscopic and macroscopic aspects of the interaction between a wave and a living organism. Modeling and dosimetry techniques have been steadily improved in recent years. The long-standing question of biological effects, traditionally formulated in terms of West $v.$ East or thermal $v.$ specific effects, could be settled soon. Indeed, a few points of agreement have been emerging, e.g., on the nonlinear, window-like nature of the interactions, the importance of hotspots and geometrical resonances, and the relevance of certain phenomena such as transmembrane flux modulation that are not thermally mediated. Safety standards tend to evolve in a parallel way, so that the present inconsistencies between national standards should eventually disappear and be replaced by a global consensus. In the medical field, antitumoral hyperthermia, used alone or in conjunction with conventional treatment, is a fast-spreading technique. The lst chapter traces back its historical evolution and describes its present state of development, the existing hardware, and the preliminary clinical results obtained. Form a different prospective, it is noted that the possibility of specific interactions could open new therapeutic avenues for immune response enhancement.

The whole book has been conceived as a bibliographic synthesis and is intended to be as comprehensive as possible. Most of the 4000 references ar drawn from English sources. They originate mostly from North America, but there are also important contributions from the European Community. Sweden, Poland, and Japan too have produced an appreciable quantity of publications. Soviet science and industry have been largely inaccessible so far and are under-represented.

I should like to apologize in advance for any errors and inaccuracies, and extend my sincere thanks to all those who have made possible the completion of this work, especially my wife, Beatriz.

Palo Alto (California),
October 1991.

Part I

Microwaves

Chapter 1

Electromagnetism and Radiation

The purpose of this chapter is to place thermal applications of microwaves in their electromagnetic, historical, and industrial contexts. Medical applications are very specific and will be treated separately in Part IV, Chapter 4.

Section 1.1 of this chapter is dedicated to the electromagnetic spectrum and its regulation and administration, including governing organizations and frequency bands allocated to industrial scientific, medical, and domestic uses of microwaves. Sections 1.2 to 1.5 describe the relevant electromagnetic phenomena. They summarize the enormous technical progress in telecommunications and radar, which has made possible the use of microwaves. Finally, Section 1.6 examines briefly the market for industrial microwave applications, thus serving as an introduction to Part II, Chapter 1 and Part III, Chapter 3.

1.1 Electromagnetic spectrum, ISM bands

Microwaves, or hyperfrequency waves, form a continuous electromagnetic spectrum that extends from low-frequency alternating currents to cosmic rays (Table 1.1) There are no essentially different types of electromagnetic radiation, and the
division into bands arises only from historical or physiological factors (visible light), modes of production, and specific
properties of the radiation considered. In this continuum, the
radio-frequency range is conventionally divided into bands by
decades of frequency or wavelength (Table 1.2). Bands 9, 10,
and 11 (300 MHz–300 GHz) constitute the microwave (hyperfrequency) range that is limited on the low-frequency side by
the HF and on the high-frequency side by the far infrared.
At these frequencies, wavelengths are often on the same order
of magnitude as the dimensions of circuit components, which
means that the discrete approximation cannot be used in calculations. Traditionally, the microwave domain is subdivided
into bands designated by the letters P, L, S, X, K, Q, V, and
W (Table 1.3).

Table 1.1

The electromagnetic spectrum

Region	Frequencies		Wavelength		
Audiofrequencies	30 -	30×10^3 Hz	10 Mm	-	10 km
Radiofrequencies	30×10^3 -	3×10^{11} Hz	10 km	-	1 mm
Infrared	3×10^{11} -	4.1×10^{14} Hz	1 mm	-	730 nm
Visible	4.1×10^{14} -	7.5×10^{14} Hz	730 nm	-	400 nm
Ultraviolet	7.5×10^{14} -	10^{18} Hz	400 nm	-	0.3 nm
X rays		$>10^{17}$ Hz	< 3 nm		
γ rays		$>10^{20}$ Hz	< 3 pm		
Cosmic rays		$>10^{20}$ Hz	< 3 pm		

The HF and ultrahigh frequency bands constitute a natural
resource managed by the following international organizations
[206], [25]:

Table 1.2

Frequency bands [245]

Band		Designation	Frequency limits		
4	VLF	very low frequency	3 kHz	-	30 kHz
5	LF	low frequency	30 kHz	-	300 kHz
6	MF	medium frequency	300 kHz	-	3 MHz
7	HF	high frequency	3 MHz	-	30 MHz
8	VHF	very high frequency	30 MHz	-	300 MHz
9	UHF	ultra high frequency	300 MHz	-	3 GHz
10	SHF	super high frequency	3 GHz	-	30 GHz
11	EHF	extremely high frequency	30 GHz	-	300 GHz

Table 1.3

Microwave bands [219]

Designation	Frequency limits, GHz			Corresponding wavelength, cm		
P	0.225	-	0.39	133.3	-	76.9
L	0.39	-	1.55	76.9	-	19.3
S	1.55	-	5.20	19.3	-	5.77
X	5.20	-	10.90	5.77	-	2.75
K	10.90	-	36.00	2.75	-	0.834
Q	36.00	-	46.00	0.834	-	0.652
V	46.00	-	56.00	0.652	-	0.536
W	56.00	-	100.00	0.536	-	0.300

(a) *The International Telecommunication Union* (ITU), created in 1865 at the Paris Conference as the International Telegraph Union, became the ITU in 1932 and was reorganized in 1947 as a specialist institution of the United Nations with headquarters in Geneva. The ITU organises world and regional conferences on the management of radio communication (CAMR or CARR), which assign parts of this spectrum to different services. The *regulation of radio communication* (RR) is derived from these conferences after agreement between countries. The ITU has permanent bodies, including the *International Frequency Registration Board* (IFRB) that oversees conformation to the rules of attribution, and the *International Consultative Committees* (CCI) that coordinate technical studies and give recommendations. These committees include *the International Telegraph and Telephone Consultative Committee* (CCITT), created in 1956 by the fusion of the CCIF (telephone, created in 1923) and of the CCIT (telegraph, created in 1926), and the *International Radio Consultative Committee* (CCIR), created in 1927. This latest project comprises a working group (GTI1/4) dedicated to the *industrial, scientific, and medical* (ISM) radiation levels.

(b) *The International Union for Radio Science* (URSI), founded in 1919 as the *International Union of Radiotelegraphic Science*, under the presidency of General Gustave Ferrié (1868–1932).

(c) *The International Electrotechnical Commission* (IEC), founded as the outcome of the St. Louis Missouri electricity congress of 1904, under the presidency of William Thomson (Lord Kelvin; 1824–1907). The IEC monitors compatibility and manages the *French Electrotechnical Committee* (CEF), created in 1907, and the *Special International Committee for Radioelectrical Perturbations* (CISPR). This body is divided into six subcommittees that make recommendations.

At the national level [25], the allocation of frequencies to different users is the responsibility of two organizations originating from the *Telecommunication Coordination Committee* (CCT) set up by Decree 64800 of 29 July 1964 and by the Regulation of 7 December 1964: the *Mixed Frequency Commission* (CMF) oversees the distribution of bands among civil users, the PTT, military forces, and civil aviation, and the *Frequency Allocation Committee* (CAF).

Radio Regulations (RR) [245] divide the world into three regions: Region one covers Europe, Africa, the Arabian peninsula, Turkey, the Middle East except Iran, the Siberian part of the USSR, and Mongolia. Region two covers the two Americas and Region three the rest of the world.

Section 1.14, Article 16 of RR defined ISM use as follows: "The setting up of equipment or installations conceived to produce or use, in a confined space, radiofrequency energy for industrial, scientific, medical, domestic and all analogous applications other than telecommunication".

Several frequency bands are reserved for these purposes. A

full list is given in Table 1.4 in which notes a and b refer to the following:

(a) "The use of this frequency band for ISM applications is subject to specific authorization by the authority concerned in agreement with other authorities whose radio communication services may be affected".

(b) "Radio communication services operating in this band must accept adverse interference that may occur from these applications [ISM]".

Table 1.4

ISM bands [245]

Band	Central frequency	RR Article	Regions	Note
433.05 - 434.79 MHz	433.92 MHz	661	1 except: D, A, FL P, CH, YU	a
		662	D, A, FL P, CH, YU	b
902 - 928 MHz	915 MHz	707	2	b
2 400 - 2 500 MHz	2 450 MHz	752	1, 2, 3	b
5 725 - 5 875 MHz	5 800 MHz	806	1, 2, 3	b
24 - 24.25 GHz	24.125 GHz	881	1, 2, 3	a
61 - 61.5 GHz	61.25 GHz	911	1, 2, 3	a
122 - 123 GHz	122.5 GHz	916	1, 2, 3	a
244 - 246 GHz	245 GHz	922	1, 2, 3	a

The services partaking with or without priorities in the frequency band allocation to ISM applications are radio broadcasting; radar; stationary, mobile, and amateur radio; satellite transmission; and, in higher frequency bands, satellite exploration of the Earth, intersatellite communication, and space research.

Frequencies in excess of 10 GHz are not used in practice

because of their low penetration range and because the utility of RF sources and technologies declines rapidly with frequency.

Finally, we note that in the United Kingdom, the 915-MHz band is actually centered at 896 MHz (886–906 MHz), whereas in Albania, Bulgaria, Hungary, Poland, Romania, Czechoslovakia, and the USSR, the 2450-Hz band is actually centered at 2375 Hz (2325–2425 Hz).

1.2 Electromagnetism

According to Peregrinus, the science of magnetism began with Pierre de Maricourt who, around 1269, identified the north and south poles of a magnet. Athanasius Kircher (1601–1680) demonstrated that the two poles were of equal magnitude, and John Mitchell (1724–1793) showed that the magnetic attraction varied inversely with the square of the distance between magnetic poles. Electricity began with Stephen Gray (1670–1736) and, particularly, Jean-Theophile Desaguliers (1683–1744) who showed that there were two types of material, namely, conductors and insulators. Humphry Davy (1778–1829) measured the conductivity of metals by means of the pile invented by Allessandro Volta (1745–1827); Georg Simon Ohm (1787–1854) explained the laws of conduction (1826); Joseph Henry (1797–1878) discovered self-induction (1832); and James Prestcott Joule (1818–1889) found the relation between the electric current and the heat dissipated in a metal (1841). In electrostatics, Benjamin Franklin (1706–1790), Joseph Priestley (1733–1804), John Robinson (1739–1805), Henry Cavendish (1731–1810), and, particularly, Charles-Augustin de Coulomb (1736–1806) established the laws of attraction and repulsion between charges by analogy with the laws of gravitation and magnetism. Siméon-Denis Poisson (1781–1840) showed that a conductor was an equipotential body in electrostatic equilibrium and calculated the charge distribution on the surface of two neighboring spheres. Karl-

Friedrich Gauss (1777–1855) demonstrated the flux theorem (1813).

The connection between electricity and magnetism became apparent in 1819 when Christian Oersted (1777–1851) observed that the magnetic needle oriented itself perpendicularly to a wire carrying an electric current. Pierre-Simon de Laplace (1749–1827), Jean-Baptiste Biot (1774–1862), and Félix Savart (1791–1841) formulated the theoretical basis of this phenomenon, but it was André-Marie Ampère (1775–1836) who created a new science based on the work of Oersted. He elucidated the laws that governed the production of magnetic field by electric current and determined the forces acting between two conductors carrying a current (1825). The existence of an induced current in a circuit placed in an alternating magnetic field, predicted in 1824 by Jean-François Arago (1786–1853), was demonstrated in 1831 by Michael Faraday (1791–1867), who also introduced the concept of a dielectric, defined as a medium in which electric induction can take place. This opened the way for the development of propagation theory.

The theory of these phenomena suffered from contradictions due to the postulated existence of the aether until James Clerk Maxwell (1831–1879) developed a theory that eliminated the need for it. According to Maxwell, electromagnetic waves consist of two fields that are at right angles to each other and vary in both space and time. They propagate in dielectric media, and their speed in vacuum is that of light. Maxwell's equations were published in 1873 in his *Treatise on Electricity and Magnetism* and constitute the basis of electromagnetism (Part I, Section 2.2). Their experimental verification, first attempted in 1879 by Hermann von Helmholtz (1821–1894) and the Prussian Academy, was completed in 1886–87 as a result of the work of Heinrich-Rudolf Hertz (1857–1894) [86],[53].

1.3 Radio broadcasting

Alternating currents were being produced in 1851 by means
of a coil, designed by Heinrich Daniel Ruhmkorff (1803–1877),
in which an iron core placed along the axis of a solenoid became magnetized during the passage of a current, and tripped
a circuit breaker which interrupted the current. As soon as
magnetization is lost, the circuit breaker is released, the current flows again, and a new cycle begins (Fig. 1.1). The
result is that a low-frequency alternating current is induced
in the secondary winding [47]. Elihu Thompson and Nikola
Tesla reached 400 kHz by means of the periodic discharge of
a Leyden jar through a coil [59]. The spirals of Knochenhauer
inspired Hertz to conceive the idea of the first transmitter-
receiver. The transmitter was a half-wave dipole with a short
gap of a few millimeters in the middle, terminated by two small
copper spheres. A potential difference across this gap caused
a discharge and an HF alternating current, accompanied by a
spark at each cycle. The charge was displaced from one side
to the other, and some of the energy of the system was lost
by radiation. A frequency of 430 MHz was reached with this
electric oscillator. The receiver was a simple resonant circuit
in the form of a rectangular loop with an adjustable spark
gap (Fig. 1.3). The induced current created sparks, the size
of which gave an approximate indication of the field strength.
By adding parabolic zinc mirrors and a large ashphalt prism
(Fig. 1.4) to this arrangement, Hertz was able to observe propagation, deflection, refraction, and the effects of polarization.
He was able to detect the maxima and minima of a stationary
wave (Fig. 1.5) and verified that the speed of propagation was
that of light. He declared that optics was a branch of electricity
[47], [56], [22], [193], [24], [111].

The spark gap was minituarized by Augusto Righi (1850–
1920) and this enabled 15 GHz to be reached with very good
power output [24]. As a detector, the spark gap was replaced
by the coherer or iron filing tube invented around 1884–1886

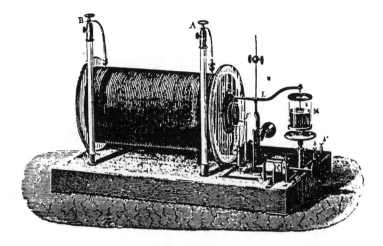

Figure 1.1 Ruhmkroff's coil [47].

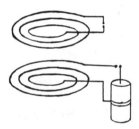

Figure 1.2 Knochenhauer's spirals powered by a Leyden jar [111].

by Calzecchi Onesti, described in 1890 by Edouard Branly (1844–1940), and improved by Aleksandr Stepanovich Popov (1859–1906) around 1895 [207]. A number of innovations were introduced by Guglielmo Marconi (1874–1937), who used the coherer in his first demonstration of communication by wireless in 1894 in Bologne and in 1896 in England (obtaining a range of 4 km and then 14 km at 1.2 GHz). He founded, with William Preece, the Wireless Telegraph and Signal Company, which later became Marconi Wireless Company. The patent for the concept of the transmitter-receiver was awarded

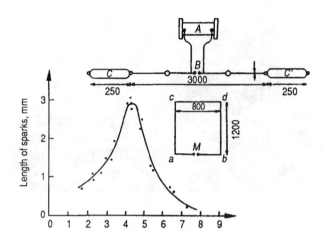

Figure 1.3 Hertz's experiment: dipole source and detecting loop. Demonstration of a resonance [111].

in 1898 to Oliver Dodge [215]. In 1899, Marconi transmitted to Branly a message across the English Channel, and the British Army used a 50-m vertical antenna during the Boer War. The double-masted HMS *Thetis* was equipped in 1900. On 12 December 1901, Marconi achieved the first transatlantic connection between Cornwall and Newfoundland, and in 1905 he drew the first polar diagram of an antenna. On 27 May 1905, wireless communication enabled the Japanese to destroy the Russian Baltic fleet at Tsushima.

The coherer was superseded by the electrolytic detector, the germanium crystal, and the triode, invented in 1906 by Lee de Forest (1873–1961) and improved by Irving Langmuir (1881–1957), using the principle of the diode detector of John Ambrose Fleming (1894–1945). Thermionic emission by hot filaments had been demonstrated in 1883 by Thomas Alva Edison (1847–1931). By 1915, AT&T was in a position to establish a telephone connection over 300 km and then 8000 km at 170 kHz. Marconi transmitted over 500 km at 3 and 20 MHz with

Figure 1.4 Hertz's experimental set-up for electric waves [47].

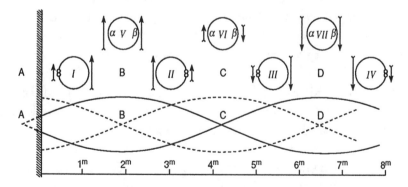

Figure 1.5 Measurements on a stationary wave in the Herz's experiment [24].

the help of electronic oscillators designed in 1919 by Heinrich Georg Barkhausen (1881–1956). In 1923–24, signals transmitted from Cornwall were received all over the world [193]. Microwaves, neglected for a while, saw a comeback in 1930 when E. Pierret and C. Gutton received a very clear radio transmission at 2 GHz over 7 km, using TMC-metal lamps, quarterwave antennas, and cylindroparabolic reflectors [160]. The following year, Marconi demonstrated over-the-horizon transmission with 'microwaves', the word having been introduced by André Clavier (born in 1901) of the (LMT) society, who was responsible for duplex telephone communication at 1.7 GHz over 51 km between Dover and Calais for the Western Electrical Company. [29], [30], [31], [77], [3], [24].

1.4 Electromagnetic detection

The detection of metallic masses by electromagnetic reflection was first used in 1904 by Christian Hülsmeyer [13]. The theory was established in 1917 by Nikola Tesla (1856–1943). In 1922, A. H. Taylor and L. C. Young of the Aircraft Radio Laboratory, rediscovered the phenomenon by observing the distortion of hertzian signals by a passing ship. The first systematic trials were made by C. Gutton and E. Pierret and, in England, E. V. Appleton and A. F. Barnett used the electromagnetic echo to evaluate the distance to the ionosphere (1927). In 1930, A. H. Taylor and L. A. Hyland observed an echo from aircraft, but the true initiators of the *electromagnetic detection* (DEM) or *radio detection and ranging* (RADAR) were Maurice Ponte (1902–1987) and his colleagues at the CSF, and H. Gutton and R. Warneck, who in November 1934 equipped the *Oregon* with a detection system operating at 375 and 1875 MHz, and the *Normandy* in August 1935 [192].

In England, Robert Alexander Watson-Watt (born in 1892) of the National Physical Laboratory obtained results that were thought so positive by an evaluation committee presided over by Sir Henry Thomas Tizard that the British Air Ministry undertook in December 1935 the construction of the first five 'hertzian beacons'. The 80-m high towers constituted the first metropolitan *radio detection finding* (RDF) chain. The set-up operated at 11.5 MHz and then at 25 MHz. It played an important part in the Battle of Britain. On 4 September 1937, an airborne radar (*airborne search for surface vessels* (ASV) on board the Anson K-6260) detected for the first time the buildings of the Royal Navy [13], [9]. In the United States, the first trial took place in 1937 on the destroyer *Leary* and then in 1938 on the steamer *New York*, provided with meter-wave equipment. In Germany radar (Wurzburg de Telefunken) was declared operational in April 1940 [13]. Italy, the Soviet Union, Japan, The Netherlands, and Canada were all carrying out research in the greatest secrecy. World War II resulted in considerable advances in decimeter and centimeter power sources, klystrons, and magnetrons.

The magnetron, conceived in 1921 by Albert Wallace Hull, is a tube with electric and magnetic fields at right angles to each other, and is fundamentally different from grid tubes. It was discovered independently in Czechoslovakia, Germany, Japan (K. Okabe), the United States (model MS-30 at 3.3 GHz, 1929–1931, Westinghouse), England, and France. The peripheral distribution of resonant cavities was introduced in 1939 by John Turton Randall, Henry A.H. Boot, and J. Sayers of the University of Birmingham (Fig. 1.6). In April 1940, Eric C. S. Megaw, of the General Electric Company Ltd. at Wembley, used a permanent magnet to shorten the anode block. Maurice Ponte constructed the first oxide-coated cathode magnetron, the M-16 of the French Radioelectric Society, which he presented on 9 May 1940 to the researchers at Wembley [217], [24], (*cf.* Part I, Section 4.2.1).

The klystron, invented in 1935 by Oscar Heil and A. Arsenjewa-Heil of the C. Lorenz AG Company in Berlin, is a linear tube based on the velocity modulation of an electron beam. The first implementation of this idea by Russell

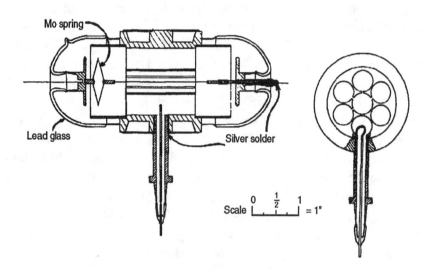

Figure 1.6 First multi-cavity magnetron designed by J. T. Randall and H. A. H. Boot (21 February 1940) [24].

and Sigurd Varian and David L. Webster of Stanford University dates from 19 August 1937. Industrial production began at San Carlos (California) in collaboration with Sperry Gyroscope Company (William G. Cooke, and Joseph L. Caldwell) and with Charles V. Litton, maker of glass and vacuum pumps. A continuously-pumped 750-MHz tube was tested in low visibility landing and a second 3-GHz tube rapidly acquired worldwide notoriety. The Sperry factory opened in New York in December 1940 [83], [213], [212], [24] (*cf.* Part I, Section 4.2.2).

The other fundamental discovery was that of guided propagation. It was made between 1887 and 1889 by William Strutt (Lord Rayleigh, 1842–1919) and Arnold Sommerfeld (1868–1951). The directional coupler was patented in 1927 by H. A. Affel. In the field of coaxial cables , the N-type connector was made in 1942 by Paul Neill, and the use of polyethylene as a dielectric began in 1943 [199], [194], [23], [32], [153], [159] (*cf.* Part I, Section 2.9).

1.5 Thermal applications

In radio broadcasting and in radar, the radio wave carries a signal (i.e., information). Thermal applications, on the other hand, use the radio wave as a carrier of energy. Apart from the Kassner's patents [103], [104], [105], these applications derive primarily from techniques developed for radar applications. Military magnetrons were in fact designed to produce *pulsed waves* (PW). In 1943–1944, W. C. Brown and P. Derby of Raytheon created the *continuous wave* (CW) magnetron. The prototype QK44 (American patent No. 2463524 of 8 March 1949) delivered 100 W at 3 GHz, and was modifed as RK5609 for hyperthermia (Part IV, Section 4.1.1). It was finally extended to 1 kW as the reference device designated QK65 [158].

The microwave oven was born as American patent No. 2495 429 of 24 January 1951. The designs had been deposited since 8 October 1945 by Percy L. Spencer of Raytheon and involved two magnetrons mounted in parallel on a waveguide. K. J. Stiefel conceived in 1946 the first slotted waveguide (American patent No. 2560903). In 1947, P. W. Morse and H. E. Rivercomb of the General Electric Company proposed a prototype oven for the thawing and cooking of prepared meals [144]. The first tunnel-oven was manufactured in 1948 by the French company Thomson-Houston (French patent No. 640783). Wolfgang Schmidt produced in 1959 the first multimode cavity incorporating wave traps (German patent No. 3048 686) and, in 1963, Cryodry Corporation created the first tunnel-oven based on the slotted waveguide in which water loads absorbed the energy radiated from the slots. A decisive stage was reached in 1963 with the development by the British Company AEI of the BM 25A magnetron, delivering 25 kW at 896 MHz [158].

As Miesel [126] has pointed out, Europe was not, for once, trailing behind the United States. In 1962, the first American restaurant without a kitchen, serving only prepared meals reheated by microwave (Tad's, on 42nd Street in New York) was equipped with ovens constructed under European licence. In the early 1960s, European activities were mostly confined to large companies such as CSF, Philips, English Electric, and Atlas Elektronik. The first industrial applications were in the pasteurization of prepacked bread and the final drying of chips. These doubtful applications did not survive (Part III, Section 3.2.1), and the failure of food applications seemed inevitable. From then on, vulcanization became the principal application, and the Svenska Livsmedelsinstitute (SIK) of Goteburg was the only institute to continue with applications to food. The recovery came from a small British company, founded by former employees of Philips, CSF, and AEI, namely, Magnetronics of Leicester (UK), Industries Micro-ondes Internationales (IMI) of Epône (France), Püschner GmbH of Schwanewede bei Bremen (RFA), and Skandinaviska Processinstrument (Scanpro) that became Calorex AB at Broma (Sweden), specializing in cooking, thawing, dehydration under vacuum, pasteurization, and the production of meat paté [126].

In the United States, the US Army Research and Development Laboratory at Natick (Massachusetts) in general, and Robert V. Decareau in particular, played an important role from the early 1940s onward. Their studies of the sterilization of rations, thawing, lyophilization, dehydration, and the dielectric properties of food led to the adoption of microwaves by the military authorities [41]. The fifties were dominated by Raytheon, which established an industrial microwave division under the direction of W. C. Brown, and which was active in the field of drying of chips, dehydration under vacuum, drying of polyamide fibers, and so on. In 1966, Litton Industries Atherton Division installed two cooking ovens for poultry (50–80 kW at 2.45 GHz), and diversified its activities to the evaporation of alcohol, treatment of polyurethane foam, and, above all, final drying of chips, which remained the principal application at 700 kW (fourteen 50-kW ovens) in May 1967 [91], [158]. A number of small companies were formed, but none survived with the exception of Cryodry, which became Microdry Corporation of San Ramon, California.

In Japan, industrial applications of microwaves began in 1968. They were aimed at the disinfestation of grain (34 appliances totalling 550 kW were in existence in 1971), mildew inhibition, and freeze drying of rice cakes and of products based on seaweed and eggs [101].

Despite the disappointment born out of the failure of the first applications, Ernie Okress organized in 1964, in Florida, the *Microwave Power Applications Symposium* under the auspices of the IEEE. This meeting gave birth to a specialist body, the *International Microwave Power Institute* (IMPI), officially created in 1966 at the University of Alberta in Edmonton (Canada). IMPI aims to promote all microwave applications, with the exception of telecommunications and radar, by means of annual symposia, specialist journals (*Journal of Microwave Power* and *Microwave Energy Applications Newsletter*), and the promotion of industrial and academic research.

1.6 Microwaves in industry

The development of the industrial microwave power market was slow and weak in absolute terms. For example, in 1984 the whole of the food industry had 19 MW of UHF installed worldwide [44]. Compared with the 5.1 MW of UHF in use in 1978 [18], the growth in the years 1978–1984 was on the order of 2.5 MW per year or, on average, an annual increase of 20%. In addition, a number of industrial installations built by users for their own processes were not included in this statistic. In absolute terms, the 19 MW of industrial microwave power are equivalent to 32 000 600-W domestic ovens, which is derisory by comparison to the 7 300 000 units in domestic service in 1984 in the United States, where they showed the highest growth rate in the electric domestic appliance sector between 1976 and 1986. The saturation level of the market, which was 40% in 1984, and expected to exceed 70% by 1990 [121], [145]. The reluctance of industry to use microwave power was not confined to the food sector. In market terms, the use of microwaves in ISM applications is marginal by comparison to its use in the domestic sector.

One questions the reluctance of industry to use microwaves in view of their universally recognised advantages, namely,

- speed of heating, better efficiency, savings in space and manpower
- ease of operation, instantaneous on and off operations
- ease of adaptation to existing on-line operations, possibile combination with other thermal processes
- no energy loss by radiation, improved efficiency and working conditions, possibility of eliminating the need for temperature control of the environment
- better thermal efficiency compared with traditional processes
- improved product quality

The response of industry is heavily influenced by the size of the investment required, which is the order of $15,000$–$20,000$ F ($750) per kW against $\sim 2,500$ F ($125) per kW for a domestic appliance and $\sim 10^6$ F ($50,000$) for a typical industrial appliance. Industry's response is also governed by natural reluctance to change, particularly in the food industry. In addition, there are several negative processes that can intervene unfavorably. They include the following:

- conflicts related to patent rights (i.e., depending on whether the manufacturer remains the owner of patents) the client may or may not benefit from guarantees;
- bad or insufficient specification on the part of the client, often for reasons related to industrial secrecy or misunderstanding by the manufacturer;
- inadequate management of production (i.e., insufficient

training of the work force) and shortage of spare parts.

Microwaves do not provide a universal solution to all problems, but should be considered whenever all other processes fail to solve an industrial problem, in which case the advantages of microwaves become unique and offer considerable savings as compared to other existing processes [131]. A necessary (but not sufficient) condition then is that the product to be treated must be of high quality, so that the cost of UHF treatment can be justified. The study of efficiency should include a detailed analysis of the traditional process, of which the user often has only a very vague idea.

Investment in a UHF system must be evaluated by classical cost effectiveness techniques [48], [93], [51]. If D_0 is the capital investment and $A_1, A_2, ..., A_i, ..., A_n$ are the yields achieved in n years, we can define the following criteria:

• the time required to recoup the investment, j, such as that

$$\sum_{i=1}^{j} A_i = D_0 \tag{1.1}$$

• the rate at which investment is recouped

$$r = \frac{1}{nD_0} \sum_{i=1}^{n} A_i \tag{1.2}$$

• the relative increase in actual value with interest k

$$g = \frac{1}{D_0} \sum_{i=1}^{n} \frac{A_i}{(1+k)^i} \tag{1.3}$$

• the theoretical yield rate, m, such that

$$\sum_{i=1}^{n} \frac{A_i}{(1+m)^i} = D_0 \tag{1.4}$$

The use of these techniques presuppose that the annual yields A_i are known year by year, so that the method can be used only for UHF systems for which there are sufficient economic data. An analysis of microwave cooking of bacon by Edgar [51] will be given in Part III, Section 1.2.3. For a new application, the setting up of the process includes a laboratory feasibility study, followed by an experimental study on a pilot plant in order to determine the industrial parameters, and, finally, the controlled production of a quantity of the product to be used in market research d[131]. In all cases, the evaluation must take into account a number of side issues such as gain in quality, savings in manpower, improved working conditions, minimum heat dissipation, savings in ventilation and/or air-conditioning, improved pollution control, and so on [221]. The study must remain sufficiently critical so that any application contravening good economic sense is stopped at the source.

References

1. AGOSTO W. Microwave power. *IEEE Spectr.*, 13, n° 7, 1976, pp. 24-25.

2. ALAMI R. An appraisal of potential applications of microwave energy and the governing factors for implementation. In : Radiant Wave Electroheat Workshop, Toronto, 2 Dec. 1985, pp. 77-89.

3. ALPHANDERY M. Dictionnaire des inventeurs français. Paris, 1962, 371 p. (Collection Seghers).

4. ASSINDER I. Microwaves for food processing. *Trans. IMPI*, 2, 1973. Industrial applications of microwave energy.

5. BADGER G. Microwave heating. Machine Design. 1st Oct. 1970, pp. 88-93.

6. BARBER H., DJIAN R. Techniques et applications de l'électrothermie industrielle. In : Conférence mondiale de l'énergie, Cannes, 5-11 oct. 1986, 32 p.

7. BARBER H., HARRY J. Electro-heat : electric power for industrial heating processes. *IEE Proc.*, 126, n° 11 R, *IEE Rev.*, Nov. 1979, pp. 1126-1148.

8. BARTHOLIN G. Micro-ondes. Industries alimentaires. *Conserv. Agric.*, 17, n° 1, 1974, pp. 24-27. (C.D.I.U.P.A., n° 69477).

9. BARTON D. A half century of radar. *IEEE Trans. Microw. Theory Tech.*, MTT-32, n° 9, 1984, pp. 1161-1170.

10. BASSOLE P. Protection des radiocommunications contre la pollution électromagnétique. In : Compte-rendu du Symposium international sur les applications énergétiques des micro-ondes, IMPI-CFE. 14, Monaco, 1979. Résumés, pp. 213-216.

11. BAUSCH L., DUFFNER P. Consumer and commercial microwave ovens. *Trans. IMPI*, 5, 1975. Future impact of microwave ovens on the food industry.

12. BEDROSIAN K. The necessary ingredients for successful microwave applications in the food industry. *J. Microw. Power*, 8, n° 2, 1973, pp. 173-178.

13. BEKKER C. Radar. Paris, Ed. France-Empire, 1960, 313 p.

14. BELLAVOINE R. Communication personnelle. SFAMO, 0160 RB/BG, 4 mars 1986, 3 p.

15. BENGTSSON N. Radio-frequency heating applications in the European food industry. *Microw. Energy Appl. Newsl.*, 2, n° 4, 1969, pp. 3-6. (C.D.I.U.P.A., n° 30584).

16. BENGTSSON N. Mikrovagsvarmning inom livsmedelstekniken. 1. Principer och berakningsmetoder. *Livsmed. tek.*, 16, n° 8, 1974, pp. 348-351. (C.D.I.U.P.A., n° 79340).

17. BENGTSSON N., OHLSSON T. Microwave heating in the food industry. *Proc. IEEE*, 62, n° 1, 1974, pp. 44-55.

18. BENGTSSON N., OHLSSON T. Application of microwave and high-frequency heating in food processing. In : LINKO P., MALKKI Y., OLKKO J., LARINKARI J. Food process engineering, Vol. 1. London, Applied Science Publishers, 1980.

19. BERTEAUD A. Activités en micro-ondes au CNRS de Thiais. *Econ. Progrès Electr.*, n° 13, janv.-fév. 1985, pp. 3-4.

20. BERTEAUD A., LEPRINCE P., PRIOU A. Mission au Japon de représentants du Club Microondes EDF, avril 1985, 12 p.

21. BERTEAUD A., LEPRINCE P., PRIOU A. Compte-rendu de la mission au Japon de représentants du Club Microondes EDF (avril 1985). *Econ. Progrès Electr.*, n° 16, juil. 1985, pp. 3-6.

22. BROWN W. The history of power transmission by radio waves. *IEEE Trans. Microw. Theory Tech.*, MTT-32, n° 9, 1984, pp. 1230-1242.

23. BRYANT J. Coaxial transmission lines, related two-conductor transmission lines, connectors, and components : a US historical perspective. *IEEE Trans. Microw. Theory Tech.*, MTT-32, n° 9, 1984, pp. 970-983.

24. BRYANT J. The first century of micro-
waves-1886 to 1986. *IEEE Trans.
Microw. Theory Tech.*, MTT-36, n° 5,
1988, pp. 830-858.

25. CHASPOUL P. Utilisation du spectre
des fréquences radioélectriques.
Tech. Ing. Electron., E 1009, mars
1982, 13 p.

26. CHATEAUMINOIS M. Rayonnements électromagnétiques : trois techniques
industrielles pour le chauffage ou
l'activation de réaction. *J. Fr.
Electrothermie*, 16 juil. 1986,
pp. 39-45.

27. CHEF M. Risques découlant de la
pollution électromagnétique dans les
bandes de fréquences attribuées à
l'aéronautique. In : Compte-rendu
du Symposium international sur les
applications énergétiques des micro-
ondes, IMPI-CFE. 14, Monaco, 1979.
Résumés, pp. 217-221.

28. CHEN H., SHEN Z., FU C., WU D. The
development of microwave power applications in China. *J. Microw. Power*,
17, n° 1, 1982, pp. 11-15.

29. CLAVIER A. Micro-ray radio. *Electr.
Commun.*, July 1931, pp. 20-21.

30. CLAVIER A. Production and utilization
of micro-rays. *Electr. Commun.*, July
1933, pp. 3-11.

31. CLAVIER A., GALLANT L. The Anglo-
French micro-ray link between Lympne
and St. Inglevert. *Electr. Commun.*,
Jan. 1934, pp. 222-229.

32. COHN S., LEVY R. History of microwave passive components with particular attention to directional couplers. *IEEE Trans. Microw. Theory
Tech.*, MTT-32, n° 9, 1984, pp. 1046-
1054.

33. COLTON R. Radar in the United States
Army. *Proc. IRE*, 33, Nov. 1945, pp.
740-753.

34. COPSON D. Microwave energy in food
procedures. *IRE Trans. Prof. Group
Med. Electron.*, PGME-4, 1956, pp.
27-35.

35. COPSON D. Microwave heating. 2nd ed.
Westport, Avi Publishing Co., 1975,
615 p. (C.D.I.U.P.A., n° 93299).

36. CORNU A. Cours de physique. Ecole
Polytechnique, 2ème div., 1899-1900,
401 p.

37. CZARNOWSKI L. Zastosowanie w przemy-
śle spozywczym nagrzewania dielek-
trycznego wysokiej czestotliwości.

Przem. Spoz., 24, n° 11, 1970, pp.
468-473. (C.D.I.U.P.A., n° 30700).

38. CZIGANY S., GYORI A. Measurement
methods for industrial applications
of microwaves. In : Proceedings of
the IMPI Symposium, Waterloo (Can.),
1975, pp. 133-136. (Paper 8.2.).

39. DECAREAU R. Microwave energy in food
processing applications. *CRC Crit.
Rev. Food Technol.*, 1, n° 2, 1970,
pp. 194-224.

40. DECAREAU R. The impact of microwaves
on the food market. *Microw. Energy
Appl. Newsl.*, 3, n° 6, 1970, pp.
10-16. (C.D.I.U.P.A., n° 30593).

41. DECAREAU R. Microwave research and
development at the US Army Natick
Research and Development Laboratories.
J. Microw. Power, 17, n° 2, 1982,
pp. 127-135.

42. DECAREAU R. Microwaves in food processing. *Food Technol. Australia*,
36, n° 2, 1984, pp. 81-86.

43. DECAREAU R. Microwaves in the food-
processing industry. New York,
Academic Press, 1985, 234 p.

44. DECAREAU R. Microwave food processing
equipment throughout the world. *Food
Technol.*, 40, n° 6, 1986, pp. 99-105.
(C.D.I.U.P.A., n° 214710).

45. DECAREAU R. Future research developments in microwave technology. Non
Publ., sans réf. s.d., 12 p.

46. DECAREAU R., PETERSON R. Microwave
processing and engineering. Weinheim,
VCH Verlagsgesellschaft, 1986, 224 p.

47. DESBEAUX E. Physique populaire. Paris,
Ed. Ernest Flammarion, Ca 1900, 835 p.

48. DUMONTIER J. Eléments d'économie.
Paris, Ecole Polytechnique, Promotion 1972, 503 p.

49. DUNN D. Microwave power systems -
an introduction. In : Proceedings of
the IMPI Symposium, Stanford, 1967.
Abstracts pp. 2-4.

50. DUTESCU F., NOVACEANU M. Incalzirea
cu micro- si macrounde. *Ind. Aliment.*,
23, n° 2, 1972, pp. 95-99.
(C.D.I.U.P.A., n° 42475).

51. EDGAR R. The economics of microwave
processing in the food industry.
Food Technol., 40, n° 6, 1986, pp.
106-112. (C.D.I.U.P.A., n° 214710).

52. EDIN B. Communication personnelle.
5 mars 1986.

53. ELLIOTT R. The history of electro-magnetics as Hertz would have known it. *IEEE Trans. Microw. Theory Tech.*, MTT-36, n° 5, 1988, pp. 806-823.

54. EULER G. Mikrowellenanwendungen für die Industrie. Valvo GmbH, D., 1971, 48 p.

55. EULER G. Messungen und Entwicklungs-gesichtspunkte für Mikrowellengeräte mit Dauerstrichmagnetrons. Valvo Bauelemente für die gesamte Elektro-nik, Valvo GmbH, D., 1978, 75 p.

56. FABRY M. Cours de physique. Ecole Polytechnique, 1ère div., 1929-1930, 478 p.

57. FAILLON G., COUASNARD C., MALONEY E. New uses of microwave power in the food industry. *J. Microw. Power*, 12, n° 1, 1977, pp. 79-86.

58. FAILLON G., MALONEY E. New uses of microwave power in the food industry. In : Proceedings of the IMPI Sympo-sium, Leuven, 1976. (Paper 6.B.1.). Abstract in *J. Microw. Power*, 11, n° 2, 1976, pp. 210-211.

59. FERNET E. Traité de physique élémen-taire. Paris, Ed. G. Masson, 12ème éd, 1893, 848 p.

60. FONTEYNE J. Brouillage de la radio-diffusion par les appareils ou installations ISM. In : Compte-rendu du Symposium international sur les applications énergétiques des micro-ondes, IMPI-CFE. 14, Monaco, 1979. Résumés pp. 222-225.

61. FREEDMAN G. The future of microwave power in industrial applications. In : Proceedings of the IMPI Sympo-sium, Ottawa, 1972, pp. 10-12. (Paper 1.2.).

62. FREEDMAN G. The future of microwave power in industrial applications. *J. Microw. Power*, 7, n° 4, 1972, pp. 353-365.

63. FREEDMAN G. The future of microwave heating equipment in the food indus-tries. *J. Microw. Power*, 8, n° 2, 1973, pp. 161-166.

64. FREY J. The growth of microwave sys-tems and applications. In : American Public Health Association, Atlantic City (N.Y.), 1972, pp. 79-95.

65. GARDIOL F. The microwave oven - an energy saver or an infernal machine. *Bull. Assoc. Suisse Electr.*, 67, n° 8, 1976, pp. 388-393.

66. GEDDES J. Seek new uses for micro-wave. *Food Eng.*, 39, n° 4, 1967, pp. 62-65. (C.D.I.U.P.A., n° 67/283).

67. GERLING J. Microwave processing equipment - a modular approach. *Food Technol.*, 22, n° 1, 1968, pp. 106-109. (C.D.I.U.P.A., n° 01384)

68. GERLING J. High power applications. In : Proceedings of the IMPI Sympo-sium, Loughborough, 1973, abstract 1 p. (Paper 1.A 1/1).

69. GERLING J. Microwave heating patents issued/microwave ovens sold. *Microw. World*, 2, n° 1, 1981, p. 17.

70. GERLING J. A listing of microwave patents associated with industrial processing and domestic/commercial microwave ovens. Gerling Lab., report n° 83-017, Aug. 1983.

71. GERLING J. Microwaves in the food industry : promise and reality. *Food Technol.*, 40, n° 6, 1986, pp. 82-83. (C.D.I.U.P.A., n° 214710).

72. GILLES P. Microwave food applications in the United Kingdom : domestic and commercial microwave ovens. *J. Microw. Power*, 8, n° 2, 1973, pp. 129-132.

73. GOLDBLITH S. Microwaves and process design. *Process Biochem.*, 4, n° 7, 1969, pp. 37-38. (C.D.I.U.P.A., n° 25970).

74. GOLDBLITH S. Radio-frequency energy-the theory of its interaction with foodstuffs, and possible applications in food processing. *Microw. Energy Appl. Newsl.*, 5, n° 6, 1972, pp. 3-11. (C.D.I.U.P.A., n° 57771).

75. GOLDBLITH S., DECAREAU R. An annota-ted bibliography on microwaves. Their properties, production, and applications to food processing. Cambridge, the Massachusetts Insti-tute of Technology, 1973, 366 p. (C.D.I.U.P.A., n° 62683).

76. GUERGA M. Le chauffage hyperfré-quentiel. Applications aux industries agricoles. In : Journée du séchage de l'A.F.E.T.I.A., 13 nov. 1968. 1968, pp. 1-6. (C.D.I.U.P.A., n° 29743)

77. GUIERRE M. Les ondes et les hommes-histoire de la radio. Paris, Ed. René Julliard, 1951, 258 p.

78. GUILLEMAIN A. Le magnétisme et l'électricité. I : Phénomènes magné-tiques et électriques. Paris, Ed. Hachette et Cie, 1980, 272 p.

79. HAMID M. Applications of microwaves to the agricultural industry. Industrial enterprise fellowship program report, Manitoba Research Council, sans réf., Dec. 1970, 49 p.

80. HAMID M. Microwave ovens - continuing special section. *J. Microw. Power*, 9, n° 2, 1974, pp. 49-50.

81. HAMID M., RZEPECKA M., STUCHLY S. Optimum frequency of microwave heating by radiometry. In : Proceedings of the IMPI Symposium, Ottawa, 1972, Résumé pp. 55-56. (Paper 4.1.)

82. HARVEY A. Industrial, biological and medical aspects of microwave radiation. *Proc. IEEE Bull.*, 107, n° 36, 1960, pp. 557-566.

83. HEIL A., HEIL O. Une nouvelle méthode pour engendrer des ondes électromagnétiques courtes non amorties de forte intensité. (En allemand). *Z. Phys.*, 95, Jul. 1935, pp. 752-773.

84. HENOCH B. Industrial applications of microwaves. In : 12th European microwave Conference, Helsinki, Sept. 13-17, 1982, pp. 47-53.

85. HERBAUGH R. ISM microwaves : some facts, some finances, some fears, and the future. *Microw.*, March 1979, pp. 43-52.

86. HOWARD A. Chambers's dictionary of scientists. London, W & R Chambers, Ltd., 1951, 500 p.

87. HUANG H. Thirty years of microwaves in China. In : IEEE Microwave Theory and Techniques Symposium digest, A-2, pp. 2-6.

88. ISHII T. Theoretical basis for decision to microwave approach for industrial processing. In : Proceedings of the IMPI Symposium, Milwaukee, 1974, 5 p. (Paper A.5-1).

89. ISHII T. Theoretical basis for decision to microwave approach for industrial processing. *J. Microw. Power*, 9, n° 4, 1974, pp. 355-359.

90. JEPPSON M. Consider microwaves. *Food Eng.*, n° 11, 1964, pp. 93-94.

91. JEPPSON M. The evolution of industrial microwave processing in the United States. *J. Microw. Power*, 3, n° 1, 1968, pp. 29-38.

92. JOLION M. Applications industrielles des micro-ondes. *Rev. Gén. Electr.*, n° 11, nov. 1981, pp. 810-815.

93. JOLLY J. Financial techniques for comparing the monetary gain of new manufacturing processes such as microwave heating. *J. Microw. Power*, 7, n° 1, 1972, pp. 5-16.

94. JOLLY J. Industrial microwave power adoption rate and the diffusion of technical information. *J. Microw. Power*, 8, n° 3-4, 1973, pp. 337-356.

95. JOLLY J. The nature of the firms manufacturing microwave products in the USA. In : Proceedings of the IMPI Symposium, Loughborough, 1973, 1 p. Résumé. (Paper 1.A.2/1).

96. JOLLY J. Economics and energy utilisation aspects of the application of microwaves : a tutorial review. *J. Microw. Power*, 11, n° 3, 1976, pp. 233-245.

97. JOLLY J. Industrial engineer awareness of industrial microwave power. *J. Microw. Power*, 14, n° 1, 1979, pp. 15-20.

98. JONES P. Developments in dielectric heating. In : Radiant Wave Electroheat Workshop, Toronto, 2 Dec. 1985, pp. 47-62.

99. KALAFAT S., KROGER M., DECAREAU R. Microwave heating of foods - use and safety considerations. *Crit. Rev. Food Technol.*, 4, n° 2, 1973, pp. 141-151. (C.D.I.U.P.A., n° 64129).

100. KASE Y. Microwave oven in Japan. *IMPI Newsl.*, 3, n° 2, 1975, pp. 2-4.

101. KASE Y., OGURA K. Microwave power applications in Japan. *J. Microw. Power*, 13, n° 2, 1978, pp. 115-123.

102. KASHYAP S. Industrial microwave research applications at National Research Council. In : Radiant Wave Electroheat Workshop, Toronto, 2 Dec. 1985, pp. 111-123.

103. KASSNER E. Process for altering the energy content of dipolar substances. US Patent n° 2 089 966, 17 Aug. 1937.

104. KASSNER E. Apparatus for the generation of short electromagnetic waves. US Patent n° 2 094 602, 5 Oct. 1937.

105. KASSNER E. Apparatus for generating and applying ultra-short electromagnetic waves. US Patent n° 2 109 843, 1 March 1938.

106. KEITLEY R. Some possibilities of heating by centimetric power. *J. IRE*, 9, March 1949, pp. 97-121.

107. KENYON E. Microwave energy applications to food processing. *Chim. Ind. - Génie Chim.*, 105, n° 15, 1972, p. C 35.

108. KIEREBINSKI C. Wybrane zastosowania mikrofal w przemysle spozywczym. *Przem. Spoz.*, 26, n° 1, 1972, pp. 6-10. (C.D.I.U.P.A., n° 41861).

109. KINN T., MARCUM J. Possible uses of microwaves for industrial heating. *Prod. Eng.*, 18, Jan. 1947 pp. 137-140.

110. KIRK D. Microwave ovens. In : SMITH R. Microwave heating for the food industry. Sypplementary notes for short course. IMPI, European Chapter. Univ. Bradford, 1975, pp. 1-15.

111. KRAUS J. Heinrich Hertz-theorist and experimenter. *IEEE Trans. Microw. Theory Tech.*, MTT-36, n° 5, 1988, pp. 824-829.

112. LANDON R. Microwave energy - a growing production tool. *Automation*, n° 9, 1968, pp. 52-55.

113. LE POMMELEC G. Les applications des micro-ondes dans les industries alimentaires et pharmaceutiques. Publication Sté IMI-SA, 1977, 30 p.

114. LILEN H. Fours à micro-ondes. *Ind. Electron. Microélectron.*, 151, 15 fév. 1972, pp. 27-30.

115. LORENZ K. Microwave cooking at high altitudes. *Microw. Energy Appl. Newsl.*, 9, n° 1, 1976, pp. 3-6. (C.D.I.U.P.A., n° 93958).

116. LORENZ K., DILSAVER W. Microwave heating of food materials at various altitudes. *J. Food Sci.*, 41, n° 3, 1976, pp. 699-702. (C.D.I.U.P.A., n° 95717).

117. Mac CONNELL D. Impact of microwaves on the future of the food industry. *J. Microw. Power*, 8, n° 2, 1973, pp. 123-127.

118. Mac CONNELL D. Energy consumption : a comparison between the microwave oven and the conventional electric range. In : Proceedings of the IMPI Symposium, Milwaukee, 1974, pp. 1-4. (Paper A.6.6.).

119. Mac CONNELL D. Energy consumption : a comparison between the microwave oven and conventional electric range. *J. Microw. Power*, 9, n° 4, 1974, pp. 341-347.

120. Mac LACHLAN A. Method of measurement of radiofrequency interference from microwave ovens. In : Proceedings of the IMPI Symposium, Loughborough, 1973, 2 p. (Paper 5.B.3.)

121. MARKOV N. The microwave oven market. An overview. *Microw. World*, 6, n° 1, 1985, pp. 6-10.

122. MAXWELL J. A dynamical theory of the electromagnetic field. *Trans. Royal Soc.*, 155, 8 Dec. 1864, pp. 459-512.

123. MEIKELJOHN B. The potential for application of dielectric technologies in Ontario. In : Radiant Wave Electroheat Workshop, Toronto, 2 Dec. 1985, pp. 91-109.

124. MEISEL N. Les micro-ondes dans l'industrie alimentaire. In : Les industries alimentaires. Massy-Douai, ENSIA, 1971, pp. 51-54.

125. MEISEL N. Nouveau domaine d'application des micro-ondes dans l'industrie alimentaire. *Hranit. Prom.*, 20, n° 1, 1971, pp. 10-14. (C.D.I.U.P.A., n° 56494).

126. MEISEL N. Microwave applications to food processing and food systems in Europe. *J. Microw. Power*, 8, n° 2, 1973, pp. 143-148.

127. MEISEL N. Applications industrielles des micro-ondes dans les industries alimentaires. *Rev. Gén. Froid*, 67, n° 1, 1976, pp. 9-20. (C.D.I.U.P.A., n° 92131).

128. MEISEL N. Nuove applicazioni delle microonde nell'industria alimentare. *Ind. Aliment.*, 15, n° 128, 1976, pp. 90-93. (C.D.I.U.P.A., n° 95624).

129. MEISEL N. Applications des micro-ondes dans les industries agricoles et alimentaires : séchage-décongélation. In : Colloque sur les applications industrielles des hautes fréquences et des micro-ondes. Lyon, Univ. Claude Bernard. Résumés, 2ème partie, 1977, 13 p.

130. MEISEL N. Microwave is gaining ground in the food processing industry. *S.A. Food Rev.*, 1980, p. 43.

131. MEISEL N. Les micro-ondes. Savoir... et...bien faire. *Ind. Aliment. Agric.*, 101, n° 4, 1984, pp. 259-264. (C.D.I.U.P.A., n° 221917).

132. MEISEL N. Les micro-ondes dans les industries agro-alimentaires, bilan 1974-1984. *Ind. Aliment. Agric.*, 101, n° 10, 1984, pp. 929-932. (C.D.I.U.P.A., n° 193861).

133. MEISELS M. Industry warming to microwave power. *Microw.*, n° 5, 1968, pp. 10-16.

134. MEREDITH R. Applications at 900 MHz. *Trans. IMPI*, 2, 1973, Industrial applications of microwave energy.

135. MEREDITH R. Industrial microwave heating : the thaw begins. *Electr. Rev.*, 212, n° 19, May 1983, pp. 31-32.

136. MEREDITH R. Communication personnelle, RJM/GLM. 27 fév. 1986, 1 p.

137. MEREDITH R. Recent advances in industrial microwave processing in the 896/915 MHz frequency band. Leicester, Magnetronics Ltd., Non publié, sans réf., s.d., 12 p.

138. METAXAS A., MEREDITH R. Industrial microwave heating. IEE power engineering series n° 4. Stevenage, Peter Peregrinus Ltd., 1983, 357 p.

139. MINETT P. Radio frequency and microwave heating save labour and energy. *Elect. Rev.*, 197, n° 23, 1975, pp. 742-744.

140. MINETT P., WITT J. Radio frequency and microwaves. *Food Process. Ind.*, 45, n° 532, 1976, pp. 36-41. (C.D.I.U.P.A., n° 92563).

141. MOBLEY M. FCC equipment type approval program with reference to microwave tests. *J. Microw. Power*, 6, n° 4, 1971, pp. 297-303.

142. MOBLEY M. Revisions of Federal Communications Commission type-approval test procedures. In : Proceedings of the IMPI Symposium, Loughborough, 1973, 2 p. (Paper 4.B.6).

143. MOBLEY M. Interference measurements on radiofrequency devices. In : Proceedings of the IMPI Symposium, Waterloo (Can.), 1975, pp. 130-132. (Paper 8.1.).

144. MORSE P., RIVERCOMB H. UHF heating of frozen foods. *Electron.*, 20, Oct. 1947, pp. 85-89.

145. MURRAY M. Facts and features - an insider's report. *Microw. World*, 7, n° 1, 1986, pp. 13-16.

146. NICOLAS-GROTTE M. Utilisation des micro-ondes dans les industries alimentaires. Univ. Paris, Thèse de Doctorat, mention Sciences, 1972, 13 p. (C.D.I.U.P.A., n° 43175)

147. NIKOL'SKIJ V. Electrodynamique et propagation des ondes radio-électriques. Moskva, Izdatel'stvo Mir, 1982, 573 p.

148. ODA S. Ontario Hydro's R and D program on radiant wave technologies. In : Radiant Wave Electroheat Workshop, Toronto, 2 Déc. 1985, pp. 33-45.

149. OGURA K., KASE Y. Microwave power applications in Japan. *J. Microw. Power*, 13, 1978, pp. 115-123.

150. ØJELID G. Ten years development in the use of microwave heating in Scandinavia. In : Proceedings of the IMPI Symposium, Boston, 1969, Résumé, p. 9. (Paper B.4.).

151. ØJELID G. Ten years development in the use of microwave heating in Scandinavia. *J. Microw. Power*, 4, n° 1, 1969, pp. 29-35.

152. OKRESS E., BROWN W., MORENO T., GOUBEAU G., HEENAN N. Microwave power engineering. *IEEE Spectr.*, 1, n° 10, 1964, pp. 76-100.

153. OLINER A. Historical perspectives on microwave field theory. *IEEE Trans. Microw. Theory Tech.*, MTT-32, n° 9, 1984, pp. 1022-1045.

154. Ó MEARA J. What can be expected from microwave energy. *Trans. IMPI*, 1, 1973. Microwave power in industry.

155. Ó MEARA J. Why did they fail ? (A backward look at microwave applications in the food industry). *J. Microw. Power*, 8, n° 2, 1973, pp. 167-172.

156. ORFEUIL M. Electrothermie industrielle, fours et équipements électriques industriels. Paris, Ed. Dunod, 1981, 800 p.

157. OSEPCHUK J. Basic principles of microwave ovens. *Trans. IMPI*, 4, 1975. Future impact of microwave ovens on the food industry.

158. OSEPCHUK J. A history of microwave heating applications. *IEEE Trans. Microw. Theory Tech.*, MTT-32, n° 9, 1984, pp. 1200-1224.

159. PACKARD K. The origin of waveguides : a case of multiple rediscovery. *IEEE Trans. Microw. Theory Tech.*, MTT-32, n° 9, 1984, pp. 961-969.

160. PIERRET E. Sur les ondes hertziennes ultra-courtes (10 à 18 cm) et leurs applications. In : Compte-rendu de la 55ème session de l'Association française pour l'Avancement des Sciences, Nancy, 1931, pp. 114-116.

161. PIETERMAAT P. Le chauffage par micro-ondes. *Inter-Electron.*, 23, n° 6, 1968, pp. 50-63.

162. POITEVIN J., CHEF M. Les problèmes de sécurité posés par l'emploi des hautes fréquences et des micro-ondes. In : Colloque sur les applications industrielles des hautes fréquences et des micro-ondes, Lyon, 1977. Lyon, Univ. Claude Bernard, 1977, pp. 1-7, résumés (complément).

163. POITEVIN J., CHEF M. Commentaires sur la protection contre les brouillages des services aéronautiques. Organisation de l'Aviation Civile Internationale, Bureau Europe. In : Colloque sur les applications industrielles des hautes fréquences et des micro-ondes, Lyon, 1977. Lyon, Univ. Claude Bernard, 1977, pp. 8-16, résumés (complément).

164. POLLAK G., FOIN L. Comparative heating efficiencies of a microwave oven and a conventional electric oven. *Food Technol.*, 14, n° 9, 1960, pp. 454-457.

165. POTTER N. Food irradiation and microwave heating. In : POTTER N. Food science. 2nd ed. Westport, Avi Publishing Co., Inc., 1973, pp. 298-328. (C.D.I.U.P.A., n° 71876).

166. PROCTOR B., GOLDBLITH S. Electromagnetic radiation fundamentals and their applications in food technology. In : MRAK E., STEWART G. Advances in Food Research, Vol. 3, New York, Academic Press, 1951, pp. 119-196.

167. PÜSCHNER H. Microwave heating technique in Europe. In : Proceedings of the IMPI Symposium, Stanford, 1967, Résumé p. 5.

168. PÜSCHNER H. Mikrowellenwärme : Anwendungsbeispiele in der Ernährungsindustrie. *Ernahr.wirtsch.*, n° 1, 1973, pp. 16-23. (C.D.I.U.P.A., n° 53835).

169. RITALY J. Applications énergétiques des micro-ondes : réflexions d'un utilisateur dans l'industrie. *Rev. Gén. Electr.*, n° 11, 1981, pp. 847-851.

170. ROSEN C. Effects of microwaves on food and related materials. *Food Technol.*, 26, n° 7, 1972, pp. 36-55. (C.D.I.U.P.A., n° 47428).

171. SALE A. Microwaves for food processing. In : SMITH R. Microwave heating for the food industry. Supplementary notes for short course. IMPI, European chapter. Univ. Bradford, 1975, pp. 16-34.

172. SALE A. A review of microwaves for food processing. *J. Food Technol.*, 11, n° 4, 1976, pp. 319-329. (C.D.I.U.P.A., n° 97988).

173. SASAKI T. Applications of microwave power in Japan. *J. Microw. Power*, 3, n° 2, 1968, pp. 85-91.

174. SASAKI T. Application of microwave power in Japan. In : Proceedings of the IMPI Symposium, Boston, 1969, Résumé p. 10. (Paper B.5.).

175. SASAKI T. Growth of Japanese microwave industry. In : Proceedings of the IMPI Symposium, Monterey, 1971, (Paper 9.2.).

176. SASAKI T., KASE V. Growth of the microwave oven industry in Japan. *J. Microw. Power*, 6, n° 4, 1971, pp. 283-290.

177. SCHIFFMANN R. A breakthrough in industrial microwave applications marketing. *J. Microw. Power*, 5, n° 2, 1970, pp. 141-142.

178. SCHIFFMANN R. Microwaves and the food industry in the 1970's. *J. Microw. Power*, 8, n° 2, 1973, pp. 119-121.

179. SCHIFFMANN R. The applications of microwave power in the food industry in the United States. *J. Microw. Power*, 8, n° 2, 1973, pp. 137-142.

180. SCHIFFMANN R. Microwave challenge : today's heat processing. *Food Eng.*, 47, n° 11, 1975, pp. 72-76. (C.D.I.U.P.A., n° 97808).

181. SCHIFFMANN R. The future potential of microwaves in the food industry. In : SMITH R. Microwave heating for the food industry. Supplementary notes for short course. IMPI, European chapter. Univ. Bradford. 1975, pp. 42-57.

182. SCHIFFMANN R. An update on the applications of microwave power in the food industry in the United States. *J. Microw. Power*, 11, n° 3, 1976, pp. 221-224.

183. SCHIFFMANN R. Microwave power applications in the world : present and future, opening address, 1979 IMPI Conference. *J. Microw. Power*, 14, n° 3, 1979, pp. 197-200.

184. SCHIFFMANN R. Le chauffage microondes dans l'industrie aux Etats-Unis. *Rev. Gén. Electr.*, n° 11, 1981, pp. 802-804.

185. SCHIFFMANN R. Microwave applications in the food industry. In : Radiant Wave Electroheat Workshop, Toronto, 2 Dec. 1985, pp. 125-142.

186. SCHIFFMANN R. Food product development for microwave processing. *Food Technol.*, 40, n° 6, 1986, pp. 94-98. (C.D.I.U.P.A., n° 214710).

187. SCHILZ W. Reduction of pollution in industrial microwave systems. In : Compte-rendu du Symposium international sur les applications énergétiques des micro-ondes, IMPI-CFE, 14, Monaco, 1979 (résumés p. 242).

188. SEEBAUER W., FRIEDRICH W. Thermische Verfahren mit Mikrowellen und Infrarotstrahlung. *Mühle Mischfuttertech.*, 111, n° 38, 1974, pp. 571-573. (C.D.I.U.P.A., n° 76277).

189. SHUTE R. Microwave power in the food industry. In : SMITH R. Microwave heating for the food industry. Supplementary notes for short course. IMPI, European chapter. Univ. Bradford, 1975, pp. 58-68.

190. SHUTE T. Microwave heating. *Food Process. Ind.*, 45, n° 532, 1976, pp. 41-43. (C.D.I.U.P.A., n° 92564).

191. SILCOX H., SPENCER M., WISCOMB L. Microwave research bibliography 1970-83. CAS Res. Comm., Cornell Univ., IMPI Publications, Vienna (USA), 111 p.

192. SKOLNIK M. Fifty years of radar. *Proc. IEEE*, 73, n° 2, 1985, pp. 182-197.

193. SOBOL H. Microwave communications - an historical perspective. *IEEE Trans. Microw. Theory Tech.*, MTT-32, n° 9, 1984, pp. 1170-1181.

194. SOMMERFELD A. Fortpflanzung elec-trodynamischer Wellen an einem zylindrischen Leiter. *Ann. Phys. Chem.*, 67, Dez. 1899, pp. 233-290.

195. SPASH D. Frequency allocations for industrial process heating applications. In : Compte-rendu du Symposium international sur les applica-tions énergétiques des micro-ondes, IMPI-CFE, 14, Monaco, 1979, Résumés, pp. 234-237.

196. SPENCER P. Means for treating food-stuffs. US Patent n° 2605383, July 1952.

197. SPRENG H. Mikrowellen in der Lebens-mittelbereitung. *Sci. Technol. Aliment.*, 6, n° 3, 1973, pp. 77-85. (C.D.I.U.P.A., n° 61376).

198. STEPHANSEN E. Economics of indus-trial use of microwave energy. *Trans. IMPI*, 1, 1973, Microwave power in industry.

199. STRUTT, lord RAYLEIGH J. On the passage of electric waves through tubes, or the vibrations of dielec-tric cylinders. *Phil. Mag.*, XLIII, 1897, pp. 125-132.

200. STUCHLY M., STUCHLY S. Industrial, scientific, medical and domestic applications of microwaves. *IEE Proc.*, 130, n° 8, part A, Nov. 1983, pp. 467-503.

201. SÜSSKIND C. Heinrich Hertz : a short life. *IEEE Trans. Microw. Theory Tech.*, MTT-36, n° 5, 1988, pp. 802-805.

202. SUZUKI T., OSHIMA K. Microwave applications to food in Japan : industrial applications. *J. Microw. Power*, 8, n° 2, 1973, pp. 149-159.

203. TAIEB B. SERMO Electronique : rècher-che et innovation tous azimuts. *Créez !* juin 1984, 3 p.

204. TAPE N. Application of microwave energy in food manufacture. *Can. Inst. Food Technol. J.*, 3, n° 2, 1970, pp. 39-43.

205. THOUREL B. Microwaves in the food and agricultural industries. *Elec-trón. Fís. Apl.*, 16, n° 3, 1973, pp. 537-539.

206. THUE M. Organisations internatio-nales dans le domaine des télécom-munications. *Tech. Ing. Electron.*, E40, juin 1976, 12 p.

207. TURPAIN A. 1- Un premier appareil récepteur de TSF, 2- Sur les ori-gines de la TSF. In : Compte-rendu de la 47ème session de l'Association française pour l'Avancement des Sciences, Bordeaux, 1923, pp. 300-311.

208. URBAIN W. Food processing with microwave energy - trends and potentials. In : Proceedings of the IMPI Symposium, Alberta, 1969, pp. 143-144.

209. URBAIN W. Some thoughts on the problems of microwave heating and food processing. *J. Microw. Power*, 4, n° 2, 1969, pp. 59-61.

210. VAN LOOCK W. Microwave power applications within the European Community. In : Proceedings of the IMPI Symposium, Chicago, 1985, pp. 78-89.

211. VAN MIEGHEM J. L'équation aux dérivées partielles des ondes électromagnétiques. In : Compte-rendu de la 56ème session de l'Association française pour l'Avancement des Sciences, Bruxelles, 1932, pp. 139-141.

212. VARIAN D. From "The inventor and the pilot", Russell and Sigurd Varian. *IEEE Trans. Microw. Theory Tech.*, MTT-32, n° 9, 1984, pp. 1248-1263.

213. VARIAN R., VARIAN S. A high-frequency oscillator and amplifier. *J. Appl. Phys.*, 10, May 1939, pp. 321-327.

214. VOSS W., TINGA W. A materials evaluation technique for microwave power processing. In : Proceedings of the IMPI Symposium, Stanford, 1967, Résumé p. 45.

215. WEISS E. Les merveilles des sciences et de l'industrie Paris, Ed. Hachette, 2 vol., 1926, 352 p. et 363 p.

216. WHITE J. How to plan, execute and extrapolate industrial feasibility studies : helpful suggestions for executive analysis. In : VOSS W., ISHII T. Transactions of the IMPI, 3, Edmonton, 1974.

217. WILTSE J. History of millimeter and submillimeter waves. *IEEE Trans. Microw. Theory Tech.*, MTT-32, n° 9, 1984, pp. 1118-1127.

218. WINCOTT P. Applications at 2450 MHz. *Trans. IMPI*, 2, 1973, Industrial applications of microwave energy.

219. YOUNG E. The new Penguin dictionary of electronics. Harmondsworth, Penguin Book, 1979, 618 p.

220. Agro-alimentaire, éclosion de procédés. *Usine Nouv.*, 1979, p. 72.

221. Applications industrielles des micro-ondes. Publication Thompson CSF, Div. Tubes Electroniques, réf. APH 6211, juin 1980, 13 p.

222. Ce que l'on pensait des applications industrielles des micro-ondes il y a une dizaine d'années. Texte intégral d'une étude réalisée par un cabinet spécialisé à la demande du Ministère de l'Industrie. *Econ. Progrès Electr.*, n° 11, sept.-oct. 1984, pp. 8-12.

223. Continuous microwave processing systems for the food processing industry : IMI Gigavac/Gigatron. *Food Technol. Aust.*, n° 8, 1977, p. 321.

224. FCC test procedure for microwave ovens submitted for type approval under part 18. Federal Communications Commission, *Bull. OCE 20*, Oct. 1974, 5 p.

225. Frequency allocations and radio treaty matters ; general rules and regulations, Sept. 1972. In : Rules and Regulations, Vol. II. (part 2). Federal Communications Commission, Washington, 1972, pp. 7-84.

226. General aspects of radio interference suppression. British Standard Code of Pratice, CP 1006, 1955, elaborated by a common committee of the Institution of Electrical Engineers and the British Standards Institution, ref. GR7, 60 p.

227. Guide for submission of information on microwave heating equipment required pursuant to 21 CFR 1002.10, April 1971. US Dep. Health, Educ. Welf., Bur. Radiol. Health, réf. OMB n° 57 R 0068, 7 p.

228. Industrial, scientific and medical equipment. In : Rules and regulations. Vol. 2, (part 18). Washington, Federal Communication Commission, 1972, pp. 123-139.

229. Information pertaining to FCC rules governing operation of industrial heating equipments. FE Bull. n° 1, revised. Washington, Federal Communications Commission, Field Eng. Bur., 1, 1963, 16 p.

230. Le chauffage par micro-ondes, homogène et instantané. *Usine Nouv.*, n° 2, 1979, pp. 78-86.

231. Les micro-ondes en grandes lignes. *Surgél.*, fév. 1985, pp. 43-45.

232. Les utilisations des micro-ondes s'étendent grâce à IMI. *Agra-Aliment.*, n° 710, pp. E et T, 1-3.

233. L'expérience microondes à votre service. Publication Microondes Energie, Association pour la valorisation des microondes, s.d., 2 p.

234. Microwave and R.F. heating in the food industry. In : Proceedings of a symposium held in London. 10, 1971. Leatherhead, BFMIRA, 1971, pp. 1-36. (C.D.I.U.P.A., n° 58144).

235. Microwave heating. Leicester, Publication Magnetronics ltd., 8 p.

236. Our business card... CALOREX. Danderyd, Publication Calorex AB, 1986, 6 p.

237. Perturbations radioélectriques, valeurs limites CISPR et recueil des valeurs limites nationales. Commission Electrotechnique Internationale, Publication CISPR, n° 9, 75 p.

238. Position paper on frequency allocations for industrial process heating applications. Bull. BNCE/-/77/1, British National Committee for Electro-heat, 1977, 16 p.

239. Premier complément à la publication 4 du CISPR (1ère édition 1967) : Spécification de l'appareillage de mesure du CISPR pour les fréquences comprises entre 300 MHz et 1000 MHz. Commission Electrotechnique Internationale, Publication CISPR, n° 4A, 1975, 3 p.

240. Proposed IMPI position paper on frequency allocation matter. *Microw. Energy Appl. Newsl.*, 9, n° 6, Nov.-Dec. 1976, pp. 14-16.

241. Questions sur les fours à microondes. *Aliment.*, 79, 1979, pp. 79-84.

242. Rayonnement Systèmes, pour tous vos problèmes concernant les applications industrielles et biomédicales des HF et des micro-ondes 2450 MHz. Publication Sté Rayonnements-Systèmes, sans réf. 1986, 13 p.

243. RAYTEK Générateurs micro-ondes. Notice G1, 1987, 2 p.

244. Reference data for radio engineers. 3rd ed. Indianapolis, Howard W. Sams and Co., Inc., ITT, 1979, 1332 p.

245. Règlement des radiocommunications. 2 vol. Genève, Ed. Union Internationale des Télécommunications, 1982, révision 1985.

246. Selectional list of British standards : electrical engineering. British Standards Institution, June 1975, 27 p., p. 23.

247. SERMO Electronique. Plaquette de présentation, s.d. 13 p.

248. Spécification de l'appareillage de mesure CISPR pour les fréquences comprises entre 300 et 1000 MHz. Commission Electrotechnique Internationale. Publication CISPR, n° 4, 1967, 37 p.

249. Specification for radio-interference measuring apparatus for the frequency range 0.015 MHz to 1000 MHz. British Standard 727, 1967, UDC 621.317.7 : 621.391.82. British Standards Institution, ref. GR 5, 25 p.

250. Survey of selected industrial applications of microwave energy. US Dep. Health, Educ. Welf., Bur. Radiol. Health, DEP/ 70-10, 1970, 67 p.

251. The wireless telegraphy (control of interference from radio-frequency heating apparatus) regulations 1971. British Telegraphs, Statutory Instruments, n° 1675, 1971, 8 p. (Règlement entré en vigueur le 21.10.1972).

252. Utilisation des rayonnements dans les industries alimentaires et agricoles. Paris, APRIA, 1974, 103 p. (APRIA, Paris, Symposium MATERA, 1974, Paris). (C.D.I.U.P.A., n° 85065).

Chapter 2

The Laws of Radiation

Microwave radiation obeys the laws of electromagnetism. Its interactions with matter and its applications will now be summarized. Our description of the processes of interaction (Part I, Chapter 3 and Part IV, Chapter 1) will rest on the concepts of field, permittivity, propagation medium, boundary conditions, reflection and transmission at interfaces, and dielectric losses. The design of industrial applicators (Part I, Chapter 4) relies on the theory of cavities, guided propagation, and antennas, as well as the idea of a stationary wave. The study of emissions, the effects on organisms, the safety limits, and the design of therapeutic applicators involve spherical and cylindrical waves, emissions from slots, open ends, and horns, and the notion of an irradiation zone.

Despite its relatively difficult mathematical formulation, this chapter should be regarded as a necessary introduction to what will follow.

2.1 Basic definitions

An electromagnetic wave is a propagation phenomenon that requires no material support and involves an electric and a magnetic field, each of which is a function of time.

The basic properties of electromagnetic radiation can be deduced from Coulomb's law. The magnitude of the electrostatic attraction between two charges q and q' of different sign, separated by a distance r is given by

$$F = \frac{1}{4\pi\epsilon_0}\frac{qq'}{r^2} \tag{2.1}$$

where the permittivity of free space (or air) is $\epsilon_0 = 8.854188 \times 10^{-12}$ F m^{-1} [L^{-3}M^{-1}T^{4}I^2]

In a medium other than free space, the permittivity ϵ' is larger than ϵ_0. The relative permittivity is defined by

$$\epsilon'_r \overset{\mathrm{d}}{=} \frac{\epsilon'}{\epsilon_0} \tag{2.2}$$

The charge q' perturbs the surrounding space by producing an electric field $\vec{E}$ (in V m^{-1} [L M T^{-3}I^{-1}]), proportional to the Coulomb force

$$\vec{F} \overset{\mathrm{d}}{=} q\vec{E} \tag{2.3}$$

We also define the electric induction $\vec{D}$ (in C m^2 [L^{-2}T I]), which is a function of the properties of the medium:

$$\vec{D} \overset{\mathrm{d}}{=} \epsilon'\vec{E} \tag{2.4}$$

A moving charge q' creates a magnetic field $\vec{H}$ (in A m^{-1} [L^{-1}I]) that corresponds to the magnetic induction $\vec{B}$ (in Tesla [M T^{-2}I^{-1}])

$$\vec{B} \overset{\mathrm{d}}{=} \mu\vec{H} \tag{2.5}$$

where μ [L M T^{-2}I^{-2}] is the magnetic permeability of the medium. In space (or air), $\mu_0 = 4\pi 10^{-7}$ Hm^{-1}

The charge q entering the magnetic field with the velocity $\vec{v}$ experiences a force

$$\vec{F} = q\left(\vec{v} \times \vec{B}\right) \tag{2.6}$$

When both bound and moving charges are present in the medium, they are described by charge densities ρ (in Cm^{-3} $[L^{-3}i]$) and $\vec{J}$ (in A m^{-3} $[L^{-3}I]$) Both are functions of space and time, and obey the continuity equation

$$\text{div}\vec{J} + \frac{\partial\rho}{\partial t} = 0. \tag{2.7}$$

The quantities $\vec{J}$ and $\vec{E}$ are related by Ohm's law

$$\vec{J} = \sigma\vec{E} \tag{2.8}$$

where σ (in $\Omega^{-1}m^{-1}$ $[L^{-3}M^{-1}T^3I^2]$) is the conductivity of the medium.

The fields are functions of time. In practice, they are generally *sinusoidal* functions with frequency f, period $T = 1/f$, and angular frequency (pulsatance) $\omega = 2\pi f$. The following complex notation can then be used:

$$E(x, y, z, t) = \text{Re}[E(x, y, z)e^{j\omega t}]$$

$$e^{j\omega t} = \cos\omega t + j\sin\omega t, \quad j = \sqrt{-1} \tag{2.9}$$

$$\frac{d}{dt}e^{j\omega t} = j\omega e^{j\omega t}$$

2.2 Maxwell's equations

The eight variables $\vec{E}, \vec{D}, \vec{H}\ \vec{B}, \rho, \vec{J}$ (or σ), ϵ', and μ are related by Maxwell's equations, which are fundamental to electromagnetism. These equations are:

$$\text{curl}\vec{H} = \vec{J} + \frac{\partial \vec{D}}{\partial t} = \sigma\vec{E} + \frac{\partial \vec{D}}{\partial t} \qquad (\text{Maxwell} - \text{Ampere law})$$

(2.10)

$$\text{curl}\vec{E} = -\frac{\partial \vec{B}}{\partial t} \qquad (\text{Maxwell} - \text{Faraday law}) \qquad (2.11)$$

$$\text{div}\vec{D} = \rho \qquad (2.12)$$

$$\text{div}\vec{B} = 0 \qquad (2.13)$$

For a sinusoidal wave, (2.10) and (2.11) become

$$\text{curl}\,\vec{H} = (\sigma + j\omega\epsilon')\vec{E} \qquad (2.14)$$

$$\text{curl}\,\vec{E} = -j\omega\mu\vec{H} \qquad (2.15)$$

The asymmetry between electric and magnetic terms is due to the fact that there are no magnetic charges.

Combining (2.10) and (2.13), we obtain

$$\text{div}\left(\vec{J} + \frac{\partial \vec{D}}{\partial t}\right) = 0 \qquad (2.16)$$

where the total current $\vec{j} + \partial\vec{D}/\partial t$ is the sum of the conduction current and the *displacement current* that is present only in an alternating field, and propagates without material support. The function of an antenna is to transform the conduction current into a displacement current and vice versa.

2.3 Propagation equation

Consider a medium containing no charges ($\rho = 0$). The following vector identity will be used for the electric field:

$$\operatorname{curl} \operatorname{curl} \vec{A} = \operatorname{grad} \operatorname{div} \vec{A} - \Delta \vec{A} \qquad (2.17)$$

In view of (2.4) and (2.12), we have

$$\operatorname{curl} \operatorname{curl} \vec{E} = \operatorname{grad} \frac{\rho}{\epsilon'} - \Delta \vec{E} = -\Delta \vec{E} \qquad (2.18)$$

According to (2.11)

$$\operatorname{curl} \operatorname{curl} \vec{E} = -\operatorname{curl} \frac{\partial \vec{B}}{\partial t}$$

and, using (2.5) and (2.10),

$$\operatorname{curl} \operatorname{curl} \vec{E} = -\frac{\partial}{\partial t} \operatorname{curl} \mu \vec{H} = -\mu \frac{\partial}{\partial t} \left(\vec{J} + \frac{\partial \vec{D}}{\partial t} \right)$$

$$\operatorname{curl} \operatorname{curl} \vec{E} = -\mu \frac{\partial}{\partial t} \left(\sigma \vec{E} + \epsilon' \frac{\partial \vec{E}}{\partial t} \right) \qquad (2.19)$$

Finally, comparing (2.18) and (2.19), we obtain the propagation equation for $\vec{E}$ (the equation for $\vec{H}$ is obtained similarly):

$$\Delta \vec{E} = \mu \sigma \frac{\partial \vec{E}}{\partial t} + \mu \epsilon' \frac{\partial^2 \vec{E}}{\partial t^2} \qquad (2.20)$$

$$\Delta \vec{H} = \mu \sigma \frac{\partial \vec{E}}{\partial t} + \mu \epsilon' \frac{\partial^2 \vec{H}}{\partial t^2} \qquad (2.21)$$

If there is no conduction current, we have ($\sigma = 0$)

$$\Delta \vec{E} - \mu \epsilon' \frac{\partial^2 \vec{E}}{\partial t^2} = 0 \qquad (2.22)$$

$$\Delta \vec{H} - \mu \epsilon' \frac{\partial^2 \vec{H}}{\partial t^2} = 0 \qquad (2.23)$$

In a sinusoidal field, (2.20) to (2.23) become

$$\Delta \vec{E} - j\omega\mu(\sigma + j\omega\epsilon')\vec{E} = 0 \qquad (2.24)$$

$$\Delta \vec{H} - j\omega\mu(\sigma + j\omega\epsilon')\vec{H} = 0 \qquad (2.25)$$

$$\Delta \vec{E} + \omega^2 \mu\epsilon' \vec{E} = 0 \qquad (2.26)$$

$$\Delta \vec{H} + \omega^2 \mu\epsilon' \vec{H} = 0 \qquad (2.27)$$

These equations describe wave propagation and are the analogues of the Helmholtz law for mechanical vibrations. The propagation constant is defined by

$$\gamma^2 \stackrel{\mathrm{d}}{=} j\omega\mu(\sigma + j\omega\epsilon') \qquad (2.28)$$

and the wave number by

$$k \stackrel{\mathrm{d}}{=} \omega\sqrt{\mu\epsilon'} \qquad (2.29)$$

If $\sigma = 0$, then $\gamma = jk$. Equation (2.28) may be expressed as:

$$\gamma^2 = -\omega^2\mu\left(\epsilon' - j\frac{\sigma}{\omega}\right) \qquad (2.30)$$

$$\gamma^2 = -\omega^2\mu(\epsilon' - j\epsilon'') \qquad (2.31)$$

where

$$\epsilon'' \stackrel{\mathrm{d}}{=} \frac{\sigma}{\omega} \qquad (2.32)$$

ϵ'' is the loss factor of the medium defined in terms of the complex permittivity

$$\epsilon \stackrel{\mathrm{d}}{=} \epsilon' - j\epsilon'' \qquad (2.33)$$

and the loss angle δ

$$\tan \delta \overset{d}{=} \frac{\epsilon''}{\epsilon'} = \frac{\sigma}{\omega \epsilon'} \qquad (2.34)$$

and γ is a complex quantity given by

$$\gamma^2 = -\omega^2 \mu \epsilon = -\omega^2 \mu \epsilon'(1 - j \tan \delta) \qquad (2.35)$$

$$\gamma = \alpha + j\beta \qquad (2.36)$$

which enables us to rewrite (2.28) in the form of a dispersion relation:

$$\alpha^2 - \beta^2 + \omega^2 \mu \epsilon' + j(2\alpha\beta - \omega\mu\sigma) = 0 \qquad (2.37)$$

Separating real and imaginary parts, we have

$$\alpha = \omega \sqrt{\frac{\mu \epsilon'}{2} \left(\sqrt{1 + \tan^2 \delta} - 1 \right)} \qquad (2.38)$$

$$\beta = \omega \sqrt{\frac{\mu \epsilon'}{2} \left(\sqrt{1 + \tan^2 \delta} + 1 \right)} \qquad (2.39)$$

In a medium of low conductivity ($\tan \delta \ll 1$), these equations become

$$\alpha = \frac{\omega \sqrt{\mu \epsilon'}}{2} \tan \delta \qquad (2.40)$$

$$\beta = \omega \sqrt{\mu \epsilon'} = k \qquad (2.41)$$

where α and β are, respectively, the attenuation coefficient and the phase angle. Since wave velocity is given by

$$v = \lambda f = \frac{1}{\sqrt{\mu \epsilon'}} \qquad (2.42)$$

we obtain

$$k = \frac{\omega}{v} = \frac{2\pi}{\lambda} = \beta \qquad (2.43)$$

where λ is the wavelength.

The phase plane of the wave is the plane perpendicular to its direction of propagation. The polarization of a wave is defined by the orientation of the vectors $\vec{E}$ and $\vec{H}$ on this plane. For a fixed orientation, the polarization is said to be rectilinear (horizontal or vertical, depending on the position of $\vec{E}$). However, in general, $\vec{E}$ and $\vec{H}$ rotate together in the phase plane, each describing a circle or, more generally, an ellipse. The polarization is then circular or elliptic.

2.4 Plane wave

A wave propagating in free space, a long distance from the source, is defined as a plane or *Transverse Electromagnetic* (TEM) wave when $\vec{E}$ and $\vec{H}$ lie in the phase plane, perpendicular to the direction of propagation (the z direction). The components E_z and H_z are both equal to zero, and

$$\text{curl } \vec{E} = -\frac{\partial E_y}{\partial z}\hat{x} + \frac{\partial E_x}{\partial z}\hat{y} \tag{2.44}$$

$$\text{curl } \vec{H} = -\frac{\partial H_y}{\partial z}\hat{x} + \frac{\partial H_x}{\partial z}\hat{y} \tag{2.45}$$

Maxwell's equations then take the form

$$-\frac{\partial H_y}{\partial z} = \sigma E_x + \epsilon'\frac{\partial E_x}{\partial t}, \qquad \frac{\partial H_x}{\partial z} = \sigma E_y + \epsilon'\frac{\partial E_y}{\partial t}$$

$$\frac{\partial E_y}{\partial z} = \mu\frac{\partial H_x}{\partial t}, \qquad \frac{\partial E_x}{\partial z} = -\mu\frac{\partial H_y}{\partial t} \tag{2.46}$$

and the propagation equation becomes

$$\frac{\partial^2 E_x}{\partial z^2} = \mu\sigma\frac{\partial E_x}{\partial t} + \mu\epsilon'\frac{\partial^2 E_x}{\partial t^2} \tag{2.47}$$

There are similar relations for E_y, H_x and H_y.

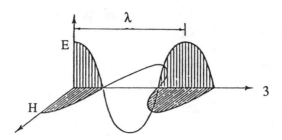

Figure 2.1 Propagation of a plane wave [2].

A possible solution is

$$E = E_0 \left[\Gamma e^{(\alpha+j\beta)z} + e^{-(\alpha+j\beta)z} \right] e^{j\omega t} \qquad (2.48)$$

$$H = H_0 \left[\Gamma e^{(\alpha+j\beta)z} + e^{-(\alpha+j\beta)z} \right] e^{j\omega t}$$

where Γ is the reflection coefficient, the wave being the sum of a propagating and a reflected wave. In an infinite medium, there is no reflection, and (2.48) becomes

$$E = E_0 e^{-\alpha z} e^{j(\omega t - \beta z)} \qquad (2.49)$$

$$H = H_0 e^{-\alpha z} e^{j(\omega t - \beta z)}$$

The propagation equation therefore becomes

$$(\alpha + j\beta)^2 E_x = j\omega\mu\sigma E_x - \omega^2 \mu\epsilon' E_x$$

which is equivalent to (2.37).

Maxwell's equations may now be written as

$$(\alpha+j\beta)H_y = (\sigma+j\omega\epsilon')E_x, \qquad -(\alpha+j\beta)E_y = j\omega\mu H_x \quad (2.50)$$

$$-(\alpha + j\beta)H_x = (\sigma + j\omega\epsilon')E_y, \qquad (\alpha + j\beta)E_x = j\omega\mu H_y$$

It can readily be shown that

$$E_x H_x + E_y H_y = 0 \qquad (2.51)$$

This shows that $\vec{E}$ and $\vec{H}$ are mutually perpendicular (Fig. (2.1)).

The ratio E/H, which is equal to E_0/H_0, is the wave impedance η (in Ω [$L^2 M T^{-3} I^{-2}$]):

$$\eta = \sqrt{\frac{E_x^2 + E_y^2}{H_x^2 + H_y^2}} \tag{2.52}$$

Squaring (2.50), we have

$$\eta = \frac{\alpha + j\beta}{\sigma + j\omega\epsilon'} \tag{2.53}$$

whereby (2.28),

$$\eta = \sqrt{\frac{j\omega\mu}{\sigma + j\omega\epsilon'}} \tag{2.54}$$

When the conductivity is zero,

$$\eta = \sqrt{\frac{\mu}{\epsilon'}} \tag{2.55}$$

so that, in the case of air,

$$\eta_0 = \sqrt{\frac{\mu_0}{\epsilon_0}} = 120\,\pi \simeq 377\,\Omega \tag{2.56}$$

2.5 Spherical and cylindrical waves

The propagation equation has a number of solutions, the simplest of which is the homogeneous plane wave (Section 2.4), but such waves exist only in theoretical situations or as local approximations.

Equations (2.24) and (2.25) may be written in spherical coordinates (r, θ, φ) [44], [26] and the solutions expressed in terms of Bessel, Neumann, and spherical Hankel functions. Thus, in a lossless medium ($\sigma = 0$)

$$I_n(\beta r) \overset{\mathrm{d}}{=} \sqrt{\frac{\pi}{2\beta r}} J_{n+1/2}(\beta r) \tag{2.57}$$

$$\mathcal{N}_n(\beta r) \overset{\mathrm{d}}{=} \sqrt{\frac{\pi}{2\beta r}} N_{n+\frac{1}{2}}(\beta r) \tag{2.58}$$

$$\mathcal{H}_n^{(1)}(\beta r) \overset{\mathrm{d}}{=} I_n(\beta r) + j\mathcal{N}_n(\beta r) \tag{2.59}$$

$$\mathcal{H}_n^{(2)}(\beta r) \overset{\mathrm{d}}{=} I_n(\beta r) - j\mathcal{N}_n(\beta r) \tag{2.60}$$

where $J_{n+\frac{1}{2}}$ and $N_{n+\frac{1}{2}}$ are Bessel and Neumann functions of order $n + \frac{1}{2}$. At long distances from the source,

$$\mathcal{H}_n^{(1)}(\beta r) \sim (-j)^{n+1} \frac{e^{j\beta r}}{\beta r} \tag{2.61}$$

$$\mathcal{H}_n^{(2)}(\beta r) \sim j^{n+1} \frac{e^{j\beta r}}{\beta r} \tag{2.62}$$

When $n = 0$, (2.62) describes a wave propagating away from the source. The factor r^{-1} is typical for spherical waves [44], [57].

When (2.24) and (2.25) are written in cylindrical coordinates (r, θ, z) [57], [26], the solutions are combinations of Hankel functions of order one or two. By analogy with (2.48), for $n = 0$ we have

$$E_z = E_0 \left[\Gamma H_0^{(1)}(\beta r) + H_0^{(2)}(\beta r) \right] e^{j\omega t} \tag{2.63}$$

where

$$H_0^{(1)}(\beta r) \overset{\mathrm{d}}{=} J_0(\beta r) + j N_0(\beta r) \tag{2.64}$$

$$H_0^{(2)}(\beta r) \overset{\mathrm{d}}{=} J_0(\beta r) + j N_0(\beta r) \tag{2.65}$$

In an infinite medium the reflected wave tends to zero, and only the propagating term remains:

$$\vec{E} = E_0 H_0^{(2)}(\beta r)\hat{z} \qquad (2.66)$$

$$\vec{H} = \frac{E_0}{j\omega\mu}\frac{d}{dr}H_0^{(2)}(\beta r)\hat{\varphi} \qquad (2.67)$$

At long distances from the source $(\beta r \to \infty)$

$$H_0^{(2)}(\beta r) \sim j^{\frac{1}{2}}\sqrt{\frac{2}{\pi\beta r}}e^{-j\beta r} \qquad (2.68)$$

where the factor $r^{-1/2}$ is typical for cylindrical waves. More generally, the field at long distances varies as

$$j^\nu \frac{e^{j\beta r}}{(\beta r)^\nu} \qquad (2.69)$$

where $\nu = 1$ for a spherical wave, $\nu = \frac{1}{2}$ for a cylindrical wave, and $\nu = 0$ for a plane homogeneous wave [20].

2.6 Propagation media

The two extreme types of media are the perfect conductor and the perfect dielectric.

The conductivity σ of a perfect conductor is infinite. Since the conduction current $\vec{J}$ is finite, Ohm's law (2.8) requires that $\vec{E}$ is zero, which implies that $\vec{H}$ and $\vec{J}$ are also zero. The wave cannot propagate, and the current exists only in the form of a surface current $\vec{J_s}$.

The conductivity of a real conductor is high, that is

$$\sigma \gg \omega\epsilon' \qquad (2.70)$$

but not infinite, and the displacement current is negligible in comparison to the conduction current; namely

$$\sigma E \gg \frac{\partial D}{\partial t} \tag{2.71}$$

The dispersion relation (2.38) - (2.39) gives

$$\alpha \simeq \beta \simeq \sqrt{\frac{\omega \mu \sigma}{2}} \tag{2.72}$$

and the wave propagates with a velocity

$$v \simeq \frac{\omega}{\beta} = \sqrt{\frac{2\omega}{\mu \sigma}} \tag{2.73}$$

The penetration depth is defined by

$$d \stackrel{\mathrm{d}}{=} \frac{1}{\alpha} = \sqrt{\frac{2}{\omega \mu \sigma}} \tag{2.74}$$

When $z = d$, the field amplitude is reduced by the factor of e^{-1} as compared with the initial values E_0, H_0. In a good conductor, d is generally less than 1 μm, indicating that, in a metal, the wave is confined to the surface layer (this is the *skin effect*).

The conductivity and dielectric loss of a perfect dielectric are both zero, meaning

$$\sigma = 0$$

$$\epsilon'' = 0, \quad \epsilon \text{ is real}$$

The dispersion relation now gives

$$\alpha = 0$$

$$\beta = \omega \sqrt{\mu \epsilon'} \tag{2.75}$$

and the wave propagates without attenuation with velocity

$$v = \frac{1}{\sqrt{\mu \epsilon'}} \qquad (2.76)$$

and a wavelength

$$\lambda = \frac{v}{f} = \frac{1}{f\sqrt{\mu \epsilon'}} \qquad (2.77)$$

In free space, the propagation velocity is that of light:

$$v_0 = c = \frac{1}{\sqrt{\mu_0 \epsilon_0}} \simeq 3 \times 10^8 \,\mathrm{m/s} \qquad (2.78)$$

In a dielectric,

$$v = \frac{c}{\sqrt{\mu_r \epsilon'_r}} \qquad (2.79)$$

The optical refractive index of the medium is defined by

$$n = \frac{c}{v} = \sqrt{\mu_r \epsilon'_r} \qquad (2.80)$$

A real dielectric has a very small but finite conductivity, which implies that it contains very few moving charges. Its dielectric losses can be represented by the dielectric conductivity σ_d

$$\sigma_d = \omega \epsilon'' = \omega \epsilon' \tan \delta \qquad (2.81)$$

hence the propagation equations (2.24)–(2.25) with

$$\gamma^2 = j\omega\mu(\sigma_d + j\omega\epsilon') \qquad (2.82)$$

The permeability of common dielectric materials is the same as that of free space: $\mu = \mu_0$.

The concept of penetration depth remains valid. From (2.38),

$$d \stackrel{\mathrm{d}}{=} \frac{1}{\alpha} = \frac{1}{\omega} \left[\frac{\mu \epsilon'}{2} \left(\sqrt{1 + \tan^2 \delta} - 1 \right) \right]^{-\frac{1}{2}} \qquad (2.83)$$

For low-loss materials, (2.38) and (2.39) may be simplified, since

$$\sqrt{1 + x^2} \simeq 1 + \frac{x^2}{2}$$

Hence

$$\alpha \simeq \frac{\omega}{2}\sqrt{\mu\epsilon'}\left(\tan\delta + \frac{\sigma}{\omega\epsilon'}\right) \tag{2.84}$$

$$\alpha \simeq \frac{1}{2}\sqrt{\frac{\mu}{\epsilon'}}\left(\sigma_d + \sigma\right)$$

$$\alpha \simeq \frac{\eta}{2}\left(\sigma_d + \sigma\right) \tag{2.85}$$

$$\beta \simeq \omega\sqrt{\mu\epsilon'} \tag{2.86}$$

When conduction losses dominate,

$$\frac{\sigma}{\omega\epsilon'} \gg \tan\delta, \qquad \alpha \simeq \frac{\sigma}{2}\sqrt{\mu\epsilon'} \tag{2.87}$$

When the dielectric losses dominate,

$$\frac{\sigma}{\omega\epsilon'} \ll \tan\delta, \qquad \alpha \simeq \omega\frac{\tan\delta}{2}\sqrt{\mu\epsilon'} \tag{2.88}$$

The wavelength in a medium is always shorter than in free space:

$$\lambda = \frac{2\pi}{k} \simeq \frac{2\pi}{\omega\sqrt{\mu\epsilon'}} = \frac{1}{f\sqrt{\mu\epsilon'}} = \frac{\lambda_0}{\sqrt{\mu_r\epsilon_r'}} \tag{2.89}$$

The transmitted power is given by

$$P(z) = P_0 e^{-2\alpha z} \tag{2.90}$$

Energy is deposited in the medium by mechanisms that will be described in Part I, Chapter 3.

2.7 Boundary conditions

An abrupt change in the medium is described by a discontinuity in the parameters ϵ', μ, σ. The normal component of $\vec{B}$ and the tangential component of $\vec{E}$ are continuous across the interface, and so are the normal component of $\vec{D}$ and the tangential component of $\vec{H}$, namely

$$\left(\vec{B}_2 - \vec{B}_1\right) . \hat{n} = 0, \qquad \left(\vec{D}_2 - \vec{D}_1\right) . \hat{n} = \rho_s \qquad (2.91)$$

$$\left(\vec{E}_2 - \vec{E}_1\right) \times \hat{n} = 0, \qquad \left(\vec{H}_2 - \vec{H}_1\right) \times \hat{n} = -\vec{J}_s$$

where $\hat{n}$ is a unit vector normal to the interface, and ρ_s and $\vec{J}_s$ are, respectively, the charge density and the current density on the discontinuity.

When medium 1 is a dielectric (ϵ', μ) and medium 2 is a metal (σ), and since we know that the fields do not propagate in a metal, we have

$$\vec{H} . \hat{n} = 0, \qquad \vec{E} . \hat{n} = \frac{\rho_s}{\epsilon'} \qquad (2.92)$$

$$\vec{E} \times \hat{n} = \vec{0}, \qquad \vec{H} \times \hat{n} = -\vec{J}_s$$

which show that $\vec{E}$ is normal to the metallic surface and that $\vec{H}$ is tangential to it. The magnetic field and the current flowing at the interface have the same absolute value.

2.8 Reflection and transmission

Consider two lossless dielectric media, separated by an interface. When a wave propagating in medium 1 reaches medium 2 at normal incidence, it is partially transmitted and partially reflected. We shall use i, r, and t to label incident, reflected, and transmitted waves, respectively. The tangential components of the field given by (2.52) and (2.91) are now related as follows:

$$E_i = \eta_1 H_i \qquad (2.93)$$

$$E_r = -\eta_1 H_r \qquad (2.94)$$

$$E_t = \eta_2 H_t \qquad (2.95)$$

$$E_i + E_r = E_t \qquad (2.96)$$

$$H_i + H_r = H_t \qquad (2.97)$$

Combining these expressions, we have

$$H_i + H_r = \frac{1}{\eta_1}(E_i - E_r) = H_t = \frac{1}{\eta_2}(E_i + E_r) \qquad (2.98)$$

where

$$\frac{E_r}{E_i} = \frac{\eta_2 - \eta_1}{\eta_2 + \eta_1} \overset{\text{d}}{=} \Gamma \qquad (2.99)$$

$$\frac{E_t}{E_i} = \frac{2\eta_2}{\eta_2 + \eta_1} \overset{\text{d}}{=} T \qquad (2.100)$$

$$\frac{H_r}{H_i} = -\frac{E_r}{E_i} = \frac{\eta_i - \eta_2}{\eta_1 + \eta_2} = -\Gamma \qquad (2.101)$$

$$\frac{H_t}{H_i} = \frac{\eta_1}{\eta_2}\frac{E_t}{E_i} = \frac{2\eta_1}{\eta_1 + \eta_2} = \frac{\eta_1}{\eta_2}T \qquad (2.102)$$

Since the permeability of commonly used dielectrics is not significantly different from that of free space (i.e., $\mu_1 \simeq \mu_2 \simeq \mu_0$), the above equations may be rewritten in terms of permittivity alone:

$$\frac{E_r}{E_i} = \frac{\sqrt{\epsilon'_1} - \sqrt{\epsilon'_2}}{\sqrt{\epsilon'_1} + \sqrt{\epsilon'_2}} \qquad (2.103)$$

$$\frac{H_r}{H_i} = \frac{\sqrt{\epsilon'_2} - \sqrt{\epsilon'_1}}{\sqrt{\epsilon'_1} + \sqrt{\epsilon'_2}} \qquad (2.104)$$

$$\frac{E_t}{E_i} = \frac{2\sqrt{\epsilon'_1}}{\sqrt{\epsilon'_1} + \sqrt{\epsilon'_2}} \qquad (2.105)$$

$$\frac{H_t}{H_i} = \frac{2\sqrt{\epsilon'_2}}{\sqrt{\epsilon'_1} + \sqrt{\epsilon'_2}} \qquad (2.106)$$

In the case of oblique incidence from medium 1 to medium 2, we have the Snell-Descartes law

$$\frac{\sin \theta_1}{\sin \theta_2} = \sqrt{\frac{\epsilon'_2}{\epsilon'_1}} \qquad (2.107)$$

$$\theta_3 = \theta_1 \qquad (2.108)$$

where θ_1 is the angle of incidence and θ_2, θ_3 the angles of transmission and reflection (Fig. 2.2).

Two special cases will now be considered (i.e., *horizontal polarization* (HP) and *vertical polarization* (VP)) depending on whether the electric or magnetic field is parallel to the interface (Fig.2.2).

It can be shown [26] that the reflected and incident fields are related by

$$\frac{E_r}{E_i} = \frac{\cos \theta_1 - \sqrt{(\epsilon'_2/\epsilon'_1) - \sin^2 \theta_1}}{\cos \theta_1 + \sqrt{(\epsilon'_2/\epsilon'_1) - \sin^2 \theta_1}} \quad \text{(HP)} \qquad (2.109)$$

$$\frac{E_r}{E_i} = \frac{(\epsilon'_2/\epsilon'_1) \cos \theta_1 - \sqrt{(\epsilon'_2/\epsilon'_1) - \sin^2 \theta_1}}{(\epsilon'_2/\epsilon'_1) \cos \theta_1 + \sqrt{(\epsilon'_2/\epsilon'_1) - \sin^2 \theta_1}} \quad \text{(VP)} \qquad (2.110)$$

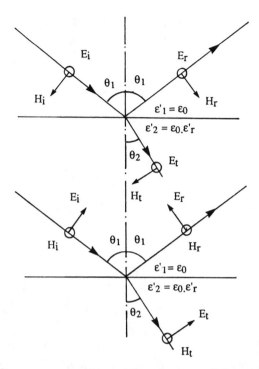

Figure 2.2 Plane wave incident obliquely on a dielectric interface; $\odot$ represents a vector perpendicular to the plane of the paper and pointing towards the reader.

$$\frac{E_t}{E_i} = \frac{2\cos\theta_1}{\cos\theta_1 + \sqrt{(\epsilon_2'/\epsilon_1') - \sin^2\theta_1}} \quad \text{(HP)} \qquad (2.111)$$

$$\frac{E_t}{E_i} = \frac{2\sqrt{\epsilon_2'/\epsilon_1'}\cos\theta_1}{(\epsilon_2'/\epsilon_1')\cos\theta_1 + \sqrt{(\epsilon_2'/\epsilon_1')\cos\theta_1 - \sin^2\theta_1}} \quad \text{(VP)}$$

$$(2.112)$$

Vertical polarization exhibits an interesting feature: There is no reflection when θ_1 is equal to the so-called Brewster angle, which may be obtained by equating to zero to the numerator of (2.110):

$$\theta_{1B} = \tan^{-1}\sqrt{\frac{\epsilon_2'}{\epsilon_1'}} \qquad (2.113)$$

At normal incidence, reflection can be minimized by interfacing the two media with an intermediate layer of thickness $\lambda/4$ with $\lambda = 1/f\sqrt{\epsilon'}$ and impedance $\eta = \sqrt{\mu/\epsilon'}$. The total reflection coefficient

$$\Gamma = \Gamma_1\Gamma_2 = \frac{\eta - \eta_1}{\eta + \eta_1}\frac{\eta_2 - \eta}{\eta_2 + \eta}$$

must then be a minimum, which is obtained for

$$\eta = \sqrt{\eta_1\eta_2} \tag{2.114}$$

2.9 Guided propagation

Two types of waveguide can be used to channel and transmit microwaves, namely, guides with simply connected cross sections (rectangular, circular, elliptic, and so on), or multiply connected (coaxial) guides with dimensions depending on the wavelength of the propagating field. Multiple reflections of the propagating wave at the walls of the waveguide produce a certain distribution of fields inside the guide and of conduction currents on the walls, giving rise to a *mode of guided propagation*.

The plane wave (Section 2.4) corresponds to the TEM mode. It can be shown that this mode will propagate in the coaxial guide, but cannot propagate in the simply connected (cavity) guide. In the latter, it is either the electric or the magnetic field that is perpendicular to the direction of propagation. Two modes, namely, *transverse electric* (TE) and *transverse magnetic* (TM) are found to propagate. In these modes, a single field component (H_z and E_z, respectively) is parallel to the waveguide axis. This component satisfies the equation of propagation

$$\frac{\partial^2 V_z}{\partial x^2} + \frac{\partial^2 V_z}{\partial y^2} + \left(\beta_0^2 - \beta_g^2\right) V_z = 0 \tag{2.115}$$

where V_z represents E_z in TM and H_z in TE, β_0 the propagation constant in free space, and β_g, the propagation constant
in the waveguide in the z direction. We now put

$$\beta_0^2 - \beta_g^2 \overset{\text{d}}{=} \beta_c^2 \tag{2.116}$$

where β_c is the propagation constant in the xOy plane. The
corresponding wavelengths are defined by

$$\lambda_0 = \frac{c}{f} = \frac{2\pi}{\beta_0}, \qquad \lambda_g = \frac{2\pi}{\beta_g}, \qquad \lambda_c = \frac{2\pi}{\beta_c} \tag{2.117}$$

and λ_c is the cut-off wavelength. There are three possible cases.
When $\lambda_0 < \lambda_c$,

$$\lambda_g = \frac{\lambda_0 \lambda_c}{\sqrt{\lambda_c^2 - \lambda_0^2}} > \lambda_0 \tag{2.118}$$

The wave propagates without attenuation (except for resistive
losses in the wall) with phase velocity

$$v_g = \frac{\omega}{\beta_g} = f\lambda_g > c \tag{2.119}$$

When $\lambda_0 = \lambda_c$, the waveguide is cut-off, λ_g is infinite, and
propagation occurs only on the cross section with a velocity

$$\frac{1}{\sqrt{\mu\epsilon'}}$$

Finally, when $\lambda_0 > \lambda_c$, the quantities λ_g, β_g, and v_g are imaginary, and the wave is said to be *evanescent* and suffers rapid
exponential attenuation. The waveguide cut-off frequency is
$f_c = c/\lambda_c$, and propagation is not possible for $f_0 < f_c$.

Applying the boundary conditions at the wall of the waveguide to the propagation equation (2.115) for a rectangular
waveguide of cross section $a \times b$ in the Oxy plane, we have
respectively.

TE:

$$\frac{\partial H_z}{\partial x}(0,y) = \frac{\partial H_z}{\partial x}(a,y) = \frac{\partial H_z}{\partial y}(x,0) = \frac{\partial H_z}{\partial y}(x,b) = 0$$

(2.120)

TM:

$$E_z(0,y) = E_z(a,y) = E_z(x,0) = E_z(x,b) = 0 \qquad (2.121)$$

The solutions have a discrete form, in which each pair of integers (m,n) generates a solution, TE_{mn} or TM_{mn}, yielding the following expressions for the fields:

TE_{mn} modes:

$$E_x = j\frac{\omega\mu}{\beta_c^2}\frac{n\pi}{b}H_0\cos\frac{m\pi x}{a}\sin\frac{n\pi y}{b} \qquad (2.122)$$

$$E_y = -j\frac{\omega\mu}{\beta_c^2}\frac{m\pi}{a}H_0\sin\frac{m\pi x}{a}\cos\frac{n\pi y}{b} \qquad (2.123)$$

$$E_z = 0 \qquad (2.124)$$

$$H_x = j\frac{\beta_g}{\beta_c^2}\frac{m\pi}{a}H_0\sin\frac{m\pi x}{a}\cos\frac{n\pi y}{b} \qquad (2.125)$$

$$H_y = j\frac{\beta_g}{\beta_c^2}\frac{n\pi}{b}H_0\cos\frac{m\pi x}{a}\sin\frac{n\pi y}{b} \qquad (2.126)$$

$$H_z = H_0\cos\frac{m\pi x}{a}\cos\frac{n\pi y}{b} \qquad (2.127)$$

TM_{mn} modes:

$$E_x = -j\frac{\beta_g}{\beta_c^2}\frac{m\pi}{a}E_0\cos\frac{m\pi x}{a}\sin\frac{n\pi y}{b} \qquad (2.128)$$

$$E_y = -j\frac{\beta_g}{\beta_c^2}\frac{n\pi}{b}E_0\sin\frac{m\pi x}{a}\cos\frac{n\pi y}{b} \qquad (2.129)$$

$$E_z = E_0 \sin \frac{m\pi x}{a} \sin \frac{n\pi y}{b} \qquad (2.130)$$

$$H_x = j\frac{\omega \epsilon'}{\beta_c^2} \frac{n\pi}{b} E_0 \sin \frac{m\pi x}{a} \cos \frac{n\pi y}{b} \qquad (2.131)$$

$$H_y = -j\frac{\omega \epsilon'}{\beta_c^2} \frac{m\pi}{a} E_0 \cos \frac{m\pi x}{a} \sin \frac{n\pi y}{b} \qquad (2.132)$$

$$H_z = 0 \qquad (2.133)$$

The general expressions for the fields are as follows:

$$\bar{E}(x,y,z) = (E_x \hat{x} + E_y \hat{y} + E_z \hat{z})e^{-j\beta_g z} \qquad (2.134)$$

$$\bar{H}(x,y,z) = (H_x \hat{x} + H_y \hat{y} + H_z \hat{z})e^{-j\beta_g z} \qquad (2.135)$$

It can be shown that, for a (m,n) mode,

$$\beta_c = \frac{2\pi}{\lambda_c} \sqrt{\left(\frac{m\pi}{a}\right)^2 + \left(\frac{n\pi}{b}\right)^2} \qquad (2.136)$$

from which

$$\lambda_c = \frac{2}{\sqrt{\left(\frac{m}{a}\right)^2 + \left(\frac{n}{b}\right)^2}} \qquad (2.137)$$

It must be noted that the TM_{0n} and TM_{m0} cannot exist unlike their equivalent TE modes.

The fundamental mode of propagation is defined as having the shortest cut-off wavelength. This is the TE_{10} mode for which $\lambda_c = 2a$.

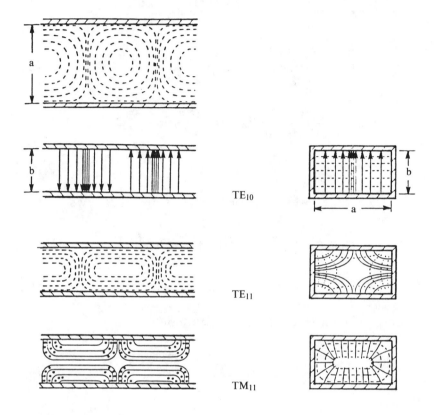

Figure 2.3 Electric and magnetic field lines in a rectangular guide for the first modes [26]: − electric field, - - - magnetic field.

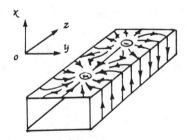

Figure 2.4 Surface currents on guide walls for the TE_{10} mode [1].

2.10 Stationary wave

The propagation of a wave along a waveguide is more efficient when reflections toward the source are negligible. The extreme case of total mismatch, or short circuit, occurs when a metal plate closes the waveguide. A reflected wave then appears and is superimposed on the incident wave. This results in a stationary wave in which zeros and maxima alternate at intervals of $\lambda_g/4$ (Fig. 2.4).

In the general case, the load is an impedance in which the resistive part represents the absorption of energy. The reflected-wave intensity is lower than the incident-wave intensity. Zero values are then no longer observed on the stationary wave, and instead, minima and maxima alternate at the same intervals as before. Moreover, the stationary pattern is shifted along the waveguide, depending on the capacitive or inductive characteristics of the impedance (Fig. 2.5).

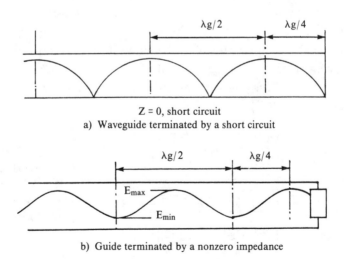

Z = 0, short circuit
a) Waveguide terminated by a short circuit

b) Guide terminated by a nonzero impedance

Figure 2.5 Stationary wave: (a) waveguide terminated by a short circuit, (b) waveguide terminated by a non-zero load.

The *standing wave ratio* (SWR) is given by

$$\tau = \frac{1 + \Gamma}{1 - \Gamma} = \frac{E_{\max}}{E_{\min}} \qquad (2.138)$$

where Γ (2.19) ranges between 0 and 1, and τ between 1 and infinity. Impedance matching techniques (Part I, Section 4.3.2.1) can be used to bring τ as near as possible to unity.

2.11 Electromagnetic cavities

2.11.1 Resonant modes

A cavity is defined as a volume enclosed by a conducting wall. Depending on their dimensions as compared to the wavelength, cavities are said to be *resonant* or *oversized*, respectively.

The electromagnetic energy trapped in a cavity is reflected by its walls and takes the form of stationary waves. In waveguides, the possible frequencies constitute a discrete series, or characteristic frequencies of the cavity. The corresponding special configurations, TE_{mnp} and TM_{mnp}, are the cavity modes. The cavity behaves as a filter that transmits only certain frequencies (Fig. 2.6).

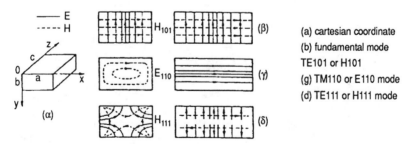

Figure 2.6 Rectangular cavities: (α) cartesian coordinates, (β) fundamental TE_{101} or H_{101} mode, (γ) TM_{110} or E_{110} mode, (δ) TE_{111} or H_{111} mode

Proceeding by analogy with (2.122)–(2.133), the solutions for a rectangular cavity of dimensions $a \times b \times L$ (Ox, Oy, Oz) can be shown to be as follows:

TE_{mnp}:

$$E_x = j\frac{\omega\mu}{\beta_r^2}\frac{n\pi}{b}H_0 \cos\frac{m\pi x}{a} \sin\frac{n\pi y}{b} \sin\frac{p\pi z}{L} \qquad (2.139)$$

$$E_y = -j\frac{\omega\mu}{\beta_r^2}\frac{m\pi}{a}H_0 \sin\frac{m\pi x}{a} \cos\frac{n\pi y}{b} \sin\frac{p\pi z}{L} \qquad (2.140)$$

$$E_z = 0 \qquad (2.141)$$

$$H_x = -\frac{1}{\beta_r^2}\frac{m\pi}{a}\frac{p\pi}{L}H_0 \sin\frac{m\pi x}{a} \cos\frac{n\pi y}{b} \cos\frac{p\pi z}{L} \qquad (2.142)$$

$$H_y = -\frac{1}{\beta_r^2}\frac{n\pi}{b}\frac{p\pi}{L}H_0 \cos\frac{m\pi x}{a} \sin\frac{n\pi y}{b} \cos\frac{p\pi z}{L} \qquad (2.143)$$

$$H_z = H_0 \cos\frac{m\pi x}{a} \cos\frac{n\pi y}{b} \sin\frac{p\pi z}{L} \qquad (2.144)$$

TM_{mnp}:

$$E_x = -\frac{1}{\beta_r^2}\frac{m\pi}{a}\frac{p\pi}{L}E_0 \cos\frac{m\pi x}{a} \sin\frac{n\pi y}{b} \sin\frac{p\pi z}{L} \qquad (2.145)$$

$$E_y = -\frac{1}{\beta_r^2}\frac{n\pi}{b}\frac{p\pi}{L}E_0 \sin\frac{m\pi x}{a} \cos\frac{n\pi y}{b} \sin\frac{p\pi z}{L} \qquad (2.146)$$

$$E_z = E_0 \sin\frac{m\pi x}{a} \sin\frac{n\pi y}{b} \cos\frac{p\pi z}{L} \qquad (2.147)$$

$$H_x = j\frac{\omega\epsilon'}{\beta_r^2}\frac{n\pi}{b}E_o \sin\frac{m\pi x}{a} \cos\frac{n\pi y}{b} \cos\frac{p\pi z}{L} \qquad (2.148)$$

$$H_y = -j\frac{\omega\epsilon'}{\beta_r^2}\frac{m\pi}{a}E_o \cos\frac{m\pi x}{a} \sin\frac{n\pi y}{b} \cos\frac{p\pi z}{L} \qquad (2.149)$$

$$H_z = 0 \qquad (2.150)$$

where E_0 and H_0 are equal to twice the corresponding values in the waveguide, and β_r is the same as the wave number k_r:

$$\beta_r = \sqrt{\left(\frac{m\pi}{a}\right)^2 + \left(\frac{n\pi}{b}\right)^2 + \left(\frac{p\pi}{L}\right)^2} \qquad (2.151)$$

$$\beta_r \stackrel{d}{=} \frac{2\pi}{\lambda_r} = \omega_r\sqrt{\mu\epsilon'} \qquad (2.152)$$

where ω_r and λ_r are the angular frequency and wavelength of the resonant mode (m, n, p), respectively, and

$$f_r = \frac{\omega_r}{2\pi} = \frac{1}{2\sqrt{\mu\epsilon'}}\sqrt{\left(\frac{m}{a}\right)^2 + \left(\frac{n}{b}\right)^2 + \left(\frac{p}{L}\right)^2} \qquad (2.153)$$

$$\lambda_r = \frac{2}{\sqrt{\left(\frac{m}{a}\right)^2 + \left(\frac{n}{b}\right)^2 + \left(\frac{p}{L}\right)^2}} \qquad (2.154)$$

In a TM mode, only p can be zero, and in a TE mode, only p cannot be zero.

The fundamental resonant mode has the smallest wave number, which means that if L is the shortest side, we have

$$\min \beta_r = \beta_{r110} = \pi\sqrt{\frac{1}{a^2} + \frac{1}{b^2}} \qquad (2.155)$$

which defines the TM_{110} mode.

The subscripts m, n, and p indicate the number of stationary-wave maxima along Ox, Oy, and Oz. The fundamental mode is that for which there are no field variations along the shortest side.

Consider a resonant cavity in the fundamental mode. As its dimensions are increased, new modes appear and superimpose on the fundamental mode. The resonant cavity therefore becomes an oversized or multimode cavity. The number of modes present increases very rapidly with the dimensions [36], [37].

Relatively laborious calculations show that the final result is simple: the number of modes is given by

$$N(f) = \frac{8\pi}{3} abL \frac{f^3}{c^3} - (a + b + L)\frac{f}{c} + \frac{1}{2} \qquad (2.156)$$

where f is the excitation frequency. This expression is correct for $\beta \min(a, b, L) \geq \pi$. For example, Figure 2.7 shows the number of modes generated in a cavity of $4.57 \times 3.05 \times 2.74$ m^3 as a function of excitation frequency.

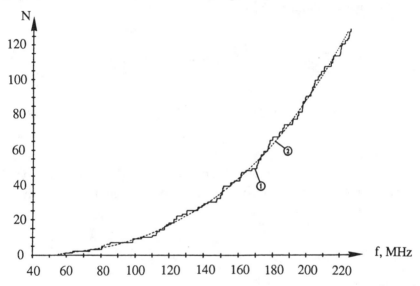

Figure 2.7 Number of modes generated in a cavity, as given by (2.156) [41].

2.11.2 Energy balance

Multiplying (2.10) and (2.11) by $\vec{E}$ and $\vec{H}$, respectively, and subtracting one from the other, we have

$$\vec{H}.\text{curl}\,\vec{E} - \vec{E}.\text{curl}\,\vec{H} = -\vec{E}.\frac{\partial \vec{D}}{\partial t} - \vec{H}.\frac{\partial \vec{B}}{\partial t} - \vec{J}.\vec{E} \qquad (2.157)$$

Applying the following identity to the first term

$$\text{div}\,(\vec{A} \times \vec{B}) = \vec{B}.\text{curl}\,\vec{A} - \vec{A}.\text{curl}\,\vec{B} \qquad (2.158)$$

and evaluating the integral in (2.157) over the volume, we have

$$\iiint_V \operatorname{div}(\vec{E} \times \vec{H}) dV = -\iiint_V \left(\vec{E}.\frac{\partial \vec{D}}{\partial t} + \vec{H}.\frac{\partial \vec{B}}{\partial t} \right) dV$$

$$-\iiint_V \vec{J}.\vec{E} dV$$

(2.159)

Finally, if S is the surface bounding the volume V, The Gauss–Ostrogradskii theorem gives

$$\iint_S \vec{E} \times \vec{H}.d\vec{S} = -\iiint_V \left(\vec{E}.\frac{\partial \vec{D}}{\partial t} - \vec{H}.\frac{\partial \vec{B}}{\partial t} \right) dV$$

(2.160)

$$-\iiint_V \vec{J}.\vec{E} dV$$

The terms of this equation have the dimensions of power $[\mathrm{L^2 M\, T^{-3}}]$ (i.e., the second term on the right represents the instantaneous power dissipated in the volume V). The first term may be rewritten as

$$-\iiint_V \left(\epsilon' E \frac{\partial E}{\partial t} + \mu H \frac{\partial H}{\partial t} \right) dV$$

or

$$-\frac{1}{2} \iiint_V \left(\epsilon' \frac{\partial E^2}{\partial t} + \mu \frac{\partial H^2}{\partial t} \right) dV$$

or

$$-\frac{1}{2} \frac{\partial}{\partial t} \iiint_V \left(\epsilon' E^2 + \mu H^2 \right) dV$$

where the integrand $\frac{1}{2}(\epsilon' E^2 + \mu H^2)$, which has the dimension $\mathrm{L^{-1} M\, T^{-2}}$, is the sum of the electric and magnetic energy densities stored in the volume V. The triple integral gives the total energy in the volume, and the time derivative term gives the rate of change of this energy. Finally, the term $-\iint_S (\vec{E} \times \vec{H}) d\vec{S}$

represents the energy flux entering the volume V. Equation (2.160) is thus seen to describe the conservation of energy.

The vector

$$\vec{\Pi} = \vec{E} \times \vec{H} \tag{2.161}$$

expressed in $\mathrm{W\,m^{-2}}$ $[MT^{-3}]$ is known as the Poynting vector and is a measure of the energy flux per unit area at a given point.

For an isolated volume

$$- \iint_S \vec{E} \times \vec{H}.d\vec{S} = 0$$

we are thus left with

$$\iiint_V \vec{J}.\vec{E}dV = - \iiint_V \left(\vec{E}.\frac{\partial \vec{D}}{\partial t} + \vec{H}.\frac{\partial \vec{B}}{\partial t} \right) dV \stackrel{d}{=} -\frac{\partial W}{\partial t} \tag{2.162}$$

where

$$W = \iiint_V \left(\epsilon \frac{E^2}{2} + \mu \frac{H^2}{2} \right) dV \tag{2.163}$$

$W = W_E + W_H$ is the sum of electric and magnetic energy stored in the system. In the sinusoidal case, W_E and W_H oscillate about their mean values:

$$\bar{W}_E = \frac{1}{4} \iiint_V \epsilon' E^2 dV$$

and

$$\bar{W}_H = \frac{1}{4} \iiint_V \mu H^2 dV \tag{2.164}$$

At resonance,

$$\bar{W}_E = \bar{W}_H \tag{2.165}$$

and

$$W = 2\bar{W}_E = 2\bar{W}_H \tag{2.166}$$

In complex notation, the average energy stored is

$$\bar{W}_c = \frac{1}{4}\left(\epsilon'\vec{E}.\vec{E}^* + \mu\vec{H}.\vec{H}^*\right) \tag{2.167}$$

and the average Poynting vector is

$$\bar{\Pi} = \frac{1}{2}\text{Re}\left(\vec{E}^* \times \vec{H}\right) = \frac{1}{2}\text{Re}\left(\vec{E} \times \vec{H}^*\right) \tag{2.168}$$

For a plane wave (Section 2.4)

$$\Pi = EH \tag{2.169}$$

From (2.55)

$$\frac{E}{H} = \eta = \sqrt{\frac{\mu}{\epsilon'}}$$

$$\Pi = \eta H^2 = \frac{E^2}{\eta} = \frac{1}{2}\left(\eta H^2 + \frac{E^2}{\eta}\right) \tag{2.170}$$

$$\Pi = \frac{1}{2\sqrt{\mu\epsilon'}}\left(\epsilon' E^2 + \mu H^2\right) \tag{2.171}$$

and, from (2.42),

$$\Pi = |\vec{E} \times \vec{H}| = EH = \frac{v}{2}\left(\epsilon' E^2 + \mu H^2\right) \tag{2.172}$$

2.11.3 Power loss in the walls

As mentioned in Section 2.6, an electromagnetic wave will penetrate slightly a metal wall, generating currents and giving rise to Joule energy losses.

Consider an air-metal interface yOz, where the x-axis points into the metal. The fields can now be expressed as $\vec{E}(x) = \vec{E}_0 e^{-\gamma x}$ and $\vec{H}(x) = \vec{H}_0 e^{-\gamma x}$. Hence it can be shown that

$$\hat{x} \times \vec{H}_0 = -\frac{1}{\eta}\vec{E}_0 \qquad \hat{x}.\vec{H}_0 = 0 \tag{2.173}$$

$$\hat{x} \times \vec{E}_0 = \eta \vec{H}_0 \qquad \hat{x} . \vec{E}_0 = 0$$

where

$$\eta = \sqrt{\frac{j\omega\mu}{\sigma + j\omega\epsilon}}$$

is the intrinsic impedance of the metal. For a good conductor,

$$\eta = \sqrt{\frac{j\omega\mu}{\sigma}} = (1 + j)\sqrt{\frac{\omega\mu}{2\sigma}} \tag{2.174}$$

The relations given by (2.173) constitute the Leontovich boundary conditions.

The energy flux per unit area that penetrates the conductor is given by the x component of the Poynting vector:

$$\vec{\Pi}_x = \frac{1}{2}\left(\vec{E}_0 \times \vec{H}_0^*\right)$$

From (2.173)

$$\vec{\Pi}_x = \frac{1}{2}\left[\left(-\eta\hat{x} \times \vec{H}_0\right) \times \vec{H}_0\right] = \frac{1}{2}\eta H_0^2 \hat{x} \tag{2.175}$$

so that, using (2.74) and (2.174), we have

$$\vec{\Pi}_x = \frac{1+j}{2\sigma d} H_0^2 \hat{x} \tag{2.176}$$

The power is obtained by integrating over the cavity surface S:

$$P = \frac{1+j}{2\sigma d} \iint_S H_0^2 dS \tag{2.177}$$

and the average power loss is

$$\bar{P}_d = \text{Re} P = \frac{1}{2\sigma d} \iint_S H_0^2 dS \tag{2.178}$$

2.11.4 Quality factor

Consider a lossy cavity, exhibiting either dielectric (ϵ'') or conduction (σ) losses: Energy is absorbed and free oscillations in the cavity are replaced by damped oscillations. It can be shown that the wave number β must be real, so that, since ϵ is complex, this is possible only if the frequencies are complex:

$$f = f' + jf'' \tag{2.179}$$

where f' is the eigenfrequency of the oscillation, and f'' is the absorption coefficient. The fields take a damped form: sinusoidal

$$e^{-2\pi f'' t} \cos(2\pi f' t + \phi)$$

For low losses $f' \gg f''$ and

$$f \simeq f_0 \left(1 + j\frac{f''}{f_0}\right) \tag{2.180}$$

where f_0 is the frequency of the undamped oscillation ($f_0 \simeq f'$).

The quantity Q is the socalled *quality factor*, which is large for low losses:

$$Q \overset{\mathrm{d}}{=} \frac{f_0}{2f''} \tag{2.181}$$

The fields vary as

$$e^{j2\pi f_0(1+j/2Q)t} = e^{-\pi f_0 t/Q} e^{j2\pi f_0 t} \tag{2.182}$$

This is a quasiperiodic process, characterized by pseudo-period $T' = 1/f'$ and time constant $T'' = 1/f''$ defined as the time in which the field amplitude falls from its initial value by a factor of e.

The energy is proportional to the square of the field; that is it varies as

$$e^{-2\pi f_0 t/Q} = e^{-2\pi t/QT'}$$

After a time T', it is reduced to a fraction $e^{-2\pi/Q}$ of its initial value, namely

$$W(t+T') = W(t)e^{-2\pi/Q} \simeq W(t)\left(1 - \frac{2\pi}{Q}\right) \qquad (2.183)$$

$$W(t+T') - W(t) \simeq \frac{2\pi}{Q}W(t) \qquad (2.183)$$

The missing energy has been dissipated in the cavity. If $\bar{P}_d$ is the average power dissipated, we have

$$\frac{2\pi}{Q}W(t) \simeq \bar{P}_d T'$$

or

$$Q \simeq \omega_0 \frac{W(t)}{\bar{P}_d} \qquad (2.184)$$

The Q factor is equal to ω_0 times the ratio of the localized energy and the dissipated power.

If the system is isolated, (2.164) and (2.166) give

$$W = 2\bar{W}_E = 2\bar{W}_H = \frac{1}{2}\iiint_V \mu H^2 dV$$

and if the losses are exclusively due to conduction in the wall, (2.178) becomes

$$\bar{P}_d = \frac{1}{2\sigma d}\iint_S H_0^2 dS$$

and (2.184) gives

$$Q = 2\pi f_0 \mu\sigma d\frac{\iiint_V H^2 dV}{\iint_S H_0^2 dS} \qquad (2.185)$$

We can now use this expression to calculate Q for different modes

$$Q_{mnp}^{\mathrm{TM}} = \frac{\eta_0}{4}\sigma d\frac{abL\beta_{xy}^2\beta_r}{b\beta_x^2(a+L) + a\beta_y^2(b+L)}$$

$$Q_{mnp}^{\text{TE}} = \frac{\eta_0}{4} \sigma d \frac{abL\beta_{xy}^2 \beta_r^3}{bL\left(\beta_{xy}^4 + \beta_y^2\beta_z^2\right) + aL\left(\beta_{xy}^4 + \beta_x^2\beta_z^2\right) + ab\beta_{xy}^2\beta_z^2}$$

$$\text{(2.186)}$$

where

$$\beta_x = \frac{m\pi}{a}, \quad \beta_y = \frac{n\pi}{b}, \quad \beta_z = \frac{p\pi}{L}$$

$$\beta_{xy} = \sqrt{\left(\frac{m\pi}{a}\right)^2 + \left(\frac{n\pi}{b}\right)^2}$$

and so on. For an oversized cavity, the composite quality factor $\bar{Q}$ is defined as the average of Q for all the modes present. Details of the calculation can be found in [36] and [37]. The result is

$$\bar{Q} = \frac{V}{S d} \frac{3}{2} \frac{1}{1 + \frac{3\pi}{8\beta}\left(\frac{1}{a} + \frac{1}{b} + \frac{1}{L}\right)} \tag{2.187}$$

where V and S are the volume and surface area of the internal cavity. This formula is valid if $\beta \min(a, b, L) \geq \pi$.

If P_1, P_2, and P_3 are the conduction, dielectric, and radiation losses, we have

$$Q_1 = \omega_0 \frac{\bar{W}}{P_1}, \quad Q_2 = \omega_0 \frac{\bar{W}}{P_2}, \quad Q_3 = \frac{\bar{W}}{P_3}$$

where

$$\bar{P}_d = P_1 + P_2 + P_3$$

and

$$\frac{1}{Q} = \frac{1}{Q_1} + \frac{1}{Q_2} + \frac{1}{Q_3} \tag{2.188}$$

Additional information on cavities can be found in the literature [22], [4], [38], [39], [12], [15], [14], [40], [43], [36], [37], [55], [41], [2].

2.12 Radiation sources

2.12.1 Characteristics

The shape and intensity of the radiation is a function of the characteristics of the source (i.e., the emitting antenna). They are polarization (Section 2.3), radiation pattern, gain, and directivity [29], [25], [26], [11], [51], [17], [45], [46].

The power radiated per unit area in a given direction is given by (2.170):

$$\Pi = \frac{E^2}{\eta_0}$$

the units being Wm^{-2}. We can also define the intensity of radiation $\phi(\theta, \varphi)$

$$\phi(\theta, \varphi) = r^2 \Pi = \frac{r^2 E^2}{\mu_0} \tag{2.189}$$

where r, θ, φ are the polar coordinates. The total radiated power is

$$P_r = \int \phi d\Omega \tag{2.190}$$

and the average power per unit solid angle is

$$\frac{P_r}{4\pi} \stackrel{\mathrm{d}}{=} \bar{\phi} \tag{2.191}$$

where $\bar{\phi}$ is the radiation intensity due to an isotropic antenna emitting an amount of power P_r. The directional gain is defined as

$$g_d(\theta, \varphi) \stackrel{\mathrm{d}}{=} \frac{\phi(\theta, \varphi)}{\bar{\phi}} = 4\pi \frac{\phi(\theta, \varphi)}{P_r} \tag{2.192}$$

The directivity D is the maximum directional gain. The equivalent area of an antenna is defined by

$$A \stackrel{\mathrm{d}}{=} \frac{g_d \lambda^2}{4\pi} \tag{2.193}$$

The *radiation pattern* is the angular distribution of radiated energy measured by a receiving antenna that moves over a sphere centered on the source. It has a number of lobes when its polar diagram is cut by different planes containing the optical axis (e.g., the E and H planes containing the electric and magnetic field, respectively). The principal lobe is characterized by its width at 3 dB, which is equivalent to the angular separation of the directions in which the radiated power is half that along the axis of the lobe. The secondary lobes are measured in terms of the relative amplitude (dB_c) with respect to the principal lobe. A high gain indicates high concentration of radiation within a narrow principal lobe.

2.12.2 Radiation from a slot

A parallel-edge slot in a thin infinite conducting plane constitutes an antenna that is the reciprocal of the simple dipole. In particular, the radiation patterns are identical if $\vec{E}$ and $\vec{H}$ are interchanged: The pattern due to a thin slot is omnidirectional in the plane perpendicular to the slot (E plane), and directional in the parallel plane (H plane). The radiated magnetic field is analogous to the electric field of a dipole:

$$H_\theta = H_0 \frac{\cos\left(\frac{\beta l}{2}\cos\theta\right) - \cos\frac{\beta l}{2}}{\sin\theta} e^{-j\beta r} \qquad (2.194)$$

where l is the length of the slot and θ the angular position $(\vec{O}z, \vec{O}M)$, $\vec{O}z$ being parallel to the slot [46]. Resonance occurs when the length is a multiple of $\lambda/2$, or half the perimeter of a large slot.

When the length l of a slot of width l' is very large as compared with the wavelength, it can be shown that the radiated field at a large distance from the source is given by

$$E = E_0 l' \sqrt{\frac{\beta}{2\pi r}} \qquad (2.195)$$

(i.e., it decreases as $r^{-1/2}$), which is typical for a cylindrical wave [20].

For a finite length, the field varies approximately as

$$E = E_0 ll' \frac{\beta}{2\pi r} \qquad (2.196)$$

and decreases as r^{-1}, which is typical for a spherical wave.

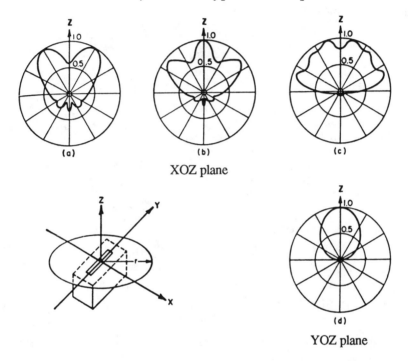

Figure 2.8 Polar radiation pattern from a half-wave slot in a circular plate of radius r: (a) $r = \lambda$, (b) $r = 3\lambda/2$, (c),(d) $r = 5\lambda$ [25].

In practice, the conductor in which the slot is cut is not infinite, and its size affects the radiation pattern. Figure 2.8 shows the change in the radiation pattern of a half-wave slot as a function of the diameter of a circular conductor.

The thickness E is also relevant. The transfer of power through a slot is defined by a transmission coefficient

$$T_p = \frac{P_t}{P_i} = \frac{4\mathrm{Re}Y}{l'\eta_0 |Y_t|^2} \qquad (2.197)$$

where Y is the complex admittance of the aperture

$$Y = \frac{1}{\eta_0 \lambda_0} \left(\pi - 2j \ln \frac{2\pi C l'}{\lambda_0} \right) \qquad (2.198)$$

and Y_t is the complex transfer admittance

$$Y_t = 2Y \cos \frac{2\pi E}{\lambda_f} + j \left(\frac{1}{l'\eta_0} + l'\eta_0 Y^2 \right) \sin \frac{2\pi E}{\lambda_f} \qquad (2.199)$$

where C' is a constant of approximately 0.2 and λ_f is the wavelength in the slot, which for TE_{10} can be found from (2.118) [21]:

$$\lambda_f = \frac{2l\lambda_0}{\sqrt{4l^2 - \lambda_0^2}} \qquad (2.200)$$

2.12.3 Radiation of an aperture

Each point on an illuminated plane aperture behaves as a source of spherical waves, and the resultant wave radiated by the aperture is the sum of these secondary waves [2]. The fields may be calculated using Shchelkunov's equivalence principle in which the distributions of the fields $\vec{E}$ and $\vec{H}$ on the aperture are represented by the following electric and magnetic surface current densities:

$$\bar{J} = \hat{n} \times \bar{H} \qquad (2.201)$$

$$\bar{M} = -\hat{n} \times \bar{E} \qquad (2.202)$$

where $\hat{n}$ is the unit vector normal to the aperture.

The field at a point M in space is given by Kottler's equations

$$E_M = \frac{1}{j4\pi\omega\epsilon'} \iint_S \left[\beta^2 \left(\hat{n} \times \bar{H} \right) \psi + \right.$$
$$\left. + \left[\left(\hat{n} \times \bar{H} \right).\nabla \right] \nabla \psi + j\omega\epsilon' \left(\hat{n} \times \bar{E} \right) \times \nabla \psi \right] dS \qquad (2.203)$$

$$H_M = -\frac{1}{j4\pi\omega\mu} \iint_S \left[\beta^2 \left(\hat{n} \times \bar{E} \right) \psi + \right.$$
$$\left. + \left[\left(\hat{n} \times \vec{E} \right).\nabla \right] \nabla\psi - j\omega\mu' \left(\hat{n} \times \vec{H} \right) \times \nabla\psi \right] dS \qquad (2.204)$$

$$\psi = \frac{e^{-j\beta r}}{r} \qquad (2.205)$$

$$\text{grad}\,\psi = \nabla\psi = \left(j\beta + \frac{1}{r} \right) \frac{e^{-j\beta r}}{r} \hat{r}_1 \qquad (2.206)$$

where r is the distance between the point M and a point P on the aperture, and $\hat{r}_1$ the unit vector along PM. The special case of a large circular aperture is discussed in [13] (see also Table 2.1).

Table 2.1

Parameters of the radiation pattern of a circular aperture of radius a in polar coordinates (r, θ) [25]

Distribution $(0 \leq r \leq 1)$	Directivity	3-dB aperture of main lobe (deg)	Position of the first zero (deg)	First side lobe (dBc)	Relative gain
$f(r) = 1$	$\pi a^2 \, (J_1(u)/u)$	58.9 ($\lambda/2a$)	69.8 ($\lambda/2a$)	-17.6	1.00
$f(r) = 1 - r^2$	$2\,\pi a^2 (J_2(u)/u^2)$	72.7 ($\lambda/2a$)	93.6 ($\lambda/2a$)	-24.6	0.75
$f(r) = (1-r^2)^2$	$8\,\pi a^2 (J_3(u)/u^3)$	84.3 ($\lambda/2a$)	116.2 ($\lambda/2a$)	-30.6	0.56

$u = (2\pi a/\lambda) \sin\theta$

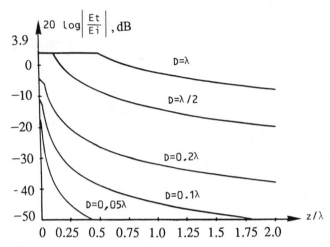

Figure 2.9 Power transmitted by a circular aperture of diameter D [9].

If the linear dimensions of an aperture are small in comparison with λ, for example, if the diameter is less than $\lambda/2$, the only field component that is transmitted is rapidly attenuated along the z direction (Fig. 2.9).

An important special case is that of an open waveguide: The tangential fields over the aperture are, in the TE_{10} rectangular mode,

$$E_{y0} = E_0 \sin \frac{\pi x}{a} \tag{2.207}$$

$$H_{x0} = -\frac{E_{y0}}{\eta_0} \sqrt{1 - \left(\frac{\lambda}{\lambda_c}\right)^2} \tag{2.208}$$

When the frequency of the transmitted wave is close to the cut-off frequency $\lambda_c = 2a$,

$$\sqrt{1 - \left(\frac{\lambda}{\lambda_c}\right)^2} \to 0 \quad \text{as} \quad \lambda \to \lambda_c \tag{2.209}$$

and the characteristic impedance of the waveguide

$$Z(\text{TE}) = \frac{\eta_0}{\sqrt{1 - \left(\frac{\lambda}{\lambda_c}\right)^2}} \tag{2.210}$$

becomes very large as compared with η_0, we have an impedance mismatch, and a set of stationary waves is established in the waveguide [26].

If the point under consideration is chosen so that the largest value of λ does not give rise to parasitic reflections, for example, $\lambda = a = 2b$, then the equivalent area is approximately $8S/\pi^2$ or

$$A = \frac{4\lambda^2}{\pi^2} \qquad (2.211)$$

and, by (2.193), the maximum gain is

$$G = \frac{4\pi A}{\lambda^2} = \frac{16}{\pi} \simeq 5.1 \qquad (2.212)$$

2.12.4 Radiation from a horn

The matching and gain of an open waveguide is improved by widening the opening to form a horn. Depending on whether the widening takes place in the E or H plane, or in both, the result is known as sectoral E or H plane or pyramidal horn; with aperture $a \times b', a' \times b$, and $a' \times b'$, respectively, where a and b are the large and small sides of the waveguide, increased to a' and b'. If L is the edge length of the horn, the gain is given by

$$G_E = \frac{64La}{\pi\lambda b'} \left[C^2 \left(\frac{b'}{\sqrt{2\lambda L}} \right) + S^2 \left(\frac{b'}{\sqrt{2\lambda L}} \right) \right] \qquad (2.213)$$

for an E plane horn; and

$$G_H = \frac{4\pi Lb}{\lambda a'} \left\{ [C(u) - C(v)]^2 + [S(u) - S(v)]^2 \right\} \qquad (2.214)$$

for an H plane horn, and

$$G = \frac{\pi}{32} \left(\frac{\lambda}{b'} G_H \right) \left(\frac{\lambda}{a'} G_E \right) \qquad (2.215)$$

for a pyramidal horn:

$$u = \frac{1}{\sqrt{2}} \left(\frac{\sqrt{\lambda L}}{a'} + \frac{a'}{\sqrt{\lambda L}} \right), \quad v = \frac{1}{\sqrt{2}} \left(\frac{\sqrt{\lambda L}}{a'} - \frac{a'}{\sqrt{\lambda L}} \right)$$

(2.216)

$$C(x) = \int_0^x \cos\left(\frac{\pi q^2}{2} \right) dq, \quad S(x) = \int_0^x \sin\left(\frac{\pi q^2}{2} \right) dq$$

(2.217)

where the latter are the Fresnel's integrals [51].

The magnitude of the field radiated by an E-plane horn is:

$$|E|^2 = \frac{2L}{\lambda} \left(\frac{2aE_{y0}}{\pi r} \right) \left[C^2 \left(\frac{b'}{\sqrt{2\lambda L}} \right) + S^2 \left(\frac{b'}{\sqrt{2\lambda L}} \right) \right]$$

(2.218)

at a distance r on the axis, where E_{y0} is the field on the aperture [26].

To ensure an equiphase distribution of emitted fields, the horn edge is chosen so that

$$L \geq \frac{a'^2}{2\lambda} \quad \text{or} \quad \frac{b'^2}{2\lambda}$$

(2.219)

with aperture angles such that

$$\sin\theta \leq \frac{\lambda}{2b'}$$

(2.220)

$$\sin\phi \leq \frac{\lambda}{2a'}$$

(2.221)

2.12.5 Radiation zones

The space in front of a radiating aperture is conventionally divided into three zones, depending on the approximations that can be made:

(a) the near or Rayleigh zone, in which no simplification is possible. In this zone, the radiated energy is confined to a tubular beam with the radiating aperture as its base.

(b) the intermediate or Fresnel zone characterized by

$$r = PM \simeq R = OM \tag{2.222}$$

and by

$$\hat{n}.\hat{r}_1 \simeq \hat{n}.\hat{r}_0 \tag{2.223}$$

where O is the center of the aperture and $\hat{r}_0$ is the unit vector along OM; no approximation is possible for $e^{-j\beta r}$; the beam begins to diverge in this zone.

(c) the far or Fraunhofer zone, in which an approximation for $e^{-j\beta r}$ is possible and the energy spreads in all directions, with the power falling as r^{-2} [57], [51], [48].

If D is the large dimension of the aperture, (e.g., the diameter in the case of a circular aperture) the Rayleigh zone extends to a distance $D^2/2\lambda$ for a uniform distribution over the aperture, and to a shorter distance for a nonuniform distribution. For example, for a parabolic distribution, it extends to a distance $0.61D^2/2\lambda$. The positions of the Fresnel and Fraunhofer zones are shown in Figure 2.10.

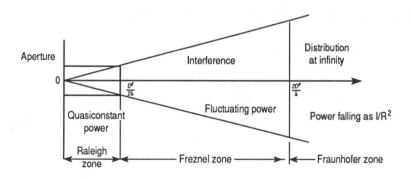

Figure 2.10 Radiation zones of an aperture [51].

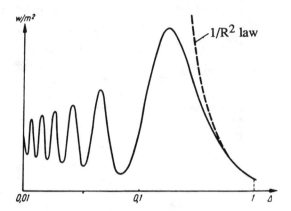

Figure 2.11 Radiation pattern in the Fresnel zone [51].

The radiated power as a function of the distance from an aperture illuminated by a uniform field is given by [51]. Figure 2.11 shows that there is a fluctuation in the Fresnel zone, followed by the r^{-2} region, and beyond $r = 2D^2/\lambda$ (far zone) by the r^{-1} variation, typical for spherical waves.

Appendix

Ax, Ay, Az are the components of $\vec{A}$ along the directions of the unit vectors $\hat{x}, \hat{y}, \hat{z}; \partial/\partial x; \partial/\partial y; \partial/\partial z$ and $\partial^2/\partial x^2; \partial^2/\partial y^2; \partial^2/\partial z^2$ are the first and second partial derivatives.

Nabla operator:

$$\vec{\nabla} \stackrel{\mathrm{d}}{=} \frac{\partial}{\partial x}\hat{x} + \frac{\partial}{\partial y}\hat{y} + \frac{\partial}{\partial z}\hat{z}$$

Divergence of a vector:

$$\mathrm{div}\,\vec{A} \stackrel{\mathrm{d}}{=} \vec{\nabla}.\vec{A} = \frac{\partial Ax}{\partial x} + \frac{\partial Ay}{\partial y} + \frac{\partial Az}{\partial z}$$

Curl of a vector:

$$\text{curl } \vec{A} \stackrel{d}{=} \nabla \times \vec{A} = \left(\frac{\partial A_z}{\partial y} - \frac{\partial A_y}{\partial z} \right) \hat{x} + \left(\frac{\partial A_x}{\partial z} - \frac{\partial A_z}{\partial x} \right) \hat{y}$$
$$+ \left(\frac{\partial A_y}{\partial x} - \frac{\partial A_x}{\partial y} \right) \hat{z}$$

Gradient of a scalar function

$$\text{grad } a \stackrel{d}{=} \vec{\nabla} a = \frac{\partial a}{\partial x} \hat{x} + \frac{\partial a}{\partial y} \hat{y} + \frac{\partial a}{\partial z} \hat{z}$$

Laplacian applied to a scalar function

$$\Delta a \stackrel{d}{=} \nabla^2 a = \frac{\partial^2 a}{\partial x^2} + \frac{\partial^2 a}{\partial y^2} + \frac{\partial^2 a}{\partial z^2}$$

Laplacian applied to a vector function

$$\Delta \vec{A} \stackrel{d}{=} \nabla^2 \vec{A} = \Delta A_x . \hat{x} + \Delta A_y . \hat{y} + \Delta A_x . \hat{z}$$

References

1. BERTEAUD A. Les hyperfréquences. Paris, P.U.F., 1976, 125 p. (Que sais-je ? n° 1643).

2. BERTEAUD A. Hyperfréquences. (Cours polycopié). Institut supérieur d'électronique de Paris, 224 p. (non daté).

3. BERTIN M., FAROUX J., RENAULT J. Electromagnétisme. Tome 4. Paris, Bordas-Dunod, 1984, 270 p.

4. BOUDOURIS G. Cavités électromagnétiques. Paris, Dunod, 1971, 166 p.

5. BOUDOURIS G., CHENEVIER P. Circuits pour ondes guidées. Paris, Bordas, 1975, 300 p.

6. BRADY M. Standard rectangular waveguide constants. London, Transcripta Books, 1972, 97 p.

7. BROGLIE L. de. Problèmes de propagations guidées des ondes électromagnétiques. 2ème éd. Paris Gauthier-Villars, 1951, 118 p.

8. CASTAGNETTO L. Lignes. Propagation. Rayonnement. Toulouse, ENSEEIHT, 1963, 3 tomes, 450 p.

9. CATHEY W. Approximate expressions for field penetration through circular apertures. *IEEE Trans. Electromagn. Compat.*, EMC-25, n° 3, Aug. 1983, pp. 339-345.

10. CHEUNG W. Microwaves made simple : principles and applications. London, Adtech Books Co. Ltd., 1985, 400 p.

11. COLLIN R., ZUCKER F. Antenna theory. 2 vol. New York, McGraw-Hill Book Co., 1969.

12. COMBE R. Les cavités parallélépipédiques. Cours d'hyperfréquences. Chapitre VI. Univ. Clermont-Ferrand, UER de Sciences Mathématiques et Physiques, avril 1975, 30 p.

13. COMBES P. Etude de la zone de Rayleigh des ouvertures circulaires par les formules de Kottler et la théorie géométrique de la diffraction. Thèse de Doctorat d'Etat, Toulouse, Univ. Paul Sabatier, 25 sept. 1978, 318 p.

14. CORONA P., LATMIRAL G., PAOLINI E. Performance and analysis of a reverberating enclosure with variable geometry. *IEEE Trans. Electromagn. Compat.*, EMC-22, n° 1, Feb. 1980, pp. 2-5.

15. CORONA P., LATMIRAL G., PAOLINI E., PICCIOLI L. Use of a reverberating enclosure for measurements of radiated power.in the microwave range. *IEEE Trans. Electromagn. Compat.*, EMC-18, n° 2, May 1976, pp. 54-59.

16. CRAMPAGNE M. Adaptation en hyperfréquence. (Cours polycopié). Institut national polytechnique de Toulouse, ENSEEIHT, 1976, 72 p.

17. DUBOST G., ZISLER S. Antennes à large bande. Paris, Masson, 1976, 337 p.

18. GARDIOL F. Traité d'électricité. Vol. III : Electromagnétisme ; Vol. XIII : Hyperfréquences. St. Saphorin (Suisse), Georgi, 1977.

19. GARDIOL F. Introduction to microwaves. London, Adtech Book Co. Ltd., 1984, 425 p.

20. GUTIÉRREZ R. Communication personnelle. 4 sept. 1987.

21. HARRINGTON R., AUCKLAND D. Electromagnetic transmission through narrow slots in thick conducting screens. *IEEE Trans. Antennas Propag.*, AP-28, n° 5, Sept. 1980, pp. 616-622.

22. HAUCK H. Design considerations for microwave oven cavities. *IEEE Trans. Ind. Gen. Appl.*, IGA-6, n° 1, Jan.-Feb. 1970, pp. 74-80.

23. HAYT W. Engineering electromagnetics. Tokyo, McGraw-Hill Kogakusha Ltd., International Student Edition, 1958, 1976, 435 p.

24. HURAUX C. Les isolants. Paris, P.U.F., 1968, pp. 5-50. (Que sais-je ? n° 1300).

25. JASIK H. Antenna engineering handbook. New York, McGraw-Hill Book Co., 1961, 1009 p.

26. JORDAN E., BALMAIN K. Electromagnetic waves and radiating systems. 2nd ed. Englewood Cliffs (New Jersey), Prentice-Hall inc., 1968, 753 p.

27. KAHAN T. Les ondes hertziennes. Paris, P.U.F., 1974, 127 p. (Que sais-je ?).

28. KING R., MIMNO H., WING A. Transmission lines, antennas and wave guides. New York, McGraw-Hill Book Co., 1945 et New York, Dover Publications Inc., 1965, 347 p.

29. KRAUS J. Antennas. New York, McGraw-Hill Book Co., 1950, 552 p.

30. KRETZSCHMAR J. Maximum obtainable electric field for TM modes on hollow cylindrical cavities with rectangular and elliptical cross-section. In : Proceedings of the IMPI Symposium, Ottawa, 1972, pp. 157-162. (Paper 9.4.).

31. LANDAU L., LIFSIC E. Elektrodinamika sploŝnyx sred. Moskva, Nauka, 1959. Trad. fr. : Electrodynamique des milieux continus. Moskva, Mir, 1969, 536 p.

32. LAVERGHETTA T. Microwave measurements and techniques. London, Artech, s.d., 394 p.

33. LEFEUVRE S. Hyperfréquence. Paris, Dunod Univ., 1969, 160 p.

34. LEFEUVRE S. Technique des microondes. Institut national polytechnique de Toulouse, ENSEEIHT, 1977, s.d., 187.p.

35. LEFEUVRE S. Ondes électromagnétiques. (Cours polycopié). Institut national polytechnique de Toulouse, ENSEEIHT, s.d., 150 p.

36. LIU B., CHANG D., MA M. Eigenmodes and the composite quality factor of a reverberating chamber, PB 84-128016. Boulder, US National Bureau of Standards, August 1983, 47 p.

37. LIU B., CHANG D., MA M. Design considerations of reverberating chambers for electromagnetic interference measurements. In : Proceedings of IEEE International Symposium on Electromagnetic compatibility, Washington, 1983, pp. 508-512.

38. LUYPAERT P., SCHOONAERT D. The use of cavities, synthesized with nonseparable solutions of Helmholtz wave equation, for heating applications. In : Proceedings of the IMPI Symposium, Milwaukee, 1974, 4 p. (Paper B5).

39. LUYPAERT P., SCHOONAERT D. On the attenuation and Q factor of waveguides and cavities synthesized with nonseparable solutions of Helmholtz wave equation. In : Proceedings of the IMPI Symposium, Waterloo (Can.), 1975, pp. 235-240. (Paper 14.4.).

40. LUYPAERT P., SCHOONAERT D. On the synthesis of waveguides and cavities realized with nonseparable solutions of Helmholtz wave equation. *IEEE Trans. Microw. Theory Tech.*, MTT-$\underline{23}$, n° 12, 1980, pp. 1061-1064.

41. MA M., KANDA M., CRAWFORD M., LARSEN E. A review of electromagnetic compatibility/interference measurement methodologies. In : Proceedings of the IEEE, $\underline{73}$, n° 3, March 1985, pp. 388-411.

42. MARCUVITZ N. Waveguide handbook. Massachusetts Institute of Technology, Vol. 10, New York, McGraw-Hill Book Co., 1951.

43. NIKOL'SKIJ V. Electrodinamika i rasprostranjenije radiovoln. Moskva, Nauka, 1978. Trad fr. : Electrodynamique et propagation des ondes radioélectriques, Moskva, Mir, 1982, 573 p.

44. REITZ J., MILFORD F. Foundations of electromagnetic theory. Reading (Massachusetts, USA), Addison-Wesley Publishing Co. Inc., 1960, 387p.

45. ROUBINE E., BOLOMEY J. Antennes. 1 : Introduction générale, Paris, Masson, 1978, 202 p.

46. ROUBINE E., DRABOWITCH S., ANCONA C. Antennes. 2 : Applications. Paris, Masson, 1978, 233 p.

47. SAAD S. Microwave engineer's handbook. Dedham (Massachusetts), Artech House Inc., 1971, 192 p. et 209 p.

48. SIMÓ-PONS R. Rayonnement des ouvertures. (Cours polycopié). Toulouse, ENSEEIHT, 1975, 44 p.

49. STRATTON J. Electromagnetic theory. New York, McGraw-Hill Book Co., 1949.

50. THOUREL L. Propagation des ondes électromagnétiques. (Cours polycopié). Toulouse, Ecole nationale supérieure de l'aéronautique et de l'espace, 1970, 138 p.

51. THOUREL L. Les antennes. 2ème éd., Paris, Dunod, 1971, 461 p.

52. THOUREL L. Cours d'hyperfréquence. (Cours polycopié). Toulouse, Ecole nationale supérieure de l'aéronautique et de l'espace, 1972, 263 p.

53. THOUREL L. Théorie des lignes. (Cours polycopié). Toulouse, Ecole nationale supérieure de l'aéronautique et de l'espace, 1976, 68 p.

54. THOUREL L. Réseaux et filtres en hyperfréquences. (Cours polycopié). Toulouse, Ecole nationale supérieure de l'aéronautique et de l'espace, 1976, 138 p.

55. TURNER R., VOSS W., TINGA W., BALTES H. On the counting of modes in rectangular cavities. *J. Microw. Power*, 19, n° 3, 1984, pp. 199-208.

56. VASSALLO C. Electromagnétisme classique dans la matière. Paris, Dunod, 1983, 272 p.

57. WALTER C. Traveling-wave antennas. New York, Dover Publications, inc., 1965, 429 p.

Chapter 3

Microwaves and Matter

Commonly used electromagnetic radiation (infrared, visible light) has relatively short wavelengths and therefore very limited penetration of matter. Microwaves have wavelengths on the order of a few centimeters or tens of centimeters (i.e., of the same order of magnitude as the size of the objects to be heated)., because Their penetration depth is therefore of the same order. The release of heat is instantaneous, so that thermal phenomena of conduction, convection, and radiation play only a secondary role in temperature equilibrium.

The release of heat involves a mechanism that is specific to the frequency range used, namely, dipolar rotation accompanied by intermolecular friction, and hysteresis between the applied field and the induced electric response. The molecules of the material must possess a sufficiently high dipole moment, which limits efficient microwave heating to high-loss dielectrics, a category in which water is of major importance.

3.1 Dielectric polarization

The common feature of dielectric materials is their ability to store electric energy. This is accomplished by the displacement of positive and negative charges under the effect of the applied electric fields and against the force of atomic and molecular attraction. The mechanism of charge displacement (i.e., polarization) depends on the type of dielectric material and the frequency of the applied field. There are four main types of dielectric polarization. They have very similar qualitative effects but appear at very different frequencies. The microscopic elements that are involved in this are at the level of a complete zone of the material, the molecule, the ion, and the atom, respectively. In all cases, the electric equilibrium is disturbed because the applied field causes the spatial separation of charges of opposite sign. With an alternating field, the frequency determines the dominant type of polarization.

Space charge polarization gives rise to low-frequency response (VLF, LF). It occurs when the material contains free electrons whose displacements are restricted by obstacles, such as grain boundaries. When an electric field is applied, the electrons accumulate on the obstacle, and the resulting charge separation polarizes the material. Entire regions of the material become either positive or negative. This type of polarization is fundamental in semiconductor electronics.

Polarization by dipole alignment occurs at higher frequencies (HF, microwaves) and also at the molecular level. It lies at the basis of dielectric heating.

Ionic polarization takes place at infrared frequencies; it is due to the separation of positive and negative ions in the molecule.

Electronic polarization occurs at very high frequencies close to the ultraviolet region. The atomic nucleus is positive and fixed in the matrix of the dielectric material. The negative electronic cloud surrounding it is displaced in the direction of the applied field.

In practice, these phenomena may overlap and it is not always easy to establish strict barriers between them.

3.2 Polarization by dipole alignment in a static field

3.2.1 Polar and nonpolar media

In a polar dielectric, the constituent molecules are neutral, but their centers of positive $(+q)$ and negative $(-q)$ charges do not geometrically coincide. This asymmetry is then responsible for a permanent electric dipole $\vec{p}$ (in C m [LTI]):

$$\vec{p} = q\vec{l} \tag{3.1}$$

where $\vec{l}$ is the relative position vector of the positive and negative charges (Fig. 3.1). A medium in which $\vec{l}$ is zero is defined as nonpolar.

3.2.2 Induced dipole moment

Nonpolar molecules have no permanent electric dipole moment, but may acquire an induced moment by molecular deformation in an applied electric field.

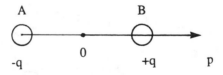

Figure 3.1 Molecular dipole moment [120].

The polarizability α' (in F m^2 [M^{-1} T^4 I^2]) is a measure of the dipole moment induced by a unit field. In other words,

$$\vec{p} = \alpha' \vec{E}_{\text{loc}} \tag{3.2}$$

where $\vec{E}_{loc}$ is the local field near the molecule.

At the macroscopic level, alignment by the field is opposed by thermal agitation. A statistical equilibrium is then established in which, at a given temperature and in a given field, the number of aligned molecule per unit volume N ($[L^{-3}]$) remains constant.

The total dipole moment is characterized by the polarization $\vec{P}$ (in $C\,m^{-2}$ $[L^{-2}T\,I]$)

$$\vec{P} = N\alpha'\vec{E}_{loc} \tag{3.3}$$

or, in function of the applied field:

$$\vec{p} \overset{d}{=} \chi\vec{E} \tag{3.4}$$

where χ (in $F\,m^{-1}$ $[L^{-3}M^{-1}T^4I^2]$) is the dielectric susceptibility of the medium.

In free space, the relationship between induction and electric field is given by (2.4)

$$\vec{D} = \epsilon_0\vec{E}$$

and in a dielectric it becomes (Fig. 3.2)

$$\vec{D} = \epsilon_0\vec{E} + \vec{P} \tag{3.5}$$

or

$$\vec{D} = (\epsilon_0 + \chi)\vec{E} \tag{3.6}$$

or

$$\vec{D} = \epsilon'\vec{E} \tag{3.7}$$

hence:

$$\chi = \epsilon' - \epsilon_0 \tag{3.8}$$

If we define the relative susceptibility χ_r by

$$\chi_r = \frac{\chi}{\epsilon_0} \tag{3.9}$$

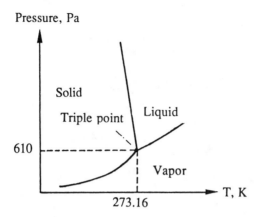

Figure 3.2 Dielectric block placed between the plates of a capacitor. D is a function of the free charges $\pm q$, and is the same everywhere; E is weaker in the dielectric, where polarization charges produce an opposite field (P. Lorrain and D. Corson, *Electromagnetic Fields and Waves*, W.H. Freeman and Co., San Franciso, 1970, p 107).

we have

$$\chi_r = \epsilon'_r - 1 \tag{3.10}$$

It is useful to establish a relationship between the permittivity ϵ', which is a characteristic of the medium, and the polarizability α', which is a characteristic of the molecule. The local field $\vec{E}_{loc}$ is different from the applied field $\vec{E}$, except for low pressure gases for which, from (3.3), (3.4), and (3.10),

$$\frac{N\alpha'}{\epsilon_0} = \epsilon'_r - 1 \tag{3.11}$$

For other media, it can be shown that the local, or Mosotti, field is

$$\vec{E}_{\text{loc}} = \vec{E} + \frac{\vec{P}}{3\epsilon_0} = \vec{E}\left(1 + \frac{\chi}{3\epsilon_0}\right)$$
$$= \frac{\epsilon' + 2\epsilon_0}{3\epsilon_0}\vec{E} = \frac{\epsilon'_r + 2}{3}\vec{E} \tag{3.12}$$

The polarization (3.3) now becomes

$$\overline{P} = N\alpha'\frac{\epsilon'_r + 2}{3}\overline{E}$$

Hence, using (3.10),

$$\frac{N\alpha'}{3\epsilon_0} = \frac{\epsilon'_r - 1}{\epsilon'_r + 2} \tag{3.13}$$

which is the Clausius - Mosotti formula

3.2.3 Permanent dipole moment

In a polar medium, the applied field produces a couple on the permanent dipole moment $\vec{p}$ of each molecule, which is given by

$$\vec{C} = \vec{p} \times \vec{E} \tag{3.14}$$

This couple tends to align the dipole moment with the field. Each individual moment contributes to an average moment

$$\overline{p} = p\overline{\cos\theta} \tag{3.15}$$

where $\overline{\cos\theta}$ represents the average cosine of the angle between $\vec{p}$ and $\vec{E}$.

It can be shown [27] that

$$\overline{p} = p\left(\tanh^{-1}\frac{pE_{\text{loc}}}{kT} - \frac{kT}{pE_{\text{loc}}}\right) = pL\left(\frac{pE_{\text{loc}}}{kT}\right) \tag{3.16}$$

where

$$L(u) \stackrel{\mathrm{d}}{=} \tanh^{-1} u - \frac{1}{u}$$

is the Langevin function.

If the field is small as compared to the thermal agitation energy, we have

$$L(u) \simeq \frac{u}{3}$$

where

$$\bar{p} \simeq \frac{p^2 E_{\mathrm{loc}}}{3kT} \tag{3.17}$$

and the polarization of the medium is

$$p \simeq \frac{Np^2 E_{\mathrm{loc}}}{3kT} \tag{3.18}$$

The field E_{loc} can be calculated from the Onsager model in which the molecule is represented by a point at the center of a cavity of molecular dimensions, volume $\frac{1}{N}$ and radius $(3/4\pi N)^{\frac{1}{3}}$. The local field produced by E in the empty cavity is the Onsager field:

$$\vec{E}_{\mathrm{cv}} = \frac{3\epsilon_r'}{2\epsilon_r' + 1} \vec{E} \tag{3.19}$$

The difference between the Mosotti field (3.12) and the Onsager field is a reaction field due to the enclosed dipole moment. This field is given by

$$\vec{E}_R = \left(\frac{\epsilon_r + 2}{3} - \frac{3\epsilon_r}{2\epsilon_r + 1} \right) \vec{E} = \frac{2}{3} \frac{(\epsilon_r - 1)^2}{2\epsilon_r + 1} \vec{E} \tag{3.20}$$

The dipole moment contained by the cavity is

$$p = VP = \frac{4}{3}\pi a^3 P = \frac{4}{3}\pi a^3 \epsilon_0 (\epsilon_0 - 1)E$$

Hence

$$E = \frac{3p}{4\pi a^3 \epsilon_0 (\epsilon_r - 1)}$$

so that (3.20) becomes

$$E_R = \frac{p}{4\pi a^3 \epsilon_0} 2 \frac{\epsilon_r - 1}{2\epsilon_r + 1} = \frac{2}{3} \frac{pN}{\epsilon_0} \frac{\epsilon_r - 1}{2\epsilon_r + 1} \qquad (3.21)$$

The field E_R is parallel to the dipole and does not affect it. The only force is the couple due to the Onsager field. Hence, using (3.18),

$$\epsilon_0 (\epsilon_r - 1) E = \frac{Np^2}{3kT} \frac{3\epsilon_r}{2\epsilon_r + 1}$$

and

$$p^2 = \frac{kT\epsilon_0}{N} \frac{(\epsilon_r - 1)(2\epsilon_r + 1)}{\epsilon_r} \qquad (3.22)$$

3.3 Dipole alignment polarization in an alternating field

So far, the fields were considered to be static. In (3.7) and elsewhere, ϵ' represents the permittivity in a static field, which will now be denoted by ϵ_S.

In an alternating field, the orientation of a dipole varies cyclically with the period T of the field. At low frequencies, the dipoles readily synchronize their orientations with the field, but, as the frequency increases, the inertia of the molecule and the binding forces become dominant. The medium becomes tetanized [24], and the dipolar polarization ceases to contribute to the dielectric properties. The permittivity tends toward the optical permittivity ϵ_∞, given by

$$\epsilon_\infty = n^2 \qquad (3.23)$$

where n is the optical index of the medium.

To account for this phenomenon, the relationships established for the static field are modified so that (3.11) and (3.17) become

$$\epsilon_{rs} - 1 = \frac{N}{\epsilon_0} \left(\alpha' + \frac{p^2}{3kT} \right) \tag{3.24}$$

$$\epsilon_{r\infty} - 1 = \frac{N\alpha'}{\epsilon_0} \tag{3.25}$$

from which the contribution of the permanent dipoles to the static permittivity is

$$\epsilon_{rs} - \epsilon_{r\infty} = \frac{Np^2}{3kT\epsilon_0} \tag{3.26}$$

The Clausius-Mosotti expression (3.13) becomes

$$\frac{N}{3\epsilon_0} \left(\alpha' + \frac{p^2}{3kT} \right) = \frac{\epsilon_{rs} - 1}{\epsilon_{rs} + 2} \tag{3.27}$$

$$\frac{N}{3\epsilon_0} \alpha' = \frac{\epsilon_{r\infty} - 1}{\epsilon_{r\infty} + 2} \tag{3.28}$$

and (3.22) becomes

$$p^2 = \frac{9kT\epsilon_0}{N} \frac{(\epsilon_{rs} - \epsilon_{r\infty})(2\epsilon_{rs} + \epsilon_{r\infty})}{\epsilon_{rs}(\epsilon_{r\infty} + 2)^2} \tag{3.29}$$

3.4 Dielectric relaxation

3.4.1 Hysteresis

Relaxation is a general phenomenon [136] due to the delayed response of a system to external forces.

Dielectric relaxation occurs when the electric field that induces polarization in a dielectric is removed. The material takes a certain time to return to molecular disorder, and the polarization subsides exponentially with time constant τ, or relaxation time. This constant is the time necessary for the polarization of the material to fall by a factor of e^{-1} (to 36.79% of its value in the field.)

If P_1 and P_2 are the polarization components due to deformation and dipole alignment, respectively, then in the alternating field

$$E = E_0 e^{j\omega t} \tag{3.30}$$

P_2 tends exponentially to its maximum value

$$P_2 = (P - P_1)\left(1 - e^{-\frac{t}{\tau}}\right) \tag{3.31}$$

According to (3.4) and (3.10)

$$P = (\epsilon_{rs} - 1)\,E \tag{3.32}$$

$$P_2 = (\epsilon_{r\infty} - 1)\,E \tag{3.33}$$

From (3.31):

$$\frac{dP_2}{dt} = \frac{P - P_1 - P_2}{\tau} = \frac{(\epsilon_{rs} - \epsilon_{r\infty})}{\tau} E_0 e^{j\omega t} - \frac{P_2}{\tau} \tag{3.34}$$

for which the solution is

$$P_2 = \frac{(\epsilon_{rs} - \epsilon_{r\infty})E}{1 + j\omega\tau} \tag{3.35}$$

The complex nature of P_2/E implies phase difference, or hysteresis, between the applied field and the polarization, similar to that between the applied field and induction $\vec{D}$.

3.4.2. Debye equation

The complex permittivity [Part I, (2.33)] $\epsilon = \epsilon' - j\epsilon''$ is given by Debye's equation [27], which has the same form as (2.35) :

$$\epsilon = \epsilon_\infty + \frac{\epsilon_S - \epsilon_\infty}{1 + j\omega\tau} \tag{3.36}$$

Hence, separating real and imaginary parts, we have

$$\epsilon' = \epsilon_\infty + \frac{\epsilon_S - \epsilon_\infty}{1 + \omega^2\tau^2} \tag{3.37}$$

$$\epsilon'' = \frac{(\epsilon_S - \epsilon_\infty)\omega\tau}{1 + \omega^2\tau^2} \tag{3.38}$$

where ϵ' is the dispersion factor and ϵ'' the dissipation or loss factor.

The loss factor has a maximum:

$$\epsilon''_{\max} = \frac{\epsilon_S - \epsilon_\infty}{2} \tag{3.39}$$

when $\omega = 1/\tau = \omega_r$, where ω_r is the relaxation frequency. This is not a resonance but an exponential process. For example, $\epsilon''_{\max}/2$ is obtained very far away from $\epsilon''_{\max}$ at $\omega\tau = 2 \pm \sqrt{3}$.

3.4.3 Intermolecular bonds

The weak attractive forces between neighboring molecules contribute to the mechanism responsible for dielectric heating. They give rise to different types of intermolecular bonds.

Hydrogen bonds occur between hydrogen-containing dipoles and an electronegative element, such as oxygen in the water molecule. The tetrahedral structure of ice is due to bonds of this type.

Van der Waals bonds are involved in three types of mechanism: (1) attraction between permanent dipoles that is inversely proportional to the temperature (Keesom force); (2) attraction between a permanent dipole and a nonpolar molecule in which the permanent dipole creates an induced moment

(Debye force); and (3) attraction between nonpolar molecules when momentary fluctuations in the electron pattern give rise to instantaneous dipoles (London force). Such Van der Waals bonds give the medium a certain viscosity [119].

To take the effect of these bonds into account, a correlation parameter g is introduced in the formulas. This parameter depends on the interaction between the given molecule and neighboring molecular layers. The expression given by (3.29) is now replaced by the Kirkwood-Fröhlich equation [39]

$$p^2 = \frac{9kT\epsilon_0}{gN} \frac{(\epsilon_{rs} - \epsilon_{r\infty})(2\epsilon_{rs} + \epsilon_{r\infty})}{\epsilon_{rs}(\epsilon_{r\infty} + 2)^2} \tag{3.40}$$

3.4.4 Relaxation time

The molecular dipole tends to rotate when it experiences the couple (3.14)

$$\vec{C} = \vec{p} \times \vec{E}$$

due to the applied field. Its angular velocity $d\theta/dt$ is proportional to the couple:

$$C = \zeta \frac{d\theta}{dt} \tag{3.41}$$

The coefficient ζ is the frictional factor that depends on the size of the molecule and its intermolecular bonds. For a spherical molecule of radius a, rotating in a liquid of viscocity ξ (in Pa.s [$L^{-1} MT^{-1}$]), Stokes' law gives

$$\zeta = 8\pi\xi a^3 \tag{3.42}$$

from which Debye [27] deduced the following expression:

$$\tau = \frac{1}{\omega_r} = \frac{\zeta}{2kT} = \frac{4\pi a^3 \xi}{kT} \tag{3.43}$$

from which it can be seen that the relaxation frequency increases as the size of the molecule decreases, the temperature increases, and the viscocity decreases. The molecular relaxation time τ' is in general different from the macroscopic relaxation time τ

$$\tau' = A\tau$$

where $A = (\epsilon_{r_\infty} + 2)/(\epsilon_{r_s} + 2)$ in Debye's theory and $2(\epsilon_{r_s} + \epsilon_{r_\infty})/3\epsilon_{r_\infty}$ in the Powles-Glarum treatment of the Onsager field.

3.4.5 Debye and Cole-Cole diagrams

The graphical representation of ϵ' and ϵ'' as a function of frequency constitutes the Debye plot of the material (Fig. 3.3).

When ϵ'' is plotted as a function of ϵ' in the complex plane, (3.36) shows that the result is a semicircle whose center is at $(\epsilon_S + \epsilon_\infty)/2$ on the horizontal axis. This semicircle is the Cole-Cole plot of the material (Fig. 3.4).

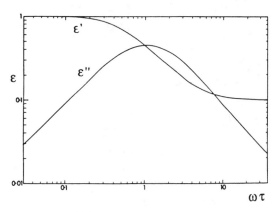

Figure 3.3 Debye plot [50].

In practice, the relaxation spectrum of liquids and solids is often flatter and more extended than predicted by Debye's equation, and the center of the semicircle lies below the horizontal axis. This is described mathematically by introducing the distribution parameter α in the Debye equation.

The modified formula is known as the Cole-Cole equation

$$\epsilon = \epsilon_\infty + \frac{\epsilon_S - \epsilon_\infty}{1 + (j\omega\tau)^{1-\nu}} \tag{3.44}$$

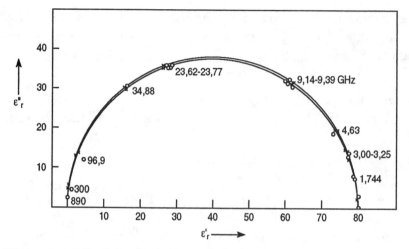

Figure 3.4 Cole-Cole plot for water [50]: $+$ – Debye's equation, $\times$ – Cole-Cole's equation; $\odot$ – experimental data.

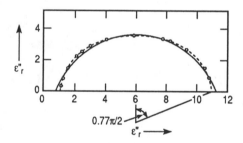

Figure 3.5 Cole-Cole plot (3.74) with $\alpha = 0.23$ [50].

where

$$\nu \in [0,1]$$

The parameter ν is usually less than 0.3 except for polymers, and $\nu\pi/2$ is the angle between the horizontal axis and the radius of the circle joining the center to the intersection with the horizontal axis (Fig. 3.5).

Separating the real and imaginary parts of (3.44) [50], we

obtain

$$\epsilon' = \epsilon_\infty + \frac{(\epsilon_S - \epsilon_\infty)\left[1 + (\omega\tau)^{1-\nu}\sin\frac{\nu\pi}{2}\right]}{1 + 2(\omega\tau)^{1-\nu}\sin\frac{\nu\pi}{2} + (\omega\tau)^{2(1-\nu)}} \qquad (3.45)$$

$$\epsilon'' = \frac{(\epsilon_S - \epsilon_\infty)(\omega\tau)^{1-\nu}\cos\frac{\nu\pi}{2}}{1 + 2(\omega\tau)^{1-\nu}\sin\frac{\nu\pi}{2} + (\omega\tau)^{2(1-\nu)}} \qquad (3.46)$$

The relaxation time no longer has a unique value, but it is distributed around the central value τ_0 according to a function

$$f(\tau) = \frac{\sin\nu\pi}{2\pi\cosh\left[(1-\nu)\ln\frac{\tau}{\tau_0}\right] - \cos\nu\pi} \qquad (3.47)$$

3.5 Different types of dielectrics

3.5.1 Permittivity measurements

As previously noted, the complex permittivity $\epsilon = \epsilon' - j\epsilon''$ is sufficient to describe dielectric behavior. Several methods have been developed for the determination of ϵ' and ϵ'' or ϵ' and $\tan\delta$. Some employ transmission lines and the others use resonant cavities into which the sample is introduced and the resulting perturbation measured. Cavity techniques are generally more accurate (less than 1% and 5% error in ϵ' and ϵ'', respectively).

The description of these techniques is outside the scope of this text. The results obtained for different frequencies and temperatures for a large variety of materials are available in published tables and graphs (e.g. the work performed at MIT [52], [54]), and to more recent data by [2], [10], [17], [18], [19], [23], [29], [32],[35], [44], [45], [49], [57], [62], [64], [66], [68], [69, 70], [88],[91],[92], [93], [95], [96], [98], [103, 104], [109], [110], [116],[121], [122], [123], [124], [129], [139], [145], [147], [148], and [149].

3.5.2 Lowloss dielectrics

Certain compounds have a very regular structure and, hence, low molecular polarity, so that they exhibit practically no dielectric loss. Examples include

polyethylene

$$(-CH_2-CH_2-)_n$$

polypropylene

$$\begin{bmatrix} CH_3 \\ | \\ -CH - CH_2 \end{bmatrix}_n$$

and polytetrafluoroethylene (or PTFE)

$$(-CF_2-CF_2-)_n$$

Theoretically, polyethylene has no dipole moment. At 3 GHz and 25°C, the values of ϵ' and $\tan \delta$ for these materials are, respectively 2.26 and 0.0003 for PE, 2 and 0.0002 - 0.0005 for PP, and 2.1 and 0.00015 for PTFE.

Certain mineral compounds such as silicon ($-O$-$Si R_2$-O-$Si R_2$-$)_n$, alumina, ($Al_2 O_3$), silica ($Si O_2$), and so on have extremely low dielectric losses. For example, for silica, $\epsilon'_r = 3.78$ and $\tan \delta = 0.00006$.

3.5.3 Aqueous dielectrics

Because of its highly asymmetric configuration, the water molecule has an exceptionally high polarity which makes it an ideal material for microwave heating.

The two hydrogen atoms in water form an angle of 105° when joined to the oxygen atom (Fig. 3.6). The distance O-H is 0.096 nm. The resulting dipole moment is directed along the bisector of the angle HOH and is estimated to be 0.62×10^{-29} C m.

The water in aqueous media can take a number of different forms. It can appear as ice. It can also be free water in cavities or capillaries, or spread on a solid surface, with very different

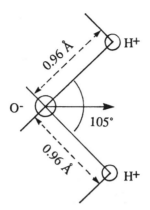

Figure 3.6 The water molecule [86].

properties from those of pure water. It can be bound water, whose properties are not well determined but are thought to be intermediate between those of liquid water and ice. Finally it can be water of crystallization or water of constitution in a number of ionic complexes in which water molecules are hydrogen-bound to negative ions.

At 3 GHz, liquid water has a permittivity ϵ'_r of 80 at 1.5°C and 52 at 95°C. Tangent δ varies over the same temperature range from 0.31 to 0.047. The permittivity ϵ'_r of ice at this frequency is 3.20 and $\tan \delta = 0.0009$. This means that there is a difference of three or four orders of magnitude between absorption in liquid water and ice.

Aqueous materials have a number of relaxations that correspond to different states of water: a few kHz for ice and water of crystallization and about 10 GHz and a few THz for free water. Bound water may exhibit a number of relaxations between those of ice and free water (Fig. 3.7).

Free water, for which maximum absorption occurs at about 10 GHz, exhibits nevertheless significant losses in the commonly used ISM band (915 MHz and 2.45 GHz).

There is an extensive literature on this subject [50], [6]), [12], [14], [26], [38], [59], [60], [65], [71], [72, 73], [87], [90],

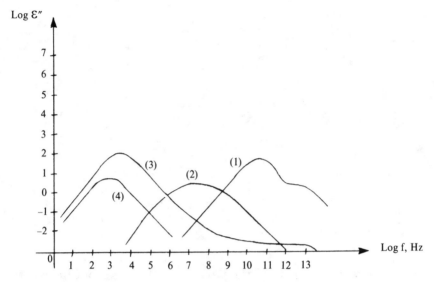

Figure 3.7 Dielectric losses associated with different forms of water: 1 - free water, 2 - bound water, 3 - ice, 4 - water of crystalization [50].

[132], [153].

3.5.4 Mixtures

When a component i is uniformly dispersed in the form of spherical particles in a continuum c, the permittivity ϵ_m of the mixture is given by the Bruggeman equation

$$1 - v_i = \frac{\epsilon_i - \epsilon_m}{\epsilon_i - \epsilon_c} \left(\frac{\epsilon_c}{\epsilon_m} \right)^{1/3} \tag{3.48}$$

where v_i is the volume fraction of the component i $(v_c + v_i = 1)$. For discoidal particles

$$1 - v_i = \frac{\epsilon_i - \epsilon_m}{\epsilon_i - \epsilon_c} \frac{2\epsilon_i + \epsilon_c}{2\epsilon_i + \epsilon_m} \tag{3.49}$$

and for needle-like particles

$$1 - v_i = \frac{\epsilon_i - \epsilon_m}{\epsilon_i - \epsilon_c} \left(\frac{\epsilon_i + 5\epsilon_c}{\epsilon_i + 5\epsilon_m} \right)^{2/3} \tag{3.50}$$

which is valid for large v_i.

Looyenga's equation can be obtained from (3.48), (it was also obtained by Landau and Lifshitz [75]) :

$$\epsilon_m = \left(v_c \epsilon_c^{1/3} + v_i \epsilon_i^{1/3} \right)^3 \tag{3.51}$$

The behavior of particular material is better described by the quadratic law

$$\epsilon_m = \left(v_c \sqrt{\epsilon_c} + v_i \sqrt{\epsilon_i} \right)^2 \tag{3.52}$$

Theories of mixtures were discussed and developed in [30], [31], [33], [34],[68], [99], [105], [106], [124], [142], [144], [146], and [152].

3.5.5 Saline solutions and biological constituents

The dielectric properties of water are greatly modified by the addition of salts. Figure 3.8 shows an example of this. The salt molecule is surrounded by bound water molecules, and the number of the latter is known as the hydration number. For such mixtures,

$$\epsilon = \epsilon_\infty + \frac{\epsilon_S - \epsilon_\infty}{1 + j^{1-\nu}(\omega\tau)^{1-h}} \tag{3.53}$$

$$h = 1 + \frac{1}{\epsilon''} \left[\frac{d\epsilon'}{d(2\pi f t)} \right]_{2\pi f t = 1} \times \cos\frac{\nu\pi}{2}$$

The behavior of biological molecules and of biological tissue is particularly important because of the many applications of these materials in food and agriculture, as well as in medical diathermy. There is an extensive literature on this subject that will be referred to in the following section.

The dipole moment of a protein molecule is given by Oncley's equation

$$p^2 = \frac{2kT\epsilon_S M(\epsilon_S - \epsilon_\infty)}{N_0 C(1 + \psi)} \tag{3.54}$$

where M is the molecular mass, N_0 is Avogadro's number, C is the protein concentration (in kg m^{-3}) and ψ a parameter

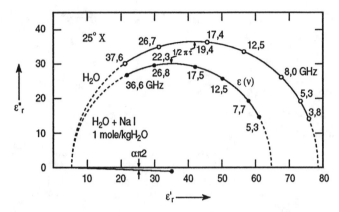

Figure 3.8 Cole-Cole diagram for water (○) and a solution of NaI (●) of 1 mol/l [50].

representing the orientation of water molecules in the vicinity of the protein molecule. The latter is in fact surrounded by two layers of bound water.

At microwave frequencies, the protein molecule with its layers of bound water is too large for its dipole alignment to contribute to the permittivity. A protein solution is therefore equivalent to a nonpolar solution in water.

Amino acids have significantly higher relaxation frequencies. In practice, three relaxations can be observed. They correspond to the amino acid itself, the adjacent bound water, and the free water [50].

The dielectric behavior of biological materials at lower frequencies will be treated in greater detail in Part IV, Section 1.1.

3.6 Heat generation

Microwave heating of a polar dielectric is due to the dissipation of part of the energy of the electromagnetic field. This energy dissipation occurs over the entire relaxation frequency range and is a maximum at the frequency at which ϵ'' passes its maximum.

The molecular mechanisms involved are complex. They can be described as frictional phenomena in which the rotation of the dipoles is hindered by intermolecular bonds, resulting in hysteresis between applied field and polarization.

At the macroscopic level, the power dissipated as heat in a volume v is proportional to the electromagnetic power penetrating this volume (i.e., to the square of the local electric field

$$\frac{P_d}{V} = \sigma_d E_{\text{loc}}^2 \tag{3.55}$$

where σ_d is defined in Part I, (2.81)). Hence

$$\begin{aligned} \frac{P_d}{V} &= \omega \epsilon_0 \epsilon_r'' E^2 = 2\pi f \epsilon_0 \epsilon_r'' E_{\text{loc}}^2 \\ &= 2\pi f \epsilon_0 \epsilon_r' \tan \delta E_{\text{loc}}^2 \end{aligned} \tag{3.56}$$

substituting for the constants, we have (in W m^{-3})

$$\begin{aligned} \frac{P_d}{V} &= 55.6325 \times 10^{-12} f \epsilon_r'' E_{\text{loc}}^2 \\ &= 55.6325 \times 10^{-12} f \epsilon_r' \tan \delta E_{\text{loc}}^2 \end{aligned} \tag{3.57}$$

Optimum heating is obtained when ϵ'' is maximum, not when $\tan \delta$ is maximum; this is due to the fact that ϵ' decreases as resonance approaches.

The local field in the material is different from the incident field. If the radiation propagates along the z direction [Part I, (2.90)], we have

$$P(z) = P_0 e^{-2\alpha z}$$

and, over a distance d_1,

$$\int_0^{d_1} e^{-2\alpha z} dz = \frac{1}{2\alpha}\left(1 - e^{-2\alpha d_1}\right)$$

so that

$$\frac{P_d(d_1)}{V} = 55.6325 \times 10^{-12} f\epsilon_r'' E_0^2 \frac{1}{2\alpha}\left(1 - e^{-2\alpha d_1}\right) \qquad (3.58)$$

This expression is valid for a cylinder of unit cross section and height d_1. The problem is more complex for other shapes. For example, for a spheroid with semi-axes a and b, the second term in (3.57) must be multiplied by a factor [143]

$$F \simeq 1 + \frac{1}{\pi}\frac{a}{b}\left(1 - \frac{b^2}{a^2}\right)^{-b/2a} \qquad (3.59)$$

The problem is also more complex for inhomogeneous materials (different media separated by interfaces), anisotropic materials (permittivity a function of direction), and when permittivity varies significantly with temperature. This last point will be discussed further in Section 3.7.

The heat generated in the medium is transmitted in the material by conduction, convection, and radiation. The heat flux transmitted by conduction in the x direction is given by (in W $[L^2MT^{-3}]$)

$$q_\Lambda = -\Lambda s \frac{dT}{dx} \qquad (3.60)$$

where Λ is the thermal conductivity (in $W\,m^{-1}K^{-1}$ $[LMT^{-3}\Theta^{-1}]$) and s the flux cross sectional area.

Radiation occurs only at the surface of the material. From Stefan's law, the radiated heat flux is (in W)

$$q_r = \sigma_S e S T^4 \qquad (3.61)$$

where σ_S is Stefan's constant (5.67×10^{-8} W m^{-2}K^{-4} [$MT^{-3}\Theta^{-4}$]), S is the radiating area, and e is the emissivity coefficient.

Convection occurs only in liquids and the corresponding heat flux is (in W)

$$q_c = \alpha_c s \Delta T \qquad (3.62)$$

where ΔT is the temperature differential, s the cross sectional area, and α_c is the heat transfer coefficient in (W m^{-2}K^{-1} [$MT^{-3}\Theta^{-1}$]).

Consider a medium in which heat transfer occurs by conduction alone. Microwave heating of the medium can be described by the standard equation of heat transfer, including a term for internal heat release. For a slab

$$\frac{\partial^2 T}{\partial x^2} + \frac{\partial^2 T}{\partial y^2} + \frac{\partial^2 T}{\partial z^2} + \frac{q(x,y,z)}{\Lambda} = \frac{1}{a}\frac{\partial T}{\partial t} \qquad (3.63)$$

where a is the thermal diffusivity (in m$^2 s^{-1}$ [$L^2 T^{-1}$])

$$a \stackrel{\mathrm{d}}{=} \frac{\Lambda}{\rho C_p} \qquad (3.64)$$

where ρ is the density, C_p the specific heat at constant pressure (in J kg^{-1}K^{-1} [$L^2 T^{-2}\Theta^{-1}$]), and $q(x,y,z)$ is the power per unit volume (in W m^{-3}) generated by microwaves at (x,y,z):

$$q = \frac{P_d}{V} \qquad (3.65)$$

For a cylindrical or spherical object, we have

$$\frac{\partial^2 T}{\partial z^2} + \frac{\partial^2 T}{\partial r^2} + \frac{1}{r}\frac{\partial T}{\partial r} + \frac{q(z,r)}{\Lambda} = \frac{1}{a}\frac{\partial T}{\partial t} \qquad (3.66)$$

and

$$\frac{\partial^2 T}{\partial r^2} + \frac{2}{r}\frac{\partial T}{\partial r} + \frac{q(r)}{\Lambda} = \frac{1}{a}\frac{\partial T}{\partial t} \qquad (3.67)$$

respectively. For a slab of thickness d_1 in which $T(x_0, 0)$ is the temperature at x_0 at time $t = 0$, the temperature at x at time t is

$$T(x,t) = \frac{1}{2\pi} \int_0^{d_1} T(x_0, 0) \sqrt{\frac{\pi}{at}} \left[e^{-(x-x_0)^2/4at} \right] dx_0 \qquad (3.68)$$

and if the power generated is $P(x) = P_0 e^{-2\alpha x}$

$$T(x,t) = \frac{1}{2\pi} \int \int_0^{d_1} T(x_0, 0) \sqrt{\frac{\pi}{at}} \left[e^{-(x-x_0)^2/4at} \right] e^{-2\alpha x} dx_0 \, dx$$

$$(3.69)$$

Microwave heatings can be computer modeled using the above equations. In general, the time is discretized in successive intervals Δt and the material in slices Δx to simulate the $T(x)$ curve by a series of segments [143].

3.7 Thermal runaway

Many solid products have very low dielectric losses, but these losses increase with temperature. For example, many metal oxides that are almost lossless at low temperatures, become very lossy at temperatures above a few hundred degrees. Their $\tan \delta$ increases rapidly with temperature, and the phenomenon assumes an avalanche character. It is therefore possible to fuse silica by microwaves provided it can be pre-heated to about 400°C (Fig. 3.9).

Temperature stabilization is possible only if heat can be removed at a sufficiently high rate, or by limiting the microwave power [127]. Roussy et al. [127] have proposed an analysis of thermal runaway in a granular material exchanging heat by convection. They represented ϵ_r'' by Taylor series

$$\epsilon_r''(T) = \epsilon_{r0}'' + \epsilon_{r1}''(T - T_0) + \epsilon_{r2}''(T - T_0)^2 \qquad (3.70)$$

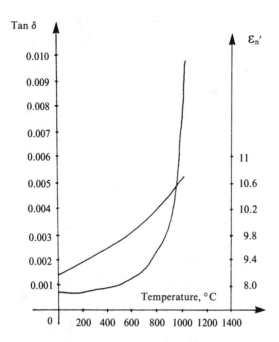

Figure 3.9 Dielectric loss in alumina as a function of temperature [154]

where T_0 is the initial temperature. The equation of conservation of energy then becomes

$$mC_p \frac{\partial T}{\partial t} + q_c = \epsilon_r'' P_i \qquad (3.71)$$

where m is the mass, q_c the heat removed by convection (8.6) with $\Delta T = T - T_0$, and P_i is proportional to the incident power in free space. The solution of (3.71) is

$$\int_0^T \frac{1}{n_0 + n_1 u + n_2 u^2} \, du = t \qquad (3.72)$$

with

$$n_0 = \frac{\epsilon_{r0}'' P_i}{mC_p}, \qquad n_1 = \frac{\epsilon_{r1}'' P_i - q_c}{mC_p}, \qquad n_2 = \frac{\epsilon_{r2}'' P_i}{mC_p}$$

It can be shown that when $n_2 < n_1^2/4n_0$, T tends to a finite limit, and the system is stable; when $n_2 > n_1^2/4n_0$, the system is unstable; and when $n_2 = n_1^2/4n_0$, the system is critical: $T(t)$ is a hyperbola whose horizontal asymptote determines the critical temperature T_c, where

$$T_c - T_0 = \frac{-2n_0}{n_1} = \sqrt{\frac{n_0}{n_2}} = \sqrt{\frac{\epsilon_{r0}''}{\epsilon_{r2}''}} \qquad (3.73)$$

T_c is a characteristic of the material and is the highest possible stable temperature irrespective of the conditions of irradiation and cooling.

In solids and compact materials, thermal equilibrium is established by conduction. Depending on the depth of penetration of the wave, the thermal behavior may be stable in parts of the material and unstable in the others, particularly near the centre.

For example, for (EPDM) (Part II, Section 2.2) (i.e., an ethylene-propylene copolymer) for which
$C_p = 1394$
$q_c = 0.0273\Delta T$
$\epsilon_{r0}'' = 0.42$
$\epsilon_{r1}'' = 0$
$\epsilon_{r2}'' = 8.22 \times 10^{-6}$
$T_c = 244$ degrees C

References on thermal runaway include [21], [117], [126], [27], [39], [41], [52, 53], [76], [77], [94], [97], [77], [94], [118], [125], [130], [135], [136], [141], [143], [160] and [161].

References

1. AARON M., GRANT E., YOUNG S. The dielectric properties of some amino-acids, peptides, and proteins at decimetre wavelengths. In : Molecular relaxation processes. Proceedings of the Chemical Society Symposium. Aberystwyth, 1965. London, Academic Press, 1966, pp. 77-82.

2. AKYEL C., BOSISIO R., CHAHINE R., BOSE T. Measurement of the complex permittivity of food products during microwave-power heating cycles. *J. Microw. Power*, 18, n° 4, 1983, pp. 355-365.

3. ALTSCHULER H. Dielectric constant. In : SUCHER M., FOX J. Handbook of microwave measurements, Vol. 2. Brooklyn, Polytechnic Press, 1962.

4. ARDENNE M. von, GROSS O., OTHERBEIN G. Dispersionsmessungen im Gebiet der dezimeterwellen. *Phys. Z.*, 39, 1936, p. 533.

5. ATTEMA E., De HAAN C., KRUL L. Modelling of microwave heating processes. In : Proceedings of the IMPI Symposium, Ottawa, 1978, pp. 77-79. (Paper 8.3.). (C.D.I.U.P.A., n° 143425).

6. BARKER J. Monte-Carlo studies of the dielectric properties of water-like models. *Mol. Phys.*, 26, n° 3, 1973, pp. 789-792.

7. BAUER E., MAGAT M. Contribution à la théorie de dispersion diélectrique dans les liquides. In : Polarisation de la matière. Colloques Internationaux du CNRS. Paris, CNRS, 1949, pp. 10-13.

8. BECKER G. AUTLER S. Water vapor absorption of electromagnetic radiation in the centimeter wavelength range. *Phys. Rev.*, 70, 1946, pp. 5-6.

9. BENGTSSON N. Dielektrisk varmning inom livsmedelstekniken -matning av livsmedels dielektriska egenskaper samt forsok med upptining och pastorisering. Thesis Ph. D., Göteborg, Chalmers Univ., 1971, 165 p. *D.A.I.* n° 73-31,315. (C.D.I.U.P.A., n° 78801).

10. BENGTSSON N., RISMAN P. Dielectric properties of foods at 3 GHz as determined by a cavity perturbation technique. II. Measurements on food materials. *J. Microw. Power*, 6, n° 2, 1971, pp. 107-123.

11. BENNETT R., HALL G. Measurements of the complex permittivity of liquids at UHF under high pressure conditions. In : BENNETT R., HALL G. Dielectric materials, measurements and applications. Cambridge, 1975, pp. 147-150.

12. BERTEAUD A. Le rôle de l'eau dans les transferts énergétiques. Colloque international des journées d'électronique de Toulouse, 7-11 mars, II-1, 1977, 9 p.

13. BERTEAUD A. Electro-magnetic waves and living systems. In : Symposium international sur les applications énergétiques des micro-ondes, IMPI - CFE. 14, 1979, Monaco, pp. 29-34.

14. BONDARENKO P. The relationship between the moisture content and electrical properties of liquid dielectrics. *Elektrotehn.*, 1, 1973, pp. 38-40.

15. BÖTTCHER C. The polarizability of ions in solution. In : Polarisation de la matière. Colloques internationaux du CNRS. Paris, CNRS, 1949, pp. 69-72.

16. BÖTTCHER C. The polarizability and the internal field. In : Polarisation de la matière. Colloques internationaux du CNRS. Paris, CNRS, 1949, pp. 73-75.

17. BOURGOIN C., VOLF E., JOLY M. Mesure de la permittivité complexe en micro-ondes sur les liquides fortement polaires ; application à l'eau et au diméthylsulfoxyde. *J. Phys. D.*, 5, n° 3, 1972, pp. 589-600.

18. BUCHANAN T. The dielectric properties of some long-chain fatty acids and their methyl esters in the microwave region. *J. Chem. Phys.*, 22, n° 4, 1954, p. 578.

110 Part I - Microwaves

19. BUCKLEY F., MARYOTT A. Tables of dielectric dispersion data of pure liquids and dilute solutions. National Bureau of Standards, Circulation 589, 1958.

20. BUDO A., FISCHER E., MIGAMOTO S. Einfluss der Molekülform auf die elektrische Relaxation. *Phys. Z.*, 11, 1939, p. 337.

21. CHAHINE R., BOSE T. AKYEL C., BOSISIO R. Computer-based permittivity measurements and analysis of microwave power absorption instabilities. *J. Microw. Power*, 19, n° 2, 1984, pp. 127-134.

22. CHAMBERLAIN J., AFSAR M., DAVIES G., HASTED J., ZAFAR M. High-frequency dielectric processes in liquids. *I.E.E.E. Trans. Microw. Theory Tech.*, MTT-22, 12, 1974, pp. 1028-1032.

23. CHARREYRE-NEEL M., THIEBAUT J., ROUSSY G. Permittivity measurements of materials during high-power microwave irradiation and processing. *J. Phys.*, E.17, 1984, pp. 678-682.

24. COPSON D. Theory of microwave heating. In : COPSON D. Microwave heating. 2nd ed. Westport, Avi Publishing Co., 1975, pp. 1-34. (C.D.I.U.P.A., n° 93299).

25. DALBERT, MAGAT M., SURDUT A. Dispersion diélectrique dans les alcools normaux. In : Polarisation de la matière. Colloques internationaux du CNRS, Paris, CNRS, 1949, pp. 14-21.

26. DAVIDSON D. Dielectric relaxation and rotational disorders in ices. In : Molecular relaxation processes. Proceedings of the Chemical Society Symposium. Aberystwyth, 1965. London, Academic Press, 1966, pp. 15-19.

27. DEBYE P. Polar molecules. New York, Chemical Catalog Co., 1929 et New York, Dover Publications, 1945.

28. DEBYE P. Hochfrequenzverluste und Molekülstruktur. *Phys. Z.*, 35, 1934, p. 101.

29. DELBOS G., BOTTREAU A., MARZAT C., SALEFRAN J. Microwave dielectric relaxation of aqueous solutions of dextran. *J. Microw. Power*, 13, n° 1, 1978, pp. 69-75.

30. DE LOOR G. Dielectric properties of mixtures containing water. *J. Microw. Power*, 3, n° 2, pp. 67-73.

31. DE LOOR G. Dielectric properties of heterogenous systems containing water. In : Proceedings of the IMPI Symposium, Boston, 1969, (abstract p. 31), (Paper F2).

32. DE LOOR G., MEIJBOOM F. The dielectric constant of foods and other materials with high water contents at microwave frequencies. *J. Food Technol.*, 1, 1966, pp. 313-322.

33. DUBE D. Study of Landau-Lifshitz-Looyenga's formula for dielectric correlation between powder and bulk. *J. Phys. D. : Appl. Phys.*, 3, 1970, pp. 1648-1652.

34. DUBE D., YAVADA R., PARSHAD R. A formula for correlating dielectric constants of powder and bulk. *J. Phys. D. : Appl. Phys.*, 9, 1971, pp. 719-721.

35. ELEY D., PETHIG R. Microwave dielectric and Hall effect measurements on biological materials. *Discuss. Faraday Soc.*, 51, 1971, pp. 164-175.

36. FISCHER E. Dielektrische Relaxationsuntersuchungen im Hinblick auf die molekulare und intermolekulare Struktur von Dipolflüssigkeiten. *Phys. Z.*, 21, 1939, p. 645.

37. FOSTER K., AYYASWAMY P., SUNDARARAJAN T., RAMAKRISHNA K. Heat transfer in surface-cooled objects subject to microwave heating. *I.E.E.E. Trans. Microw. Theory Tech.*, MTT-30, n° 8, 1982, pp. 1158-1166.

38. FREYMANN M., FREYMANN R. Influence de la température sur l'adsorption ultra-hertzienne de l'eau de cristallisation et de l'eau d'adsorption. *Comptes Rendus Acad. Sci.*, 2 B2, 1951, p. 1096.

39. FRÖHLICH H. Theory of dielectrics. London, Oxford Univ. Press, 1949.

40. GANDAR J. Über die mechanische und dielektrische Zerkleinerung von Getreide- und Ölsaaten. *Fette Seifen Anstrichm.*, 70, n° 6, 1968, pp. 433-439.

41. GAY H. Structure et propriétés électroniques de la matière. In : Propriétés diélectriques de la matière. Tome II, livre V, Toulouse, ENSEEIHT, 1970, 33 p.

42. GEVERS M., DU PRE F. Power factor and temperature coefficient of solid (amorphous) dielectrics. *Trans. Faraday Soc.*, 43, 1947, pp. 47-55.

43. GILES P., MOORE E., BOUNDS L. Investigation of heat penetration of food samples at various frequencies. *J. Microw. Power*, 5, n° 1, 1970, pp. 40-43.

44. GOLDBLITH S. Dielectric properties of foods and their importance in processing with microwave energy. *Microw. Energy Appl. Newsl.*, 7, n° 3, 1974, pp. 9-14.

45. GOLDBLITH S., WANG D. Dielectric properties of foods. U.S. Army Natick Development Center, Technical Report TR 76-27-FEL, 1975.

46. GRANT E., KEEFE S., TAKASHIMA S. The dielectric behaviour of aqueous solutions of bovine serum albumin from radio wave to microwave frequencies. *J. Phys. Chem.*, 72, n° 13, 1968, pp. 4373-4380.

47. HAKIM R. Conduction, breakdown and dielectric properties of insulating liquids. *Dig. Lit. Dielectr.*, 35, 1971, pp . 208-248.

48. HAMID M., BHARTIA P., STUCHLY S. Microwave energy in processing liquids. Progress report n° 4, Microwave applications project, Industrial Enterprise Fellowship Program Report, Manitoba Research Council, March 1972, 56 p.

49. HAMID M., RZEPECKA M., STUCHLY S. Routine permittivity measurements of granular substances at microwave frequencies. Progress report n° 6, Microwave applications project, Industrial Enterprise Fellowship Program Report, Manitoba Research Council, May 1972, 18 p.

50. HASTED J. Aqueous dielectrics. London, Chapman & Hall, 1973, 302 p.

51. HENNELLY E., HESTON W., SMYTH C. Microwave absorption and molecular structure in liquids. III. Dielectric relaxation and structure in organic halides. *J. Am. Chem. Soc.*, 70, 1948, p. 4102.

52. HIPPEL A. von. Dielectrics-and waves. 1st ed. Cambridge, M.I.T. Press, 1966, 284 p.

53. HIPPEL A. von. Dielectric materials and applications. 1st ed. Cambridge, M.I.T. Press, 1966, 438 p.

54. HO W., HALL W. Measurements of dielectric properties of sea water and NaCl solutions at 2,65 GHz. *J. Geophys. Res.*, 78, n° 27, pp. 6301-6315.

55. HOFFMANN J. Dielectric relaxation in molecular crystals : multiple site models. In : Molecular relaxation processes. Proceedings of the Chemical Society Symposium. Aberystwyth, 1965. London, Academic Press, 1966, pp. 47-60.

56. HONZOWA T. Heating of ions by modulated microwaves. *Phys. Lett. A*, A 40, n° 4, 1972, p. 335.

57. HU C. Online measurements of the fast-changing dielectric constant in oil shale due to high-power microwave heating. *I.E.E.E. Trans. Microw. Theory Tech.*, MTT-27, n° 1, 1979, pp. 38-43.

58. ISHII T. Theory of continuing heating after microwave radiation is turned off. In : Proceedings of the IMPI Symposium, Chicago, 1985, pp. 141-146.

59. ITAGAKI K. Dielectric properties of dislocation-free ice. *Bull. Am. Phys. Soc.*, 21, n° 6, 1976, p. 896.

60. JOHARI G. Dielectric properties of ice VII and VIII and phase boundary between ice VI and VII. *J. Chem. Phys.*, 61, n° 10, 1974, pp. 4292-4300.

61. JORDAN E., BALMAIN K. Electromagnetic waves and radiating systems. 2nd ed. Englewood Cliffs, Prentice Hall, 1968, 753 p.

62. KAMATH K. Variation of dielectric properties of some vegetable oils in liquid-solid transition phase. *Indian J. Technol.*, 9, n° 8, 1971, p. 312.

63. KAUZMANN W. Dielectric relaxation as a chemical rate process. *Rev. Mod. Phys.*, 14, 1942, p. 12.

64. KENT M. Complex permittivity of protein powders at 9,4 GHz as a function of temperature and hydration. *J. Phys. D. : Appl. Phys.*, 5, 1972, pp. 394-409.

65. KENT M., MEYER W. Dielectric relaxation of adsorbed water in microcrystalline cellulose. *J. Phys. D. : Appl. Phys.*, 16, 1983, pp. 915-925.

66. KENT M., MEYER W. Complex permittivity spectra of protein powders as a function of temperature and hydration. *J. Phys. D. : Appl. Phys.*, 17, 1984, pp. 1687-1698.

67. KOSTENKO E. Absorption d'énergie par processus de relaxation dans les diélectriques (en russe). *Elektr. Tehnol.*, 4, 1971, p. 135.

68. KRASZEWSKI A. Prediction of the dielectric properties of two-phase mixtures. *J. Microw. Power*, 12, n° 3, 1977, pp. 215-222.

69. KRASZEWSKI A. A model of the dielectric properties of wheat at 9.4 GHz. In : Proceedings of the IMPI Symposium, Ottawa, 1978, pp. 28-30. (Paper 4.2.). (C.D.I.U.P.A., n° 143425).

70. KRASZEWSKI A. A model of the dielectric properties of wheat at 9.4 GHz. *J. Microw. Power*, 13, n° 4, 1978, pp. 293-296.

71. KRASZEWSKI A., KULINSKI S., MATUSZEWSKI M. Dielectric properties and a model of biphase water suspension at 9.45 GHz. *J. Appl. Phys.*, 47, n° 4, 1976, pp. 1275-1277.

72. LAGOURETTE B. Study of the dielectric properties of disperse microcrystals of ice near melting temperature. I. Experimental results. *J. Phys.*, 37, n° 7-8, 1976, pp. 945-954.

73. LAGOURETTE B. Study of dielectric properties of disperse microcrystals of ice near melting temperature. II. Discussions and interpretation. *J. Phys.*, 37, n° 7-8, 1976, pp. 955-964.

74. LANDAU L., LIFSHITZ E. Electrodynamics of continuous media, 8, Course of theoretical physics. Oxford, Pergamon Press, 1960, pp. 45-47.

75. LANDAU L., LIFSIC E. Elektrodinamika splošnyh sred. Moskva, Nauka, 1959. Trad. fr. : Electrodynamique des milieux continus. Moskva, Mir, 1969, 536 p.

76. LEFEUVRE S. Dipôles et polarité. *Rev. Gén. Electr.*, 11, nov. 1981, p. 792.

77. LEFEUVRE S. Propriétés caractéristiques du chauffage micro-ondes. *Rev. Gén. Electr.*, 11, nov. 1981, pp. 793-801.

78. LENTZ R. On the microwave heating of saline solutions. *J. Microw. Power*, 15, n° 2, 1980, pp. 107-111.

79. LE PETIT J., DELBOS G., BOTTREAU A. DUTUIT Y., MARZAT C., CABANAS R. Dielectric relaxation of emulsions of saline aqueous solutions. *J. Microw. Power*, 12, n° 4, 1977, pp. 335-340.

80. LOEB H., YOUNG G., QUICKENDEN P., SUGGETT A. New methods for measurement of complex permittivity up to 13 GHz and their application to the study of dielectric relaxation of polar liquids including aqueous solutions. *Ber. Bunsenges. Phys. Chem.*, 75, 1971, pp. 1155-1165.

81. LOOYENGA H. Dielectric constants of heterogeneous mixtures. *Physica*, 31, 1965, pp. 401-406.

82. MAGALINSKIJ V. Théorie élémentaire de la perméabilité diélectrique des matériaux polaires (en russe). *Fiz. Tverd. Tela*, 17, n° 7, 1975, pp. 2135-2136.

83. MALECKI J. Theory of non linear dielectric effects in liquids. *J. Chem. Soc. Faraday Trans.*, 2, 72, 1976, pp. 104-112.

84. MALLEMANN R. de. Les modalités de la polarisation électro-optique et la corrélation des phénomènes. In : Polarisation de la matière. Colloques internationaux du CNRS. Paris, CNRS, 1949, pp. 144-156.

85. MANOURY M., ROCHAS J., GUILLEMAUT H., CHANAL B. Interaction eau-cellulose : application au chauffage diélectrique. *Rev. Gén. Electr.*, 11, nov. 1981, pp. 839-846.

86. MASSIEU G. Chimie minérale. Paris, Baillière et fils, 1966, 381 p.

87. MASSZI G. The dielectric characteristics of bound water. *Acta Biochim. Biophys. Acad. Sci. Hung.*, 5, n° 3, 1970, pp. 321-331.

88. METAXAS A. Properties of materials at microwave frequencies. Transactions of the IMPI, 2. Industrial Applications of Microwave Energy, 1973.

89. METAXAS A., MEREDITH R. Industrial microwave heating. Stevenage, Peter Peregrinus Ltd., 1983, IEEE Power Eng. Series n° 4, 357 p.

90. MEYER W., SCHILZ W. Microwave absorption by water in organic materials. Dielectric materials, measurements and applications. Publ. IEE, 177, 1979, pp. 215-219.

91. MIDDLEHOEK J., BÖTTCHER C. The dielectric behaviour of the four straight-chain heptanols. In : Molecular relaxation processes. Proceedings of the Chemical Society Symposium. Aberystwyth, 1965. London, Academic Press, 1966, pp. 69-75.

92. MOHSENIN N. Electromagnetic radiation properties of foods and agricultural products. New York, Gordon and Breach Science Publishers, 1983, 672 p. (C.D.I.U.P.A., n° 205642).

93. MUDGETT R. Electrical properties of foods in microwave processing. *Food Technol.*, 36, n° 2, 1982, pp. 109-115. (C.D.I.U.P.A., n° 165665).

94. MUDGETT R. Microwave properties and heating characteristics of foods. *Food Technol.*, 40, n° 6, 1986, pp. 84-93. (C.D.I.U.P.A., n° 214710).

95. MUDGETT R., GOLDBLITH S., WANG D., WESTPHAL W. Prediction of dielectric properties in solid foods at ultrahigh and microwave frequencies. In : Proceedings of the IMPI Symposium, Waterloo (Can.), 1975, pp. 272-274. (Paper 16.4.).

96. MUDGETT R., GOLDBLITH S., WANG D., WESTPHAL W. Dielectric behavior of a semi-solid food at low, intermediate and high moisture contents. *J. Microw. Power*, 15, n° 1, 1980, pp. 27-36.

97. MUDGETT R., SCHWARTZBERG H. Microwave food processing : pasteurization and sterilization - a review. American Institute of Chemical Engineers, Symposium Series 78, n° 218, 1982, 11 p.

98. MUDGETT R., SMITH A., WANG D., GOLDBLITH S. Prediction of the relative dielectric loss factor in aqueous solutions of non fat dried milk through chemical simulation. *J. Food Sci.*, 36, n° 6, 1971, pp. 915-918. (C.D.I.U.P.A., n° 41453).

99. MUDGETT R., WANG D., GOLDBLITH S. Prediction of dielectric properties of oil-water and alcohol-water mixtures at 3000 MHz, at 25° C, based on pure components properties. *J. Food Sci.*, 39, n° 3, 1974, pp. 632-635. (C.D.I.U.P.A., n° 73346).

100. MULLER F., SCHMELZER C. Dielektrisches Verhalten in Zusammenhang mit dem polaren Aufbau der Materie. *Ergeb. exakten Nat.-wiss.*, 25, 1951, p. 359.

101. NAJIM M., MATHEAU J., LEFEUVRE S. Microwave Kerr effect on polar liquids. *Appl. Phys. Lett.*, 21, n° 8, 1972, pp. 399-400.

102. NEKRASOV L., RIKENGLAZ L. Chauffage par micro-ondes d'un diélectrique avec amortissement en température (en russe). *Ž. Teh. Fiz.*, 43, n° 4, 1973, pp. 694-697.

103. NELSON S. Electrical properties of agricultural products. *Trans. Am. Soc. Agric. Eng.*, 16, n° 2, 1973, pp. 384-400.

104. NELSON S. Dielectric properties of agricultural products. In : Proceedings of the IMPI Symposium, Loughborough, 1973, 2 p. (Paper 6B1/1-2).

105. NELSON S. Observations on the density dependence of dielectric properties of particulate materials. *J. Microw. Power*, 18, n° 2, 1983, pp. 143-152.

106. NELSON S. Density dependence of the dielectric properties of wheat and whole-wheat flour. *J. Microw. Power*, 19, n° 1, 1984, pp. 55-64.

107. OGURA K., NAMBA I., ARAI S., ITOH K. A study of X-band heating. *J. Microw. Power*, 14, n° 3, 1979, pp. 269-273.

108. OHLSSON T., BENGTSSON N., Microwave heating profiles in foods. *Microw. Energy Appl. Newsl.*, 4, n° 6, 1971, pp. 3-8.

109. OHLSSON T., BENGTSSON N., RISMAN P. Dielectric data of foods at 915 MHz and 2800 MHz as a function of temperature - a comparison. In : Proceedings of the IMPI Symposium, Loughborough, 1973, 7 p. (Paper 3A2/1-7).

110. OHLSSON T., BENGTSSON N., RISMAN P. The frequency and temperature dependence of dielectric food data as determined by a cavity perturbation technique. *J. Microw. Power*, 9, n° 2, 1974, pp. 129-145.

111. OHLSSON T., HENRIQUES M., BENGTSSON N. Dielectric properties of model meat emulsions at 900 and 2800 MHz in relation to their composition. *J. Food Sci.*, 39, n° 6, 1974, pp. 1153-1156. (C.D.I.U.P.A., n° 78231).

112. OHLSSON T., RISMAN P. Temperature distribution of microwave heating : spheres and cylinders. *J. Microw. Power*, 13, n° 4, 1978, pp. 303-309.

113. OSTER G. The dielectric properties and structure of liquids and solutions. In : Polarisation de la matière. Colloques internationaux du CNRS. Paris, CNRS, 1949, pp. 22-27.

114. PACE W. Dielectric properties of certain foodstuffs at selected frequencies and at different temperatures. *J. Microw. Power*, 2, n° 3, 1967, p. 98. (Thesis summary).

115. PACE W., WESTPHAL W., GOLDBLITH S. Dielectric properties of commercial cooking oils. *J. Food Sci.*, 33, n° 1, 1968, pp. 30-36. (C.D.I.U.P.A., n° 1892).

116. PEREIRA R. Permittivity of some dairy products at 2450 MHz. *J. Microw. Power*, 9, n° 4, 1974, pp. 277-288.

114 Part I - Microwaves

117. PIQUEMAL P., THIEBAUT J., BENNANI
A., ROUSSY G. Etude de l'emballe-
ment thermique des matériaux irra-
diés par micro-ondes. In : Comptes-
rendus du Colloque sur les Applications énergetiques des rayonnements.
Toulouse, Univ. Paul Sabatier, 12
juin 1986. Toulouse, ENSEEIHT, pp.
119-121.

118. PRIOU A. Interactions avec les milieux diélectriques, applications
industrielles des micro-ondes. In :
Ensemble des travaux, doctorat
d'Etat, Univ. Paul Sabatier, Toulouse, 1981, Chap. II, pp. 367-377.

119. PROVOST P. Chimie générale. Paris,
Masson et Cie, 1968, 364 p.

120. RAVAILLE M. Chimie générale. Paris,
Baillière et fils, 1968, 370 p.

121. RISMAN P., BENGTSSON N. Dielectric
properties of foods at 3 GHz as
determined by a cavity perturbation
technique : I. Measuring technique.
J. Microw. Power, 6, n° 2, 1971,
pp. 101-106.

122. RISMAN P., OHLSSON T. Dielectric
constants for high loss materials.
J. Microw. Power, 8, n° 2, 1973,
pp. 185-188.

123. ROEBUCK B., GOLDBLITH S. Dielectric
properties at microwave frequencies
of agar gels. Similarity to the dielectric properties of water. J. Food
Sci., 40, n° 5, 1975, pp. 899-902.
(C.D.I.U.P.A., n° 92670).

124. ROEBUCK B., GOLDBLITH S., WESTPHAL W.
Dielectric properties of carbohy-
drate-water mixtures at microwave
frequencies. J. Food Sci., 37, n° 2,
1972, pp. 199-204.
(C.D.I.U.P.A., n° 43586).

125. ROUSSY G. Dielectric materials. In :
Symposium international sur les
applications énergétiques des micro-
ondes, IMPI-CFE. 14, 1979, Monaco.
Tome 2, pp. 15-20.

126. ROUSSY G., BENNANI A., THIEBAUT J.
Temperature runaway of microwave-
irradiated materials. In : Proceedings of the IMPI Symposium, Chicago,
1985, pp. 97-99.

127. ROUSSY G., MERCIER A., THIEBAUT J.,
VAUBOURG J. Temperature runaway of
microwave-heated materials, study
and control. J. Microw. Power, 20,
n° 1, 1985, pp. 47-51.

128. ROUSSY G., THIEBAUT J. Calorimetric
microwave loss measurements of granular materials. J. Microw. Power,
19, n° 3, 1984, pp. 181-186.

129. RZEPECKA M., PEREIRA R. Dielectric
properties of dairy products at
2450 MHz. In : Proceedings of the
IMPI Symposium, Loughborough, 1973,
3 p. (Paper 3 A5/1-3).

130. SCAIFE B. The role of dipole-dipole
coupling in dielectric relaxation.
In : Molecular relaxation processes.
Proceedings of the Chemical Society
Symposium. Aberystwyth, 1965. London,
Academic Press, 1966, pp. 15-19.

131. SCHWAN H. Molecular response characteristics to ultra-high frequency
fields. In : Proceedings of the 2nd
Tri-Service Conference on Biological
effects of microwave energy. Rome,
New York, 1958, p. 33.

132. SCHWAN H. Electrical properties of
bound water. Ann. N.Y. Acad. Sci.,
125, art. 2, 1965, pp. 344-354.

133. SCHWARZ G., SAITO M., SCHWAN H.
On the orientation of non-spherical
particles in an alternating electrical field. J. Chem. Phys., 43, 1965,
pp. 3562-3569.

134. SMYTH C. Dielectric constant and
molecular structure. In : Chemical
Catalog Co., New York, 1931.

135. SMYTH C. Microwave absorption and
molecular structures in liquids.
In : Polarisation de la matière.
Colloques internationaux du CNRS.
Paris, CNRS, 1949, pp. 5-9.

136. SMYTH C. Dielectric relaxation times.
In : Molecular relaxation processes.
Proceedings of the Chemical Society
Symposium. Aberystwyth, 1965. London,
Academic Press, 1966, pp. 1-13.

137. SMYTH C. The absorption of microwaves by molecules. In : Proceedings
of the IMPI Symposium, Ottawa, 1972,
pp. 60-62. (Paper 4.3.).

138. SRIVASTA G., MATHUR P., TRIPATHI N.,
DAWAR A. Effect of frequency dependence of viscosity on dielectric
relaxation in liquids. Indian J.
Pure Appl. Phys., 10, n° 9, 1972,
pp. 696-697.

139. STUCHLY S. Dielectric properties of
some granular solids containing
water. J. Microw. Power, 5, n° 2,
1970, pp. 62-68.

140. STUCHLY S., HAMID M. Physical parameters in microwave heating processes. J. Microw. Power, 7, n° 2,
1972, pp. 117-137.

141. STUCHLY M., STUCHLY S. Industrial, scientific, medical and domestic applications of microwaves. *IEE Proc.*, 130, partie A, n° 8, nov. 1983, pp. 467-503.

142. TAYLOR L. Dielectric properties of mixtures. *I.E.E.E. Trans. Antennas Propag.*, AP-13, n° 6, 1965.

143. THOUREL L. Comportement de la matière dans un champ de haute fréquence. Toulouse, Centre d'Etudes et de Recherches (C.E.R.T.-O.N.E.R.A.), 1976, 54 p.

144. TINGA W. Multiphase dielectric theory applied to cellulose mixtures. Thesis Ph. D., Univ. Alberta, Edmonton, Dep. Electr. Eng., 1969.

145. TINGA W., NELSON S., Dielectric properties of materials for microwave processing-tabulated. *J. Microw. Power*, 8, n° 1, 1973, pp. 23-65.

146. TINGA W., VOSS W., BLOSSEY D. Generalized approach to multiphase dielectric mixture theory. *J. Appl. Phys.*, 44, 1973, p. 3897.

147. TO E. Dielectric properties of meats. Thesis.Ph. D., Massachusetts Inst. Technol., Cambridge, 1974.

148. TO E., GOLDBLITH S., WANG D., DECAREAU R. Dielectric properties of foodstuffs. In : Proceedings of the IMPI Symposium, Milwaukee, 1974, 2 p. (Paper B3-2/1-2).

149. TO E., MUDGETT R., WANG D., GOLBLITH S., DECAREAU R. Dielectric properties of food materials. *J. Microw. Power*, 9, n° 4, 1974, pp. 303-315.

150. TRAN V. Variations of dielectric properties of rice with bulk densities. In Proceedings of the IMPI Symposium, Chicago, 1985, p. 181.

151. TRAN V., STUCHLY S., KRASZEWSKI A. Dielectric properties of selected vegetables and fruits 0.1-10.0 GHz. *J. Microw. Power*, 19, n° 4, 1984, pp. 251-258.

152. VAN BEEK L. Dielectric behavior of heterogeneous systems. In : BIRKS J. Progress in dielectrics. 7. London, Heywood Books, 1967, pp. 69-114.

153. VANT M., GRAY R., RAMESEIER R., MAKIOS V. Dielectric properties of fresh and sea ice at 10 and 35 GHz. *J. Appl. Phys.*, 45, n° 11, 1974, pp. 4712-4717.

154. WESTPHAL W., SILS A. Dielectric constant and loss data. Technical Report AFML-TR-72-39 (Air Force materials laboratory, Air Force systems command, Wright-Patterson Air Force base), Dayton (Ohio), 1972, 224 p.

155. WHEELER H., LODWIG R. The thermal dielectric quotient for characterizing dielectric heat conductors. *IEEE Trans. Microw. Theory Tech.*, MTT-29, n° 11, 1981, pp. 1231-1233.

156. WHITE J. Physics of microwave heating. In : Proceedings of the IMPI Symposium, Stanford, 1967, p. 44. (Abstracts).

157. WHITE J. The intensive and extensive microwave properties of materials. In : Proceedings of the IMPI Symposium, Boston, 1969, p. 30, (Abstracts), (Paper F-1).

158. WHITE J. Why materials heat. Transactions of the IMPI, 1, Microwave power in industry. 1973.

159. Digest of the literature on dielectrics. Vol. XIX (1965), Vol. XXII (1959). National Academy of Science, National Research Council, Washington.

160. Initiation à la technique des hyperfréquences. Thomson-Houston, Groupe Electronique, Ca. 1970, réf. 1066 456, Chap. V. Propagation des ondes électromagnétiques dans des milieux homogènes et isotropes, 57 p.

161. Molecular relaxation processes. Proceedings of the Chemical Society Symposium. Aberystwyth, 1965. London, Academic Press, 1966, 41 papers, 303 p. (Chemical Society Special Publication n° 20).

Chapter 4

Generators and applicators

4.1 Introduction

Irrespective of their use in industrial, scientific, medical, or simply domestic applications, microwave appliances have two main components: a microwave generator with its electric power supply and an applicator. The latter may be an antenna for direct irradiation (hyperthermia) or a resonant or oversized cavity in which the treatment takes place. The connection between the two components is assured by a waveguide or a coaxial cable, depending on the power transmitted. When the applicator is a cavity, it is coupled to the transmission line by an antenna-type junction.

4.2 Microwave generators

The use of solid-state oscillators in microwave appliances has been the subject of a number of patents since the seventies [14], [18]. The power outputs available with silicon transistors are on the order of 100 W at 950 MHz and 15 W at 2.45 GHz [63]. These power levels are likely to double in the near future, which should lead to an increase in the use of solidstate sources, particularly in medical applications and in domestic ovens.

At present, the production of the high powers required for most ISM applications requires the use of vacuum tubes. There are two types of tube: linear beam tubes (type O) and cross field tubes (type M) in which the electron beam follows a curved trajectory under the combined effect of mutually orthogonal electric and magnetic fields.

The *klystron* and the *traveling wave tube* belong to the former category and the *magnetron* to the latter. The high-power klystron can provide tens of kilowatts at 2.45 GHz, but is rarely used for practical and economic reasons. The traveling wave tube is used in medical diathermy equipment. As for the magnetron (Part I, Section 1.4), it has always been the source used in radar and in microwave heating.

4.2.1 The magnetron

The magnetron (Fig. 4.1) is a circular symmetric tube containing a hollow cylindrical anode with a directly or indirectly heated tungsten cathode along its axis. In the case of an indirectly heated filament, the latter is wound on the cathode.

The radial separation between the anode and cathode defines the highly evacuated region of interaction. A number of resonant cavities of different shape (Fig. 4.2) are placed in the curved interior of the anode block.

A constant potential difference is applied between the anode and cathode. It amounts to several kilovolts for an electrode separation of a few millimeters. Permanent magnets or an electromagnet produce a magnetic field parallel to the axis of the tube.

Figure 4.1 Photograph of a magnetron [213].

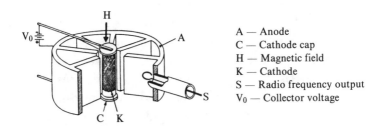

A — Anode
C — Cathode cap
H — Magnetic field
K — Cathode
S — Radio frequency output
V_0 — Collector voltage

Figure 4.2 Diagram of a magnetron [4].

Electrons emitted by the cathode are accelerated by the electric field and follow radial trajectories on their way toward the anode. The magnetic field bends these trajectories into almost cycloidal paths (Fig. 4.3). The critical magnetic induction B_c

is defined as a function of the applied potential V_0 by

$$V_0 = \frac{e}{8m} B_c b^2 \left(1 - \frac{a^2}{b^2}\right)^2 \tag{4.1}$$

where a and b are the radii of the cathode and anode, respectively, and e and m the charge and mass of the electron at rest ($e = 1.6021 \times 10^{-19}$C, $m = 1.6748 \times 10^{-27}$kg), so that

$$V_0 = 1.1957 \times 10^7 B_c b^2 \left(1 - \frac{a^2}{b^2}\right)^2 \tag{4.2}$$

For magnetic inductions greater than B_c, the electrons cannot reach the anode and form a space-charge cloud that rotates in the interaction region. As B increases, the space-charge cloud comes closer to the cathode. The anode and its cavities constitute a periodic structure that interacts with the electron cloud. Electrons passing a cavity undergo acceleration or deceleration, depending on the direction of the local RF field (if the latter is already established). Accelerated electrons take energy from the RF field and move back toward the anode. They heat the cathode and contribute to the emission of secondary electrons by it. Decelerated electrons give up energy to the RF field and continue on their way toward the anode.

An anode with N cavities (N being even) has $N/2 + 1$ resonant frequencies that correspond to phase differences of $0, 2\pi/N, 4\pi/N, ..., \pi$. When the angular frequency ω_e of the electrons is such that $(\omega/\omega_e)(2\pi/N)$ is equal to one of the above phase differences plus or minus $2k\pi$, an oscillation of angular frequency ω is found to occur. A phase difference of π is usually selected to give the best output. Because of the acceleration and deceleration to which they are subjected, the electrons tend to form bunches that pass in front of a resonant cavity in the time equal to the RF period (i.e., $2\pi/\omega$). This fundamental mode of operation is known as the π mode [52].

On average, the RF energy required to send an electron back toward the cathode is less than the kinetic energy given up by

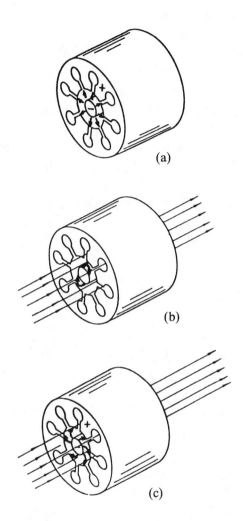

Figure 4.3 Electron trajectory in the anode block of a magnetron [200]:
a–electric field only, *b*–magnetic field only, *c*–superposition of the two
orthogonal fields.

an electron on reaching the anode. The process is globally
stable, the RF oscillations in the cavities are self-sustaining,
and the excess RF energy is extracted by a coupling loop and

sent to the coaxial exit of the tube where it takes the form of a TEM wave (Fig. 4.4).

Most magnetrons have short metal straps that connect half the intercavity poles. Their function is to separate the different resonant modes. The same result can be obtained by designing the cavities in the form of short and long strips known as a rising sun structure.

The numerical modeling of magnetrons has been studied by a number of authors [20], [52], [65], [87], [92], [159].

A magnetron is usually characterized by its operating chart in which the abscissa represents the anode current and the ordinate the anode potential. Lines of equal magnetic field are also plotted on the chart together with output power and efficiency.

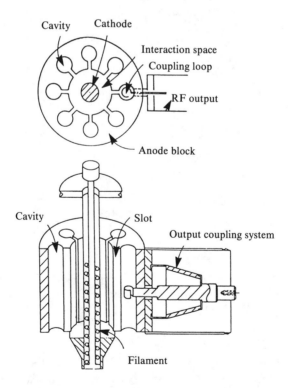

Figure 4.4 Output coupling of a magnetron [200].

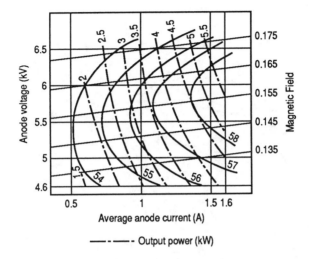

Figure 4.5 Operating chart of the MCF 1327 magnetron [212]: dot-dash curves-power output (kW), solid lines-efficiency (percent).

For example, the magnetron in Figure 4.5 would operate with an efficiency of 58% for an anode potential, current, and magnetic field of 6.1 kV, 1.42 A, and 0.163 T, respectively.

The output impedance, characterized by the *voltage standing wave ratio* (VSWR) and the phase, determines the oscillation frequency and the output power. The phase represents the distance, measured in fractions of $\lambda_g/2$, between the first minimum of the standing wave and a plane of reference specified by the manufacturer for the tube output. The Rieke diagram of the magnetron (Fig. 4.6) shows the frequency and power output under different output conditions. Optimum operating conditions are not generally obtained for a matched output impedance ($\tau = 1$) [170].

The efficiency of the magnetron can be quite high (i.e., on the order of 60–65%). The remaining 35–40% contributes to the heating of the cathode or is eliminated through Joule power dissipation in the anode where it is removed by means of radiator fins or a circulating water jacket system.

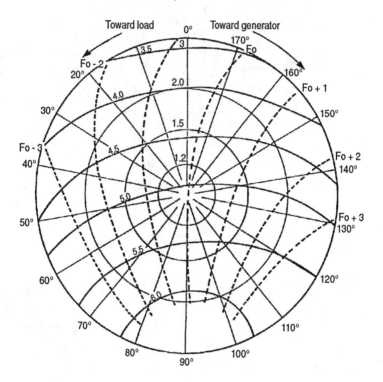

Figure 4.6 Rieke diagram of magnetron F 1123 [209]. Concentric circles: VSWR = 1 at the center and ∞ on the periphery; angular scale: 360° corresponds to $\lambda/2$; dashed line: operating frequency in MHz ($F_0 = 2450$ MHz); solid line: power output in kW.

There is a technical limit to the power output of a tube at a given frequency. The higher the frequency, the higher the obtainable power. At the ISM frequencies of 915 and 2450 MHz, there are magnetrons capable of delivering 100 kW and 6 kW, respectively. Examples include the 2450-MHz Philips

tubes YJ 1600 (5 kW) and YJ 1540 (1.3 kW) [207], [208] and
the Thomson-TTE's F 1123(5 kW), MCF 1165 (5 kW), MCF
1166 (1.2 kW), MCF 1327 (5 kW), and TH 3094 (6 kW) [209,
210, 211, 212, and 213]. At 915 MHz, there is the M 1287 tube
of the New Japan Radio Company [162]. The last magnetron
delivers 100 kW RF and remains stable up to 10 kW. The
cathode runs at a temperature of 2340 K and is subjected to
1 kW of recoil bombardment. The 35 kW to be removed from
the anode requires a water supply of 42 liters per minute.

From the electrical point of view, the magnetron has a
current-voltage characteristic similar to that of a diode (Fig.
4.7). Since its internal resistance in the vicinity of the operating point is small, a change in the potential of the secondary
feeder results in a very large change in the anode current. For
example, for the F 1123 tube [136], a 10% increase in voltage
results in an increase of 2.6 A in the current. This means that
stabilized current supplies have to be employed. Tube power
supplies are discussed extensively in the literature [7], [135],
[136], [162], [166], [167].

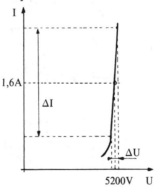

Figure 4.7 Current-voltage characteristic of the Thomson-CSF F1123
magnetron [136].

4.2.2 Klystron and TWT

Although the klystron and the TWT are linear beam tubes,
they also depend on the speed modulation and bunching of the
electrons.

The klystron (Fig. 4.8) consists of a series of resonant cavities bounded by grids and separated from each other by drift tubes. The beam produced by the electron gun (cathode, focusing electrode, and electromagnet) is accelerated by a steady potential difference into the first resonant cavity into which the driving RF signal is injected and produces the RF potential at the entrance to the first drift tube. This potential creates the alternating electric field that accelerates the electrons in one half-cycle and retards them in the other. This is how the speed modulation of the beam is achieved. Bunches of accelerated and decelerated electrons are thus formed in the first drift tube. The modulated beam induces RF power in the second resonant cavity and an electric field is established at the entrance of the second drift tube, which results in a second modulation of the speed of the beam, and so on. Finally, the RF energy amplified at the frequency of the driving signal is extracted in the last cavity [199].

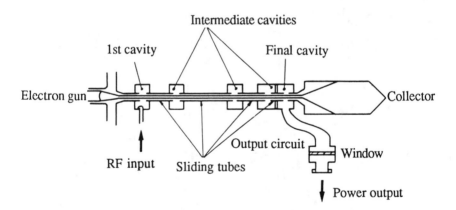

Figure 4.8 Schematic cross section of a klystron [199].

The efficiency of the klystron can be very high, e.g., 65% for the Thomson-CSF TH 2054 tube (five cavities, 50 kW, 2.45 GHz).

Further details about the klystron can be found in the literature [61] ([52] may also be consulted).

In the *traveling wave tube* (TWT), a beam of electrons is focused by magnets in a periodic nonresonant structure of such dimensions that the beam is forced to interact with a traveling wave. The periodic structure is a helix or a series of cavities capable of supporting the propagation of a slow wave at, say, 20% of the speed of light. The accelerating potential is adjusted to ensure that the speed of the electrons is close to that of the slow wave. Accelerated electrons catch up with the decelerated ones ahead of them, thus producing electron bunches. The modulated electron beam induces in the periodic structure an additional wave whose amplitude follows the transfer of energy from the electrons. The amplified RF energy is extracted at the end of the tube, and the electrons are drained by one or more collectors (Fig. 4.9). The efficiency of the TWT is on the order of 30% and the tube behaves like a wideband device. The interaction of electrons and electromagnetic waves is discussed in greater detail in [91].

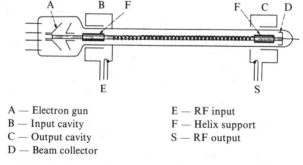

A — Electron gun   E — RF input
B — Input cavity   F — Helix support
C — Output cavity   S — RF output
D — Beam collector

Figure 4.9 Schematic cross section of a helical TWT [4].

4.2.3 RF energy transmission

The RF energy generated by the tube or tubes must be transmitted to an applicator. Coaxial cables are used for power transmission up to a few hundred watts, and matched waveguides beyond that. At 2.45 GHz, the following standard sizes may be used: WR 284 (72.14 × 34.04 mm² section), WR

340 (86.36 × 43.18 mm²), and WR 430 (109.22 × 54.61 mm²), whose guided wavelength λ_g is 144.3, 172.7, and 218.4 mm, respectively, for a free-space wavelength of 122.4 mm. At 915 MHz, the usual size is WR 975 (247.65 × 123.83 mm²) whose guided wavelength is 1258.1 mm for a free-space wavelength of 327.9 mm.

Figure 4.10 shows an example of how the Thomson-CSF F 1123 magnetron can be coupled to waveguides WR 430 and WR 340 (109.22 × 54.61 mm and 86.36 × 43.18 mm, respectively).

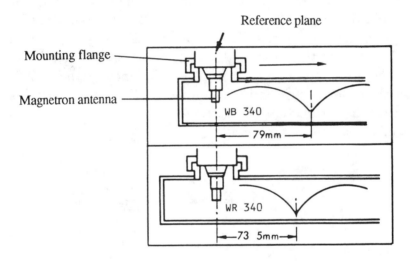

Figure 4.10 Magnetron F 1123 to waveguide coupling [209].

4.3 Applicators

4.3.1 Different types

A microwave applicator is a device designed to ensure, under optimum condition, the transmission of electromagnetic energy from the source to the material to be treated. Its design depends on the nature, shape, and dimensions of the material. It also depends on the frequency, RF power, and the nature of the process (continuous or batch). The design is subject to a number of constraints, for example, impedance matching, uniform irradiation of the product, safety aspects, industrial environmental factors, and so on.

For high-volume materials, the applicator is usually a multimode cavity whose linear dimensions are large compared to those of the material and the wavelength. In continuous mode processing, the cavity becomes a microwave tunnel. Several examples of this type of applicator will be given later.

Energy coupling in an oversized cavity becomes totally inadequate when small-volume samples are being treated (e.g., a yarn threading at high speed or a liquid pouring into a small-diameter dielectric tube.) Resonant cavities (Part I, Section 2.11.4) are then used instead. The local field amplitudes are proportional to the square of the Q-factor and are therefore much greater than those of the exciting field. Under resonance conditions, heating is very rapid. Circular or ellipsoidal cavities, excited to resonate in a mode with an axial maximum, (e.g., TM_{01p}) are the most commonly used. Descriptions of such applicators can be found in [76], [139], [108], [113], [88], [145], [147], [152], [153], [160], [178], [180], and [196]. A spherical cavity for the treatment of droplets is described in [45].

Figure 4.11 shows an applicator for yarns or liquid products [145]. A grooved rectangular cavity, in which the TE_{113} mode is excited at 2.45 GHz and produces 1.5 kW, is described in [172]. It is capable of fusing alumina or glass rods 30 mm in diameter in 3 minutes.

An unusual cavity, capable of functioning efficiently outside resonance, is reported in [180]. It consists of a cylindrical waveguide excited in the TM_{01} mode and internally divided into cavities by metal disks. RF is introduced at one extremity, the other being terminated by a water load (Fig. 4.12). Finally, the applicator described in [152] is a cylindrical cavity excited in the TM_{02p} mode and designed for products that can be pumped through it. This mode gives an axial maximum and a minimum at a certain radial distance that can be adjusted

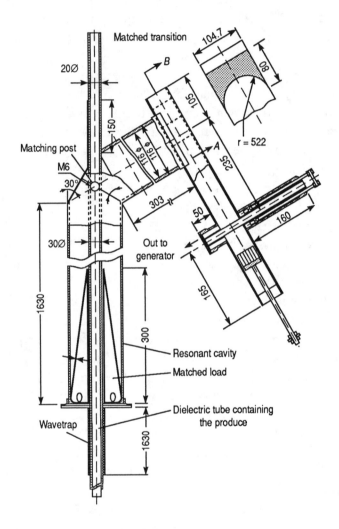

Figure 4.11 Resonant cylindrical applicator [145].

so that it occurs at the walls of the tube containing the sample in order to avoid overheating.

The heating produced in resonant cavities is essentially uni-directional and inconvenient for thin sheet-like materials. Such materials are better heated in a slotted waveguide with a slot

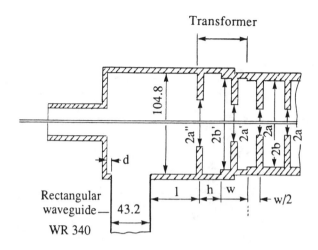

Figure 4.12 A resonant applicator for yarns and similar material [180].

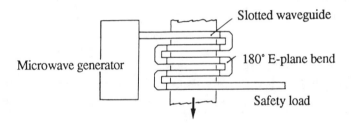

Figure 4.13 Meander waveguide applicator for sheet type material [199].

runing along the axis of each long side, and thus not perturbing the lines of current in the fundamental mode (Part I, Chapter 2, Fig. 2.3), or folded according to a meander pattern (Fig. 4.13). In the TE_{10} mode, the electric field is a maximum in the middle of the long side (Part I, Chapter 2, Fig. 2.2), and the folded meander structure ensures the entire width of the sheet crosses the maximum-field zone. The waveguide may contain internal ridges that improve uniformity (Fig. 4.14). Further information on this subject can be found in [38], [40], and [41].

A similar system consisting of N identical resonant cavities

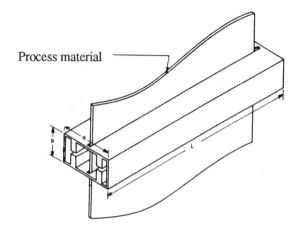

Process material

Figure 4.14 Ridged waveguide applicator for sheet-type material [41].

has been patented [12]. The sheet to be treated passes through the cavities separated by $(1/N)\lambda_g/2$ from each other in order to spread the effect of the field maxima along the entire width.

4.3.2 Design constraints

The design of an applicator must follow certain basic rules, the most important of which are listed in the following section.

4.3.2.1 Impedance matching

The energy reflected by the load must be reduced to a minimum in order to ensure optimum transfer of energy to the treated material and at the same time to protect the generator.

The load seen from the generator consists of the transmission line, the antenna, the irradiation cavity, and the material to be treated. These impedances are mismatched and multiple reflections occur at their interfaces. A variety of microwave components is used for impedance matching, including:

- mobile piston short circuits (Fig. 4.15) before the magnetron exit.

- capacitive or inductive waveguide impedance elements in the

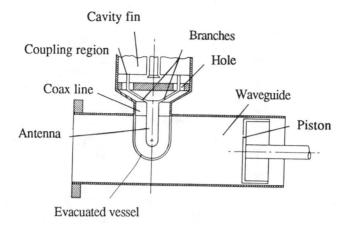

Cavity fin

Branches

Coupling region

Hole

Coax line

Waveguide

Antenna

Piston

Evacuated vessel

Figure 4.15 Piston-type short circuit [161].

form of thin obstacles inserted into the waveguide (apertures, irises, posts, or tuning screws); Figure 4.16 shows the
equivalent circuits of these obstacles.

- quarter-wave transformers in the form of waveguide sections
of length $\lambda_g/4$ and characteristic impedance Z_t chosen so
that (see Part I, Section 3.2.2):

$$Z_t = \sqrt{Z_1 Z_2} \tag{4.3}$$

where Z_1 and Z_2 are the characteristic impedances of the
components to be matched and

$$Z(\text{TE}) = \frac{\eta_0}{\sqrt{1 - \frac{\lambda_0^2}{\lambda_c^2}}} \tag{4.4}$$

$$Z(\text{TM}) = \eta_0 \sqrt{1 - \frac{\lambda_0^2}{\lambda_c^2}} \tag{4.5}$$

For instance, in the fundamental TE_{10} mode, for which
$\lambda_c = 2a$, a WR 340 waveguide ($a = 86.36\,\text{mm}$, so that
$Z_1 = 534.54\,\Omega$) can be matched to a WR 430 waveguide
($a = 109.22\,\text{mm}$, so that $Z_2 = 455.24\,\Omega$) by an intermediate
section with $Z_t = 493.30\,\Omega$, so that $a = 94.93\,\text{mm}$.

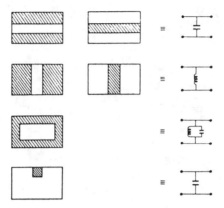

Figure 4.16 Waveguide inserts [161].

In general, the aim is to have the VSWR as low as possible (e.g., less than 1.15) over the entire frequency band of the tube. The design of the applicator, and in particular of the cavity coupling system, plays a very important role to the extent that none of the impedance matching techniques mentioned in the last section can remedy a faulty design.

In certain cases, it is practically impossible to achieve impedance matching over a broad band (e.g., when the material to be treated exhibits very low losses or when it contains conductive parts, or when a resonant cavity is used). The tube must then be protected by an isolator that consists of a ferrite circulator and a matched load. The circulator is a three-port device inserted between two sections of the guide. The incident wave is transmitted without loss, but the reflected wave is diverted to a third port. This effect is due to the use of ferrites (i.e., ferrimagnetic materials of the form MO, Fe_2O_3) where M is a divalent metal. They have isotropic dielectric properties, but anisotropic magnetic permeability [174]. The reflected energy thus diverted is dissipated in the matched load which is generally a water load. At 2450 MHz, and even more so at 915

MHz, the isolator is a bulky and costly component, so that it is difficult to envisage its use in an industrial oven. References on this subject (at 2.45 GHz) include [44], [201], [95], [132], [221], and (at 915 MHz) [95], [132], and [221].

Impedance matching to resonant cavities presents specific problems. In fact, operation is then closely related to loading changes that result in field-strength changes inside the cavity. To maintain resonance accurately, the frequency must be continuously tuned in accordance with the variation in field strength. References in this subject include [39], [42], [55], [60], [108], [114], [125], [140], [144], [186], and [196].

4.3.2.2 Field homogeneity

The design of resonant cavities is not constrained by special demands imposed on field homogeneity because the field by its very nature is concentrated at certain well-defined points.

A large oversized cavity or reverberating chamber (Part I, Section 2.1) produces, at least in principle, a homogeneous and isotropic electric field [23], [22], [98], [99], but this model is not valid for all multimode cavities (e.g., domestic microwave ovens whose dimensions are not large enough compared to the wavelength) and in which the field varies considerably from point to point. The spatial scale of field gradients thus created inside the cavity is usually on the order of the dimensions of the product, so that there is serious risk of local overheating, especially since the heating rate does not allow thermal conduction to establish equilibrium.

The design problem depends on whether the applicator is of the continuous or static type. For a nonstatic applicator, it is sufficient to have a nearly uniform field throughout the width of the conveyor belt and to rely on a statistical homogenization as the product passes through the applicator. In the case of a static applicator, the field must be uniform in the two directions in the plane of irradiation. When the thickness of the product to be treated is on the same order as the wavelength, field homogeneity becomes a three-dimensional problem.

The classical (though not very elegant) solutions make use

of turntables and metallic helical devices, commonly known as mode stirrers, in order to produce statistical field homogenization by moving the product or by periodically modifying the energy distribution. A more original solution is to use multifrequency sources and to ensure uniformity by the superposition of fields at different frequency [106]. This solution is, however, limited by the frequency allocation to microwave heating.

A better approach is to design the cavity and the radiating sources as functions of the product to be treated. This involves the construction of models and the use of trial and error in achieving minimum variation in the monitored quantity, usually the temperature. A variety of techniques can be used for this purpose. For example, a number of small containers filled with water may be placed inside the cavity, and their temperature or loss of weight by evaporation may be measured after the irradiation. Samples of the product, or of agar gel or gelatine/sugar mixtures whose dielectric and thermal properties are close to those of biological materials, may be treated in similar fashion. Simple qualitative visualization of the field is often sufficient and easily produced using heat-sensitive paper, liquid crystals, silica gel, or photocopier toner [130], [188]. Further information may be found in [101], [11], [15], [69], [84], [89], [96], [128], [129], [148], [164], [189], [190], and [193].

Theoretical models of multimode cavities are given in [47], [54], [64], [73], [80], [86], [94], [137], [145], [146], [176], and [185].

4.3.2.3 Shielding

Shielding is a serious constraint with important financial consequences. Radiation escaping from microwave equipment should not exceed certain well-defined limits in order to avoid causing electromagnetic interference on the one hand, and becoming a biological hazard on the other. Biological safety considerations will be discussed fully in Chapter 3 of Part IV. Radio interference is mostly due to harmonics and out-of-band noise, except when use is made of nonallocated frequency (e.g., 915 MHz in Region 1) or a restricted frequency (e.g., 434 MHz) in which cases fundamental emission is the main concern (Part I, Section 1.1).

The legal requirements (RR 18-3, 1815, §10) are that all practical measures must be taken to ensure that radiation from industrial, scientific, and medical appliances is reduced to a minimum and that, outside the frequency bands allocated to these applications, the level of radiation should not cause interference likely to affect radio communication, especially radio navigation and other security services.

Leaking radiation usually takes the form of spherical or cylindrical waves, depending on whether it originates from a point source or a slot, and its amplitude decreases with distance as $1/r$ and $1/\sqrt{r}$, respectively. In the frequency domain, radiation is usually concentrated in a narrow frequency band and its harmonics $(2f_0, 3f_0, \ldots)$ [31], but can also cause serious wideband perturbations. According to [81], at a distance of 1 km from a radio telescope, a simple domestic oven can increase by a factor of 12 the background noise in the emission band of neutral hydrogen atoms (1.4 GHz).

Table 4.1 shows the limits of radiation emission recommended by CISPR. It is interesting to note that these limits are not confined to the 2450 MHz and 5800 MHz bands (Part I, Chapter 1, Table 1.4).

Table 4.1

CISPR emission limits [81]

Frequencies	Limiting field values at :		
	30 m (in μV m⁻¹)	100 m (in μV m⁻¹)	300 m (in μV m⁻¹)
30-470 MHz			
TV bands	30	50	
Outside TV bands	500		
470-1000 MHz			
TV bands	100		200
Outside TV bands	500		
1-18 GHz	57 dBpW equivalent radiated power above that of a half wave dipola.		

The classification of ISM equipments was established by CAMR in 1979 and was adopted in Europe on the recommendation of the Council of the European Community. There are three categories [111]:

Class A: Equipments whose maximum leakage is weak enough for the authorization of installation without restriction everywhere except certain designated sensitive zones; **Class B**: Equipments with low level of leakage, which may be installed anywhere; **Class C**: Equipments that may have to be tested prior to use.

In the European Community, certification by a member state is usually recognised and accepted by the other members.

The shielding of static applicators is a relatively simple problem that often amounts to shielding a door, as in the classical case of the domestic microwave ovens. The design of the door seal incorporates quarter-wave structures known as radiation traps, bringing about a maximum electric field and a minimum current at the points of escape. If necessary, the energy traps can be filled with absorbing material (Fig. 4.17).

Some equipment has observation windows or other apertures that must be shielded by glass with embedded fine metal mesh. Contacts can be made RF-tight by metal braids or absorbing material (see [26, 27]).

The shielding of open-ended applicators presents more subtle problems. When the dimensions of the apertures are small, metal tubes under cut-off can provide very efficient shielding. For a rectangular tube with cross section $a \times b$, the cut-off condition for the fundamental mode is (Part I, Section 2.9)

$$\lambda_0 > \lambda_c(\text{TE}_{10}) = 2a \qquad (4.6)$$

or

$$a < \frac{\lambda_0}{2}$$

(i.e., $a < 6.12\,\text{cm}$ at $2.45\,\text{GHz}$). For a circular tube of radius r, the condition is

$$\lambda_0 > \lambda_c(\text{TE}_{11}) = \frac{2\pi r}{B_{11}} \qquad (4.7)$$

where B_{11} is the first root of the Bessel function J_1, so that

$$\lambda_0 > \frac{2\pi r}{1.841} = 3.413r$$

and

$$r < \frac{\lambda_0}{3.413}$$

(i.e., $r < 3.59$ cm at 2.45 GHz).

These dimensions must be scaled down in accordance with the permittivity of the material transported through the tubes, (i.e., by the factor of $\sqrt{\epsilon_r'}$ [90]).

In general, it is possible to reduce leakages to very low levels by means of quarter-wave traps, metal flaps, dielectric serpentines with water circulation, air-cooled absorbing material, and so on. For example, the microwave oven described in [35] has two tunnel traps in series. The first attenuates the central frequency and the second contains a ferromagnetic material to attenuate out-of-band radiation.

In all cases, two rules must be followed to reduce the level of leakage [75]: (a) the load must be well matched to ensure maximum energy transfer to it and (b) great care must be taken in the design of the cavity, particularly at the joints, in order to avoid disturbing the lines of current.

Further information may be found in [3], [8], [13], [19], [24], [43], [62], [82], [107], [109], [115], [116], [117], [141], [142, 143], [157], [182], [197], [203], and in the publications mentioned in Part IV, Chapter 3.

4.3.2.4 Materials

The metals used for enclosures and for waveguides must be nonmagnetic and as good conductors as possible. The surface resistivity is found to be a good measure of ohmic losses as previously discussed in connection with the expressions for d and Q [Part I, (2.67) and (2.122)]. The ideal materials would be silver-plated copper and brass. In practice, stainless steel is used almost exclusively except for certain components that

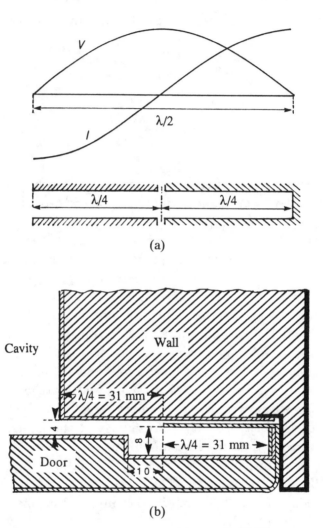

Figure 4.17 Principle (a) and design (b) of a $2.45\,\mathrm{Ghz}$, quarter-wave radiation trap [21].

carry high current (e.g., magnetron antennas). The resistivity of stainless steel is $90\,\mu\Omega\,\mathrm{cm}$, which may be compared with 1.72 for copper and 1.62 for silver.

It is imperative that supports, conveyor belts, partitions, water seals, joints, and so on be made of very low loss dielectric

materials (see Part I, Section 3.7 and [67], [68], [70], [183]). In this respect, CW high power is subject to more stringent conditions than pulsed radars. The list of acceptable materials is in fact very restricted. Essentially, it consists of the following:
Polyethylene whose dielectric properties are remarkably independent of temperature. It has good mechanical properties, including exceptionally high flexibility at low temperatures. Polyethylene shows no reaction to most common acids, but it is flammable and cannot be used above 90° (fusion point: 105° C). It is also sensitive to the effect of corona discharge.

Polypropylene has properties similar to those of polyethylene. It is very susceptible to oxidation, and for this reason it usually contains antioxidants that may degrade its dielectric properties.

Polytetrafluorethylene (PTFE or Teflon) needs machining. It is totally inert, nonflammable, and retains its solidity and shape up to 260° C. In addition, PTFE has extremely suitable dielectric properties and often constitutes the best possible choice of material despite some disadvantages, such as its tendency to flow under stress and to degrade in the presence of plasmas. In comparison with other materials, PTFE is rather expensive.

Silicones are frequently used in the form of rubber (for water seals) and as grease for joints because of their temperature-independent dielectric properties.

Conveyor belts for tunnel ovens and internal partitions are often made of silicone glass, fiber composites, or PTFE glass fiber.

4.3.2.5 Electric arcing and corona discharges

An electric discharge is likely to occur when metallic parts within the microwave applicator are close to one another in the region where the field is very high. The discharge carries the risk of deterioration for the material, the applicator, and the generator [72]. At atmospheric pressure, the breakdown field is on the order of 10^6 V m^{-1}. At 2.45 GHz and for 700 W RF power, the field in WR 284, WR 340, and WR 430 waveguides does not exceed 57, 20, and $14.6\,\mathrm{kVm^{-1}}$, respectively. In the cavity of a domestic microwave oven, the field does not exceed $5300\,\mathrm{Vm^{-1}}$, which excludes all risks of arcing and dielectric breakdown except between sharp angled metal pieces.

Assuming that a thin metal plate is placed in the xz plane at a distance x from the cavity wall, the induced charged density on the surface is

$$\rho_s = A\epsilon_0 \frac{1}{\sqrt{x}} \ (\mathrm{in\ C\,m^{-2}}[L^{-2}TI]) \tag{4.8}$$

where A is a constant. The electric field near the plate in the y direction is

$$E_y = \frac{\rho_s}{\epsilon_0} = A\frac{1}{\sqrt{x}} \tag{4.9}$$

which is valid for a plate of zero thickness. In practice, there is a risk of arcing if the thickness is less than 0.05 mm.

The risk of discharge can be almost completely eliminated by taking a few relatively simple precautions in the design of the oven. For instance, sharp angles and metal objects must be avoided, quarter-wave field traps can be used to reduce the current to zero at the level of all critical points, and, finally, all soldered surfaces must be smooth and the enclosure must be free of all traces of iron filings or dust.

Discharges constitute much more of a problem in an enclosure under vacuum. The Paschen curves for different gases (Fig. 4.18) show that the breakdown point decreases rapidly with pressure, reaching a minimum at around 133 Pa, and then increases again as the pressure continues to drop. The critical breakdown field is $7\,\mathrm{kVm^{-1}}$ at 950 MHz and $18\,\mathrm{kVm^{-1}}$ at 2450 MHz.

In addition to arcing, dielectric breakdown in vacuum leads to the formation of plasmas that will be treated very briefly in the following section.

A plasma is essentially an electrically neutral ionized gas whose charged particles interact with electric and magnetic

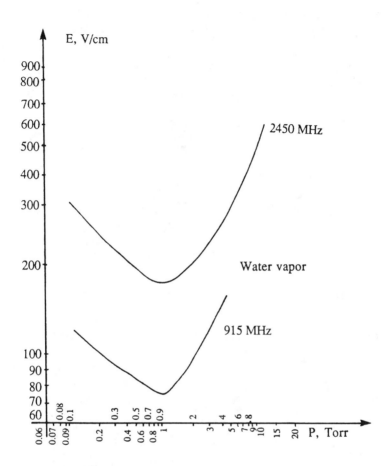

Figure 4.18 Breakdown field as a function of pressure [1].

fields [78]. Free electrons ionize atoms, create more free electrons and are accelerated by the field. When the rate of generation of electrons exceeds their loss by recombination, conduction in metal walls, or other processes, a chain reaction is established and plasma appears as a luminous haze whose color depends on the gases present in the vessel.

If N is the plasma density, expressed as the number of free electrons per cubic meter, ω is the angular frequency $2\pi f$ of the applied field, e and m are the charge and rest mass of the electron, and ξ its mean free path, the force exerted on an electron in the absence of collisions can be expressed as

$$m\ddot{\xi} = -eE \qquad (4.10)$$

The conductivity of the plasma is

$$\sigma = \frac{-jNe^2}{m\omega} \qquad (4.11)$$

and its permittivity is

$$\epsilon'_r = 1 - \frac{\omega_p^2}{\omega^2} \qquad (4.12)$$

where the angular frequency of the plasma is

$$\omega_p = \frac{e^2 N}{\epsilon_0 m} \qquad (4.13)$$

The propagation constant [Part I, equation (2.34)] is

$$\gamma^2 = -\omega^2 \mu \epsilon_0 \epsilon'_r \qquad (4.14)$$

giving

$$\gamma = j\frac{\omega}{c}\sqrt{1 - \frac{\omega_p^2}{\omega^2}} \qquad (4.15)$$

When $\omega < \omega_p$, γ is real and the plasma behaves as a metal. When $\omega > \omega_p$, γ is imaginary and the plasma behaves as a dielectric of relative permittivity less than unity.

A more sophisticated model [93], [124] takes collisions into consideration. The equation of motion of the electron now becomes

$$m\ddot{\xi} + \nu m\dot{\xi} = -eE \qquad (4.16)$$

where ν is the collision frequency (i.e., the number of collisions between electrons and heavy particles per second). Equation (4.16) yields

$$\xi = \frac{eE}{m\omega(\omega - j\nu)} \qquad (4.17)$$

The conductivity and permittivity are

$$\sigma = \frac{Ne^2}{m}\frac{\nu - j\omega}{\nu^2 + \omega^2} \tag{4.18}$$

$$\epsilon_r = 1 - \frac{\omega_p^2}{\omega^2 + \nu^2} - j\frac{\nu}{\omega}\frac{\omega_p^2}{\nu^2 + \omega^2} \tag{4.19}$$

If the plasma is highly ionized ($\nu \ll \omega_p$), the attenuation and propagation coefficients may be calculated from the real and imaginary parts of γ shown in Figure 4.19. The plasma behaves like a metal with

$$\sigma = \frac{Ne^2}{m\nu} \tag{4.20}$$

when $\omega < \nu$.

It behaves like a cut-off waveguide for $\nu < \omega < \omega_p$ and like a dielectric for $\omega > \omega_p$.

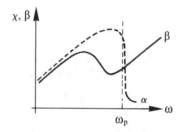

Figure 4.19 Absorption and propagation coefficients in plasma [93].

Plasma generation may be very detrimental in RF heating applications when the field is applied in vacuum. The plasma absorbs some of the RF power, degrades the quality factor Q of resonant cavities, and initiates a rapid deterioration in dielectric components, thus introducing a potentially dangerous impedance mismatch. A study of this phenomenon [16] has shown that the absorbed fraction of RF power is a maximum at a pressure approximately half that at which the plasma is extinguished. The latter is on the order of 6700 Pa for O_2 and N_2 and 20 kPa for Ar and He below 250 W at 900 MHz. Figure 4.20 shows the power absorbed by a plasma in air at 900 MHz as a function of pressure and incident power.

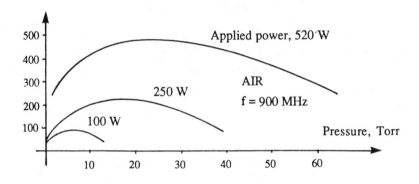

Figure 4.20 Power absorption in plasma [34].

It is difficult to protect microwave applicators from the effect of plasmas because microwaves produce higher concentrations N than any other type of excitation, even in the absence of electrodes. Microwave plasma can sustain themselves at pressures that would otherwise be considered too high for plasmas. References to this subject include [37], [48], [49], [50], [53], [66], [77], [100], [105], [101], [110], [163], [171], and [175].

4.3.2.6 Environmental constraints

Industrial microwave sources and applicators must often be designed for very hostile environments. In particular, casings may have to be dust- and water-proof as well as resistant to corrosion by salt and acid, which may necessitate the use of stainless steel and inert plastic material. In addition, they may have to withstand wide variations in temperature, shocks, and vibrations. They must be sturdy and equipped with safety cut-outs to guard against misuse.

4.4 Conclusion

We have shown that microwave sources and applicators are complex technical components, often used in harsh environments where they should display the same degree of sturdiness as traditional equipments.

In addition, the constraints specific to microwaves (impedance matching, field homogeneity, choice of material, risk of discharge) and shielding regulations involve design concepts that are often difficult to realise and always costly. However, it will be shown later that despite (and partly because of) these disadvantages, microwaves are an extremely interesting - and in some cases the only - alternative in many fields.

References

1. ARCHIERI C. Déshydratation par micro-ondes. Evaporation sous vide. Sublimation sous vide. *Bios*, $\underline{8}$, n° 1, 1977, pp. 10-21. (C.D.I.U.P.A., n° 102774).

2. AREF M., BRACH E., TAPE N. A pilot-plant continuous process microwave oven. *Can. Inst. Food Technol. J.*, $\underline{2}$, n° 1, 1969, pp. 37-41.

3. AZOULAY A. Perturbations électromagnétiques et protection des télécommunications. *Tech. Ing., Electron.*, E 1510, sept. 1984, 12 p.

4. BABILLON C. Emetteurs radioélectriques, caractéristiques. *Tech. Ing., Electron.*, E 7700, mars 1979, 19 p.

5. BAEYER von H..The effect of silver plating on attenuation at microwave frequencies. *Microw. J.*, avril 1960, pp. 47-50.

6. BARBER H. Microwave generators and applicators. *Trans. IMPI*, $\underline{2}$, 1973, Industrial application of microwave energy.

7. BARJON M. Construisez vous-même un générateur micro-ondes délivrant 800 W utiles à 2450 MHz. *Econ. Progrès Electr.*, n° 14, mars-avril 1985, pp. 23-25.

8. BASSOLE P. Protection des radiocommunications contre la pollution électromagnétique. In : Symposium international sur les applications énergétiques des micro-ondes, IMPI-CFE. $\underline{14}$, 1979, Monaco. Résumés, pp. 213-216.

9. BAUDRAND H. Interactions progressives entre électrons et ondes électromagnétiques. (Cours polycopié) Institut National Polytechnique de Toulouse, ENSEEIHT, 1977, 113 p.

10. BENGTSSON N., LYCKE E. Experiments with a heat camera for recording temperature distribution in foods during microwave heating. *J. Microw. Power*, $\underline{4}$, n° 2, 1969, pp. 48-54.

11. BERNTSEN W., DAVID B. Determining the electric field distribution in a 1250 watt microwave oven and its effect on portioned food during heating.

Microw. Energy Appl. Newsl., $\underline{8}$, n° 4, 1975, pp. 3-10. (C.D.I.U.P.A., n° 91974).

12. BERTEAUD A., CLEMENT R., MERLET C., LECLERCQ C. Procédé et dispositif de traitement par micro-ondes de produits en feuilles. Brevet français n° 2 523 797, A1, 82 04398, 16 mars 1982.

13. BIALOD D. TOURAINE . Implications techniques des règlements sur les perturbations électromagnétiques des appareils HF-UHF. In : Comptes-rendus des Journées d'Etudes de la SEE, champs électromagnétiques dans le proche environnement des équipements industriels micro-onde et des antennes. Univ. Toulon et Var, 19-20 janv.1984, réf. 84 25 23, 7 p.

14. BICKEL S. Solid-state microwave-oven power source. U.S. Patent n° 3953702, 27 april 1976.

15. BOBENG B., DAVID B. Identifying the electric field distribution in a microwave oven : a practical method for food service operators. *Microw. Energy Appl. Newsl.*, $\underline{8}$, n° 6, 1975, pp. 3-6. (C.D.I.U.P.A., n° 91015).

16. BOSISIO R., WEISSFLOCH C., WERTHEIMER M. The large-volume microwave plasma generator : a new tool for research and industrial processing. Document commercial. Québec, Cie Physique Appliquée, 1972, 58 p.

17. BRACEWELL R. Charts for resonant frequencies of cavities. In : Proceedings of the IRE (Waves and Electrons), $\underline{35}$, August 1947, pp. 830-841.

18. CHANG K. Solid-state microwave heating apparatus. U.S. Patent n° 3 691 338, 12 Sept. 1972.

19. CHEF M. Risques découlant de la pollution électromagnétique dans les bandes de fréquences attribuées à l'aéronautique. In : Comptes-rendus du Symposium international sur les applications énergétiques des micro-ondes, IMPI-CFE, $\underline{14}$, 1979, Monaco. Résumés, pp. 217-221.

20. COLLINS G. Microwave magnetrons. New York, McGraw Hill Book Co., 1948.

21. COPSON D. Design and development of microwave freeze-dryers. In : COPSON D. Microwave heating. 2nd ed. Westport, AVI Publishing Co., 1975, pp. 117-139. (C.D.I.U.P.A., n° 93299).

22. CORONA P., LATMIRAL G., PAOLINI E. Performance and analysis of a reverberating enclosure with variable geometry. *IEEE Trans. Electromagn. Compat.*, EMC-22, n° 1, Feb. 1980, pp. 2-5.

23. CORONA P., LATMIRAL G., PAOLINI E., PICCIOLI L. Use of a reverberating enclosure for measurements of radiated power in the microwave range. *IEEE Trans. Electromagn. Compat.*, EMC-18, n° 2, May 1976, pp. 54-59.

24. CZIGANY S., GYORI A. Measurement methods for industrial applications of microwaves. In : Proceedings of the IMPI Symposium, Waterloo (Can.), 1975, pp. 133-136 (Paper 8.2.).

25. DAY J. Construction for tuning microwave heating applicator. U.S. Patent n° 3 590 202, June 1971.

26. DE BRUYNE R., VAN LOOCK W. New class of microwave shielding materials. *J. Microw. Power*, 12, n° 2, 1977, pp. 145-154.

27. DE BRUYNE R., VAN LOOCK W. Microwave shielding properties of thin stainless steel needles embedded in rubber. *J. Microw. Power*, 12, n° 4, 1977, pp. 361-367.

28. DECAREAU R. Microwave heating control techniques. *Microw. Energy Appl. Newsl.*, 5, n° 5, 1972, pp. 3-7. (C.D.I.U.P.A., n° 054237).

29. DECAREAU R. Notes on some European microwave equipments. *Microw. Energy Appl. Newsl.*, 9, n° 2, 1976, pp. 7-10. (C.D.I.U.P.A., n° 95884).

30. DEHN R. Microwave heating apparatus with improved couplers and solidstate power source. U.S. Patent n° 4 006 338, 1 Feb. 1977.

31. DEPARIS G., PERRICHON P. Perturbations apportées aux réceptions de radiodiffusion. *Tech. Ing., Electron.*, E1520, juin 1984, 11 p.

32. DERBY P. Electron discharge devices. U.S. Patent n° 2 463 524, 8 March 1949.

33. DILLS R., HUNT R., FITZMAYER L. Microwave oven with dual feed excitation system. U.S. Patent n° 4 458 126, 3 July 1984.

34. DORMAN F., Mac TAGGART F. Absorption of microwave power by plasmas. *J. Microw. Power*, 5, n° 1, 1970, pp. 4-16.

35. DUDLEY K., EVES E., STONE W. Leakage suppression tunnel for conveyorized microwave oven. U.S. Patent n° 4 488 027, 11 Dec. 1984.

36. DUNN D. Slow wave couplers microwave dielectric heating systems. *J. Microw. Power*, 2, n° 1, 1967, pp. 7-20. (C.D.I.U.P.A., n° 46989).

37. DYMŠIC B., KORECKI J. Température d'une colonne de plasma libre dans un champ micro-ondes à haute pression (en russe). *Opt. Spektrosk.*, 33, n° 1, 1972, p. 32.

38. EL-SAYED E., ABDEL HAMID T. Use of sheath helix slow-wave structure as an applicator in microwave heating systems. *J. Microw. Power*, 16, n° 3-4, 1981, pp. 283-288.

39. EL-SAYED E., FARGHALY S. Influence of load location in mode tuning of microwave ovens. *J. Microw. Power*, 18, n° 2, 1983, pp. 197-207.

40. EL-SAYED E., HASHEM A. Wave propagation in rectangular waveguides with symmetrically placed tapered ridges. *J. Microw. Power*, 19, n° 1, 1984, pp. 35-46.

41. EL-SAYED E., HASHEM A. Ridged waveguide applicators for uniform microwave heating of sheet material. *J. Microw. Power*, 19, n° 2, 1984, pp. 111-117.

42. EL-SAYED E., MORSY M. Use of transmission-line matrix method in determining the resonant frequencies of loaded microwave ovens. *J. Microw. Power*, 19, n° 1, 1984, pp. 65-71.

43. FONTEYNE J. Brouillage de la radiodiffusion par les appareils ou installations ISM. In : Comptes-rendus du Symposium international sur les applications énergétiques des micro-ondes, IMPI-CFE. 14, 1979, Monaco, Résumés pp. 222-225.

44. FORTERRE G., FOURNET-FAYAS C., PRIOU A. A 50 kW CW S-band ferrite circulator. *J. Microw. Power*, 13, n° 1, 1978, pp. 65-68.

45. GARELIS E. Initial heating rate of a droplet in a spherical microwave cavity. *Phys. Fluids*, 17, n° 11, 1974, pp. 2002-2008.

150 Part I - Microwaves

46. GERLING J. Microwave heating patents. *J. Microw. Power*, 7, n° 2, 1972, pp. 143-153.

47. GERLING J. Applicators and their design. *Trans. IMPI*, 1, 1973. Microwave Power in Industry.

48. GOULD L. Handbook on breakdown of air in wave guide systems. In : Microwave high power breakdown energy. U.S. Navy Dep., Bur. Ships, Electron. Div., Contract NOBSR 63295, 1956.

49. GOULD W., KENYON E. Corona breakdown and electric field strength in microwave freeze-drying. In : Proceedings of the IMPI Symposium, Den Haag, 1970.

50. GOULD J., KENYON E. Gas discharge and electric field strength in microwave freeze-drying. *J. Microw. Power*, 6, n° 2, 1971, pp. 151-167.

51. GRIEMSMANN J. Oversized waveguides. *Microw.*, 2, Dec. 1963, pp. 20-31.

52. GUENARD P. Tubes pour hyperfréquences. *Tech. Ing., Electron.*, E 760, Dec. 1975, 21 p.

53. GWAL A. Reflection of microwave through laboratory plasma. *Int. J. Electron.*, V/I, 33, n° 1, 1972, pp. 91-95.

54. HAMID M. Optimization of microwave power applicators. In : Proceedings of the IMPI Symposium, Leuven, 1976, (Paper 2B-3) and *J. Microw. Power*, 11,, n° 2, 1976, pp. 185-186.

55. HAMID M., BHARTIA P., KASHYAP S., STUCHLY S. Tuning, coupling and matching of microwave heating applicators operating at high-order modes. *J. Microw. Power*, 6, n° 3, 1971, pp. 221-228.

56. HAMID M., BHARTIA P., STUCHLY S. Microwave power applicators for liquid materials. Progress report n° 9, Microwave applications project, Industrial enterprise fellowship program report, Manitoba research council, Sept. 1972, 31 p.

57. HAMID M., MOHSEN A., KASHYAP S., BOERNER W. Field distribution in multilayered dielectric-loaded rectangular waveguides. *Proc. IEE*, 117, n° 4, 1970, pp. 709-712.

58. HAMID M., RZEPECKA M., STUCHLY S. Applicators for microwave treatment of granular materials in agricultural technology. Progress report n° 3, Microwave applications project, Industrial enterprise fellowship program report, Manitoba research council, Dec. 1971, 56 p.

59. HAMID M., RZEPECKA M., STUCHLY S. Applicators for microwave treatment of granular materials in agriculture. Performance and design criteria. IMPI Technical Report, TE 72-1, s.d., 58 p.

60. HAMID M., STUCHLY S., BHARTIA P. Tuning and excitation of a prolate spheroidal cavity resonator for microwave drying. *J. Microw. Power*, 6, n° 3, 1971, pp. 213-220.

61. HAMILTON D., KNIPP J., KUPER J. Klystrons and microwave triodes. New York, Dover Publications, 1966, 533 p. (1st ed., McGraw Hill Book, 1948).

62. HARADA A., OHARA I., OGURO T. Studies on microwave leakage suppression of magnetrons. In : Proceedings of the IMPI Symposium, Waterloo (Can.), 1975, pp. 137-139. (Paper 8.3.).

63. HARRIS J., MALLINGER M., GOOCH G. Design state-of-the-art assessed (for solid state) at RF frequencies. *Microw. Syst. News*, August 1985, pp. 79-96.

64. HAUCK H. Design considerations for microwave oven cavities. *IEEE Trans. Ind. Gen. Appl.*, IGA-6, n° 1, 1970, pp. 74-80.

65. HINKEL K. Magnetrons. Eindhoven, N.V. Philips Gloeilampenfabrieken, 98 p. (Philips Technische Bibliothek).

66. HIPPEL A. Von. Der elektrische Durchschlag in Gasen und festen Isolatoren. *Ergeb. exakten Nat.-Wiss.*, 14, 1935, p. 79.

67. HIPPEL A. von. Dielectrics and waves. (1st ed., 1954). Cambridge (USA), M.I.T. Press, 1966, 284 p.

68. HIPPEL A. von. Dielectric materials and applications. (1st ed. 1954). Cambridge (USA), M.I.T. Press, 1966, 438 p.

69. HIRATSUKA A., INOUE H., TAKAGI T. Observations of electric field intensity patterns in microwave ovens. *J. Microw. Power*, 13, n° 2, 1978, pp. 189-191.

70. HOWE H. Dielectric material development. *Microw. J.*, 21, n° 11, 1978, pp. 39-40.

71. HUANG H. Microwave apparatus for rapid heating of threadlines. *J. Microw. Power*, 4, n° 4, 1969, p. 289.

72. ISHII T. Theoretical analysis of arcing structure in microwave ovens. *J. Microw. Power*, 18, n° 4, 1983, pp. 337-344.

73. ISHII T., YEN Y., KIPP R. Improvement of microwave power distribution by the use of the first-order principle of geometrical optics for scientific microwave oven cavity. *J. Microw. Power*, 14, n° 3, 1979, pp. 201-208.

74. ISHITOBI Y., TOGAWA M. A practical diagram for computing long-line effects for CW magnetron. *J. Microw. Power*, 13, n° 2, 1978, pp. 131-137.

75. JACOMINO J. Mesure du rayonnement des équipements micro-onde et hautes fréquences. In : Comptes-rendus des Journées d'Etudes de la SEE, Champs électromagnétiques dans le proche environnement des équipements industriels micro-onde et des antennes, Univ. Toulon et Var, 19-20 janv. 1984, réf. 842525, 7 p.

76. JOHNSON R., SMITH F. Resonant-cavity microwave applicator. U.S. Patent n° 3 597 566, August 1971.

77. JOHNSTON D. Bibliography II : Microwave plasmas. *J. Microw. Power*, 5, n° 1, 1970, pp. 17-22.

78. KADOMCEV B. Kollektivnye javljenija v plazmje. Moskva, Nauka, 1976. Trad. Fr. :.Phénomènes collectifs dans les plasmas. Moskva, Mir, 1979, 240 p.

79. KADONO K., OSHIMA K. OGURA K. Waveguide accessories for microwave heating. Sans réf., s.d., 22 p.

80. KASE Y. Methods for uniform cooking in microwave oven. In : Proceedings of the IMPI Symposium, Milwaukee, 1974, 3 p. (Paper A6.5/1-3).

81. KASHYAP S. Emissions from ISM sources. *J. Microw. Power*, 18, n° 2, 1983, pp. 153-161.

82. KASHYAP S., HUNT F. Radiation emitted by microwave ovens in the 30-10000 MHz range. Rapport NRC-ERB/943, Div. Génie Electr., Conseil Nat. Rech., Canada, May 1982.

83. KASHYAP S., WYSLOUZIL W. Automatic control of non-resonant dryers. In : Proceedings of the IMPI Symposium, Waterloo (Can.), 1975, pp. 197-200. (Paper 12.6.)

84. KASHYAP S., WYSLOUZIL W. Methods for improving heating uniformity of micro-wave ovens. *J. Microw. Power*, 12, n° 3, 1977, pp. 223-230.

85. KAWAGUCHI T. Analysis of magnetron operation in microwave oven. In : Proceedings of the IMPI Symposium, Chicago, 1985, pp. 126-132.

86. KERREGEIS J., GOULD J., KENYON E. Microwave power absorption in a multimode cavity. In : Proceedings of the IMPI Symposium, Ottawa, 1977, p. 115. (Paper 7.6.)

87. KOBAYASHI D. A new analysis of magnetrons. Reports of the Electrical Comm. Lab. Nippon, 7, n° 4, 1959, pp. 100-115.

88. KRETZSCHMAR J., PIETERMAAT F. Concentrated microwave heating in elliptical waveguides and cavities. In : Proceedings of the IMPI Symposium, Monterey, 1971. (Paper 8.4.).

89. KUMPFER B. A simple system of microwave pattern measurement. *Microw. Energy Appl. Newsl.*, 5, n° 4, 1972, pp. 10-11. (C.D.I.U.P.A., n° 48752).

90. LASZLO T., NEWMAN T. The application of the cutoff-tube principle to the end seal of microwave conveyor systems. *J. Microw. Power*, 15, n° 3, 1980, pp. 173-176.

91. LAVOLTE L. Procédé de chauffage par pertes diélectriques et dispositif de mise en oeuvre. Brevet Français n° 2 500 708 (8103418), 20 fév. 1981.

92. LEBLOND A. Les tubes hyperfréquences. Paris, Masson & Cie, 1972, 309 p.

93. LEFEUVRE S. Hyperfréquences. Paris, Dunod.Univ., 1969, 160 p.

94. LEIDIGH W., STOLTE W., RHUDY R. The applicator simulator - An economical means to optimum process design. In : Proceedings of the IMPI Symposium, Monterey, 1971. (Paper 3.4.).

95. LEPPIN J. Thermal design of ferrite isolators for industrial microwave equipment. *J. Microw. Power*, 9, n° 3, 1974, pp. 251-261.

96. LEVINE H., MOORE R. Microwave oven test load evaluation and determination of microwave energy distribution. Report BRH/DEP 70.23, US Dep. Health, Education and Welfare, Rockville, 1970.

97. LEWIS R., WHITE J. Heating apparatus. U.S. Patent n° 3 461 261, August 1969.

98. LIU B., CHANG D., MA M. Design consideration of reverberating chambers for electromagnetic interference measurements. In : Proceedings of the International IEEE Symposium on Electromagnetic compatibility, Washington, 1983, pp. 508-512.

99. LIU B., CHANG D., MA M. Eigenmodes and the composite quality factor of a reverberating chamber. U.S. Dep. Commerce Publication, National Bureau of Standards, T.N. 1066, August 1983, PB 84-128016, 54 p.

100. LLEWELLYN JONES F. Electrical discharges. Reports on Progress in Physics, $\underline{16}$, 1953, p. 216.

101. LOEB L. Fundamental processes of electrical discharge in gases. New York, Wiley & Sons Inc. 1939.

102. LONNE J., PATAY L. Applicateur hyperfréquence perfectionné. Brevet français n° 2 500 218 (81 03495), 19 fév. 1981.

103. LUYPAERT P., VAN DE CAPELLE A. Power applications of nonconventional cavities. In Proceedings of the IMPI Symposium, Ottawa, 1978, pp. 68-69. (Paper 7.3.) (C.D.I.U.P.A., n° 143425).

104. MA M., KANDA M., CPAWFORD M., LARSEN E. A review of electromagnetic compatibility/interference measurement methodology. In : Proceedings of the IEEE, $\underline{73}$, n° 3,1985, pp. 388-411.

105. Mac DONALD A. Microwave breakdown in gases. New York, John Wiley & Sons, 1966.

106. Mac KAY A., TINGA W., VOSS W. Controlled heating microwave ovens. U.S. Patent n° 4 196 332, 1st april 1980.

107. Mac LACHLAN A. Method of measurement of radiofrequency interference from microwave ovens. In : Proceedings of the IMPI Symposium, Loughborough, 1973, 2 p. (Paper 5.B.3.).

108. MAIER L., SLATER J. Field strength measurements in resonant cavities. J. Appl. Phys., $\underline{23}$, n° 1, 1952, pp. 68-77.

109. MARTIN H., TABOR F. Radio-frequency emission characteristics and measurement procedures of incidental radiation devices and ISM equipment. Rapport FAA. RD-72-80, Sept. 1972.

110. MEEK J., CRAGGS J. Electric breakdown of gases. Oxford, Clarendon Press, 1953.

111. MEISEL N. Les micro-ondes. Savoir... et... bien faire. Ind. Aliment. Agric., $\underline{101}$, n° 4, 1984, pp. 259-264. (C.D.I.U.P.A., n° 188055).

112. METAXAS A. Design of a TM_{010} resonant cavity as a heating device at 2.45 GHz. J. Microw. Power, $\underline{9}$, n° 2, 1974, pp. 123-128, et Electricity Council Research Centre M 767.

113. METAXAS A. Rapid heating of liquid foodstuffs at 896 MHz. J. Microw. Power, $\underline{11}$, n° 2, 1976, pp. 105-115.

114. MIHRAN T. Oven mode tuning by slab dielectric loads. IEEE Trans. Microw. Theory Tech., MTT-$\underline{26}$, n° 6, 1978, pp. 381-387.

115. MOBLEY M. FCC equipment type approval program with reference to microwave tests. J. Microw. Power, $\underline{6}$, n° 4, 1971, pp. 297-303.

116. MOBLEY M. Revisions of Federal Communications Commission type-approval test procedures. In : Proceedings of the IMPI Symposium, Loughborough, 1973, 2 p. (Paper 4.B.6.).

117. MOBLEY M. Interference measurements on radiofrequency devices. In : Proceedings of the IMPI Symposium, Waterloo (Can.), 1975, pp. 130-132. (Paper 8.1.).

118. NAMBA I., TASHIRO N., KUME M., KAWAGUCHI T., AZUMA T. Low-voltage magnetron for microwave ovens. J. Microw. Power, $\underline{16}$, n° 3-4, 1981, pp. 257-261.

119. NAMBA I., TASHIRO N., NAKAI K., SAITO H. A new magnetron (2 M 172 A) from Toshiba. J. Microw. Power, $\underline{19}$, n° 3-4, 1984, pp. 209-220.

120. NELSON R. A magnetron oscillator for dielectric heating. J. Appl. Phys., $\underline{18}$, 1947, pp. 356-361.

121. NELSON R. Magnetrons for dielectric heating. Electr. Eng., $\underline{70}$, n° 7, 1951, pp. 627-633.

122. NELSON R. Industrial magnetrons for dielectric heating. Electron., $\underline{25}$, 1952, pp. 104-109.

123. NELSON R. Tubes for dielectric heating at 915 megacycles. 1st. part. Trans. AJEE, $\underline{71}$, n° 1, 1952, pp. 72-80.

124. NIKOL'SKIJ V. Elektrodinamika i rasprostanjenije radiovoln. Moskva, Nauka, 1978. Trad. Fr. : Electrodynamique et propagation des ondes-radio-électriques. Moskva, Mir, 1982, 574 p.

125. NOBUE T., KUSUNOKI S. Microwave oven having controllable frequency microwave power source. U.S. Patent n° 4 415 789. 15 Nov. 1983.

126. OGURA K., TAKAHASHI H., KOINUMA T., TASHIRO N. Design problems in magnetrons for microwave ovens. *J. Microw. Power*, 13, n° 4, 1978, pp. 357-361.

127. OGURA K., TAKAHASHI H., NAMBA I. Input impedance measurement of microwave ovens. *J. Microw. Power*, 13, n° 2, 1978, pp. 183-188.

128. ÖHLSSON T. Temperature distribution in microwave oven heating : experiments and computer simulation. In : Proceedings of the IMPI Symposium, Leuven, 1976. (Paper 1B.2a). (Abstracts in *J. Microw. Power*, 11, n° 2, 1976, pp. 178-179).

129. OHLSSON T. Temperature distribution in microwave oven heating : a comparison of mapping methods and influence of sample parameters. In : Proceedings of the IMPI Symposium, Leuven, 1976. (Paper 1B.2b). (Abstracts in *J. Microw. Power*, 11, n° 2, 1976, pp. 180-181).

130. OHLSSON T. Methods for measuring temperature distribution in microwave ovens. *Microw. World*, 2, n° 2, March-April 1981, pp. 14-17, and SIK publication, n° 346.

131. OHTANI T. Hybrid microwave heating apparatus. U.S. Patent n° 3 867 607, 18 Feb. 1985.

132. OKADA F., OHWI K., MORI M., YASUDA M. A 100 kW waveguide Y-junction circulator for microwave power systems at 915 MHz. *J. Microw. Power*, 12, n°3, 1977, pp. 201-207.

133. OKATSUKA H., TANIGUCHI K., KONNO M., NARITA R. Microwave heating apparatus with solid-state microwave oscillating device. U.S. Patent n° 4 504 718, 12 March 1985.

134. OSEPCHUK J Life begins at forty : microwave tubes. *Microw. J.*, Nov. 1978, pp. 51-60.

135. PELLISSIER J. Conception et réalisation pratique d'un générateur modulaire délivrant 0,8 kW à 2450 MHz. *Econ. Progrès Electr.*, n° 1, janv.-fév. 1982, pp. 16-19.

136. PELLISSIER J. Réalisation pratique d'une alimentation régulée en courant pour magnétron. *Econ. Progrès Electr.*, n° 7, janv.-fév. 1983, pp. 20-21.

137. PETERS W. Der Hochfrequenzstrahlungs-herd. *Heiz.- Lueftung-Haustech.*, 9, n° 4, 1958, pp. 85-91 et 9, n° 5, 1958, pp. 110-114.

138. PETERS W. Technische Probleme und praktische Verwendungsmöglichkeiten des Hochfrequenzstrahlungsherds. *Elektr.-verwert.*, 33, n° 5, 1958, pp. 111-119.

139. PIETERMAAT F., KRETZSCHMAR J. Le chauffage par micro-ondes avec cavité à section ellipsoïdale, réf. n° 412, s.d., 4 p.

140. POINSOT A., CHANUSSOT J., JUREK R. Asservissement d'un générateur hyperfréquence sur la fréquence de résonance d'une cavité. *Econ. Progrès Electr.*, n° 12, nov.-déc. 1984, pp. 22-25.

141. POITEVIN J. Perturbations radioélectriques produites par les appareils industriels, scientifiques et médicaux. Etat actuel de la législation française et perspectives d'évolution. *Rev. Gén. Electr.*, n° 11, 1981, pp. 805-809.

142. POITEVIN J., CHEF M. Les problèmes de sécurité posés par l'emploi des hautes fréquences et des micro-ondes. In : Colloque sur les applications industrielles des hautes fréquences et des micro-ondes, Lyon, 1977. Lyon, Univ. Claude Bernard, 1977, pp. 1-7, résumés (complément).

143. POITEVIN J., CHEF M. Commentaires sur la protection contre les brouillages des services aéronautiques. Organisation de l'Aviation Civile Internationale, Bureau Europe. In : Colloque sur les applications industrielles des hautes fréquences et des micro-ondes, Lyon, 1977. Lyon, Univ. Claude Bernard, 1977, pp. 8-16, résumés (complément).

144. PRAKASH V., ROBERTS J. Perturbation of a resonant microwave cavity by select alcohol vapors. *J. Microw. Power*, 21, n° 1, 1986, pp. 45-50.

145. PÜSCHNER H. Wärme durch Mikrowellen. N.V. Philips Gloeilampenfabrieken, Eindhoven, 1964, 325 p. (Philips Technische Bibliothek).

146. PÜSCHNER H. Design consideration for high power microwave conveyor belt systems. In : Proceedings of the IMPI Symposium, Boston, 1969. (Paper G1). (Abstract p. 34).

147. RENGARAJAN S., LEWIS J. Quality factor of elliptical cylindrical resonant cavities. *J. Microw. Power*, 15, n° 1, 1980, pp. 53-57.

148. RINGLE E., DAVID B. Measuring electric field distribution in a microwave oven. *Food Technol.*, 29, n° 12, 1975, pp. 46-54. (C.D.I.U.P.A., n° 90658).

149. RISMAN P. A commercial microwave oven using a near field applicator. *J. Microw. Power*, 9, n° 2, 1974, pp. 163-167.

150. RISMAN P. Microwave energy control by humidity sensing. In : Proceedings of the IMPI Symposium, Leuven, 1976, (Paper 1B1) and in *J. Microw. Power*, 11, n° 2, 1976, pp. 177-178.

151. RISMAN P. A microwave applicator for drying food samples. *J. Microw. Power*, 13, n° 4, 1978, pp. 297-301.

152. RISMAN P., OHLSSON T. Theory for and experiments with a TM_{02n} applicator. *J. Microw. Power*, 10, n° 3, 1975, pp. 271-280.

153. ROCHAS J. Etude et réalisation de cavités électromagnétiques en vue du traitement thermique d'éléments filiformes. Thèse d'Université. Univ. Lyon I, 1980, 126 p.

154. RZEPECKA M., STUCHLY S., HAMID M. Applicators for microwave treatment of granular materials in agricultural technology. In : Proceedings of the IMPI Symposium, Ottawa, 1972, pp. 163-164. (Paper 9.5.).

155. SAITO H., MINO N. Improved magnetron with very low-level harmonic radiation. In : Proceedings of the IMPI Symposium, Chicago, 1985, pp. 133-134.

156. SCHAUBERT D., WITTERS D., HERMAN W. Spatial distribution of microwave oven leaks. *J. Microw. Power*, 17, n° 2, 1982, pp. 113-119.

157. SCHILZ W. Reduction of pollution in industrial microwave systems. In : Comptes rendus du Symposium international sur les applications énergétiques des micro-ondes, IMPI-CFE. 14, Monaco, 1979. (Résumés p. 242).

158. SCHMIDT W. Mikrowellengeneratoren mit abgeschlossenem Arbeitsraum zur dielektrischen Erwärmung von Nahrungsmitteln und Industrieprodukten. *Elektron. Rundsch.*, 12, n° 11, 1958, pp. 390-393, 12, n° 12, 1958, pp. 417-420 et 13, n° 1, 1959, pp. 13-16.

159. SCHMIDT W. Dauerstrichmagnetrons und ihre Anwendung im Mikrowellengerät. *Philips Tech. Rundsch.*, 21, n° 12, 1959/60, pp. 341-355.

160. SCHOONAERT D., LUYPAERT P. A new type of microwave applicator operating in the TM_{221} mode. In : Proceedings of the IMPI Symposium, Milwaukee, 1974, pp. 1-3. (Paper B5.1).

161. SHEN Zh. CHEN K., CHENG R., LI Ch. Design of the axial output structure of the CW magnetron. *J. Microw. Power*, 17, n° 2, 1982, pp. 151-160.

162. SHIBATA C., AKIOKA T., SATO Y., TAMAI H. 100 kW, 915 MHz CW magnetron for industrial heating application. *J. Microw. Power*, 13, n° 1, 1978, pp. 59-64.

163. SKINNER J., BRADY J. Effect of gas flow on the microwave dielectric breakdown of O_2. *J. Appl. Phys.*, 34, n° 4, 1963, pp. 975-978.

164. SLUCE P. Simple method for determining energy distribution in a microwave oven. *Non-Ioniz. Radiat. 1*, n° 3, 1969, p. 131.

165. SOULIER J. Nouveau dispositif pour le traitement thermique de matières en poudres ou en grains. Brevet européen n° 0 036 362, 23 sept. 1981, (FR 80 05607, 13 mars 1980).

166. STUCHLY M., STUCHLY S. Industrial, scientific, medical and domestic applications of microwaves. *Proceedings of the IEE*, 130, part A, n° 8, Nov. 1983, pp. 467-503.

167. STUPP E., FELLOWS M. Power supply for a magnetron. U.S. Patent n° 4 481 447, 6 Nov. 1984.

168. SUCHER M. Measurements of Q. In : SUCHER M., FOX J. Handbook of microwave measurements. Brooklyn, Polytechnic Press.

169. SWIFT J. Notes on a feasibility study of some microwave applicators of potential use to the food industry. In : SMITH R. Microwave heating for the food industry. Suppl. notes for short course, IMPI, European Chapter. Bradford, Univ. Bradford, 1975, pp. 35-41.

170. TAKAHASHI H., NAMBA I., AKIYAMA K. Magnetron Rieke-diagram plotting and application. *J. Microw. Power*, 14, n° 3,1979, pp. 261-267.

171. TETENBAUM S., WEISS J. Microwave breakdown of water vapor. In : Proceedings of the IMPI Symposium, Ottawa, 1978, pp. 57-59. (Paper 6.2). (C.D.I.U.P.A., n° 143425).

172. THIEBAUT J., BERTEAUD A., ROUSSY G. A new microwave resonant applicator. *J. Microw. Power*, 14, n° 3, 1979, pp. 217-222.

173. THIEBAUT J., ROCHAS J., MANOURY M., ROUSSY G. Control of the fields and the hysteresis heating process in a microwave resonant applicator. *J. Microw. Power*, 17, n° 3, 1982, pp. 187-194.

174. THOUREL L. Les ferrites en microondes. (Cours polycopié) Univ. Toulouse, ENSEEIHT, 1968, 144 p.

175. TRUMP J. Insulation strength of high pressure gases and of vacuum. In : HIPPEL A. von. Dielectric materials and applications. New York, John Willey and Sons, 1954, pp. 147-156.

176. TURNER R., VOSS W., TINGA W., BALTES H. On the counting of modes in rectangular cavities. *J. Microw. Power*, 19, n° 3, 1984, pp. 199-208.

177. TWISLETON J. Twenty-kilowatt 890 Mc/S continuous-wave magnetron. *Proc. IEE*, 111, n° 1, 1964, pp. 51-56.

178. VAN KOUGHNETT A. A microwave applicator for filamentary materials. *J. Microw. Power*, 7, n° 1, 1972, p. 17.

179. VAN KOUGHNETT A., DUNN J. Doubly-corrugated chokes for microwave heating systems. *J. Microw. Power*, 8, 1973, pp. 101-110.

180. VAN KOUGHNETT A., DUNN J., WOODS L. A Microwave applicator for heating filamentary materials. *J. Microw. Power*, 9, n° 3, 1974, pp. 195-204.

181. VAN LOOCK W., DE WAGTER C. Ring applicator with resonant slots. In : Proceedings of the IMPI Symposium, Chicago, 1985, pp. 94-96.

182. VAUTRIN J. Susceptibilité aux divers parasites industriels des systèmes et plus particulièrement des systèmes électroniques. Etude INRS n° 2.1/6.1.18 sur 5 ans, début 1985.

183. VENE J. Les plastiques. 8ème édition. Paris, P.U.F., 1976, 121 p. (Que sais-je ?).

184. VERWELL J. Magnetrons. *Philips Tech. Rev.*, 14, n° 2, 1952, pp. 44-58.

185. VOSS W. Applicator principles. In : Proceedings of the IMPI Symposium, Monterey, 1971. (Paper 8.1.).

186. WALDRON R. Perturbation theory of resonant cavities. *Inst. Electr. Eng.*, Part 107, n° 12, 1960, pp. 272-274.

187. WALKER C., HUIZINGA A., VOSS W., TINGA W. A system for controlled microwave heating of small samples. *J. Microw. Power*, 11, n° 1, 1976, pp. 29-32.

188. WASHISU S., FUKAI I. A simple method for indicating the electric field distribution in a microwave oven. *J. Microw. Power*, 15, n° 1, 1980, pp. 59-61.

189. WATANABE M., SUZUKI M., OHKAWA S. Analysis of power density distribution in microwave ovens. *J. Microw. Power*, 13, n° 2, 1978, pp. 173-181.

190. WHITE J. Measuring the strength of the microwave field in a cavity. *J. Microw. Power*, 5, n° 2, 1970, pp. 145-147.

191. WHITE J. The Smith chart : an endangered species ? *Microw. J.*, Nov. 1979, pp. 49-54.

192. WHITE J., TINGA W. Measuring microwave power. *Trans. IMPI*, 1, Microwave power in industry, 1973.

193. WILHELM M., SATTERLEE L. A 3-dimensional method for mapping microwave ovens. *Microw. Energy Appl. Newsl.*, 4, n° 5, 1971, pp. 3-5. (C.D.I.U.P.A., n° 40540).

194. WYSLOUZIL W., KASHYAP S. A cascading rotary microwave processor. In : Proceedings of the IMPI Symposium, Chicago, 1985, pp. 175-177.

195. WYSLOUZIL W., KASHYAP S. A cascading rotary microwave processor. Ottawa, Div. Electr. Eng., Nat. Res. Council Canada, non publié, s.d., 4 p.

196. WYSLOUZIL W., VAN KOUGHNETT A. Automated matching of resonant microwave heating system. *J. Microw. Power*, 8, n° 1, 1973, pp. 89-100.

197. YAMAMOTO K., KURONUMA H., KOINUMA T., TASHIRO N. Experimental study of noise generated by magnetrons for microwave ovens. *J. Microw. Power*, 16, n° 3-4, 1981, pp. 271-276.

198. YOUNG E. The new "Penguin" dictionary of electronics. Hardmonsworth, Penguin Books, 1980, 618 p.

199. Applications industrielles des microondes. Publ. Thomson-CSF, Division Tubes Electroniques, Juin 1980, 13 p. (Réf. APH 6211).

200. Chauffage micro-ondes. La Radiotechnique Compelec, 1978, 44 p. (Publication technique, réf. 4720-06).

201. Circulateurs-isolateurs hyperfréquences. La Radiotechnique Compelec, 1978, 16 p. (Publication technique, réf. 53927-03).

202. Dauerstrichmagnetrons : Industrielle Anwendungen. Hamburg, Valvo GmbH, 1966, 160 p.

203. FCC document on notice of proposed rule making in the matter of overall revision of part 18 concerning ISM equipment. Docket 20718, Washington, Sept. 1978.

204. Générateur micro-onde 5 kW/2450 MHz. Publ. Thomson CSF, Division Tubes Electroniques, août 1982, 4 p. (Notice TEH 4974).

205. Klystrons de grande puissance pour chauffage industriel en hyperfréquence. Publ. Thomson CSF, Division Tubes Electroniques, juin 1982, 6 p. (Notice TH 2054-TH 2054 A).

206. Le magnétron OM 75, source élémentaire pour la réalisation de générateur de grande puissance. *Econ. Progrès Electr.*, n° 4, juil.-août 1982, pp. 23-26.

207. Magnétron à ondes entretenues YJ 1540. Documentation provisoire, Philips/RTC, Janv. 1986, 7 p.

208. Magnétron à ondes entretenues YJ 1600. Documentation provisoire, Philips/RTC, janv. 1986, 9 p.

209. Magnétron F 1123, 5kW/2450 MHz. Publ. Thomson CSF, Division Tubes Electroniques, oct. 1980, 7 p. (Notice TEH 4808).

210. Magnétron MCF 1165. Publ. Thomson CSF, Division Tubes Electroniques, Sept. 1980, 5 p. (Notice TEH 4801).

211. Magnétron MCF 1166. Publ. Thomson CSF, Division Tubes Electroniques, sept. 1980, 5 p. (Notice TEH 4802).

212. Magnétron MCF 1327. Publ. Thomson CFS, Division Tubes Electroniques, sept. 1980, 7 p. (Notice TEH 4803).

213. Magnétron TH 3094. Publ. Thomson CSF, Division Tubes Electroniques, sept. 1980, 6 p. (Notice 4804).

214. Matériaux magnétiques, céramiques piézoélectriques. La Radiotechnique Compelec, 1978, 68 p. (Publication technique n° 3).

215. Microwave heating. Document commercial, Magnetronics Ltd., sans réf., s.d., 8 p.

216. Microwave heating for agriculture-sterilising, drying, cooking, blanching, heating. Publication Magnetronics Ltd., ref. AGR 60 E, s.d., 4 p.

217. Microwave power generators. Document commercial, Magnetronics Ltd., réf. B40-100 e, s.d., 6 p.

218. TH 2054, klystron de grande puissance pour chauffage industriel en hyperfréquence. Document commercial Thomson CSF, Division Tubes Electroniques, reproduit dans *Econ. Progrès Electr.*, n° 13, janv.-fév. 1985, pp. 22-26.

219. Tubes for microwave equipment. Electron tubes, part 2. Philips data Books, 1971, 724 p.

220. Tubes pour télécommunications et tubes industriels-circulateurs et isolateurs. La Radiotechnique Compelec, 1978, 46 p. (Publication technique n° 5).

221. 915 MHz, 500 kW CW circulator, CU H 214. Raytheon Co., Special Microwave Devices Operation. *Microw. Syst. News Commun. Technol.*, 16, n° 4, 1986, p. 166.

Part II

Industrial applications

Chapter 1

Drying

Many industrial processes involve one or more stages of product drying, which is often very costly. Microwaves offer what is sometimes the most viable alternative in terms of energy efficiency.

We shall consider different examples of industrial microwave drying. Most of them will be examined at their experimental stage because of the secrecy that invariably colors the reporting of industrial processes.

1.1 Humidity and drying

Humid materials occur in a variety of physicochemical forms (e.g., pastes, colloids, suspensions, and porous solids containing an absorbed liquid, usually water). In the case of a porous solid, water binding involves both the adsorption and capillarity mechanisms [7]. Adsorption occurs at the solid/liquid interface in pores in which Van der Waals forces (Part I, Section 3.4.3) cause the adhesion of the first layer of water molecules

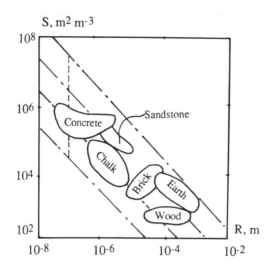

Figure 1.1 Type of porous material as a function of both the radius of the pores (R) and specific surface area (S) [7].

to which other molecules may get attached. The adsorption efficiency of a material is characterized by its specific surface area S which is a measure of the interface area per unit volume of the product. The phenomenon of capillarity governs the interaction when the porous structure contains liquid water. The liquid spreads over the pore surface, forming concave interfaces with air. The pressure difference across the interface (i.e., the capillary pressure acting on the liquid) is given by

$$P_c = \frac{2\sigma \cos \theta}{r} \tag{1.1}$$

where σ is the surface tension ($[MT^{-2}]$, $N\,m^{-1}$), r is the radius of the capillary tube, and θ is the angle of contact which is a measure of the wettability of the surface. For water at 20°C, we have $\sigma = 73 \times 10^9\,N\,m^{-1}$.

The quantity of water that can bind to a porous material is therefore a function of the specific adsorption area and of the pore radius. Porous solids may thus be divided into the two main categories indicated in Figure 1.1.

Materials falling into region A in Figure 1.1 are said to be hygroscopic and have the ability to bind large quantities of water by adsorption. They have high specific surface areas and pore radii below a threshold of 10^{-7} m. The drying or dessication of wet material requires the transfer, without boiling, of water from the material to the ambient medium. This involves mostly the removal of free water, in contrast to dehydration in which all forms of bound water are also removed. We have already noted that microwaves have little effect on some bound-water components (Part I, Section 3.5.3).

1.2 Drying kinetics

Drying is regulated by two fundamental mechanisms, namely, the transfer of heat and of mass. Consider a material at a temperature T, with mass water content m, exposed to a current of air at a temperature T_a greater than T and with partial vapor pressure P_a that is by definition less than the saturation vapor pressure $P_{vs}(T_s)$. A rise is observed in the surface temperature T_s and, consequently, in the saturated vapor pressure at the surface, $P_{vs}(T_s)$. The pressure gradient $P_{vs}(T_s) - P_{vs}(T_a)$ gives rise to the flow of vapor between the surface and the air current. This initial evaporation reduces the water content m_s of the superficial layers and results in an increase in the capillary pressure P_{cs} at the surface, with the capillary pressure gradient causing the rehumidification of the superficial layers. The rate of evaporation is constant and remains so as long as the capillary suction ensures surface rehumidification at a rate sufficient to compensate for the evaporation. The superficial layers eventually dry up, the diffusion of vapor slows down, and the rate of drying falls [7].

1.3 Microwave drying

In the conventional process, the two drying mechanisms function in opposite directions, i.e., the temperature gradient (heat transfer) and the humidity gradient (mass transfer) operate in opposition [59]. The net effect is usually the formation of a crusty surface that is thermally insulating and thus impermeable to the transmission of heat and hindering capillarity. The advantages of deep energy deposition within the material are thus very evident.

Microwave drying is governed by equations incorporating mass and energy transfer terms as well as microwave energy propagation and boundary conditions. The first two equations are

$$\frac{\partial m}{\partial t} = \text{div} \left(D_m \, \text{grad} \, m + D_T \, \text{grad} \, T \right) \qquad (1.2)$$

$$\frac{\partial T}{\partial t} = \frac{1}{\rho C} \left[\text{div} \left(\Lambda \, \text{grad} \, T \right) + A \frac{\partial m}{\partial t} + q \right] \qquad (1.3)$$

where m is the water content, t is the time, T is the temperature, Λ is the thermal conductivity, ρ is the density, and C is the heat capacity (Part I, Section 3.6). The quantity q is the power per unit volume released by the wave in accordance with equation (3.57) of Part I. The definition of A and of the isothermal and nonisothermal mass diffusion coefficients D_m and D_T is given in [8].

Analysis of the above equations shows that by combining microwaves with traditional heat source we can control the temperature gradient and achieve uniform drying [59]. The only difficulty is in determining the optimum time for starting microwave irradiation.

1.4 Paper and printing industries

1.4.1 Paper

Drying is a very important process in the manufacture of paper and cardboard. On average, 1.5 litres of water must be evaporated per kilogram of dry product. This operation consumes a great deal of energy and accounts for approximately 10% of the production cost [15], [89]. The drying process has, in addition, an important qualitative effect because its parameters determine the mechanical properties of the final product, the quality of its surface, and its permeability to air [36]. Conventional drying incorporates mechanical and thermal stages. The level of humidity is first reduced from 96–98% to 86–92% by means of a metal mesh set in horizontal translational motion, which releases water but retains solid constituents in a fibrous mass. The water content of this mass is then reduced in cylindrical presses to 65%. The remaining water is removed by evaporation by pressing the felt-supported paper against a series of heated drums. The evaporation process is often aided by a stream of air that serves to accelerate mass transfer, assists in the uniformity of drying over the entire width of the paper, and eliminates the layer of saturated air that tends to stagnate at the surface [15].

The final humidity is between 5 and 10%, but a uniform result is difficult to achieve after the cylinders are dried, and this often necessitates additional drying to 3–4%, followed by rewetting. This operation does not make economic sense and has an undesirable and irreversible effect on the quality of the product [37, 38].

Trials performed since 1967 by EIMAC, a subsidiary of Varian Associates, using a 100-kW klystron operating at 2.45 GHz, have demonstrated the advantages of microwave drying, which include three-dimensional uniformity of humidity distribution, no need for overdrying, selective microwave heating of the water component, and a 30% reduction in the length of the dryers. The applicator consists of a series of slotted waveguides and employs a stream of air at 104° [38]. There is a clear economic advantage of an initial humidity level of 15–20%.

There seems to be a large number of successful industrial processes involving microwaves [89]. For example, a Shanghai enterprise [24] has achieved a radical saving of time and a 50% energy saving by employing a microwave drying process at 2450 and 915 MHz.

The use of microwaves in this field is, however, in strong competition with RF (27.12 MHz), which is better in controlling the humidity profile [36].

Microwaves can also help in the drying of paper coatings [36] that consist of a very thin layer of mineral particulates (kaolin or calcium carbonate), the effect of which is that a microporosity is substituted for the macroporosity of the paper. Microwave drying tests on binding agents have demonstrated improved drying efficiency and better coating cohesion. A hybrid drying process involving both microwaves and forced convection ensures precise control of the migration of bonding material.

1.4.2 Printing inks

A special issue of the *Coventry Standard*, released in April 1967, was probably the first publication to benefit from microwave drying applied to printing inks. The dryer manufactured by Eden, Fisher, and Hirst was installed on a Halley-Aller press. It delivered 25 kW at 2.45 GHz. The temperature of the paper emerging from the press did not exceed 80° as compared with the 200–300° produced by the naked flame or a stream of hot air for an equivalent production rate of 300 m/min. The advantages were quite important: no surface overheating and hence no risk of darkening or cracking of paper, and no need for protection from high temperatures and for vapor and fume extractors. The risk of fire was thus greatly diminished. Since no water was lost from the paper during the process, ordinary offset paper could be used, and this resulted in a saving on the order of £4 per ton (1967).

Printing quality was judged to be superior and the colors

deeper and well defined. The user pointed to additional advantages related to the start up and shut down speed, ease of access, and economy of space [11]. The capital outlay (about £10,000 in 1967) was considered to be on the same order as that of a traditional installation. Cost savings achieved by using the microwave drying process were estimated at about 30% [65].

However, the inks employed must be specially formulated to have high dielectric loss. Such inks have been developed for lithography, offset printing, and typography.

Microwaves can also be used in conjunction with infrared drying [82]. The rapid deposition of ink on the surface and the low paper penetration result in good quality colors without the additional need for powders or suspensions normally vaporized on inks.

A microwave applicator consisting of N rectangular cavities, $(1/N)\lambda_g/2$ apart, through which a sheet of inked paper can circulate at high speed, is described in the patent specification given in [5].

1.4.3 Glued products

An interesting application of microwaves to the bulk processing of paper bags is described in [1],[2]. The bags contain on average 10 g of water-based adhesive with 14% of dry matter. Inadequate drying of the adhesive produces water migration into the paper, causing loss of strength. The complete drying cycle requires 21 days of immobilization, for which large storage facilities are necessary; more problems may arise from handling and the possible sticking of the bags to one another. Microwaves bring about the usual advantages of selective water heating, ease of control, action in depth, and lower temperatures. Pilot studies showed that a 10-m long, 96-kW microwave unit was capable of treating 200 bags per minute. Assuming that the capital outlay is written off after three years, the cost amounts to about 1% of the unit price of the bags.

1.5 Leather and textile industries

1.5.1 Leathers

The treatment of skins includes two stages of drying after tanning and tinting. The latter stage creates bottlenecks that result in delays in delivery.

The British company Gomshall Tanneries is known to have used microwave conditioning as early as 1976 [101]. It employed three 25-kW microwave units operating at 896 MHz for drying leather after tanning. Leather was first treated in a steam-heated dryer operating under vacuum, which reduced the skin water content to 35%. At this stage, the distribution of water was not uniform. The microwave dryer then reduced the water content to 25% in a very uniform way, thus facilitating subsequent processes. A prototype applicator with a 1.83-m polypropylene trellis shelf enabled the treatment of 360 sheepskins per hour at 12 kW microwave power. More recently, the microwave laboratory of ENSEEIHT [77] produced a prototype measuring $1 \times 1 \times 3$ m and employing a vertical airstream. This results in a time saving of between 30 and 50%. Shrinkage is reduced and a highly uniform distribution of humidity is achieved, thus avoiding overdrying and rehumidification.

1.5.2 Tufts and yarns

A new process for the manufacture of tufts and yarns without torsion has been set up at the *Mulhouse Center for Textile Research* (CRTM). Production rates of 150 to 200 m/min can be reached. Fiber cohesion is obtained by means of an adhesive rather than the conventional torsion technique. The finished products exhibit exceptionally high degrees of body suppleness, comfort, and thermal insulation. The process ensures a 12-% saving of material as compared with the traditional process [21], [22].

The manufacturing problem in this process is the fragility of the wet and sticky yarns and tufts. In fact they emerge from the production head at the rate of 2 to 3 m s^{-1} and must be deposited, without being subjected to any constraint, on a horizontal conveyor belt. In conventional hot-air drying, the stream of air must be quite weak to avoid the deterioration of the material. This implies longer drying time, which in turn necessitates a longer dryer with yields of between 6 to 7% only, or 10 to 12 kW/l of evaporated water [16].

Infrared drying is very irregular and causes yarn breaking during loading on bobbins. On the other hand, the first microwave trials carried out in 1977 by the IMI-SA Company suggested that a satisfactory solution was at hand [14]. A specific applicator was then developed in the form of a multimode stainless-steel cavity with an active length of 70 cm and a fiberglass/PTFE conveyor belt running through the system. The points of entry and exit were equipped with wave traps and air-cooled absorbing plates. Microwave power was fed through the bottom wall via a power divider and two waveguides, each with three helical antennas (Fig. 1.2). The power generator delivered 5 kW at 2.45 GHz from an 11-kVA power supply. Water vapor was removed through an extraction hood.

Table 1.1

Drying of polyester/cotton tufts [21]

Speed of conv. belt m/min	Rate of evap. of water, l/hr	Mass of water in cavity, g	Energy consumption, kWh/kg evaporated
1.27	1.39	7.4	3.6
0.81	1.67	10.1	3.0
0.38	1.71	21.0	2.9

Table 1.1 lists some typical results obtained at CRTM. The product in question is in the form of polyester-cotton tufts of 0.32 g/m. The yield is 37% and is far superior to that obtained with traditional dryers [16]. The performance remains quite acceptable even at high speed, and the quality of the dried product is judged to be excellent. Because of the fast treatment (30 to 60 s), the material has no time to change color or degrade

Figure 1.2 Microwave applicator designed by CRTM/IMI-SA Company for the treatment of textile tufts and yarns (CRTM document).

in any other way, there is no sticking at the point of contact (which is a persistent problem in hot-air dryers), and there is no overdrying.

1.5.3 Dyeing and finishing

The final thermal treatments after dyeing and finishing are usually performed on hanks, bobbins, or fabric. This may lead to a certain degree of color migration and, hence, to defects and nonuniformity of the product. Microwaves may provide a solution if a resonant cavity is used [40].

Preliminary trials at the Institut Textile de France and the Claude Bernard University in Lyon, using a rectangular TE_{01n} cavity, gave interesting results for cotton yarns threading at 100 m/min. The excess humidity of 30–100% was reduced to 2 and 20%, respectively. The use of circular and elliptic cavities was reported in [40]. In the latter case, all the energy in the cavity can in theory be concentrated on one focal axis by exciting the desired mode by an antenna placed along the other focal axis. A frequency in excess of 2.45 GHz, for example 10 GHz, is best suited to the process, and an automated RF tuning may prove essential.

Another possible application [13] is the microwave fixation of colors and dressings on cotton fabrics as a substitute for vapor. The resulting evaporation without liquid displacement avoids problems associated with color migration.

Folded, slotted waveguides have been used in continuous dyeing processes since the 1970s. For example, a 2.5-kW applicator operating at 2.45 GHz has been used for the dyeing of wool and polyamid, authorizing speeds of fixation 25 to 50 times higher and acid coloration yields on the same order as in traditional techniques (90–93%). In the case of viscose, the process was even more efficient, with speeds nearly 200 times higher and yields of 95% instead of 80. With acrylics, the speed of fixation may be 40 to 50% higher and the yield of basic components comparable to that of traditional vapor processing.

The efficiency of microwave dyeing of wool with reactive colorants can be improved by the addition of metabisulfite or sodium sulfate [28].

In 1972, the British firm Smith and Company produced a 2.5-kW applicator operating at 2.45 GHz and capable of dyeing a strip of fabric 25 cm wide at 7 m/min.

A cylindrical TM_{010} cavity operating at 2.45 GHz and delivering 1.5 kW has been used in combination with a hot-air dryer [69]. The tuning of the cavity is not significantly affected by the type of fabric and dye or the temperature distribution, and small changes due to inevitable mismatches can be easily compensated.

Professor H. Needles of the University of California has contributed some interesting work on dye fixation and fabric finishing [53]. He has noted that with anthraquinone-substituted dyes on polyester fibers, microwaves gave a much better finished product as compared with the conventional treatment, and a much better diffusion of color into fibers. The last characteristic might be due to some specific effect of microwaves whereby they reduce the molecular aggregates of the dye. The process is not only more rapid and more stable than traditional techniques, but is also very effective on thick materials. A continuous process in which microwaves are used for dye fixation was patented in 1984 [55].

The Japanese company Ichikin has developed an industrial applicator (Apolotex) measuring 8.75 m, 2.50 m, and 3.10 m in length, depth, and height, respectively. The applicator is fed by two 5-kW, 2.45-W sources through four waveguide outputs. About 20 units are working worldwide, including a unit at Intessa at Como (Italy). The cost of this applicator was just under $200,000 in 1984. It was originally designed for the chemical treatment of Pad Roll type fabric, in partial replacement of hot-air treatment. It is used in fabric dyeing and finishing, in bleaching with oxygenated water, in cleaning and deglueing of cotton fabric, and in surface finishing of polyesters [17].

1.5.4 Tufted carpets

This application was studied by the Electricity Council and the University of Leeds in Britain and was the subject of several publications [71], [72], [73], [50], [70].

The production of tufted carpets involves fixing the pile on a polypropylene base before applying a backing of jute, latex, or foam. The printing of the carpet can take place before or after the application of the backing.

Drying of the carpets consumes vast amounts of energy. For example, the power necessary to remove the last 10% of humidity from a 4-m wide carpet of $2 \, kg/m^2$, transported at $5 \, m/min$, is on the order of 200 kW.

Metaxas et al. at the Electricity Council in Britain have suggested the use of traditional techniques of vacuum extraction and hot-air drying to bring the humidity levels from 160% down to 30 to 50%, followed by microwave drying to reach a humidity level of 1.8%. They proposed the use of a TE_{10n}-mode double applicator consisting of two resonant cavities, designed so that the stationary-wave troughs of one chamber coincide with the crests in the other cavity. The $\lambda/4$ difference between the cavities thus ensures uniform heating over the entire width of material. This principle was used to build a prototype delivering 25 kW with hot air circulating between the two cavities (Fig. 1.3). This prototype was placed at the end of a carpet-drying chain, where humidity levels were between 5 and 20%. The chain was traveling at between 1 and 6 m/min. This stage of microwave drying reportedly is capable of increasing the output by 20%.

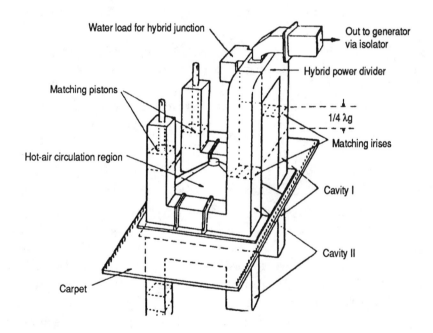

Water load for hybrid junction
Out to generator via isolator
Hybrid power divider
Matching pistons
1/4 λg
Matching irises
Hot-air circulation region
Cavity I
Cavity II
Carpet

Figure 1.3 A two-cavity microwave dryer for tufted carpets [50].

1.6 Construction

1.6.1 Wood and plywood

Decareau has described [27] the use in Finland of a 25-kW microwave system developed by Magnetronics for the drying of pine and birch floors. In China, a microwave drying installation processes wooden frames to a very high standard with loss rates of only 2% as compared with 22% in hot-air drying. Wood can indeed be readily dried by microwaves provided the process is slow enough to avoid the creation of internal strains that affect product quality. This type of drying is also an efficient means of insect control [94].

The Boise Cascade Plywood Mill Company of Yakima, (Washington) has achieved excellent results by combining traditional and microwave dryers of plywoods (50 kW at 950 MHz). These pioneering experiments with plywood were not, however, exploited on an industrial scale. Nevertheless, microwave processing of plywood avoids the need for overdrying and ensures uniformity and the absence of bending [63].

1.6.2 Plaster, concrete, and ceramics

Industrial plaster tiles measure $50 \times 60 \text{ cm}^2$ with a thickness of 4, 6, 7, or 10 cm. They are usually dried in 50-m tunnel ovens, which takes about 36 hours for the 10-cm tiles. The mass loss is about 18%. Lefeuvre [58], [59] has developed a combination oven with 8 kW of conventional power (hot-air radiators) and a 3-kW microwave unit operating a 2.45 GHz with a U-shaped waveguide applicator exciting coaxial elements with radiating slots.

Cober Electronics of Stamford, Connecticut has designed 24-kW, 2.45-GHz microwave tunnel ovens for the treatment of ceramics, providing excellent uniformity of penetration and heating (Fig. 1.4). Microwave ceramic dryers are also produced by the Italian company OMAC Srl at Pratissolo di Scandiano (Reggio Emilia). Very little information is available about the users of these dryers, but it may be noted that the Mori company of Madena uses the Cober dryer and that the Verona-based company Siti, world leader in ceramics, has designed and built the mechanical and thermal parts of a dryer producing $7 \times 5\,\text{kW}$ in which the microwave part is supplied by the Compagnie Française des Microrayonnements [18].

Figure 1.4 Microwave tunnel dryer for ceramics from Cober Electronics Inc.[56].

The Polish company Digicom produces a microwave applicator consisting of an electromagnetic horn of $900\,\text{cm}^2$ base fed by a 2.45-GHz, 750-W magnetron via a cylindrical antenna extending into the horn. This applicator, called Emibeam, was specially designed for drying walls (on which it had a fungicidal effect) [97]. However, the safety of this equipment can be called into question.

1.7 Foundries

The drying of foundry moulds requires a considerable amount of energy because of the very poor heat transfer properties of the materials to be dried. This is a classic situation in which the use of microwaves is more advantageous than gas dryers. The advantages include ease of use, cleanliness, instant energy deposition, savings in workshop space and time (by a factor of up to 10), the absence of fumes and heat loss, and savings in energy and in manpower. Moreover, uniform drying ensures the least possible distortion of dimensions, better surface quality, and greater hardness [99].

Cober Electronics (Fig. 1.5) offers a large range of drying equipment operating at 2.45 GHz: batch driers BPH18A (1.7 m³, 18 kW, 230 kg capacity); BPH30A (13.8 m³, 30 kW, 900 kg capacity); tunnel ovens CCO, CCD, CCM, and CCE, producing 30, 60, 120, and 180 kW and capable of removing 32, 64, 128, and 256 l/hr of water. The CCO models have an aperture of 710 × 150 mm² whilst models CCD and CCM have an aperture of 710 × 710 mm² and an automatic door-closing device at the entrance and exit. The CCM model has a conveyor belt going in and out of the oven through wave traps. Unfortunately no information is available about the users of these devices.

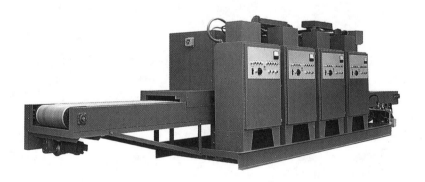

Figure 1.5 Microwave drying tunnel for foundry material, Cober Electronics [56].

1.8 Rubbers and plastics

1.8.1 Drying of polymers

Polymers are mostly nonpolar materials (Part I, Section 3.5.2), in contrast to the solvents to be evaporated. Microwave drying is therefore recommended particularly because traditional techniques tend to damage the products and produce changes in molecular weight. Moreover, certain materials such as halogenated rubbers decompose under the influence of heat and give off corrosive gases.

Trial drying experiments have been performed on polyethylene, polypropylene, *butyl rubber* (BR), and polyvinyl chloride. Samples of between 5 and 100 g were exposed to between 100 and 1, 000 W of microwave energy at 2.45 GHz [96]. In the case of polyvinyl chloride, humidity could be reduced from 18% to less than 1 ppm, and better quality films of higher molecular weight material (10,000 as compared with 8,000 in traditional drying) were produced. A hybrid microwave/hot-air system used for drying butyl rubber to less than 1 ppm from 5% humidity resulted in an energy saving of 33%.

1.8.2 Photographic film and magnetic tape

Photographic film is made of polyesters and acetates that have very low dielectric loss. In contrast, wet emulsions have high loss values. This is a classic situation in which microwaves can be used for the selective heating.

The process was developed by Production, a subsidiary of Houston Fearless Corporation of Los Angeles and by Reeves Electronics Corporation of Chicago. The first applicator was in the form of a slotted, folded waveguide that gave only mediocre results. The second applicator had radiating apertures and achieved the extraction of 5 l/hr of water from 16 and 35 mm films transported at 54 m/min [30], [31].

Microwave drying offers savings of space, dust-free operation, uniform drying, no accidental burns, no water pockets under dried emulsions, no need for overdrying, energy savings, and no environmental heating. Moreover, in case of breakdown, only about 10 m of film is lost instead of 100 to 1000 m because the microwave dryer is much shorter than the conventional one.

A similar application was reported in China [24]. In this case, magnetic tapes are dried in a hot-air/microwave oven operating at 2.45 GHz. The drying tunnel is only 8 m as compared with 78 m for the traditional infrared oven. The drying time is 14 s instead of 175 s, and the solvent evaporates at the rate of 1.12g/Whr as compared with 0.2.

1.9 Pharmaceutical industry

The manufacture of granular pharmaceutical compounds consists of four stages: (1) the primary materials are weighed and mixed until a homogeneous powder is obtained; (2) the dry mass is moistened with aqueous and alcoholic solvents to produce a soft paste used to make granules; (3) the wet granules are spread on trays; and (4) the granules are dried in a hot-air oven.

The process is not continuous between these stages, which can be costly in terms of manpower. The drying in particular presents a serious interruption of the production chain, and conventional continuous dryers will require a very heavy infrastructure because of their size.

Trials of microwave drying of pharmaceutical products have been performed using an LMT oven producing 1.5 kW at 2.45 GHz [57]. The product tested was a mixture of senna powder, sucrose, coco, and excipients in respective proportions of 4%, 85%, 5%, and 6%. The product took 10 min. to dry, which is approximately 2% of the usual time. Six months later, the quality of the dried product was no different from that of conventionally dried controls. A large number of trials performed by the IMI-SA Company using a Gigatron tunnel oven have confirmed the superior quality of microwave drying for this type of product.

An important achievement in this field is the rotating oven developed jointly by Professor Pellissier, ADERME, and the Rayonnements Systèmes Company for the Boiron Labs of Ste. Foy lès Lyon. This oven was adapted from a hot-air conventional sugar-coating machine by the addition of a microwave array fed by twenty 800-W, 2.45-GHz magnetrons protected by a radome. The array can readily be removed for drying. The oven rotates around the horizontal axis while the microwave element remains fixed parallel to the axis. The design has necessitated the development of a crown-shaped wave trap.

The product is a homeopathic granular preparation obtained by the successive coating of sugar grains with a concentrated syrup. The initial load consists of 400 kg of sugar grains, which becomes 800 kg after the first coating. Half of this is removed and the process is repeated with the remaining 400 kg. The process is repeated 21 times before the final diameter of 4 mm is reached. During these operations, the product is continuously stirred by blades to improve thermal transfer. At the end of each sequence, water is removed from the syrup by drying in hot air (100 kW) and by microwave heating (16 kW), thus raising the temperature of the product to between 40° and 70° C.

The use of microwaves has led to a 50% improvement in productivity. The user is said to appreciate the flexibility of the technique, the precision of control, and the ease with which the whole installation can be supervised. The process is also thought to be economically interesting [99], [75], [105].

1.10 Drying of tobacco

The use of microwaves for drying cigarettes was reported in several studies in the late sixties [47], [48], [45], [46]. The purpose of the treatment was to increase the volume of tobacco or, more precisely, the cigarette filling factor, without increasing the quantity of tobacco used.

At first, the cigarettes were treated before packaging directly on the monorail-transported trays. They were stacked in eighty rows of fifty layers. A 4-kW microwave generator produced a core temperature of 60° C at the rate of 2,500 cigarettes per minute. However, the treatment was accompanied by a significant loss of water, reduced filling factor at the extremity of each cigarette, and a degradation of quality and aroma. It was soon realised that the loss of humidity should not exceed 1.5%. Having then introduced conveyor belts into the production chain [46], it became possible to build a microwave cavity around it, forming a tunnel of $450 \times 550 \times 400\,mm^3$, delivering 4.5 kW at 2.45 GHz, followed by a cooling area that consumed 3.5 kW in forcing air at the rate of $16\,m^3/min$. The speed of the conveyor belt was between 0.3 and 3 m/min or a maximum of 3,000 pieces per minute. This time, the results were quite acceptable, provided the temperature at exit of the oven was between 55 and 65°C. The humid-air cooling limits the water loss to less than 1.5%, and the process yields a significant improvement in the cigarette filling factor along with an insecticidal effect.

Magnetronics have developed a microwave tunnel with 2×30 kW at 896 MHz and a PVC-terylene belt carrying a tobacco bed 600 mm wide and 100 mm thick, preheated by hot air. The temperature at the end of the tunnel is $95 \pm 2°C$. The use of PTFE is strictly forbidden in the tobacco industry [103]. PVC-terylene, a poor dielectric, could be used instead only because of the very high dielectric loss of wet tobacco [103].

1.11 Regeneration of zeolites

Zeolites, or tectosilicates, are silicates with a special three-dimensional structure, characterized by the fact that the water they contain is not structural water but is weakly bound to the crystal lattice. This means that zeolites play an important part in the purification of gases by adsorption. Once saturated with humidity, the crystal must be regenerated by heating, which eventually leads to the degradation of its properties.

Research at the University of Nancy has been concerned with the use of microwaves in this very specific drying process. For instance, a study of the 13 X zeolite (sodium silicoaluminate, crystallizing in the cubic lattice) was reported in [85], [86], [87]. The unit cell measures 0.25 nm and contains 192 $Si - Al\,O_4$ octahedra, forming eight cuboctahedra or sodalite cages, 0.7 nm in diameter, linked by hexagonal prisms. This structure defines a 26-hedral supercage, 1.3 nm in diameter. The supercages are joined to each other and are in contact with the exterior via channels 0.74 nm in diameter (Fig. 1.6). Absorbed water spreads in cages and supercages at the rate of 3 and 15 molecules per unit, respectively. At the macroscopic level, the material can be described as consisting of 30 to 50-μ crystals, aggregated in spherical granules of 2, 4, and 8 mm in diameter, or 223-mm long rods, bound by an inert medium.

Dehydration by microwaves cannot be explained in terms of temperature alone. Its characteristics are completely different

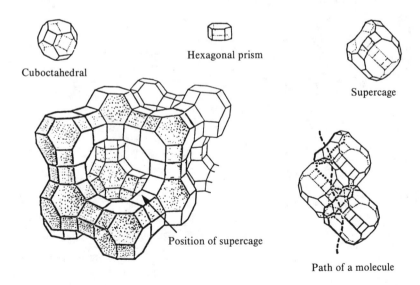

Cuboctahedral

Hexagonal prism

Supercage

Position of supercage

Path of a molecule

Figure 1.6 The structure of zeolite [85].

from classical dehydration. In particular, the process occurs in two separate stages, which could be explained by the fact that adsorbed water in supercages has a rotational mobility close to that of free water, whilst water molecules in sodalite cages are more hindered [85].

Other research in this field [34] has been concerned with molecular sieves regenerated by extraction of adsorbed ethanol. The combination of microwaves (750 W at 2.45 GHz) and a carrier gas (CO_2 or N) has been reported as giving excellent results.

References

1. BELLAVOINE R. Tunnel de traitement micro-ondes et son application au séchage de produits encollés. Brevet français n° 2 501 450 (81 01615), 9 mars 1981.

2. BELLAVOINE R. Séchage de colle sur papier par micro-ondes, application au séchage sur sac papier. *Econ. Progrès Electr.*, n° 1, 1982, pp. 5-8.

3. BELTON J. Microwave-accelerated dye fixation. In : International dyer and textile printer, 143, 1970, pp. 662-664.

4. BERTEAUD A. Absorption électromagnétique dans les matériaux. In : Les micro-ondes. 1980, 19 p. (Cycle de formation permanente, CNRS/ENSCP).

5. BERTEAUD A., CLEMENT R., MERLET C., LECLERCQ C. Procédé et dispositif de traitement par micro-ondes de produits en feuille. Brevet français n° 2 523 797 (82 04398), 16 mars 1982.

6. BERTEAUD A., MATHIEU A., MANOURY M., ROCHAS J. Perspectives de traitement d'éléments filiformes par micro-ondes. In : Comptes-rendus du Symposium international sur les applications énergétiques des micro-ondes, IMPI-CFE. 14, Monaco, 1979, pp. 122-124.

7. BORIES S. Thermodynamique et cinétique du séchage. In : Les micro-ondes. 1980, 19 p. (Cycle de formation permanente, CNRS/ENSCP).

8. BORIES S., LE POURHIET A. Transferts couplés de chaleur et de masse en milieu poreux. Modélisation mathématique - simulation numérique - application au séchage par micro-ondes. In : Comptes rendus du Symposium international sur les applications énergétiques des micro-ondes, IMPI-CFE. 14, Monaco, 1979, pp. 105-108.

9. BRADY M. Loss measurements of wet textiles at 9 GHz. Teknisk Notat E-35, ref. 176 jobb 123, Forvarets Forskningsinstitut, juil. 1964, 9 p.

10. BREWER R. Microwave drying techniques, Hirst microwave drying techniques, British Printer, Hirst Microwave Industries publ. Ltd., s.d., 6 p.

11. BROWN J. Microwave drying - another advance in Coventry colour. In : Coventry Evening Telegraph, Coventry Standard, special n°, April 1967, 8 p.

12. BURKHOLDER H., FANSLOW G., BLUHM D. Recovery of ethanol from a molecular sieve using dielectric heating. Univ. Iowa, Ames lab. USDOE, s.d., 13 p. (ref. 13-4785).

13. CHABERT J. Possibilités d'applications des techniques de chauffage par micro-ondes, dans l'industrie textile. Publication du Centre de Recherches Textiles de Mulhouse, 1975, 14 p.

14. CHABERT J. Application des techniques de chauffage haute fréquence et micro-ondes au séchage et aux traitements des fils textiles en écheveaux ou des mèches de gros titre. In : Comptes-rendus du Colloque sur les applications industrielles des hautes fréquences et des micro-ondes, Lyon, 1977, 8 p.

15. CHABERT J. Séchage de papier et de fibres textiles. Comité français d'électrothermie. Journées d'études et d'informations électro-industrielles des régions Bourgogne et Franche-Comté, Dijon, 1978, 23 p.

16. CHABERT J. Microwaves in the textile industry : drying and processing. In : Comptes-rendus du Symposium international sur les applications énergétiques des micro-ondes, IMPI-CFE. 14, Monaco, 1979, pp. 83-91.

17. CHABERT J. Séchage et traitements textiles par chauffage haute fréquence et micro-ondes. *Econ. Progrès Electr.*, n° 10, Juin-Juil. 1984, pp. 8-12.

18. CHABERT J. Communication personnelle. 12 fév. 1986.

19. CHABERT J., VIALLIER P. Microwave drying of rovings of textile fibers. Publication du Centre de Recherches Textiles de Mulhouse, s.d., 4 p.

20. CHABERT J., VIALLIER P., MEISEL N. Application des micro-ondes au séchage en continu de mèches et fils de gros titre. In : Comptes-rendus du Symposium international sur les applications énergétiques des micro-ondes, IMPI-CFE. 14, Monaco, 1979, pp. 115-117.

21. CHABERT J., VIALLIER P., MEISEL N. Microwave drying of untwisted rovings and yarns in the textile industry. In : Proceedings of the IMPI Symposium, Toronto, 1981, pp. 61-63.

22. CHABERT J., VIALLIER P., MEISEL N. Séchage par micro-ondes des mèches et fils encollés sans torsion dans l'industrie textile. *Rev. Gén. Electr.*, n° 11, 1981, pp. 823-825.

23. CHARREYRE-NEEL M. Station d'irradiation et de mesure de la permittivité des diélectriques en micro-ondes. Application à l'étude de la déshydratation des zéolites et du séchage du papier. Thèse Doct.-Ing., Univ. Nancy I, 1983.

24. CHEN H., SHEN Z., FU C., WU D. The development of microwave power applications in China. *J. Microw. Power*, 17, n° 1, 1982, pp. 11-15.

25. CLEMENTS R. Improvements in the carpet and paper industries. In : Comptes-rendus de la Conférence Internationale d'Electrothermie (U.I.E.), Brighton, 402, 1968.

26. DAWSON T. What are the textile possibilities for microwave heating ? *Text. Month*, Feb. 1972, pp. 47-50.

27. DECAREAU R. Future research developments in microwave technology. Non publ., s.d., 12 p.

28. DELANEY M. Pad batch microwave dyeing of wool. *Text. Chem. Color.*, 4, n° 5, 1972, pp. 119-122.

29. DE WAGTER C., DE POURCA M., VAN LOOCK W. Microwave heating of laminated materials. In : Proceedings of the IMPI Symposium, Toronto, 1981, pp. 225-227.

30. EBERSOL E. Microwaves speed film drying. *Microw.*, May 1967, pp. 14-15.

31. EBERSOL E. Microwave film drying saves costs, improves the product. *Microw.*, Dec. 1967, p. 16.

32. EVANS D., SKELLY K. Application of microwave heating in dye fixation. *J. Soc. Dyers Colour.*, 88, n° 12, 1972, pp. 429-433.

33. EZZAIDI M. Simulateur de séchage micro-onde à commande numérique. Thèse de Doctorat 3ème cycle, EEA, Inst. Nat. Polytech. Toulouse, n° 170, 1983.

34. FANSLOW G., BURKHOLDER H., BLUHM D. Microwave-assisted desorption of alcohol from a molecular sieve. In : Proceedings of the IMPI Symposium, Toronto, 9-12 June 1981.

35. FORSTER E. Microwave drying process for synthetic polymers. U.S. Patent n° 3 434 220, 25 March 1969.

36. GALLAND G. Les utilisations des hautes fréquences et des micro-ondes dans l'industrie papetière. In : Comptes-rendus du Colloque sur les Applications industrielles des hautes fréquences et des micro-ondes, Lyon, 1977, 16 p.

37. GOERZ D., JOLLY J. The economic advantages of microwave energy in the paper industry. In : Proceedings of the IMPI Symposium, Stanford, 1967, p. 6.

38. GOERZ D., JOLLY J. The economic advantages of microwave energy in the paper industry. *J. Microw. Power*, 2, n° 3, 1967, pp. 87-94.

39. GOHEL H., METAXAS A. Microwave drying of nylon tufted carpets. I. Dielectric property and Q measurements. In : Proceedings of the IMPI Symposium, Ottawa, 1978. (C.D.I.U.P.A., n° 143425)

40. GROZELLIER M. Perspectives des traitements de fils textiles par micro-ondes. In : Comptes-rendus du Colloque sur les Applications industrielles des hautes fréquences et des micro-ondes, Lyon, 1977, 9 p.

41. HAMID M. Microwave applications in the printing industry. Part I : Microwave drying of book-binding glue. Progress report n° 7, Microwave applications project, Industrial enterprise fellowship program report, Manitoba Research Council, May 1972, 6 p.

42. HAMID M. Microwave paint dryer. Brevet canadien n° 1038 458, 12 Sept. 1978.

43. HAMID M., STUCHLY S., BHARTIA P., MOSTOWY N. Microwave drying of leather. Progress report n° 1, Microwave applications project, Industrial enterprise fellowship program report, Manitoba Research Council, Nov. 1971, 15 p.

44. HAMID M., STUCHLY S., BHARTIA P., MOSTOWY N. Microwave drying of leather. *J. Microw. Power*, 7, n° 1, March 1972, pp. 43-49.

45. HIROSE T., ABE I., IWAHASHI M., SUZUKI T., OSHIMA K., OKAKURA T. Microwave heating of cigarettes. In : Proceedings of the IMPI Symposium, Ottawa, 1972, pp. 212-213.

46. HIROSE T., ABE I., IWAHASHI M., SUZUKI T., OSHIMA K., OKAKURA T. Microwave heating of multi-layered cigarettes. J. Microw. Power, 13, n° 2, 1978, pp. 125-129.

47. HIROSE T., ABE I., MATSUOKA K., SUZUKI T., OSHIMA K., OKAKURA T. Improving the filling ability of tobacco by microwave heating. In : Proceedings of the IMPI Symposium, Den Haag, 1970.

48. HIROSE T., ABE I., SAITO K., SUZUKI T., OSHIMA K., OKAKURA T. Some effects of microwave heating to tobacco. In : Proceedings of the IMPI Symposium, Monterey, 1971.

49. HIROSE T., SUGIE R., OSHIMA K., AMANO K. The characterisiics of microwave on tobacco shred. In : Proceedings of the IMPI Symposium, Edmonton, 1969.

50. HOLME I., METAXAS A. Microwave drying of nylon tufted carpets. III. Field trials. J. Microw. Power, 14, n° 4, 1979, pp. 367-382.

51. HUANG H. A microwave apparatus for rapid heating of threadlines. J. Microw. Power, 4, n° 4, 1969, pp. 288-293.

52. HULLS P. Development of the industrial use of dielectric heating in the United Kingdom. J. Microw. Power, 17, n° 1, 1982, pp. 29-38.

53. ISRAEL S., CHABERT J., JOLION M., LEFEUVRE S., ROUSSY G. Compte-rendu de mission Etats-Unis-Club Microondes, 1981, 66 p.

54. KASHYAP S., DUNN J. A waveguide applicator for sheet materials. IEEE Trans. Microw. Theory Techn., 24, n° 2, 1976, pp. 125-126.

55. KAWAGUCHI B. Apparatus for development and fixation of dies with a printed textile sheet by application of microwave emanation. U.S. Patent n° 4 425 718, 17 Jan. 1984.

56. KRIEGER B. Communication personnelle. Stamford (USA), Cober Electronics, 28 déc. 1987.

57. LABRADOR J., LAVIEC J., LORTHIOIR-POMMIER J. Sur l'utilisation en pharmacie industrielle des ondes hyperfréquences. Application au séchage des granulés. Prod. Probl. Pharm., sept. 1971, pp. 622-633.

58. LEFEUVRE S. Séchage de carreaux de plâtre par chauffage hybride conventionnel plus micro-ondes. In : Comptes-rendus du Colloque sur les applications industrielles des hautes fréquences et des micro-ondes. Lyon, 1977, 10 p.

59. LEFEUVRE S. Technologie du séchage micro-onde. In : Les micro-ondes. 1980, 12 p. (Cycle de formation permanente, CNRS/ENSCP).

60. LE GOFF R. A survey of the existing applications. In : Comptes-rendus du Symposium international sur les applications énergétiques des micro-ondes, IMPI-CFE. 14, Monaco, 1979, pp. 77-81.

61. LEIDIGH W., STEPHANSEN E., STOLTE W. Some experimental and analytical aspects of microwave paper drying. In : Proceedings of the IMPI Symposium, Edmonton, 1969, G6, pp. 138-139.

62. LIGHTSEY G., GEORGE C., RUSSELL L. Low-temperature processing of plastics utilizing microwave heating techniques. In : Proceedings of the IMPI Symposium, Memphis, 1986. (Abstract in J. Microw. Power, 21, n° 2, 1986, pp. 86-87).

63. LOFDAHL C. Microwave veneer redrying : a general report. Cryodry Corp. 4 Jan. 1968, 45 p.

64. MARCE D. Quelques aspects de la thermique dans les industries du bois, du papier et des textiles. Sté Française des Thermiciens, 1985, pp. 1-16.

65. MATZ B., KEARNEY D., KRIEGER B. Microwave drying of refractory-coated foam patterns used in the evaporative casting process. In : Proceedings of the IMPI Symposium, Memphis, 1986. (Abstract in J. Microw. Power, 21, n° 2, 1986, pp. 84-85).

66. MEISEL N. Le chauffage par hyperfréquence dans l'industrie. Document I.M.I.-S.A., s.d., 12 p.

67. MEREDITH R., METAXAS A. Microwave carpet dryer. British patent n° 6531/77, case n° 211. Feb. 1978.

68. METAXAS A. Microwave dye fixation. I. Design of equipment. Electricity Council Research Centre. International Report, ECRC/M877, 1975.

69. METAXAS A. Microwave dye fixation.
II : Dye fixation tests. Electricity
Council Research Centre. Internal
Report, ECRC/P.1123, 1978.

70. METAXAS A. The future of electrical
techniques in the production of printed tufted carpets. *J. Microw. Power*,
16, n° 1, 1981, pp. 43-55.

71. METAXAS A., CATLOW N., EVANS D. Microwave assisted dye fixation. *J. Microw.
Power*, 13, n° 4, 1978, pp. 341-350.

72. METAXAS A., MEREDITH R. Microwave
drying of nylon tufted carpets. II :
Electrical characterization of a modified TE_{10n} resonant cavity. *J.
Microw. Power*, 13, n° 4, 1978, pp.
315-320.

73. METAXAS A., MEREDITH R., HOLME I.
Boost your shift production with
microwaves. *Carpet Rev. Wkly*, 19
July 1978.

74. MOORE D. The rapid drying of print
by microwave energy. *J. Microw.
Power*, 3, 1968, pp. 158-165.

75. PELLISSIER J. Communication personnelle. 30 janv. 1987.

76. PETIT J. Application de l'énergie des
micro-ondes au problème pratique du
séchage des matériaux poreux. In :
Les micro-ondes. 1980, 16 p. (Cycle
de formation permanente, CNRS/ENSCP).

77. PHILIPPE A., SOUSTELLE J. Optimisa-
tion du séchage du cuir par un appoint
d'énergie microondes sur un séchoir
pompe à chaleur intégrant un dispositif automatique de manutention des
peaux. In : Comptes-rendus du Colloque sur les Applications énergétiques
des rayonnements. Toulouse, Univ.
Paul Sabatier, ENSEEIHT, 12 juin
1986, pp. 56-61.

78. PRIOU A. Ensemble des travaux. Thèse
de Doctorat d'Etat. Toulouse, Univ.
Paul Sabatier, 1981, 699 p. §3.1.,
pp. 423-437.

79. RESCH H. Über die Holztrocknung mit
Mikrowellen. Holz als Roh- und Werk-
stoff, n° 9, 1968, pp. 317-324.

80. RITALY J., FILDERMAN R., RICCO M.
Séchage de tresses chargées de con-
ducteurs métalliques et leur revête-
ment filmogène. In : Comptes-rendus
du Symposium international sur les
applications énergétiques des micro-
ondes, IMPI-CFE. 14, Monaco, 1979,
pp. 112-114.

81. RITALY J., LESPINATS C., MINET M.
Installation de séchage industriel
à micro-ondes. Brevet français
n° 2 522 798 (82 03 620), 4 mars 1982.

82. RITCHIE R., Mac FARLANE J. Microwave
drying of sheet work. Publication
Hirst Microwave Industries Ltd.,
s.d., reprint from *Print. Trade
J.*, Dec. 1970, 2 p.

83. ROCHAS J., MANOURY M., BERTEAUD A.,
GUILLEMANT H. Procédé et dispositif
pour le traitement thermique d'éléments filiformes. Brevet français
ANVAR n° 79-15083, 7 juin 1979.

84. ROUSSEAU M., EXCELL P. Modification
of the microwave breakdown threshold
by the presence of radioactive material. *J. Microw. Power*, 17, n° 3,
1982, pp..231-235.

85. ROUSSY G. Déshydratation micro-onde
des zéolites comme exemple d'apport
d'énergie sélectif et non équivalent
à un chauffage classique. In : Les
micro-ondes. 1980, 20 p. (Cycle de
formation permanente, CNRS/ENSCP).

86. ROUSSY G., CHENOT P. Selective energy supply to adsorbed water and non-
classical thermal process during
microwave dehydration of zeolite.
In : Proceedings of the IMPI Symposium, Toronto, 1981, pp. 24-26.

87. ROUSSY G., CHENOT P. Déshydratation
par micro-ondes des zéolites. Exemple d'apport d'énergie sélectif et
non équivalent à un chauffage classique. *Rev. Gén. Electr.*, n° 11,
1981, pp. 816-822.

88. ROUSSY G., COLIN P., THIEBAUT J.,
BERTRAND G., WATELLE G. Microwave
enhancement of evaporation of a
polar liquid. II : Theoretical
approach. *J. Therm. Anal.*, 28, 1983,
pp. 49-57.

89. ROUSSY G., THIEBAUT J., CHARREYRE-
NEEL M. A chemical-physical model
for describing microwave paper
drying. *J. Microw. Power*, 19, n° 4,
1984, pp. 243-250.

90. RZEPECKA-STUCHLY M. Microwave energy
in foam-mat dehydration process. *J.
Microw. Power*, 11, 1976, pp. 255-266.

91. SMALL H., HATCHER J., LYONS D. Microwave drying of latex carpet backing.
J. Microw. Power, 7, n° 1, 1972,,
pp. 29-34.

92. THIEBAUT J., COLIN P., ROUSSY G.
Microwave enhancement of evaporation
of a polar liquid. I : Experimentation.
J. Therm. Anal., 28, 1983, pp. 37-47.

93. THOUREL L. Applicateurs pour matériaux en nappes. Publication du CERT/DERMO, ONERA, 1975, 18 p.

94. THOUREL L., MEISEL N. Les applications industrielles du chauffage par micro-ondes. EDF-Journées d'informations électro-industrielles, Paris, Publication EDF, nov. 1974, 12 p.

95. VAN KOUGHNETT A., WYSLOUZIL W. Microwave dryer for ink lines. *J. Microw. Power*, 7, 1972, pp. 347-351.

96. VASILAKOS N., MAGALHÃES F. Microwave drying of polymers. *J. Microw. Power*, 19, n° 2, 1984, pp. 135-144.

97. WIETESKA A., ŁUKJANIK J., PIWOŃSKI K. Drying building constructions under the "Emibeam" system electromagnetic heating equipment. Publication Sté DIGICOM-POLAND, s.d., 4 p.

98. WILLIAMS N. Moisture levelling in paper, woods, textiles, and other mixed dielectric sheets. *J. Microw. Power*, 1, 1966, pp. 73-80.

99. ZAHOUANI M. Etude d'une barre rayonnante fournissant 40 kW micro-ondes et destinée à un four tournant rotatif. In : Comptes-rendus du Colloque sur les Applications énergétiques des rayonnements. Toulouse, Univ. Paul Sabatier, ENSEEIHT, 12 juin 1986, 10 p.

100. COBER Microwave... a breakthrough for the foundry. Publication COBER Electronics, Inc., s.d., 8 p.

101. Microwave conditioning of leather. Case History, The Electricity Council, EC 3608/12/76, 1976, 1 p.

102. Mise au point d'un ensemble générateur-applicateur adapté au séchage par micro-ondes des rubans, mèches et fils. Rapport technique. Mulhouse, Centre de Recherches Textiles, 1978, 50 p.

103. Recent advances in industrial microwave processing in the 896/915 MHz frequency band. Leicester, Publication Magnetronics Ltd., s.d. 12 p.

104. RF heating system speeds drying of print webs. *Electron. Weekly*, 16 June 1965.

105. Séchage de granulés - 50 % de gain de productivité par dopage électrique. Publication EDF-Industrie, CINELI, 4ème trim. 1986, 4 p. (réf. 1-45-2-07/86).

106. Utilisation des micro-ondes dans l'industrie du plâtre. Publication du CERT/DERMO, ONERA, s.d., 21 p.

Chapter 2

The treatment of elastomers

A large number of industrial applications involving the use of rubber have become possible since the discovery of certain physicochemical treatments, including rubber vulcanization.

Vulcanization requires very high product temperatures, as do the preheating of vulcanizable mixtures and the devulcanization of rubber for recycling. Elastomers are poor thermal conductors that require complex and costly heating procedures. Microwaves may therefore have a considerable role to play in this field as a carrier of energy that is independent of the thermal properties of the material.

Industrial concerns are particularly secretive about details of their production processes, so that it is rather difficult to evaluate the role of microwaves. There is, however, little doubt that the treatment of elastomers is an area of successful industrial penetration by microwaves, which can advantageously replace certain traditional processes as well as provide new applications (e.g., the vulcanization of thick or complex sections) and the production of high-quality recycled rubber.

2.1 Macromolecules and microwaves

2.1.1 Principles of interaction

Macromolecules are formed by the polymerization of a number of elementary molecular units called monomers. The final macromolecule is known as a homopolymer or a copolymer, depending on whether the elementary molecular units are identical or belong to two or more distinct types.

Macromolecules can be linear monochains, with or without ramifications, or three-dimensional structures. Vulcanized rubbers have a nonisotropic three-dimensional structure that results from the reaction of linear macromolecules with small polyfunctional molecules.

When the basic structure of the polymer is not totally symmetric, the macromolecule may have a permanent dipole moment that is the vector sum of the elementary dipole moments with the appropriate formfactor [14]. However, the molecular weight is such that the relaxation frequency is extremely low, far below the microwave region, and so it is unlikely that dielectric losses due to this dipole moment can be observed at microwave frequencies.

2.1.2 Relaxation mechanisms

The dielectric behavior that results from the molecular structure just described is far more complex than that observed for small molecules. There are two relaxation regions that correspond to two distinct mechanisms, namely, α relaxation that occurs at low frequencies and is due to microbrownian motion within the chain, and β relaxation that occurs at higher frequencies and is due to dipole reorientation (lateral subgroups or chain segments), as well as torsional movement of the chains [14], [31].

These relaxation processes are influenced by a number of other factors, including the following:

- The manner in which the dipoles are bonded to the chain (i.e., polymers with lateral dipolar elements that are long and flexible) are subject to α relaxation that is very temperature and frequency sensitive, whereas polymers with short, rigid dipolar elements, which are more or less integrated into the chain, undergo β relaxation whose frequency is very temperature sensitive.

- Crystallinity, which reduces β relaxation and shifts the α relaxation toward the high-temperature region, increases.

- Cross-linking, which does not affect β relaxation but, because of the reduction in the mobility of the chains, reduces the α relaxation and shifts it toward higher temperatures.

- Plastification, which facilitates chain movement and thus reduces the α relaxation temperature.

- The presence of moisture, which induces a particular relaxation process known as γ at low temperatures and high frequencies.

Microwaves produce an increase in temperature and, hence, an increase in the β relaxation frequency, bringing it closer to the irradiation frequency. The dielectric losses thus increase, which leads to a further temperature increase, and so on, with the risk of thermal runaway. At the same time, the α relaxation frequency increases to the extent that it approaches the β relaxation frequency [14].

2.1.3 Dielectric properties of elastomers

Elastomers are linear macromolecules with high intrinsic flexibility and very few side chains that can lead to intermolecular interactions. The chains can thus wind on themselves in a disorderly manner. A tensile force can considerably extend the elastomer by unwinding its macromolecules, which subsequently return to their original state. The untreated natural product will, however, retain a residual irreversible linear expansion. Untreated products are also too plastic and too temperature sensitive to be of any practical use, and must be submitted to vulcanization procedures. The mechanical properties may be further improved by the incorporation of certain fine substances known as fillers. The principal material in this category is carbon black.

Elastomers do not exhibit uniform dielectric behavior. Some have very low dielectric losses and do not react to microwaves. This is so for the two important elastomers, namely, *natural rubber* (NR) and *butadiene-styrene rubber* (SBR). Other low-loss rubbers include *ethylene-propylene rubber* (EPDM) and *polynorbornene* (PNB), both of which are white, and also *butyl rubber* (BR).

In contrast, other natural elastomers exhibit high dielectric losses (e.g., *chlorobutadiene rubber* (CR) or *nitrile butadiene rubber* (NBR)).

The dielectric losses of these products can be found in the literature [8]. Some values are reproduced below to highlight these differences.

$$CR: \quad \epsilon_r'' = 1363.0 \times 10^{-4}$$
$$NBR: \quad \epsilon_r'' = 441.5 \times 10^{-4}$$
$$NR: \quad \epsilon_r'' = 64.5 \times 10^{-4}$$
$$SBR: \quad \epsilon_r'' = 107.0 \times 10^{-4}$$
$$BR: \quad \epsilon_r'' = 53.8 \times 10^{-4}$$
$$IIR: \quad \epsilon_r'' = 21.0 \times 10^{-4}$$

The curves shown in Figure 2.1 [8] illustrate the opposite behavior of the two families (i.e., polar and nonpolar elastomers).

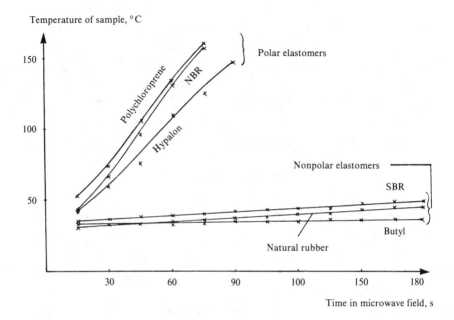

Figure 2.1 Heating of crude rubber [8].

2.2 Vulcanization

The first products made from crude rubber were quite unsatisfactory and lost their shape rapidly because of their changing
mechanical properties. In 1839, a major discovery by Charles
Goodyear laid the foundations for the development of rubber as
an industrial product. He found that crude rubber treated with
sulfur at temperatures in excess of its fusion point underwent
a transformation that considerably improved its mechanical
and thermal properties. This process, known as *vulcanization*,
creates intermolecular cross linkage that anchors the structure
and ensures that the new properties of the material become
permanent.

Other vulcanizing agents were discovered later. They include sulfur chloride, selenium, polynitrobenzene, and ionizing radiation. For synthetic elastomers that cannot be vulcanized by sulfur, the alternative is to use organic peroxides such as benzoyl peroxide [17].

The discovery that certain organic compounds could accelerate the process was another important step. In 1906, Oenslager reported the accelerating effect of aniline and of other organic compounds, such as thiocarbanilide and hexamethylene tetramine. There is, indeed, a very extensive and varied range of accelerating agents, some of which take the vulcanization time to below one minute. Most accelerators produce an optimum effect in the presence of other substances called activators (e.g., metallic oxides and organic fatty acids).

Different vulcanization procedures are employed, depending on the formulation of the final product. For example, the crude mixture can be placed in a mould that is itself placed between the plates of a press. The plates can then be heated directly or placed in an autoclave.

This classical procedure ceases to be useful in the case of extruded sections because it involves the interruption of a continuous process (i.e., extrusion) in order to perform an intermittent process (i.e., vulcanization). The first to resolve this problem [21] were cable manufacturers who used vapor-filled pipes for continuous vulcanization under pressure. Unfortunately, this procedure is not suitable for unstructured sections that do not possess the mechanical resilience that cables have by virtue of their metal core.

Research in this field was then directed toward pulsed hot-air or infrared tunnels, hot-bath (oil or salt) vulcanization, fluidized beds (sand or glass beads), and microwaves.

There are two main obstacles to continuous vulcanization, namely, the poor thermal conduction of rubber and the necessity for vulcanization under pressure as a means of avoiding the formation of porosities. Actually, all rubber mixtures that are not vulcanized under pressure have porosities due to air or water inclusions, the liberation of gas by the elastomer itself, and gas-yielding decomposition of certain additives. This difficulty has been removed by the development of degasing extruders.

As far as heat transfer is concerned, the most interesting techniques are those employing the fluidized bed, the salt bath, and microwaves. Finally, the salt bath technique, in preference to the fluidized bed, has become the main competitor of microwaves. It may therefore be useful to give a brief description of this technique [21].

A heated stainless steel basin contains the eutectic mixture of salts, usually 53% of potassium nitrate, 40% of sodium nitrite, and 7% of sodium nitrate. This mixture melts at 141°C, so that at the vulcanization temperatures of 200 to 250°C it possesses enough fluidity to penetrate the interstices of the mouldings. The heating of the rubber is thus initiated by contact and is activated by the displacement of the moulding. The transfer of heat is approximately 50 times better than in air. The length of the basin is between 7 and 15 m. The extrusion rate is of the order of 25 m/min for small section mouldings of approximately 50 g/m.

The salt mixture has high density (on the order of 1.9) and is maintained by its own weight in intimate contact with the rubber. This ensures very good heat transfer by conduction and, hence, an acceptable thermal efficiency. The temperatures reached are steady and readily adjustable. Almost any type of mixture can be vulcanized in the salt bath. However, the procedure is only suitable for relatively thin sections of less than 20 to 25 mm. Furthermore, the shape of sections cannot be too complicated because the salt bath tends to press the moulding against a steel carrier rail, which has the effect of deforming complex shapes. Other disadvantages relate to the salt carried by the moulding as it leaves the bath. The salt consumption may well reach 4 to 5 kg/hr. As a consequence, the level of the bath is reduced, thus lowering the pressure exerted on the rubber in production and giving rise to variations in the section of the moulding. This produces a change in product quality. The clinging salt must be removed, sometimes by hand brushing. Vulcanization of white rubbers brings about some discoloration. Finally, the process consumes significant amounts of energy in the two hours necessary to set up the required temperature, and there are significant losses of energy by radiation.

Another process competing with microwaves is known as the shearing head technique [6]. A rotor mixes the material at the extruding head, initiating vulcanization by heating. At the exit of the drawing plate, a simple hot-air tunnel completes the reaction.

2.3 Microwave vulcanization

2.3.1 Formulation of mixtures

We have already noted that rubbers exhibit a wide range of dielectric properties. For some, the dielectric loss factor is high enough to cause rapid heating on microwave irradiation. This is the case for the nitriles and chloroprenes. For others, the rise in temperature is quite negligible, so that it is necessary to use polar materials to increase the loss factor.

2.3.1.1 Addition of carbon black

Carbon black has a very high dielectric loss factor. It is a very common additive to rubbers because it also improves their mechanical properties [8]. The heating efficiency depends on the type of carbon black used. For example, the heating rates for NR loaded with different carbon black varieties are shown in Figures 2.2 and 2.3 [8].

2.3.1.2 Addition of light-colored fillers

Although less effective than carbon black, light-colored fillers such as clay, magnesium carbonate, certain types of calcium carbonate or silicate, and aluminum silicates are widely used in the rubber industry [17].

The natural substances (chalk, kaolin, talc, barite) have very little effect on microwave heating. Synthetic loads (silica and

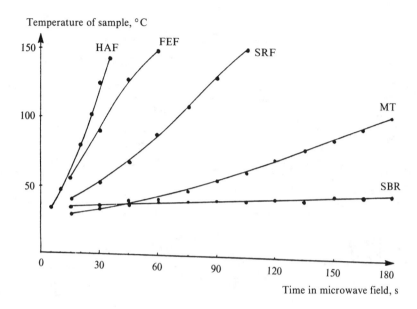

Figure 2.2 Heating of SBR loaded with different types of carbon black (45 parts)[8].

silicates) increase dielectric losses, but not sufficiently to allow continuous vulcanization by microwaves (Fig. 2.4). Aluminum silicate gives a very poor result [25, 8, 7].

2.3.1.3 Mixtures of polar and nonpolar elastomers

When filling with polar constituents is undesirable, a possible solution may be to introduce 5 to 10% of a polar elastomer to the nonpolar rubber to be heated by microwaves. For instance, it is possible to incorporate NBR or *chloroprene rubber* (CR) in *natural rubber* (NR) or in SBR.

The 10% limit is adopted for two reasons. One is the higher price due to the use of a more expensive substance (CR or NBR), and the other is the change in properties of the final mixture[8] . Figure 2.5 shows the increase in microwave heating for NR mixed with NBR.

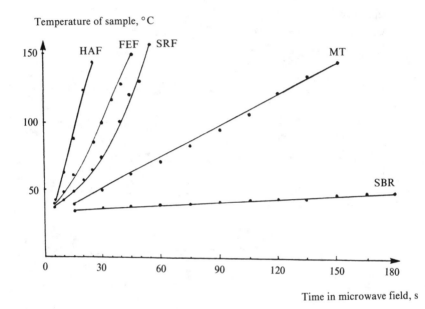

Figure 2.3 Heating of SBR loaded with different types of carbon black (80 parts) [8].

2.3.1.4 Addition of polar activators

Certain polar additives such as *diethylene glycol* (DEG) or *triethanol amine* (TEA) may enhance sensitivity to microwaves when used in the proportion of ten parts per hundred parts of the nonpolar filler [8].

A document [39] published by the IMI-SA Company gives a number of Bayer formulations for microwave vulcanizable mixtures.

2.3.2 Advantages and disadvantages of microwave vulcanization

The degree of vulcanization is a function of the temperature reached and of the time for which it is maintained. Microwaves cause a rapid increase in temperature (in a few seconds) throughout the mass of the mouldings. Once this tem-

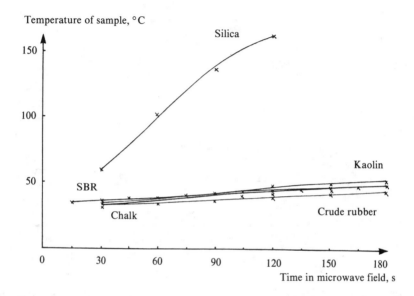

Figure 2.4 Heating of a mixture of natural rubber, chalk, kaolin, and silica (50 parts)[8].

perature is reached, it must be maintained for the time necessary for vulcanization by the traditional technique, usually a hot-air tunnel that prevents the loss of heat to the surroundings.

The main disadvantage of microwaves is the restriction to polar mixtures. Mouldings of large cross section can be vulcanized rapidly and uniformly by bulk absorption of energy [24]. The overheating associated with classical processing is avoided and quality is improved by the absence of internal strains due to the simultaneous action of heat and pressure [16]. There are no contraindications for the vulcanization of complex shapes, and shape stability during vulcanization guarantees excellent three-dimensional precision [21]. In the case of porous mouldings, the porosity obtained is uniform and the

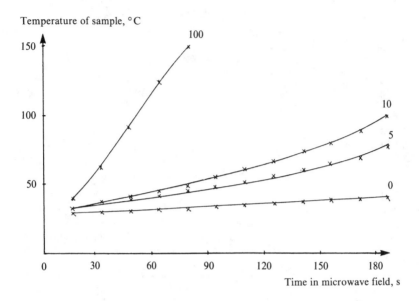

Figure 2.5 Heating of a mixture of natural rubber and nitrile rubber (0, 5,10, and 100 parts) [8].

skin well formed [24]. The surface of the moulding requires no pre-treatment or post-vulcanization cleaning. However, it has been reported [24] that the surface can be made sticky by oxidation after microwave vulcanization with peroxide.

The technique is usually judged to give excellent results in terms of efficiency, reliability, and adaptability to the industrial environment. The British company William Warne, a user since 1970, estimated [35] that an energy saving of up to 30% (as compared to the salt-bath or fluidized beds) was achieved. They also stressed reliability and the economic and environmental advantages of not using salt.

Figure 2.6 shows the yield of microwave and salt-bath vulcanization as a function of the diameter of the moulding, demonstrating unequivocally that large-diameter mouldings can only be vulcanized by microwaves [8].

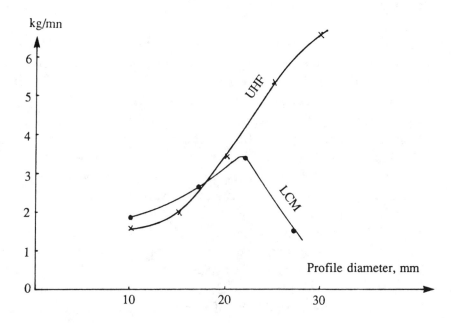

Figure 2.6 Yield as a function of the cross section of the moulding. Comparison between microwave and salt-bath vulcanization [40].

2.3.3 Materials available

The first patent on dielectric vulcanization by high-frequency heating was submitted in 1935 by Dufour and Leduc [39]. The first industrial installations operating in the microwave band were introduced in the 1960s. At that time, high-power microwave generators had become available and were reasonably priced. Moreover, the mechanism of vulcanization and the action of chemical accelerators had become much better understood [21].

The first industrial installations were manufactured in Europe by Hertz-Four and Bewe-Plast in France, Elliott in the United Kingdom, and Troester and Berstoff in Germany [21]. For example, the applicator jointly designed by Elliott and Dunlop [16] delivered 32 kW at 2.45 GHz, and used 21 magnetrons mounted in a tunnel incorporating wave traps. The dimensions of this tunnel were [16] height 2.3 m, width 3.3 m, and length 11.6 m.

Two irradiation techniques were employed concurrently: a circular waveguide, used as the vulcanization canal (Hertz-Four, Elliott, Berstoff), and a meander waveguide containing the vulcanization canal (Troester, Bewe-Plast).

More recently, the German company Johannes Menschner Maschinenfabrik produced industrial equipment operating at 2.45 GHz. For example, the MWG 100 K model delivers 12 kW of microwave power and comes complete with a conveyor-belt-type delivery mechanism moving at the rate of 15 m/min maximum in a microwave cavity 5.9 m long, followed by a hot-air tunnel 7.1 m long. This assembly is used in the vulcanization of rubber moulds of large cross section (up to $180 \times 120 \, \text{mm}^2$) [33]. Modern equipment has also been developed by Hermann Berstoff Maschinenbau GmbH in Germany. Its model PH 20-200.100 delivers 36 kW of microwave power in an 8.75-m microwave cavity followed by a 6.6-m hot-air tunnel. The whole assembly is capable of vulcanizing rubber moulds of $200 \times 100 \, \text{mm}^2$ at the rate of between 1 and 30 m/min. Figure 2.7 shows the main features of the Berstoff microwave vulcanization system [40].

The Vulcatron 10, manufactured by the French company IMI-SA, delivers between 3 and 10 kW. It is 2.8 m long, with an adjustable speed of between 0.75 and 45 m/min and a cross section of $55 \times 55 \text{mm}^2$, followed by a 15-kW hot-air tunnel 6.6 m long, and a 4.18-m cooling tunnel [39]. Other makes that must be mentioned are Techmo (14 kW) and Remy Electronique.

In the United States, Raytheon-Radarline produces a range of microwave vulcanization units (e.g., model QMP 1816 that delivers 10 to 20 kW and maximum output of 450 kg/hr for rubber mouldings of 130×130 or $230 \times 20 \text{mm}^2$) [13].

A number of new applicators were developed in the 1980s, including a multimode cavity for the continuous vulcanization of rubber moulds or rubber tubes [1], a microwave vulcanization tunnel patented by Kai and Ito [15] and using sectoral horns and radiating devices, a microwave tunnel incorporating an air cushion to prevent the deformation of rubber mouldings [20], a resonant tunnel patented by Beck [3] with alternating TE and TM zones, and, finally, a multimode tunnel fed by coupling slots and incorporating microwave components capable of transforming it into a resonant cavity with an axial field maximum [4].

2.4 Thawing and preheating of rubber

The British company Magnetronics produced for Simon Carves of Stockport a microwave applicator for thawing rubber bales before mixing. This applicator is fed by thirty 1.5-kW magnetrons and is capable of raising the core temperature of the bales from −18 to +32°C at the rate of 1.5 ton/hr [34].

The Dunlop-Angus Belting Group Company uses in its factory in Cardiff (Wales) a microwave unit for preheating rubber strips on a continuous basis. The equipment was developed by the Rotax-Lucas Aerospace Company and can raise the temperature of the product to 90°C prior to treatment in a Rotocure unit. The latter ensures the continuous vulcanization under pressure between a vapor-heated cylinder and a stainless steel sheet heated by infrared radiation. The microwave preheating unit delivers 13.5 kW from nine 1.5-kW magnetrons at maximum power per unit of 27 kW [36].

Batch vulcanization under pressure can be accelerated if preceded by a microwave preheating stage. This initial stage raises the temperature to between 75 and 105°C, which reduces considerably the vulcanization time [22]. The procedure is used in the manufacture of tyres for trucks by Avon Tyres (UK). The mouldings are preheated to 90°C in approximately ten minutes in Rotax microwave ovens delivering 1.5 to 3 kW at 2.45 GHz. In the second stage, the preheated rubber is placed in a heated

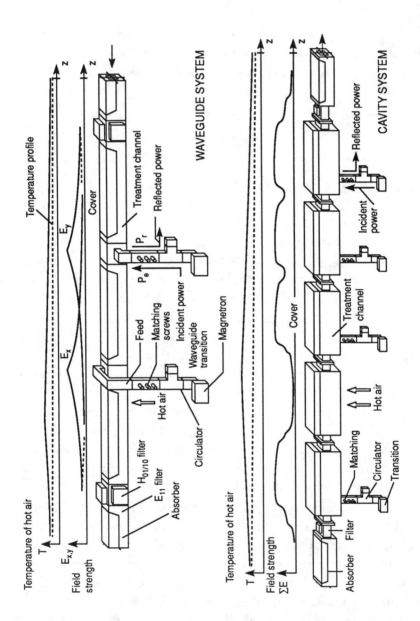

Figure 2.7 The main features of the Berstoff microwave vulcanization system [40].

press for both vulcanization and moulding. For batches of 30 kg, the processing time is reduced from 2 hr to 1 hr 10 min as compared with traditional preheating, with a 40% energy saving.

In the United States, the Raytheon Company recommends the use of preheaters for large metal-reinforced tyres. The use of this equipment results in a saving of between 40 and 60% in heating time, with the potential to increase the yield of the presses by a factor of ten. There are also secondary advantages that result from the reduction in stock and reject levels, as well as better adherence of rubber to metal when such combinations are used [13].

A robotic microwave preheating unit developed by Raytheon is described in [9]. It comprises an enclosure of $4.5 \times 4.5 \times 3.6 \mathrm{m}^3$ with a rotating platform and a mechanical arm for sample handling. Two radiating antennas are positioned so as to ensure uniform irradiation of the tyre. The microwave power delivered is 2×50 kW at 915 MHz.

2.5 Microwave devulcanization

The recycling of rubber requires the breaking of bridging bonds created by vulcanization. Substantial heating is needed to achieve devulcanization. The difficulty is that the devulcanization and depolymerization temperatures are not very different (350 and 370°C, respectively for EPDM). This small difference can be explained by the similarity between the atomic links to be broken (S-S and C-S links in devulcanization and C-C bonds in depolymerization):

$$\begin{array}{ll} \text{S-S:} & 272\,\mathrm{kJ/mole} \\ \text{S-C:} & 301\,\mathrm{kJ/mole} \\ \text{C-C:} & 347\,\mathrm{kJ/mole} \end{array}$$

Clearly, the main advantages of microwaves are the uniformity of heating and the precise control of temperature. However, the procedure has been implemented only for EPDM, according to the American patent 4,104,205, submitted by the Gerling Moore Company. Research in this area has been active since 1971. An industrial unit was installed in 1977 in the Goodyear factories at Lincoln (Nebraska). It consisted of a 70-kW oven operating at 2.45 GHz with 28 magnetrons of 2.5 kW mounted on a multimode cavity of $1.20 \times 1.20 \times 8.40 \, m^3$. Rubber pieces of approximately $1 \, mm^3$ in volume entered the cavity in a dielectric tube 200 mm in diameter. The distribution of energy was optimized by an empirical adjustment of feeding waveguides. The processing took approximately 2 min, and the yield reached 280 kg/hr.

Devulcanization was followed by the extrusion of rubber, and the extruded artefacts were pressed to form a sheet. Recycled rubber was then combined with new rubber in the proportion of 15 to 20%.

Tests performed on a mixture of 80% butyl rubber and 20% EPDM, which was used on the internal surfaces of tyres, gave positive results.

By 1981, a devulcanization line had been functioning for four years at Goodyear. The total investment was estimated to be $10 million and the running cost $0.33 per kg. The Goodyear company considered these figures to be very satisfactory, the additional advantages including the elimination of pollution, energy saving, and ease of integration of microwaves in a continuous processing line. The mechanical properties of rubber were better preserved as compared with classical devulcanization techniques at comparable or lower cost [13].

In France, Roussy, Thiebaut, and Zoulalian of the University of Nancy are currently conducting research in this field.

References

1. BECK M. Applicateur à micro-ondes, notamment pour la vulcanisation ou la réticulation en continu de profilés, de tuyaux ou de câbles à base de caoutchouc. Brevet français n° 2 478 930 (80 06 224), 20 mars 1980.

2. BECK M. Applicateur rotatif pour le traitement thermique ou thermochimique par micro-ondes d'éléments granulaires en matière polaire. Brevet français n° 2 519 224 (81 24 529), 30 déc. 1981.

3. BECK M. Installation pour le traitement thermique ou thermochimique en continu, par micro-ondes, de matières au moins en partie polaires. Brevet français n° 2 521 810 (82 02 517), 16 fév. 1982.

4. BECK M. Applicateur à micro-ondes, à densité d'énergie ajustable, destiné au traitement d'objets au moins en partie polaires. Brevet français n° 2 548 507 (83 10 636), 28 juin 1983.

5. BUECHE F. Electrical properties of carbon black in an SBR-wax matrix. *J. Polym. Sci.*, 11, 1973, pp. 1319-1330.

6. BOUTEVILLE A. La vulcanisation du caoutchouc par micro-ondes. *Rev. Gén. Electr.*, 11, nov. 1981, pp. 834-838.

7. CHANET P., MONCEL M. Rôle de l'humidité des mélanges dans le chauffage diélectrique par ultra-haute-fréquence. *Rev. Gén. Caoutch. Plast.*, 49, n° 4, 1972, pp. 309-314.

8. CHEVILLON G. Les applications des micro-ondes dans l'industrie du caoutchouc. In : Comptes-rendus du Colloque sur les Applications industrielles des hautes-fréquences et des micro-ondes. Lyon, 1977, 10 p.

9. EDGAR R. Robotics in large-scale curing processes. In : Proceedings of the IMPI Symposium, Memphis, 1986. (Abstract in *J. Microw. Power*, 21, n° 3, 1986, pp. 194-195).

10. FIX S. Microwave devulcanization of rubber. *Elastom.*, 112, n° 6, 1980, pp. 38-40.

11. HULLS P. Development of the industrial use of dielectric heating in the United-Kingdom. *J. Microw. Power*, 17, n° 1, 1982, pp. 29-38.

12. IPPEN J. Formulation for continuous vulcanization in microwave heating systems. *Rubber Chem. Technol.*, 44, 1971, pp. 294-308.

13. ISRAEL S., CHABERT J., JOLION M., LEFEUVRE S., ROUSSY G. Compte-rendu de mission Etats-Unis, Club micro-ondes, 1981, 66 p.

14. JULLIEN H. Interaction entre micro-ondes et macromolécules. In : Les micro-ondes. 1980, pp. 12-32. (Cycle de formation permanente, CNRS/ENSCP).

15. KAI T., ITO M. Appareil de chauffage à micro-ondes. Brevet français n° 2 525 063 (83 05 696), 7 avril 1983.

16. KIRSZENBERG J. Préchauffage par micro-ondes. *Rev. Gén. Caoutch. Plast.*, 46, n° 12, 1969, pp. 1453-1454.

17. LE BRAS J. Le caoutchouc. Paris, PUF, 1969, 126 p. (Collection Que sais-je ?).

18. LE GOFF R. A survey of the existing applications. In : Comptes-rendus du Symposium international sur les applications énergétiques des micro-ondes, IMPI-CFE. 14, Monaco, 1979, pp. 77-81.

19. LUYPAERT T., REUSENS G. Microwave applicator for continuous vulcanization. In : Proceedings of the IMPI Symposium, Chicago, 1985, pp. 100-102.

20. LUYPAERT P., REUSENS P. A new microwave applicator for the continuous vulcanization of rubber. In : Proceedings of the IMPI Symposium, Memphis, 1986. (Abstract in *J. Microw. Power*, 21, n° 2, 1986, p. 75).

21. MEISEL N. Vulcanisation en continu d'élastomères extrudés en bain de sel ou par hyperfréquences. *Rev. Gén. Caoutch. Plast.*, 48, n° 4, 1971, pp. 397-401.

22. MEISEL N. Le chauffage par micro-ondes dans l'industrie. Document IMI-SA, s.d., 12 p.

23. MINETT P. Microwave and RF heating in plastics and rubber processing. *Plast. Rubber*, 1, 1976, pp. 197-200.

24. ÖTTNER C. Le comportement d'élastomères et de mélanges dans un champ électrique à hyperfréquence. *Rev. Gén. Caoutch. Plast.*, 46, 1969, pp. 973-979.

25. ÖTTNER C. Vulcanisation continue dans le champ d'ultra-haute fréquence. Informations BAYER, sans réf., s.d., pp. 43-58.

26. SCHWARZ H., BOSISIO R., WERTHEIMER M., COUDERC D. Microwave curing of synthetic rubbers. *J. Microw. Power*, 8, 1973, pp. 303-322.

27. SHUTE R. Industrial microwave systems for the rubber industry. *J. Microw. Power*, 6, 1971, pp. 193-206.

28. TERSELIUS B., RANBY B. Cavity-perturbation measurements of the dielectric properties of vulcanizing rubber and polyethylene compounds. *J. Microw. Power*, 13, n° 4, 1978, pp. 327-335.

29. THOUREL L., MEISEL N. Les applications industrielles du chauffage par micro-ondes. EDF, Journées d'informations électro-industrielles, Paris, Publication EDF, nov. 1974, 12 p.

30. VALOT H. Aperçu sur la structure et les propriétés des macromolécules. In : Les micro-ondes. 1980, 32 p. (Cycle de formation permanente, CNRS/ENSCP).

31. WILLIAMS G. The complex dielectric constant of dipolar polymeric compounds as a function of frequency, temperature and pressure. In : Molecular relaxation processes. Proceedings of the Chemical Society Symposium. Aberystwyth, 1965. London, Academic Press, 1966, pp. 21-31.

32. Cinquième forum électro-industriel national plastique-caoutchouc, 25-26 mai 1983, Avignon, C.F.E.

33. Menschner UHF Ultrahochfrequenz Erwärmungsanlagen. Publication de la Sté Johannes Menschner Maschinenfabrik, s.d., 26 p. (alle,, fran., espa., angl.).

34. Microwave power generators/45 kW-2450 MHz special-purpose microwave oven. Publ. Magnetronics Ltd., s.d., 6 p. (ref. B40-100e).

35. Microwave rubber curing, case history. The Electricity Council, EC 3349/1/76, 1976, 1 p.

36. Microwave rubber preheating, case history. The Electricity Council, EC 3436/1/76, 1976, 1 p.

37. Microwave speed production of conveyor belts. News Information, the Electricity Council, 18 March, 1976, 7 p.

38. Radarline - Continuous vulcanizing unit. Raytheon company, Rubber processing equipment, 1975, 3 p.

39. Systèmes intégrés à conception modulaire pour la vulcanisation en continu par effet diélectrique de profilés extrudés - Vulcatron. Publication Sté IMI-SA, s.d., 4 p.

40. Vulcanisation en continu avec installation UHF et bain salin. Publication Sté Hermann Berstorff Maschinenbau GmbH, s.d., 26 p.

Chapter 3

Miscellaneous applications

The aim of this chapter is to note and briefly describe the numerous industrial applications of microwaves other than drying and the treatment of rubber. Most of these applications are in different stages of development and occupy only a minor position on the industrial production scene, but this situation may well change in the future. It is difficult to be exhaustive in recording these applications because research in this subject is being very actively pursued, and this steadily extends the range of potential applications.

3.1 Polymerization

3.1.1 Thermosetting and thermoplastic polymers

Plastic materials fall into one of two main families: thermosetting and thermoplastic polymers. The former, obtained by the polycondensation reaction, set irreversibly under the effect of heat (reticular reaction); and the latter, obtained by simple polymerization, soften when heated.

Resins such as polyesters, polyurethanes, epoxies, phenolics, amino resins, and so on, are classified as thermosetting materials. Thermoplastics include polyolefines, vinyls, polyamides, polycarbonates, and so on.

3.1.2 Microwave reticulation of thermosetting resins

The microwave reticulation of thermosetting resins is the subject of numerous research products, so we may expect interesting future developments, particularly for composite materials based on an epoxy matrix for the space and aeronautic industries and the manufacture of films.

3.1.2.1 Polyesters

Polyesters are formed by the condensation at 200°C of a mixture of a polyacid and a polyalcohol in which one of the two components is unsaturated. This is followed by the dissolution of the resulting polyester in a vinyl monomer, usually styrene. Copolymerization occurs between 70° and 120°C in the presence of a catalyst whose role is to open the double bonds in the polyester and the monomer [25], [74].

The utilization of microwaves in copolymerization was reported in [32]. An unsaturated polyester dissolved at 70°C in 30% styrene was heated by microwaves in the presence of a radical primer, in this case, per-2-ethylhexoate tertiobutyl. The temperature/time curves show that the conversion of styrene begins as soon as microwaves are turned on, reaching very high rates (97%) with an abrupt release of heat. It is clear that polymerization occurs in two stages, namely, the addition of styrene to the double bond of the polyester chains and a very exothermic polymerization of the styrene.

Polyesters are widely used for the production of reinforced materials. The French company SFAMO has produced an oven operating at 2.45 GHz and delivering between 42 and 84 kW. It was designed for the reticulation of resin coatings on glass fibers (Fig. 3.1). This equipment has already been exported to Germany, Sweden, Finland, Turkey, and South Korea. It is capable of processing tubes of internal and external diameters in the ranges of 13–300 mm and 40–500 mm, respectively, and can easily be incorporated in a production line after a silica fusing furnace [6].

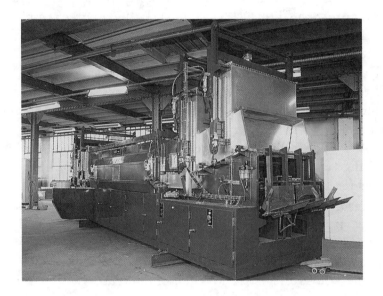

Figure 3.1 The EMO oven built by SFAMO for the treatment of glass-resin fibers [6].

3.1.2.2 Polyurethanes

Polyurethanes are formed when polyalcohols react with polyisocyanates. Depending on the nature of the alcohol, the structure can be linear and hence made thermoplastic or three-dimensional [74].

The mixture proposed in [32] consisted of dicyclohexyl methane di-isocyanate and polyoxypropylene glycol and polyoxypropylene glycerol, with tin dibutyl laurate as catalyst. The temperature/time curves obtained under microwave irradiation at 2.45 GHz show a rise in temperature of the resin, followed by polymerization. This releases energy which reinforces the energy of the microwaves. In the final stage, the polymerized product is not responsive to microwave heating, and eventually loses heat by convection. Depending on the microwave power used (30, 40, 60, or 70 W), the maximum temperature is reached after 9 to 23 min.

There are several reports on the microwave reticulation of polyurethane films [13], [38], [11], [9]. For example, on polyurethane laquer with isocyanate fixed with phenol and deposited in layers of 100 to 300 μm, the reticulation reaction will always occur in less than five minutes at 2.45 GHz, and will result in remarkably hard finishings. Possible applications include laquering of glass bottles, impregnation of corrugated cardboard (50 g of resin for 100 g of cardboard) at the rate of 100 kg per hour, with excellent energy efficiency achieved by using ten 1-kW modules, and, finally, coating of fiber-optic cables for telecommunications.

A resonant cavity was proposed for this application in [13]. The cavity makes use of the thermal runaway properties of certain materials such as alumina (Part I, Section 3.7). It operates in the TM_{010} mode and incorporates a coaxial dielectric element whose dielectric loss factor increases with temperature, namely, an alumina sleeve on a quartz tube. The cavity Q-factor is considerably reduced, and tuning over the feed frequency is achieved simply by shifting the sleeve. The alumina heats up, the dielectric losses rise, and the Q-factor is reduced still further. The equilibrium temperature is reached in a few seconds (e.g., 800 to 1,000°C in 15 to 20 s with 300-W RF). The fiber is then introduced along the axis of the cavity and the temperature of its coating rises, partly due to the infrared radiation from the alumina and partly due to direct heating by the microwaves. The thermal stability of this setup is remarkably good, the energy uptake is very rapid, and the process as a whole is energy efficient. The fiber is processed at the rate of up to 30 to 50 m/min, and its mechanical properties are superior to those obtained with traditional techniques.

A microwave application specific to polyurethanes is the simulation of spontaneous scorching of foam. Polyurethane foams contain haloalkylophosphates as fireproofing materials. They can produce scorching of the material during or just after polymerization, causing a certain degradation of the aesthetic quality of the product, its commercial value, and, eventually, its physical properties. There is also a risk of spontaneous combustion of the recently manufactured blocks.

The evaluation of such fireproofing products relies on simulation techniques that are difficult to implement and generally show poor correlation with industrial production data. Only microwaves result in satisfactory simulation of the scorching process, and even of spontaneous combustion. Tests have shown that the technique is very fast and the results correlate well with industrial processes [63].

3.1.2.3 Epoxy resins

Ethoxyline resins or epoxies are obtained by condensing epichlorohydrin with bisphenol. Reticulation is stimulated by drying agents such as diacide anhydride or polyamines [25], often in the presence of phenols that accelerate the reaction. These resins are ideal for the manufacture of composite materials used in the aerospace industry in combination with carbon fiber, fiberglass and kevlar. Although most of the research is concerned with polymerization agents such as *tetraethylene-pentamine* (TEPA) at ambient temperatures, treatments employing microwaves are equally promising [65, 47].

The processing of carbon-fiber-epoxy bobbins 600 mm long and 200 mm in diameter [62] results in a gain of time and energy by a factor of 3 to 10. In the case of mouldings with metallic inserts, the gain is between 5 and 15, respectively. There is also an improvement by 15 and 20% in mechanical properties. CNET is currently working on the development of a rapid thin-layer polymerization technique for the aeronautic and building industries, and on new types of material produced by microwave polymerization [61]. In Japan, the Fuji Heavy Industries use microwaves (2.45 GHz) in industrial-scale polymerization of epoxy matrices loaded with carbon or kevlar and destined for the aeronautic industry [12].

3.1.2.4 Amino resins

The polycondensation of formol H-CHO and an aldehyde, in particular urea NH_2-CO-NH_2, gives an amino or urea-formaldehyde resin:

$$\left[\begin{array}{l} \quad\quad\quad NH - CO - NH_2 \\ H - CH \\ \quad\quad\quad N - CO - NH_2 \\ H - CH \\ \quad\quad\quad N - CO - NH_2 \\ \quad\quad\quad\quad \ldots\ldots \end{array} \right]_n$$

Urea-formaldehyde glues are used, among other applications, in the manufacture of glued laminates and certain types of honeycomb structures. An applicator for the microwave gluing of chevron butt joints in frames was recently developed [56], [57], [58], [59]. It is used for sections of $35 \times 70\,mm^2$, covered in urea-formaldehyde glue. The applicator is in the form of a cross, and the joint to be glued is inserted in a guide in such a way that the glued zone is exposed to 800-W, 2.45 GHz radiation from the orthogonal guide.

The Applications Electriques Company of Mont-de-Marsan (France) has two working industrial applicators for gluing plywood at Langon and Bergerac [54].

Other applicators include a device for sealing cartons [39], [78], [79], which employs a short-circuited guide (WR 284) that heats an area of $6\,cm^2$, and an applicator based on hybrid thick-film technology that includes tuning rods. The latter applicator is the more efficient of the two and may be used to glue a $5\,cm^2$ surface in a quarter of a second with 800 W of applied power. A prototype of this applicator has processed cardboard at the rate of 20 m/min.

3.1.3 Thermoplastic polymers

Thermoplastic polymers have a linear chain structure produced by a simple polymerization reaction without prior condensation. This reaction is catalyzed by a variety of chemical and physical agents.

Professor E. Stahel of the State University of North Carolina has used microwaves to polymerize plastic material spread in a thin layer on paper. This product was subsequently used in the manufacture of filters for car engines. He used a 20-kW Gerling-Moore applicator incorporating a 60-cm wide cavity fed by slotted guides. The conditions under which he achieved the polymerization of styrene and methyl methacrylate suggested that an industrial application was possible [36].

3.2 Fusion

3.2.1 Dewaxing of casting moulds

Wax moulding is one of the oldest procedures used in foundries. It is employed in the manufacture of foundry pieces with excellent mechanical precision and consists of the following manufacturing stages:

- making wax or urea models,
- arranging the models in batches on a wax or urea tree,
- creating a ceramic shell on the tree with a batch dipping technique,
- dewaxing, generally in an autoclave,
- baking shells at temperatures between 900 and 1, 100°C [81].

Baking shells eliminates the vitrification products that appear during dewaxing as a result of the decomposition of urea. It also results in some sintering that increases their resistance in preparation for the pouring of metal.

In practice, the two operations of dewaxing and baking are carried out simultaneously. When the material of the model (wax or urea) is thus raised to very high temperatures, it decomposes and cannot be recovered.

The use of microwaves for dewaxing can improve this procedure very considerably. The first trials in this area were reported from Bucharest [73]. A 1-kW, 2.45-GHz oven was used to melt the wax in 7 min. Synthetic waxes (50% styrene, 50% paraffin) showed a slower start, but a faster total melting period. Since, in general, waxes have low dielectric loss factors, trials were performed on waxes loaded with 3% of graphite but failed to show any noticeable improvement in their properties.

The first report of an industrial application of this procedure appeared in 1979 [51], [81]. It was developed by the Swiss company Precision-Fondeurs, which, with Thomson-CSF/DTE, designed a 5.5-kW, 2.45-GHz dewaxing oven incorporating a simple cavity of $450 \times 450 \times 500 \, mm^3$. Inverted moulding trees were mounted on a rotating tube through which the molten wax, or urea was let out.

This very simple setup produced complete dewaxing in a few minutes. The wax or urea, previously lost was recovered at the rates of 100 and 80%, respectively. The saving in material was estimated at tonnes per annum. The saving in energy at 1 kWh per piece at the rate of 300,000 pieces per annum amounted to 300 MWh per annum. Investment was recovered in approximately eight months. Additional benefits included the absence of toxic fumes, uniformity of heating, and, consequently a smaller number of broken pieces. The subsequent baking temperature of the shells was reduced from 1,000 to 900°C and the baking time was reduced from 90 to 45 min, so that the energy consumption became 2.3 kWh per piece instead of 3.2 kWh. The user was reported to be very satisfied and considered increasing the capacity of the microwave installation. A similar application was reported from the United Kingdom [35].

3.2.2 Viscous materials in metal containers

Certain thermoplastic materials, solid at ambient temperature, need to be heated in their containers to reach a pouring consistency. They are often bad heat conductors that require long and costly treatment.

An elegant solution is to use the metal container as a multimode cavity whilst applying microwaves through a window. Measures must be taken to ensure the continuous evacuation of the molten material because its dielectric loss factor is higher than that of the solid, so that it tends to absorb the energy preferentially and to overheat.

A number of trials by IMI-SA with congealed or crystallized materials in metallic containers have established the efficacy of this procedure, for which the SERMO Electronique Company of Metz also produces the necessary hardware. A semiportable microwave unit has also been developed for this application [35].

Microwaves can also be used to improve the pumping of bitumen from road delivery tankers. A microwave system [34] has been designed and optimized to heat bitumen in resevoirs 3 m long and 1.5 m in diameter. Uniform heating is achieved by inducing a circularly polarized field (Part I, Section 2.3), which produces a slow variation of the stationary wave pattern. The applicator is fed by a 30-kW source at 896 MHz with an energy efficiency of 48%.

3.2.3 Oil and shale oil

Sandy oil formations in Athabasca (Alberta, Canada) constitute a considerable reserve that amounts to 890 billion barrels of oil (i.e., approximately twice the conventional world oil reserves). They consist of a grain matrix of mostly quartz with some kaolinite and illite, the interstatial space being occupied by water and tar. Tar is recovered by heating on-site to between 50 and 100°C. The traditional techniques of conduction or hot-gas injection are not very satisfactory, and this has led to the interest in microwaves [20], [75]. For large-scale applications, HF seems more appropriate [45].

Microwaves could also be used for the pyrolization of shales [22], the gasification of coal [3], and the liquifaction of heavy oil prior to pumping [31].

3.2.4 Road repairs

An unusual application of microwaves involves the heating of asphalt up to its melting point until cracks and gaps at the joints between sections become filled. A radiating horn applicator, producing localized longitudinal heating, was developed in 1974 at the University of Montreal [21]. In 1981, the American company, Microdry, produced a prototype combining 100-kW microwaves at 2.45 GHz with a hot-air system [36]. The actual equipment consumes 1.2 MW of electric power [68]. The temperature rise is about 77°C. The cost of the operation (microwaves, hot air, profiling, compacting) is estimated at $12 per ton at 1986 prices.

The Applications Electriques Company of Mont-de-Marsan (France) has developed for the Central Laboratory of the Ponts et Chaussées a "Lomos"-type equipment (Part III, Section 5.2), generating 24 kW of microwave power at 2.45 Ghz for joint repairs and for the repair of tar covered roadways [54].

3.2.5 Defrosting of soil

In certain cold climates it may be useful to defrost soils for civil engineering work in the winter, or simply for road and pavement maintenance.

The University of Manitoba has developed an applicator for this purpose, which consists of an H-plane sectoral horn and a corrugated flange [33].

3.3 Consolidation

3.3.1 Hardening of foundry mouldings

The most common foundry procedure is sand moulding in which the moulds are made of silica sand with natural (clay) or synthetic binding material. The sand is pressed around the model, and the amount of pressure applied ensures a more or less accurate reproduction. The mould is then thoroughly baked, thus rendering it immune to handling and pouring of metal.

Preliminary trials of microwave baking were performed in 1977 at the Bucharest Polytechnic [73] and led to the following classification of binding materials according to their dielectric loss factors:

- high-polarity materials: sodium silicate, Covasil, phenol-formaldehyde and urea-formaldehyde resins;
- medium-polarity materials: molasses, colophony;
- low-polarity materials: all oil byproducts.

The most appropriate support during baking is an aluminum plate protected by a low-loss dielectric material (i.e., polystyrene, epoxy resin, rubber, or silicon), which heats up sufficiently to prevent the condensation of humidity, whilst actually absorbing a very small fraction of microwave energy.

Moulds and cores made of 6% sodium silicate, hardened by microwaves, showed a resistance to compression between 530 and $1,020 \, \text{N cm}^{-2}$, the minimum value of $530 \, \text{daN cm}^{-2}$ being reached in one minute. This may be compared with the resistance of moulds hardened using CO_2, which is on the order of 260 to $280 \, \text{N cm}^{-2}$. The proportion of sodium silicate could thus be reduced to just 4%. The baking time was also reduced by a factor of 10, and the quality of all cores was improved in all cases. A 60-kW industrial application has been working in the United Kingdom since 1968 [35]. The potential market is very important, but the current economic climate does not favor the exploitation of this technique in an industry hit by recession.

3.3.2 Fast-setting concrete

The setting of concrete is a chemical reaction between constituents that results in the formation of crystals of calcium hydrosilico aluminate. Irradiation with microwaves triggers the reaction throughout the sample and accelerates it significantly, as shown by trials with concrete contained in cellular moulds [44], [70].

3.3.3 Sintering of ferrites and ceramics

Ferrites (Part I, Section 4.3.2.1) constitute a class of ferromagnetic dielectric materials usually with the spinel structure Fe_2O_3, MO, where M is a divalent metal (e.g. nickel, magnesium, maganese, zinc, or copper) or a combination of divalent metals. Recently it has become possible to obtain garnet structures $5\,Fe_2O_3, 3TO_3$, where T is either a rare earth element or yttrium.

In a magnetic field, this material has an anisotropic magnetic permeability that is tensorial in nature whilst its dielectric permittivity is a complex scalar.

When a linearly polarized electromagnetic wave penetrates this material at right angles to the direction of magnetization, it exhibits a phenomenon analogous to the Faraday effect in optics: the fields **E** and **H** split into two components with opposite circular polarizations and different propagation constants. When the components recombine on leaving the medium, the planes of polarization of **E** and **H** are found to have rotated by a certain angle [7], [10].

This special interaction between wave and matter is exploited in the design of microwave components providing asymmetric propagation (e.g., circulators) used to match the RF source irrespective of the impedance of the system [7].

Different manufacturing procedures are used to produce single-crystal and polycrystalline materials. The former are ordered arrangements of atoms and are obtained by controlled melting and cooling. The latter consist of very fine grains, each of which behaves as a single crystal, but without crystalline continuity between one grain and another. They are produced by a technique similar to that of traditional ceramics and based on sintering. Their hardness is such that they are subsequently machined with diamond [10].

The sintering procedure starts with a powder of well-defined granular size and a number of organic additives acting as plasticizers, lubricants, and binders. All the additives are eventually destroyed by thermal treatment. The powder is put into shape by high-pressure moulding, followed by baking at a temperature that ensures the consolidation with the required density. The final product then has practically zero porosity.

Sintering can be performed with microwaves if the material has a high enough dielectric loss factor, which is indeed the case for raw ferrites. Extremely high temperatures can thus be achieved in very short time intervals.

Some very encouraging trials have been performed by the IMI-SA and by others, but the process does not appear to have been adopted on an industrial scale [8], [10].

Very recently, a number of research projects have involved ceramic sintering [55], and it was concluded that the idea had interesting possibilities. A vacuum cavity, 1 m long and 760 mm in diameter, was fed by a gyrotron delivering 200 kW CW at 28 GHz, with the alumina samples placed on a tantalum platform. The difference between the dielectric loss factors of the two materials was sufficient to ensure the preferential heating of the alumina, and the thermal runaway (Part I, Section 3.7) led very rapidly to uniformly high temperatures on the order of 1600°C [42].

The use of even higher frequencies has been reported [46] (i.e., 60 GHz, with 400–700 W fed to a cylindrical cavity in TE_{02} mode) producing a maximum field of 20 kV/m. As in the previous applications, temperatures on the order of 1600°C are reached very rapidly. The crystallographic laboratories of the University of Caen are working on the sintering of perovskites (barium titanate, $BaTiO_3$) at 2.45 GHz. This material is used mainly for type II ceramic capacitors. The product is mixed with lithium fluoride (LiF) and a complex fluoride of lithium and barium. After sintering at 1,000°C or higher, the product should exhibit constant dielectric properties that are independent of temperature [59].

This procedure seems to have reached the industrial stage in the United States where the Cober Electronics Company produces a 6-kW oven (Fig. 3.2) for the sintering of ceramics to 1,000°C [43].

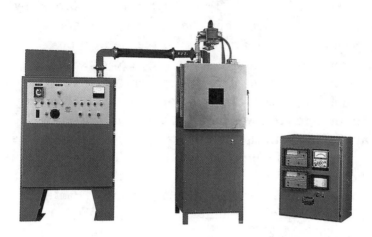

Figure 3.2 The Cober Electronics oven for the sintering of ceramics [43].

3.4 Emulsification

A procedure in which microwaves are used to produce a stable mixture of two immiscible liquids is discussed in [60]. It is essential, however, that the two liquids have different dielectric loss factors. This technique can be used to emulsify water in benzene, using a TE_{101} resonant cavity. Water droplets have a high dielectric loss factor and explode in benzene under the influence of the microwave radiation. However, this procedure, which may induce and favor certain chemical reaction, has not been exploited on an industrial scale.

3.5 Crushing

The use of microwaves for crushing rocks and concrete is an old application that has not seen recent advances. Concrete contains capillary water as well as water of crystallization, whereas rocks contain at least water of crystallization. In addition, these are materials with significant dielectric loss factors. It is therefore possible to use microwave irradiation to produce rapid heating that leads to the development of internal stresses that in turn generate microscopic fractures. The material thus fractured readily falls to pieces. Special "hammers" incorporating a 5-kW, 2.45-kW microwave generator and connected by a flexible cable to a radiating horn were developed in the United States. These devices can be used to fracture $1\,m^2$ of concrete 20 cm thick in five minutes [72].

3.6 Purification of coal

The combustion of coal creates pollution problems because it contains sulfur and, especially, pyrite single crystals (cubic FeS_2). High-gradient magnetic separation can be used to eliminate these single crystals, but the procedure is riddled with problems and difficulties due to the weak magnetism of the pyrites. The microwave method was investigated by Iowa State University for the United States Department of Energy. Microwaves were used to increase the magnetic susceptibility of the pyrites by selectively heating the two materials in a mass of coal. The technique relies on the fact that the ratio of energy absorbed by the two constituents is between 1.9 and 2.6. There is in addition a certain amount of conversion of pyrites into pyrotine (Fe_7S_8), and a 1% conversion rate is sufficient to increase the magnetic permeability by two orders of magnitude.

There are several published reports on this application [27], [28], [50], [15], [49], [16]. However, the gain in permeability remains relatively low and not really sufficient to obtain a significant improvement in magnetic sorting. An additional difficulty is the extreme nonuniformity of the permittivity of different types of coal as a function of frequency.

Another application to coal was developed in collaboration with the General Electric Company of Philadelphia [17], [18], [26], [37], [80]: Microwaves were used for the purification of coal as a means of reducing its ash and sulfur contents. An experimental applicator, known as the Ames Laboratory *Experimental Microwave Applicator System* EMAS was developed and consists of several resonant cavities (6 kW, TM_{01}) and an oversized cavity (3 kW, 2.45 GHz). The microwave irradiation of between 30 seconds and 3 minutes takes place between an alkaline solution treatment and acid water rinsing. It is carried out in an inert argon and nitrogen atmosphere. The relative reduction in ash and sulfur content is 97 and 53%, respectively, for a single treatment, and 98 and 66% when coal goes twice through EMAS. Although the mechanisms exploited in this treatment are not fully understood, the procedure is nevertheless considered to be fully optimized [26].

3.7 Nuclear waste treatment

Microwaves are being used to solidify droplets of radioactive waste in solution [19]. To comply with subsequent treatment, the solidified droplets must be perfectly spherical and uniform in size. Traditional techniques require the use of precipitation from a liquid bath, which is not convenient for two reasons, namely,

(*a*) the liquid acting as the energy carrier is contaminated and becomes itself a radioactive waste, and

(*b*) the solidified droplets must be cleaned of the residual bath liquid by a rinsing solution that becomes radioactive waste too.

Microwaves require no physical contact and provide an interesting solution to this problem. In the experiments at the Technical University of Vienna [19], the solutions to be treated were mixtures of a metal salt, resorcinol, and formaldehyde. The exothermic condensation reaction between the last two components is greatly accelerated by the presence of the metal salt solution (uranylnitrate). The vertical microwave cavity used had a square cross section, varying between 10×10 and

$14 \times 14 \, \text{mm}^2$; it was between 150 and 1000 mm long and was excited at 14 GHz in the TE_{10p} mode by a TWT amplifier of 250 W driven by a Gunn diode oscillator (Part I, Section 4.2.2). The solutions were cooled and mixed immediately before they were introducted into the cavity in the form of droplets whose temperature should be raised very rapidly to approximately 100°C to accelerate the chemical reaction of solidification. Approximately 5 drops were introduced per second, and heating occurred along the droplet trajectory inside the cavity. The optimum cavity dimensions were found to be $10 \times 10 \times 300 \, \text{mm}^3$, which meant that a droplet traversed the cavity in 0.25 seconds. With 80 to 90 W of RF power, 100% hardened spherical drops with fine-grained interior were produced.

An analogous procedure was developed for the Toshiba Company [52]. It involved the treatment of an uranyl-nitrate solution introduced through a tube into a 16-kW microwave oven. The first stage of the reaction is the evaporation of water and then of azeotropic $HNO_3/5H_2O$, resulting in $UO_2(NO_3)_2$. The second phase begins when the temperature reaches 350°C and produces porous UO_3 as the final product. Another equipment, installed at the Power Reactor and Nuclear Fuel Development Corporation (P.N.C., Japan), treats uranium-plutonium solutions with nitric acid to produce their oxides, which are then reduced to granular U + Pu.

Another study was performed by Professor Pellissier of the University of Lyon for the French Atomic Energy Authority in which an applicator generating 3 kW at 2.45 GHz was evaluated for the treatment of contaminated materials. These materials are reduced to grains and imbedded in glass cylinders. Glass can be melted at approximately $1,000$°C by HF induction if it is preheated to trigger the reaction. The microwave treatment can also be applied with the help of a suitable high-loss material that heats the glass by contact and triggers the chain reaction that simultaneously increases the temperature and the dielectric loss factor [59].

3.8 Cellulosic waste treatment

The saccharification of cellulosic waste plays an important part in the utilization of biomasses. It aims to separate the three constituents lignine, hemicellulose, and cellulose, the last having a large number of potential industrial uses. The process of enzymatic hydrolysis has a number of advantages over acid hydrolysis, but is more difficult to achieve because of the presence of lignine and of the crystallinity of cellulose, which require several types of pretreatment. A traditional pretreatment process, applicable to hardwood only, involves heating in the presence of water, which liberates acetic acid and provokes the autohydrolysis of lignine and hemicellulosic carbohydrates.

The Institute of Wood Research at Kyoto and Osaka University (Japan) has for many years been active in the development of a microwave technique for the pretreatment of biomass in the presence of water [1, 53, 12]. For this purpose, they developed a 5-kW applicator operating at 2.45 GHz under a pressure of between 2.5 and 2.8 MPa. Trials performed on sugar cane wastes, rice straw, and resinous shavings have demonstrated a clear improvement in enzymatic activity by a factor of about 1.6 for rice straw after treatment at 170°C for 5 min and a factor of 3.2 for sugar cane wastes after treatment at 200°C for 5 min. The optimum temperature is probably between 223 and 228°C.

References

1. AZUMA J., TANAKA F., KOSHIJIMA T. Enhancement of enzymatic susceptibility of lignocellulosic wastes by microwave irradiation. *J. Ferment. Technol.*, 62, n° 4, 1984, pp. 377-384. (C.D.I.U.P.A., n° 194747).

2. BADOT J., COLOMBAN P. Production de hautes températures par chauffage micro-ondes. In : Les micro-ondes. 1980, 44 p. (Cycle de formation permanente CNRS/ENSCP)

3. BALANIS C. Electromagnetic techniques in the development of coal-derived energy sources - a review. *J. Microw. Power*, 18, n° 1, 1983, pp. 45-54.

4. BALANIS C., RICE W., SMITH N. Microwave measurements of coal. *Radio Sci.*, 11, 1976, pp. 413-418.

5. BELICAR J. Peroxyde organique pour la rétification des polymères par micro-ondes. *Econ. Progrès Electr.*, n° 7, janv.-fév., 1983, pp. 8-10.

6. BELLAVOINE R. Communication personelle. 4 mars 1986.

7. BERTEAUD A. Les hyperfréquences. Paris, P.U.F., 1976, 125 p. (Collection Que sais-je ?).

8. BERTEAUD A. Les applications des micro-ondes à la production de hautes températures. In : Comptes-rendus du Colloque sur les applications industrielles des hautes fréquences et des micro-ondes. Lyon, 1977, 6 p.

9. BERTEAUD A. Activités en micro-ondes au CNRS de Thiais. *Econ. Progrès Electr.*, 13, janv.-fév. 1985, pp. 3-4.

10. BERTEAUD A. Hyperfréquences. Cours de 3ème année, Institut Supérieur d'Electronique de Paris, s.d., 224 p.

11. BERTEAUD A., JULIEN H., VALOT H. Les micro-ondes appliquées aux macromolécules filmogènes. *Rev. Gén. Electr.*, 11, nov. 1981, pp. 826-829.

12. BERTEAUD A., LEPRINCE P., PRIOU A. Mission au Japon de représentants du club micro-ondes EDF. Publication du Club Micro-ondes EDF, avril 1985, 12 p.

13. BERTEAUD A., MATHIEU A., MANOURY M., ROCHAS J. Perspectives de traitement d'éléments filiformes par micro-ondes. In : Comptes-rendus du Symposium international sur les applications énergétiques des micro-ondes, IMPI-CFE. 14, Monaco, 1979, pp. 122-124.

14. BERTEAUD A., VALOT H., JULLIEN H. Procédé et appareil pour le traitement par micro-ondes de revêtements sur substrats. Brevet français ANVAR n° 79-14709, 8 juin 1979.

15. BLUHM D., FANSLOW G., NELSON S. An analysis of dielectric heating and conductive cooling of pyrite in coal. In : Proceedings of the IMPI Symposium, Toronto, 1981, pp. 59-60.

16. BLUHM D., FANSLOW G., NELSON S. Enhanced magnetic separation of pyrite from coal after microwave heating. (Manuscrit révisé pour publication dans *IEEE Trans. Magn.*, 16 July 1985, 13 p. ref. IS-J-579).

17. BLUHM D., MARKUSZEWSKI R., RICHARDSON C., FANSLOW G., DURHAM K., GREEN T. Microwave interactions with coal components. In : Proceedings of the IMPI Symposium, Chicago, 26-28 August 1985, 3 p.

18. BLUHM D., RICHARDSON C., MARKUSZEWSKI R., FANSLOW G., DURHAM K., GREEN T. Effect of microwave irradiation on the desulfurization and deashing of caustic-treated coal. In : Proceedings of the IMPI Symposium, Memphis, 1986. (Abstract in *J. Microw. Power*, 21, n° 2, 1986, p. 75.).

19. BONEK E., KNOTIK K., LEICHTER P., MAGERL G., ROHRECKER L. Microwave hardening of free-falling radioactive droplets. In : Proceedings of the 12th European Microwave Conference, Helsinki, 13-17 Sept. 1982, pp. 610-614.

20. BOSISIO R., CAMBON J., CHAVARIE C., KLVANA D. Experimental results on the heating of Athabasca tar sand samples with microwave power. *J. Microw. Power*, 12 , 1977, pp. 301-307.

226 Part II - Industrial applications

21. BOSISIO R., SPOONER J., GRANGER J. Asphalt road maintenance with a mobile microwave power unit. *J. Microw. Power*, 9, 1974 pp. 381-386.

22. BRIGGS W., LEWIS J., TRANQUILLA J. Dielectric properties of New Brunswick oil shale. *J. Microw. Power*, 18, n° 1, 1983, pp. 75-82.

23. BUTTS J., LEWIS J., STEWARD F. Microwave heating of New Brunswick oil shale. *J. Microw. Power*, 18, n° 1, 1983, pp. 37-43.

24. CHAN TANG T., BOSISIO R., DALLAIRE R. Microwave heating of iron ore pellets. *Can. Electr. Eng. J.*, 4, 1979, pp. 37-40.

25. DUFLOS J., DESJEUX J. Les plastiques renforcés. Paris, P.U.F., 1977, 124 p. (Collection Que sais-je ?).

26. FANSLOW G. Communication personnelle. 13 mars 1986.

27. FANSLOW G., BLUHM D., SIMPSON R. Dielectric heating in mixtures containing coal and pyrite. In : Comptes-rendus du Symposium international sur les applications énergetiques des micro-ondes, IMPI-CFE. 14, Monaco, 1979, pp. 133-135.

28. FANSLOW G., BLUHM D., NELSON S. Dielectric heating in mixtures containing coal and pyrite. *J. Microw. Power*, 15, n° 3, 1980, pp. 187-191.

29. FORD J., PEI D. High-temperature chemical processing via microwave absorption. *J. Microw. Power*, 2, n° 2, 1967, pp. 61-64.

30. FROSINI V., BUTTA E., CALAMIA M. Dielectric behavior of some polar high polymers at ultra-high frequencies (microwaves). *J. Appl. Polymer Sci.*, 11, 1967. p. 527-551.

31. GILL H. The electrothermic system for enhancing oil recovery. *J. Microw. Power*, 18, n° 1, 1983, pp. 107-110.

32. GOURDENNE A., MAASSARANI A., MONCHAUX P., AUSSUDRE S., THOUREL L. Quelques aspects de la polymérisation par chauffage diélectrique micro-ondes. In : Comptes-rendus du Symposium international sur les applications énergétiques des micro-ondes, IMPI-CFE. 14, Monaco, 1979, pp. 127-129.

33. HAMID M. Microwave thawing of frozen soil. *J. Microw. Power*, 17, n° 3, 1982, pp. 167-174.

34. HEARD N. Microwave heating of bitumen in road delivery tankers. In : Proceedings of the IMPI Symposium, Memphis, 1986. (Abstract in *J. Microw. Power*, 21, n° 2, 1986, pp. 76-78).

35. HULLS P. Development of the industrial use of dielectric heating in the United Kingdom. *J. Microw. Power*, 17, n° 1, 1982, pp. 29-38.

36. ISRAEL S., CHABERT J., JOLION M., LEFEUVRE S., ROUSSY G., Compte-rendu de mission. Etats-Unis, Club Micro-ondes, 1981, 66 p.

37. JACOBS I., ZAVITSANOS P., GOLDEN J. Tracking pyritic sulfur in the microwave desulfurization of coal. *J. Appl. Phys.*, 53. n° 3, 1982, pp. 2730-2732.

38. JULLIEN H. Interactions entre micro-ondes et macromolécules. In : Les micro-ondes. 1980, pp. 12-32. (Cycle de formation permanente, CNRS/ENSCP).

39. KASHYAP S., WYSLOUZIL W. Single-sided microwave applicator for sealing cartons. U.S. Patent n° 4 160 144, 3 July 1979.

40. KETE R., THOUREL L., VIALLIER P., JANNOT M. Traitements d'ennoblissement thermiques et chimiques des textiles par micro-ondes. In : Comptes-rendus du Symposium international sur les applications énergétiques des micro-ondes, IMPI-CFE. 14, Monaco, 1979, pp. 118-121.

41. KIMREY H., KAUFFMAN J., EL-SHIEKH A. A study of microwave texturing of polyester yarns. In : Proceedings of the IMPI Symposium, Toronto, 1981, pp. 52-54.

42. KIMREY H., WHITE T., BIGELOW T., BECHER P. Initial results of a high-power microwave sintering experiment at ORNL. In : Proceedings of the IMPI Symposium, Memphis, 1986. (Abstract in *J. Microw. Power*, 21, n° 2, 1986, pp. 81-82).

43. KRIEGER B. Communication personnelle. Stamford (USA), COBER Electronics, inc. 28 déc. 1987.

44. LE CORRE H. Prise rapide du béton par micro-ondes. *Econ. Progrès Electr.*, n° 1, jan-fév. 1982, p. 15.

45. Mac PHERSON R., CHUTE F., VERMEULEN F. The electromagnetic flooding process for in-situ recovery of oil from Athabasca oil sand. *J. Microw. Power*, 21, n° 3, 1986, pp. 129-147.

46. MEEK T., BROOKS M., BLAKE R.,
    PETROVIC J., JORY H., MILEWSKI J.
    Microwave processing of ceramics.
    In : Proceedings of the IMPI Sympo-
    sium, Memphis, 1986. (Abstract in
    J. Microw. Power, 21, n° 3, 1986,
    pp. 193-194).

47. MONCHAUX P. Etude thermodynamique
    des interactions micro-ondes/résines
    epoxydes : aspects théoriques et
    pratiques. Thèse de Doctorat d'Ingé-
    nieur, Univ. Paul Sabatier, Toulouse,
    Physique et structure des solides
    n° 809, 1982.

48. MUNOZ M. Chaudière à micro-ondes
    pour la production d'un fluide chaud
    à usage domestique, industriel ou de
    chauffage de locaux, et procédé mis
    en oeuvre par cette chaudière.
    Brevet français n° 2 521 809 (8202460),
    12 fév. 1982.

49. NELSON S., BECK-MONTGOMERY S.,
    FANSLOW G., BLUHM D. Frequency-depen-
    dence of the dielectric properties
    of coal. Part II. *J. Microw. Power*,
    16, n° 3-4, 1981, pp. 319-326.

50. NELSON S., FANSLOW G., BLUHM D. Fre-
    quency-dependence of the dielectric
    properties of coal. *J. Microw. Power*,
    15, n° 4, 1980, pp. 277-282.

51. NENIN J. Fonderie en cire perdue :
    les avantages des micro-ondes. *Usine
    Nouv.*, n° 45, 1979, pp. 138-139.

52. OGURA K., TOISHI T., MURANAKA T.,
    TARUTANI K., SASAKI S. Microwave
    heating equipment for nuclear fuel
    recycle. In : Proceedings of the IMPI
    Symposium, Toronto, 1981, pp. 17-18.

53. OOSHIMA H., ASO K., HARANO Y.,
    YAMAMOTO T. Microwave treatment of
    cellulosic materials for their enzy-
    matic hydrolysis. *Biotechnol. Lett.*,
    6, n° 5, 1984, pp. 289-294.
    (C.D.I.U.P.A., n° 188956).

54. PATAY L. Communication personnelle.
    8 fév. 1986.

55. PEKARSKY A. Microwave application in
    ceramic industry. In : Proceedings
    of the Radiant Wave Electroheat
    Workshop, Toronto, 2 Dec. 1985, pp.
    73-76.

56. PELLISSIER J. Fabrication en continu
    par un procédé de collage utilisant
    des micro-ondes de profilés en maté-
    riau diélectrique. Brevet français
    n° 2 502 884 (81 06053), 24 mars 1981.

57. PELLISSIER J. Les micro-ondes, une
    nécessité pour l'aboutage automatique
    de chevrons. *Rev. Bois*, nov. 1982,
    pp. 19-24.

58. PELLISSIER J. Pénétration des tech-
    niques H.F. et micro-ondes dans
    l'industrie du bois. *Econ. progrès
    Electr.*, n° 14, mars-avril 1985,
    pp. 9-11.

59. PELLISSIER J. Communication person-
    nelle. 30 janv. 1987.

60. PELLISSIER J., JACOMINO J. Utilisa-
    tion des rayonnements électromagné-
    tiques à des fins "émulsives". In :
    Comptes-rendus du Symposium interna-
    tional sur les applications énergé-
    tiques des micro-ondes, IMPI-CFE.
    14, Monaco, 1979, pp. 125-126.

61. PRIOU A. Communication personnelle.
    Sept. 1986.

62. PRIOU A., AUSSUDRE S. Intérêt de
    l'utilisation du rayonnement micro-
    onde dans les procédés et fabrica-
    tion de matériaux composites. Publi-
    cation DERMO, Centre d'Etudes et de
    Recherches de Toulouse/ONERA, 1986,
    5 p.

63. REALE M., JACOBS B. A rapid protec-
    tive test for urethane foam scorch.
    J. Cell. Plast., sept.-oct. 1978,
    pp. 273-275.

64. RICHARDSON C., MARKUSZEWSKI R.,
    DURHAM K., BLUHM D. Effect of caustic
    and microwave treatment on clay
    minerals associated with coal. Ames,
    Univ. Iowa, lab. dep. Earth Sci.,
    non publ., s.d., 10 p.

65. SALERNO P., WILSON L., HALPIN B.
    Application of microwave heating to
    the curing of fiberglass-epoxy lami-
    nates. In : Proceedings of the South-
    eastcon 1977 on Imaginative Enginee-
    ring through Education and Experience,
    Williamsburg, 1977, pp. 504-506.

66. SEN P., CHEW W. The frequency-depen-
    dent dielectric and conductivity
    response of sedimentary rocks. *J.
    Microw. Power*, 18, n° 1, 1983, pp.
    95-105.

67. SIRKIS M., CONRAD G. Heating of low-
    resistivity semiconductor wafers and
    ribbons with traveling microwaves.
    In : Proceedings of the IMPI Sympo-
    sium, Memphis, 1986. (Abstract in
    J. Microw. Power, 21, n° 2, 1986,
    pp. 83-84).

68. SMITH F. Microwave methods enable
    energy savings in restoration of
    highway pavements. In : Proceedings
    of the IMPI Symposium, Memphis, 1986.
    (Abstract in *J. Microw. Power*, 21,
    n° 2, 1986, pp. 79-81).

69. TEFFAL M. Interactions des micro-ondes (2,45 GHz) avec quelques composés hydroxyles. Applications aux réactions de polymérisation. Thèse de 3ème Cycle. Univ. Paul Sabatier, Toulouse, Physique des solides n° 2693, 1982.

70. TERESCHENKO A., KONONOV A., KONTAR A. On the application of microwave energy for acceleration of concretes hardening. In : Comptes-rendus du Symposium international sur les applications énergétiques des micro-ondes. IMPI-CFE. 14, Monaco, 1979, p. 136. (Résumé seul).

71. THIEBAUT J., VAUBOURG J., ROUSSY G. Control of ore flotation by electromagnetic irradiation. In : Proceedings of the IMPI Symposium, Chicago, 1985, pp. 90-93.

72. THOUREL L., MEISEL N. Les applications industrielles du chauffage par micro-ondes. EDF-Journées d'informations électro-industrielles, Paris, Publication EDF, nov. 1974, 12 p.

73. TRISPA I., DRAGOI F., COSNEANU C., TARASESCU M. Etuvage des noyaux par micro-ondes. Fonderie, n° 365, mars 1977, pp. 93-97. Mémoire n° 33 présenté au 43ème Congrès International de fonderie, Bucureşti, 5-10 sept. 1976.

74. VENE J. Les plastiques. Paris, P.U.F., 1976, 121 p. (Collection Que-sais-je ?).

75. WALL E. Interaction of microwave energy with fuel precursors. J. Microw. Power, 18, n° 1, 1983, pp. 31-36.

76. WESTBROOK W., JEFFERSON R. Dissolution of samples by heating with a microwave oven in a teflon vessel for instrumental analyses. J. Microw. Power, 21, n° 1, 1986, pp. 25-32.

77. WESTPHAL W., SILS A. Dielectric constant and loss data. Technical report AFML-TR-72-39. U.S. Air Force Materials Lab., Air Force Systems Command, Wright-Patterson Air Force Base, Ohio, 1972.

78. WYSLOUZIL W., KASHYAP S. Single-sided applicator for microwave heating. Brevet canadien n° 1167 531, 15 mai 1984.

79. WYSLOUZIL W., KASHYAP S. Single-sided microwave applicators for sealing cartons. J. Microw. Power, 20, n° 4, 1985, pp. 267-272.

80. ZAVITSANOS P., GOLDEN J., BLEILER K. Coal desulfurization by a microwave process. Technical progress report, Philadelphia, General Electric Co., Feb.-May 1981, Jan. 1982, May 1982, Sept. 1982, Dec. 1982, March 1983.

81. Fonderie à la cire perdue : une utilisation du four à micro-ondes. L'électricité dans l'industrie, Industrie de la Fonderie, 1979, Industrie-Electricité Informations, réf. 241, 4 p.

82. Procédé perfectionné pour la fabrication, par moulage, de lentilles de contact, et dispositif de mise en oeuvre. Sté Medicornea S.A. Brevet français n° 2 523 505 (82 04663), 17 mars 1982.

Part III

Applications in the food industry

Chapter 1

Cooking

Historically, it is in the field of food cooking that the thermal applications of microwaves took off. Applications currently in use include the cooking of poultry, bacon, and meatloaves; the preparation of sprats; and the frying of fritters and doughnuts, as well as the reheating of prepared meals.

The effect of microwave cooking on different animal and vegetable products will be examined in this chapter, with particular emphasis on four points, namely, hygiene, nutrition, sensory perception, and the advantages and disadvantages of each industrial process. There will also be a brief mention of the digestibility of microwave-cooked food.

1.1 Mechanisms

It is clear from from Table 1.1 that lean products have high permittivity values at positive temperatures. By contrast, fatty materials have remarkably low permittivity and therefore absorb very little electromagnetic energy. For example the ratio of energies absorbed by steak and smoked lard is found to be 96:1.

Figure 1.1 shows that the loss factor $\epsilon_r'' = \epsilon'_r \tan \delta$ varies smoothly with temperature. For most products, the loss of water during heating leads to a slight reduction in the loss factor as a function of temperature [1].

The dielectric properties of food products are not only functions of their water content, but also of the physical and chemical state of this water (Fig. 1.2). In common with most biological materials, food products contain monolayers of water strongly bound to organic macromolecules that account for approximately 7% of dry weight [214]; subsequent layers of water are bound to a lesser and lesser extent. Pockets of free water may also be found within the structure of certain products, and microwaves exert their full effect on this fraction of water.

In general, the dielectric behavior of food products follows the two-phase mixture model (Part I, Section 3.5.4): a liquid phase containing free water and dissolved ions, and a solid phase containing bound water, dissolved ions, lipids, proteins, and coloidal carbohydrates [300], [213], [214], [303]. The model is in reasonably good agreement with experimental results.

There is an extensive literature on the dielectric properties of food products [100], [141], [234], [125], [27], [258], [128], [139], [217], [226], [227], [127], [299], [300], [130], [212–214], [210], [1], [207], [211], [302], and, more particularly, the properties of meat [133], [308], [316], [297], [228], [298], dairy products [265], [242], fatty materials [124], [235], [152], fruits and vegetables [235], [303], [301], and fish [153], [154], [155].

Microwave cooking obeys the classic heat equation

$$W = mC\Delta T = P\Delta t$$

Table 1.1

Permittivity of food products at 3 GHz [141], [316], [297]

Product	Temperature	Moisture content, %	ε'_r	tan δ
Cooked beef	20°C	67.8	41.6	0.31
Raw beef	20°C	73.8	48.3	0.28
	60°C	73.8	39.5	0.29
Steak	25°C		40.0	0.3
Smoked bacon	25°C		2.50	0.05
Corn oil	25°C		2.53	0.06
Soya oil	25°C		2.51	0.06
Cooked cod	20°C		46.5	0.26
Milk	20°C	7.3	51.0	0.59
Egg white	25°C		35.0	0.5
Potato	25°C	78.9	81.0	0.38 [316]
	25°C	79.5	57.3	0.27 [297]

where W is the energy in joules, m is the mass in kilograms, C is the heat capacity in $\mathrm{J\,kg^{-1}\,K^{-1}}$, Δt is the rise in temperature in K, P the power in watts, and Δt is the time interval in seconds.

Values of C are on the order of 2.9–3.5 kJ kg K^{-1} for lean meat, 1.3–2.6 for fatty meat, 1.4 for butter, 2.1 for cheese, 2.8–3.6 for fish, 3.5–3.8 for shell fish, and 2.0 for oil (the figure for water is 4.2 in the same units [201], [223]).

Microwave heating is, by its very nature, nonuniform for many reasons:

- nonuniformity of the electric field in the applicator, which is an important feature of domestic microwave ovens, but much less so in industrial ovens, and, particularly, the specially designed microwave tunnels such as the BS Gigatron made by IMI-SA (Fig. 1.3).
- heterogeneity of the product, whose structure contains interfaces separating regions of high and low dielectric loss.
- the shape of the product and, in particular, cylindrical and

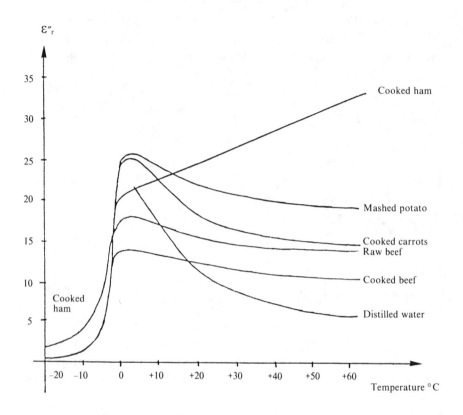

Figure 1.1 Dielectric loss factor as a function of temperature [27].

spherical interfaces that may give rise to axial or central hot spots as described in [229]; this mechanism is responsible for eggs exploding under microwave irradiation.

• sharp edges that produce very inconvenient local overheating [223].

Microwave heating is generally uniform over the entire thickness of the product if it does not exceed about 1 to 1.5 penetration depths as defined in Part I, formula (2.83), which is approximately 3 cm for steak at 2.45 GHz. The frequencies of 915 and 896 MHz, allowed on the American continent and in the United Kingdom, respectively, are evidently more appro-

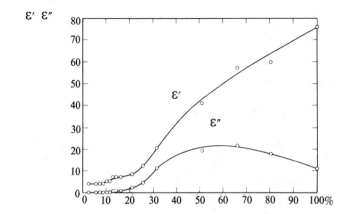

Figure 1.2 Permittivity of freeze-dried and rehydrated potatoes as a function of water content [214].

priate for cooking products with large cross sections and high dielectric loss factors [220].

A numerical simulation of the microwave heating of meat [225] has shown that important thermal gradients appeared in large joints. The study also highlighted the importance of conductivity in the evolution of thermal equilibrium after microwave irradiation. It is therefore recommended that domestic users of the microwave oven should allow food to stand after cooking until the temperature equilibrium is reached [11, 266]. The latency period is the time interval during which the temperature at the initially coldest point continues to increase.

1.2 Animal products

1.2.1 Red meat

Although the main industrial applications of microwave cooking involve poultry, bacon, and meatloaves, there is an extensive literature on the cooking of red meat in domestic microwave ovens, and on its bacteriological, nutritional, and sensory evaluation.

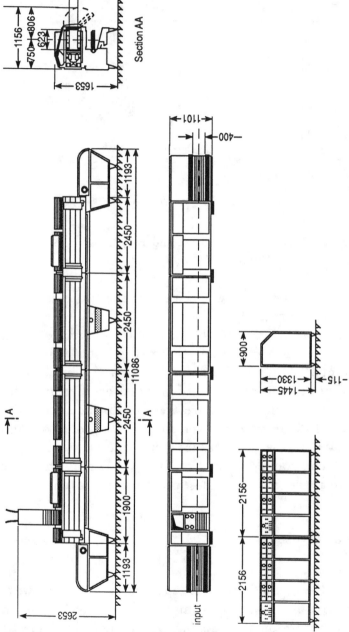

Figure 1.3 The Gigatron BS cooking tunnel operating at $60\,\mathrm{kW}$ RF, $2.45\,\mathrm{GHz}$ (Courtesy of IMI-SA).

The texture of meat is a function of the mechanical rigidity of collagen of the connective tissue. Collagen is a fibrous protein whose molecule consists of a triple helix formed by three polypeptide chains interlinked with hydrogen bonds and covalent bonds. Cooking induces profound changes: muscle fibers disintegrate as the connective tissues supporting them lose solid consistency and collagen transforms into gelatine. This transformation occurs in three stages, namely, denaturation of the protein due to the breaking of hydrogen bonds; thermal contraction as the temperature approaches 58°C, which shortens the fibers by about 75%; and, finally, solubilization, which occurs at temperatures between 60 and 80°C and leads to the progressive destruction of the structure of the collagen [106].

Microwave cooking phenomena fall within this general framework but with their own specific characteristics that originate mainly from deep heat deposition, selective heating that depends on the permittivity of tissues, and, above all, the short cooking time. The end result has often been judged to be qualitatively inferior to that achieved by conventional cooking. However, it is practically impossible to get an objective evaluation from the experimental data available on this subject. This is primarily because different studies relate to different cooking conditions, and also because reported differences between the methods of cooking are quite marginal anyway. Thus, for beef, tenderness was considered to be identical in three recent studies [105], [121], [238] and inferior in other studies [208], [159]. Taste is generally judged to be the same [159], [121], [257], [238] and juiciness was also generally considered to be identical [159], [105], [238]. In contrast, microwave cooking gave inferior results in one study [121]. It is equally difficult to find a general trend in the assessment of color and even of weight loss during microwave cooking, which is judged to be similar to that of conventional cooking in [159], better in [121], [238], and inferior in [67]. Losses by dripping increase and those by evaporation decrease, and, in general, both are less important

Table 1.2

Cooking losses for broiled and microwaved beef

Cooking losses, %	Microwaves	Broiling
Dripping	15.4 ± 5.9	3.3 ± 0.8
Evaporation	10.8 ± 0.7	33.5 ± 3.1
Total	26.2 ± 6.1	36.9 ± 3.0

in microwave cooking (Table 1.2).

Microwave cooking is generally shown to be satisfactory in studies of the retention of proteins, amino acids, phosphorus, potassium [159], and nitrogenous products [22]. The comparison ceases to be clear-cut regarding the retention of vitamins B1 and B2 (thiamine and riboflavin), which have been reported [159] to be 25 and 90%, respectively, against 19 and 81% in conventional cooking. Less significant differences were reported in [238](Table 1.3).

Table 1.3

Retention of B1 and B2 vitamins by beef

		Microwave convection oven	Forced air convection oven	Conventional oven	Raw meat
Moisture content	(%)	59	60	60	70
Fat	(%)	20	20	19	18
Thiamin (wet base)	(mg hg⁻¹)	0.11	0.11	0.12	0.09
Thiamin (dry base)		0.54	0.57	0.57	0.77
Thiamin retention	(%)	70	77	76	
Riboflavin (wet base)	(mg hg⁻¹)	0.23	0.19	0.22	
Riboflavin (dry base)		1.03	1.06	0.97	1.28
Riboflavin retention	(%)	83	76	81	

Speed is cited as the main advantage of microwave cooking [202], [224], [121], [146]. A number of studies report on the retention of iron [259], [314], [150], [249], [274], [28], [293]. The retention of thiamine by pork is the only parameter mentioned as being significantly improved by microwave cooking [55], all other sensory and nutritional aspects having been found identical in microwave and conventional cooking [2], [55], [209], [296].

In respect to hygiene, we note that pork meat has been the subject of a controversy related to the destruction of *Trichinella spiralis* by heat. This parasite, otherwise known as roundworm, is a nematode. Its larva becomes encysted in pork meat and develops further in the human intestine after ingestion of partially cooked infected pork. Trichinosis is the illness resulting from this infestation, and it may occasionally prove fatal. To avoid this danger, an American directive (Food Service Sanitation, 1978 [332] and United States Department of Agriculture 1981 [348]) imposed a final cooking temperature of at least 76.7°C at every point in the pork meat. A number of studies [326], [163], [322], [323] suggest that microwave cooking gives trichinela larva a higher chance of survival, probably because of uneven heating and short cooking time [324]. For example, Zimmermann reported the presence of *Trichinella* in 50 roasts from a sample of 189 after they were cooked in five different microwave ovens by 39 different cooking procedures. The temperature of one sample, in which the larvae survived, was tested at five different places and found to exceed 76.7°C. Other systematic studies by the same author [326, 322, 323, 324, 325] were concerned with meatloaves and roast pork. He recommends cooking at lower powers and for longer times to ensure better temperature uniformity.

This problem should not, however, be overrated as there are no reported clinical cases that are directly attributable to microwave cooking. The simple rule is that longer cooking at lower microwave powers with a final temperature of at least 77°C after ten minutes standing time, should be enough to ensure the safety of the product. Further information can be found in [221], [173], [15], [11], [266], [341], and [346].

1.2.2 Poultry

The first industrial installation for the microwave cooking of chicken was developed in 1966 at Ocoma Foods of Berryville, Arkansas. The microwave equipment consisted of two parallel microwave and vapor tunnels, one used for wings and thighs and the other for legs and wings. These tunnels were equipped with 2.5-kW modules operating at 2.45 Ghz, delivering a total of 80 kW of microwave power for the first and 50 kW for the second, and allowing a processing rate of 1140 kg/hr [89], [90], [92], [233]. Preliminary results [279] showed that fast cooking gave poor results in terms of tenderness and, paradoxically, higher weight loss as compared to steam cooking. Finally, a cooking time of 8–15 min resulted in excellent product quality, and the weight loss, mostly due to the evaporation of water and its condensation on the cold tunnel walls, was greatly reduced in the presence of steam. It was concluded that the combination of microwaves and saturated vapor gave much better results than either of these two forms of energy separately. It was better than conventional steam cooking, which is time consuming, uneconomic, and results in mediocre quality [89].

Other installations have since been developed and built for 915 MHz or 2.45 GHz. Some of the latter equipments are powered by 30-kW klystrons [89].

The quality of the finished product suggests that microwaves are less efficient in destroying harmful bacteria such as *Straphylococcus aureus, Salmonella typhimurium, and Clostridium prefringens*. For example, microwave cooking of infested turkey, followed by a 30-min temperature equalization time at 10% power, is not sufficient for the complete destruction of *S. typhimurium or S. Aureus* [3]. Additional information on this subject can be found in [31], [60], [69], [13], and [177]. It should be noted, however, that, as in case of trichinella in pork meat, the cooking was performed in domestic ovens and that the uniformity of heating in such ovens was, by design, inferior to that of industrial applicators.

The oxidation of lipids is another important qualitative aspect of the microwave cooking of poultry.

Fatty acids are characterized by the presence of a carboxyl group at the end of the chain. The acids are said to be saturated if they contain no double bonds (e.g., butyric acid $CH_3 - CH_2 CH_2 - CO_2 H$). If they contain double bonds, they are known as unsaturated fatty acids: either monounsturated like the oleic acid $CH_3 - (CH_2)_7 - CH = CH(CH_2)_7 - CO_2H$ or polyunsaturated like the linoleic acid $CH_3 - (CH_2)_4 - CH = CH - CH_2 - CH = CH - (CH_2)_7 - CO_2H$

Saturated fatty acids are now much less common in foods because of their effect on blood cholesterol levels. This change has occurred mostly at the expense of red meat: the annual consumption of beef in the United States has fallen from 42 to 34 kg per capita between 1976 and 1985, and there has been a consequent rise in the consumption of fish and poultry, which are rich in polyunsaturated fatty acids. However, the increased consumption of the latter is associated with a new health risk originating from the lability of the double bonds, which results in the loss of a hydrogen atom, followed by a series of transformations [245] with the fixation of two oxygen atoms and the breakup of the molecule into several fragments of which the *malodialdehyde* (MDA) molecule, $O = CH - CH_2 - CH = O$, is highly reactive and carcinogenic. Moreover, MDA damages the proteins and membranes by creating the fluorescent structure $-N = C - C = C - N$ known as the Schiff base.

The two successive phases of auto-oxidation can be evaluated by testing for malodialdehyde, using thio-2-barbituric acid, and for flourescent products, respectively. Normally, auto-oxidation occurs during metabolism, but it may also start before ingestion, when the frozen product is stored, and during cooking [245, 175, 246, 174]. Comparative studies of cooking by convection and by microwaves [174] have shown that the MDA levels are not significantly different, but the levels of flourescent products are noticeably lower for microwaves, indicating a lower degree of auto-oxidation. Microwave cooking of poultry is therefore thought to be safer than conventional cooking (Table 1.4).

The sensory quality of poultry meat has often been judged to be superior in microwave cooking. For example, tastings of poultry cooked using microwaves, a fryer, and a conventional frying pan were found to favor microwaves [71]. Other studies

Table 1.4

Relative content of fluorescent products from chicken breast (A)
and legs (B) for frozen, cooked, and reheated chilled samples

	Conservation prior to cooking	Content prior to cooking	Content after cooking	Content after 4 days' refrigeration	Content after reheating by microwave/convection
A	2 days	1.14 ± 0.04	1.43 ± 0.07	1.96 ± 0.07	2.08 ± 0.06
			1.16 ± 0.04	1.39 ± 0.04	1.46 ± 0.05
	3 months	1.20 ± 0.05	1.51 ± 0.06	2.38 ± 0.08	2.43 ± 0.07
			1.32 ± 0.06	1.68 ± 0.06	1.71 ± 0.05
	6 months	1.31 ± 0.06	1.85 ± 0.07	2.48 ± 0.06	2.65 ± 0.07
			1.45 ± 0.06	1.76 ± 0.07	1.85 ± 0.06
B	2 days	1.49 ± 0.06	1.68 ± 0.08	2.95 ± 0.11	3.02 ± 0.08
			1.54 ± 0.06	2.07 ± 0.07	2.28 ± 0.06
	3 months	1.57 ± 0.06	2.28 ± 0.07	2.97 ± 0.09	3.18 ± 0.07
			1.61 ± 0.05	2.28 ± 0.09	2.35 ± 0.08
	6 months	1.87 ± 0.07	2.54 ± 0.09	3.06 ± 0.08	3.35 ± 0.10
			1.96 ± 0.07	2.60 ± 0.08	2.75 ± 0.07

have led to similar conclusions [200], [37], but more recent evaluations [109] suggest that microwave cooking results in inferior tenderness and in a degradation of color and taste. At best, organoleptic parameters (taste, appearance, and juiciness) are thought to be identical, but the tenderness of the meat is usually judged to be inferior in the case of microwaves[193]. However, differences arising from the method of cooking seem very marginal [56], [58], [290].

In contrast, nutritional properties are better preserved by microwaves. For example, the retention of pyridoxine (vitamin B6) is 92.5%, as against 88.4% in conventional cooking [318]. The retention of proteins and of mineral salts is also better [247].

The energy efficiency of microwave cooking is far superior to conventional techniques. For example, five minutes of microwave cooking are equivalent to to 20–30 minutes in water and 25–50 minutes in steam [200]. The energy efficiency is on the order of $300\,\mathrm{W\,hr\,kg^{-1}}$ for microwaves, as against 730 and $640\,\mathrm{W\,hr\,kg^{-1}}$ in conventional and convection cooking at $190°\mathrm{C}$ [109]. Another study has revealed that the energy consumption ratios are 3:1 for conventional and microwave cooking and 5:1 for convection and microwave cooking [193]. Further information may be found in [29], [12].

1.2.3 Bacon and fat

The main industrial application of microwaves in the food industry is probably the precooking of bacon in the United States, which uses 3 MW of microwave power at 915 MHz. Raytheon and Microdry are the main American equipment manufacturers in this field.

Although the microwave cooking of bacon was the subject of a patent in 1967 [149], and was mentioned in some publications in the 1970s [171], [241], its commercial importance is relatively recent and is still perceived as somewhat risky.

The advantages of microwave cooking are, however, very clear: the finished product is of a much higher quality with respect to color, appearance, taste, nutritional properties, and the remarkable absence of curling. This cooking method results in very considerable savings of energy (30 to 50%) and of raw material (35%). Moreover, fat can be recovered and wastes and smells minimized, which make antipollution equipment no longer necessary [91].

The economics of the microwave cooking of bacon has been modeled [107] in comparison with the production of an equal quantity of uncooked but simply cured bacon. The installation considered had a 400-kW microwave oven fed by eight 50-kW tubes operating at 915 MHz and amounting to a total investment of $1,125,000. The annual 3,630 tonnes of raw material are now cooked by microwaves, resulting in the recovery and sale of 1,630 tonnes of rendered fat. Table 1.5 gives an example of the evaluation of the annual cash flow generated by the two types of production. Despite the costs of installation, depreciation, interest rates, running costs (mainly tube replacement, but also additional manpower, energy, packaging, and maintenance relative to the uncooked product) there is a net advantage in favor of the cooked product, which can be recovered from the first year of production. The investment (Part I, Section 1.6) gives over four years a theoretical return of $m = 58\%$ (Table 1.5).

Table 1.5
Economics of microwave cooking of bacon [107]

	Base	Microwaves	Differential cashflow Year 1	Differential cashflow Year 2
Fixed costs				
Linear depreciation over 10 years	0	112 500	(112 500)	(112 500)
Transport & installation	0	50 000	(50 000)	0
Loan interest	0	96 000	(96 000)	0
General & sales costs	100 000	366 400	(266 400)	(266 400)
Variable costs				
Labor ($16 per hr)	256 000	1 088 000	(832 000)	(832 000)
Raw materials				
($1.3 per kg)	4 800 000	4 800 000	0	0
Refrigeration	10 000	10 000	0	0
Power ($0.05kW h^{-1}, 600kW)	5 000	120 000	(115 000)	(115 000)
Tube replacement ($2.88 per hr)	0	92 160	(92 160)	(92 160)
Packaging	10 900	88 000	(77 100)	(77 100)
Maintenance	5 000	20 000	(15 000)	(15 000)
Total costs	5 186 900	6 843 060	(1 656 160)	(1 510 160)
Revenues				
Uncooked				
($3.3 per kg, 3 630t)	12 000 000	0	(12 000 000)	(12 000 000)
Cooked				
($0.08 per slice, 44 000 slices/h)	0	14 080 000	14 080 000	14 080 000
Fat ($0.44 per kg)	0	720 000	720 000	720 000
Margin			1 143 840	1 289 840
Tax (46%)			526 166	593 326
Net profit			617 674	696 514

Microwaves also give a safer product in the sense that they do not involve carcinogenic compounds, particularly *nitrosopyrrolidine* (NPYR) which is produced at high temperatures when nitrites, used as preservatives, combine with hydroxyproline in bacon [241], [292]. Nitrite-free bacon is not without its risk because high-temperature cooking may produce new mutagen compounds, probably N-substituted amines through the Maillard reaction [206]. Although the etiological importance of these compounds is unknown, it is *a priori* prudent to avoid them, and microwave cooking is the only practical way of doing so.

The rendering of fat is another potential application for microwaves despite inherent difficulties due to the low dielectric loss factor of fats. Trials performed about ten years ago on a microwave rendering unit in Provo, Utah produced high-quality fats. Further details can be found in a patent [30] and elsewhere [90, 92, 343].

1.2.4 Meat patties

Meat patties destined for the frozen-food market are usually fried in oil at temperatures between 150 and 200°C. This results in a very fatty product and requires frequent changes of oil. Inevitable exchanges between meat fats and oil tend to degrade taste flavor when meat patties of different types are being processed, and bond breaking within the fat molecule can also affect the taste and the shelf life of the product.

A new processing technique incorporating a microwave cooking stage has been the subject of a collaborative study by three companies: Indra AB of Helvingborg (Sweden), Skandinaviska Processinstrument AB of Stockholm (which has since become Calorex AB), and the National Industrial Manufacturing Company of Burlingame in California. The meat patties are pressed into shape, topped with a portion of margarine, introduced into a browning unit between two hot aluminum panels, and then finished off in a microwave cooking oven. The $6 \times 1.3 \times 1\,\text{m}^3$ microwave tunnel contains three slow-wave applicators positioned above and below the conveyor belt and fed by a 30-kW

2.45-GHz source. Hot air is circulated to prevent condensation. The internal temperature of the meat patties reaches a very uniform 77°C, which is an optimum figure for food hygiene with minimum cooking loss.

After a trial period of one year, Indra AB revealed that the new technique was more cost effective, saved a lot of fat, and the processing time was much shorter than in conventional deep-fat frying. The elimination of oil vapor and the lower ambient temperature create a better working environment. Finally, the quality of the product, including shape, taste, and fat content, is significantly improved [218], [92].

The bacteriological state of meat patties cooked or reheated by microwaves has been the subject of several studies [42], [104], [72]. They highlight the need for research into more reproducible and reliable levels of hygiene. Studies of the rates of survival of *E. coli* in hamburgers cooked by microwaves [72], [122] show significant differences between the center and the periphery (0.7 and 30% survival rates, respectively). Other studies [74], [268] have found no differences between patties reheated by conduction, convection, and microwaves. In contrast, there have been reports that conventional cooking was better at destroying bacteria [62], [176], [320]. In general, microwave meat patties look better than the conventional product, but have a very similar taste [267]. Vitamin retention is comparable in the case of thiamine, and better in the case of riboflavin and niacin [143], [144].

1.2.5 Fish

The weight of whiting fillets cooked in the microwave oven is approximately 15% higher than in conventional cooking. The protein content is also about 15% higher, and the skin does not deteriorate and is readily removed.

Similar results have also been obtained with mackerel fillets for which the lower cooking losses compensate for losses due to mechanical filleting. Frozen fish cooked in the microwave oven is more tender and suffers no alteration in flesh or skin. The same can be said for sardine, sprat, and herring [330].

Coated fish is not usually suitable for microwave cooking except when specially treated to give a product of similar sensory quality as fried fish [179]. This treatment involves replacing water in batter with hydrogenated soya oil, using cornflower, and cooking at low microwave power. The same product frozen is judged to be better when reheated by microwaves. Fish is probably one of the products that gain most from this type of cooking: its flavor is significantly improved. At worst, the sensory quality is judged to be identical to that of conventional cooking [78], [195].

The different types of cooking produce no difference in the nutritional value in terms of proteins, lipids, fatty acids, and metallic trace elements [123], [321]. The speed, energy efficiency, and cooking output are much better in the case of microwaves [78], [123].

Several studies have reported on bacteriological quality in microwave cooking [18], [53]. In the case of shellfish, numerous microwave cooking trials on prawns have shown that microwave cooking significantly improves bacteriological quality but does not reduce cooking losses [330].

An interesting application was set up in 1979 by the IMI-SA and by Bjelland of Stavanger (Norway) for the precooking or coagulation of sprats and sprats in open metal tins. These collaborative studies by the two companies have led to the development of a prototype microwave oven - the Gigamet - that is integrated in a fully automated line for the production of sprats in tins.

The Gigamet (Figs. 1.4 and 1.5) is a stainless steel system incorporating two parallel tunnels of cross section slightly larger than the size of the tins. The tins are transported one after another on two conveyor belts. Microwave irradiation along the tunnel is produced by helical coils coupled to a WR 248 waveguide. The microwave power consumption is 24 kW for 2,800 tins (105 g each) per hour, or 36 kW for 3,600 tins per hour.

This application constitutes a novel technique because in-

Figure 1.4 Gigamet 24 tunnel for the coagulation of sprats (IMI-SA document).

dustrial microwave applications have tended until recently to exclude metal parts from the cavity to eliminate the risk of arcing and to avoid reflections from the base of the tins. There is also greater risk of temperature nonuniformity because of stationary waves set up inside the tins (Part I, Section 4.3.2). However, the oven is so designed that arcing is avoided and reflections are relatively insignificant. The result is that precooking in tins is definitely more homogeneous than batch cooking, with no overheating of tails and fins. Users claim [110] that the product has better organoleptic quality, the flesh and the skin are not torn, and the reduction in cooking losses translates into weight gain of up to 10%. The oven is thought to be reliable, hygenic, easy to clean, and very easy to use. The saving in energy is on the order of 40% of conventional processing.

1.2.6 Dairy products

Dairy products are well-suited to microwave processing, but

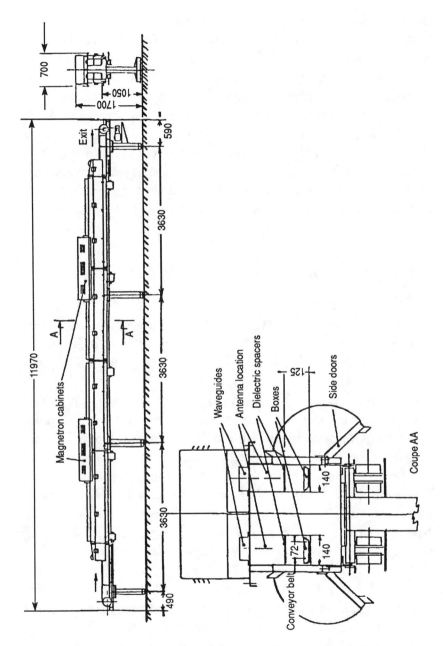

Figure 1.5 The 24-kW Gigamet 24 tunnel for the coagulation of sprats at 2.45 GHz (IMI-SA documentation).

there do not seem to be any industrial applications in this field. There have been trials on curd for the manufacture of mozzarella [330], [44]. The curd is cut into pieces of approximately 25 mm and must be heated to about 60–80°C to acquire the necessary elasticity. This operation is normally carried out using hot water. In pilot studies, the Gigatron 45 B IMI (Fig. 1.3) was used to treat 90 kg of curd per hour. The curd expanded and boiled slowly, reaching 78.8°C with excellent elasticity. The bacteriological quality achieved in microwave processing can be reached with hot water processing but only at the price of considerable loss of fatty material.

1.3 Vegetable products

1.3.1 Vegetables

Sensory comparison between potatoes cooked conventionally (boiled, oven at 200°C) and by microwaves shows no significant evidence in favor of either cooking method [108], [197], [158], [312], [267], [162]).

The American diet had been generally deficient in green vegetables, but the the use of microwave ovens in catering has led to a rise in demand by the public [283]. Indeed, like fish, green vegetables are very well suited to microwave cooking.

The retention of chlorophyl has been found to be superior in five out of six vegetables after microwave cooking (as compared to conventional cooking). Vitamin C (ascorbic acid) is also better retained in microwave cooking (e.g., 92% against 84% in cauliflower, 89% against 79% in spinach, 69% against 67% in broccoli, and 74% against 71% in peas) [38]. Figure 1.6 [45] illustrates the retention of vitamins in cabbage as a function of cooking method; microwaves are clearly more advantageous. Other authors have reported similar results [187].

A comparative study of different methods of cooking cabbage [192] has been carried out to determine whether the production of certain aromas was purely thermal or whether it

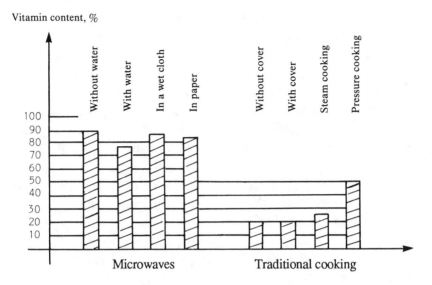

Figure 1.6 Vitamin content of cabbage as a function of the method of cooking [45].

was also a function of time. Comparison of dry microwave cooking with conventional cooking in water revealed that the former method resulted in the rapid production of allyl isothiocyanate and allyl cyanide in quantities that were respectively lower and higher than in water. These two compounds are produced from the same precursor, the first by the action of an enzyme and the second by simple thermal degradation. The conclusion of this study was that microwaves inactivate enzymes much more rapidly than conventional cooking, which may explain why vegetables cooked by microwaves retain their sensory qualities better.

1.3.2 Cereals and soya

In the first reported trials on the precooking of rice, 7 kg of wet rice in a polythene container was heated to 100°C in four minutes. The product did not stick and was easier to handle than precooked rice. The authors [147] proposed a type of industrial applicator for the treatment of rice at 50% humidity at final temperature of between 65 and 100°C. The energy requirement was estimated at 102 J/g. Assuming a microwave thermal efficiency of 80%, a 40-kW microwave tunnel would thus be capable of treating 1400 kg/hr.

The use of microwaves was reported to have significantly improved the quality of grains and beans (soyabean in particular) [131]. The applicator was a cylinder under air pressure, using an Archimedean screw to transport the product at 600 kg/hr. The power source was a 36-kW, 2.45 GHz generator. The use of microwaves should increase the nutritional value of cooked soyabeans [35], [51]. For example, it was reported in [318] that two to three minutes of microwave irradiation caused no harm to fatty acids in soyabeans and produced a product of high nutritional value. Finally, it was shown that wheat starch heated to 75°C [126], [328] showed a lower weight loss and less particle shrinkage as compared with convection heating, but at the price of a certain degradation of microscopic texture.

1.3.3 Roasting

Roasting or torrefaction is the process in which a product is submitted to a certain degree of calcination in order to extract an aromatic constituent. The conventional method cannot be automated and often gives a nonuniform product. There are also problems related to process control (e.g., hazlenuts need 140°C for at least 15 to 30 minutes without the temperature ever exceeding 150°C so that there is no loss of aromatic compounds). Because of the low permittivity of fatty materials, the microwave roasting of hazlenuts required a resonant applicator. The applicator designed and developed by the University of Bordeaux, which is fed by 16 magnetrons of 800 W at 2.45 GHz is an example. Its base is $1.2 \times 1.2 \, m^2$ at a height of 3 m, and the system is capable of processing 90 kg of hazlenuts per hour. The final temperature is only 130°C, but the results are very satisfactory. They include low migration of oils, no abrasion, manageable loss of volatile products, and excellent organoleptic quality [340].

Another interesting application [186] concerns contaminated

peanuts. The contaminants are certain metabolites of the mushroom *Aspergillus flavus* known as aflatoxines. Their carcinogenic effect on the liver is particularly virulent. Conventional roasting - five hours at 150°C - destroys 80% of the aflatoxines. Exposure to microwave for 16 minutes at 1.6 kW or five minutes at 3.2 kW (2.45 GHz), produces a final temperature in excess of 150°C and destroys 95% of the toxins. The residual amount does not exceed 5ng g^{-1} as compared with the standard of 15ng g^{-1}. The roasting process is in itself very satisfactory. For an output of 370 kg/hr, the cost of the process was estimated at 0.015 dollars per kilogram for an investment of $100,000 and a 50-kW microwave oven. The authors of this report were considering the construction of a pilot plant.

1.4 Catering

An increasing number of organizations, including many American hospitals, make use of meals prepared or semiprepared in central kitchens and refrigerated in portions to be reheated at the point of consumption. The reheating process is rapid and is conveniently performed in microwave tunnels, but it is important to ensure its bacteriocidal efficacy. This topic has been the subject of a large number of publications, for example, [66], [42], [72], [73], [74], [75], [120], [284], [339].

It was suggested in [117] that microwaves were less efficient than pulsed hot air or immersion for a given final temperature of 75°C. In contrast, another study [268] compared microwave reheating of meat patties, mashed potatoes, and frozen peas and concluded that microwave ovens used in hospitals produced no significant bacteriological difficulties as compared with reheating by conduction or convection.

Different conclusions were reached in a study of 120 samples of 360 g of mashed potatoes infested with 105–107 cells per gram, frozen, and then reheated with 1.5 kW of microwave power at 2.45 GHz. The results showed that certain microorganisms (e.g., *Salmonella cubana* and *Staphylococcus aureus* were totally destroyed in three minutes whilst others (e.g., *Bacillus cereus* and *Clostridium perfringens*) survived longer than four minutes. Considering that after 3 minutes and 30 seconds the temperature was 83°C, and the product started to dry up, it was concluded that the reheating instruction specified by the manufacturers allowed a higher risk of survival of pathogenes than was encountered in conventional reheating.

In Sweden, a survey [230] showed that fear of poor hygiene together with certain cooking difficulties (dehydration, high temperature gradients, burning at the edges, rapid cooling, high residual water content) were enough to explain why microwaves are not extensively used in Swedish hospitals. It appears, however, that the problem may be resolved by the optimization of cooking procedures [230]. In so far as hygiene is concerned, United States authorities have established a standard known as *Hazard Analysis Critical Control Point Model* (HACCP), which imposes a final temperature of between 74 and 77°C for the optimal conservation of microbiological, sensory, and nutritional qualities [32], [33], [34], [268].

A large number of manufacturers produce microwave reheating systems for communal use. Two examples are, the *Automatic Food Supply* (AFS) of the Swedish Rejler Company whose first prototype was installed in the late sixties at the West Jersey Hospital System of Camden (New Jersey), and the oven manufactured by Enersyst and featuring a conveyor belt that passes through microwave cavities fed with 600 W of microwave power. The latter oven is used at the United Hospitals of St. Paul (Minnesota) to prepare 300 meals per hour. The advantages of this type of installation are quite significant. For example, the AFS system reduces manpower by 60%, the time taken in the preparation of trays by 80%, and the washing up time by 75%. In addition, the separation of meal production from service facilitates the management of periods of high activity, an increase in the shelf life of meals, and a reduction in waste. In contrast to the Swedish survey, frozen meals reheated by microwaves [230] were judged superior to meals prepared in the conventional manner [88], [102], [204].

1.5 Baking

1.5.1 Bread

The first reported use of microwaves for bread baking is due to N. Chamberlain of the Flour Milling and Baking Research Association at Chorleywood (UK). The absence of crust made it evident that microwaves could be successfully used only in a combined process involving conventional baking. Chamberlain [49] obtained good results with 896-MHz in the case of soft bread. The loaves cooked by microwaves after cooling and slicing had a significantly lower level of enzymatic activity and were therefore less sticky, less humid, and more robust than conventionally cooked bread. In the meantime, it was shown in the United States [182] that microwaves were not suitable for bread baking on an industrial scale. In a comparative study of microwaves and conventional baking [304], the levels of amino acids, and in particular lysine, were found to be identical in the two types of cooking. Conventional cooking gave a product of larger volume and better texture, but microwaves increased the protein efficiency.

The industrial plant described in [198] is based on a combination oven consuming 55 kW/hr. It can produce 50% more bread than an electric oven with the same power consumption and three times the ouput of a gas oven of similar thermal consumption.

Two French companies, Pavailler S.A. of Bourg-lès-Valence and the Remy Company of Neuilly-Plaisance have patented the design of a multimode cavity fed by 27 generators and supplied with a vertical rotating basket holding 216 baguettes of 250 g each on 18 shelves. This 6-kW prototype can bake 1000 baguettes per hour [253], [59].

A combination oven, using microwaves and infrared, is described in [251] for the manufacture of breads and cakes. The cavity is provided with a metal grid which stops microwaves but allows the passage of infrared generated by hot steel sheets at the top and bottom of the oven.

There is an extensive literature on the subject [23], [94], [95], [165], [289], [262], [50], [239], [334], [338].

1.5.2 Doughnuts

The production of doughnuts, very popular in the United States, constitutes one of the most successful commercial applications of microwaves. The rising dough has a porous structure of very low thermal conductivity and crusting of the surface also prevents the penetration of heat in conventional cooking. The first reported use of microwaves for the cooking of doughnuts is described in [271]. The structure of the doughnuts was found to be very uniform. They rose in four minutes and reached a final temperature of between 40 and 50°C as compared with the 25–45 minutes at between 50 and 60°C normally necessary in conventional cooking. These results led the American company DCA of Jessup (Maryland) - which has since abandoned its microwave activities [140] - to develop microwave tunnels operating at 2.45 GHz and producing 4800 units per hour at 11 kW. The processing time is four minutes [333], [291].

Doughnuts are often made from flour that is relatively low in protein and high in α-amylase. This flour is normally available in the United Kingdom and Canada and gives the doughnut surface a very porous structure. The microwaves have the favorable effect of activating the α-amylase, but the cooking time is too short for an impermeable crust to form and to retain the gases produced by the yeast. Special conditioning ingredients and emulsifiers must therefore be employed [90], [269].

Microwaves can also be used to improve the frying of doughnuts. Conventional deep frying in oil tends to tear the crust and to leave some of the interior uncooked and flat. The combination of microwaves and deep-fat frying gives a new product that is more tender and stable, with 25% less fat for the same volume, but at the price of a new recipe patented together with the process itself [333], [269]. A number of other publications may be consulted on this subject [261], [264], [270], [263], [184], [305].

1.6 Digestibility of foods cooked by microwaves

This concluding section is devoted to some of the very few investigations of the metabolism of microwave-cooked products.

The University of Maryland [135] has reported that chicken fed on microwave-treated soyabeans had a higher rate of growth, higher gain in weight, and lower degree of hypertrophy of the pancreas. The report also suggests that there is an optimum microwave treatment time beyond which the beneficial effect disappears. Thus, for 1 kg treated in a 650-W oven at 2.45 GHz, the optimum time is between 9 and 12 minutes. Under the same conditions, the digestability of soyabean protein ingested by mice is 73, 84, 87, and 81% for microwave treatment time of 0, 9, 12, and 15 minutes, which tends to confirm that there is an optimum treatment time [134].

Asparagus officinalis boiled and cooked by microwaves [137] showed no difference in digestability by mice. Finally, another study [286] reported higher digestibility of protein mixtures cooked by extrusion to the detriment of microwaves. This beneficial effect of extrusion cooking may be due to the breakup of cellular structure by the heat and pressure of the process.

Finally, a clinical study performed at the Kiev Institute of Commerce and Industry [167] examined a total of 148 patients aged 18 to 66 years, 86 of whom were on a diet of microwave-cooked meals. No evidence was found of adverse effects on gastric pH, blood flow, and a number of other characteristics of the gastric mucus membrane. Histological examination of a sample obtained by biopsy presented no functional or morphological modifications. The conclusion of this study was that there were no obstacles to the use of microwaves for hospital catering.

References

1. AKYEL C., BOSISIO R., CHAHINE R., BOSE T. Measurement of the complex permittivity of food products during microwave-power heating cycles. *J. Microw. Power,* 18, n° 4, 1983, pp. 355-365.

2. ALBRECHT R., BALDWIN R. Sensory and nutritive attributes of roast pork reheated by one conventional and two microwave methods. *J. Microw. Power,* 17, n° 1, 1982, pp. 57-61.

3. ALEIXO J., SWAMINATHAN B., JAMESEN K., PRATT D. Destruction of pathogenic bacteria in turkeys roasted in microwave ovens. *J. Food Sci.,* 50, n° 4, 1985, pp. 873-880. (C.D.I.U.P.A., n° 203105).

4. ANDRE M. Emploi du chauffage par haute et hyper fréquences dans les industries de cuisson. In : ANDRE M. La cuisson. Paris, Centre Technique de l'Union, 1967, pp. 31-44. (C.D.I.U.P.A., 23002).

5. ANG C., CHANG C., FREY A., LIVINGSTON G. Effects of heating methods on vitamin retention in six fresh or frozen prepared food products. *J. Food Sci.,* 40, n° 5, 1975, pp. 997-1003. (C.D.I.U.P.A., n° 92694).

6. ANG C., LIVINGSTON G. Nutrient implications of microwave cooking and heating of frozen foods. *Trans. IMPI,* 5, Frozen foods and microwaves make cents, 1975.

7. ARMBRUSTER G. Ascorbic acid content of fruits and vegetables after microwave and conventional methods of cooking. In : Proceedings of the IMPI Symposium, Ottawa, 1978, pp. 103-104. (Paper 10.5.). (C.D.I.U.P.A., n° 143425).

8. ARMBRUSTER C., ECROYD L. The use of pulsed heating periods in 915 MHz and 2450 MHz cooking of meat. In : Proceedings of the IMPI Symposium, Loughborough, 1973. (Paper 3A.4/1-2).

9. ARMBRUSTER G., HAEFELE C. Quality of foods after cooking in 915 MHz and 2450 MHz microwave appliances using plastic film covers. *J. Food Sci.,* 40, n° 4, 1975, pp. 721-723. (C.D.I.U.P.A., n° 85600).

10. ARMSTRONG D., STANLEY D. Utilization of microwave heating in production of soymilk protein. *Cereal Sci. today,* 19, n° 9, 1974, p. 415.

11. BAKANOWSKI S., ZOLLER J. Endpoint temperature distributions in microwave and conventionally cooked pork. *Food Technol.,* 38, n° 2, 1984, pp. 45-51. (C.D.I.U.P.A., n° 186438).

12. BAKER R., DARFLER J., REHKUGLER G. Electrical energy used and time consumed when cooking foods by various home methods : chickens. *Poult. Sci.,* 60, n° 9, 1981, pp. 2062-2070. (C.D.I.U.P.A., n° 161224).

13. BAKER R., POON W., VADEHRA D. Destruction of Salmonella typhimurium and Staphylococcus aureus in poultry products cooked in a conventional and microwave oven. *Poult. Sci.,* 62, n° 5, 1983, pp. 805-810. (C.D.I.U.P.A., n° 83-4541).

14. BALDWIN R. Microwave cookery for meats. In : Proceedings of the 30th Annual Recipr. Meat Conf. of the Am. Meat Soc., Assoc. Nat. Livestock and Meat Board, Chicago, 1977, pp. 131-136.

15. BALDWIN R. Microwave cooking : an overview. *J. Food Prot.,* 46, n° 3, 1983, pp. 266-269. (C.D.I.U.P.A., n° 83-3219).

16. BALDWIN R., BRANDON M. Browning of meats cooked by microwaves. *Microw. Energy Appl. Newsl.,* 6, n° 5, 1973, pp. 3-5. (C.D.I.U.P.A., n° 65701).

17. BALDWIN R., CLONINGER M., FIELDS M. Growth and destruction of Salmonella typhimurium in egg white foam products cooked by microwaves. *Appl. Microbiol.,* 16, n° 12, 1968, pp. 1929-1934. (C.D.I.U.P.A., n° 21976).

18. BALDWIN R., FIELDS M., POON W., KORSCHGEN B. Destruction of Salmonellae by microwave heating of fish with implications for fish products. *J. Milk Food Technol.,* 34, n° 10, 1971, pp. 467-470. (C.D.I.U.P.A., n° 40376).

19. BALDWIN R., KORSCHGEN B., RUSSELL M., MABESA L. Proximate analysis free amino-acid, vitamin and mineral content of microwave cooked meat. *J. Food Sci.*, 41, n° 4, 1976, pp. 762-765. (C.D.I.U.P.A., n° 97936).

20. BALDWIN R., RUSSELL M. Microwave cooking of prebrowned steaks. *Microw. Energy Appl. Newsl.*, 4, n° 3, 1971, pp. 3-5. (C.D.I.U.P.A., n° 40538).

21. BALDWIN R., SNIDER S. Factors influencing spatter of carp during microwave cooking. *Microw. Energy Appl. Newsl.*, 4, n° 4, 1971, pp. 9-11. (C.D.I.U.P.A., n° 40539).

22. BALDWIN R., TETTAMBEL J. Nitrogen content of rib-eye steaks heated by microwaves and a conventional method. *Microw. Energy Appl. Newsl.*, 7, n° 3, 1974, pp. 3-4. (C.D.I.U.P.A., n° 74388).

23. BALDWIN R., UPCHURCH R., COTTERILL O. Ingredient effects on meringues cooked by microwaves and by baking. *Food Technol.*, 22, n° 12, 1968, pp. 1573-1576.

24. BARKER H. Microwaves for cooking and heating. *Electron. Power*, May 1981, pp. 401-402.

25. BENGTSSON N., JAKOBSSON B. Cooking of meat patties for freezing. A comparison of conventional methods and their combination with microwave heating. *Microw. Energy Appl. Newsl.*, 7, n° 6, 1974, pp. 3-10. (C.D.I.U.P.A., n° 79362).

26. BENGTSSON N., OHLSSON T. Microwave cooking and ready meals. In : Comptes rendus du 6ème Symposium européen sur les aliments. London, Soc. Chem. Ind., 1975, pp. 9-10 (angl.) et pp. 11-12 (alle.).

27. BENGTSSON N., RISMAN P. Dielectric properties of foods at 3 GHz as determined by a cavity perturbation technique. II. Measurements on food materials. *J. Microw. Power*, 6, n° 2, 1971, pp. 107-123.

28. BERG P., MARCHELLO M., ERICKSON D., SLANGER W. Selected nutrient content of beef Longissimus muscle relative to marbling class, fat status, and cooking method. *J. Food Sci.*, 50, n° 4, 1985, pp. 1029-1033. (C.D.I.U.P.A., n° 203116)

29. BERRY J., STADELMAN W., PRATT D., SWEAT V. Estimating cooking times and meat yields from roasted turkeys. *J. Food Sci.*, 45, n° 3, 1980, pp. 629-631. (C.D.I.U.P.A., n° 144049).

30. BIRD L. Rendering of material such as meat. U.S. Patent n° 4168 418, 1979.

31. BLANCO J., DAWSON L. Survival of Clostridium perfringens on chicken cooked with microwave energy. *Poult. Sci.*, 53, n° 5, 1974, pp. 1823-1830. (C.D.I.U.P.A., n° 80559).

32. BOBENG B., DAVID B. HACCP models for quality control of entree production in foodservice systems. *J. Food Prot.*, 40, n° 9, 1977, pp. 632-638. (C.D.I.U.P.A., n° 113517).

33. BOBENG B., DAVID B. HACCP models for quality control of entree production in hospital foodservice systems. I. Development of hazard analysis critical control points models. *J. Am. Diet. Assoc.*, 73, n° 5, 1978, pp. 524-529. (C.D.I.U.P.A., n° 125838).

34. BOBENG B., DAVID B. HACCP models for quality control of entree production in hospital foodservice systems. II. Quality assessment of beef loaves utilizing HACCP models. *J. Am. Diet. Assoc.*, 73, n° 5, 1978, pp. 530-535. (C.D.I.U.P.A., n° 125839).

35. BORCHERS R., MANAGE L., NELSON S., STETSON L. Rapid improvement in nutritional quality of soybeans by dielectric heating. *J. Food Sci.*, 37, n° 2, 1972, pp. 333-334. (C.D.I.U.P.A., n° 43619).

36. BOWERS J., FRYER B., ENGLER P. A research note : vitamin B6 in pork muscle cooked in microwave and conventional ovens. *J. Food Sci.*, 39, n° 2, 1974, pp. 426-427. (C.D.I.U.P.A., n° 70724).

37. BOWERS J., HEIER M. Microwave cooked turkey : heating patterns, eating quality, and histological appearance. *Microw. Energy Appl. Newsl.*, 3, n° 6, 1970, pp. 3-6. (C.D.I.U.P.A., n° 30592).

38. BOWMAN F., BERG E., CHUANG A., GUNTHER M., TRUMP D., LORENZ K. Vegetables cooked by microwave vs. conventional methods. Retention of reduced ascorbic acid and chlorophyll. *Microw. Energy Appl. Newsl.*, 8, n° 3, 1975, pp. 3-8. (C.D.I.U.P.A., n° 85792).

39. BOWMAN F., PAGE E., REMMENGA E., TRUMP D. Microwave vs. conventional cooking of vegetables at high altitude. *J. Am. Diet. Assoc.*, 58, n° 5, 1971, pp. 427-433. (C.D.I.U.P.A., n° 35493).

40. BROCKHUIZEN S., SCHILENBURG A. Tot dusverre Verkregen Resultaten met het hoogfrequent bakken van brood. Bakkerij Wetenschap, 1950.

41. BUCHANAN T. The dielectric properties of some long-chain fatty acids and their methylesters in the microwave region. *J. Chem. Phys.*, 22, n° 4, 1954, p. 578.

42. BUNCH W., MATTHEWS M., MARTH E. Fate of Staphylococcus aureus in beef-soy loaves subjected to procedures used in hospital chill foodservice systems. *J. Food Sci.*, 42, n° 2, 1977, pp. 565-566. (C.D.I.U.P.A., n° 106346).

43. BURDICK D., COX N., THOMSON J., BAILEY J. Heating by microwave, hot air, and flowing steam to eliminate inoculated Salmonella from poultry feed. *Poult. Sci.*, 62, n° 9, 1983, pp. 1780-1785. (C.D.I.U.P.A., n° 83-8044).

44. CADEDDU S. L'utilizzazione delle tecniche a microonda nella produzione di formaggio mozzarella. *Latte*, 8, n° 3, 1983, pp. 181-183. (C.D.I.U.P.A., n° 83-7409).

45. CAMPBELL C., TUNG Y., PROCTOR B. Microwave vs. conventional cooking. I. Reduced and total ascorbic acid in vegetables. *J. Am. Diet. Assoc.*, 34, 1958, pp. 365-370.

46. CARLIN A., MOTT C., CASH D., ZIMMERMANN W. Destruction of trichina larvae in cooked pork roasts. *J. Food Sci.*, 34, n° 2, 1969, pp. 210-212. (C.D.I.U.P.A., n° 14253).

47. CARLIN F., ZIMMERMANN W., SUNDBERG A. Destruction of trichina larvae in beef-pork loaves cooked in microwave ovens. *J. Food Sci.*, 47, n° 4, 1982, pp. 1096-1118. (C.D.I.U.P.A., n° 171976).

48. CARPENTER Z., ABRAHAM H., KING G. Tenderness and cooking loss of beef and pork. I. Relative effects of microwave cooking, deep-fat frying, and oven broiling. *J. Am. Diet. Assoc.*, 53, n° 4, 1968, pp. 353-356. (C.D.I.U.P.A., n° 12219).

49. CHAMBERLAIN N. Microwave energy in the baking of bread. *Br. Bak.*, 20 July, 1973, pp. 14-16.

50. CHAMBERLAIN N. Microwave energy in the baking of bread. *Food Trade Rev.*, 43, n° 9, 1975, pp. 8-12. (C.D.I.U.P.A., n° 61828).

51. CHEN X., BAU H., GIANNANGELI F. DEBRY G. Evaluation de l'influence de la cuisson par les micro-ondes sur les propriétés physico-chimiques et nutritionnelles de la farine entière de soja. *Sci. Aliments*, 6, n° 2, 1986, pp. 257-272. (C.D.I.U.P.A., n° 213416).

52. CHEN T., CULLOTTA J., WANG W. Effects of water and microwave energy pre-cooking on microbiological quality of chicken parts. *J. Food Sci.*, 38, n° 1, 1973, pp. 155-157. (C.D.I.U.P.A., n° 56395)

53. CHEN H., WU C., LIN H., CHUNG C. Effect of microwave and open flame heating on microbial counts and weight-loss of eel. *Bull. Jpn. Soc. Sci. Fish.*, 42, n° 4, 1976, pp. 405-410. (C.D.I.U.P.A., n° 97642).

54. CHENG H., BALDWIN R. Quality of pork cooked by stir-fry and two microwave procedures. In : Proceedings of the IMPI Symposium, Chicago, 1985, PP. 3-5.

55. CHENG H., BALDWIN R. Quality of pork cooked by stir-fry and two microwave procedures. *J. Microw. Power*, 20, n° 4, 1985, pp. 261-265.

56. CIPRA J., BOWERS J. Flavor of micro-wave and conventionally reheated turkey. *Poult. Sci.*, 50, n° 3, 1971, pp. 703-706. (C.D.I.U.P.A., n° 037720).

57. CIPRA J., BOWERS J., HOOPER A. Pre-cooking and reheating of turkey. Effects of microwave vs. conventional methods. *J. Am. Diet. Assoc.*, 58, n° 1, 1971, pp. 38-40. (C.D.I.U.P.A., n° 35486).

58. CORNFORTH D., BRENNAND C., BROWN R., GODFREY D. Evaluation of various methods for roasting frozen turkeys. *J. Food Sci.*, 47, n° 4, 1982, pp. 1108-1112. (C.D.I.U.P.A., n° 171978).

59. COTTE J. Communication personnelle, 27 août 1986.

60. CRAVEN S., LILLARD H. Effect of microwave heating and precooked chicken on Clostridium perfringens. *J. Food Sci.*, 39, n° 1, 1974, pp. 211-212. (C.D.I.U.P.A., n° 68900).

61. CREMER M. Food quality and energy use - microwave vs. convection ovens. *Ohio Rep.*, 68, n° 2, 1983, pp. 21-23. (C.D.I.U.P.A., n° 83-5737).

62. CREMER M. Sensory quality of spaghetti with meat sauce after varying hol-ding treatments and heating in insti-tutional microwave oven and convec-tion oven. *J. Food Sci.*, 48, n° 6, 1983, pp. 1579-1582. (C.D.I.U.P.A., n° 183146).

63. CREMER M., CHIPLEY J. Time and tempe-rature, microbiological, and sensory assessment of roast beef in a hospital foodservice system. *J. Food Sci.*, 45, n° 6, 1980, pp. 1472-1477. (C.D.I.U.P.A., n° 150420).

64. CREWS G., GOERTZ G. Moisture and microwave effects on selected characteristics of turkey pectoral muscles. *Poult. Sci.*, 52, n° 4, 1973, pp. 1496-1500. (C.D.I.U.P.A., n° 66981).

65. CROSS G., FUNG D. The effect of microwaves on nutrient value of foods. *CRC Crit. Rev. Food Sci. Nutr.*, 16, n° 4, 1982, pp. 355-381. (C.D.I.U.P.A., n° 168682).

66. CULKIN K., FUNG D. Destruction of Escherichia coli and Salmonella typhimurium in microwave-cooked soups. *J. Milk Food Technol.*, 38, n° 1, 1975, pp. 8-15. (C.D.I.U.P.A., n° 81399).

67. CULLARS-SAMUEL B., LOVINGOOD R. Microwave/convection versus electric range ovens : tradeoffs in energy use, time, and food quality. *J. Microw. Power*, 21, n° 1, 1986, pp. 1-8. (C.D.I.U.P.A., n° 221936).

68. CULLOTTA J., CHEN T. Hot water and microwave energy for precooking chicken parts : effects on yield and organoleptic quality. *J. Food Sci.*, 38, n° 5, 1973, pp. 860-869. (C.D.I.U.P.A., n° 63258).

69. CUNNINGHAM F. The effect of brief microwave treatment on numbers of bacteria in fresh chicken patties. *Poult. Sci.*, 57, n° 1, 1978, pp. 296-297. (C.D.I.U.P.A., n° 121994).

70. CUNNINGHAM F. Influence of microwave radiation on psychotrophic bacteria. *J. Food Prot.*, 43, n° 8, 1980, pp. 651-655. (C.D.I.U.P.A., n° 148135).

71. CUNNINGHAM F., LOHMEYER P. Use of microwaves for cooking or heating processed poultry items. *Microw. Energy Appl. Newsl.*, 6, n° 1, 1973, pp. 3-5. (C.D.I.U.P.A., n° 58837).

72. DAHL C., MATTHEWS M., MARTH E. Cook/chill foodservice system with a microwave oven : aerobic plate counts from beef loaf, potatoes and frozen green beans. *J. Microw. Power*, 15, n° 2, 1980, pp. 95-105.

73. DAHL C., MATTHEWS M., MARTH E. Fate of Staphylococcus aureus in beef loaf, potatoes and frozen and canned green beans after microwave-heating in simulated cook/chill hospital foodservice system. *J. Food Prot.*, 42, n° 12, 1980, pp. 916-932. (C.D.I.U.P.A., n° 152577).

74. DAHL C., MATTHEWS M., MARTH E. Cook/chill foodservice system with a microwave oven : injured aerobic bacteria during food product flow. *Europ. J.*

75. DAHL C., MATTHEWS M., MARTH E. Survival of Streptococcus faecium in beef loaf and potatoes after microwave-heating in a simulated cook/chill foodservice system. *J. Food Prot.*, 44, n° 2, 1981, pp. 128-136. (C.D.I.U.P.A., n° 157869).

76. DAHL-SAWYER C., JEN J., HUANG D. Cook/chill foodservice systems with conduction, convection and microwave reheat subsystems. Nutrient retention in beef loaf, peas, and potatoes. *J. Food Sci.*, 47, n° 4, 1982, pp. 1089-1095. (C.D.I.U.P.A., n° 171974).

77. DANLEY A., STEINGRABER M. Microwave oven popcorn popper, steamer and roaster. U.S. Patent n° 4477 705, 16 Oct. 1984.

78. DAVIDOVICH L., PIGOTT G. The use of microwave power in the fabrication of "kamaboko" (fish cake). *J. Microw. Power*, 17, n° 4, 1982, pp. 335-340.

79. DAVIS D., PRATT D., REBER E., KLOCKOW R. Microwave cooking in meal management. *J. Home Econ.*, 63, n° 2, 1971, pp. 97-99.

80. DAWSON L., SKELLY J. Stability of microwave precooked chicken during frozen storage. *Poult. Sci.*, 51, n° 5, 1972, p. 1799.

81. DAWSON L., SISON E. Stability and acceptability of phosphate-treated and precooked chicken pieces reheated with microwave energy. *J. Food Sci.*, 38, n° 1, 1973, pp. 161-164. (C.D.I.U.P.A., n° 56397).

82. DECAREAU R. Utilization of microwave cookery in meat processing. In : Proceedings of the 20th Annual Reciproc. Meat Conference of the American Meat Science Association, Lincoln, Univ. Nebraska, 14-16 July 1967.

83. DECAREAU R. Utilization of microwave cookery in meat processing. *Microw. Energy Appl. Newsl.*, 1, n° 1, 1968, pp. 3-4, 10.

84. DECAREAU R. The browning reaction and the microwave oven. *Microw. Energy Appl. Newsl.*, 6, n° 3, 1973, pp. 3-5. (C.D.I.U.P.A., n° 62300).

85. DECAREAU R. Microwave meat roasting. *Microw. Energy Appl. Newsl.*, 7, n° 4, 1974, pp. 3-5. (C.D.I.U.P.A., n° 78348).

86. DECAREAU R. Nutritional aspects of food service. *Microw. Energy Appl. Newsl.*, 7, n° 5, 1974, pp. 3-5. (C.D.I.U.P.A., n° 79361).

87. DECAREAU R. Do microwaves make food more nutritious ? *Microw. Energy Appl. Newsl.*, 10, n° 2, 1977, pp. 13-19. (C.D.I.U.P.A., n° 107475).

88. DECAREAU R. Conveyor microwave ovens. *J. Foodserv. syst.*, 2, 1982, pp. 127-137.

89. DECAREAU R. Microwaves in food processing. *Food Technol. Aust.*, 36, n° 2, 1984, pp. 81-86. (C.D.I.U.P.A., n° 201725).

90. DECAREAU R. Microwave food processing equipment throughout the world. *Food Technol.*, 40, n° 6, 1986, pp. 99-105. (C.D.I.U.P.A., 214710).

91. DECAREAU R. Future research developments in microwave technology. Non publ., n.d., 12 p.

92. DECAREAU R. Microwave food-processing equipment throughout the world. Food Engineering Laboratory, U.S. Army Natick Res. and Dev. Center, Natick, Non publ., n.d., 15 p.

93. DECAREAU R., MUDGETT R. Microwaves in the food-processing industry. (Food science and technology, a series of monographs). New York, Academic Press Inc., 1985, 247 p.

94. DECAREAU R., PETERSON R. Potential applications of microwave energy to the baking industry. *J. Microw. Power*, 3, n° 3, 1968, pp. 152-157.

95. DECAREAU R., PETERSON R. Potential applications of microwave energy to the baking industry. In : Proceedings of the IMPI Symposium, Boston, 1969, Abstract p. 41. (Paper H3).

96. DEETHARD D., COSTELLO W., SCHNEIDER K. Effect of electronic, convection and conventional oven roasting on acceptability of pork loin roasts. *J. Food Sci.*, 38, n° 6, 1973, pp. 1076-1077. (C.D.I.U.P.A., n° 066654).

97. DEHNE L., BÖGL W. Einfluss der Mikrowellenerwärmung auf den Nährwert tierischer Lebensmittel, eine Übersicht. Teil II. *Fleischwirtsch.*, 63, n° 7, 1983, pp. 1206-1211. (C.D.I.U.P.A., n° 83-6442).

98. DEHNE L., BÖGL W., GROSSKLAUS D. Einfluss der Mikrowellenerwärmung auf den Nährwert tierischer Lebensmittel. Eine Übersicht. Teil I. *Fleischwirtsch.*, 63, n° 2, 1983, pp. 231-237. (C.D.I.U.P.A., n° 83-1834).

99. DELANY E., VAN ZANTE H., HARTMAN P. Microbial survival in electronically heated foods. *Microw. Energy Appl. Newsl.*, 1, n° 3, 1968, pp. 11-14.

100. DE LOOR G., MEIJBOOM F. The dielectric constant of foods and other materials with high water contents at microwave frequencies. *J. Food Technol.*, 1, 1966, pp. 313-322.

101. DESSEL M., BOWERSOX E., JETER W. Bacteria in electronically cooked foods. *J. Am. Diet. Assoc.*, 37, n°9, 1960, pp. 230-233.

102. DREUILLET J. Les micro-ondes en restauration professionnelle. *Surgél.*, n° 223, 1984, pp. 18-21. (C.D.I.U.P.A., n° 190662).

103. DREW F., RHEE K. Fuel consumption by cooking appliances. *J. Am. Diet. Assoc.*, 72, n°1, 1978, pp. 37-44.

104. DREW F., RHEE K. Microwave cookery of beef patties. *J. Am. Diet. Assoc.*, 74, n° 6, 1979, pp. 652-656. (C.D.I.U.P.A., n° 132805).

105. DREW F., RHEE K., CARPENTER Z. Effects on energy use and quality of top round beef roats. Cooking at variable microwave power levels. *J. Am. Diet. Assoc.*, 77, n° 4, 1980, pp. 455-459. (C.D.I.U.P.A., n°149388).

106. DUMONT B. La viande de boeuf : structure et tendreté. *Pour la Science*, juin 1986, pp. 88-96.

107. EDGAR R. The economics of microwave processing in the food industry. *Food Technol.*, 40, n° 6, 1986, pp. 106-112. (C.D.I.U.P.A., n° 214710).

108. EHEART M., GOTT C. Conventional and microwave cooking of vegetables : ascorbic acid and carotene retention and palatability. *Food Technol.*, 19, n° 5, 1965, pp. 185-188.

109. EHRCKE L., RASDALL J., GIBBS S., HAYDON F. Comparison of eight cooking treatments for energy utilization, product yields, and quality of roasted turkey. *Poult. Sci.*, 64, n° 10, 1985, pp. 1881-1886. (C.D.I.U.P.A., n° 206038).

110. EIK C. Microwave precooking of sardines in open metal cans. In : Comptes rendus du Symposium International IMPI-CFE sur les Applications énergétiques des micro-ondes, Monaco, 1979, Résumé pp. 77-78.

111. ELFIMOV V., ŽARINOV A., LAZAREV A., NALETOV N., ROGOV I., FOMIN A. Modifications de la structure histologique du tissu musculaire au cours de son traitement dans un champ de micro-ondes (en russe). *Mjasnaja Ind. SSSR*, 43, n° 7, 1972, pp. 32-34. (C.D.I.U.P.A., n° 59973).

112. ESSARY E. Influence of microwave heat on bone discoloration. *Poult. Sci.*, 38, 1959, pp. 525-529.

113. EVANS K. The processing of meat and baked products using microwaves. In : Proceedings of the Symposium on Microwave and RF heating in the food industry. London, Zoological Society, 1971.

114. FENTON F. Research on electronic cooking. *J. Home Econ.*, 49, 1957, pp. 709-716.

115. FERGUSSON J. Exciting products from microwave ovens. *Food Eng. Intern.*, 1, n° 9, 1976, pp. 24-26. (C.D.I.U.P.A., n° 098773).

116. FETTY H. Microwave baking of partially baked products. In : Proceedings of the 42nd Annual Meeting of the American Society of Bakery Engineering, 1966, pp. 144-152.

117. FOURNAUD J., LE POMMELEC G. Qualité bactériologique des plats cuisinés. *Rev. Gén. Froid*, 68, n° 11, 1977, pp. 703-707. (C.D.I.U.P.A., n° 113773).

118. FOURNAUD J. LE POMMELEC G. Qualité bactériologique des plats cuisinés. *Rev. Gén. Froid*, 68, n° 12, 1977, pp. 741-745. (C.D.I.U.P.A., n° 114994).

119. FRANCOIS D. Coagulation of fish products prior to canning. In : Proceedings of the IMPI Symposium, Leuven, 1976. (Paper 6B.6.).

120. FRUIN J., GUTHERTZ L. Survival of bacterias in food cooked by microwave oven, conventional oven and slow cookers. *J. Food Prot.*, 45, n° 8, 1982, pp. 695-705. (C.D.I.U.P.A., n° 170956).

121. FULTON L., DAVIS C. Roasting and braising beef roasts in microwave ovens. *J. Am. Diet. Assoc.*, 83, n° 5, 1983, pp. 560-563. (C.D.I.U.P.A., n° 183420).

122. FUNG D., CUNNINGHAM F. Effect of microwaves on microorganisms in foods. *J. Food Prot.*, 43, n° 8, 1980, pp. 641-650. (C.D.I.U.P.A., n° 148134).

123. GALL K., OTWELL W., KOBURGER J., APPLEDORF H. Effects of four cooking methods on the proximate, mineral and fatty acid composition of fish fillets. *J. Food Sci.*, 48, n° 4, 1983, pp. 1068-1074. (C.D.I.U.P.A., n° 83-7606).

124. GANDAR J. Über die mechanische und dielektrische Zerkleinerung von Getreide- und Ölsaaten. *Fette Seifen Anstrichm.*, 70, n° 6, 1968, pp. 433-439.

125. GILES P., MOORE E., BOUNDS L. Pork heating in a 200 MHz chamber ; investigation of heat penetration of food samples at various frequencies. *J. Microw. Power*, 5, n° 1, 1970, pp. 37-43.

126. GOEBEL N., GRIDER J., DAVIS E., GORDON J. The effects of microwave energy and convection heating on wheat starch granule transformations. *Food Microstruct.*, 3, 1984, pp. 73-82.

127. GOLDBLITH S. Dielectric properties of foods and their importance in processing with microwave energy. *Microw. Energy Appl. Newsl.*, 7, n°3, 1974, pp. 9-14.

128. GOLDBLITH S., DECAREAU R. An annotated bibliography on microwaves. Cambridge, M.I.T. Press, 1973, 356 p. (C.D.I.U.P.A., n° 062683).

129. GOLDBLITH S., TANNENBAUM S., WANG D. Thermal and 2450 MHz microwave energy effect on the destruction of thiamine. *Food Technol.*, 22, n° 10, 1968, pp. 64-66. (C.D.I.U.P.A., n° 5133).

130. GOLDBLITH S., WANG D. Dielectric properties of foods. U.S. Army Natick Development Center, Technical Report TR 76-27-FEL, 1975.

131. GOLDSMID A. Microwave-baked potatoes in large-scale food preparation. *J. Am. Diet. Assoc.*, 51, n° 6, 1967, p. 536.

132. GORAKHPURWALLA H. Continuous microwave grain cooker. U.S. Patent n° 3 626 838, Dec. 1971.

133. GRANT E., KEEFE S., TAKASHIMA S. The dielectric behaviour of aqueous solutions of bovine serum albumin from radio wave to microwave frequencies. *J. Phys. Chem.*, 72, n° 13, 1968, pp. 4373-4380.

134. HAFEZ Y., MOHAMED A., HEWEDY F., SINGH G. Effects of microwave heating on solubility, digestibility and metabolism of soy protein. *J. Food Sci.*, **50**, n° 2, 1985, pp. 415-423. (C.D.I.U.P.A., n° 199523).

135. HAFEZ Y., SINGH G., Mac LELLAN M., MONROE-LORD L. Effects of microwave heating on nutritional quality of soybeans. *Nutr. Rep. Int.*, **28**, n° 2, August 1983, pp. 413-421. (C.D.I.U.P.A., n° 186685).

136. HAMID M. Microwave roasting apparatus. Brevet Canadien n° 1061867, 4 Sept. 1979.

137. HARBERS C., HARBERS L. Asparagus tissue changes by cooking methods and digestion in rats as observed by scanning electron microscopy. *Nutr. Rep. Int.*, **27**, n° 1, Jan. 1983, pp. 97-101. (C.D.I.U.P.A., n° 83-2784).

138. HASSOUN Y. Oven energy use compared. *Microw. World*, **3**, n° 3, 1982, pp. 13-16. (C.D.I.U.P.A., n° 170152).

139. HASTED J. Aqueous dielectrics. London, Chapman & Hall, 1973, 302 p.

140. HEINMULLER C. Communication personnelle, 20 fév. 1986.

141. HIPPEL (A. von) Dielectric materials and applications. 1st ed., Cambridge (USA), M.I.T, Press, 1966, 438 p.

142. HLEBNIKOV V., ABALDOVA V., BLINOVA T. Vlijanije tenpa nagreva mjasnyh farševy izdelij SVČ - energiej na dostiženije ih gotovnosti. *Vopr. Pitan.*, n° 1, 1976, pp. 63-65. (C.D.I.U.P.A., n° 95059).

143. HOFFMAN C., ZABIK M. Effects of microwave cooking/reheating on nutrients and food systems : a review of recent studies. *J. Am. Diet. Assoc.*, **85**, n° 8, 1985, pp. 922-926. (C.D.I.U.P.A., n° 204377).

144. HOFFMAN C., ZABIK M. Current and future foodservice applications of microwave cooking/reheating. *J. Am. Diet. Assoc.*, **85**, n° 8, 1985, pp. 929-933. C.D.I.U.P.A., n° 204378).

145. HOSTETLER R., DUTSON T. Investigations of a rapid method for meat tenderness evaluation using microwave cookery. *J. Food Sci.*, **43**, n° 2, 1978, pp. 304-306. (C.D.I.U.P.A., n° 117596).

146. HSIEH J., BALDWIN R. Storage and microwave reheating effects on lipid oxidation of roast beef. *J. Microw. Power*, **19**, n° 3, 1984, pp. 187-194. (C.D.I.U.P.A., 203147).

147. HUXSOLL C., MORGAN A. Microwaves for quick-cooking rice. *Cereal Sci. Today*, **13**, n° 5, 1968, pp. 203-206. (C.D.I.U.P.A., n° 3286).

148. ISHITOBI Y., TOGAWA M. Browning of foodstuffs by heating at 10 GHz. In : Proceedings of the IMPI Symposium, Ottawa, 1978, pp. 8-10. (Paper 1.3.)

149. JEPPSON M. Process for cooking bacon with microwave energy. U.S. Patent n° 3 321 314, May 1967, 6 p.

150. JOHNSTON M., BALDWIN R. Influence of microwave reheating on selected quality factors of roast beef. *J. Food Sci.*, **45**, n° 6, 1980, pp. 1460-1462. (C.D.I.U.P.A., n° 150417).

151. KAHN L., LIVINGSTON G. Effect of heating method on the thiamine retention in fresh or frozen prepared foods. *J. Food Sci.*, **35**, n° 4, 1970, pp. 349-351.(C.D.I.U.P.A., n° 28546).

152. KAMATH K. Variation of dielectric properties of some vegetable oils in liquid-solid transition phase. *Indian J. Technol.*, **9**, n° 8, 1971, p. 312.

153. KENT M. Complex permittivity of white fish meal in the microwave region as a function of temperature and moisture content. *J. Phys. D. : Appl. Phys.*, **3**, 1970, pp. 1275-1283.

154. KENT M. Microwave dielectric properties of fish meal. *J. Microw. Power*, **7**, n° 2, 1972, pp. 109-116.

155. KENT M. Complex permittivity of fish meal : a general discussion on temperature, density, and moisture dependence. *J. Microw. Power*, **12**, n° 4, 1977, pp. 341-345.

156. KIEREBINSKI C. Temperature changes in meat heated by radiowaves (2450 MHz) and in hot water. *Roczn. Technol. Chem. Żywn.*, **15**, 1969, pp. 171-187 et *J. Sci. Food Agric.*, Abstracts, **21**, n° 2, 1970, i-78.

157. KIRK D., HOLMES A. The heating of foodstuffs in a microwave oven. *J. Food Technol.*, **10**, n° 4, 1975, pp. 375-384. (C.D.I.U.P.A., n° 86544).

158. KLEIN L. Potatoes in relation to magnesium fertilization, sprouting, microwave baking and product development. Thesis doct. Ph. D., Cornell Univ., 1982, 174 p. (C.D.I.U.P.A., n° 83-7886).

159. KORSCHGEN B., BALDWIN R. Moist heat microwave and conventional cooking of round roasts of beef. *J. Microw. Power*, 13, n° 3, 1978, pp. 257-262.

160. KORSCHGEN B., BALDWIN R. Comparison of methods of cooking beef roasts by microwaves. *J. Microw. Power*, 15, n° 3, 1980, pp. 169-172.

161. KORSCHGEN B., BALDWIN R., SNIDER S. Quality factors in beef, pork and lamb cooked by microwaves. *J. Am. Diet. Assoc.*, 70, n° 6, 1976, pp. 635-640.

162. KOSTIĆ S., TOSOVIĆ T. Uticaj mikro talasnog zagrevanja na termicki kalo i senzorna svojstva nekih proizvoda. *Hrana Ishrana*, 26, n° 3-4, 1985, pp. 101-102. (C.D.I.U.P.A. n° 208057).

163. KOTULA A., MURRELL K., ACOSTA-STEIN L., LAMB L., DOUGLASS L. Destruction of Trichinella spiralis during cooking. *J. Food Sci.*, 48, n° 3, 1983, pp. 765-768. (C.D.I.U.P.A., n° 83-7431).

164. KOTULA A., MURRELL K., ACOSTA-STEIN L., TENNENT I. Influence of rapid cooking methods on the survival of Trichinella spiralis in pork chops from experimentally infected pigs. *J. Food Sci.*, 47, n° 3, 1982, pp. 1006-1007. (C.D.I.U.P.A., n° 171084).

165. KRIEMS P., MOLLER B. Ist das Backen mit Hochfrequenz noch aktuell ? *Back. Konditor*, 20, n° 11, 1972, pp. 330-333. (C.D.I.U.P.A., n°50370).

166. KUZNECOVA Z., BOL'ŠAKOV A., ROGOV I. Variations dans la composition des lipides des volailles pendant le traitement dans un champ électro-magnétique UHF (en russe). *Mjasn. Ind. SSSR*, 42, n° 3, 1971, pp. 36-37.

167. KVITNICKIJ M., ŠLYKOV I., JULIN A., TARASENKO I. Kliniko-instrumental' nyje issledovanija piščevaritel' nogo apparata pri ispol'zovanii V lečebnom pitanii kulinarnyh izdelij, prigotovlennyh v SVC-pečat. *Vopr. Pitan.*, n° 1, 1984, pp. 18-21. (C.D.I.U.P.A., n° 186919).

168. KYLEN A., CHARLES V., MacGRATH B., SCHLETER J., WEST L., VAN DUYNE F. Microwave cooking of vegetables : ascorbic acid retention and palatability. *J. Am. Diet. Assoc.*, 39, 1961, pp. 321-326.

169. KYLEN A., Mac GRATH B., HALLMARK E., VAN DUYNE F. Microwave and conventional cooking of meat. *J. Am. Diet. Assoc.*, 45, n° 2, 1964, pp. 139-145.

170. LACEY B., WINNER H., Mac LELLAN M., BAGSHAWE K. Effects of microwave cookery on the bacterial counts of food. *J. Appl. Bacteriol.*, 28, n° 2, 1965, pp. 331-335.

171. LATRONICA A., ZIEMBA J. Microwaves precooked bacon. *Food Eng.*, 44, n° 4, 1972, pp. 62-64. (C.D.I.U.P.A., n° 44371).

172. LEON-CRESPO F., OCKERMAN H. Thermal destruction of microorganisms in meat by microwave and conventional cooking. *J. Food Prot.*, 40, 1977, pp. 442-444.

173. LEON-CRESPO F., OCKERMAN H., IRVIN K. Effect of conventional and microwave heating on Pseudomonas putrefaciens, Streptococcus faecalis and Lacto-bacillus plantarum in meat tissues. *J. Food Prot.*, 40, n° 8, 1977, pp. 588-591. (C.D.I.U.P.A., n° 113507).

174. LESZCZYNSKI D. Lipid oxidation in meat as related to cooking and health. *Microw. World*, 7, n° 2, March-April 1986, pp. 9-14.

175. LESZCZYNSKI D., MOORE H. Lipid oxidation products in chicken meat cooked in microwave and convection ovens. In : Proceedings of the IMPI Symposium, Chicago, 1985, pp. 6-10.

176. LIN W., SAWYER C. Microwave cooking : survival of Staphylococcus aureus, Escherichia coli, and aerobic bacteria in beef loaves. In : Proceedings of the IMPI Symposium, Chicago, 1985, pp. 11-19.

177. LINDSAY R., KRISSINGER W., FIELDS B. Microwave vs. conventional oven cooking of chicken : relationship of internal temperature to surface contamination by Salmonella typhi-murium. *J. Am. Diet. Assoc.*, 86, n° 3, 1986, pp. 373-374. (C.D.I.U.P.A., n° 210084).

178. LIVINGSTON G. Nutrition aspects of microwave cooking. *Microw. Energy Appl. Newsl.*, 12, n° 5, 1979, pp. 4-9. (C.D.I.U.P.A., n° 139096).

179. LOPEZ-GAVITO L., PIGOTT G. Effects of microwave cooking on textural characteristics of battered and breaded fish products. *J. Microw. Power*, 18, n° 4, 1983, pp. 345-353. (C.D.I.U.P.A., n° 191435).

180. LORENZ K. Microwave cooking at high altitudes. *Microw. Energy Appl. Newsl.*, 9, n° 1, 1976, pp. 3-6. (C.D.I.U.P.A., n° 93958).

181. LORENZ K., DECAREAU R. Microwave heating of foods. Changes in nutrient and chemical composition. *Critical Rev. Food Sci. Nutr.*, 7, n° 4, 1976, pp. 339-370. (C.D.I.U.P.A., n° 95407).

182. LORENZ K., CHARMAN E., DILSAVER W. Baking with microwave energy. *Food Technol.*, 27, n° 12, 1973, pp. 28-36. (C.D.I.U.P.A., n° 66460).

183. LORENZ K., DECAREAU R. Microwave heating of foods. Changes in nutrient and chemical composition. *Critical Rev. Food Sci. Nutr.*, 7, n° 4, 1976, pp. 339-370. (C.D.I.U.P.A., n° 95407).

184. LORENZ K., DILSAVER W., STARKJOHANN R. Proofing and baking with microwave energy. *Microw. Energy Appl. Newsl.*, 8, n° 5, 1975, pp. 3-10. (C.D.I.U.P.A., n° 89936).

185. LOVINGOOD R. Energy use of microwave ovens versus conventional cooking methods. *Microw. World*, 4, n° 2, 1983, pp. 10-12. (C.D.I.U.P.A., n° 184002).

186. LUTER L. WYSLOUZIL W., KASHYAP S. The destruction of aflatoxins in peanuts by microwave roasting. *Can. Inst. Food Sci. Technol. J.*, 15, n° 3, 1982, pp. 236-238.

187. MABESA L., BALDWIN R. Flavor and color of peas and carrots cooked by microwaves. *J. Microw. Power*, 13, n° 4, 1978, pp. 321-326.

188. MABESA L., BALDWIN R. Ascorbic acid in peas cooked by microwaves. *J. Food Sci.*, 44, n° 3, 1979, pp. 932-933. (C.D.I.U.P.A., n° 131949).

189. Mac LENNAN H. Ready foods. In : Proceedings of the IMPI Symposium, Boston, 1969, abstract pp. 39-40. (Paper H2).

190. Mac LEOD G. Microwave heating of food and its effect on flavour. Part 1. *Food Process. Ind.*, 41, n° 485, 1972, pp. 27-28. (C.D.I.U.P.A., n° 44386).

191. Mac LEOD G. Microwave heating of food and its effect on flavour. Part 2. *Food Process. Ind.*, 41, n° 486, 1972, pp. 51-53. (C.D.I.U.P.A., n° 44391).

192. Mac LEOD A., Mac LEOD G. Effects of variation of cooking methods on the flavor volatiles of cabbage. *J. Food Sci.*, 35, n° 6, 1970, pp. 744-750. (C.D.I.U.P.A., n° 30399).

193. Mac NEIL M., PENFIELD M. Turkey quality as affected by ovens of varying energy costs. *J. Food Sci.*, 48, n°3, 1983, pp. 853-855. (C.D.I.U.P.A., n° 83-7440).

194. MADDUX R., OLSEN C. The application of microwaves for the processing of poultry. In : Proceedings of the IMPI Symposium, Stanford, 1967, abstract p. 17.

195. MADEIRA K., PENFIELD M. Turbot fillet sections cooked by microwave and conventional heating methods : objective and sensory evaluation. *J. Food Sci.*, 50, n° 1, 1985, pp. 172-177. (C.D.I.U.P.A., n° 197385).

196. MADSON R., CORDARO J., KOLLER R., VOELKER G. Effect of microwaves on bacteria in frozen foods. US Government Reports - Topical Announcements, US Dep. Commerce Publication, 25 May 1971.

197. MAGA J., TWOMEY J. Sensory comparison of four potato varieties baked conventionally and by microwaves. *J. Food Sci.*, 42, n° 2, 1977, pp. 541-542. (C.D.I.U.P.A., n° 106334)

198. MANGIN M., LEFEUVRE S. Emploi des micro-ondes dans les fours de boulangerie. In : Colloque International des Journées d'Electronique de Toulouse, 7-11 mars 1977. Communication IV-5.

199. MATTHEWS E. Microwave ovens : effect on food quality and safety. *J. Am. Diet. Assoc.*, 85, n° 8, 1985, pp. 919-921. (C.D.I.U.P.A., n° 204376).

200. MAY K. Applications of microwave energy in preparation of poultry convenience foods. *J. Microw. Power*, 4, n° 2, 1969, pp. 54-59.

201. MEISEL N. Apport d'énergie instantanée au coeur des produits alimentaires, les micro-ondes. *Ind. Aliment. Agric.*, 91, n° 9-10, 1974, pp. 1203-1214. (C.D.I.U.P.A., n° 076032).

202. MEISEL N. Les hyperfréquences au service des techniques alimentaires. *Ind. Aliment. Agric.*, 78, n° 95, 1978, pp. 997-1003. (C.D.I.U.P.A., n° 124554).

203. MEISEL N. Tunnel à micro-ondes de précuisson de sardines en boîtes métalliques ouvertes, type "Gigamet". Publication de la Société IMI, 1979.

204. MEISEL N. Les micro-ondes dans les industries agro-alimentaires. *Ind. Aliment. Agric.*, 101, n° 10, 1984, pp. 929-932. (C.D.I.U.P.A., n° 193861).

205. MEREDITH R. Recent advances in industrial microwave processing in the 896/915 MHz frequency band. Document Magnetronics Ltd., s.d., 12 p.

206. MILLER A., BUCHANAN R. Reduction of mutagen formation in cooked nitrite-free bacon by selected cooking treatments. *J. Food Sci.*, 48, n° 6, 1983, pp. 1772-1775. (C.D.I.U.P.A., n° 183409).

207. MOHSENIN N. Electromagnetic radiation propertie of foods and agricultural products. New York, Gordon and Breach Science Publishers, 1983, 672 p. (C.D.I.U.P.A., n° 205642).

208. MONTGOMERY T. Microwave and conventional cookery of hot and cold processed pork loins. Thesis Doct. Ph. D., Texas Technical Univ., 1975, 135 p. (Dissertation Abstracts International n° 76-7377). (C.D.I.U.P.A., n° 95244).

209. MONTGOMERY T., RAMSEY C., LEE R. Microwave and conventional precooking of hot and cold processed pork loins. *J. Food Sci.*, 42, n° 2, 1977, pp. 310-315. (C.D.I.U.P.A., n° 106280).

210. MUDGETT R. Electrical properties of foods in microwave processing. *Food Technol.*, 36, 1982, pp. 109-115.

211. MUDGETT R. Microwave properties and heating characteristics of foods. *Food Technol.*, 40, n° 6, 1986, pp. 84-98. (C.D.I.U.P.A., n° 214710).

212. MUDGETT R., GOLDBLITH S., WANG D., WESTPHAL W. Prediction of dielectric properties in solid foods at ultrahigh and microwave frequencies. In : Proceedings of the IMPI Symposium, Waterloo (Can.), 1975, pp. 272-274. (Paper 16.4.).

213. MUDGETT R., GOLDBLITH S., WANG D., WESTPHAL W. Prediction of dielectric properties in solid foods of high moisture content at ultrahigh and microwave frequencies. *J. Food Process. Preserv.*, 1, n° 2, 1977, pp. 119-151. (C.D.I.U.P.A., n° 115601).

214. MUDGETT R., GOLDBLITH S., WANG D., WESTPHAL W. Dielectric behavior of a semi-solid food at low, intermediate and high moisture contents. *J. Microw. Power*, 15, n° 1, 1980, pp. 27-36.

215. MYERS S., HARRIS N. Effect of electronic cooking of fatty acids in meats. *J. Am. Diet. Assoc.*, 67, n° 3, 1975, pp. 232-234. (C.D.I.U.P.A., n° 87654).

216. NELSON S. Electrical properties of agricultural products. In : Proceedings of the IMPI Symposium, Loughborough, 1973, 2 p. (Paper 6B1/1-2).

217. NELSON S. Electrical properties of agricultural products - A critical review. *Trans. ASAE*, 16, n° 2, 1973, pp. 384-400.

218. NILSSON K. Belt grill and microwave oven improve prepared meat patties. *Quick Frozen Foods Int.*, avril 1975, pp. 62-65.

219. NYKVIST W., DECAREAU R. Microwave meat roasting. In : Proceedings of the IMPI Symposium, Waterloo (Can.), 1975, p. 69. (Paper 4.2.).

220. NYKVIST W., DECAREAU R. Microwave meat roasting. *J. Microw. Power*, 11, n° 1, 1976, pp. 3-24.

221. OCKERMAN H., CAHILL V., PLIMPTON R., PARRETT N. Cooking inoculated pork in microwave and conventional ovens. *J. Milk Food Technol.*, 39, n° 11, 1976, pp. 771-773. (C.D.I.U.P.A., n° 102197).

222. OCKERMAN H., LEON-CRESPO F., CAHILL V., PLIMPTON R., IRVIN K. Microorganism survival in meat cooked in microwave ovens. *Ohio Rep.*, 62, 1977, p. 38. (C.D.I.U.P.A., n° 111578).

223. OHLSSON T. Fundamentals of microwave cooking. *Microw. World*, 4, n° 2, 1983, pp. 4-9. (C.D.I.U.P.A., n° 184001).

224. OHLSSON T., ASTROM A. Sensory and nutritional quality in microwave cooking. *Microw. World*, 3, Nov.-Dec. 1982, pp. 15-16.

225. OHLSSON T., BENGTSSON N. Microwave heating profiles in foods. *Microw. Energy Appl. Newsl.*, 4, n° 6, 1971, pp. 3-8. (C.D.I.U.P.A., n° 42759).

226. OHLSSON T., BENGTSSON N., RISMAN P. Dielectric data of foods at 915 MHz and 2800 MHz as a function of temperature - a comparison. In : Proceedings of the IMPI Symposium, Loughborough, 1973, 7 p. (Paper 3A2/1-7).

227. OHLSSON T., BENGTSSON N., RISMAN P. The frequency and temperature dependence of dielectric food data as determined by a cavity perturbation technique. *J. Microw. Power*, 9, n° 2, 1974, pp. 129-145.

228. OHLSSON T., HENRIQUES M.., BENGTSSON N. Dielectric properties of model meat emulsions at 900 and 2800 MHz in relation to their composition. *J. Food Sci.*, 39, n° 6, 1974, pp. 1153-1156. (C.D.I.U.P.A., n° 78231).

229. OHLSSON T., RISMAN P. Temperature distribution of microwave heating : spheres and cylinders. *J. Microw. Power*, 13, n° 4, 1978, pp. 303-309.

230. OHLSSON T., THORSELL .' Problems microwave reheating of chilled foods. *J. Foodserv. Syst.*, 3, n° 1, 1984, pp. 9-16 et *Svenska Livsmedelsinst. SIK*, n° 386.

231. ÖJELID G. Cooking at one-fifth of the time. In : Proceedings of the IMPI Symposium, Loughborough, 1973. (Paper 5B7/1-2).

232. Ö MEARA J. Microwave processing and its applications in the poultry industry. In : Poultry and egg further processing conference. 6. Columbus, 1967, Ohio State Univ., 1967, pp. 35-41.

233. OSEPCHUK J. A history of microwave heating applications. *IEEE Trans. Microw. Theory Tech.*, MTT-32, n° 9, 1984, pp. 1200-1224.

234. PACE W. Dielectric properties of certain foodstuffs as selected frequencies and at different temperatures. *J. Microw. Power*, 2, n° 3, 1967, p. 98, (résumé de thèse).

235. PACE W., WESTPHAL W., GOLDBLITH S. Dielectric properties of commercial cooking oils. *J. Food Sci.*, 33, n° 1, 1968, pp. 30-36. (C.D.I.U.P.A., n° 1892).

236. PACE W., WESTPHAL W., GOLDBLITH S., VAN DYKE D. Dielectric properties of potatoes and potato chips. *J. Food Sci.*, 33, n° 1, 1968, pp. 37-42. (C.D.I.U.P.A., n° 1893).

237. PAYTON J., BALDWIN R. Comparison of top round steaks cooked by conventional, forced-air and microwave-convection ovens. In : Proceedings of the IMPI Symposium, Chicago, 1985, pp. 24-27.

238. PAYTON J., BALDWIN R. Comparison of top round steaks cooked by microwave-convection, forced-air convection and conventional ovens. *J. Microw. Power*, 20, n° 4, 1985, pp. 255-259.

239. PEI D. Microwave baking, new developments. *Baker's Dig.*, 56, n° 1, 1982, pp. 8-32. (C.D.I.U.P.A., n° 166536)

240. PENNER K., BOWERS J. Flavor and chemical characteristics of conventionally and microwave heated pork. *J. Food Sci.*, 38, n° 4, 1973, pp. 552-554. (C.D.I.U.P.A., n° 61972).

241. PENSABENE J., FIDDLER W., GATES R., FAGAN J., WASSERMAN A. Effect of frying and other cooking conditions on nitrosopyrrolidine formation in bacon. *J. Food Sci.*, 39, n° 2, 1974, pp. 314-316. (C.D.I.U.P.A., n° 70694).

242. PEREIRA R. Permittivity of some dairy products at 2450 MHz. *J. Microw. Power*, 9, n° 4, 1974, pp. 277-288.

243. PERILLO G. Applicazione delle microonde nell'industria conserviera. *Ind. Aliment.*, 15, n° 130, 1976, pp. 87-89. (C.D.I.U.P.A., n° 98858).

244. PETERSON A., FOERSTNER A. Evaluation of microwave oven cooking performance. *Microw. Energy Appl. Newsl.*, 4, n° 1, 1971, pp. 3-8. (C.D.I.U.P.A., n° 42760).

245. PIKUL J., LESZCZYNSKI D., BECHTEL P., KUMMEROW F. Effects of frozen storage and cooking on lipid oxidation in chicken meat. *J. Food Sci.*, 49, n° 3, 1984, pp. 838-843. (C.D.I.U.P.A., n° 190462).

246. PIKUL J., LESZCZYNSKI D., KUMMEROW F. Oxidation products in chicken meal after frozen storage, microwave and convection oven cooking, refrigerated storage, and reheating. *Poult. Sci.*, 64, n° 1, 1985, pp. 93-100. (C.D.I.U.P.A., n° 203058).

247. PROCTOR V., CUNNINGHAM F. Composition of broiler meat as influenced by cooking methods and coating. *J. Food Sci.*, 48, n° 6, 1983, pp. 1696-1699. (C.D.I.U.P.A., n° 183397).

248. PROCTOR B., GOLDBLITH S. Radar energy for rapid food cooking and blanching, and its effects on vitamin content. *Food Technol.*, 2, 1948, p. 95.

249. RAY E., BERRY B., LOUCKS L., LEIGHTON E., GARDNER B. Effects of electrical stimulation and conditioning periods upon pre-rigor beef samples cooked with a microwave oven. *J. Food Prot.*, 46, n° 11, 1983, pp. 954-964. (C.D.I.U.P.A., n° 183336).

250. REAM E., WILCOX E., TAYLOR F., BENNETT J. Microwave vs. conventional cooking method : tenderness of beef roasts. *J. Am. Diet. Assoc.*, 65, n° 2, 1974, pp. 155-160. (C.D.I.U.P.A., n° 74248).

251. REMY R. Four perfectionné, à rayonnements dits "micro-ondes" et "infrarouges" combinés, notamment pour la
boulangerie et la pâtisserie. Brevet
français n° 2 531 828 (82 08130),
11 mai 1982.

252. REMY R. Communication personnelle.
29 août 1986.

253. RENE J., COTTE F. Four de cuisson
pour la pâte à pain. Brevet français
n° 2 494 562 (80 25497), 27 nov. 1980.

254. RICHARDSON S. Convenience foods and
home-prepared foods heated with an
electric range and a microwave oven.
Thesis Doct. Ph. D., Virginia Polytech. Inst. State Univ., 1982, 251 p.

255. RICHARDSON S., PHILLIPS J., AXELSON
J., LOVINGOOD R., PEARSON J.,
SALTMARCH M. Energy required to heat
convenience and home-prepared foods
with an electric range and a microwave oven. *J. Microw. Power*, 19,
n° 2, 1984, pp. 89-95.
(C.D.I.U.P.A., n° 198482).

256. RICHARDSON S., PHILLIPS J., AXELSON
J., LOVINGOOD R., PEARSON J.,
SALTMARCH M. Total and active time
required to prepare convenience and
home-prepared foods with an electric
range and a microwave oven. *Home
Econ. Res. J.*, 14, 1985, pp. 21-28.

257. RIFFERO L., HOLMES Z. Characteristics of pre-rigor pressurized versus
conventionally processed beef cooked
by microwaves and by broiling. *J.
Food Sci.*, 48, n° 2, 1983, pp. 346-
374. (C.D.I.U.P.A., 83-3629).

258. RISMAN P., BENGTSSON N. Dielectric
properties of foods at 3 GHz as
determined by a cavity perturbation
technique : I. Measuring technique.
J. Microw. Power, 6, n° 2, 1971,
pp. 101-106.

259. ROBERTS P., LAWRIE R. Effects on
bovine L. dorsi muscle of conventional and microwave heating. *J.
Food Technol.*, 9, n° 3, 1974, pp.
345-356. (C.D.I.U.P.A., n° 76102).

260. ROSEN C. Effects of microwaves on
food and related materials. *Food
Technol.*, 26, n° 7, 1972, pp. 36-55.
(C.D.I.U.P.A., n° 047428).

261. ROTH H., SCHIFFMANN R. The application of microwave energy to the
frying of chemically leavened doughnuts. In : Proceedings of the IMPI
Symposium, Monterey, 1971. (Paper
2.2.).

262. RUNTAG T. Continuous high frequency
dielectric equipment and technology
for baking bread. *Acta Aliment.
Acad. Sci. Hung.*, 3, n° 4, 1974,
pp. 349-356. (C.D.I.U.P.A., n° 77742).

263. RUNTAG T., DEMECZKY M. Investigations
into the possibilities of application
of high frequency electric fields to
the baking of dough. *Acta Aliment.
Acad. Sci. Hung.*, 2, n° 1, 1973, pp.
3-16. (C.D.I.U.P.A., n° 55845).

264. RUSSO J. Microwaves proof donuts.
Food Eng., 43, n° 4, 1971, pp. 55-58.
(C.D.I.U.P.A., n° 35205).

265. RZEPECKA M., PEREIRA R. Dielectric
properties of dairy products at
2450 MHz. In : Proceedings of the
IMPI Symposium, Loughborough, 1973,
3 p. (Paper 3A5/1-3).

266. SAWYER C. Post-processing temperature rise in foods : conventional hot
air and microwave ovens. *J. Food
Prot.*, 48, n° 5, 1985, pp. 429-434.
(C.D.I.U.P.A., n° 202111).

267. SAWYER C., NAIDU Y. Sensory evaluation of cook/chilled products reheated by conduction, convection
and microwave radiation. *J. Foodserv. Syst.*, 3, n° 2, 1984, pp. 89-
106. (C.D.I.U.P.A., n° 194431).

268. SAWYER C., NAIDU Y., THOMPSON S.
Cook/chill foodservice systems :
microbiological quality and endpoint temperature of beef loaf,
peas and potatoes after reheating
by conduction, convection and microwave radiation. *J. Food Prot.*, 46,
n° 12, 1983, pp. 1036-1043.
(C.D.I.U.P.A., n° 185681).

269. SCHIFFMANN R. Food product develop-
ment for microwave processing. *Food
Technol.*, 40, n° 6, 1986, pp. 94-98.
(C.D.I.U.P.A., n° 214710).

270. SCHIFFMANN R., ROTH H., STEIN E.,
KAUFMAN H., HOCHHAUSER A., CLARK F.
Application of microwave energy to
doughnut production. *Food Technol.*,
25, n° 7, 1971, pp. 58-62.
(C.D.I.U.P.A., n° 37246).

271. SCHIFFMANN R., STEIN E., KAUFMAN H.
The microwave proofing of yeast-rai-
sed doughnuts. *Bakers' Dig.*, 45,
n° 1, 1971, pp. 55-61.
(C.D.I.U.P.A., n° 34199).

272. SCHILLER E., PRATT D., REBER D.
Lipids changes in egg yolks and cakes
baked in microwave ovens. *J. Am.
Diet. Assoc.*, 62, n° 5, 1973, pp. 529-
533. (C.D.I.U.P.A., n° 62048).

273. SCHLUMPF K. Microwaves for cooking appliances. *Elektrizitätsverwert.*, *42*, n° 12, 1972, pp. 408-410.

274. SCHRICKER B., MILLER D. Effects of cooking and chemical treatment on heme and nonheme iron in meat. *J. Food Sci.*, *48*, n° 4, 1983, pp. 1340-1349. (C.D.I.U.P.A., n° 83-7518).

275. SCHRUMPF E., CHARLEY H. Texture of broccoli and carrots cooked by microwave energy. *J. Food Sci.*, *40*, n° 5, 1975, pp. 1025-1029. (C.D.I.U.P.A., n° 92701).

276. SISON E., DAWSON L., PAYNE E. Acceptance of precooked frozen chicken reheated by microwave energy. *Poult. Sci.*, *52*, n° 1, 1973, pp. 28-34. (C.D.I.U.P.A., n° 60092).

277. SISON E., DAWSON L., PAYNE E. Acceptability of microwave reheated precooked chicken after packaging, freezing and storing. *Poult. Sci.*, *52*, n° 1, 1973, pp. 70-73. (C.D.I.U.P.A., n° 60094).

278. SMITH D. A system for doubling sales value of poultry. *Broiler Ind.*, n° 2, 1967, pp. 60-67.

279. SMITH D. Industrial microwave cooking comes of age : chicken cooking in operation 2 1/2 years. *Microw. Energy Appl. Newsl.*, *2*, n° 3, 1969, pp. 9-10. (C.D.I.U.P.A., n° 23849).

280. SMITH D. Food-finishing microwave tunnel utilizes jet impingement and infrared sensing for process control. *Food Technol.*, *40*, n° 6, 1986, pp. 113-116. (C.D.I.U.P.A., n° 214710).

281. SMITH D., DECAREAU R., GERLING J. Microwave cooking for further processing. *Poult. Meat*, *3*, n° 4, 1966, pp. 65-66.

282. SMITH F. How processing with microwave heat affects food qualities. *Food Prod. Dev.*, *11*, n° 1, 1977, p. 60. (C.D.I.U.P.A., n° 105117).

283. SNYDER O. Increasing the quality of vegetables in foodservices operations. *Microw. Energy Appl. Newsl.*, *9*, n° 2, 1976, pp. 3-7. (C.D.I.U.P.A., n° 95883).

284. SNYDER P., MATTHEWS M. Effects of microwave cooking, and reheating on the microbiological quality of menu items. In : Proceedings of the IMPI Symposium, Chicago, 1985, pp. 20-23.

285. SPITE G. Microwave-inactivation of bacterial pathogens in various controlled frozen food compositions and in a commercially available frozen food product. *J. Food Prot.*, *47*, n° 6, 1984, pp. 458-462. (C.D.I.U.P.A., n° 189057).

286. SRIHARA P., ALEXANDER J. Effect of heat treatment on nutritive quality of plant protein blends. *Can. Inst. Food Sci. Technol.*, *17*, n° 4, 1984, pp. 237-241. (C.D.I.U.P.A., n° 206359).

287. STAGGS R. Cooking individual foods with microwave energy : meat. In : Proceedings of the Consumer Microwave Oven Systems Conference, Ithaca, Cornell Univ., 1970, pp. 24-25.

288. STARRAK G., Techniques in microwaving meat. *Microw. World*, *3*, n° 6, 1982, pp. 10-14. (C.D.I.U.P.A., n° 83-239)

289. STEIN E. Application of microwaves to bakery production. In : Proceedings of the American Society of Bakery Engineers. 48th. Chicago, 1972. Chicago, American Society of Bakery Engineers, *14*, 1972, pp. 46-54. (C.D.I.U.P.A., n° 51188).

290. STEINER J., JOHNSON R., TOBIN B. Effect of microwave and conventional reheating on flavor development in chicken breasts. *Microw. World*, *6*, n° 2, 1985, pp. 10-12. (C.D.I.U.P.A., n° 203145).

291. STUCHLY M., STUCHLY S. Industrial, scientific, medical and domestic applications of microwaves. *IEE Rev.*, *IEE Proc.*, *130*, Pt A, n° 8, Nov. 1983, pp. 467-503.

292. SUZUKI J., BALDWIN R., KORSCHGEN M. Sensory properties of dextrose - and sucrose - cured bacon : microwave and conventionally cooked. *J. Microw. Power*, *19*, n° 3, 1984, pp. 195-197. (C.D.I.U.P.A., n° 203148).

293. TAKI G. Cooking meat in the microwave oven. *Microw. World*, *6*, n° 4, 1985, pp. 11-13. (C.D.I.U.P.A., n° 204145).

294. THAMER D., TOWN R., ROBE K. Rapid cook-and-freeze increases fried chicken yield. *Food Process.*, *32*, n° 1, 1971, pp. 14-15. (C.D.I.U.P.A., n° 35240).

295. THOMAS M., BRENNER S., EATON A., CRAIG V. Effect of electronic cooking on nutritive value of foods. *J. Am. Diet. Assoc.*, *25*, n° 1, 1949, pp. 39-44.

296. THOMAS M., DECAREAU R., ATWOOD B. Thiamin and riboflavin content of flake-cut formed pork roasts. *J. Microw. Power*, 17, n° 1, 1982, pp. 83-87.

297. TINGA W., NELSON S. Dielectric properties of materials for microwave processing-tabulated. *J. Microw. Power*, 8, n° 1, 1973, pp. 23-65.

298. TO E. Dielectric properties of meats. Thesis Ph. D., Massachusetts Institute of Technology, Cambridge, 1974.

299. TO E., GOLDBLITH S., WANG D., DECAREAU R. Dielectric properties of foodstuffs. In : Proceedings of the IMPI Symposium, Milwaukee, 1974, 2 p. (Paper B3-2/1-2).

300. TO E., MUDGETT R., WANG D., GOLDBLITH S., DECAREAU R. Dielectric properties of food materials. *J. Microw. Power*, 9, n° 4, 1974, pp. 303-315.

301. TRAN V. Variations of dielectric properties of rice with bulk densities. In : Proceedings of the IMPI Symposium, Chicago, 1985, p. 181.

302. TRAN V., STUCHLY S. Dielectric properties of beef, chicken and salmon at radiowaves and microwaves from 100 MHz to 2400 MHz. In : Proceedings of the IMPI Symposium, Memphis, 1986. (Abstract in *J. Microw. Power*, 21, n° 2, 1986, p. 116).

303. TRAN V., STUCHLY S., KRASZEWSKI A. Dielectric properties of selected vegetables and fruits 0.1-10.0 GHz. *J. Microw. Power*, 19, n° 4, 1984, pp. 251-258. (C.D.I.U.P.A., n° 200436).

304. TSEN C., REDDY P., GEHRKE C. Effects of conventional baking, microwave baking, and steaming on the nutritive value of regular and fortified breads. *J. Food Sci.*, 42, n° 2, 1977, pp. 402-406. (C.D.I.U.P.A., n° 106301).

305. TURNBROUGH J., BALDWIN R. Enhancing nutritive value and appearance of microwave baked muffins. In : Proceedings of the IMPI Symposium, Chicago, 1985, pp. 28-31.

306. UPCHURCH R., BALDWIN R. Guar gum and triacetin in meringues and a meringue product cooked by microwaves. *Food Technol.*, 22, n° 10, 1968, pp. 107-108. (C.D.I.U.P.A., n° 5141).

307. VADEHRA D. Cooking poultry with microwave energy. In : Proceedings of the Consumer Microwave Oven System Conference. Ithaca, Cornell Univ., 1970, pp. 26-29.

308. VAN DYKE D., WANG D., GOLDBLITH S. Dielectric loss factor of reconstituted ground beef : the effect of chemical composition. *Food Technol.*, 23, n° 7, 1969, pp. 84-86. (C.D.I.U.P.A., n° 12928).

309. VAN ZANTE H. Techniques for electronic cooking research. *J. Home Econ.*, 51, 1959, pp. 454-460.

310. VAN ZANTE H. Some effects of microwave cooking power upon certain basic food components. In : Proceedings of the IMPI Symposium, Stanford, 1967, abstract pp. 39-40 et *Microw. Energy Appl. Newsl.*, 1, n° 6, 1968, pp. 3-9.

311. VAN ZANTE H., JOHNSON S. Effect of electronic cookery on thiamine and riboflavin in buffered solutions. *J. Am. Diet. Assoc.*, 56, n° 2, 1970, pp. 133-135. (C.D.I.U.P.A., n° 18374).

312. VARO P., VEIJALAINEN K., KOIVISTOINEN P. Effect of heat treatment on the dietary fibre contents of potato and tomato. *J. Food Technol.*, 19, n° 4, 1984, pp. 485-492. (C.D.I.U.P.A., n° 191589).

313. VETSUYPENS J., VAN LOOCK W. Microwave treatment of grains. In : Proceedings of the IMPI Symposium, Memphis, 1986. (Abstract in *J. Microw. Power*, 21, n° 2, 1986, p. 110).

314. VORIS H., VAN DUYNE F. Low wattage microwave cooking of top round roasts : energy consumption, thiamine content and palatability. *J. Food Sci.*, 44, n° 5, 1979, pp. 1447-1454. (C.D.I.U.P.A., n° 135885).

315. WALRADT J., LINDSAY R., LIBBEY L. Popcorn flavor : identification of volatile compounds. *J. Agric. Food Chem.*, 18, n° 5, 1970, pp. 926-928. (C.D.I.U.P.A., n° 25647).

316. WESTPHAL W., SILS A. Dielectric constant and loss data. Technical Report AFML-TR-72-39, Air Force Materials Laboratory, Air Force Systems Command, Wright-Patterson Air Force Base, Dayton, 1972, 224 p.

317. WING R., ALEXANDER J. Effect of microwave heating on vitamin B6 retention in chicken. *J. Am. Diet. Assoc.*, 61, n° 6, 1972, pp. 661-664. (C.D.I.U.P.A., n° 53085)

318. WING R., ALEXANDER J. The value of microwave radiations in the processing of full-fat soybeans. *J. Inst. Can. Sci. Technol. Aliment.*, 8, n° 1, 1975, pp. 16-18. (C.D.I.U.P.A., n° 80205).

319. WISMER . Backen mit hochfrequenz und Infrarot "System Reforma". *Zucker Susswaren Wirtsch.*, 22, n° 3, 1969, pp. 110-113. (C.D.I.U.P.A., n° 21759).

320. WRIGHT-RUDOLPH L., WALKER H., PARRISH F. Survival of Clostridium perfringens and aerobic bacteria in ground beef patties during microwave and conventional cookery. *J. Food Prot.*, 49, n° 3, 1986, pp. 203-206. (C.D.I.U.P.A., n° 211100).

321. YOWELL K., FLURKEY W. A research note : effect of freezing and microwave heating on proteins from codfish fillets : analysis by SDS polyacrylamide gel electrophoresis. *J. Food Sci.*, 51, n° 2, 1986, pp 508-509. (C.D.I.U.P.A., n° 211154).

322. ZIMMERMANN W. Evaluation of microwave cooking procedures and ovens for devitalizing Trichinae in pork roasts. *J. Food Sci.*, 48, n°3, 1983, pp. 856-899. (C.D.I.U.P.A., n°83-7441).

323. ZIMMERMANN W. An approach to safe microwave cooking of pork roasts containing Trichinella spiralis. *J. Food Sci.*, 48, n° 6, 1983, pp. 1715-1722. (C.D.I.U.P.A., n° 183401).

324. ZIMMERMANN W. Power and cooking time relationships for devitalization of Trichinae in pork roasts cooked in microwave ovens. *J. Food Sci.*, 49, n° 3, 1984, pp. 624-826. (C.D.I.U.P.A., n° 190459).

325. ZIMMERMANN W. A research note. Microwave recooking of pork roasts to attain 76,7°C throughout. *J. Food Sci.*, 49, n° 3, 1984, pp. 970-974. (C.D.I.U.P.A., n° 190472).

326. ZIMMERMANN W., BEACH P. Efficacy of microwave cooking for devitalizing Trichinae in pork roasts and chops. *J. Food Prot.*, 45, n° 5, 1982, pp. 405-409. (C.D.I.U.P.A., n° 169974).

327. ZIPRIN Y., CARLIN A. Microwave and conventional cooking in relation to quality and nutritive value of beef and beef-soy loaves. *J. Food Sci.*, 41, n° 1, 1976, pp. 4-8. (C.D.I.U.P.A., n° 93672).

328. ZYLEMA B., GRIDER J., GORDON J., DAVIS E. Model wheat starch systems heated by microwave irradiation and conduction with equalized heating times. *Cereal Chem.*, 62, n° 6, 1985, pp. 447-453. (C.D.I.U.P.A., n° 207773).

329. Applications industrielles des micro-ondes. Publ. Thomson CSF, Div. Tubes Electroniques, réf. APH 6211, juin 1980, 13 p.

330. Divers comptes-rendus d'essais. Société I.M.I.-S.A., 1975 à 1980.

331. Federal Register, Office of Federal Register, National Archives and Record Service, General Services Adm., Washington, 35, n° 194, 1970, pp. 15642-15643, 38, n° 151, 1973, pp. 21262-21264,, 40, n° 64, 1975, pp. 14750-14753.

332. Food service sanitation manual. Public Health Service, FDA, US Dep. of Health, Education, and Welfare, Publ. FDA 78-2081, Div. Retail Food Protection, 1978, Washington, 76 p.

333. 4-minutes donut proof time...fastest ever devised ! Yours... with a DCA microwave proofer. New York, DCA Food Industries Inc., s.d., 4 p.

334. Hope of new breakthrough in bread technology. *Food Manuf.*, 46, n° 7, 1971, pp. 46-47.

335. Les micro-ondes dans l'agro-alimentaire : un outil performant. Situation technologique. Cahier ARIST, ARIST Aquitaine, juillet 1984, 237 p.

336. Ligne automatique d'emboîtage et de pré-cuisson de sardines. *Surgél.*, n° 164, 1978, pp. 167-169.

337. Microwave cooking chicken parts ; installed on production line basis. *Quick Frozen Foods*, n° 7, 1966, pp. 119-120.

338. Microwave methods with bakery products. *Food Process. Ind.*, 45, n° 532, 1976, p. 44. (C.D.I.U.P.A., n° 92565).

339. Microwave ovens : effects on food quality and safety. *J. Am. Diet. Assoc.*, 85, n° 8, 1985, pp. 919-921. (C.D.I.U.P.A., n° 204376).

340. Nouvel applicateur micro-ondes destiné au traitement au défilé de produits en grains. In : Comptes-rendus du Colloque sur les Applications Energétiques des Rayonnements, Univ. Paul Sabatier, Toulouse, 12 juin 1986, ENSEEIHT, pp. 16-19.

341. Pork microwave cooking. Des Moines, National Pork Producers Council, 1977, 31 p.

342. Prepared foods-computerized pasta plant. *Prep. Foods*, 153, n° 9, 1984, pp. 87-89. (C.D.I.U.P.A., n° 192171).

343. Production near for microwave rendering system. *Render*, Oct. 1983, pp. 13-14.

344. Qualitätsverbesserung durch Mikrowellenkochen. *Ind. Obst- Gemüseverwert.*, 59, n° 1, 1974, pp. 20-21. (C.D.I.U.P.A., n° 66517).

345. "Radyne" electronic baking oven type OC 160. Wokingham, Radyne Ltd., s.d., 4 p.

346. Recommandations for microwaving pork. Chicago, National Live Stock and Meat Board, 1981, 4 p.

347. The effect of microwave energy on the nutrients in foods. *Microw. Energy Appl. Newsl.*, 2, n° 2, 1969, pp. 3-5. (C.D.I.U.P.A, n° 23847).

348. USDA advises consumers to microwave pork to uniform 170 degrees Fahrenheit. News release. Washington, USDA, 1981.

349. Weaver speeds chicken processing with giant microwave system. *Quick Frozen Foods*, 34, n° 7, 1972, p. 49. (C.D.I.U.P.A., n° 52126).

Chapter 2

Thawing and tempering

The thawing of frozen food by microwaves suffers from three fundamental difficulties: the low dielectric loss factor of frozen products, the relatively large amount of energy required to produce the change of state, and the superficial thermal runaway effect.

The first of these has a positive aspect whereby it allows excellent penetration of the product by microwaves, and the residual low-loss factor is usually sufficient to initiate the thawing process. The second difficulty may be disregarded if final temperatures of $-2°C$ to $-5°C$ are acceptable. Such temperatures are ideal for post-thawing industrial operations. The third difficulty occurs mostly at high frequencies (2.45 GHz) and can be readily and elegantly resolved by cooling the surface of the product before or during thawing.

The process offers such quantitative and qualitative advantages that, despite the high investment required, it is one of the most successful applications of microwaves in the food industry.

2.1 Conventional thawing

The solidification and melting of biological products are slow processes. The time required for either of them to take place is given by the formula established in 1913 by Max Planck:

$$t = \frac{2a\rho L_F}{T_1 - T_0} \left(\frac{1}{2h} + \frac{2a}{8\Lambda} \right) \tag{2.1}$$

where ρ is the density ($kg\, m^{-3}[L^{-3}M]$), L_F is the heat transferred during the isothermal process (i.e., the latent heat of fusion ($J\, kg^{-1}[L^2T^{-2}]$)), $2a$ is the thickness of the product, $T_1 - T_0$ is the difference between the ambient temperature and the freezing point, h is the heat transfer coefficient of the surface ($W\, m^{-2}K^{-1}[MT^{-3}\Theta^{-1}]$, and Λ is the thermal conductivity ($W\, m^{-1}K^{-1}[LMT^{-3}\Theta^{-1}]$).

For a chunk of beef ($\rho = 1150, L_F = 1.9 \times 10^5, \Lambda = 1.465$) 20 cm thick ($a = 0.1$) at $-16°C$ (T_0) in an ambient temperature of $+4°C$ (T_1), we find from (2.1) that $t = 630\, min$ if we neglect the first term in the formula.

During freezing, the external layer cools first and thus acquires better thermal conductivity than the unfrozen remainder of the object, so that internal heat can readily leave the material as the frozen layer increases in thickness. In contrast, in the case of thawing, the superficial layer warms up first and hinders heat transfer, which explains why, in practice, thawing is slower than freezing [38], [1].

Conventional thawing, which relies on air or water, employs a fluid as a heat carrier (e.g., a saturated air stream circulating at $300\, m\, min^{-1}$ or a stream of water at 18 to $20°C$), circulating at speeds of 0.3 to $1.5\, m\, min^{-1}$.

Thawing with warm air has a number of disadvantages. In particular, there is a reduction in the heat transfer coefficient h at the surface (due to condensation from the saturated air), and very large volumes of air are required because of its remarkably low heat capacity. Dripping losses are quite significant; the fats tend to get rancid, and surface oxidation gives rise to a degree of browning that results in a deterioration of quality.

The advantages of thawing in a water bath rely on the excellent thermal transfer coefficient of the surface. The major disadvantages include the deterioration in the appearance of the product, the loss of material by dissolution, the problem of disposal of the effluent, the difficulty of cleaning, and the risk of contamination [42].

A concurrent process, thawing under vacuum, was developed by the APV Company of Evreux (France) in collaboration with Electricitè de France and has been described in a number of publications [38], [1], [20], [25]. The product is introduced into an evacuated enclosure containing boiling water. Water vapor condenses on the surface of the product and gives up its latent heat. The quality of the resulting product may be judged from the data reported by ISTPM of Nantes [25] on the thawing of sardines and tuna fish. The weight loss was quite significant (i.e., 10.8% for blocks of sardines as against 4% in air thawing to 4°C, and 4.3% for 3.5 kg of tuna fish as compared with 2.3% in the case of thawing by air). Organoleptic tests performed on sardines showed certain defects of appearance, but satisfactory internal quality. The process ensures better preservation of aroma, taste, and consistency of the cooked fish by comparison with thawing by water or air. The bacteriological quality is slightly better.

2.2 Mechanisms of microwave thawing

2.2.1 Dielectric properties of frozen products

Ice is a polymorphic material that occurs in a variety of allotropic configurations, all of which have densities that are lower than the density of water. Figure 2.1 shows the temperature and pressure domains for the different types of ice. At atmospheric pressure, there is only ordinary ice (ice I), which has a tetrahedral structure formed by the hydrogen bonds (Fig. 2.2). This allows the water molecules to retain their identity with an HOH angle of 109.4° instead of the 104.5° in the liquid phase. The average separation between the two neighboring oxygen atoms is 0.276 nm and there is a hydrogen atom in between. Each oxygen atom has four neighboring hydrogen atoms (i.e., two at approximately 0.1 nm) to which it is convalently bonded, and two at 0.176 nm, to which it is hydrogen bonded [64, 30, 51].

This rigid configuration (Fig. 2.3) has anisotropic dielectric properties (a 14% difference between $\epsilon_\perp$ and $\epsilon_\parallel$), with most of the relaxation confined to the kilohertz region (Fig. 2.4). The mechanism of this dispersion is thought to be the propagation of molecular orientation defects referred to as Bjerrum defects. The residual dielectric losses at high frequencies ($\epsilon'_r = 3.20$, $\tan \delta = 0.0009$ at 3 GHz) may be explained in terms of the superposition of two asymptotic tails: the low-frequency relaxation tail and the submillimetric absorption band tail [51].

We note that, in the S band, the dielectric loss factor $\epsilon''_r = \epsilon'_r \tan \delta$ is lower by four orders of magnitudes than that of water ($\epsilon'_r = 80$, $\tan 0.31$), which seems to make microwave thawing impossible.

Actually, the freezing point of water is significantly impeded by the presence of solutes and by capillary phenomena [51] to the extent that there is still liquid water even at $-40°C$ [44], [79]. Frozen meat can thus be seen as a two-phase mixture consisting of a solid matrix of ice and biological material, sur-

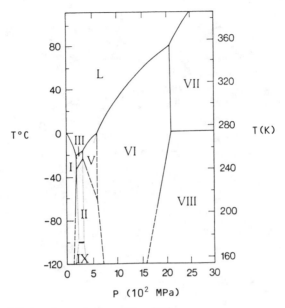

Figure 2.1 Allotropic varieties of ice [51].

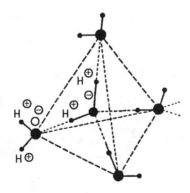

Figure 2.2 Basic structure of ice I below $180°$C [64].

rounded by monomolecular layers of tightly bound water and free liquid water saturated with dissolved salt [80]. The dielec-

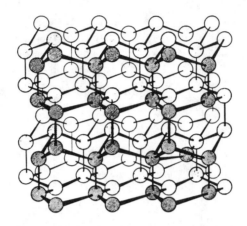

Figure 2.3 Oxygen atoms in the structure of ice I [51].

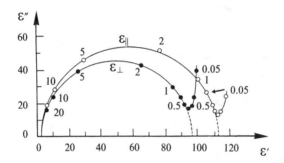

Figure 2.4 Cole-Cole diagram of monocrystalline ice I at $-10.9°$C [51].

tric activity of the mixture is much higher than that of pure ice, but remains much lower than that of the same material at temperatures above zero. For example, the loss tangent of frozen, raw, or cooked beef at $-20°$C [117] is 0.1 at 3 GHz. At the same frequency, the corresponding figures are: 0.1 for beef fat at $-10°$C, 1.2 for pork meat at $-15°$C, 0.1 for cooked cod at $-20°$C, 0.08 for carrots at $-20°$C, and 0.12 for peaches at $-15°$C. Tables 2.1 and 2.2 list some dielectric data reported in [79], [83] for frozen beef.

The permittivity of frozen food is also discussed in [7], [85], [8], [120], [77], [83], [57], [58].

Table 2.1
Permittivity of frozen beef

T_0	f, MHz	ε'_r	ε''_r	tan δ
-20°C	300	5.4	0.97	0.18
	915	4.8	0.54	0.11
	2 450	4.4	0.51	0.12
-40°C	300	3.9	0.34	0.09
	915	3.6	0.21	0.06
	2450	3.5	0.13	0.04

Table 2.2
Permittivity of frozen beef

	T_0	f, MHz	ε'_r	ε''_r	tan δ
Lean beef 73.7% moisture content 1.3% fat	-20°C	450	5.4	1.08	0.2
		900	4.8	0.82	0.171
	-10°C	450	6.81	2.39	0.351
		900	6.2	1.84	0.297
Fatty beef 72% moisture content 5% fat	-20°C	450	5.03	0.78	0.155
		900	4.5	0.62	0.138
	-10°C	450	7.11	1.82	0.256
		900	6.1	1.71	0.28

2.2.2 Thermal runaway

The low dielectric loss factor of frozen food causes thermal-runaway problems (Part I, Sec. 3.7). Microwaves are exponentially attenuated in matter, so that surface layers receive more energy and heat up faster than the interior of the medium. Since their dielectric loss factor increases with temperature, this promotes a higher rate of heating and a lower penetration depth. Pockets of free water and points that reach temperatures of around 0°C absorb energy selectively because of their high dielectric loss factor, and they promote the thermal runaway that produces a rapid rise in temperature. Meanwhile, the cold interior receives practically no energy and remains frozen.

The lower ISM frequencies (434 and 896/915 MHz) are better suited to the thawing process because they allow better penetration [Part I, (2.83)], and also because the loss factor ϵ_r'' decreases with frequency (Tables 2.5 and 2.6). It was reported [116] that the thawing of chunks of meat 18 cm thick at 2.45 GHz created temperature gradients that were twice as high as those observed at 915 MHz. It was observed that the standard deviation (or the variance) of the temperature decreased with increasing average temperature and passed through a minimum (i.e., a state of maximum uniformity) at a certain slightly negative temperature. It then began to increase again under the influence of local runaway heating. This phenomenon is illustrated in Figure 2.5, which was obtained for thawing chunks of meat at 915 MHz, in the Gigatron 25 SL IMI. It is therefore preferable to limit the process of microwave thawing to simple tempering to −2°C or −3°C.

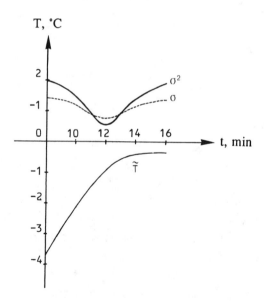

Figure 2.5 Average temperature (T), standard deviation (σ), and variance (σ^2) of the temperature of meat during microwave thawing [123].

2.2.3 Energy limitations

The second reason for favoring tempering as compared with thawing derives from the shape of the curves in Figure 2.6, which represent the energy necessary to reach a temperature T starting from $T_0 = -20°C$. This energy is made up of two terms:

$$W = m[C(T - T_0) + L_F] \qquad (2.2)$$

where m is the mass, C is the specific heat capacity $(J \, kg^{-1}K^{-1}[L^2T^{-2}\Theta^{-1}])$, and L_F is the latent heat of fusion $(J \, kg^{-1}[L^2T^{-2}])$.

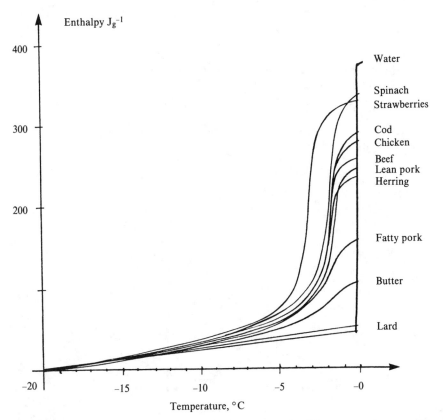

Figure 2.6 Energy required for thawing between $T_0 = -20°C$ and T (IMI-SA documentation).

For ice, $C = 2093 \, \mathrm{J \, kg^{-1} \, K^{-1}}$ and $L_F = 332\,368.4 \, \mathrm{J \, kg^{-1}}$. For frozen food, the transition is less sharp, but a point of inflection is always observed in the vicinity of 0°C, which corresponds to a change of state. It is therefore energetically unsound to cross the threshold of fusion if it is not absolutely necessary, which is generally the case because subsequent operations such as the cutting of meat can be performed at −2°C or even less; a temperature of −4°C is sufficient for the mechanical boning of shoulders of lamb [69]. A significant amount of ice is still present at the end of the tempering process (Fig. 2.7). It acts as a cold reservoir for the temperature stabilization of possible overheated regions, and enables the product to be held for several hours without special storage precautions. This is of considerable advantage in retail sales [93–101], [45], [42].

The importance of a satisfactory choice of the final temperature is illustrated by the output of the Gigatron 30 BSF microwave tunnel for the treatment of frozen beef at −20°C, which varies from 800 to 1,400 kg hr^{-1} as the final temperature varies from −2°C to −4°C.

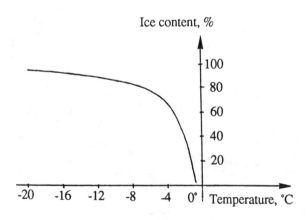

Figure 2.7 Proportion of ice as a function of temperature for lean meat and fish [45].

2.2.4 Surface cooling

Paradoxically, large temperature variations during thawing or tempering at high frequencies (2.45 GHz) can be avoided by using surface cooling. For a one-dimensional frozen medium of thickness $2a$, the standard heat transfer equation is (Part I, Sec. 3.6)

$$\frac{\partial^2 T}{\partial x^2} = \frac{\rho C_p}{\Lambda} \frac{\partial T}{\partial t} \tag{2.3}$$

where ρ is the density, C_p is the specific heat capacity at constant pressure ($\mathrm{J\,kg^{-1}K^{-1}}$ [$L^2 T^{-2} \Theta^{-1}$]), and Λ is the thermal conductivity ($\mathrm{Wm^{-1}K^{-1}}$ [$LMT^{-3}\Theta^{-1}$]). The material is initially at temperature T_0 and is placed in an ambient temperature $T_1 < T_0$. We now substitute $\theta = T - T_1$ and $\theta_0 = T_0 - T_1$ and seek a separable solution in the form $\theta(x,t) = f(t)g(x)$. Equation (2.3) becomes

$$\frac{\Lambda}{\rho C_p} f(t) \frac{d^2 g(x)}{dx^2} = g(x) \frac{df(t)}{dt}$$

where

$$\frac{\Lambda}{\rho C_p} \frac{g''(x)}{g(x)} = \frac{f'(t)}{f(t)}$$

The two sides of this equation are functions of two different variables and must therefore be equal to the same constant, say, $-\omega^2$. This is always true if

$$f(t) = C e^{-\omega^2 t} \tag{2.4}$$

$$g(x) = A \cos \omega \sqrt{\frac{\rho C_p}{\Lambda}} x + B \sin \omega \sqrt{\frac{\rho C_p}{\Lambda}} x \tag{2.5}$$

The general solution is a superposition of these two particular solutions:

$$\theta(x,t) = \sum_{n=1}^{\infty} e^{-\omega_n^2 t} \left[a \omega_n \cos \omega_n \sqrt{\frac{\rho C_p}{\Lambda}} x + b \omega_n \sin \omega_n \sqrt{\frac{\rho C_p}{\Lambda}} x \right] \tag{2.6}$$

The pulsatances of the Fourier series are given by

$$\omega_n \sqrt{\frac{\rho C_p}{\Lambda}} x = \frac{2n\pi}{T}$$

$$\omega_n = \frac{2n\pi}{T} \sqrt{\frac{\Lambda}{\rho C_p}}$$

Since the problem is confined to x in $[-a, +a]$, it may be assumed that θ is a periodic function with the period $4a$:

$$\theta(x,0) = \theta_0 \qquad \text{for} \qquad a \in \,]-a, +a[$$

$$\theta(x,0) = -\theta_0 \qquad \text{for} \qquad a \in \,]+a, +3a[$$

Hence

$$\omega_n = \frac{n\pi}{2a} \sqrt{\frac{\Lambda}{\rho C_p}} \qquad (2.7)$$

Since θ is an even function, we have

$$b_n = 0$$

and the coefficients a_n are given by

$$a_n = \frac{2}{T} \int_T f(x) \cos \frac{2n\pi x}{T} dx \qquad (2.8)$$

$$a_n = \frac{1}{2a} \int_{-a}^{+a} \theta_0 \cos \frac{n\pi x}{2a} dx + \frac{1}{2a} \int_{a}^{3a} -\theta_0 \cos \frac{n\pi x}{2a} dx \qquad (2.9)$$

and

$$a_n = \frac{4\theta_0}{n\pi} \sin \frac{n\pi}{2} \qquad (2.10).$$

Hence

$$\theta(x,t) = \sum_{n=1}^{\infty} \exp\left(-\frac{n^2 \pi^2}{4a^2} \frac{\Lambda t}{\rho C_p}\right) \frac{4\theta_0}{n\pi} \sin \frac{n\pi}{2} \cos \frac{n\pi x}{2a} \qquad (2.11)$$

$T(x,t) = \theta(x,t) + T_1$ is the temperature reached by a point at a distance x after a time t of exposure to the cold. For example, consider a chunk of lean beef 20 cm thick ($a = 0.1$), initially at $T_0 = -16°C$, cooled down to $T_1 = -60°C$ ($\theta_0 = 44\,K$) for 10 min ($T = 600\,s$). For this type of material, $\Lambda = 1.465\,W\,K^{-1}m^{-1}$, $C_p = 1674\,J\,kg^{-1}K^{-1}$, $\rho = 1150\,kg\,m^{-3}$, and (2.11) gives the temperature profile of Figure 2.8 (curve a).

The temperature dependence of permittivity and loss factor ($\epsilon'_r, \epsilon''_r$) is known [120] and is shown in Figure 2.9.

The RF power is attenuated exponentially in the material according to the expression

$$\exp[-2\alpha(a - x)]$$

where $a - x$ is the distance from the surface and α is the attenuation coefficient [Part I, (2.83)]:

$$\alpha = \omega\sqrt{\frac{\mu\epsilon'}{2}\left(\sqrt{1 + \tan^2\delta} - 1\right)} \qquad (2.12)$$

which simplifies to [Part I, (2.88)]

$$\alpha \simeq \frac{\omega}{2}\sqrt{\mu\epsilon'}\tan\delta \qquad (2.13)$$

for small values of $\tan\delta$. The use of this approximation in the above example introduces an error of less than 1% for the highest value of $\tan\delta$ encountered (i.e., 0.250 at $-1°C$).

It is readily seen that (2.13) may be written as

$$\alpha \simeq \pi f\sqrt{\mu_0\epsilon'_0}\sqrt{\epsilon'_r}\tan\delta \qquad (2.14)$$

Hence at 3 GHz

$$\alpha \simeq 31.430\sqrt{\epsilon'_r}\tan\delta \qquad (2.15)$$

The values of ϵ'_r and $\tan\delta$ (Fig. 2.9) are such that $\sqrt{\epsilon'_r}\tan\delta$ is given with sufficient precision by the approximate expression

$$\sqrt{\epsilon'_r}\tan\delta = 10^{(-0.257 + 0.027T)} \qquad (2.16)$$

where T is in °C.

The sample is discretized into parallel layers of thickness $\Delta x = 0.01\,\mathrm{m}$. The microwave power is P_0 at $x = a = 0.1\,\mathrm{m}$, and we have (at the indicated depths)

$$P_1 = P_0 \exp\left(-2\alpha_1 \Delta x\right) \quad \text{at } 0.09$$

$$P_2 = P_1 \exp\left(-2\alpha_2 \Delta x\right) \quad \text{at } 0.08$$

$$P_i = P_{i-1} \exp\left(-2\alpha_i \Delta x\right) \quad \text{at } (1-i)/100$$

so that

$$P_i = P_0 \exp\left[-2\Delta x(\alpha_1 + \alpha_2 + \cdots + \alpha_i)\right] \tag{2.17}$$

The power dissipated in each layer is the difference between the power entering and leaving

$$\Delta P_i = P_i - P_{i-1} = P_0 [\exp(-2\Delta x \sum_1^i \alpha_j) - \exp(-2\Delta x \sum_1^{i-1} \alpha_j)] \tag{2.18}$$

The application of (2.11), (2.16), (2.15), and (2.18) gives the power dissipated as heat which, over a time interval Δt, results in a temperature rise

$$\Delta T_i = \frac{\Delta t \Delta P_i}{m_i C_p} \tag{2.19}$$

where m_i is the mass of the layer under consideration. After a time interval Δt, T_i becomes $T_i + \Delta T_i$ and the calculation is iterated. For example, when $1\,\mathrm{kW}$ of power is applied to a block of 20 layers of $1\,\mathrm{kg}$, and a time step of $\Delta t = 2\,\mathrm{min}$ is employed, the resulting temperature evolution $(a) \rightarrow (b) \rightarrow (c) \rightarrow (d)$ of Figure 2.8 is obtained for an initial profile (a).

It is evident that the outer layers remain cold and that the warm inner region tends to migrate towards the surface. It is interesting to compare this process with a chunk of meat

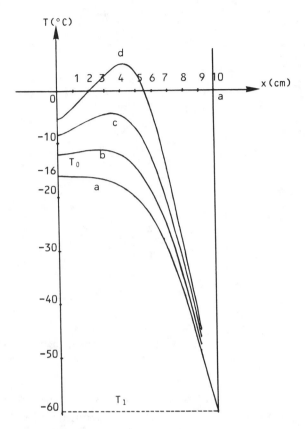

Figure 2.8 Calculated temperature profiles for microwave thawing after surface cooling: (a) cold, (b) after 2 min, (c) after 4 min, (d) after 6 min microwave exposure.

heated in the same manner, but not subjected to prior surface cooling (Fig. 2.10). In the latter case, the core remains cold, but the outer layers experience fast thermal runaway.

This relatively simple model takes into consideration neither the three-dimensional aspect of cooling and irradiation nor thermal conduction between layers. Despite these shortcomings, it highlights the effect of cooling, the inversion of the temperature gradient, and the elimination of thermal-runaway heating at the surface. A proper choice of duration and tem-

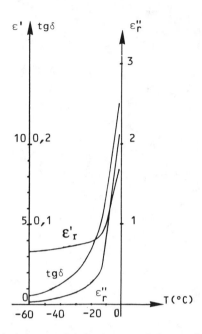

Figure 2.9 Permittivity and loss factor of frozen beef as functions of temperature [120].

perature of cooling, and microwave power input, would cause the curves of Figure 2.8 to flatten in the vicinity of $-2°$C, which corresponds to uniform thawing over the whole thickness of the block.

Microwave thawing has been the subject of a number of theoretical studies [116], [13], [2], [90]. Curves very similar to those of Figure 2.8 are reported in [89]. Further information may be found in [39], [22], [82].

2.3 Available equipment

2.3.1 896 and 915 MHz

A microwave tunnel, the QMP 1679 D, developed by Raytheon is described in [23]. It is used to temper beef in standard 27-kg loads with an output of 1,000 to 3,000 kg hr^{-1}

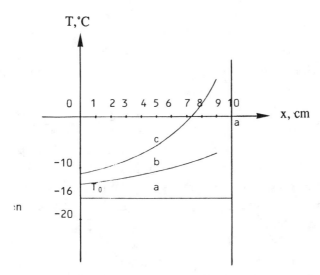

Figure 2.10 Calculated temperature profiles for microwave thawing without surface cooling: (a) - $T = T_0$, (b) - after 2 min, (c) - after 3 min.

depending on the microwave power (25, 50, and 75 kW). The final temperature of $-2.5°C$ is obtained after 18, 10, or 7 min, depending on the power supplied. The equipment is easily installed in a continuous process (e.g., in a meat-patty production line that starts with frozen blocks of meat) [129]. Raytheon has also put forward a batch oven (QMP 1765) that incorporates seven 650-W generators that enable 30 kg of beef at $-18°C$ to reach the temperature of $0°C$ in five minutes.

In the United Kingdom, APV Parafreeze of Thetford (Norfolk) manufactures microwave tunnels of 30 to 120 kW with an output of 1 to $4\,\mathrm{t\,hr}^{-1}$ for tempering blocks of meat and 1.5 to $7\,\mathrm{t\,hr}^{-1}$ for tempering butter. Microwave energy (896 MHz) is fed into the tunnel by source horns via waveguides. Magnetronics of Leicester [127], [132] has developed 896/915-MHz equipment with a microwave efficiency of close to 85%. This equipment is used for the tempering of butter by Anchor Foods of Swindon who have three 120-kW tunnels each. Four sides of blocks of $271 \times 270 \times 410\,\mathrm{mm}^3$ can be irradiated. The product passes from -10 or $-14°C$ to $5°C$ in five minutes at the rate of $7\,t\,hr^{-1}$. The space occupied by the plant is no more than $1/6$ of the traditional tempering equipment. For meat, the output is $1.5\,t\,hr^{-1}$ for 30-kW microwaves applied to standard blocks at an initial temperature of $-18°C$.

In 1977, Professor L. Thourel and his team at CERT/ONERA in Toulouse, France, studied a system for thawing beef in a cold chamber. The ambient temperature could be adjusted between -30 and $4°C$ with an air stream flowing at $3\,m\,s^{-1}$. The microwave system incorporated a 25-kW generator operating at $915\,MHz$, and a network of waveguides and T-dividers distributing the energy between four slotted-waveguide applicators, each facing a radiating aperture. The piece to be treated rotated around its own axis at 1.5 revolutions per minute and thus received energy alternately from the two radiating sources. At maximum power, the output was estimated to be $400\,kg\,hr^{-1}$, the homogeneity was excellent, and the whole piece reached $-2°C$ to $0°C$ with just a few points at $10°C$ [116], [41], [91], [89]. Unfortunately, this equipment remained at the prototype stage, possibly because of problems associated with regulations governing the use of a frequency not allowed in Region 1 (Part I, Sec. 1.1).

Similar problems have been encountered by Gigatron 25 SL, manufactured by IMI-SA. This is a static applicator capable of handling $150\,kg$ of meat in standard cartons. The tilting door is equipped with hydraulic jacks to support the moving trolley of $1460 \times 1180\,mm^2$ during loading. The source is a 25-kW generator supplying energy via a T-divider, the radiation being distributed equally from the top and bottom of the cavity. The emission inside the cavity is through radiating slots cut in the large side of bent WR 975 waveguides. Good impedance matching makes a circulator useless, which is an advantage because, in this frequency range, circulators are extremely cumbersome and costly. Excellent temperature uniformity is obtained at around $-2°C$ (Fig. 2.5) without runaway heating.

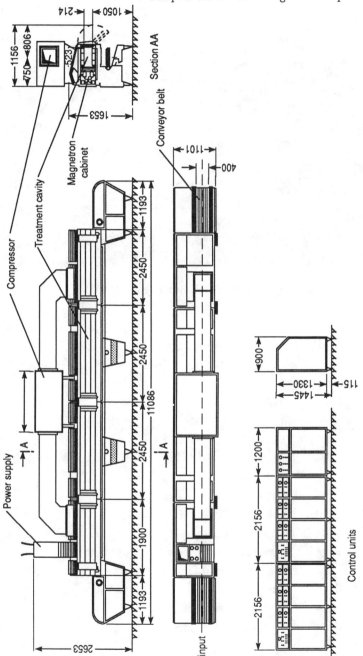

Figure 2.11 The Gigatron 60 BSF (Courtesy of IMI-SA).

Figure 2.12 The Gigatron 60 BSF(Courtesy of IMI-SA).

2.3.2 2.45 GHz

The Gigatron BSF, manufactured by IMI-SA, is the only thawing tunnel operating in the S-band. It was developed in the early seventies and is based on 15-kW modules using 5-kW Philips magnetrons. Bidirectional irradiation is produced by transverse waveguides with radiating dipoles placed alternatively at the top and bottom of the cavity. The cross section of the cavity is $500 \times 210 \, \text{mm}^2$. The cavity is separated from waveguides and magnetrons by windows of laminated fiberglass, and is totally accessible along the whole length of the tunnel. The product circulates on a fiberglass-PTFE conveyor belt. Surface cooling is produced by a compressor that circulates air at $-20°$C in a closed circuit. A system of wavetraps, including polypropylene meanders carrying water, attenuates radiation at the exit apertures, thus reducing leakages to below the safety level (Figs. 2.11 and 2.12).

The output of the Gigatron BSF with power consumption P is given by

$$D = \frac{P}{W} \tag{2.20}$$

where W is the energy required by the product at an initial temperature of T_0 to reach a final temperature T (Fig. 2.6). For example, for beef tempering from -20 to $-4°C$ ($W = 75\,kJ\,kg^{-1}$) in a 30-kW tunnel,

$$D = 0.4\,kg\,s^{-1} = 1440\,kg\,hr^{-1}$$

The Gigatron BSF conforms to the standards of the French Veterinary Authorities and to those of the United States Departments of Health (USDHEW), Agriculture (USDA), and the Federal Communications Commission (FCC) [68], [69], [70], [71], [74], [109], [63], [130], [54], [43], [42], [130], [136], [137].

The first units were installed in 1974 at Gunner Dafgard of Kallby (Sweden) for the thawing of poultry and in Japan by the Narasaki Sangyo Company for thawing whole salmon and blocks of fish [73]. This equipment is currently in service in many countries. In France, users include Buitoni, Findus, Gourault of Blois, and SICA-CANA of Ancenis [76], [35]. Table 2.3 lists examples of Gigatron BSF being used in industrial production lines.

As far as batch ovens are concerned, it is worth mentioning the prototype developed in 1977 by the Electricity Applications Department of Electricitè de France (DER) at Les Renardieres (Seine-et-Marne). The system combines microwaves with a liquid-nitrogen pulverization technique based on an original electrostatic procedure. Liquid-nitrogen droplets are ionized by a current on the order of mA in a 90-kV potential difference in the pulverization vessel. The particles thus acquire kinetic energy that takes them to the vicinity of the product to which they adhere by electrostatic attraction. The saving in nitrogen is 20% to 30% as compared with simple pulverization [3], [13], [15].

Table 2.3
Examples of industrial microwave thawing
(based on an IMI document)

Pork belly e = 6 cm	Tempering to -6°C	Mechanical shaping	Mechanical slicing
		Ditto	Ditto
Beef, chicken, etc e ~ 15 cm	Tempering to -4°C	Mechanical dicing	Refrosting or pre-cooking
		Mincing with cutter	Shaping or mixing
Fish e ~ 8 cm	Tempering to -2°C	Separation and washing	Mechanical filleting
Prawns e ~ 5 cm	Tempering to -2°C	Separation and pickling	Pre-cooking
Peaches	Tempering to -2°C	Peeling Stoning Mashing	

e : Thickness of frozen block

2.4 Advantages of microwave processing

2.4.1 Industrial aspects

The main advantage of microwave tempering is speed: It takes minutes rather than hours or tens of hours, so that stocks are no longer immobilized during the thawing stage; the process is very adaptable, with batch thawing becoming possible, and the loss of material, if any, being very limited.

Another decisive advantage over classical procedure is that the microwave tunnel can be easily inserted into a production chain for continuous processing. For example, a user of Gigatron BSF has reported that the tempering, slicing, conditioning, and refreezing operations necessary in meat processing can be completed in less than 30 min [81]. Losses by exudation are greatly diminished and often quite negligible, which means that the yield is improved and products can be processed in their packaging.

In conventional thawing, the cold room is maintained at 4°C and the additional objects (trolleys, trays, and shelves) present significant surface areas that are unnecessarily immobilized and are potential sources of contamination. There are no equivalent disadvantages in microwave tempering, which requires only about one tenth of the surface area compared to the pulsed air system [70] and ensures a remarkably low level of bacteriological risk.

In the microwave treatment, the optimum temperatures of, say, −5 to −3°C that are required for slicing can be reached with great precision and uniformity. When this is followed by refreezing, the saving in liquid nitrogen can be quite appreciable as compared with conventional thawing in which the surface temperature is positive [81]. Finally, savings in manpower achieved by the reduced manual and maintenance tasks, and low running costs, make the use of microwaves an economically interesting solution despite the initial capital investment in equipment that has to be made [63], [89], [35], [130], [133], [137].

2.4.2 Qualitative aspects

The quality of finished products is significantly better after microwave treatment than in conventional thawing. The use of microwaves produces a significant reduction in losses by exudation. This is attributed, without being fully understood, to the formation of fewer ice needles. For beef, the losses amount to about 5% in conventional thawing [89], [35]. They are lower by a factor of two to three for microwaves [105], and may be less than 1% [89]. The reduction in exudation is associated with better preservation of texture and appearance to the extent that it is often difficult to distinguish between the final product and the product still frozen [72], [89], [34], [35], [122], [130], [133]. Blocks of dairy products [123], prawns [12], frozen vegetables, and fruit exhibit similar properties [88], [4]. In the case of fruit, exudation can reach 8% to 10% in conventional processing and is accompanied by the contraction of pulp, which eventually becomes spongy; nothing of the sort occurs in microwave tempering [12].

Microwave tempering eliminates another major inconvenience of conventional processes, namely, the browning of the surface by oxidation, which is relevant to the processing of meat [42] as well as curd, dairy products generally [123], and peaches [87].

Products thawed by microwaves have very high bacteriological quality because of the speed and cleanliness of the process. There may also be a more specific bacteriocidal effect because, in certain cases, the bacterial density in the product is lower after thawing than before it was frozen [42]. This observation is particularly important for products that should be refrozen [12], [89], [35]. Organoleptic and nutritional properties are also greatly improved in comparison with classical procedures. For example, work performed at the US National Marine Fisheries Service has shown [12] that microwave processing leads to higher protein content in fish and also to improved sensory quality as indicated by organoleptic tests. Similar results have been obtained for sliced tuna fish and for sardines for which the improvement in quality is maintained even after canning [24], [28].

References

1. BAILEY C. Thawing methods for meat. Bristol, ARC Meat Research Institute, 1976, 15 p.

2. BAILLOT G. Contribution à l'étude de l'action des ondes électromagnétiques sur la matière. Application à la décongélation de carcasses de boeuf par micro-ondes. Thèse de Doctorat. Toulouse, Ecole Nationale Supérieure de l'Aéronautique et de l'Espace, 1978, 199 p.

3. BARBINI M., BIALOD D. Décongélation de produits agroalimentaires par micro-ondes associées à un refroidissement de surface. In : Colloque International, Journées d'Electronique de Toulouse, 7-11 mars 1977. Communication IV-3.

4. BEKE G., SEBOK A., SEBOK A., GUBIK G., SIROKMAN K. Mikrohullámu sütok alkalmazásanak tapasztalatai gyorsfagyasztott termekeknel. Huetoipar, 32, n° 2, 1986, pp. 49-63. (C.D.I.U.P.A., n° 212717).

5. BELDEROK B., VAN'T ROOT M. L'impiego delle microonde per lo scongelamento dei prodotti da forno surgelati. Ind. Aliment., 9, n° 67, 1970, pp. 81-84. (C.D.I.U.P.A., n° 26582).

6. BENGTSSON N. Décongélation électronique du poisson congelé. In : Congrès sur la technologie du poisson. Organisation de Coopération et de Développement Economiques, DAAF/FI/T/12, Diffusion restreinte, 48.832, Ta. 82.944, 1964, 17 p.

7. BENGTSSON N., MELIN J., REMI K., SODERLIND S. Measurement of the dielectric properties of frozen and de-frosted meat and fish in the frequency range 10-200 MHz. J. Sci. Food Agric., 14, 1963, pp. 592-604.

8. BENGTSSON N., RISMAN P. Dielectric properties of foods at 3 GHz as determined by a cavity perturbation technique : II. Measurements on food materials. J. Microw. Power, 6, n° 2, 1971, pp. 107-123.

9. BEZANSON A. Microwave tempering of frozen foods. Trans. IMPI, 5, Frozen Food and microwaves make cents, 1975.

10. BEZANSON A. Tempering. Microwave systems rapidly temper frozen foods for further processing. Food Technol., 30, n° 12, 1976, pp. 34-36. (Symposium on innovations in processing of refrigerated and frozen foods). In : Annual Meeting of the Institute of Food Technologists. 36, Anaheim. (C.D.I.U.P.A., n° 102996).

11. BEZANSON A., EDGAR R. Microwave tempering in the food-processing industry. Electron. Prog., 18, n° 1, 1976 (Raytheon Co.), pp. 8-12.

12. BEZANSON A., LEARSON R., TEICH W. Defrosting shrimp with microwaves. Microw. Energy Appl. Newsl., 6, n° 4, 1973, pp. 3-8. (C.D.I.U.P.A., n° 62301).

13. BIALOD D. Décongélation par micro-ondes ; essai de modélisation, et réalisation d'un prototype. Thèse de Doctorat. Toulouse, Institut National Polytechnique, ENSEEIHT, 1977, 120 p.

14. BIALOD D. Décongélation par micro-ondes. Rev. Gén. Electr., 11, nov. 1981, pp. 830-833.

15. BIALOD D., JOLION M., LE GOFF R. Microwave thawing of food products using associated surface cooling. J. Microw. Power, 13, n° 3, 1978, pp. 269-274.

16. BOSSAVIT A. Modèle numérique d'un four à micro-ondes pour la décongélation rapide de produits alimentaires. In : Comptes-rendus du Colloque International des Journées d'Electronique de Toulouse, 7-11 mars 1975. Communication IV-2.

17. CADEDDU S. L'utilizzazione delle tecniche a microonda nella produzione di formaggio mozzarella. Latte, 8, n° 3, 1983, pp. 181-183. (C.D.I.U.P.A., n° 83-7409)

18. CATHCART W., PARKER J. Defrosting frozen foods by high-frequency heat. Food Res., 11, 1946, pp. 341-344.

19. CHARPENTIER J., FOURNAUD J., SALE P., VALIN C., VIGNE C. Utilisation des micro-ondes dans l'industrie de la viande. Paris, DGRST, Action concertée, Contrat n° 70.02.232, 1971, 37 p. (C.D.I.U.P.A., n° 55646)

20. CHRISTIE R., JASON A. Vacuum thawing of foodstuffs. Aberdeen, Torry Research Station, 1976, 21 p.

21. CIOBANU A. Decongelarea produselor alimentare prin curenţi de înaltă frecvenţă. Ind. Aliment., 21, n° 5, 1970, pp. 250-254. (C.D.I.U.P.A., n° 22298).

22. CLELAND D., CLELAND A., EARLE L., BYRNE S. Prediction of rates of freezing, thawing or cooling in solids of arbitrary shape using the finite-element method. Int. J. Refrig., 7, n° 1, 1984, pp. 6-13.

23. COPSON D. Microwave heating. 2nd ed. Westport, The Avi Publishing Company, Inc., 1975, 615 p.

24. CREPEY J. Utilisation industrielle du poisson congelé. Paris, Action concertée : Technologie Agricole, Contrat n° 64, 1973, 17 p. (C.D.I.U.P.A., n° 57078).

25. CREPEY J. Décongélation par vapeur d'eau sous vide. Nantes, ISTPM, 1976, 8 p.

26. CREPEY J., ADJADJ A. La microstructure du poisson frais, congelé et décongelé. Rev. Gén. Froid, 7, n° 7-8, 1979, pp. 395-404. (C.D.I.U.P.A., n° 134265).

27. CREPEY J., MAILLARD J. La décongélation du poisson. In : Comptes-rendus du 6éme Congrès International de la Conserve. L'aliment conservé. Evolution des aspects technologiques, économiques et législatifs. Paris, 1972, pp. 315-338. (C.D.I.U.P.A., n° 66040).

28. CREPEY J., MAILLARD J. La décongélation des produits marins. Rev. Gén. Froid, 64, n° 4, 1973, pp. 359-365. (C.D.I.U.P.A., n° 60170).

29. CREPEY J., MAIREY D. Application des hautes fréquences à la décongélation de la sardine et du thon. In : Aliments congelés. Qualité des produits. Techniques de congélation (notamment cryogéniques). Paris, Institut International du Froid, n° 6, 1969, pp. 249-257. (C.D.I.U.P.A., n° 29826).

30. DAVIDSON D. Dielectric relaxation and rotational disorders in ices. In : Molecular relaxation processes.

Proceedings of the Chemical Society Symposium. Aberystwyth, 1965. London, Academic Press, 1966, pp. 33-43.

31. DECAREAU R. Microwave defrosting and heating. Microw. Energy Appl. Newsl., 3, n° 3, 1970, pp. 9-12. (C.D.I.U.P.A., n° 30589).

32. DECAREAU R. Microwave baking applications up-date. Microw. Energy Appl. Newsl., 3, n° 1, 1970, pp. 3-5. (C.D.I.U.P.A., n° 30587).

33. DECAREAU R. Multi-energy source oven and microwave thawing of packaged meat. Microw. Energy Appl. Newsl., 8, n° 2, 1975, pp. 3-6. (C.D.I.U.P.A., n° 83689).

34. DECAREAU R. Microwaves in food processing. Food Technol. Aust., 36, n° 2, 1984, pp. 81-86. (C.D.I.U.P.A., n° 201725).

35. DECAREAU R. Microwave food processing equipment throughout the world. Food Technol., 40, n° 6, pp. 99-105. (C.D.I.U.P.A., n° 214710).

36. DECAREAU R., MUDGETT R. Microwaves in the food processing industry. (Food Science and Technology, A Series of Monographs). London, Academic Press, Inc., 1985, 247 p.

37. EDGAR R. Microwave systems. Industrial microwave processing. Raytheon, Publication PT-3332, 26 April 1972, 19 p.

38. EVERINGTON D., COOPER A. Vacuum heat thawing of frozen foodstuffs. Food Trade Rev., n° 7, 1972, pp. 7-13. (C.D.I.U.P.A., n° 47442).

39. FOSTER K. AYYASWAMY P., SUNDARARAJAN T., RAMAKRISHNA K. Heat transfer in surface-cooled objects subject to microwave heating. IEEE Trans. Microw. Theory Tech., MTT-30, n° 8, 1982, pp. 1158-1166.

40. FOURNAUD J., LE POMMELEC G., POMA J. Restitution de plats cuisinés surgelés. Paris, DGRST, Contrat n° 72.7.0511, 1975, 59 p.

41. FOURNET-FAYAS C. Applicateur hyperfréquence destiné à rayonner une onde hyperfréquence sur une surface allongée. Brevet français n° 77 06 546, 1er mars 1977.

42. FRANÇOIS D. Note sur les tunnels de type Gigatron. Publication Sté IMI-SA, s.d., 10 p.

43. FRANDSEN J. Fransk landsbyslagters en gros salg skabte grundlaget for den hypermoderne Kødvirksomhed. *Levnedsmiddelbladet/Supermark.*, n° 9, 1979, pp. 6-10.

44. FUJINO K. Physics of snow and ice. In : OURA H.- Comptes-rendus de la Conférence internationale sur la Science des Basses Températures, 1ère partie. Sapporo, ILTS, 1967, pp. 633-648.

45. GAC A. Limites et contraintes des procédés de décongélation. *Rev. Gén. Froid*, 66, n° 12, 1975, pp. 1047-1049.

46. GAZAN C., GRUNDER D., VINCENT A. Utilisation des micro-ondes dans l'industrie de la viande. Paris, DGRST, Action concertée : Technologie Agricole, Contrat n° 70 02 271 , 1971, 39 p. (C.D.I.U.P.A., n° 104784).

47. GILLIAT C. Microwave tempering of frozen meat. In : Proceedings of the IMPI Symposium, Monterey, 1971, p. 1.

48. HAFNER T. Auftauen von Fleisch mit Hochfrequenzenergie. *Kältetech.-Klim.*, 23, n° 6, 1971, pp. 185-186. (C.D.I.U.P.A., n° 33559).

49. HAFNER T. Auftauen von Fleisch mittels Hochfrequenzwarme. *Alimenta*, 10, n° 6, 1971, pp. 224-225. (C.D.I.U.P.A., n° 42126).

50. HAMID M. Microwave thawing of frozen materials and applicators therefor. U.S. Patent n° 1044331, 12 Dec. 1978.

51. HASTED J. Aqueous dielectrics. London, Chapman and Hall, 1973, 302 p.

52. HEEREN H. Von. Essais et expériences de décongélation par haute fréquence. In : Aliments congelés. Qualité des produits. Techniques de congélation (notamment cryogéniques). Paris, Institut International du Froid, n° 6, 1969, pp. 259-262. (C.D.I.U.P.A., n° 29827).

53. HOUWING H. Het ontdooien van vis en visprodukten. In : Comptes-rendus de la Journées d'étude "Techniques industrielles de décongélation pour matiéres premiéres surgelées". Kortrijk, ASBL Interfreez, 1977, 7 p.

54. HUSTOFT K. Mikrobølgetining. *Naeringsmiddelind.*, 9, 1978, pp. 19-20.

55. JAMES S. Thawing meat blocks using microwaves under vacuum. In : Proceedings of the European Meeting of Meat Research Workers, 30, Bristol, 1984. Bristol, Food Research Institute, s.d., pp. 61-62. (C.D.I.U.P.A., n° 209059).

56. JOYAL M. Chaleur. Paris, Masson et Cie, 1967, 271 p. (Collection Joyal-Provost).

57. KENT M. Fish muscle in the frozen state : time dependence of its dielectric properties. *J. Food Technol.*, 10, n° 1, 1975, pp. 91-102.

58. KENT M. Time domain measurements of the dielectric properties of frozen fish. *J. Microw. Power*, 10, n° 1, 1975, pp. 37-48.

59. KENT M. Microwave attenuation by frozen fish. *J. Microw. Power*, 12, n° 1, 1977, pp. 101-107.

60. KLEPACKI J. Ustalenie optymalnej czestotliwości pola dla szybkiego rozmrażania masla pradami wielkiej czestotliwości. *Rocz. Inst. Przem. Mleczarsk.*, 13, n° 39, 1971, pp..55-70. (C.D.I.U.P.A., n 40775).

61. KLEPACKI J. Zastosowanie pradów wielkiej czestotliwości do szybkiego rozmrażania masla przechowywanego w stanie zamrożonym. *Rocz. Inst. Przem. Mleczarsk.*, 13, n° 41, 1971, pp. 39-51. (C.D.I.U.P.A., n° 40778).

62. KUMPFER B. The donut cooker - a conveyor type microwave oven for processing frozen snacks. In : Proceedings of the IMPI Symposium, Boston, 1969, résumé p. 38. (Paper H1).

63. LE POMMELEC G. Les applications des micro-ondes dans les industries alimentaires et pharmaceutiques. In : Comptes-rendus de la Journée de formation sur les Techniques récentes du traitement des produits biologiques. Dijon, I.U.T., 18 mars 1977, 30 p.

64. MASSIEU G. Chimie minérale. Paris, J.B. Baillière et Fils, 1966, 381 p.

65. MEISEL N. La décongélation des denrées alimentaires par les micro-ondes. *Ind. Aliment. Agric.*, 86, n° 9-10, 1969, pp. 1251-1257. (C.D.I.U.P.A., n° 15325).

66. MEISEL N. Nouvelles applications des micro-ondes à l'industrie alimentaire. *Ind. Aliment. Agric.*, 87, n° 9-10, 1970, pp. 1059-1067. (C.D.I.U.P.A., n° 025542).

67. MEISEL N. Décongélation des denrées alimentaires par micro-ondes. *Ind. Aliment. Agric.*, 88, n° 3, 1971, pp. 297-304. (C.D.I.U.P.A., n° 35369).

68. MEISEL N. La décongélation par micro-ondes. Résumé du principe, description d'un tunnel de décongélation industriel et résultats obtenus. Publication Sté IMI-SA, 1971, 5 p.

69. MEISEL N. Les micro-ondes complément (indispensable) du froid ? *Rev. Gén. Froid*, 62, n° 4, 1971, pp. 267-281. (C.D.I.U.P.A., n° 32902).

70. MEISEL N. Le chauffage hyperfréquentiel dans l'industrie alimentaire. *U.I.E.*, VII, N. 313, 1972, 8 p.

71. MEISEL N. Tempering of meat by microwaves. *Microw. Energy Appl. Newsl.*, 5, n° 3, 1972, pp. 3-7. (C.D.I.U.P.A., n° 48750).

72. MEISEL N. Microwave applications to food processing and food systems in Europe. *J. Microw. Power*, 8, n° 2, 1973, pp. 143-147.

73. MEISEL N. Apport d'énergie instantanée au coeur des produits alimentaires, les micro-ondes. *Ind. Aliment. Agric.*, 91, n° 9-10, pp. 1203-1214. (C.D.I.U.P.A., n° 76032).

74. MEISEL N. Utilisation des micro-ondes pour la décongélation de produits alimentaires. In : Proceedings of the IMPI Symposium, Madrid, 1977. Publication Sté IMI-S.A., 21 p.

75. MEISEL N. Les micro-ondes : savoir... et...,bien faire. *Ind. Aliment. Agric.*, 101, n° 4, 1984, pp. 259-264. (C.D.I.U.P.A., n° 188055).

76. MEISEL N. Les micro-ondes dans les industries agro-alimentaires, bilan 1974-1984. *Ind. Aliment. Agric.*, 101, n° 10, 1984, pp. 929-932. (C.D.I.U.P.A., n° 193861).

77. METAXAS A., Properties of materials at microwave frequencies. Vol. 2. Industrial course. *Trans. IMPI*, Univ. Loughborough, 1973, pp. 19-47.

78. MORSE P., REVERCOMB H. UHF heating of frozen foods. *Electronics*, 20, n° 1, 1947, pp. 85-89.

79. MUDGETT R., MUDGETT D., GOLDBLITH S., WANG D., WESTPHAL W. Dielectric properties of frozen meats. In : Proceedings of the IMPI Symposium, Ottawa, 1978, pp. 37-39. (Paper 4.4.).

80. MUDGETT R., MUDGETT D., GOLDBLITH S., WANG D., WESTPHAL W. Dielectric properties of frozen meats. *J. Microw. Power*, 14, n° 3, sept. 1979, pp. 209-216.

81. NENIN J. Le chauffage par micro-ondes, homogène et instantané. *Usine Nouv.*, suppl. février 1979, pp. 78-86.

82. OHLSSON T. Low-power microwave thawing of animal foods. In : ZEUTHEN P., CHEFTEL J., ...Thermal processing and quality of foods. Barking, Elsevier Applied Science Publishers Ltd., 1984, pp. 579-584. (C.D.I.U.P.A., n° 197863).

83. OHLSSON T., BENGTSSON N., RISMAN P. The frequency and temperature dependence of dielectric food data as determined by a cavity perturbation technique. *J. Microw. Power*, 9, n° 2, 1974, pp. 129-145.

84. OHLSSON T., THORSELL U. Problems in microwave reheating of chilled foods. SIK-Svenska Livsmedelsinstitutet, SIK n° 386, and *J. Foodserv. Syst.*, n° 3, 1984, pp. 9-16.

85. PACE W. Dielectric properties of certain foodstuffs at selected frequencies and at different temperatures. *J. Microw. Power*, n° 3, 1967, p. 98. (Résumé de thèse).

86. PELLISSIER J. La décongélation et la déshydratation des produits alimentaires par micro-ondes. In : Comptes-rendus du Colloque sur l'Electricité dans les Industries de Conservation, Avignon, 13-14 nov. 1984, pp. 83-100.

87. PHAN P. Microwave thawing of peaches : a comparative study of various thawing treatments. *J. Microw. Power*, 12, n° 4, 1977, pp. 261-271.

88. PHAN P., DRI D. Décongélation des pêches par micro-ondes, étude comparative de divers traitements de décongélation. In : Réfrigération, congélation, entreposage et transport : aspects biologiques et techniques. I.I.F., Commissions C2, D1 et D2, Budapest, 1978. Paris, Institut International du Froid, 1978, pp. 211-215. (C.D.I.U.P.A., n° 135281).

89. PRIOU A. Ensemble des travaux. Thèse de doctorat d'Etat, Toulouse, Univ. Paul Sabatier, 1981, 699 p. (§ 3.2., pp. 437-479).

90. PRIOU A., DEFICIS A., GIMONET E. Microwave thawing of large pieces of beef. In : Proceedings of the IMPI Symposium, Ottawa, 1978, p. 61. (Paper 6.4.). (C.D.I.U.P.A., n° 143425).

91. PRIOU A., FOURNET-FAYAS C., DEFICIS A., GIMONET E. Microwave thawing of large pieces of beef. *J. Microw. Power*, 12, n° 1, 1977, p. 49.

92. RAO M., NOVAK A., Thermal and microwave energy for shrimp processing. *Marine Fish. Rev.*, 37, n° 12, 1975, pp. 25-30.

93. RIEDEL L. Enthalpie und spezifische Wärme von Fetten und Ölen im Schmelzbereich. *Kältetech.*, **8**, n° 3, 1956, pp. 3-4.

94. RIEDEL L. Kalorimetrische Untersuchungen über das Gefrieren von Fleisch. *Kältetech.*, **9**, n° 2, 1957, pp. 38-40.

95. RIEDEL L. Enthalpie - Konzentrations - Diagram für Eiklar. *Kältetech.*, **9**, n° 11, 1957, pp. 11-12.

96. RIEDEL L. Enthalpie - Konzentrations - Diagramm für Eigelb. *Kältetech.*, **9**, n° 12, 1957, pp. 12-13.

97. RIEDEL L. Enthalpie - Konzentrations - Diagramm für Volleimasse. *Kältetech.*, **10**, n° 2, 1958, pp. 2-3.

98. RIEDEL L. Enthalpie - Konzentrations - Diagramm für Weissbrot. *Kältetech.*, **11**, n° 2, 1959, pp. 2-3.

99. RIEDEL L. Enthalpie - Konzentrations - Diagramm für das Fleisch von mageren Seefischen. *Kältetech.*, **12**, n° 4, 1960, pp. 4-5.

100. RIEDEL L. Enthalpie - Konzentrations - Diagramm für Kartoffelstärke. *Kältetech.*, **12**, n° 12, 1960, pp. 12-13.

101. RIEDEL L. Enthalpie - Konzentrations - Diagramm für Backhefe. *Kältetech.*, **20**, n° 9, 1968, pp. 9-10.

102. ROGOV I., MALJUTIN A., ROSLOVA A., MITASEVA L., MIKLAŠEVSKIJ V., DŽANGIROV A. Kačestvennye pokazateli mjasa v blokah, temperirovannogo SVČ-energiej, i gotovyh kolbas. *Mjasn. Ind. SSSR*, n° 6, 1984, pp. 34-36. (C.D.I.U.P.A., n° 190448).

103. ROSENBERG U., BOGL W. Auftauen, Trocknen, Wassergehaltsbestimmung und Enzyminaktivierung mit Mikrowellen. *Z. Lebensm.- Technol. Verfahr.*, **37**, n° 1, 1986, pp. 12-19. (C.D.I.U.P.A., n° 208632).

104. SALE M. Utilisation des micro-ondes pour la décongélation des viandes. *Rev. Gén. Froid*, **64**, n° 4, 1973, pp. 345-354.(C.D.I.U.P.A., n°60168).

105. SALE P., VALIN C., VIGNE C. Décongélation par hyperfréquence de viandes en morceaux. Theix, INRA, Station Recherches sur la viande, s.d., 6 p.

106. SANDERS H. Dielectric thawing of meat and meat products. *J. Food Technol.*, **1**, 1966, pp. 183-192.

107. SCHLUSSELBURG G. Auftauen von Fleisch mittels Hochfrequenz. *Fleischwirtsch.*, **54**, n° 4, 1974, pp. 672-680. (C.D.I.U.P.A., n° 72167).

108. SKIERKOWSKI K. Postęp w wykorzystaniu grzejnictwa mikrofalowego w technologii żywnośći. *Przem. Spoz.*, **39**, n° 8-9, 1985, pp. 289-291. (C.D.I.U.P.A., n° 212712).

109. SLATER L. Hormel creates the self-renewing plant. *Food Eng.*, **46**, n° 12, 1974, pp. 55-58 plus un appendice : Some notes on microwave tempering, 1 p. (C.D.I.U.P.A., n° 79013).

110. SMITH D. Microwave heating for the meat industry. *Microw. Energy Appl. Newsl.*, **5**, n° 4, 1972, pp. 3-5. (C.D.I.U.P.A., n° 48751).

111. STUCHLY M., STUCHLY S. Industrial, scientific, medical and domestic applications of microwaves. *IEE Proc.*, **130**, part A, n° 8, nov. 1983, pp. 467-503.

112. SUARDET R. Thermodynamique. Paris, J.B. Baillière et fils, 1965, 320 p.

113. SWIFT J., TUOMY J. Evaluation of microwave tempering of meat for use in central food preparation facilities. *Microw. Energy Appl. Newsl.*, **11**, n° 1, 1978, pp. 3-10.

114. TERJUNG F. Untersuchungen über das Auftauen von tiefgefrorenem und gefrorenem Geflugelfleisch mit einem Mikrowellengerat. *Archiv Lebensm.-hyg.*, **25**, n° 3, 1974, pp. 62-63. (C.D.I.U.P.A., n° 69382).

115. THOUREL L., MEISEL N. Les applications industrielles du chauffage par micro-ondes. In : Comptes-rendus des Journées d'Information Electro-industrielles. Paris, EDF, nov. 1974, 12 p.

116. THOUREL L., PRIOU A., DEFICIS A., FOURNET-FAYAS C. Etude théorique de la simulation de la décongélation de quartiers de boeuf par hyperfréquences. Paris, DGRST, Rapport final n° 1/1089-MO, Convention de recherche n° 75.7.03.76, publ. CERT-ONERA, 1977, 141 p.

117. TINGA W., NELSON S. Dielectric properties of materials for microwave processing - tabulated. *J. Microw. Power*, **8**, n° 1, 1973, pp. 23-65.

118. VASILESCU S., NACEA M. Recherches expérimentales concernant l'utilisation du four à micro-ondes pour la décongélation et la remise en température des plats cuisinés surgelés. In : Comptes-rendus du Congrès International du Froid. **16**, Paris, 31 août-7 sept. 1983. Progrès dans la science et la technique du froid. Tome 3. Paris, Association pour le

Congrès International du Froid, 1983, pp. 663-669. (C.D.I.U.P.A., n° 217745)

119. WATERMAN J. Thawing large blocks of frozen whole fish. Programme Dep. Sci. Ind. Res. (G.B.) and Torry Res. Stn., Aberdeen, s.d., 2 p.

120. WESTPHAL W., SILS A. Dielectric constant and loss data. In : Technical Report AFML-TR-72-39, Air Force Materials Lab., Air Force Systems Command, Wright Patterson Air Force Base (Ohio), 1972, p. 199.

121. YOUNATHAN M., FARR A., LAIRD D. Microwave energy as a rapid-thaw method for frozen poultry. *Poult. Sci.*, _63_, n° 2, 1984, pp. 265-268. (C.D.I.U.P.A., n° 191389).

122. Décongélation par tunnel micro-ondes à 2450 MHz. Doc. IMI-SA, s.d., 3 p.

123. Divers comptes-rendus d'essais. IMI-SA, 1977-1980.

124. High-frequency thawing of food. *Microw. Energy Appl. Newsl.*, _1_, n° 2, 1968, pp. 3-12. (C.D.I.U.P.A., n° 104405).

125. Les performances des appareils de décongélation et de remise en température des plats cuisinés surgelés pour collectivités. EDF, Dir. Distribution, Div. "Techniques et Applications", 1974, pp. 1-17. (Journées d'Etudes A.F.F., Strasbourg, oct. 1974).

126. Les produits de la mer. Micro-ondes. Décongélation dans l'industrie du poisson. *Aliment.*, n° 51, 1977, pp. 64-65. (C.D.I.U.P.A., n° 103733).

127. Microwave heating. Leicester, Publ. Magnetronics Ltd, s.d., 8 p.

128. Microwave tempering improves fruit texture of Weight-Watcher's dessert. *Prep. Foods*, _153_, n° 6, 1984, pp. 78-79.

129. Microwave tempering makes possible frozen blocks to frozen patties. *Meat Ind.*, n° 3, 1975, 2 p.

130. Microwave tunnel thawer Gigatron with conveyor belt. Commercial publ. Narasaki Sangyo Co., Ltd, Tokyo-L.M.I., (engl./jap.), s.d., 12 p.

131. Ontdooien bij - 30° C in de micro-golventunnel Gigatron F. In : Comptes rendus de la Journée d'Etude sur les "Techniques industrielles de décongélation pour matières premières surgelées". Kortrijk, ASBL Interfreez, 1977, 7 p.

132. Recent advances in industrial microwave processing in the 896/915 MHz frequency band. Leicester, Publ. Magnetronics Ltd, s.d., 12 p.

133. Some notes on microwave tempering. *Food Eng.*, _46_, n° 12, 1974, p. 59.

134. Tempering med mikrobølger åbner nye muligheder ved forarbejdningen. *Livnedsmiddelbladet/Supermark.*, n° 3, 1979, pp. 24-25.

135. Tempers meat in conveyorized tunnel. *Food Eng.*, _44_, n° 11, 1972, pp. 200-202. (C.D.I.U.P.A., n° 50563).

136. Thaw at -30° C microwave tunnel Gigatron F. Publ. commerciale Sté IMI-PREMO, s.d., 4 p.

137. Tunnel de décongelation/cuisson par micro-ondes Gigatron. Publ. commerciale Sté IMI-PREMO, s.d., 2 p.

Chapter 3

Drying

We have already mentioned (Part II, Chapter 1) that the drying of industrial products is a lengthy and very energy-consuming operation. The drying of food products is no different, except that it often presents a high temperature sensitivity. In many cases, the drying of food products can be significantly improved, both quantitatively and qualitatively, by using microwaves.

In this chapter we review the application of microwaves to the drying of food products, be they at the experimental or production stage, or used for the determination of the dry matter content of food. With the exception of vacuum sublimation (microwave freeze-drying), they are all based on vaporization.

At least one of the processes described in this chapter (i.e., the final drying of potato chips) is of only historic interest, having had considerable success in the past. By contrast, the drying of pasta at atmospheric pressure, the drying of preconcentrated products by expansion under vacuum (Gigavac procedure), or the determination of dry matter, offer such advantages that they have an assured future in the food industry.

3.1 Vaporization

Vaporization (i.e., the transformation of a liquid to the gaseous state) occurs at all temperatures. Depending on local conditions, it has the characteristics of boiling or evaporation.

Evaporation is the production of vapor at the surface of a liquid. If the space above the liquid is finite, vapor is formed until an equilibrium is established between molecules in the liquid and vapor states. In equilibrium, the pressure exerted by the vapor molecules on the walls of the vessels is the saturated vapor pressure P_m at the temperature of the vessel. This limiting pressure is reached whether the vessel is evacuated or full of gas - although it is reached more slowly in the second case - provided the temperature is the same. The total pressure of the gas mixture is the sum of the partial pressure of the gas and the saturated vapor pressure of the liquid.

The saturated vapor pressure P_m depends only on the nature of the gas and on temperature. There are no rigorous theoretical expressions for P_m as a function of temperature, but there are some semiempirical expressions derived from the Clapeyron equation

$$\log P_m \simeq \alpha - \beta/T - \gamma \log T \qquad (3.1)$$

For example, for water between 0 and 200°C, the Dupré formula gives

$$\log P_m \simeq 18.185 - 2795/T - 3.868 \log T \qquad (3.2)$$

where T is in K and P_m in Pa [59, 189].

If the space above the liquid is not finite, evaporation continues as long as liquid is present and whatever the temperature. The saturated vapor pressure is then never reached: We have dry vapor whose partial pressure P_1 is always less than P_m. The evaporating flux depends on the external pressure, the temperature, the exposed surface area of the liquid, and the motion of the gaseous phase. The quantity of vapor that forms at any given time is proportional to $(1/P_0)(P_m - P_1)$ and $T_1 - T_0$, where P_1 is the partial vapor pressure, P_0 is the atmospheric pressure, T_1 is the temperature of the surface of the liquid, and T_0 is the air temperature.

Boiling is the production of vapor within the body of the liquid, which results in the formation of gas bubbles. In contrast to evaporation, boiling takes place at a particular well-defined temperature at which the saturated vapor pressure is equal to the prevailing atmospheric pressure. If the pressure does not vary, the temperature remains constant for the duration of the process. The heat supplied to the liquid is consumed uniquely, as latent heat, in the vaporization process. The subject is discussed in greater detail in [107], [57], [105], [101].

3.2 Drying at atmospheric pressure

The laws that govern heat and mass transfer during drying are described by (1.2) and (1.3) of Part II in which the diffusion coefficient D_m represents all possible mechanisms of internal moisture transfer within the material. This coefficient is a monotonic function of temperature, so that any rise in temperature must improve the drying rate $\partial m/\partial t$.

However, this improvement has limits: in the case of biological materials, the highest drying rate is not always desirable because quality may suffer at excessively high temperatures [103]. For many thermally degradable or labile components, the optimum temperature is about $60°C$, which prevents drying by boiling at normal pressure.

Conventional drying is performed in two stages. The first takes place at constant rate: water leaves the interstitial spaces and its level in large capillaries decreases, so that capillaries contract by an amount that is practically equal to the volume of evacuated water. The surface of the product remains wet because internal diffusion and mechanical collapse produce an internal influx of water that compensates for evaporation. The second stage is characterized by a decreasing rate: the level of water in fine capillary tubes is reduced, the texture of the cells gradually hardens. Contraction continually slows down, and this drying "tail" has an asymptotic limit that corresponds to the structure of the dry product.

The difficulties encountered in conventional drying are mainly associated with the increasing hardening of outer layers that act as thermal insulators for the wet inner zones. The use of microwaves instead of conventional heating can accelerate the rate of mass transfer $\partial m/\partial t$ and provide an efficient water pump for the following reasons:

- Water heating is selective and almost exclusive. This helps in the expulsion of the last fraction of liquid and produces a self-regulating effect on temperature at the end of the drying process.
- The process is independent of conduction and, in particular, is unaffected by the thermally insulating, dry outer layers.
- A temperature gradient directed from the interior to the exterior is established and favors the expulsion of water.
- The surface crust is not formed because the ambient air is not heated [119].

In practice, the use of microwaves is restricted to only one stage of the drying process, namely, predrying, intermediate stage, or final drying. In the last case, microwaves allow rapid removal of the last vestiges of water, and thus avoid the asymptotic drying tail. In general, drying systems employ microwaves and hot air over periods calculated to optimize the efficiency [83], [84], [136], [117], [116], [17].

3.2.1 Final drying of potato chips

Most publications on this subject appeared around 1967-1968, and there are hardly any after 1973. This was one of the first and very important industrial applications of microwaves

[155]. In 1970, there were twenty 50-kW units and nine 75-kW units (i.e., a total of 1.7 MW of microwave power) [138]. These 29 dryers operated mostly in Canada and the United States, with some units in Sweden and the United Kingdom [135], [52].

It is considered that the elimination of Maillard's reactions is a definite qualitative advantage of microwave drying. These reactions are triggered by temperature and occur between groups of free amino acids and reducing groups (carbonyls and aldehydes) of carbohydrates (e.g., reducing sugars such as glucose and lactose) and result in the browning of the product. Potatoes can have a high content of reducing sugars, and storage has an aggravating effect on the phenomenon. When such potatoes are used in the manufacture of chips, the last stage of the frying procedure is accompanied by Maillard's reactions that make the product unfit for consumption. It was therefore necessary to stop the frying before browning, but the chips retained too much moisture. Conventional techniques being unsuitable, microwaves were adopted as a unique solution of this problem.

The first microwave applicator developed by Magnetronics of Leicester (UK) was a cooking/drying oven for chips that incorporated an English electric valve 896-MHz, 25-kW magnetron and a radiating meander waveguide (Part 1, Section 4.3.1). In the United States, the industrial microwave division set up in the 1960s by W.C. Brown at Raytheon had, among its first clients, Frito-Lay for a 60-kW tunnel operating at 915 MHz. Other manufacturers also appeared on the market (e.g., Cryodry) which has since become Microdry in San Ramón, California, and the Atherton Division of Litton [158].

This application was, unfortunately, short-lived. Several factors contributed to its decline: firstly the weather conditions in North America were such that the proportion of sugar-rich potatoes was greatly reduced on several successive harvests, and, secondly, certain imperfections in the design of the microwave equipment resulted in variable final moisture contents; thirdly, and perhaps most importantly, the storage conditions were improved and new high-volume fryers arrived on the market [135, 52].

3.2.2 The drying of pasta

Pasta is usually dried for about 10 hours at 40°C with well-controlled humidity levels. The aim is to dry it without hardening the surface, so that good rehydration on cooking is possible [131, 132]. Rapid drying results in surface cracks, which lead to further cracking on rehydration and to the escape of starch into the cooking water.

The value of the product is thus directly linked to the quality of the drying process, which involves long drying times and, hence, large static dryers occupying significant floor space and creating serious risk of bacterial contamination. This has stimulated interest in microwave drying after it was shown that it prevents surface drying, even at final temperatures close to 100°C, and that the escaping vapor creates micropores that facilitate and speed up rehydration on cooking [131, 132].

The first industrial microwave dryer was installed in 1971 by Lipton at Bramalea near Toronto, Canada. It was designed for the drying of vermicelli for dehydrated soups [140]. In view of the cost of microwave installations, the industry then looked toward combined hot-air/microwave solutions. According to Meisel [139] and Decareau [50], the best solution involves the following sequence:

- hot-air predrying (70–80°C) of extruded pasta, which reduces the moisture content from 30% to 23% in less than 1 min and then to 18% in 35 min;
- microwave (915 MHz) and hot-air drying at 80–90°C, reducing the moisture content to 13–14% in 12 min;
- hot-air dyring at 74°C, which is a sufficiently low temperature to prevent surface cracking by thermal gradients; this results in an additional water loss of 0.5 to 1% in one hour.

The Microdry Company produces dryers based on this principle. They provide 60 kW at 915 MHz and are capable of an output of 1,800 kg hr^{-1} [224], [142]. At least 27 of these dryers are in service in the United States and in Canada [52].

The process offers several advantages:

- an average 75% saving in floor space, the average length of the tunnel being 8.2 m as against 36–48 m in conventional drying;
- an energy saving of 50 to 60%;
- saving of time: 90 min instead of 8–10 hr;
- a 60% saving on maintenance and 75% on cleaning costs (6 hr instead of 24 hr per person);
- a 20–25% saving in total manufacturing costs;
- partial sterilization of pastas (2,500 bacterial colonies per gram against 32,000 after traditional drying);
- better product quality (the yellow color of egg pasta is preserved) and better texture, so that cooking is easier [139, 142, 50, 52].

3.2.3 Miscellaneous food products

It is worth mentioning at this stage the developments at SFAMO of Plombieres-les-Bains (Vosges, France) whose tunnel ovens (AIMT, Fig. 3.1) are used to dry different products. A 270-kW tunnel is working in Brittany and is used to treat 23 different products, including meat granules and fibers destined for soups, whose moisture level is reduced from 55% to 3%. Other than the improved organoleptic quality, the most interesting advantage is the bacteriological stability. Other 60-kW tunnels operating at 2.45 GHz are used for the final drying of carrots, onions, and peas, reducing the moisture level from 14% to 4% [13].

In the drying of onion, the conventional process is accompanied by a 70% reduction in bacterial infestation. Unfortunately this is followed by a new proliferation that may reach 140% of the initial value. Trials performed by Gentry International have shown that final microwave drying reduces by 90% the initial infestation while at the same time achieving a 30% energy saving. The final product is reported to be rigorously identical in sensory quality with that obtained by conventional drying [218].

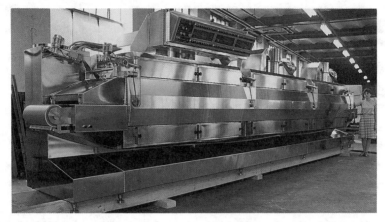

Figure 3.1 The AIMT drying tunnel manufactured by SFAMO.

Also available in Europe are the Gigatron microwave ovens manufactured by IMI-PREMO (Chapter 3, Section 2.3.2) and the equipment developed by Magnetronics. The latter evaporates $130 \, \mathrm{l \, hr^{-1}}$ of water from an initial temperature of $20°\mathrm{C}$, using $120 \, \mathrm{kW}$ of microwave power at $896 \, \mathrm{MHz}$. This means that for a material with 10% moisture content that needs to be reduced to 5%, the output is $2,000 \, \mathrm{kg \, hr^{-1}}$ [220].

Microwave drying trials have been performed on a number of products that have not reached the industrial stage. For example, spinach leaves were found to undergo extensive internal structure changes whilst their surface maintained a satisfactory aspect and their chlorophyl content was the same as in conventional drying [10]. Microwave drying of tomato paste [171] and its expansion into mousse was performed six times faster than in hot air, and at a higher temperature, but without the appearance of any trace of thermal degradation.

Animal fats can be dehydrated by microwaves three times faster than by conventional techniques. Moreover, microwave-treated fats have excellent quality and a lower and stable acidity [153]. Conclusive experiments [79], [80] have established three points in favor of microwaves in the case of oleagineous seeds: slightly better conservation, higher husking rate (94%

instead of the 60% in conventional), and almost complete destruction of sulphurous enzymes that are toxic to animals.

Finally, it is appropriate to mention a recent application to potatoes, namely, predrying before frying. In the United States, 51% of potato production is consumed in prepared meals, with french fries accounting for 47% of this. The quality of these french fries which are frozen and usually reheated in a domestic microwave oven, leaves much to be desired unless care is taken to predry the potato segments by microwaves prior to frying, which considerably improves the texture and color. The process is patented [174] and has been described in the literature [26].

3.3 Drying at low pressure

3.3.1 Freeze drying

Drying by vacuum sublimation, or lyophilization, produces products of high aromatic and nutritional value. The direct transition to the gaseous phase avoids the draining of soluble products, and the structure is generally well preserved. Freeze-drying is the ideal quality-preserving process, but it is also complex, slow, and costly.

The sublimation of ice to water vapor can take place at temperatures below 0.0098°C and pressures below 610 Pa (Fig. 3.2). In practice, the product to be dried is frozen to −40°C and held in a container at a pressure of not more than 70 Pa. Freeze drying consumes 2828 J of energy per gram of ice [78], [40], of which $334\,J\,g^{-1}$ is pent in fusion and $2494\,J\,g^{-1}$ in vaporization.

There are several sources of energy that can be used in freeze-drying, including infrared irradiation, radiating panels, heat conductors inserted into the product, and gas circulation. All these techniques rely on thermal conduction. Unfortunately, during this process, a layer of dry product with very low thermal conductivity grows while the ice/vapor interface

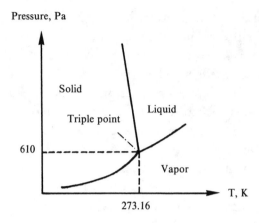

Figure 3.2 Equilibrium curves for water [189].

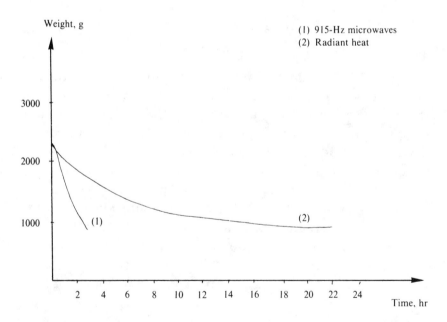

Figure 3.3 Dynamics of microwave lyophilization of meat patties and minced beef [93].

recedes. This progressively slows down the process. The difficulty may be obviated by the use of microwaves. As noted

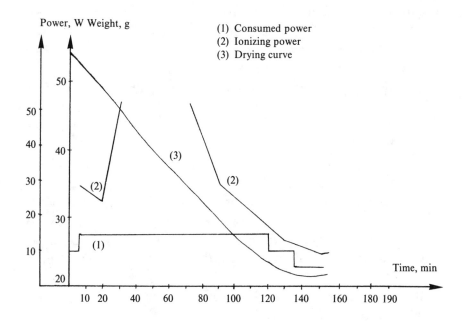

Figure 3.4 Drying curve and power used in microwave lyophilization of meat patties and minced beef [93].

in Section 3.2, frozen products have low dielectric loss factors, so that the incident wave passes through the material and is reflected many times without significant attenuation. This ensures uniform heating but requires additional protection for the source in the form of an isolator.

Microwave freeze-drying must avoid the onset of ionization phenomena (Part I, Section 4.3.2.5). Figure 4.18 in Part I shows that the low pressures required by the process are very close to the minimum of the Paschen curve, so that it is essential to keep the microwave power in the cavity low enough to prevent the formation of plasma.

There are a number of published studies on this subject [36], [37], [38], [39], [41], [47], [48], [24], [25], [86], [87]. Burke and Decareau [24] have reported that the optimum pressure for the treatment of fruit, and of high sugar content products in general, is on the order of 33 Pa. Sugar solutions have freezing points that are significantly lower than the freezing point of water, so that the use of lower pressures avoids liquefaction.

A study of microwave lyophilization was performed on an AMSCO pilot plant in which the vacuum chamber contained a resonant cavity fed by an RCA 8501 tube running at 915 MHz with adjustable output power in the range 0–500 W. The condenser was maintained at −55°C and the conclusions of the study [92], [93] were as follows:

- the thickness of the layer under treatment had little effect on the rate of drying: the same time was required to dry 6 and 25 mm of minced meat or 18 and 63 mm layers of peas.

- the saving of time is considerable in comparison with drying by radiating panels: the duration of the process is reduced by a factor of 3 and 7 for products 12 and 25 mm thick, respectively.

- the power level must be kept low enough to avoid reaching field values that can produce a discharge.

Figures 3.3 and 3.4 show the drying curves obtained for minced meat. Raw beef steaks were lyophilized in three hours at 2.45 GHz under a pressure of 200 Pa [75], [76], [77].

The first trials performed on a Microvac prototype, developed by IMI-SA, were performed in 1971. The Microvac consisted of a cylindrical cavity 1 m long and 63 cm in diameter. It was fed by a 2.5-kW magnetron operating at 2.45 GHz. The vacuum was produced by an Alcatel V 2100 pump incorporating a vapor trap. For whole raspberries loaded at $10 \, \text{kg} \, \text{m}^{-2}$, the lyophilization cycle was less than 4 hr. For a coffee extract containing 29% of dry matter loaded at 5 and $10 \, \text{kg} \, \text{m}^{-2}$, the cycle was 5 hr and 4 hr, respectively. The final product was of good quality, and the saving of time was about 66% as compared with conventional treatment.

During the same decade [135], Nestlé of Switzerland was also engaged in a study of the process. Experiments reported in the early 1970s on beef steaks 15 mm thick showed [125], [126] that the lyophilization procedure can be completed in less than 6 hr and may be further accelerated by the use of higher pressure (133 Pa instead of 27 Pa). These studies led to a model for the microwave freeze-drying, which had as its parameters the electric field (100 or 130 V cm^{-1}), the initial temperature ($-15°$C and $-25°$C), and the thermal diffusivity $a = \Lambda/\rho C_p$ (Part III, Section 2.2.4; 14 and 21 cm^2 s^{-1}). The drying time was found to be roughly inversely proportional to the square of the electric field. Neither the temperature nor the diffusivity affected the drying time.

The efficiency of heating by contact, by infrared radiation, and by microwaves in the lyophilization process was compared in 1976 [42]. It was found that microwaves and infrared radiation permit shorter cycles as compared with conduction at the same pressure (67 Pa), temperature ($-50°$C), and load (10 kg m^{-2}), and this despite lower power consumption (60 W RF). The chemical characteristics of vegetables treated by the three procedures were identical and remained so after two months of storage.

The lack of systematic studies that would permit the optimization of the process was noted in [77], [126]. In fact, microwave lyophilization is practically a dead subject and has been so for many years despite some recent studies [53], [52] and a 1986 patent [52]. We have mentioned only some of the references published in the 1960s and 1970s. Others include [70], [150], [1], [110], [111], [85], [159], [78], [81], [219], [113], [160], [40], [66], [162], [6], [199], [124], and [202].

3.3.2 Expansion in vacuum

By contrast to microwave freeze-drying, the physical mechanism involved here is not sublimation, but vaporization from the liquid phase.

Drying at low pressures has several major advantages:

- since the evaporation rate depends on $(1/P_0)(P_m - P_1)$, where P_m and P_1 are independent of the ambient pressure, the rate is inversely proportional to P_0; in perfect vacuum, vaporization would be instantaneous.
- the rate of evaporation is also proportional to the surface area S of the liquid-vapor interface, which is considerably

increased when the product foams as it expands.

- for water between 0 and 30°C, the saturated vapor pressure is given by

$$P_m \simeq 133.3T \qquad (3.3)$$

where P_m is in Pa and T in °C, and the product boils when $P_m = P_0$. For example, for $P_0 = 267$ Pa, the boiling point is $T = 20$°C, which is well outside the temperature range that may harm the product.

In conventional expansion in vacuum, the product to be dried must contain at least 60% of dry matter and is introduced into an enclosure at room temperature under low pressure: 130–460 Pa for lemon juice and 660–2700 Pa for tea and most other products. The product adheres in the form of a viscous layer to a stainless-steel conveyor belt. It first passes through a zone containing radiating panels under the conveyor belt, which cause some evaporation and the formation of a porous structure. This is followed by a heating drum on which the evaporation is much more rapid, and then by another series of radiating panels placed just above the conveyor belt, which helps to reduce the moisture content to 2–3%. At this stage, the product has an elastic consistency. It is then cooled by a cold drum placed at the other extremity of the enclosure where it is removed by a scraper. The drying cycle takes 15–120 s, depending on the product and the layer thickness. Figure 3.5 shows the main features of the dryer.

This process, developed since 1956 by Copley and Notter and, for milk, by Sinnamon, gives products that are better preserved and more soluble than those obtained by atomization or any other procedure performed at atmospheric pressure [229]. The weakness of the technique is the poor heat transfer which is not helped by the expansion under vacuum because the resulting foam has a very low thermal conductivity.

The possible improvement of this process by the use of microwaves was the subject of a collaboration between the IMI-SA, *Institute of Research Applied to Beverages* (IRAB) a subsidiary of Pernod-Ricard, and the *French Institute for Fruit*

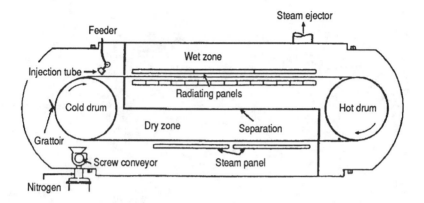

Figure 3.5 Dryer using expansion in vacuum [198].

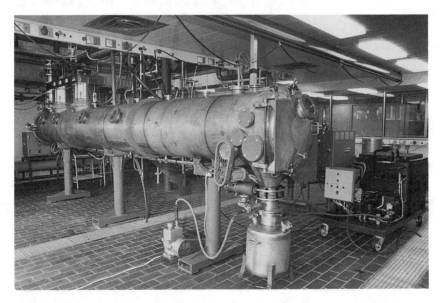

Figure 3.6 The Gigavac 7.5 microwave tunnel dryer for expansion in vacuum (courtesy of IMI-SA).

Figure 3.7 The Gigavac 50 microwave tunnel dryer for expansion in vacuum (courtesy of IMI-SA).

Research Overseas (IFAC) . The results of these efforts was the development in 1972 of a static pilot plant (the Microvac 2000, Section 3.3.1), which received the French Innovation prize in 1972 [4], [7], [146], [135], [40], [6], [145].

The Gigavac 7.5 (Fig. 3.6) is the successor to the Microvac 2000. It incorporates a cavity 6 m long and 0.63 m in diameter. It is fed by two magnetrons producing 5 and 2.5 kW and has a PTFE/fiberglass conveyor. It is capable of continuous processing at an adjustable speed of between 1 and 30 m hr^{-1}. The water extraction capacity of the equipment is 8 l hr^{-1} [147, 82, 145].

Another landmark was reached in 1977 with the development of the Gigavac 50, which is a highly optimized microwave equipment producing uniform heating with good energy efficiency (Figs. 3.7 and 3.8). The Gigavac 50 has a stainless-steel

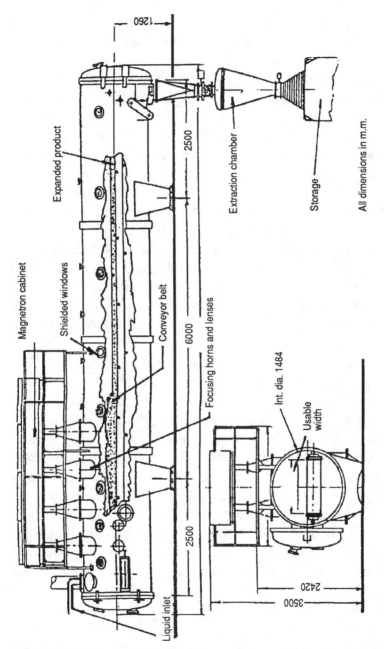

Figure 3.8 The Gigavac 50 microwave tunnel dryer for expansion in vacuum (courtesy of IMI-SA).

vacuum cavity, 1.50 m diameter and 12 m long, equipped with a number of shielded windows that are used to view the drying process. The eight 6-kW, 2.45-kHz magnetrons are placed above the enclosure and are coupled to it by a rectangular waveguide, a circulator, and a large circular horn containing a diverging dielectric lens. This assembly is isolated from the vacuum cavity by a high-density polyethylene window. A PTFE/fiberglass conveyor belt, 1.20 m wide, runs on silicon rubber guides and transports the product, which is introduced into the enclosure by three independent pumps in order to allow good control of product spreading. Two of the magnetrons are automatically controlled according to the temperature of the product via an infrared pyrometer, and the vacuum is regulated to within ± 130 Pa between two limiting values. A Redler device transports the dry product to an intermediate container. When the predetermined threshold product level is reached, the container is automatically isolated and emptied into another receptacle at atmospheric pressure.

The product is usually preconcentrated to at least 60–65% of dry matter in order to reduce the cost of the treatment and to produce a fine and homogeneous foam [96]. The pressure in the enclosure is 1–1.3 kPa. The product expands and cools down as a result. The temperature then gradually increases as the product passes beneath radiating horns, but without ever exceeding 30–40°C. It then progressively decreases in the cold zone, which is separated from the treatment zone by metal screens. After cooling, a layer of 8–10 cm of hard porous substance is obtained. It has a very low density (about $0.05 \, \mathrm{g \, cm^{-3}}$) and a residual moisture content of approximately 2%. The conveyor belt then passes under a scraper and brush, and the dry product is collected in the form of fine slithers or powder that is readily soluble.

The characteristics of the final product are remarkably constant because the irradiation system has low directivity and produces a very uniform electric field. It is designed to prevent the appearance of a high local field maximum capable of producing plasmas and arcs. In addition, the design of the treatment chamber makes systematic use of quarter-wave traps (Part 1, Section 4.3.2.3).

The rate of evaporation in the Gigavac 50 is around $49 \, l \, hr^{-1}$ for a product that has an initial dry-matter content of 63% [8, 52]. The quality of the final product depends very much on the preconcentration treatment that must preserve product structure, color, and taste. For this reason, the techniques that are particularly recommended for this process are membrane separation, predrying under vacuum with recovery of aroma, osmosis, and, particularly, cryoconcentration [145].

The Gigavac gives excellent results for concentrated fruit juices such as orange, lemon, grapefruit, pineapple, grapes, strawberries, raspberries, blackcurrent, grenadine, mango, papaya, and guava. It is also good for plant extract; herbal tea mixtures; vegetables such as onion, garlic, and asparagus; mushrooms; natural colorants; essential oils; and meat extract, as well as numerous chemical and pharmaceutical products such as proteins, enzymes, lactic fermenting agents, and so on [145], [209]. It was reported in [4], [7], [6], that nutritional quality was unaffected by the treatment, and, in particular, the proteins, sugars, and organic acids were not subjected to any alteration. It was also noted [144] that the treated fruit lost very little (if any) of vitamin C, there were no traces of hydroxymethylfurfurol, and there was no change in color. By contrast, the structure of solid products was less well preserved in comparison with freeze-drying [145].

With respect to loss of aroma [96], Table 3.1 lists the results obtained for passion-fruit juice, starting with a unique concentrate containing 28% of dry matter. It was concluded in [96] that the initial high level of dry matter, compatible with the use of microwaves, was responsible for the retention of aroma. The aroma is governed by the size of the molecule, and even the smallest aromatic molecule migrates with greater difficulty than water across dry matter, which thus plays the role of a molecular sieve. Organoleptic evaluations performed on samples of reconstituted orange juice show that the sensory qualitites are practically identical for powders obtained by freeze-drying and by microwaves. The cost of the treatment was found to be on the order of $1 per liter of evaporated water [142]. Users of Gigavac 50, or of the 10-, 15-, and 25-kW versions include Pampryl, Nestlé, Coca-Cola, Beecham, and Hoechst [142].

3.3.3 Various processes

Recent literature reports on several vacuum drying processes not involving expansion for a variety of products. For example, cotton grains with 67% moisture content were treated under vacuum and by microwaves to improve their content of free fatty acids [186].

A pilot plant at the experimental station of the American Department of Agriculture at Tifton (Georgia) is described in [50]. This plant was developed by the McDonald Douglas Aircraft Corporation of St. Louis (Missouri) and was constructed by the Aeroglide Corporation. It comprises a treatment chamber, 7 m high and 0.9 m in diameter, which is partially evacuated (3.4–6.6 kPa) and is fed by a 6-kW magnetron working at 2.45 GHz. Trials performed on soya grain led to the construction of a 150-kW, 915-MHz system designed to eliminate 2% moisture content of this product. The output is $5.9 \, \text{m}^3 \, \text{hr}^{-1}$. Finally, excellent results have been obtained in microwave dehydration of oil in a nitrogen atmosphere at 4 kPa [185].

3.4 Determination of dry content

The standard method of determining the moisture content is to heat the sample in an oven or in a water bath for a time long enough for the residual water to become negligible, and then compare the weight at the beginning and the end of the operation. This is usually a lengthy procedure because of the asymptotic character of the drying process. For example, the standard method used in the dairy industry requires 5–7 hr

Table 3.1

Retention of volatile components by fruit juice powder (passion fruit) [96]

	Atomization		
	a	b	Retention Level
Volatile substances (mg/ml)			11.0
Ethyl acetate (mm^2)	1 155	33	2.9
Methanol	75	112	150.0
Ethanol	6 622	443	6.7
Ethyl butyrate	266	19	7.2
	Freeze drying		
	a	b	Retention Level
Volatile substances (mg/ml)			33.1
Ethyl acetate (mm^2)	1 155	272	23.5
Methanol	75		
Ethanol	6 622	2 185	32.8
Ethyl butyrate	266	114	429.0
	Microwaves		
	a	b	Retention Level
Volatile reducable substances (mg/ml)			82.5
Ethyl acetate (mm^2)	950	295	31.15
Methanol	69	29	42.50
Ethanol	5 400	3 078	57.00
Ethyl butyrate	207	122	59.2

The volatile components are expressed in units of the area under the corresponding peak on the chromatograph.

(a) Raw material
(b) Raw material reconstituted from powder

for a 10 g sample. Such lengthy procedures are unsuited to the ordinary needs of modern industry [31]. There are faster procedures such as vacuum drying and the Schultz method. The former is not very accurate if the duration of the drying process is less than 2 hr and the latter suffers from systematic errors because of the small sample volumes. The only ultrafast procedure available is microwave dessication in a laboratory oven. The process is accurate and reproducible for dairy [31] and sugar products [228].

It is also worth mentioning the Labotron 600 manufactured by IMI-PREMO. This applicator has an adjustable output 0–600 W at 2.45 GHz [230] and offers the facility of vacuum drying of thermodegradable products. A 100-g sample of flour reaches 136–137°C in 6–7 min at atmospheric pressure, whilst the temperature of 60°C is not exceeded at the reduced pressure of 18 Pa.

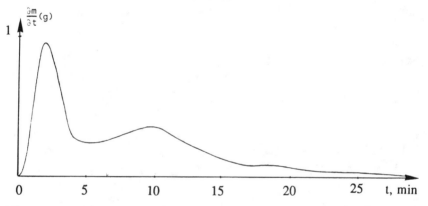

Figure 3.9 Labotron 600 drying curve for sunflower seeds [209].

Figure 3.9 shows the dehydration curve obtained for a sample of sunflower seeds. The use of the Labotron highlights some irregularities of the drying process (e.g., a slowing down, followed by an acceleration, and, finally, an asymptotic tail with a number of inflections that may be due to the evaporation of volatile products other than water or to the creation of metastable states leading to the release of vapor at certain predetermined internal pressures) [209].

The microwave drying technique is becoming widespread in the food industry. In particular, it is considered satisfactory for dairy products [31], [128], [175], [44], [11], [194] as well as for flour [45], potatoes [56], and tomato products [32]. For meat, five minutes are usually enough for samples of 30–50 g;

Table **3.2**

Measured water content of wet fish (cod) [33]

	Moisture content after drying for:	
	24 hours	**312 hours**
Air oven at 101°C	80.98 ± 0.03 %	81.16 ± 0.03 %
Air oven at 105°C	81.16 ± 0.07 %	81.31 ± 0.08 %
Vacuum oven at 70°C	80.97 ± 0.10 %	81.12 ± 0.10 %
Vacuum oven at 101°C	81.17 ± 0.16 %	
Microwave, 90 min	81.31 + 0.27%	

Table **3.3**

Measured water content after microwave drying followed by
drying with hot air [33]

	Cod	Herring	Mackerel
Microwave drying, 90 min	81.31 ± 0.27	63.55 ± 0.50•	77.58* ± 0.35
Subsequent drying with hot air (101°C)			
24 hr	81.42 ± 0.23	63.62 ± 0.50	77.59 ± 0.29
48 hr	81.45 ± 0.48	63.63 ± 0.28•	77.61 ± 0.28•
120 hr	81.46 ± 0.23		77.59 ± 0.28•
168 hr	81.48 ± 0.23	63.66 ± 0.48	77.58 ± 0.27

* 72 min • Not significant

there are no significant differences between this and the standard method [58]. In the case of fish, microwaves offer the best method to replace batch drying. Table 3.2 compares the accuracy of different existing methods. The absolute accuracy can be estimated from Table 3.3 whose data were obtained by microwave drying for 90 min, followed by hot-air drying for 24–168 hr [33]. Microwaves ensure remarkable accuracy in a very short time; in infrared drying, the same accuracy can only be obtained by successively drying eight samples for 90 min each.

Further information on this subject can be found in [19], [9], [29], [30], [163], [108], [115], [192], [88], and [23].

References

1. ALLAIRE R. Microwave heating in freeze-drying. *Trans. Am. Soc. Agric. Eng.*, 9, n° 6, 1966, pp. 752-753, 758.

2. ALLAN G. Microwave drying of solids. *J. Microw. Power*, 2, n° 3, 1967, p. 78. Thesis summary.. Electrical Engineering Department, University of Alberta (Can.).

3. ALLAN G. Microwave moving-bed dryers : a feasibility study. *J. Microw. Power*, 3, n° 1, 1968, pp. 21-28.

4. ARCHIERI C. Utilisation des micro-ondes dans la déshydratation sous vide des fruits et des extraits de fruits. Paris, Action concertée DGRST, Technologie Agricole, Contrat n° 71-72-804, 1971, pp. 1-38. (C.D.I.U.P.A., n° 084026).

5. ARCHIERI C. Déshydratation par micro-ondes. In : La lyophilisation dans les industries alimentaires. Paris, Compagnie Française d'Edition. Coll. Génie Alimentaire, 1976, pp. 137-150 (annexe 4).

6. ARCHIERI C. Déshydratation par micro-ondes. Evaporation sous vide. Sublimation sous vide. *Bios*, 8, n° 1, 1977, pp. 10-21. (C.D.I.U.P.A., n° 102774).

7. ARCHIERI C., BRICOUT J., MENORET Y. Déshydratation sous vide par micro-ondes. *Ind. Aliment. Agric.*, 88, n° 3, 1971, pp. 279-288. (C.D.I.U.P.A., n° 35367).

8. ATTIYATE Y. Microwave vacuum drying : first industrial application. *Food Eng. Int.*, 4, n° 2, 1979, pp. 30-31. (C.D.I.U.P.A., n° 126526)

9. AULENBACH D. Use of microwave oven for drying samples. In : Proceedings of the 28th Annual Purdue Industrial Waste Conference, 1973. Purdue University, Lafayette.

10. BACCI M., CHECCUCCI A., CHECCUCCI G., PALANDRI M. Biophysical and cellular effects of microwaves interacting with plant tissues. *J. Microw. Power*, 20, n° 3, 1985, pp. 153-159.

11. BARBANO D., DELLA VALLE M. Microwave drying to determine the solids content of milk and Cottage and Cheddar cheese. *J. Food Prot.*, 47, n° 4, 1984, pp. 272-278. (C.D.I.U.P.A., n° 188338).

12. BAUNACK F., POERSCH W. Choix de l'installation de séchage de plantes fourragères et produits alimentaires. *Ind. Chim. Belge*, 34, n° 12, 1969, pp. 1071-1082.

13. BELLAVOINE R. Communication personnelle. 4 mars 1986.

14. BESSER E., PIRET E. Controlled-temperature dielectric drying. *Chem. Eng. Prog.*, 51, n° 9, 1955, pp. 405-410.

15. BHARTIA P., STUCHLY S., HAMID M. Optimum combined microwave-conventional drying processes. In : Proceedings of the IMPI Symposium, Ottawa, 1972, pp. 80-82. (Paper 5.4.).

16. BHARTIA P., STUCHLY S., HAMID M. Experimental results for combinational microwave and hot air drying. *J. Microw. Power*, 8, n° 3-4, 1973, pp. 243-252.

17. BIALOD D., RAYMOND D. Les rayonnements infrarouges, hautes fréquences et UHF (micro-ondes) et le séchage. In : Comptes-rendus des journées scientifiques EDF-DER, Les Renardières (Moret-sur-Loing), 13-14 mars 1985 et In : Association française de séchage dans l'industrie et l'agriculture, mai 1985, pp. 87-102.

18. BLAU R., POWELL M., GERLING J. Results of 2450 megacycle microwave treatments in potato chip finishing. In : Proceedings of the Production and Technique Div. Meeting, Potato Chip Institute International, 2 Feb. 1965, pp. 1-8.

19. BORZANI W., PRADO A. O emprego do forno de microondas na determinação da umidade de banana, farinha de mandioca, fermento prensado e argila. *Rev. Bras. Tecnol.*, 3, n° 1, 1972, p. 25.

20. BOULANGER R., BOERNER W., HAMID A. Microwave and dielectric heating systems. *Milling*, 153, n° 2, 1971, pp. 18-28. (C.D.I.U.P.A., n° 40553).

21. BRADBURY S. Process for the dessication of aqueous materials from the frozen state. U.S. Patent n° 2513 991, 4 July 1950.

22. BRYGIDYR A., RZEPECKA M., Mac CONNELL M. Characterization and drying of tomato paste foam by hot air and microwave energy. *Can. Inst. Food Sci. J.*, 10, n° 4, pp. 313-319. (C.D.I.U.P.A., n° 118594)

23. BUONO M., ERICKSON L. Rapid measurement of Candida utilis dry weight with microwave drying. *J. Food Prot.*, 48, n° 11, 1985, pp. 958-960. (C.D.I.U.P.A., n° 208426)

24. BURKE R., DECAREAU R. Recent advances in the freeze-drying of food products. *Adv. Food Res.*, 10, 1962, pp. 1-88.

25. BURKE R., DECAREAU R. Freeze-drying of foods. In : MRAK E., STEWART G. *Adv. Food Res.*, 12, 1964, p. 1.

26. BUSHWAY A., TRUE R., WORK T., BUSHWAY R. A comparison of chemical and physical methods for treating French fries to produce an acceptable microwaved product. *Am. Potato J.*, 61, n° 1, 1984, pp. 31-40. (C.D.I.U.P.A., n° 185678)

27. CALDERWOOD D. Rice drying and storage studies. *Rice J.*, 75, n° 7, 1972, p. 63. (C.D.I.U.P.A., n° 53364).

28. CARLIER L., VAN HEE L. Microwave drying of lucerne and grass samples. *J. Sci. Food Agric.*, 22, n° 6, 1971, pp. 306-307. (C.D.I.U.P.A., n° 35535)

29. CARTER J., FLEISCHFRESSER D., ISHII T. Microwave oven techniques for biological solids determination of wastewater samples. In : Proceedings of the IMPI Symposium, Milwaukee, 1974. (Paper A3-2/1-4).

30. CARTER J., ISHII T., FLEISCHFRESSER D. Microwave oven for improved biological solids determination. *J. Microw. Power*, 11, n° 2, 1976, pp. 117-125.

31. CHEVALIER M., RETAILLEAU. Essai d'application des micro-ondes au domaine analytique en industries laitières. *Ind. Aliment. Agric.*, 87; n° 3, 1970, pp. 243-248. (C.D.I.U.P.A., n° 18261).

32. CHIN H., KIMBALL J., HUNG J., ALLEN B. Microwave-oven-drying determination of total solids in processed tomato products, collaborative study. *J. Assoc. Off. Anal. Chem.*, 68, n° 6, 1985, pp. 1081-1083. (C.D.I.U.P.A., n° 208215).

33. CHRISTIE R., KENT M., LEES A. Microwave and infra-red drying versus conventional oven drying methods for moisture determination in fish flesh. *J. Food Technol.*, 20, n° 2, 1985, pp. 117-127. (C.D.I.U.P.A., n° 200305)

34. CONLEY W., GRIGGS T. Continuous vacuum-drying of fruit juices. *Food Can.*, n° 6, 1962, 4 p.

35. COPSON D. Some aspects of orange juice processing by super voltage cathode and microwave radiations. Thesis. Cambridge (Massachusetts), Inst. Technol., 1953.

36. COPSON D. Microwave energy in food procedures. In : I.R.E. Transactions of the Prof. Group on Medical Electronics, 4, 1956.

37. COPSON D. Methods and apparatus for radio-frequency freeze-drying. U.S. Patent n° 2 859 543, 11 Nov. 1958.

38. COPSON D. Microwave sublimation of foods. *Food Technol.*, 12, n° 6, 1958, pp. 270-272.

39. COPSON D. Method and apparatus for freeze-drying. U.S. Patent n° 3 020 645, 13 Feb. 1962.

40. COPSON D. Microwave heating, 2nd ed., Westport, Avi Publishing Co., 1975, 615 p. (C.D.I.U.P.A., n° 93299).

41. COPSON D., DECAREAU R. Microwave energy in freeze-drying procedures. *Food Res.*, 22, n° 4, 1957, pp. 402-403.

42. DALL'AGLIO G., GHERARDI S., VERSITANO A. Il riscaldamento a microonde, a raggi infrarossi e per contatto nella liofilizzazione di alcuni prodotti vegetali. *Ind. Conserve*, 51, n° 4, 1976, pp. 282-289. (C.D.I.U.P.A., n° 103048).

43. DALL'AGLIO G., PORRETTA A., VERSITANO A. Impiego del riscaldamento a microonde a piastre ed a raggi infrarossi nella liofilizzazione dei gamberetti e dei calamari. *Ind. Conserve*, 46, n° 2, 1971, pp. 93-100. (C.D.I.U.P.A., n° 35340).

44. DASEN A., GRAPPIN R. Détermination rapide de l'extrait sec des fromages à l'aide d'un four à micro-ondes. *Le Lait*, 63, n° 623-624, 1983, pp. 75-84. (C.D.I.U.P.A., n° 83/1630)

45. DAVIS A., LAI C. Microwave utilization in the rapid determination of flour moisture. *Cereal Chem.*, 61, n° 1, 1984, pp. 1-4. (C.D.I.U.P.A., n° 185229)

46. DAVIS C., SMITH O., OLANDER J. Microwave processing of potato chips (3 parties). *Potato Chipper*, 25, n° 2, n° 3, n° 4, 1965, pp. 38-58, pp. 72-92, pp. 78-94.

47. DECAREAU R. How microwaves speed freeze-drying. *Food Eng.*, 33, n°8, 1961, pp. 34-36.

48. DECAREAU R. Limitations and opportunities for high frequency energy in the freeze-drying process. In : FISHER F. Freeze-drying of foods, Proceedings of a Conference, Chicago, 1961. Washington, National Academy of Sciences, National Research Council, 1961.

49. DECAREAU R. Microwave research and development at the US Army Natick Research and Development laboratories. *J. Microw. Power*, 17, n° 2, 1982, pp. 127-135.

50. DECAREAU R. Microwaves in food processing. *Food Technol. Aust.*, 36, n° 2, 1984, pp. 81-86. (C.D.I.U.P.A., n° 201725)

51. DECAREAU R. Future research developments in microwave technology. Non publ. communiqué par l'auteur, 1986, 12 p.

52. DECAREAU R. Microwave food processing equipment throughout the world. *Food Technol.*, 40, n° 6, 1986, pp. 99-105. (C.D.I.U.P.A., n° 214710).

53. DECAREAU R., MUDGETT R. Microwaves in the food-processing industry. (Food Science and Technology, a series of monographs). New York, Academic Press Inc., 1985, 247 p.

54. DEFROMONT C., GUERGA J. Applications des hyperfréquences au séchage des graines oléagineuses et des corps gras animaux. *Rev. Fr. Corps Gras*, 16, n° 12, 1969, pp. 771-780. (C.D.I.U.P.A., n° 16576).

55. DEMECZKY M. Folytonos teknologiak komló vegszárításra, fagyasztott tömbhúsok felmelegítesere es kakaoporkolesre nagyfrekvencias erőter alkalmazásaval. *Elelmiszertudomany*, 2, nos 1-2, pp. 77-89.

56. DOMAGALA A. Möglichkeiten des Einsatzes von Mikrowellen zur Bestimmung der Feuchtigkeit von gewürfelten Speisekartoffeln. *Industrielle Obst- und Gemüseverwert.*, 70, n° 8, 1985, pp. 361-366. (C.D.I.U.P.A., n° 203340)..

57. DRION C., FERNET E. Traité de physique élémentaire. Paris, Masson, 1893, 848 p.

58. EUSTACE I., JONES P. Microwave drying of meat for rapid estimation of water and fat contents. *CSIRO Food Res. Quarterly*, 44, n°2, 1984, pp. 38-44. (C.D.I.U.P.A., n° 194250).

59. FABRY M. Cours de physique. 1ère partie : thermodynamique, Chap. 1. Ecole Polytechnique, 2ème div., 1930-1931, 416 p.

60. FANSLOW G., GITTINS L., WEDIN W., MARTIN N. Power absorption and drying patterns of forage crops dried with microwave power. In : Proceedings of the IMPI Symposium, Ottawa, 1972, pp. 207-211. (Paper 11.4.).

61. FANSLOW G., GITTINS L., WEDIN W., MARTIN N. Power absorption and drying patterns of forage crops dried with microwave power. *J. Microw. Power*, 8, n° 1, 1973, pp. 83-87.

62. FANSLOW G., SAUL R. Drying field corn with microwave power and unheated air. In : Proceedings of the IMPI Symposium, Monterey, 1971. (Paper 4.3.)

63. FANSLOW G., SAUL R. Drying field corn with microwave power and unheated air. *J. Microw. Power*, 6, n° 3, 1971, pp. 229-235.

64. FINN-KELCEY P. The applications of electricity to crops drying on farms : a critical review. The British Electrical and Allied Industries Research Association. Technical Report, ref. W/T 25.

65. FITZPATRICK T., PORTER W. Microwave finishing of potato chips. Effect on the amino-acids and sugars. In : Proceedings of the 17th National Potato Utilization Conference, 24-26 July 1967, Aroostock State College, Presque Isle (Maine), USDA, ARS, 1967, 51 p.

66. FORD J., PEI D., ANG T. Microwave freeze-drying of beef. In : Proceedings of the IMPI Symposium, Waterloo (Can.), 1975, pp. 72-74. (Paper 4.6.).

67. FRITZ K. Microwave drying apparatus. U.S. Patent n° 3 276 138, 4 Oct. 1966.

68. FUENTEVILLA M. Method and apparatus for continuous freeze-drying. U.S. Patent n° 3 264 747, 9 August 1966.

69. FUTER R. Method and apparatus for treating pieces of material by microwaves. British Patent n° 1247324, 22 Sept. 1971.

70. GALL B., LA PLANTE R. Methods of application of microwave energy in industrial processes. In : FISHER F. Freeze-drying of foods. Proceedings of a Conference, Chicago, 1961. Washington, National Academy of Sciences, National Research Council, 1962, pp. 16-24.

71. GERLING J. Method and apparatus for processing potato chips with micro-wave energy. Canadian Patent n° 868 296, 13 April 1971.

72. GOLDBLITH S., PACE W. Some considerations in the processing of potato chips by microwaves. In : Proceedings of the IMPI Symposium, Standford, 1967, abstracts pp. 37-38.

73. GOLDBLITH S., PACE W. Some considerations in the processing of potato chips. *J. Microw. Power*, 2, n° 3, 1967, pp. 95-98.

74. GOMES A., LEONHARDT G., TORLONI M., BORZANI W. Microwave drying of micro-organisms : influence of the micro-wave energy and the sample thickness on the drying of yeast. *J. Microw. Power*, 10, n° 3, 1975, pp. 264-270.

75. GOULD W., KENYON E. Corona breakdown and electric field strength in micro-wave freeze-drying. In : Proceedings of the IMPI Symposium, Den Haag, 1970.

76. GOULD J., KENYON E. Gas discharge and electric field strength in microwave freeze-drying. *J. Microw. Power*, 6, n° 2, 1971, pp. 151-167.

77. GOULD J., PERRY J., KENYON E. Theoretical and experimental aspects of microwave freeze-drying. In : Proceedings of the IMPI Symposium, Monterey, 1971. (Paper 2.1.).

78. GRIMM A. A technical and economic appraisal of the use of microwave energy in the freeze-drying process. ST-3980, *RCA Rev.*, n° 12, 1969.

79. GUERGA M. Application des hyperfréquences au séchage des corps gras. *Ind. Aliment. Agric.*, 88, n° 3, 1971, pp. 289-294. (C.D.I.U.P.A., n° 35368).

80. GUERGA M. Application of microwaves for drying fats. *Microw. Energy Appl. Newsl.*, 5, n° 2, 1972, pp. 3-6. (C.D.I.U.P.A., n° 48749).

81. GUIGO E., MALKOV L., KAUHČEŠVILI E. Certaines particularités du transfert de chaleur et de masse au cours de la lyophilisation de produits poreux dans un champ ultra haute fréquence. In : Développements récents en lyophilisation. Tome I, Paris, 1969. Paris,

Institut International du Froid, 1969, pp. 75-78. (C.D.I.U.P.A., n° 17770)

82. GUILLAUDEAU J. Applications des micro-ondes : déshydratation sous vide d'extraits de fruits. *RTC Actual.*, n°28, 1975, pp. 13-14.

83. HAMID M. Experimental results for combinational microwave and hot-air drying. *J. Microw. Power*, 8, n° 3, 1973, pp. 245-252.

84. HAMID M., BHARTIA P., STUCHLY S. Experimental results for combinational microwave-hot air drying. Progress report n° 11, Microwave applications project, Industrial enterprise fellowship program report, Manitoba Research Council, March 1973, 15 p.

85. HAMMOND L. Economic evaluation of UHF dielectric vs. radiant heating for freeze-drying. *Food Technol.*, 21, n° 5, 1967, pp. 735-736 and 742-743. (C.D.I.U.P.A., n° 67/559).

86. HARPER J., CHICHESTER C. Freeze-drying. Application of dielectric heating. In : Proceedings of freeze-dehydration of food meeting, 20-21 Sept. 1960. Chicago, Research and Development Associates, Food and Container Institute, 1960.

87. HARPER J., CHICHESTER C., ROBERTS T. Freeze-drying of foods. *Agric. Eng.*, 43, n° 2, 1962, pp. 78-81, 90.

88. HAYWARD L., KROPF D. Sample position effects on moisture analyses by a microwave oven method. *J. Food Prot.*, 43, n° 8, 1980, pp. 656-663. (C.D.I.U.P.A., n° 148136)

89. HEINEMAN H., MAVIS J. Concentrazione dei succhi e delle bevande con radio-frequenza. *Ind. Aliment.*, 9, n° 66, 1970, pp. 73-76. (C.D.I.U.P.A., n° 26586).

90. HILTON B. Research in the potato chip industry. In : Proceedings of the 19th National Potato Utilization Conference, Ferris State College, Big Rapids (Michigan), 28-29 July 1969. USDA Publications, ARS 73-65.

91. HOFFMAN M. Microwave heating as an energy source for predrying of her-bage samples. *Plant Soil*, 93, n° 1, 1965, p. 145.

92. HOOVER M., MARKANTONATOS A, PARKER N. Engineering aspects of using UHF dielectric heating to accelerate the freeze-drying of foods. *Food Technol.*, 20, n° 6, 1966, pp. 811-814.

93. HOOVER M., MARKANTONATOS A., PARKER N. UHF dielectric heating in experimental acceleration of freeze-drying of foods. *Food Technol.*, 20, n° 6, 1966, pp. 807-811.

94. HOPKINS A. Radio-frequency spectroscopy of frozen biological material: dielectric heating and the study of bound water. In : Annals of the New York Academy of Sciences, 85, art. 2, 1960, pp. 714-722.

95. HUANG H., YATES R. Radio-frequency drying of fungal material and resultant textured product. *J. Microw. Power*, 15, n° 1, 1980, pp. 15-17.

96. HUET R., MURAIL M., HUBERT-BRIERRE C., BANTIGNIE M., MAIRESSE M. Rétention des arômes dans les poudres de fruits tropicaux obtenues dans un four à micro-ondes sous vide. *Fruits, Fruits d'Outre-mer*, 29, n° 5, 1974, pp. 399-405. (C.D.I.U.P.A., n° 72200).

97. HUXSOLL C., MORGAN A. Microwave dehydration of potatoes and apples. *Food Technol.*, 22, n° 6, 1968, pp. 47-72. (C.D.I.U.P.A., n° 02559).

98. HUXSOLL C., MORGAN A. Microwaves for quick-cooking rice. *Cereal Sci. Today*, 13, n° 5, 1968, pp. 203-206. (C.D.I.U.P.A., n° 03286).

99. ISHIMARU K., ABE K., Combination heating apparatus. U.S. Patent n° 4 435 628, 1984.

100. JACKSON S., RICKTER S., CHICHESTER C. Freeze-drying of fruit. *Food Technol.*, 11, n° 9, 1957, pp. 468-470.

101. JAVORSKIJ B., DETLAF A. Aide-mémoire de physique. 2ème éd., Moskva, Ed. Mir, 1975, 963 p.

102. JEPPSON M. Production drying of potato chips with microwave energy. In : Proceedings of the Production and Technique Division Meeting. Potato Chip Institute International, 2 Feb. 1965, pp. 24-27.

103. JOLLY P. Temperature controlled combined microwave-convection drying. *J. Microw. Power*, 21, n° 2, 1986, pp. 65-74.

104. JONES D., GRIFFITH G. Microwave drying of herbage. *J. Br. Grassl. Soc.*, 23, 1968, pp. 202-205.

105. JOYAL M. Chaleur. Paris, Masson et Cie, Coll. Joyal-Provost, 1967, 271 p.

106. KACZMAREK R., SOBIECH W. Mikrofalowe suszenie pieczarek. *Przem. Spoz.*, 27, n° 11, 1973, pp. 524-527. (C.D.I.U.P.A., n° 64753).

107. KÄPPELIN B. Cours élémentaire de physique. Colmar, Ed. Z. KÄPPELIN, 1833, 403 p.

108. KARLESKIND A., VALMALLE G., CHEMIN S. Méthode rapide de dosage de la teneur en huile par RMN après séchage des graines à l'aide d'un four à micro-ondes. *Rev. Fr. Corps Gras*, 23, n° 3, 1976, pp. 147-150. (C.D.I.U.P.A., n° 94073).

109. KASHYAP S., WYSLOUZIL W., Automatic control of non-resonant dryers. *J. Microw. Power*, 11, n° 1, 1976, pp. 33-39.

110. KAZANOWSKI H., PARKER W. RF accelerated freeze-drying of foods. *RCA Eng.*, 12, 1966, pp. 4-8.

111. KIMURA S., TAKI M. Dehydrated foods by freeze-drying. Effect of microwave dielectric heating on freeze-drying of foods. *Nippon Shokuhin Kogyo Gakkaishi*, 13, n° 4, 1966, pp. 141-142. (Chemical Abstracts, 64, 20517 C.).

112. KING C. The freeze-drying of foods. Cleveland, CRC Press, Chemical Rubber Co., 1971, 86 p. (C.D.I.U.P.A., n° 40954).

113. KOČERGA S., BOIM B., VOSKOBOJNIKOV V. High frequency freeze-drying equipment for food products. *Konservnaja Ovoščesušilnaja Prom.*, 4, 1971, pp. 31-32.

114. LAIGNELET B. Veränderungen der Farbtönung und anderer Qualitätsfaktoren von Teigwaren im Verlauf der Trocknung. *Getreide Mehl Brot*, 30, n° 10, 1976, pp. 277-280. (C.D.I.U.P.A., n° 99808).

115. LEE J., LATHAM S., Rapid moisture determination by a commercial-type microwave oven technique. *J. Food Sci.*, 41, n° 6, 1976, p. 1487. (C.D.I.U.P.A., n° 102166).

116. LEFEUVRE S. Possibilities of microwave heating. In : Comptes-rendus de la 10ème Conférence Européenne sur les micro-ondes, Warszawa, 1980, pp. 227-231.

117. LEFEUVRE S., PARESI A., AHMADPANAH M., MANGIN B. Compound drying : air plus electromagnetic waves. In : Comptes-rendus de la 8ème Conférence Européenne sur les micro-ondes, Paris, 1978, pp. 579-583.

118. LEONHARDT G., GOMES A., BORZANI W., TORLONI M. Microwave drying of microorganisms. 2. : The use of microwave oven for the determination of moisture content of pressed yeast. *J. Microw. Power*, 13, n° 3, 1978, pp. 235-237.

119. LE POMMELEC G. Les applications des micro-ondes dans les industries alimentaires et pharmaceutiques. In : Comptes-rendus de la Journée de formation sur les Techniques récentes de traitements des produits biologiques, I.U.T. de Dijon, 18 mars 1977, 30 p.

120. LEPVRAUD M. Utilisations nouvelles de l'électricité dans l'industrie agro-alimentaire, *Génie Rural*, 67, n° 11, 1974, pp. 475-478. (C.D.I.U.P.A., n° 79056).

121. LE VIET L. Traitements thermiques des produits granulés par micro-ondes. In : Comptes-rendus du Séminaire international sur la recherche en technologie alimentaire. La Baule, 21-23 mai 1970, pp. N1-N4. (C.D.I.U.P.A., n° 35869).

122. LIPOMA S., WATKINS H. Process for making fried chips. U.S. Patent n° 3 365 301, 23 Jan. 1968.

123. LYONS D. The drying of porous fibrous materials using microwave heating. Ph. D. Thesis, Georgia Institute of Technology, 1966.

124. MA Y., ARSEM H. Freeze-drying by a combination of microwave and radiant heat. In : Proceedings of the IMPI Symposium, Ottawa, 1978, p. 60. (Paper 6.3.). (C.D.I.U.P.A., n° 143425).

125. MA Y., PELTRE P. Mathematical simulation of a freeze-drying process using microwave energy. AIChe Symposium Series, 69, n° 132, 1973, pp. 47-54.

126. MA Y., PELTRE P. Freeze dehydration by microwave energy. *AIChe J.*, 21, n° 2, 1975, pp. 335-350.

127. MARCHART H., HOFFER H. Untersuchungen zur Schnellbestimmung von Milchtrockenmasse mittels eines Mikrowellenherdes. *Österr. Milchwirtsch.*, 30, n° 8, 1975, pp. 129-133. (C.D.I.U.P.A., n° 81604).

128. MARSCHKE R., SNOW A., TOMMERUP J. Rapid moisture testing of butter. *Austr. J. Dairy Technol.*, 34, n° 1, 1979, pp. 36-37. (C.D.I.U.P.A., n° 132394).

129. MARTINI G. L'evoluzione nelle tecniche di essiccazione delle paste alimentari. *Tec. Molit.*, 25, n° 23-24, 1974, pp. 173-180. (C.D.I.U.P.A., n° 79635).

130. MASSZI G., ORKENYI J. Microwave investigation of biological substances. *Acta Biochim. Biophys. Acad. Sci. Hung.*, 2, n° 1, 1967, pp. 69-78.

131. MAURER R., TREMBLAY M., CHADWICK E. Microwave processing of pasta. *Food Technol.*, 25, n° 12, 1971, pp. 32-37. (C.D.I.U.P.A., n° 41275)

132. MAURER R., TREMBLAY M., CHADWICK E. Use of microwave energy in drying alimentary pastes. In : Proceedings of the IMPI Symposium, Monterey, 1971. (Paper 4.2.).

133. MAURER R., TREMBLAY M., CHADWICK E. Microwave oven occupies 75 % less floor space, dries 2000 lbs pasta per hour -- cuts process time 95 %. *Food Process.*, 33, n° 1, 1972, pp. 18-19.

134. MEISEL N. Nouvelles applications des micro-ondes à l'industrie alimentaire. *Ind. Aliment. Agric.*, 87, n° 9-10, 1970, pp. 1059-1067. (C.D.I.U.P.A., n° 25542)

135. MEISEL N. Microwave applications to food processing and food systems in Europe. *J. Microw. Power*, 8, n° 2, 1973, pp. 143-147.

136. MEISEL N. Apport d'énergie instantanée au coeur des produits alimentaires. Les micro-ondes. *Ind. Aliment. Agric.*, 91, n° 9-10, 1974, pp. 1203-1214. (C.D.I.U.P.A., n° 076032)

137. MEISEL N. Applications industrielles des micro-ondes dans les industries alimentaires. *Rev. Gén. Froid*, 1, n° 1, 1976, pp. 9-20. (C.D.I.U.P.A., n° 92131).

138. MEISEL N. Essiccamento delle paste alimentari con microonde. *Tec. Molit.*, 27, n° 12, 1976, pp. 101-105. (7ème Symposium "Durum und Teigwaren", Detmold, 1976). (C.D.I.U.P.A., n° 103549).

139. MEISEL N. Mikrowellen beim Trocknen von Teigwaren. *Getreide Mehl Brot*, 30, n° 7, 1976, pp. 187-189. (C.D.I.U.P.A., n° 96749)

140. MEISEL N. Les micro-ondes au service des techniques alimentaires. *Ind. Aliment. Agric.*, 95, n° 9-10, 1978, pp. 997-1003 ; Séchage, pp. 1000-1001. (C.D.I.U.P.A., n° 124554).

141. MEISEL N. Procédé Gigavac pour la fabrication en continu de poudres de fruits instantanément solubles. In : Comptes-rendus des Journées d'études et d'informations électro-industrielles des régions Bourgogne et Franche-Comté. Dijon, 12-13 oct. 1978. Publication du Comité Français d'Electrothermie, 12 p.

142. MEISEL N. Les micro-ondes dans les industries agro-alimentaires, bilan 1974-1984. *Ind. Aliment. Agric.*, 101, n° 10, 1984, pp. 929-932. (C.D.I.U.P.A., n° 193861).

143. MEISEL N. Microwave heating in vacuum drying. In : SPICER A. Advances in preconcentration and dehydration of foods. London, Applied Science Publishers Ltd., 1975, pp. 505-509.

144. MEISEL N. Microwave vacuum drying : Gigavac process for continuous manufacture of instantly soluble fruit powders. Publication I.M.I., 11 p.

145. MEISEL N., ARCHIERI C. La déshydratation par micro-ondes dans les industries alimentaires. In : Comptes-rendus du Colloque sur les Techniques électroindustrielles de séparation et de classement, 21-22 avril 1977, Versailles. Publication du Comité Français d'Electrothermie, 1977, 20 p.

146. MENORET Y., ARCHIERI C. Utilisation des micro-ondes pour déshydrater sous vide des extraits de fraise, framboise et grenadille. Fédération Internationale des Producteurs de Jus de Fruits, 12, 1972, pp. 87-99. (C.D.I.U.P.A., n° 62650).

147. MENORET Y., ARCHIERI C. Déshydratation de fruits et jus de fruits sous vide par micro-ondes. In : Comptes-rendus du 6ème Symposium européen sur les Aliments. London, Soc. Chem. Ind., 1975, pp. 109-110.(fran.).

148. MENORET Y., ARCHIERI C. Microwave vacuum dehydration of fruits and fruit juices. In : Comptes-rendus du 6ème Symposium européen sur les aliments. London, Soc. Chem. Ind., 1975, pp. 107-108 (angl.).

149. MENORET Y., ARCHIERI C. Utilisation des micro-ondes pour déshydrater sous vide les fruits et les extraits de fruits. Document I.M.I., non daté.

150. MERYMAN H. Induction and dielectric heating for freeze-drying. In : REY L. Aspects théoriques et industriels de la lyophilisation. Paris, Hermann, 1964, pp. 183-196.

151. MERYMAN H. Principles of freeze-drying. *Annals New York Acad. Sci.*, 85, art. 2, pp. 630-640.

152. MUZZARELLI G. Impiego di un forno-tunnel a microonde per la produzione e la confezione di formaggi a pasta filata. *Latte*, 1, n° 8-9, 1976, pp. 445-447. (C.D.I.U.P.A., n° 100055).

153. NICOLAS-GROTTE M. Utilisation des micro-ondes dans les industries alimentaires. Thèse Doctorat. Paris, Fac. Sci., 1972, 14 p. (C.D.I.U.P.A., n° 043175).

154. NURY F., SALUNKHE D. Effects of microwave dehydration on components of apples. USDA, ARS-74-75, Western Util. Res. Dev. Div., Albany, 1968, 19 p.

155. OLSEN C. Promising applications of microwave energy in the food and pharmaceutical industry. Varian Industrial Microwave Operation, 1968, 22 p. Drying, pp. 15-16.

156. Ò MEARA J. Progress report on microwave drying. In : Proceedings of Production and Technique Division Meetings, Potato Chip Inst. Internat. 30 Jan.-3 Feb. 1966, Las Vegas.

157. Ò MEARA J. Finish drying of potato chips. In : OKRESS E. Microwave Power Engineering. Vol. 2. Applications. New York, Academic Press, 1968, pp. 65-73.

158. OSEPCHUK J. A history of microwave heating applications. *IEEE Trans. Microw. Theory Tech.*, MTT-32, n° 9, 1984, pp. 1200-1224.

159. PARKER W. Freeze-drying. In : OKRESS E. Microwave power engineering. Vol. 2. Applications, New York, Academic Press, 1968.

160. PASSEY A., RIEL R., BOULET M. Mathematical model for microwave freeze-drying of spherical particles. In : Congrès International du Froid. 13, Washington, 1971. Progrès dans la science et la technique du froid. Vol. 2. Westport, Avi Publishing Co., Inc., 1973, pp. 347-355. (C.D.I.U.P.A., n° 71689)

161. PELLISSIER J. Séchoir rotatif utilisant des micro-ondes. Doc. fourni par l'auteur, 1986, 6 p.

162. PELTRE P., MA Y. Application of computer simulation in the study of microwave freeze-drying. In :Proceedings of the IMPI Symposium, Waterloo (Can.), 1975, pp. 70-71. (Paper 4.5.).

163. PETTINATI J. Microwave oven method for rapid determination of moisture in meat. *J. Assoc. Official Anal. Chem.*, 58, n° 6, 1975, pp. 1186-1193. (C.D.I.U.P.A., n° 92838).

164. PORTER V., NELSON A., STEINBERG M., WEI L. Microwave finish drying of potato chips. *J. Food Sci.*, 38, n° 4, 1973, pp. 583-585. (C.D.I.U.P.A., n° 61979).

165. REVERON A., GELMAN A., TOPPS J. Microwave drying of carcass samples from experimental lambs. *Lab. Pract.*, 20, n° 12, 1971, pp. 943-945. (C.D.I.U.P.A., n° 39416).

166. RIVOCHE E. Treatment and partial dehydration of foodstuffs under vacuum. U.S. Patent n° 2 708 636, 17 May 1955.

167. ROCKWELL W., LOWE E., HUXSOLL C., MORGAN A. Apparatus for experimental microwave processing. *Food Technol.*, 21, n° 9, 1967, pp. 93-94. (C.D.I.U.P.A., n° 67/1445).

168. ROSENBERG U., BOGL W. Auftauen, Trocknen, Wassergehaltsbestimmung und Enzyminaktivierung mit Mikrowellen. *Z. Lebensm:- Technol. Verfahr.*, 37, n° 1, 1986, pp. 12-19. (C.D.I.U.P.A., n° 208632).

169. RZEPECKA M., HAMID M., Mac CONNELL M. Microwave fluidized bed dryer. In : Proceedings of the IMPI Symposium, Waterloo (Can.), 1975, pp. 297-299. (Paper 18.2.).

170. RZEPECKA-STUCHLY M. Microwave energy in foam-mat dehydration process. *J. Microw. Power*, 11, n° 3, 1976, pp. 255-266.

171. RZEPECKA-STUCHLY M., BRYGIDYR A., Mac CONNELL M. Tomato paste dehydration by hot air and microwave energy. In ; Proceedings of the IMPI Symposium, Leuven, 1976. (Paper 6B.5.).

172. SAHARA H. Methods and apparatus for continuous freeze-drying. U.S. Patent n° 3 531 871, 6 Oct. 1970.

173. SALES M., MAIA G., TELES F., STULL J., WEBER C., BERRY J., TAYLOR R. Cowpea. solids prepared by mild alkali treatment, dry roasting, and microwave heating. *Nutr. Rep. Intern.*, 30, n° 4, 1984, pp. 973-981. (C.D.I.U.P.A., n° 202342).

174. SANDERS F., Mac LAUGHLIN R. Potato segment and process for preparing frozen French fried potatoes suitable for microwave reheating. U.S. Patent n° 4 219 575. 26 August 1980.

175. SHANLEY R., JAMESON G. A study of the rapid determination of moisture in cheese by microwave heating. *Aust. J. Dairy Technol.*, 36, n° 3, 1981, pp. 107-109. (C.D.I.U.P.A., n° 162553).

176. SLATER L. Advanced technology : innovating with microwaves. *Food Eng.*, 47, n° 7, 1975, pp. 51-53. (C.D.I.U.P.A., n° 88465).

177. SMITH F. Microwave fluidized bed dryers. U.S. Patent n° 3 528 179, 15 Sept. 1970.

178. SMITH F. Microwave-hot air drying of pasta, onions, and bacon. *Microw. Energ. Appl. Newsl.*, 12, n° 6, 1979, pp. 6-14. (C.D.I.U.P.A., n° 141171).

179. SMITH F., SILBERMANN K. Method and apparatus for drying sheet materials. U.S. Patent n° 3 507 050, 21 April 1970.

180. SMITH O. Processing potato chips with microwave energy. In : Proceedings of Production and Technique Division Meetings, The Potato Chips Institute International, Feb. 1966, pp. 60-64.

181. SMITH O. Impact of microwave finishing on processing of potato chips. In : Proceedings of the 17th National Potato Utilization Conference, 24-26 July 1967. USDA, ARS, Arrostock State College, Presque Isle (Maine), 1967, pp. 89-92.

182. SMITH O. Impact of microwave processing on our industry. In : Proceedings of Production and Technique Division Meetings, The Potato Chips Institute International, 24 Jan. 1967, pp. 5-6.

183. SMITH O., DAVIS C. Use of microwaves in batch preparation of potato chips. In : Proceedings of the 15th National Potato Utilization Conference, 21-23 July 1965, p. 51.

184. SOBIECH W. Microwave-vacuum drying of sliced parsley root. *J. Microw. Power*, 15, n° 3, 1980, pp. 143-154.

185. SOURTY M., PLAISANT E. Déshydratation d'un liquide industriel. In : Comptesrendus du Colloque sur les Applications Energétiques des Rayonnements, Univ. Paul Sabatier, Toulouse, 12 juin 1986, ENSEEIHT, pp. 53-57.

186. STANLEY-ANTHONY W. Vacuum microwave drying of cotton : effect on cottonseed. *Trans. ASAE*, 26, n° 1, 1983, pp. 275-278.

336 Part III - Applications in the food industry

187 STETSON L., OGDEN R., NELSON S. Effect of radio-frequency electric fields on drying and carotene retention of chopped alfalfa. *Trans. ASAE*, 12, n° 3, 1969, pp. 407-410.

188. STUCHLY M., STUCHLY S. Industrial, scientific, medical and domestic applications of microwaves. *IEE Proc.*, 130, part A, n° 8, Nov. 1983, pp. 467-503.

189. SUARDET R. Thermodynamique. Paris, J.-B. Baillière et fils, 1965, 320 p.

190. SUCHY J., BALCAR V., KOPRIVA M. Expanzni sušeni. *Potravin. Chladiči Tech.*, 4, n° 2, 1973, pp. 49-52. (C.D.I.U.P.A., n° 60084).

191. SUNDERLAND J. An economic study of microwave freeze-drying. *J. Food Technol.*, 36, n° 2, 1982, pp. 50-56. (C.D.I.U.P.A., n° 165658).

192. TAKAHASHI Y., NAGAI A., SUGA N., CHIBA J. Measurement of total milk solids by microwave heating. *J. Microw. Power*, 13, n° 2, 1978, pp. 167-171.

193. TAPE N. The dehydration of foods and some new methods of drying. *Can. Food Ind.*, 41, n° 8, 1970, pp. 29-31 (C.D.I.U.P.A., n° 29177).

194. THOMASOW J. Mikrowellengeräte zur Wasser- Trockenmassebestimmung. Erfahrungen mit neuzeitlichen Analysenautomaten. *Dtsch. Milchwirtsch.*, 36, n° 9, 1985, pp. 263-265. (C.D.I.U.P.A., n° 197188).

195. THOUREL L., MEISEL N. Les applications industrielles du chauffage par microondes. In : Comptes-rendus des journées d'Information Electro-industrielles, Paris, EDF, nov. 1974, 12 p.

196. TOOBY G. Method for drying food. U.S. Patent n° 3 249 446, 3 May 1966.

197. TSUYKI H. SHUTO A. Foam drying techniques of food by microwave heating. XII. Foam drying method of fish protein. Raw material (E) - A lecture. *Shokuhin Kogyo*, 13, n° 20, 1970, pp. 81-88.

198. TURKOT V., ACETO N., SCHOPPET E., CRAIG J. Continuous vacuum drying of whole milk foam. *Food Eng.*, 41, n° 8, 1969, pp. 59-64. (C.D.I.U.P.A., n° 15181).

199. VAN DE CAPELLE A., LUYPAERT A. Applications of microwaves for freeze-drying processes. In : Comptes-rendus du Colloque international des Journées d'Electronique de Toulouse, 7-11 mars 1977, IV-1, 13 p.

200. VAN LEEUWEN J. Dielectric drying and heating. *Polytech. Tijd. Elektrotech. Elektron.*, 29, n° 11, 1974, pp. 363-369.

201. VAN OLPHEN G. Method and apparatus for dehydrating foods and other products. U.S. Patent n° 3 270 428, 6 Sept. 1966.

202. WEISS J., TETENBAUM S. A travelling-wave technique for microwave-aided freeze-drying. In : Proceedings of the IMPI Symposium, Ottawa, 1978, pp. 54-56. (Paper 6.1.). (C.D.I.U.P.A., n° 143425).

203. WESLEY R. Microwave drying of cottonseed and its effect on cottonseed quality. Master Thesis, Clemson Univ., August 1972.

204. WESLEY R., LYONS D., GARNER T., GARNER W. Some effects of microwave drying on cottonseed. *J. Microw. Power*, 9, n° 4, 1974, pp. 329-340.

205. WING R., ALEXANDER J. The value of microwave radiations in the processing of full-fat soybeans. *Can. Inst. Food Sci. Technol. J.*, 8, n° 1, 1975, pp. 16-18. (C.D.I.U.P.A., n° 80205).

206. WINSTON M. Microwaves and pasta drying. *Macaroni J.*, 55, n° 9, 1974, pp. 22-23. (C.D.I.U.P.A., n° 68090).

207. ZAHOUANI M. Etude d'une barre rayonnante fournissant 40 kW micro-ondes et destinée à un four tournant rotatif. In : Comptes-rendus du Colloque sur les Applications Energétiques des Rayonnements, Univ. Paul Sabatier, Toulouse, 12 juin 1986, ENSEEIHT, pp. 1-10.

208. Déshydratation sous vide au moyen des micro-ondes. *Alembal*, n° 14, 1972, p. 78. (C.D.I.U.P.A., n° 53560).

209. Divers comptes-rendus d'essais, Sté I.M.I.-S.A., 1977-1980.

210. Fresh flavor banana crystals. *Food Process. Staff. Food process.*, n° 4, 1961, 3 p.

211. Gigavac-Microwave vacuum drier with continuous conveyor belt system. Publ. commerciale Sté I.M.I.-S.A., n.d., 4 p.

212. Le procédé de séchage par micro-ondes des pâtes alimentaires est une réalité industrielle au Canada. Publication du Salon de l'Agriculture, Paris, 16 fév. 1972. (Séchage des pâtes alimentaires), 1 p. (AGRA Alimentation, n° 372, 1972)(MPP2).

213. Les micro-ondes dans l'agro-alimentaire : un outil performant. Cahier
ARIST, ARIST-Aquitaine, 1984,
pp. A44-A54. (C.D.I.U.P.A., n° 221917)..

214. Method for manufacturing a farina-
ceous product based on wheat flour.
UNILEVER, Nederlands Patent
n° 7 106 629, 16 Nov. 1971.

215. Microvac-microwave tunnel oven with
conveyor belt for the vacuum drying
of labile products. Publ. commerciale
Sté L.M.I. (I.M.I.-S.A.), 1972, 4 p.

216. Microwave dryer and sterilizer on
the way ? *Microw.*, 11, n° 2, 1972,
pp. 12-14.

217. Microwave drying cuts costs. *Food
Eng.*, 44, n° 11, 1972, pp. 78-79.
(C.D.I.U.P.A., 50552).

218. Microwave drying of onions produces
lower microbial counts. *Canner-Packer*,
n° 10, 1975, pp. 64-65.

219. Microwave freeze-drying update.
Microw. Energy Appl. Newsl., 2,
n° 5, 1969, p. 14.
(C.D.I.U.P.A., n° 23851).

220. Microwave heating. Doc. commercial
Sté Magnetronics Ltd., Leicester,
n.d., 8 p.

221. Microwave heating for potato crisps.
Food Manuf., 43, n° 10, 1968, pp.
46-47.

222. Microwave potato chips finish drying
update. *Microw. Energy Appl. Newsl.*,
2, n° 6, 1969, pp. 12-15.
(C.D.I.U.P.A., n° 30586).

223. Microwave vacuum dryer produces instant orange juice. *Microw. Energy
Appl. Newsl.*, 12, n° 3, 1979, p. 12.

224. Microwaves dry pasta. *Food Eng.*,
44, n° 4, 1972, pp. 94-96.
(C.D.I.U.P.A., n° 44379).

225. Microwaves improve color control.
Food Eng., 36, n° 8, 1964, p. 95.

226. Mikrobølge-Vakuum Tørringsanlæg.
Niro-Atomizer News Mag., 1, n° 4,
août 1979, pp. 8-9 (danois/anglais).

227. Mikrowellen in der Lebensmittel-
produktion. *Ernähr.wirtschaft*, 18,
n° 6, 1971, p. 383.

228. Note technique sur l'application des
fours à micro-ondes en sucrerie.
Document S.N.F.S./I.R.I.S., 1978,
5 p.

229. Séchoir continu sous vide. Publication Sté Zschokke-Wartmann S.A., 8 p.

230. Technical data sheet, Labotron 2000.
Publ. Sté I.M.I.-S.A., n.d., 1 p.

231. The drying method for the future.
Macaroni J., 55, n° 12, 1974, p. 8.
(C.D.I.U.P.A., n° 70822).

Chapter 4

Preservation

Preservation

The biochemical stabilization of food that is necessary for preservation relies on the elimination of proliferating parasitic micro-organisms (bacteria, yeasts, and mildew) and the inhibition of undesirable chemical reactions by inactivating their catalyzing enzymes. Under certain conditions, operations such as enzymatic inactivation, sterilization, and pasteurization may be advantageously performed with microwaves. Many such applications are being developed at present, and several have already produced interesting results and are promising for industrial uses.

4.1 Enzymatic inactivation

Enzymes are soluble, colloidal, organic catalysts produced by living beings. In contrast to mineral catalysts, enzymes have a very specific action, and each enzyme promotes only one type of reaction. The adaptation of an enzyme to a given substrate is comparable to that of a key to its lock.

All known enzymes are proteins, i.e., polypeptide chains obtained from the polymerization of amino acids

$$NH_2-CH-COOH$$
$$|$$
$$R$$

These chains, known as primary structures, are linked by hydrogen and covalent bonds into secondary structures known as subunits. The latter are assembled into the tertiary or quaternary structures that constitute the enzyme. The enzyme itself consists of two complementary parts: the active or prosthetic group that carries the active sites and is directly responsible for the catalytic effect, and the activating part of the apoenzyme that provides the protein support for the active part. The apoenzyme adsorbs substrate molecules and produces extremely high local concentrations that raise the efficiency of the active part to an enormous extent.

The Baldwin nomenclature distinguishes between four families of enzymes:

- cleaving enzymes that fragment different components and fix certain groups on molecules
- transfer enzymes that lift a fragment of the substrate and fix it on another molecule
- isomerization enzymes that favor a change from one isomeric form to another
- polymerization enzymes that accelerate the formation of chains starting with a monomer.

The name given to an enzyme derives from its specificity and, in general, consists of the name of the substrate, a semantic element that suggests the mode of action, and the suffix *-ase*. The activity of an enzyme is characterized by the amount of substrate transformed in the shortest possible time at the start of the reaction (i.e., by the initial reaction rate). This activity may be affected by a number of factors, namely, the concentrations of enzyme and substrate, pH, temperature, and external disturbances such as stirring, ultrasound, ultraviolet radiation, and so on [142].

Enzymes play an important role in the chemistry of life. They appear, among others, at the origin of many reactions that affect raw fruit and vegetables, and produce undesirable compounds whose smell, color, and structure have a deleterious effect on the quality of these products.

4.1.1 Blanching of fruits and vegetables

Blanching is a heat treatment that aims to stabilize the sensory qualities of food by inactivating certain enzymes. This inactivation is generally effective at 82°C; it is achieved in a few minutes at 88°C or in 30 s at 100°C. The process is monitored by means of a specific test that measures the activity of the enzyme to be inactivated, or by an activity test on two enzymes characterized by high thermal stability, namely, catalase and peroxidase, whose inactivation signifies prior inactivation of other enzymes that may be present.

When the surface area of the product to be treated is sufficiently large, vapor can be used to prevent aromatic losses. Blanching in water at 95–100°C is frequently used for bulky vegetables and is faster than vapor treatment [172]. The efficiency of the two procedures is a function of thermal conductivity, but the uniformity of the treatment is rarely satisfactory. For example, in the case of bulky vegetables, the surface is often overblanched whilst the center continues to contain active enzymes. Overblanching results in a softening of the product, a loss of aroma, a wasteful consumption of energy, and sometimes thermal reactions that are as harmful as enzymatic reactions. Conversely, inadequate blanching may produce tissue damage, bringing the enzyme into contact with the substrate, and, paradoxically, inducing faster degradation than in the absence of any treatment.

Compared to conventional procedures, microwaves offer several advantages. The action of microwaves on enzymes is purely thermal. This has been demonstrated by experiments performed at 2.45 GHz on glucose-6-phosphatase-dehydrogenase from human red blood cells and adenylate kinase extracted from mice liver [161], and also by experiments with lactate dehydrogenase at 3 GHz [22]. The conclusions of these studies were identical: There was no cummulative effect of irradiation, and the reaction was accelerated by the purely thermal effect of microwaves.

Microwave blanching tests have been carried out on different types of vegetables. In the case of potatoes, the aim was to inactivate polyphenyloxidase, a transfer enzyme that oxidizes polyphenols and gives rise to a brown-black polymer. Microwave treatment at 2.45 GHz can be three times faster than treatment with boiling water [35]. Structure changes induced by heat are identical, and sensory evaluation shows that there is no difference in taste. Combined microwave and boiling-water treatment is found to produce good conservation of structure. The superposition of two opposing temperature gradients, one due to conduction and the other to microwaves, produces moderate heat deposition in intermediate layers. The time required for complete inactivation is then 90 s of irradiation at 1 kW, followed by 3 min in hot water [32], [1].

Lipoxygenase is responsible for the deterioration in the quality of peas, since it produces with linoleic acid unstable hydroxides that subsequently degrade into the aldehyde hexanal [77]. Good inactivation of this enzyme was achieved in an 8-kW, 2.45-GHz IMI-SA microwave tunnel. The time inside the tunnel was 1 min for a load of 500 g m^{-1}. Sugar and soluble nitrogen losses were very low, and the deterioration in color and taste was identical to that produced in conventional blanching [157], [158]. By contrast, a sugar content of 92% against only 80% in conventional hot-water blanching has been reported for microwave-blanched peas (500 g treated for 210 s with 1, 8-kW microwaves) [171].

Peas, green beans, horseradish, and other products contain peroxidase, a transfer enzyme that lifts oxygen from the peroxides and fixes it on certain phenolic compounds; this reaction is accompanied by the development of unpleasant odours [77].

The blanching of green beans with microwaves [66] gives a product that is less green, firmer, and less tasty than in the case of water blanching. The ascorbic-acid contents are identical. Tests performed on peas, green beans, and squash showed [96] that the retention of vitamins (e.g., ascorbic acid) was approximately the same for microwave and vapor treatments. Microwaved French beans had a higher level of dry matter, a better retention of flavor, and a better structure of the pods, with hardly any skin tearing [171].

Also reported in the literature is the blanching of celery with 500 W at 915 MHz [159]. The treatment was applied for 90, 100, 110, 120, and 130 s in glass containers, and produced satisfactory inactivation of peroxidase and polyphenyloxidase with an improvement in texture and color, a slight deterioration in aroma and flavor, and a weight gain of 8%. An industrial application resulting in considerable gain in time was envisaged, but it was also noted that the energy control of the process was difficult: most of the characteristics of the product were improved as compared with conventional blanching so long as the temperature did not exceed 110°C, at which point a rapid degradation was found to occur.

Practical implementations include the experimental equipment produced by Magnetronics of Leicester (United Kingdom) for the blanching of brussel sprouts. A temperature rise up to 60°C was achieved with steam, and with microwaves beyond that. Good energy efficiency, very good adaptability to production procedures, and a reduction in effluents and mechanical damage were achieved [180], [187]. In France a 25-kW prototype operating at 915 MHz was introduced around 1980, but has been kept very confidential. Blanching of vegetables is also discussed in [113], [112], [130], [83], [49], [81], [179], [32], [17], [172], [44], [4], and [134].

For fruits, and in particular peaches, water or vapor blanching produces overcooking of the surface and a significant loss of soluble matter. By contrast, microwave treatment avoids almost completely any fluid exudation and preserves the flavor and appearance of the product whilst ensuring total enzymatic inactivation [7], [10], [8], [9]. Table 4.1 shows the percentage residual activity of the peroxidase and polyphenyloxidase as a function of irradiation time. The only problem is that it is impossible to inactivate enzymes in the skin, which is very rich in polyphenyloxidase and is cooled by ambient air. A combination of microwaves and infrared radiation, or microwaves and a hot fluid, was therefore considered in [7]. It was subsequently reported [9] that, after six months of storage, the browning of the tissue, the loss of flavor, and the degree of exudation were less than for untreated peaches.

Table 4.1

Inactivation of enzymes in peaches [9]

Varieties		Initial activity ΔOD/min	Residual % activity after		
			6 min MW	8 min MW	10 min MW
Ambergen	Peroxydase	2.3	41	32	0
	Polyphenyloxydase	7.9	52	37	0
Babygold 6	Peroxydase	1.9	12	5	1.75
	Polyphenyloxydase	5.3	8	0	0
Dixon	Peroxydase	3.5	34	15	1.5
	Polyphenyloxydase	13.3	58	2	0
Tuscan	Peroxydase	2.8	37	7	0
	Polyphenyloxydase	5.4	91	6	0

4.1.2 Inactivation of α-amylase in wheat

α-Amylase is a cleaving enzyme that breaks down polysaccharides. Its presence in wheat flour produced from germinated grains promotes the hydrolysis of starch and the appearance of dextrines during the fermentation of dough . The result is low-quality bread that is leathery and tough.

Microwave treatment of flour (1.8 kW, 2.45 GHz) has shown [3] that a 30-s exposure raised the temperature to 77°C , but this was not sufficient for satisfactory inactivation. By contrast, 60-s irradiation with a final temperature of 120°C resulted in total inactivation without dough degradation and with very low water loss. These results are in agreement with reports by IMI-SA, which suggested that a final temperature of at least 100°C was necessary [173].

4.1.3 Treatment of grains and soya beans

Soya beans contain lipoxygenase that oxidizes the unsaturated linoleic and linolenic fatty acids with the release of undesirable flavors. It also contains products that have an inhibiting effect on trypsin, a proteolytic enzyme present in the intestine. It was shown in [117] that 2.45-GHz microwaves were as efficient as vapor in reducing the activity of antitrypsin, and that this mode of heating inactivated completely the lipoxigenase. These results were later confirmed in other studies [68], [67] in which poultry were fed on soya beans treated with 650-W, 2.45-GHz radiation. This resulted in faster growth and a reduction in pancreatic hypertrophy that is generally caused by this diet. Maximum amelioration and digestibility was achieved for microwave irradiation times of 9–12 min. An adverse effect was observed after 15 min. The lipoxigenase was completely destroyed by a 90-s treatment of a 150-g sample with a moisture content of 14–28%. Soya beans with 10% moisture content were treated similarly [57] with satisfactory inactivation of antitrypsin, and the same irradiation time threshold was observed for digestbility. These later studies were performed on a rotary applicator that was previously reported in [165].

The microwave bean roaster described in [72], [73] also permits the destruction of trypsin inhibitors in soya beans, peanuts, sunflower seeds, and coffee. It uses only 66% of the energy required for the conventional treatment.

Myrosinase, an enzyme present in mustard and rape seeds, produces a very characteristic bitter taste. The traditional treatment is to maintain the temperature of the product between 94 and 105°C for 20 min. The disadvantages of this treatment include the length and volume of the ovens, and the cooked taste of the final product, which, reduced to flour, is commonly used as a protein supplement and a texture agent for meat products. A recent project by Canada Agriculture of

Ottawa was aimed at developing a microwave treatment. The product was placed in a PTFE tube that was inserted in a waveguide and submitted to 900–1500 W at 2.45 GHz, reaching a temperature of between 85 and 140°C in 13 s. Inactivation was found to be effective for temperatures in the range 105–125°C, depending on the initial moisture levels. These temperatures had no adverse effects on quality that in all cases was judged to be superior to that after conventional treatment. A more powerful applicator working at 915 MHz confirmed these results [153], [78].

In rape grains, in which efficient inactivation of myrosinase has not taken place, the hydrolytic decomposition of glucosinates can be observed and their derivatives found in the oil. Researchers from the University of Agriculture and Technology at Olsztyn (Poland) and the Institute of Animal Nutrition of Dummerstorf-Rostock (Germany) [92] compared the efficacy of 2.45-GHz microwaves, γ-rays, and hydrothermal procedures combining a fluidized bed and water vapor. Microwaves produced 90% inactivation in 210 s with much better qualitative results than the other two procedures.

4.2 Sterilization

Food products are usually infested with animal and vegetable micro-organisms, bacteria, yeasts, and mildews. The first category includes aerobic germs such as gram positive cocci (*Staphylococci*, *Streptococci*, *Bacillaceae*), gram negative cocci (enterobacterial types such as *Escherichia coli*, *Salmonella*, and *Protei*), *Pseudomonas*, and anaerobic germs such as *Clostridia*. The second category contains different phycomycetes such as mildews and ascomycetes, such as yeasts, and mildews of the classes *Penicillus* and *Aspergillus*.

Bacteria can cause serious ailments because of their virulence or toxicity. As for yeasts and mildews, they mostly affect the acceptability and the quality of the product. All these types of micro-organisms require product sterilization, which is generally achieved by conventional techniques employing heat, toxic gas, or ionizing radiation.

Heat treatment requires the product to be in contact with a fluid (i.e., air or water) at high temperature; the unit of sterilization value (s.v.) corresponds to a product held for one minute at 121°C with an attendant reduction in the microbial population by an order of magnitude. The irradiation treatment consists of exposing the product to a source of ionizing radiation equivalent to 10 kGy or $10 \, J \, g^{-1}$. For example, an absorbed dose of 3 kGy enables minced meat to be preserved for 10 days at room temperature despite the fact that it is an ideal medium for the growth of pathogenic germs. The most commonly used ionizing radiations are γ-rays from ^{60}Co and ^{137}Cs sources, electron beams generated in particle accelerators with energies below 10 MeV, and bremsstrahlung x-rays [175].

Heat treatment alters the characteristics of the product by superficial overheating, and produces noticeable degradation of organoleptic properties (the product tastes cooked); there is also a reduction in nutritional value due to the destruction of vitamins. Manufacturers are generally uncomfortable with irradiation. In gas treatment, using, for example, ethylene oxide on thermodegradable products, it seems that the desorption of the highly toxic gas is not complete. Microwaves may therefore offer an interesting alternative.

For delicate products (mostly liquids), sterilization may be replaced by more appropriate procedures that are less harsh (e.g., stassanization and pasteurization). The former procedure is not commonly used, although it ensures the destruction of 99.9% of all germs; it consists of bringing the liquid to 75°C for 15 s as it flows between two hot tin-plated copper cylinders that are only a few tenths of a millimeter apart. As for pasteurization, three techniques are presently employed:

- *low-temperature, long-time pasteurization* (LTLT) with an exposure to 62.8 to 65.6°C for 30 min at least

- *high-temperature, short-time pasteurization* (HTST) at 72°C for 15 s or 80–85°C for 5 s

- *ultrahigh-temperature pasteurization* (UHT), which involves placing the product in superheated vapor and then allowing it to expand in vacuum at 140–150°C for a fraction of a second.

Better results are obtained with fast, high-temperature techniques that capitalize on the fact that the rate of destruction of micro-organisms increases more rapidly with temperature than the rate of degradation of the characteristics of the product. Unfortunately, the low thermal conductivity of liquids makes the implementation of these techniques rather difficult [122], [121].

Sterilization was among the early suggested applications of microwaves [54], [174], [144], [139], [25], [75], [146]. This went hand in hand with the exploration of the biological effects of radiation, including the search for specific effects on cells and the development of microwave cooking. As early as 1960, it was noted [48] that germs such as *Staphyloccus aureus* and *Salmonella typhosa* were more effectively destroyed by microwaves than by conventional cooking. A vast range of products was investigated and it was found that the becteriocidal effect of microwaves was at least as good as that of conventional methods, without substantially affecting the freshness of the products. However, the main advantage of the technique was that it allowed continuous sterilization, thus advoiding the chain breaking of conventional autoclaving.

In theory, it is highly unlikely that there are any athermal effects of radiation at the common frequencies of 915 and 2450 MHz. This conclusion had already been drawn in 1967 from experimental data on *E. coli* [62] for which the destruction curve is given in Figure 4.1.

Research performed at the University of Massachusetts has shown [115], [114] that microbial survival under microwave irradiation is a function of time and temperature, and can be

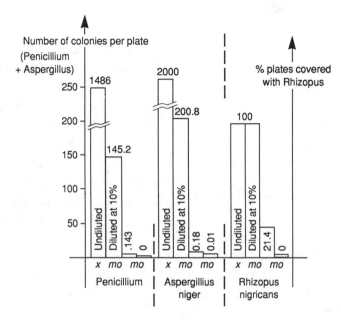

Figure 4.1 Destruction of $E.\ coli$ (2.45 GHz, 416 W per liter of solution) [62].

modeled as follows:

$$C = C_0 \exp(-Kt) \qquad (4.1)$$

$$K = K_0 \exp\left[-W_a(T - T_0)/R\,T\,T_0\right] \qquad (4.2)$$

where C is the concentration in m^{-3}; K is a thermal denaturization parameter over and above the reference temperature T_0, measured in s^{-1}; t is the time of treatment in seconds; T is the temperature in kelvins $R = 8.31441\,\mathrm{J\,mol^{-1}\,K^{-1}}$ is the gas constant; W_a is the activation energy in $\mathrm{J\,mol^{-1}}$; and K_0 is specific to the organism to be inactivated and its physico-chemical environment. This model was found to be in good agreement with experimental results. Additional information may be found in [160] and [135].

4.2.1 Animal products

In an attempt to devise a technique aiming at minimizing the weight loss suffered by ham in conventional cooking, it was found that radiofrequency (60 MHz) and microwave (2.45 GHz) pasteurization produced purely thermal effects [19]. Microwaves increased the preservation time of ham and frankfurters from 7 to 14 days, respectively, without any proliferation of surface bacterial flora.

Cooked meat and delicatessen products present conservation problems, even when packaged in plastic containers, because of contaminants introduced during handling. Conventional heating in water causes exudation and the tarnishing of packaging material. Research reported in [59] involved several thousand packets of sausages and bacon, as well as patés, in a Gigatron tunnel. Sausages previously cooked to 80°C, were found to be free from discoloration phenomena after exposure to microwaves, and the bacteriological results were better than for hot water. The results for bacon cooked at 70°C were more irregular and indicated the need for counter-pressure. As for patés, the effect of microwaves was judged good for small terrines and inadequate for large ones.

Experiments with pork meat infected with *Bacillus subtilis*, *Leuconostoc mesenteroides*, and *Pseudomonas putrefaciens* 24 hours earlier, and subsequently irradiated with microwaves to bring its temperature to 60, 68, 77, and 85°C, led to the conclusion [119] that microwaves were the most efficient means of reducing bacterial populations. This is discussed further in Part II, Section 1.2.1.

Microwave pasteurization of milk and milk products is currently a subject of renewed interest. The first experiments in this field, reviewed and reported in [115], were performed by the National Research Council of Canada and had started in the early 1970s. The development of a dual-frequency treatment led to pasteurization that was six times faster than the conventional process, the shelf-life being between 4 and 6 weeks. This was also the concept behind the design of the combined HF-microwave applicator produced in Germany [13], [14] for the pasteurization of yoghurt in containers (40 kW at 27.12 MHz and 8 kW at 2.45 GHz), with a final temperature of 60°C. Excellent bacteriocidal effects were reported in [148] in which a sample heated with microwaves to a temperature of 95°C in 15 s was found to contain only a trace of contamination as compared with an initial population of 25×10^6 cm^{-3}. More recently, researchers at the Agricultural Research Office at Bet-Dagan in Israel [107] noted that the exposure of 400 ml of milk to 700 W of 2.45-GHz radiation had an identical effect to that of LTLT pasteurization, but the rate of protein denaturation was slightly higher. Microwaves are much faster than conventional methods, but are nevertheless judged capable of reducing the number of micro-organisms and of deactivating alkaline phosphatase enzymes. Researchers in Yugoslavia [145] performed experiments at 2.45 GHz on samples of milk and yoghurt and reached identical conclusions.

Research at Cornell University in Ithaca (New York) has demonstrated that a final temperature of 55 to 60°C produced by 2.45-GHz microwaves was sufficient to ensure efficient pasteurization of milk. Counting was performed according to directives issued by the American Public Health Association (Standard Plate Count and Psychotrophic Bacterial Count) and showed very remarkable bacteriocidal rates (e.g., for psychotrophic micro-organisms that produced the proteolytic and lipolytic thermally stable enzymes that were responsible for the rancid and bitter taste of degraded milk). The shelf-life of cheese in soft packaging can be extended to between 7 and 12 days by microwave pasterization at 82°C [154].

4.2.2 Vegetable products

A technique for decontaminating mustard grains infested with *Nematospora coryli* yeast was described in [78] and [79]. Samples of mustard grains with 6.8% and 9.7% moisture content were exposed to 2.45-GHz microwaves for 13 s and reached final temperatures of 87° and 71°, respectively. In both cases, decontamination was complete without any significant effect on the level of allylisothiocyanate that is the principal aromatic compound in mustard. The concurrent procedure, namely, sterilization with ethylene oxide, took six hours, affected germination, and may have produced carcinogenic derivatives. Relatively poor results were reported in [60] and [61] on the microflora of pepper, marjoram, and paprika. In particular, there was a lowering in the level of pigments and essential oils. Pondering over these conclusions, it is interesting to note that the pasteurization of spices, ginger, medical herbs, and cocoa powder constitutes one of the main applications of the Micron tunnels produced by the Oshikiri Machinery Company of Tokyo.

Vegetables in glass containers are usually processed by HTST techniques that are deemed better from the standpoint of preservation than LTLT procedures. Researchers at St. Paul University (Minnesota) [102] sterilized broccoli and asparagus tips with microwaves without observing any significant degradation. They used using a 61-cm cubic cavity with a 30-cm diameter opening for the insertion of the sterilization container (usually an epoxy-fiberglass cylinder sealed under a pressure of 1 MPa). The treatment with the 1-kW RF took 10 min at 2.45 GHz, and the final maximum temperature was 165°C. It was concluded that microwaves offered the most promising and economically viable sterilization technique.

At the industrial level, a 30-kW microwave tunnel was installed in 1974 in Stenstorp (Sweden) by the Skandinaviska Processinstrument AB of Bromma (this company is no longer producing microwave equipment). It was used for the pasteurization of peeled potatoes after conditioning in plastic containers. The throughput was 500 kg hr^{-1} and the final temperature was 80°C [182].

In a somewhat different field, interesting results were reported on soya beans in [87]. A 675-W, 2.45-GHz oven produced a reduction of between 45% and 7% in mosaic virus infestation of beans with a moisture content of 8.5% for a temperature rise of only 20° from an initial 22°C in 160 s.

Experiments with manufactured products were largely confined to pasta, bread, wine, and fruit syrups. Researchers from OMAC Impianti SRL of Pratissolo di Scandiano (Emilia-Romagna) [30] showed that the market for fresh pasta was doomed to remain strictly local unless use was made of refrigerated semipreservation of the product previously pasteurized in hermetically sealed packaging. Stuffed pasta is invested with lactic bacteria (the *Micrococcaceae* and several species of *Bacillus*). Gnocchi in particular constitute a perfect substrate for the development of all kinds of micro-organisms. Conventional procedures (e.g. hot water, vapor, hot air) are unsatisfactory from the standpoint of damage to packaging, and this led to the successful application of microwave pasteurization. The efficacy of the method was found to be comparable to that of classical treatment. It also allowed better presentation, much shorter treatments, and lower or similar cost as compared with hot air. In view of these excellent results, it was suggested that a continuous process could be implemented.

Microwave pasteurization of bread had been the subject of trials as far back as 1946 [31], and was judged at the time to be very satisfactory. More recently [124], Litton Industries at Palo Alto (California) exposed sliced bread to 5-kW microwaves at 2.45 GHz with a transit period of approximately 2 min and a final temperature of 55-65°C. The population of *Penicillum* was found to fall from 1,846 to 0.143 colonies per plate, and the population of *Aspergillus niger* fell from 2,000 to 0.186. The slices of microwave-pasteurized bread continued to be free of contaminations after 21 days, whereas those that were treated with sodium propionate were covered in mildew after only 12 days. The lethal temperatures for mildew are 68–70°C for a 20-min exposure to heat, and microwaves could have a selective thermal effect on spores despite their small size.

Other workers [148] succeeded in preserving microwave-pasteurized breads and pastries for several months without any sign of mildew. These results were confirmed by trials perfomed by IMI-SA [173] in which microwave-treated brioches were preserved for five months whereas controls held for the same period were completely covered in mildew.

Larger scale trials were performed in 1974 [20] by SIK of Gothenburg (Sweden) and led to the development of an 80-kW, 2.45-W applicator incorporating slow-wave structures on either side of a transporting belt. The system was designed for the pasteurization of packaged brown bread. The first equipment of this type was constructed by Skandinaviska Process Instrument AB and was operating around 1974 at Jowa AG of Gränichen (Switzerland). This equipment was 11.2 m long, 1 m wide, and 1.3 m tall. The RF output power was 80 kW. Bread temperature was raised from 20 to 80°C in one to two minutes with a throughput of 2,000 kg hr^{-1}, which was enough to ensure a shelf life of two or three months without the use of a chemical inhibitor.

The Gigatron tunnel of IMI-SA and the Micron MWC tunnel of Oshikiri that were mentioned above can be adapted for this type of application. There are nine different versions of the latter applicator, ranging from 2 to 32 kW output power in 4-kW modules, with lengths between 4 and 11 m and an aperture 44 cm wide and 20 cm high [178]. Several plants of this type are currently operating in Scandanavia.

It has been suggested [43] that the microwave pasteurization of bread could develop rapidly because of the current tendency to strengthen the regulations dealing with chemical inhibitors. Further information can be found in [116] and [26].

Wine, a very complex and delicate product, was for a long time the subject of radio-frequency pasteurization [167]. About 10 years ago, the Te chnical Institute for Wine-SICAREX of Rabastens (Tarn, France) conducted a study of stabilization by 4.6 and 2.45-Ghz radiation of musts containing the winemaking yeasts *Saccharomyces*, *Torulopsis stelata*, and *Hanseniaspora uvarum*. The results were completely negative [58]. The Australian Wine Research Institute of Glen Osmond (Southern Australia), currently working on this question [110], has shown that microwave sterilization can be defined as for all the other thermal inactivation techniques in terms of duration and temperature: the destruction of yeasts and *Leuconostoc oenos* cells requires that a temperature of 90°C be maintained for at least 15 min. Microwave coagulation of proteins is another possibility that may extend the shelf life without detriment to the quality of wine [115].

Beer has also been the subject of microwave pasteurization trials, particularly in the Soviet Union [184]. Fruit juices and syrups in glass, cardboard, and polyethylene containers with metal linings and covers are unsuitable for standard pasteurization procedures. The French company Neyret-Chavin of Bourgoin-Jallieu (Isère) has collaborated with Lambda International (Val-d'Oise, France) to develop a microwave applicator 5.5 m long, 1 m wide, fed with 12 kW of microwave power at 2.45 GHz and using a conveyor belt circulating at a maximum speed of $15 \, \mathrm{m \, min^{-1}}$. The syrup or juice is at 52°C after passing through a heat exchanger, and is heated to 56°C by microwaves in 84 s. Power consumption is less than $10 \, \mathrm{Wh \, l^{-1}}$ at a cost of $0.5 per cubic meter as against $3 to $5 per cubic meter in conventional heating. The applicator can be readily incorporated in a bottling chain, and the microbial population can be reduced by three orders of magnitude. Despite the relatively low temperatures, the effect is thought to be of purely thermal origin [118], [115], [21].

4.2.3 Prepared meals

Microwave pasteurization of military rations packaged in plastic was under way in the early 1970s [90]. It was then reported that the necessary duration of treatment in a 10-kW, 2.45-GHz oven was 9–14 min. Previous research [95], [103], [85], [86] had indicated that a counter pressure was necessary to avoid the explosion of the packaging. For example, it was suggested [85], [86] that a fluid such as mineral oil with an extremely low dielectric loss factor could be used as a counterpressure medium with good microwave transmission.

The applicator developed for the US Army [90] consisted

of an 11-cm diameter pressurized fiberglass-epoxy tube fitted
with airlocks and a polypropylene conveyor belt. The tube was
placed in a cavity of $2.88 \times 0.69 \times 0.66m^{-3}$ fed by four 1.25–
2.5 kW magnetrons operating at 2.45 GHz. The pressure on
the tube did not exceed 1 MPa; the output was one ration per
minute for a transit time of 1–12 min and a final temperature
of 121°C. It is also worth mentioning at this stage the P250
Multitherm tunnel developed by Alfastar AB of Tumba (Sweden), a subsidiary of Alfa Laval and Swedish Match: It was
designed for the sterilization of products in airtight packaging.
To date it has been tested on salads in mayonnaise sauce and
fruit-based desserts. In both cases, the shelf life of the products
was raised to five or six months without any additives or loss of
quality as compared with the fresh product. The Multitherm
seems well adapted to the taste of the Scandinavian consumer
who favors semiprepared and refrigerated, high-quality meals
[155], [183], [186], [42], [43].

References

1. ABARA A., HILL M. The effect of moisture content and enzyme activity on the blanching time of potatoes. *J. Microw. Power*, 16, n° 1, 1981, pp. 31-34.

2. ALEXANDER D. Process parameters for continuous microwave sterilization. In : Proceedings of the IMPI Symposium, Loughborough, 1973, pp. 1-2. (Paper 3.A.1).

3. AREF M., NOËL J., MILLER H. Inactivation of alpha-amylase in wheat flour with microwaves. *J. Microw. Power*, 7, n° 3, 1972, pp. 215-221.

4. ARMBRUSTER G., BRINK P. Evaluation of microwave blanching of vegetables. In: Proceedings of the IMPI Symposium, Chicago, 1985, pp. 1-2.

5. ASAMI K., HANAI T., KOIZUMI N. Dielectric properties of yeast cells. *J. Membr. Biol.*, 28, n° 2-3, 1976, pp. 169-180.

6. AUSSUDRE S. Etude d'une cavité pour stérilisation de conserves. In : Colloque international des Journées d'Electronique de Toulouse, 7-11 mars 1977, IV-6.

7. AVISSE C. Etude du blanchiment des pêches aux micro-ondes. Publ. INRA, Dijon, s.d., 16 p.

8. AVISSE C., VAROQUAUX P. Microwave blanching of peaches. In : Proceedings of the IMPI Symposium, Waterloo (Can.), 1975, pp. 67-68. (Paper 4.1.).

9. AVISSE C., VAROQUAUX P. Microwave blanching of peaches. *J. Microw. Power*, 12, n° 1, 1977, pp. 73-77.

10. AVISSE C., VAROQUAUX P., DUPUY P. Utilisation des micro-ondes pour le blanchiment des pêches. *C. R. Séances Acad. Agric. Fr.*, 60, n° 10, 1974, pp. 741-749. (C.D.I.U.P.A., n° 77938).

11. AYOUB J., BERKOWITZ D., KENYON E., WADSWORTH C. Continuous microwave sterilization of meat in flexible pouches. *J. Food Sci.*, 39, n° 3, 1974, pp. 309-313. (C.D.I.U.P.A., n° 70693).

12. AZUMA J., ASAI T., ISAKA M., KOSHIJIMA T., Effects of microwave irradiation on enzymatic susceptibility of crystalline cellulose. *J. Ferment. Technol.*, 63, n° 6, 1985, pp. 529-536. (C.D.I.U.P.A., n° 209569).

13. BACH J. Verfahren und Sterilisieren von Wasserenthaltenden organischen Produkten. BRD Patent, n° 1 812 470, 18 Juni 1970.

14. BACH J. Multiple-frequency method for prolonging shelf-life of milk products. *Dtsch. Milchwirtsch.*, 28, 1976, pp. 1376-1377.

15. BALDWIN R., CLONINGER M., FIELDS M. Growth and destruction of Salmonella typhimurium in egg white foam products cooked by microwaves. *Appl. Microbiol.*, 16, n°12, 1968, pp. 1929-1934. (C.D.I.U.P.A., n° 21976).

16. BALDWIN R., FIELDS M., POON W., KORSCHGEN B. Destruction of Salmonellae by microwave heating of fish with implications for fish products. *J. Milk Food Technol.*, 34, n° 10, 1971, pp. 467-470. (C.D.I.U.P.A., n° 40376).

17. BARTHOLIN G., DESCAMPS P., BOUDET B. Problèmes posés par l'industrie de transformation des légumes. 1. Application des micro-ondes au blanchiment. 2. Recherche d'une méthode rapide de contrôle de la qualité bactériologique des matières premières. Paris, DGRST, Action concertée : Technologie Alimentaire et Agricole, Rapport de fin de Contrat n° 72-7-0827, 1974, 48 p. (C.D.I.U.P.A., n° 81866).

18. BELKHODE M., MUC A., JOHNSON D. ,Thermal and athermal effects of 2.8 GHz microwaves on three human serum enzymes. *J. Microw. Power*, 9, n° 1, 1974, pp. 23-29.

19. BENGTSSON N., GREEN W., DEL VALLE F. Radio-frequency pasteurization of cured hams. *J. Food Sci.*, 35, n° 5, 1970, pp. 681-687. (C.D.I.U.P.A., n° 29444).

20. BENGTSSON N., OHLSSON T. Microwave heating in the food industry. *Proc. IEEE*, 62, n° 1, 1974, p. 44.

358 Part III - Applications in the food industry

21. BERNARD J. Traitement micro-ondes. Application à la pasteurisation des boissons. *Bios*, 14, n° 5, 1983, pp. 19-25. (C.D.I.U.P.A., n° 83-5840).

22. BINI M., CHECCUCCI A., IGNESTI A., MILLANTA L., RUBINO N., CAMICI G., MANAO G., RAMPONI G. Analysis of the effects of microwave energy on enzymatic activity of lactate dehydrogenase. *J. Microw. Power*, 13, n° 1, 1978, pp. 95-99.

23. BLANCO J., DAWSON L. Survival of Clostridium perfringens on chicken cooked with microwave energy. *Poult. Sci.*, 53, n° 5, 1974, pp. 1823-1830. (C.D.I.U.P.A., n° 80559).

24. BOMAR M., GRUNEWALD T. Zur Frage des nichttermischen Effektes von Mikrowellen auf Mikroorganismen. *Sci. Technol. Alim.*, 5, n° 5, 1972, pp. 166-171. (C.D.I.U.P.A., n° 50095).

25. BROWN G., MORRISON W. An exploration of the effects of strong radio-frequency fields on micro-organisms in aqueous solutions. *Food Technol.*, 8, n° 8, 1954, pp. 361-366.

26. BRUMMER J., MORGENSTERN G. Massnahmen zur Schimmelbekämpfung bei Brot. VII. Mitt.: Mikrowellenbehandlung bei Brot. *Brot Backwaren*, 33, n° 1-2, 1985, pp. 26-29. (C.D.I.U.P.A., n° 197045).

27. BURDIN J., LAVERGNE E. de. Les bactéries. Paris, P.U.F., 1978, 125 p. (Collection Que sais-je ? n° 53).

28. BURG F. Mikrowellensterilisation bei Schnittbrot. *Brot Gebäck*, 3, 1968, pp. 58-60.

29. CARROLL D., LOPEZ A. Lethality of radio frequency energy upon microorganisms in liquid, buffered and alcoholic food systems. *J. Food Sci.*, 34, n° 4, 1969, pp. 320-324. (C.D.I.U.P.A., n° 15450).

30. CASTELVETRI F., MASSINI R., MIGLIOLI L., FERRARI C. Impiego delle microonde per la pastorizzazione di paste alimentari fresche. *Ind. Conserve*, 58, n° 4, 1983, pp. 211-218. (C.D.I.U.P.A., n° 192204).

31. CATHCART W. High-frequency heating produces mold-free bread. *Food Ind.*, 18, n° 6, 1946, pp. 864-865.

32. CHEN S., COLLINS J., Mac CARTY I., JOHNSTON M. Blanching of white potatoes by microwave energy followed by boiling water. *J. Food Sci.*, 36, n° 5, 1971, pp. 742-743. (C.D.I.U.P.A., n° 36470).

33. CHEN H., WU C., LIN H., CHUNG C. Effect of microwave and open flame heating on the microbial counts and weight loss of eel. *Bull. Jpn. Soc. Sci. Fish.*, 42, n° 4, 1976, pp. 405-410. (C.D.I.U.P.A., n° 97642).

34. CHIU C., TATEISHI K., KOSIKOWSKI F., ARMBRUSTER G. Microwave treatment of pasteurized milk. *J. Microw. Power*, 19, n° 4, 1984, pp. 269-272.

35. COLLINS J., Mac CARTY I. Comparison of microwave energy with boiling water for blanching whole potatoes. *Food Technol.*, 23, n° 3, 1969, pp. 337-340. (C.D.I.U.P.A., n° 09865).

36. COMOROSA S. Effect of electromagnetic field on enzymic substrates reply. *Stud. Biophys.*, 44, n° 2, 1966, p. 166.

37. COPSON D. Microwave irradiation of orange juice concentrate for enzyme inactivation. *Food Technol.*, 25, n° 8, 1971, pp. 397-399.

38. CORELLI J., GUTMANN R., KOHAZI S., LEVY J. Effects of 2.6-4.0 GHz microwave radiation on E. coli B. *J. Microw. Power*, 12, n° 2, 1977, pp. 141-144.

39. CULKIN K., FUNG D. Destruction of Escherichia coli and Salmonella typhymurium in microwave-cooked soups. *J. Milk Food Technol.*, 38, n° 1, 1975, pp. 8-15. (C.D.I.U.P.A., n° 81399).

40. CUNNINGHAM F., ALBRIGHT K. Using microwaves to reduce bacterial numbers on fresh chicken. *Microw. Energy Appl. Newsl.*, 10, n° 4, 1977, pp. 3-4.

41. DECAREAU R. Microwave research and development at the US Army Natick Research and Development laboratories. *J. Microw. Power*, 17, n° 2, 1982, pp. 127-135.

42. DECAREAU R. Future research developments in microwave technology. Non publ., s.d., 12 p.

43. DECAREAU R. Microwave food-processing equipment throughout the world. Food Eng. Lab., US Army Natick Res. Dev. Center, Non publ., s.d., 15 p.

44. DECAREAU R., MUDGETT R. Microwaves in the food processing industry. (Food Science and Technology - a series of monographs). London, Academic Press, 1985, 247 p.

45. DEFICIS, FLAMENT. Etude des effets spécifiques des micro-ondes sur les micro-organismes du lait. Paris, DGRST, Action concertée : Technologie Agricole, Rapport de fin de Contrat n° 73-7-1065, 1974, 32 p. (C.D.I.U.P.A., n° 84027).

46. DELANY E., VAN ZANTE H., HARTMAN P. Microbial survival in electronically heated foods. *Microw. Energy Appl. Newsl.*, 1, n° 3, 1968, pp. 11-14.

47. DEMEAUX M., BIDAN P. Etude de l'inactivation par la chaleur de la polyphénoloxydase du jus de pomme. *Ann. Technol. Agric.*, 15, n° 41, 1966, pp. 349-358.

48. DESSEL M., BOWERSOX E., JETER W. Bacteria in electronically cooked foods. *J. Am. Diet. Assoc.*, 37, n° 9, 1960, pp. 230-233.

49. DIETRICH W., HUXSOLL C., GUADAGNI D. Comparison of microwave, conventional and combination blanching of Brussel sprouts for frozen storage. *Food Technol.*, 24, n° 5, 1970, pp. 105-109. (C.D.I.U.P.A., n° 22267).

50. DUNAJSKI E., STECKI W. Usefulness of microwave heating for in-bottle pasteurization of liquids. *Przem. Spoz.*, 25, n° 2, 1971, pp. 80-81.

51. EDWARDS G. Effects of microwave radiation on wheat and flour : the viscosity of the flour pastes. *J. Sci. Food Agric.*, 15, 1964, pp. 108-114.

52. EVANS K., TAYLOR H. Microwaves extend shelf-life of cakes. *Food Manuf.*, 42, n° 10, 1967, pp. 50-51. (C.D.I.U.P.A., n° 00040).

53. EVANS K., TAYLOR H., Las microondas prolongan la conservación de los bizcochos. *Tecnol. Alimenti*, 2, n° 8, 1968, pp. 28-30. (C.D.I.U.P.A., n° 22797).

54. FLEMING H. Effects of high frequency fields on micro-organisms. *Electr. Eng.*, 63, n° 1, 1944, pp. 18-21.

55. FOURNAUD J., LAURET R. Action des micro-ondes sur la flore microbienne de viandes en cours de décongélation ou de produits de charcuterie emballés sous film plastique et sous vide. Paris, DGRST, Rapport de fin de contrat n° 70-02-232, s.d., 9 p.

56. FOURNAUD J., LE POMMELEC G. Qualité bactériologique des plats cuisinés. *Rev. Gén. Froid*, 68, n° 12, 1977, pp. 741-745. (C.D.I.U.P.A., n° 114994).

57. FULLER J., OWINGS W., FANSLOW G. Microwave treated soybeans as a feedstuff in poultry diets. Communication personnelle du Pr. FANSLOW, 1986, 3 p.

58. GAILLARD M. Etude de la stabilisation des moûts par l'effet non thermique des micro-ondes, essais 1978. Inst. Tech. Vin-SICAREX Sud-Ouest (Rabastens)-C.S.T., session ITV 1979, 6 p.

59. GAZAN C., GRUNDER D., VINCENT A. Utilisation des microondes dans l'industrie de la viande. Paris, DGRST, Action concertée : Technologie Agricole, Rapport de fin de contrat n° 70-02-271, 1971, 71 p. (C.D.I.U.P.A., n° 104784).

60. GERHARDT U., ROMER H. Einfluss von Mikrowellen auf Gewürze. *Fleischwirtsch.*, 65, n° 6, 1985, pp. 718-723. (C.D.I.U.P.A., n° 201458).

61. GERHARDT U., ROMER H. Wesen und Eigenschaften von Mikrowellen. *Z. Lebensm.-Technol. Verfahr.-Tech.*, 36, n° 5, 1985, pp. 309-316. (C.D.I.U.P.A., n° 204727).

62. GOLDBLITH S., WANG D. Effect of microwaves on Escherichia coli and Bacillus subtilis. *Appl. Microbiol.*, 15, n° 6, 1967, pp. 1371-1375. (C.D.I.U.P.A., n° 21961).

63. GRAY O. Method and apparatus for sterilizing. U.S. Patent n° 3 494 722, 10 Feb. 1970.

64. GRAY O. Method and apparatus for controlling micro-organisms and enzymes. U.S. Patent n° 3 494 723, 10 Feb. 1970.

65. GRAY O. Apparatus for controlling micro-organisms and enzymes. U.S. Patent n° 3 494 724, 10 Feb. 1970.

66. GULLETT E., ROWE D., HINES R. Effect of microwave blanching on the quality of frozen green beans. *Can. Inst. Food Sci. Technol. J.* 17, n° 4, 1984, pp. 247-252. (C.D.I.U.P.A., n° 206282).

67. HAFEZ Y., MOHAMED A., HEWEDY F., SINGH G. Effects of microwave heating on solubility, digestibility and metabolism of soy protein. *J. Food Sci.*, 50, n° 2, 1985, pp. 415-423. (C.D.I.U.P.A., n° 199523).

68. HAFEZ Y., SINGH G., mac LELLAN M., MONROE-LORD L. Effects of microwave heating on nutritional quality of soybeans. *Nutr. Rep. Int.*, 28, n° 2, August 1983, pp. 413-421. (C.D.I.U.P.A., n° 186685)

69. HAMID M. Sterilization of articles by microwave irradiation. Canadian Patent n° 1034 734, 18 july 1978.

70. HAMID M., BOERNER W., TONG S. Microwave irradiation of potato-waste water. *J. Microw. Power*, 5, n° 1, 1970, pp. 44-46. (C.D.I.U.P.A., n° 104740).

71. HAMID M., BOULANGER R., TONG S., GALLOP R., PEREIRA R. Microwave pasteurization of raw milk. *J. Microw. Power*, 4, n° 4, 1969, pp. 272-275.

72. HAMID M., MOSTOWY N., BHARTIA P. Microwave bean roaster. In : Proceedings of the IMPI Symposium, Milwaukee, 1974, 3 p. (Paper B5-2/1-3).

73. HAMID M., MOSTOWY N., BHARTIA P. Microwave bean roaster. *J. Microw. Power*, 10, n° 1, 1975, pp. 109-114.

74. HAMRICK P., BUTLER B. Exposition of bacteria to 2450 MHz microwave radiation. *J. Microw. Power*, 8, n° 3-4, 1973, pp. 227-233.

75. HELLER J. Effect of high-frequency electromagnetic fields on microorganisms. *Radio Electron.*, 6, 1959.

76. HELLER J., TEIXEIRA-PINTO A. A new physical method of creating chromosomal aberrations. *Nature*, 183, 1959, pp. 905-906.

77. HENDERSON H., HERGENROEDER K., STUCHLY S. Effect of 2450 MHz microwave radiation on horseradish peroxidase. *J. Microw. Power*, 10, n° 1, 1975, pp. 27-35.

78. HOLLEY R., TIMBERS G. Nematospora destruction in mustard seed by microwave and moisture treatments. *Can. Inst. Food Sci. Technol. J.*, 16, n° 1, 1983, pp. 68-75. (C.D.I.U.P.A., n° 83-1891).

80. HSIEH S. Biological effects of X-band microwave irradiations on E. coli. Thesis Ph. D., Univ. Tulane, 1974.

81. HUXSOLL C., DIETRICH W., MORGAN A. Comparison of microwave with steam or water blanching of corn-on-the-cob. 1. Characteristics of equipment and heat penetration. 2. Peroxidase inactivation and flavor retention. *Food Technol.*, 24, n° 3, 1970, pp. 84-86. (C.D.I.U.P.A., n° 19311).

82. JAYNES H. A research note. Microwave pasteurization of milk. *J. Milk Food Technol.*, 38, n° 7, 1975, pp. 386-387.(C.D.I.U.P.A., n° 85650).

83. JEPPSON M. Microwave blanch, sterilize after packaging. *Food Process.*, n° 12, 1963.

84. JEPPSON M. Aseptic canning of foods having solid or semi-solid components. U.S. Patent n° 3 437 495, 8 April 1969.

85. JEPPSON M., HARPER J. Microwave heating of substances under hydrostatic pressure. U.S. Patent n° 3 335 253, 8 August 1967.

86. JEPPSON M., HARPER J. Microwave heating of substances under hydrostatic pressure. U.S. Patent n° 3 398 251, 20 August 1967.

87. JOLICOEUR G., HACKAM R., TU J. The selective inactivation of seed-borne soybean mosaic virus by exposure to microwaves. *J. Microw. Power*, 17, n° 4, 1982, pp. 341-344.

88. KAMAT G., LASKEY J. Enzyme inactivation in vitro with 2450 MHz microwaves. In : Radiation bioeffects. Summary reports. U.S. Dep. Health, Educ. Welfare, Bureau of Radiological Health, Publ. n° BRH/DBE 70-7, 1970.

89. KENYON E. The feasibility of continuous heat sterilization of food products using microwave power. U.S. Army Natick Lab., Technical Report n° 71-8-FL, 1970.

90. KENYON E., WESTCOTT D., LA CASSE P., GOULD J. A system for continuous thermal processing of food pouches using microwave energy. *J. Food Sci.*, 36, n° 2, 1971, pp. 289-293. (C.D.I.U.P.A., n° 34594).

91. KLINGER R. Mikrowellenbehandlung von Müsliprodukten. *Lebensm.-Tech.*, 16, n° 9, 1984, pp. 446-450. (C.D.I.U.P.A., n° 192195).

92. KOZLOWSKA H., ROTKIEWICZ D., BOCK H. Inactivation of the enzyme Myrosinase in whole rapeseeds. In : Congrès International sur le Colza, 6, Paris, juin 1983, Paris, CETIOM, s.d., 2, pp. 1466-1471. (C.D.I.U.P.A., n° 191574).

93. LACEY B., WINNER H., Mac LELLAN M., BAGSHAWE K. Effects of microwave cookery on the bacterial counts of food. *J. Appl. Bacteriol.*, 28, n° 2, 1965, pp. 331-335.

94. LAMPI R. Flexible packaging in thermoprocessed foods. *Adv. Food Res.*, 23, 1977, pp. 305-428 et *Microw. Process.*, pp. 397-398. (C.D.I.U.P.A., n° 124211).

95. LANDY J. Method of sterilizing food in sealed containers. U.S. Patent n° 3 215 239, 2 Nov. 1965.

96. LANE R., BOSCHUNG M., ABDEL-GHANY M. Absorbic acid retention of selected vegetables blanched by microwave and conventional methods. *J. Food Qual.*, 8, n° 2-3, 1985, pp. 139-144. (C.D.I.U.P.A., n° 209264).

97. LANGLEY J., YEARGERS E., SHEPPARD A. Effects of microwave radiation on enzymes. In : Proceedings of the IMPI Symposium, Loughborough, 1973, 2 p. (Paper 3.B.4/1-2).

98. LECHOWICH R., BEUCHAT L., FOX K., WEBSTER F. Procedure for evaluating the effects of 2450 MHz microwaves upon Streptococcus faecalis and Saccharomyces cerevisiae. *Appl. Microbiol.*, 17, n° 1, 1969, pp. 106-110.

99. LIN C., LI C. Microwave sterilization of oranges in glass pack. *J. Microw. Power*, 6, n° 1, 1971, pp. 45-47.

100. LIPOMA S. Apparatus for continuous electromagnetic sterilization. U.S. Patent n° 3 718 082, 1971.

101. LIPOMA S. Method for continuous electromagnetic sterilization of food in a pressure zone. U.S. Patent n° 3 889 009, 1972.

102. LOH J., BREENE W. A laboratory microwave sterilizer and its possible application toward improving texture of sterilized vegetables. *J. Food Process. Preserv.*, 7, n° 2, 1983, pp. 77-92. (C.D.I.U.P.A., n° 184000).

103. LONG F., SHAW F., LISLE H. Microwave sterilization and vacuumizing of products in flexible packages and apparatus therefore. U.S. Patent n° 3 261 140, 19 July 1966.

104. LUTER L., WYSLOUZIL W., KASHYAP S. The destruction of aflatoxins in peanuts by microwave roasting. *Can. Inst. Food Sci. Technol. J.*, 15, n° 3, 1982, pp. 236-238. (C.D.I.U.P.A., n° 83-3742).

105. Mac MILLAN D. Process for pasteurizing and sealing oysters. U.S. Patent n° 3 615 726, 26 Oct. 1971.

106. MASURE M., CAMPBELL H. Rapid estimation of peroxidase in vegetable extracts. *Fruit Prod. J.*, 23, 1944, pp. 369-374.

107. MERIN U., ROSENTHAL I. Pasteurization of milk by microwave irradiation. *Milchwiss.*, 39, n° 11, 1984, pp. 643-644. (C.D.I.U.P.A., n° 194172).

108. METAXAS A. Design of equipment for microwave sterilization of stews. Publ. Electr. Council Res. Centre M 921 (G.B.), 1976.

109. MODERSOHN E. Verfahren zum sterilen Verpacken von Brot, insbesondere von geschnittenem Brot. BRD Patent n° 1 586 178, 26 Aug. 1971.

110. MONK P., STEPHENSON D. Microwave sterilisation of media and membranes. *Food Technol. Aust.*, 37, n° 1, 1985, pp. 14-15. (C.D.I.U.P.A., n° 200637).

111. MOTHIRON J., VIALARD-GOUDOU A., MAUPAS P. Use of an industrial prototype machine for drying and sterilizing pharmaceutical ampoules by microwaves at 2.45GHz. In : Proceedings of the IMPI Symposium, Leuven, 1976. (Paper 4B.3).

112. MOYER J., HOLGATE K. Cooling after water and electronic blanching. *Food Ind.*, 19, 1947, p. 1370.

113. MOYER J., STOTZ E. The electronic blanching of vegetables. *Science*, 102, n° 2638, 1945, p. 68.

114. MUDGETT R. Microwave properties and heating characteristics of foods, an overview. *Food Technol.*, 40, n° 6, 1986, pp. 84-98. (C.D.I.U.P.A., n° 214710).

115. MUDGETT R., SCHWARTZBERG H. Microwave food processing : pasteurization and sterilization - a review. American Institute of Chemical Engineers, Symposium Series, 78, n° 218, 1982, 11 p.

116. NAUMANN G. Mikrowellenpasteurisierung von Schnittbrot ohne Konservierungsstoffe. *Getreide, Mehl Brot*, 39, n° 10, 1985, pp. 306-309. (C.D.I.U.P.A., n° 191049).

117. NELSON S., POUR-EL A., PECK E. Effects of 42- and 2450-MHz dielectric heating on nutrition-related properties of soybeans. *J. Microw. Power*, 16, n° 3-4, 1981, pp. 313-318.

118. NENIN J. Le chauffage par microondes, homogène et instantané. *Usine Nouv.*, n° 2, 1979, pp. 78-86 et *Stérilis.*, pp. 84, 86.

119. OCKERMAN H., CAHILL V., PLIMPTON R., PARRETT N. Cooking inoculated pork in microwave and conventional ovens. Publication of the Ohio State Univ., Columbus. Wooster, Agric. Res. Dev. Center, 39, n° 11, 1976, pp. 771-773.

120. OGURA K. Sterilization of food by microwave heating. *J. Antibact. Antifungal Agents*, 4, n° 1, 1976, pp. 21-28.

121. OHLSSON T. Possibilities and limits of microwave sterilization. In : Symposium international de Karlsruhe "How ready are ready-to-serve foods?" 23-24 August 1977, Publication n° 309 du SIK, Göteborg, 10 p. (C.D.I.U.P.A., n° 33337).

122. OHLSSON T., BENGTSSON N. Development of a pilot plant microwave food sterilizer and results of preliminary experiments. In : Proceedings of the IMPI Symposium, Loughborough, 1973, 8 p. (Paper 3A.3).

123. OHLSSON T., BENGTSSON N. Dielectric food data for microwave sterilization processing. *J. Microw. Power*, 10, n° 1, 1975, pp. 93-108.

124. OLSEN C. Microwaves inhibit bread mold. *Food Eng.*, 37, n° 7, 1965, pp. 51-53.

125. OLSEN C., DRAKE C., BUNCH S. Some biological effects of microwave energy. *J. Microw. Power*, 1, n° 2, 1966, pp. 45-56.

126. Ó MEARA J., JOHNSON R. Application of microwaves to packaged food sterilization in a helical pump hydrostatic cooker. In : Proceedings of the IMPI Symposium, Monterey, 1971. (Paper 2.6.).

127. Ó MEARA J., TINGA W., WADSWORTH C., FARKAS D. Food sterilization in a microwave pressure retort. In : Proceedings of the IMPI Symposium, Leuven, 1976. (Paper 6B.4).

128. Ó MEARA J., WADSWORTH C., FARKAS D. Microwave heat sterilization of foods in flexible pouches. Publication of the U.S. Army Natick Res. Dev. Command, 1976.

129. PEREIRA F. On the effect of electromagnetic waves on enzyme systems. *Biochem. J.*, 238, 1935, pp. 53-58.

130. PROCTOR B., GOLDBLITH S. Radar energy for rapid food cooking and blanching, and its effects on vitamin content. *Food Technol.*, 2, 1948, pp. 95-104.

131. ROBE K. Improve flavor of pasteurized products. *Food Process. Mark.*, 27, n° 3, 1966, pp. 84-86.

132. ROBERTS P. Viability studies of ascospores and vegetative cells of Saccharomyces cerevisiae exposed to microwaves at 2450 MHz. *J. Sci. Food Agric.*, 23, n° 4, 1972, p. 544.

133. ROSEN C. Effects of microwaves on food and related materials. *Food Technol.*, 26, n° 7, 1972, pp. 36-55 (C.D.I.U.P.A., n° 47428).

134. ROSENBERG U., BOGL W. Auftauen, Trocknen, Wassergehaltsbestimmung und Enzyminaktivierung mit Mikrowellen. *Z. Lebensm.-Technol. Verfahr.*, 37, n° 1, 1986, pp. 12-19.

135. ROSENBERG U., SINELL H. Reduktion von Bacillus subtilis-sporen bei Mikrowellenerhitzung. *Archiv Lebensm.-Hyg.*, 35, n° 1, 1984, pp. 7-9. (C.D.I.U.P.A., 187898).

136. RZEPECKA M., PEREIRA R. Permittivity of some dairy products at 2450 MHz. *J. Microw. Power*, 9, 1974, pp. 277-288.

137. SAMUELS C., WIEGAND E. Radio frequency blanching of cut corn and freestone peaches. *Fruit Prod. J.*, 28, n° 2, 1948, p. 43.

138. SATO T., FUNAOKA T., SAKAMOTO M. Effects of radio wave heating on mold and preservation of marine foods. *Hokusuishi Geppo*, 24, n° 10, 1967, pp. 407-415.

139. SMITH K., GRINNEL A. Method and apparatus for preserving foodstuffs. U.S. Patent n° 2 576 862, 27 Nov. 1951.

140. STENSTRÖM L. Heating of products in electromagnetic field. U.S. Patent n° 3 809 845, 1971.

141. STENSTRÖM L. Taming microwaves for solid food stabilization. In : Proceedings of the IMPI Symposium, Ottawa, 1972, pp. 110-112. (Paper 7.4.).

142. STOLKOWSKI J. Les enzymes. 5ème éd. Paris. P.U.F., 1973, 126 p. (Collection Que sais-je ? n° 434).

143. STUCHLY M., STUCHLY S. Industrial, scientific, medical and domestic applications of microwaves. *IEE Proc.*, 130, part A, n° 8, Nov. 1983, pp. 467-503.

144. SWENSON T. Process for pasteurization and enzyme inactivity of fruits by electronic heating. U.S. Patent n° 2 476 251, 12 July 1949.

145. TABORŠAK N., BAŠIĆ V. Mogućnosti primjene mikrovalova u tehnologiji prerade mlijeka. *Mljekarstvo*, 34, n° 3, 1984, pp. 82-87. (C.D.I.U.P.A., n° 189325).

146. TEIXEIRA-PINTO A., HELLER J. The behaviour of unicellular organisms in an electromagnetic field. *Exper. Cell. Res.*, 20, 1960, pp. 548-564.

147. THOUREL L. Action des micro-ondes sur les enzymes et les micro-organismes. Publication du CERT-ONERA, Toulouse. s.d., 7 p.

148. THOUREL L., MEISEL N. Les applications industrielles du chauffage par micro-ondes. In : Journées d'informations électro-industrielles sur la stérilisation et la pasteurisation. Paris, Document EDF, 1974, pp. 5-6.

149. THOUREL L., PAREILLEUX A., PRIOU A., THOUREL B., AUGE C. Microwave specific effects on beer yeast. In : Proceedings of the IMPI Symposium, Waterloo (Can.), 1975, pp. 127-128 (Paper 7.7) et Publication du CERT-ONERA, Toulouse, 12 p.

150. THOUREL L., PAREILLEUX A., THOUREL B., AUGE C. Microwaves specific effects on beer yeast. In : Proceedings of the IMPI Symposium, Milwaukee, 1974, 2 p. (Paper A3.4).

151. THOUREL L., THOUREL B. Effet des micro-ondes sur les champignons et les microorganismes. Paris, DGRST, Rapport de fin de contrat n° 74-70821, 1975, 45 p. et Publication CERT-ONERA, Toulouse. (C.D.I.U.P.A., n° 99410).

152. THOUREL L., THOUREL B. Destruction des champignons par hyperfréquences. Paris, DGRST, Action concertée : Technologie Alimentaire et Agricole, Contrat n° 73 7 1103, 21 p. Document n° 1042. Publication du CERT-ONERA, Toulouse. s.d.

153. TIMBERS G., WYSLOUZIL W., KASHYAP S. Inactivation of myrosinase enzyme in mustard seed by microwave heating. *J. Microw. Power*, 17, n° 4, 1982, pp. 324-325.

154. TOCHMAN L., STINE C., HARTE B. Thermal treatment of Cottage cheese "in-package" by microwave heating. *J. Food Prot.*, 48, n° 11, 1985, pp. 932-938. (C.D.I.U.P.A., n° 207901).

155. UHNBOM P. Communication personnelle, 13 oct. 1986.

156. VAROQUAUX P. Applications des micro-ondes dans les industries agricoles et alimentaires : blanchiment, pasteurisation, stérilisation. In : Comptes-rendus du Colloque sur les applications industrielles des hautes fréquences et des micro-ondes, 26-27 janv. 1977, 2ème partie, Lyon, Lyon, Univ. Claude Bernard, 1977.

157. VAROQUAUX P., AVISSE C., BUSSIDAN C. Le blanchiment des pois par l'eau et par les micro-ondes. In : Comptes-rendus du 6ème Congrès International de la Conserve, Paris, 1972. 1972, pp. 267-273. (C.D.I.U.P.A., n° 66037).

158. VAROQUAUX P., AVISSE C., NADAL N., COUSIN R. La qualité des pois en conserverie. *Int. Aliment. Agric.*, 91, n° 11, 1974, pp. 1415-1422. (C.D.I.U.P.A., n° 79116).

159. VAROQUAUX P., SAUVAGEOT F., CHAPON M., DUPUY P. Utilisation de micro-ondes pour le blanchiment de lanières de céleri. *Ann. Technol. Agric.*, 28, n° 4, 1979, pp. 387-395. (C.D.I.U.P.A., n° 142517).

160. VELA G., WU J. Mechanisms of lethal action of 2450 MHz radiation on microorganisms. *Appl. Environm. Microbiol.*, 37, n° 3, 1979, pp. 550-553. (C.D.I.U.P.A., n° 131456).

161. WARD T., ALLIS J., ELDER J. Measure of enzymatic activity coincident with 2450 MHz microwave exposure. *J. Microw. Power*, 10, n° 3, 1975, pp. 314-320.

162. WEBB S., BOOTH A. Absorption of microwaves by micro-organisms. *Nature*, 222, n° 5199, 1969, pp. 1199-1200.

163. WEBB S., DODDS D. Inhibition of bacterial cell growth by 136 GHz microwaves. *Nature*, 218, 1968, pp. 374-375.

164. WYSLOUZIL W., KASHYAP S. Microwave sterilization of pea flour and pea protein concentrate. In : Proceedings of the IMPI Symposium, Leuven, 1976, (Paper 6B.3).

165. WYSLOUZIL W., KASHYAP S., A cascading rotary microwave processor. In : Proceedings of the IMPI Symposium, Chicago, 1985, pp. 175-177.

166. YANAI S., YAMADA Y., KIMURA S. Effet des ondes haute fréquence sur le nombre de bactéries contaminant l'amidon (en japonais). *J. Jpn. Soc. Starch Sci.*, 19, n° 4, 1972, pp. 192-194. (C.D.I.U.P.A., n° 58723).

167. YANG Y., JOHNSON J., WIEGAND E. Electronic pasteurization of wine. Fruit prod. *J. Am. Food Manuf.*, 26, 1947, p. 295.

168. YEARGERS E., LANGLEY J., SHEPPARD A., HUDDLESTON G. Effects of microwave radiation on enzymes. *Ann. New York Acad. Sci.*, 247, 1975, pp. 301-304.

169. YORK G. Comparative effects of laser light and microwave energy on micro-organisms. In : Proceedings of the IMPI Symposium, Stanford, 1967. (Abstracts, pp. 33-34).

170. ZIMMER L. Verlängerung der Halt-barket von Schnittbrot. *Bäcker Konditor*, 25, n° 2, 1971, pp. 44-46. (C.D.I.U.P.A., n° 30028).

171. Application des micro-ondes au blan-chiment. Compte-rendu de fin d'essais dans le cadre d'un contrat DGRST. Stn. exp. Dury. Publication du Centre Technique des Conserves de Produits Agricoles, 1974, 36 p.

172. Blanching of vegetables. *Quick Frozen Food Int.*, 22, n° 4, 1980, pp. 183-184.

173. Divers Comptes-rendus d'essais. Sté IMI-SA, 1977-1980.

174. Electronic sterilization. *Am. Miller Process.*, 73, n° 4, 1945, pp. 43-48.

175. La conservation par irradiation. Informations Thomson-CSF, n° 50, janv.-fév. 1982, pp. 6-9.

176. Method and apparatus for sterilizing. British Patent n° 1 154 752, 11 June 1969, Gray Industries.

177. Méthode et appareil pour contrôler les micro-organismes et les enzymes. Brevet belge n° 717411, 28 juin 1968 et Brevet français n° 1 571 833, 20 juin 1969, Gray industries, correspondant au Brevet américain au nom de Gray (O) n° 688 260 du 5 déc. 1967.

178. Micron, microwave multipurpose inhi-bitor and sterilizer/pasteurizer. Publication OSHIKIRI Machinery Ltd., 1985, 7 p.

179. Microwave blanching review. *Microw. Energy Appl. Newsl.*, 3, n° 1, 1970, pp. 6-7. (C.D.I.U.P.A., n° 30588).

180. Microwave heating. Leicester, Publ. Magnetronics Ltd., s.d., 8 p.

181. Microwave tunnel for heat treatment of bread. Commercial publication Company Skandinaviska Processinstru-ment AB, Bromma, s.d., 1 p.

182. Microwave tunnel for heat treatment of potatoes. Commercial publication Company Skandinaviska Processinstru-ment AB, Bromma, s.d., 4 p.

183. Multitherm unveiled. *Food Manuf.*, 58, n° 12, 1983, pp. 23-24.

184. Pasteurisation de bière. *J. Fr. Electrotherm.*, n° 1, août-sept. 1984, p. 50.

185. Pasteurization of bread with micro-waves. Commercial publication Company Skandinaviska Processinstru-ment AB, Bromma, s.d., 1 p.

186. Processing with microwaves. *Chilton's Food Eng. Int.*, 10, n° 8, 1985, pp. 46-47. (C.D.I.U.P.A., n° 206611).

187. Recent advances in industrial micro-wave processing in the 896/915 MHz frequency band. Leicester, Publ. Magnetronics Ltd., s.d., 12 p.

188. Stérilisation par micro-ondes. Publication DIEPAL-CRD, Brive, 1974, 66 p.

189. Teighphysikalische Prüfung mit HF-behandeltem und unbehandeltem Mehl. *Alimenta*, 14, n° 1, 1975, p. 22. (C.D.I.U.P.A., n° 79817).

190. The utilization of microwave ovens for blanching prior to freezing. *Quick Frozen Foods*, 26, n° 4, 1963, pp. 273-284.

Chapter 5

Miscellaneous applications

This rather general heading covers several relatively minor microwave food applications which are, nevertheless, interesting from practical and theoretical points of view. They include the germination of dormant grains, now widely used in Australia; the disinfestation of stored cereals that has been the subject of numerous studies in the past 30 years but has to compete with high-frequency techniques; and open-air crop protection. Three other applications, namely, soil treatment, wine making by microwave-induced carbonic maceration, and the opening of oyster shells are now in the early stages of development and appear to be extremely promising.

5.1 Disinfestation

Stored grain is often infested with different insects whose larvae develop at the expense of the grain, reducing its quality and leading to significant weight loss. In western climates, these parasites fall into the three categories listed below [2].

- Weevils and other insects of the order *Coleoptera*. For example, *Sitophilus granarius* drills holes in the grains in which it deposits its eggs. The larvae undergo four transformations and leave the grain after devouring two-thirds of it. Starting with a population of two, more than 800 kg of grain can disappear in six months and 1,200 kg can be infested. This insect needs a relative humidity of at least 50% that corresponds to grain moisture content of approximately 11–12% . Reproduction is impossible below 9.5%. The threshold of activity of the adult is 10°C, the threshold for mating is 12°C, for laying eggs 13°C, and for population growth 15°C. The optimum population temperature is 26–30°C, and population growth ceases above 35°C.

- Insects of the order *Lepidoptera* such as *Oryzaephilus surinamensis* and *Sitotroga cerealella* whose larvae attack corn. *Oryzaephilus s.* is carnivorous and its larvae develop and feed on grain already infested with weevils. The thresholds for its development and growth are 18°C and 21°C, respectively, and its optimum temperature is 31–34°C. These insects do not tolerate humidity levels higher than 10%.

- Insects of the order *Acarida*, which is part of the *Arthropoda*, e.g., *Acarus siro*. These insects have a very high reproduction rate and are very well aclimatized to cold and humid conditions. Their development threshold is 7°C and their optimum temperature is 21–27°C with minimum relative humidity of 65–70%.

The conventional techniques employed to fight these parasites are mostly preventative: good hygiene in warehouses and vehicles, and ventilation of stored grain for both cooling and drying. Remedial measures include fumigations and the use of insecticides. The latter have several disadvantages, including toxicity to personnel that use them, the presence of residue, the absence of any effect on the larvae that are well protected inside the grain, and the risk that resistant strains will develop.

Fumigation with hydrogen phosphide, produced by mixing aluminum phosphide pellets with amonium carbonate in the presence of moisture, gives rise to absorption of some of the gas by the grains and to the deposition of aluminum oxide particles on the surface. In all cases, chemical procedures require efficient ventilation [2], [88].

Microwave disinfestation is possible in principle because, as we have already noted, parasites do not readily tolerate high temperatures. Another necessary condition is that the radiation must penetrate the grain without significant attenuation, and that the dielectric loss factor of the insects be significantly greater than that of the medium. For example, the loss factor of *Sitophilus oryzae* (Fig. 5.1) in the range 1–10 Ghz is 0.4 at 24°C [121], [88]. The loss factor of wheat has also been measured [58] and found to be between 0.2 and 0.5 at 24°C and 9.4 GHz, for moisture contents in the range 10–19% (Fig. 5.2). Other values reported in the literature [86] are loss factors of 0.43, 0.59, 0.67, 0.78, and 1.13 at 11.7 GHz for moisture contents of 8.5, 10.0, 10.9, 12.2, and 14.7%. The curve in Figure 5.1 shows 0.25 between 1 and 10 GHz for a moisture content of 10.6%. This value is confirmed in [81] and [58]. It is thus evident that the differential absorption between the insects and the medium is only 1.6. It is difficult to evaluate the electric field in the vicinity of different parts of the the insects for a given applied field (Part I, Section 3.6), but it has been shown [91], [88] that irradiation at frequencies in excess of 1 GHz has almost no selective effect on insects.

Despite these unfavorable circumstances, some of the old prototypes are worth mentioning, e.g., a 2.45-GHz applicator delivering 1 kW and producing a 70% mortality rate with a final temperature of 55°C and 100% mortality at 65°C in the case of *Tribolium confusum* [36]. The drying and disinfestation of cereals was considered in [13] and it was found that the lethal temperature for *Tribolium confusum*, *Sitophilus granarius*, and *Cryptolestes ferrugineus* fell from 60°C to 45°C when hot wheat was held for a few moments in the container. It was also predicted that HF should be more suitable than microwaves because of faster payoff.

The advantages of HF for disinfestation alone (microwaves are better for drying combined with disinfestation) have been reported in a number publications [3], [128]. The ratio of absorption coefficients of insects and cereals in the range 10–100 MHz is between 3 and 3.5 [88], and a 100% destruction of the parasites requires an irradiation of 3 s at 39 MHz compared to 13 s at 2.45 GHz, the final temperature of the grain being much higher in the second case.

Stuart O. Nelson, who was at the University of Nebraska until 1976 and is currently at the Richard B. Russell Agricultural Center in Athens (Georgia), played a key role in this field and has contributed to it very significantly. Further information may be found in [119], [144], [27], [4], [103], [147], [68], [69], [92], [102], [149], [70], [39], [40], [64], [118], [36], [37], [38], [55], [56], [73], [74], [130], [57], [77], [96], [97], [63], [8], [61], [62], [67], [41], [140], [26], [50], [18], [19], [94], [131] and [123]. There are extensive published data on the loss factor of grains and their parasites [72], [15], [75], [99], [79], [82], [83], [84], [28], [32], [85], [86], [87].

Experiments on HF and microwave disinfestation have been carried out on bakery products [31], packaged goods [136], timber [125], [156], and cigarettes [48], [47], [46], [44], [45].

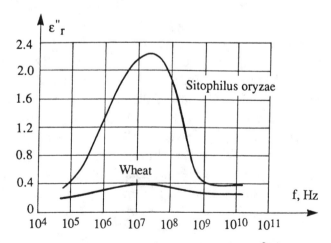

Figure 5.1 Loss factors of hard winter wheat (at $24°$C and 10.6% moisture content) and *Sitophilus oryzae* [88].

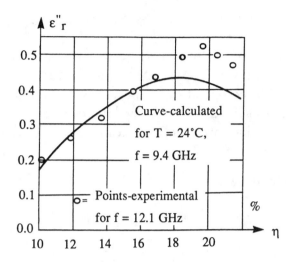

Figure 5.2 Loss factor of loose wheat as a function of moisture content [58].

5.2 Soil treatment

Vegetable tissue is very sensitive to the thermal effect of microwaves. For example, the leaves of the Mexican bean *Prosopis glandulosa Torr* heated to 46°C in a conventional oven are not harmed in any way, whereas the same temperature in a microwave oven brings about a 78% deterioration in six seconds [142]. The germination of grains of beans (*Phaseolus vulgaris*) is unaffected by heating to 27–50°C in water, but falls by 62% in the case of heating by microwaves. This may be due to preferential absorption in parts of the grain, which produces local temperatures well above the average.

The use of microwaves instead of herbicides for the destruction of unwanted seeds and parasitic plants has been under investigation since the early 1970s by the USDA Agricultural Research Center at Welasco (Texas), the University of Texas, the Phytox Corporation, and the First American Farms at Freeport (Florida) [20], [21], [141], [145], [65], [146], [143]. The aim was to destroy, before sowing, all undesirable grain and shoots. Although this material inevitably reappears as more seeds are carried and spread by the wind, useful plants will have had a head start.

The problem is complex because is is essential to reach the grain or root in a soil of variable thickness, the soil itself being of variable composition. The curves in Figure 5.3 show absorption by soil and by seeds planted at different depths. They suggest that preferential heating of the seeds is taking place.

Absorbed power density, Wm^{-3}

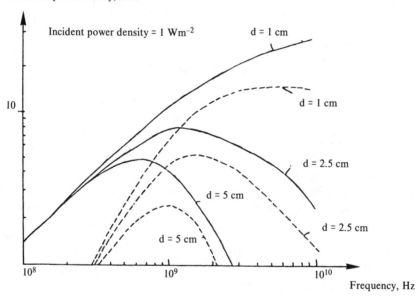

Figure 5.3 Power absorbed per unit volume of spherical seeds in soil and by the soil itself [104]. Solid line – power absorbed by seeds, dashed line – power absorbed by soil.

It is interesting to note that the loss factor ϵ_r'' is a function of the nature of the soil and the moisture content, as shown in Figure 5.4 and Table 5.1. The latter also lists the values of the reflection and transmission coefficients Γ_p and T_p at the air/soil interface (Part I, Section 2.8). The fraction of power

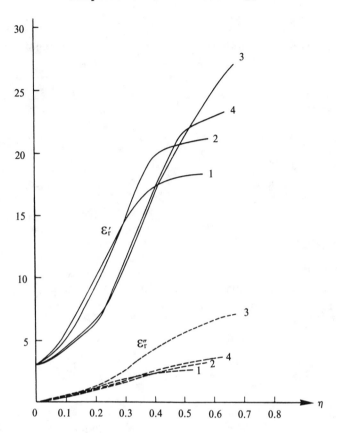

Figure 5.4 Permittivity of four types of soil at 5 GHz as a function of moisture content [138].

Solution no.	Composition, %		
	Sand	Silt	Clay
1	88.0	7.3	4.7
2	56.0	26.7	17.3
3	19.3	46.0	34.7
4	2.0	37.0	61.0

penetrating the soil is given by

$$P_t(d) = T_p P_i e^{-2\alpha d} \qquad (2.1)$$

where P_i is the incident power, d is the depth, and α is the

Table 5.1

Permittivities, reflection coefficients, and transmission coefficients
of different soils at 3 GHz [43]

Soil type	η, %	ε'_r	tan δ	ε''_r	Γ	T
Sandy soil	0	2.55	0.0062	0.016	0.053	0.947
	2.18	2.50	0.03	0.075	0.051	0.949
	3.88	4.40	0.046	0.202	0.126	0.874
	16.8	20.0	0.13	2.600	2.600	0.597
Soil rich	0	2.44	0.0011	0.003	0.048	0.952
in organic substances	2.2	3.5	0.04	0.140	0.092	0.908
(compost)	13.77	20.0	0.12	2.400	0.403	0.597
Clay	0	2.27	0.015	0.034	0.041	0.959
	20.09	11.3	0.25	2.825	2.930	0.707

attenuation coefficient [Part I, formula (2.83)]. Figure 5.5 illustrates the function $P_t(d)$ for the permittivities listed in Table 5.1. The permittivity of soils is also discussed in [30], [16],
[42], [49].

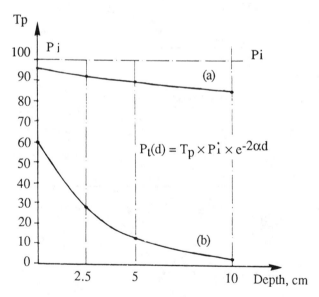

Figure 5.5 Power transmitted by soil [43]: a – dry clay soil, b – sandy
soil with 16.8% moisture content.

The energy flux $\Phi_W = tP_t(d)/S$ (in $\mathrm{J\,m^{-2}}[\mathrm{MT^{-2}}]$) required to prevent the germination of grains has been variously estimated as ranging from 100 to 1600 $\mathrm{J\,cm^{-2}}$. Olsen [104] gives a value of approximately 800 $\mathrm{J\,cm^{-2}}$ and presents the curves of Figure 5.6 which show the lethality as a function of the incident power

$$L = \int_0^{t_0} \exp[AT(t) + B]\mathrm{d}t \qquad (5.2)$$

in which A and B are constants that depend on the particular vegetable tissue, t_0 is the time required for the temperature to drop to 35°C, and $T(t)$ is the temperature of the seeds as a function of time. The irradiation is lethal if L is greater than unity.

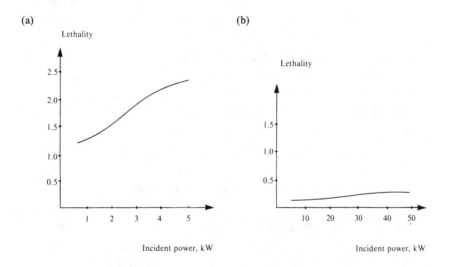

Figure 5.6 Lethality of seeds, 2 mm in radius, planted at a depth of 1 cm in organic soil (a) and mineral soil (b) irradiated with 2.45 GHz microwaves at the rate of 870 $\mathrm{J\,cm^{-2}}$; $L > 1$: lethal irradiation, $L < 1$: nonlethal irradiation [104].

We note that other studies estimate Φ_W to be $360\,\mathrm{J\,cm^{-2}}$ [65] and even $183\,\mathrm{J\,cm^{-2}}$ [143]; the latter is based on experimental results obtained at Freeport (Florida). For example, a 1-kW microwave applicator sweeping the terrain at the rate of $1.5\,\mathrm{km\,h^{-1}}$ and maintaining a constant field over 40 cm in the direction of motion, introduces 932 W in 1 s (i.e., 932 J) into a seed $1\,\mathrm{cm^2}$ in cross section, planted at a depth of 2.5 cm in soil of type (a) of Figure 5.7. In soil of type (b), the absorbed energy is only 284 J, since the absorption rates in the two soils are 93.2% and 28.4% of the incident power, respectively. The treatment be effective in soil (a) but may not be in soil (b).

The angle of incidence and the polarization are also relevant in this problem. Figure 5.7, based on (2.111) and (2.112) of Part I, shows that vertical polarization at Brewster incidence is theoretically the best choice.

The first prototype applicator for soils was built by Oceanography International of Texas [143]. Their "Zapper" could be described as a four-wheel trailer carrying four 1.5-kW generators operating at 2.45 GHz and connected by means of flexible guides to four antennas forming a square-shaped assembly, each antenna measuring $10 \times 10\,\mathrm{cm^2}$. The matching between air and soil was assured by the presence of a dielectric of intermediate permittivity [Part I, (2.114)].

The first trials with the electronic "Zapper" were carried out at Magic Valley (Idaho) and produced very good results for grass, parasitic fungi, and nematodes [150].

In France, the initial work by CERT-ONERA in Toulouse was followed in October 1977 by field trials at Pouymenjon (Landes, France). The resulting prototype was built by AUHFA (Fig 5.8) and was mounted on a Renault 656 tractor that also powered a Tractelec alternator. High-voltage transformers and rectifiers were kept in a box placed behind the seat of the driver; the magnetrons, the associated cooling devices, and the isolators were placed in a box in front of the vehicle, and two flexible waveguides came out of this box to feed the radiating horns. The installed power was 10 kW, the speed

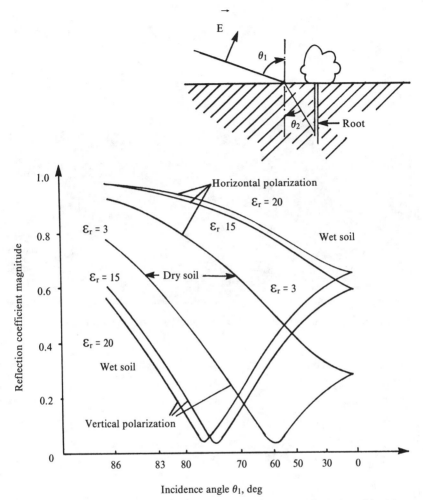

Figure 5.7 Reflection coefficient as a function of the angle of incidence [116].

was on the order of $1.5 \, \mathrm{km \, h^{-1}}$, and weeding was satisfactory for one year to a depth of 10 cm.

An operational model, the "Lomos" was constructed in 1980. A 48-kW Lomos is working at Labouheyre (Landes, France). Another version in the form of a 24-kW Lomos was used by Ponts-et-Chaussées of Nantes for road maintenance (Part II, Section 3.2.4).

Figure 5.8 Experimental setup for microwave weeding (courtesy of AUHFA).

This setup is particularly effective for annual grasses. The energy required is a function of the nature of the soil (clay soils are easier to treat than sandy soils) and of soil temperature. The germination rate remains 100% after soil treatment at $-20°C$; it is only 20% in a soil at 8°C and no germination takes place in soil at 18°C [104], [127], [129], [106], [107], [105]. The "Lomos" is capable of precise and localized irradiation that has a durable effect provided the soil is not turned over so that untreated layers get to the surface. The procedure is nontoxic, and the field strengths around the irradiation plant comply with safety standards. Of course, planting can take place immediately after the treatment.

An urban version of the Lomos is currently under study for the disinfection of sand in public parks polluted by dogs, of which at least 25% carry parasites. As many as 10% of the dogs are infested with *Toxocara* (i.e., nematodes of the order *Ascaridae* that cause toxocariosis) also known as the larva migrans syndrome. Traditional treatments of sand pits (chemical or hot steam sterilization) are neither convenient nor efficient. Microwaves, on the other hand, appear to give excellent results [11].

5.3 Germination

Having just discussed the herbicidal effects of microwaves, it may seem paradoxical to mention that they can accelerate the germination of certain seeds. There is a great deal of interest in this field, and a large number of publications is devoted to this subject [93], [52], [120], [53], [54], [98], [78], [80], [5], [89], [90], [133], [135], [132], [101], [134], [88].

Some seeds do not germinate even under optimum conditions. This can be ascribed to the impermeability of the shell, the immaturity of the embryo, the presence of inhibitors, and a lack of heat or light. These factors allow the seed to remain in the dormant state until all the conditions are favorable. For example, the seeds of *Acacia longifolia*, an Australian species, have an extremely hard impermeable shell that enables them to remain dormant for up to 50 years. Only a very small proportion - less than 10% - will therefore germinate. However, in Australia, *Acacia longifolia* is used for the reforestation after civil engineering work, which ideally requires massive and rapid germination. To achieve this, certain treatments have to be used to break the shell. They include hot baths, exposure to liquified gas, acid, or alcohol vapor, pressure, scarrification, and electromagnetic heating.

Exposure to 650-W, 2.45-GHz microwaves for about 30 s is sufficient to ensure a high rate of germination by some mechanism that is not as yet fully understood. The microwaves seem to act on the strophiola, a sensitive part located on the ventral side of the seed, which may thus become more water permeable [5], [134].

The effect of the radiation varies according to the species: clover, peas, beans, and spinach respond favorably whereas wheat, corn, and cotton are less sensitive. The positive effect on Esparto grass seeds has been maintained for 21 years without adverse reaction [88]. In view of the relatively high cost of the grain, and the small volumes to be treated, this application could be economically very viable.

5.4 Crop protection

Experiments on crop protecion by microwaves have been carried out on the McGill experimental farm at Ste. Anne de Bellevue (Québec) [9], [10]. A plot of $7.5 \times 7.5 \text{m}^2$ with 231 four-month old maize plants was exposed to an electromagnetic flux capable of protecting the crop during a cold spell. A 2.4-kW generator operating at 2.45 GHz fed an emitting horn at the top of a 2-m high tower in the northwest corner of the field. The level of irradiation in this field was well above the established standards for safety, which meant that access to the field and its environment was forbidden.

The cold spell continued for 60 hr and a gust of cold northerly wind reduced the temperatures to between $-1°C$ and $-5°C$. In addition, the storm resulted in the deposition of snow on the plants to a depth of 13 mm in certain parts. After 60 hr, 10% of the plants were dead (those that were facing the dominant wind), but, on the whole, the crop had been remarkably well protected. Because of its very low loss factor, the snow did not prevent the penetration of energy into the plants. The cost of the treatment was estimated at 180 kWh hm^{-2}.

Although this trial was apparently not repeated, it was noted in 1985 [22] that this was potentially a very interesting application. An increase in temperature by just a few degrees, produced by this technique, could prevent crop losses worth several millions of dollars.

5.5 Wine-making by carbonic fermentation

Carbonic fermentation preserves remarkably the primary aromas and creates very characteristic wines irrespective of variety. The grapes are harvested by hand and placed without pressing in the fermentation vessel. An exothermic reaction starts at the bottom of the vessel and releases carbon dioxide that eventually replaces all residual air. The reaction continues for about 20 days. To increase the profitability of the operation, the fermentation process can be accelerated by heating the grapes to 31°C, which is the temperature required to initiate the enzymatic reaction. This temperature is produced by dry vapor, the circulation of a hot gas, the heating of the vessel from the sides and the bottom, or by drenching with hot must from the bottom of the vessel. All these techniques result in very nonuniform temperatures and local overheating, both of which tend to degrade aromas, destroy enzymes, and activate bacterial lactic fermentation above 40°C.

The absence of a really satisfactory procedure led INRA of Gruissan (Aude, France) to collaborative microwave trials with the MicroRay Company of Toulouse. This resulted in a prototype using three 5-kW, 2.45-GHz generators built by the SARRE Company of Bram (Aude).

Microwave treatment does not seem to perturb the enzymatic system and, since the grapes are spherical, maximum heating occurs at the center, so that the skin and its aromas are subjected to only minimal heating. The time saved in the fermentation cycle can reach a factor of 2 or 3.

This procedure was first tried on the harvests of 1985 at Durban-Corbières. The equipment is now available in modular form and each 25-kW microwave element is capable of treating two tonnes of grapes per hour for an initial temperature of 10°C and a target temperature of 25 to 30°C. The wines obtained are judged to be remarkably good. This application seems destined for some success and is described in [148], [154], [152], [153], [151], [108], [66], [155] and [157].

5.6 Opening of oysters

In the United States, oysters are usually preserved and consumed cooked. The shelling of oysters on an industrial scale is a relatively slow and difficult operation that requires specialized manpower. This explains why about 50 patents on mechanical shelling have been put forward since 1850 without any of them being eventually used in industry.

In 1971, it was found [60] that a quick exposure to microwaves produced the opening of the shell without cooking or even killing the animal. Pilot studies by the Rock Hall Clam and Oyster Company in Chesapeake Bay showed that oysters treated at the rate of $120\,dm^3\,kWh^{-1}$ presented no chemical, bacteriological, or organoleptic peculiarity. A 5-kW, 2.45-GHz microwave tunnel incorporating an emitting horn, designed to apply maximum energy in a minimum time, was developed by the Industrial Microwave Division of Raytheon and installed in the plant of the Savage and Mears Company at Chincoteague (Virginia). It was found that the process, including the depreciation of the microwave tunnel, was more cost-effective than manual operation.

Surprisingly, there have been no other reports of this technique until its recent rediscovery by an oyster farmer, a Mr. G. Constant of Fouras (Charente Maritime, France), who studied a pilot microwave tunnel capable of opening 500 oysters per day. The drawback is that approximately 5% of the oysters get cooked. This, however, was not considered a problem by the VETRA Company of La Barre de Monts (Vendée, France), which wanted to market a terrine of oysters for which manual shelling would have been prohibitively expensive. VETRA was also considering putting on the market frozen shelled oysters [122]. Microwave processing is ideal for this type of production and will eventually achieve a degree of success provided, of course, that the new products are favorably accepted by consumers.

References

1. ANDRIEU G. Désherber dans desherbant grâce à l'appareil à micro-ondes. *Dépêche Midi*, n° 428, supplément agricole, 1980, pp. 1 et 3.

2. ANGLADE P. Les insectes et acariens des céréales stockées ; biologie et méthodes de lutte. *Bull. Anc. Elèves Ec. Meun.*, 235, 1970, pp. 23-33. (C.D.I.U.P.A., n° 016993).

3. ANGLADE P., CANGARDEL H., FLEURAT-LESSARD F. Application des O.E.M. de haute fréquence et des micro-ondes à la désinsectisation des denrées stockées. In : Comptes-rendus du Symposium International sur les applications énergétiques des micro-ondes, IMPI-CFE. 14, Monaco, 1979, pp. 67-69.

4. BAKER V., WIANT D., TABOADA O. Some effects of microwaves on certain insects which infest wheat and flour. *J. Econ. Entomol.*, 49, 1956, pp. 33-37.

5. BALLARD L., NELSON S., BUCHWALD T., STETSON L. Effects of radio-frequency electric fields on permeability to water of some legume seeds, with special reference to strophiolar conduction. *Seed Sci. Tech.*, 4, 1976, pp. 257-274.

6. BARTHAKUR N. Use of microwave radiation to study plant-environment interactions. *J. Microw. Power*, 10, n° 4, 1975, pp. 441-449.

7. BARTHAKUR N. Stomatal response to microwave induced thermal stresses. *J. Microw. Power*, 11, n° 3, 1976, pp. 247-254.

8. BENZ G. Entomologische Untersuchungen zur Entwesung von Getreide mittels Hochfrequenz. *Alimenta*, 14, n° 1, 1975, pp. 11-15. (C.D.I.U.P.A., n° 79815).

9. BOSISIO R., BARTHAKUR N. Microwave protection of plants from the cold. *J. Microw. Power*, 4, n° 3, 1969, pp. 190-193.

10. BOSISIO R., BARTHAKUR N., SPOONER J. Microwave protection of a field crop against cold. *J. Microw. Power*, 5, n° 1, 1970, pp. 47-52.

11. BOUCHET F., LEGER N. Utilisation des micro-ondes comme moyen de lutte contre la toxocarose dans les bacs à sable urbains. *Econ. Progrès Electr.*, n° 12, nov.-déc. 1984, pp. 17-21.

12. BOULANGER R., BOERNER W., HAMID M. Comparison of microwave and dielectric heating systems for the control of moisture content and insect infestations of grain. In : Proceedings of the IMPI Symposium, Edmonton, 1969, pp. 27-29. (Paper B3).

13. BOULANGER R., BOERNER W., HAMID M. Comparison of microwave and dielectric heating systems for the control of moisture content and insect infestations of grain. *J. Microw. Power*, 4, n° 3, 1969, pp. 194-208.

14. CHUGH R., STUCHLY S., RZEPECKA M. Dielectric properties of wheat at microwave frequencies. American Society of Agricultural Engineers, publ. n° 72-843, 1972.

15. CHUGH R., STUCHLY S., RZEPECKA M. Dielectric properties of wheat at microwave frequencies. *Trans. Am. Soc. Agric. Eng.*, 16, 1973, pp. 906-909.

16. CIHLAR J., ULABY F. Dielectric properties of soils as a function of moisture content. Technical report RSL 177-47, Univ. Kansas, Lawrence, Nov. 1974.

17. COARTNEY J., HAWKINS G., LARSEN D. Effects of microwaves on soil chemistry. In : Proceedings of the 27th Weed Science Society Symposium, 1974, p. 199.

18. D'AMBROSIO G., FERRARA G., TRANFAGLIA A. Disinfestation of stored foodstuffs by means of microwaves, experiments on Tenebrio molitor (L.), (Col. Tenebrionidae) and Sitophilus granarius (L.) (Col. Curculionidae). *Boll. Lab. Entomol. Agraria "Filippo Silvestri"*, 39, 1982, p. 31.

19. D'AMBROSIO G., FERRARA G., TRANFAGLIA A. Teratogenesis due to microwaves in Tenebrio molitor Coleoptera Tenebrionidae, influence of impulse modulation. *Boll. Lab. Entomol. Agraria "Filippo Silvestri"*, 39, 1982, p. 3.

20. DAVIS F., WAYLAND J., MERKLE M. Ultra-high-frequency electromagnetic fields for weed control : phytotoxicity and selectivity. *Science*, 173, 1971, pp. 535-537.

21. DAVIS F., WAYLAND J., ROBINSON C., MERKLE M. Phytotoxicity of UHF electromagnetic fields. *Nature*, 241, 1973, pp. 291-292.

22. DECAREAU R. Future research developments in microwave technology. Non publ., s.d., 12 p.

23. DIPROSE M., HACKAM R., BENSON F. Selective destruction of wild oats when mixed with cereal seeds by microwave irradiation. In : Proceedings of the IMPI Symposium, Ottawa, 1978. (Paper 7.4.).

24. FANSLOW G. Ovicidal effects of electromagnetic energy at 2.45 GHz on eggs. In : Proceedings of the IMPI Symposium, Milwaukee, 1974, 4 p. (Paper B.2.1.).

25. FANSLOW G., TOLLEFSON J., OWENS J. Ovicidal levels of 2.45 GHz electromagnetic energy for the Southern corn rootworm. *J. Microw. Power*, 10, n° 3, 1975, pp. 321-325.

26. FLEURAT-LESSARD F., LESBATS M., LAVENSEAU L., CANGARDEL H., MOREAU R., LAMY M., ANGLADE P. The biological effects of microwaves on two insects, Tenebrio molitor L. (Col. : Tenebrionidae) and Pieris brassicae L. (Lep.: Pieridae). *Ann. Zool. Ec. Anim.*, 11, 1979, p. 457.

27. FRINGS H. Factors determining the effects of radio-frequency electromagnetic fields on insects and materials they infest. *J. Econ. Entomol.*, 45, 1952, pp. 396-408.

28. FUJIWARA O., GOTOH Y., AMEMIYA Y. Characteristics of microwave-power absorption in an insect exposed to standing-wave fields. *Trans. Inst. Electron. Commun. Eng. Jpn.*, J66-B, Sept. 1983, pp. 1085-1092.

29. FUNG D., SHEREE-LIN C. Melting agar by microwave energy. *J. Food Prot.*, 47, n° 10, 1984, pp. 770-772. (C.D.I.U.P.A., n° 194676).

30. GEIGER F., WILLIAMS D. Dielectric constants of soils at microwave frequencies. NASA Report, ref. TM-X 65987, April 1972.

31. GODKIN W., CATHCART W. Effectiveness of heat in controlling insects infesting the surface of bakery products. *Food Technol.*, 3, n° 8, 1949, pp. 254-257.

32. GOTOH Y., FUJIWARA O., AMEMIYA Y. Distribution of temperature rise inside an insect due to microwave irradiation. *Trans. Inst. Electron. Commun. Eng. Jpn.*, J67-B, August 1984, pp. 928-929.

33. GREEN D., ROSENBAUM F., PICKARD W. Biological effects of microwaves on the pupae of Tenebrio molitor. U.S. Dep. Health, Education, Welfare Publication, 77-8026, 1977, pp. 253-262.

34. HAMID M. Grading agricultural products with a microwave antenna. U.S. Patent n° 4 106 340, 15 August 1978.

35. HAMID M., BADOUR S. The effects of microwaves on green algae. *J. Microw. Power*, 8, n° 3-4, 1973, pp. 267-273.

36. HAMID M., BOULANGER R. A new method for the control of moisture and insect infestations of grain by microwave power. *J. Microw. Power*, 4, n° 1, 1969, pp. 11-18 and in : *Milling*, 152, n° 5, 1970, pp. 25-32. (C.D.I.U.P.A., n° 40543).

37. HAMID M., BOULANGER R. Apparatus for the desinfestation of vegetal substances. Canadian Patent n° 880-551, 7 sept. 1971.

38. HAMID M., BOULANGER R. Microwave package for control of moisture content and insect infestation of grain. U.S. Patent n° 3 611 582, 12 Oct. 1971.

39. HAMID M., KASHYAP C., VAN CAUWENBERGHE R. Control of grain insects by microwave power. *J. Microw. Power*, 3, n° 3, 1968, pp. 126-135.

40. HAMID M., KASHYAP C., VAN CAUWENBERGHE R. Control of grain insects by microwave power. In : Proceedings of the IMPI Symposium, Boston, 1969, p. 29. (Paper E4).

41. HAMID M., KRUSH E., BOULANGER R. Electronic insect trap. U.S. Patent n° 3939662, 3 Feb. 1976.

42. HIPP J. Soil electromagnetic parameters as a function of frequency, soil density and soil moisture. *Proc. IEEE*, 62, n° 1, 1974, pp. 98-103.

43. HIPPEL A. von. Dielectric materials and applications. Cambridge (Massachusetts), M.I.T. Press, 1954, 4th ed. 1966, 438 p.

44. HIROSE T., ABE I., IWAHASHI M., SUZUKI T., OSHIMA K., OKAKURA T. Microwave heating of cigarettes. In : Proceedings of the IMPI Symposium, Ottawa, 1972, pp. 212-213.

45. HIROSE T., ABE I., IWAHASHI M., SUZUKI T., OSHIMA K., OKAKURA T. Microwave heating of multi-layered cigarettes. *J. Microw. Power*, 13, n° 2, 1978, pp. 125-129.

46. HIROSE T., ABE I., KOHNO M., SUZUKI T., OSHIMA K., OKAKURA T. The use of microwave heating to control insects in cigarette manufacture. *J. Microw. Power*, 10, n° 2, 1975, pp. 181-190.

47. HIROSE T., ABE I., SAITO K., SUZUKI T., OSHIMA K., OKAKURA T. Some effects of microwave heating to tobacco. In : Proceedings of the IMPI Symposium, Monterey, 1971.

48. HIROSE T., ABE I., SUGIE R., SUZUKI T., OSHIMA K., OKAKURA T. Killing of tobacco injurious insects by microwave heating. In : Proceedings of the IMPI Symposium, Edmonton, 1969, pp. 81-82. (Paper D.A.5.).

49. HOCKSTRA P., DELANEY A. Dielectric properties of soils at UHF and microwave frequencies. *J. Geophys. Res.*, 79, 1974, pp. 1699-1708.

50. HURLOCK E., LLEWELLING B., STABLES L. Microwaves can kill insects. *Food Manuf.*, 54, n° 8, 1979, p. 37. (C.D.I.U.P.A., n° 134649).

51. IRITANI W., WOODBURY G. Use of radio-frequency heat in seed treatment. *Res. Bull.*, n° 25, Agriculture Experiment Station, University of Idaho, Moscow.

52. JOLLY J., TATE R., Douglas fir-tree germination using microwave energy. *J. Microw. Power*, 6, n° 2, 1971, pp. 125-130.

53. KASHYAP S., LEWIS J. Microwave processing of tree seeds. In : Proceedings of the IMPI Symposium, Loughborough, 1973, 4 p. (Paper 2.A.4.).

54. KASHYAP S., LEWIS J. Microwave processing of tree seeds. *J. Microw. Power*, 9, n° 2, 1974, pp. 99-107.

55. KIRKPATRICK R., BROWER J., TILTON E. A comparison of microwave and infrared radiation to control rice weevils (Coleoptera curculionidae) in wheat. *J. Kansas Entomol. Soc.*, 45, 1972, p. 434.

56. KIRKPATRICK R., BROWER J., TILTON E. Gamma, infrared, and microwave radiation combinations for control of Rhyzopertha dominica in wheat. *J. Stored Prod. Res.*, 9, 1973, p. 19.

57. KIRKPATRICK R., BROWER J., TILTON E., BROWN G. Gamma and microwave radiation to control the rice weevil in wheat. *J. Georgia Entomol. Soc.*, 8, 1973, p. 51.

58. KRASZEWSKI A. A model of the dielectric properties of wheat at 9,4 GHz. *J. Microw. Power*, 13, n° 4, Dec. 1978, pp. 293-296.

59. LASZLO T., STEPHENS W. The effect of microwaves on the tobacco beetle. In : Proceedings of the IMPI Symposium, Milwaukee, 1974, 5 p.

60. LEARSON R., STONE W. The application of microwave energy to the shucking of oysters. Radarline-Performance/ application notes, Raytheon Co. In : Proceedings of the IMPI Symposium, Monterey, 26 May 1971, 8 p.

61. LIND H. Anlagen zur Getreideentwesung mit Hochfrequenz. *Alimenta*, 14, n° 1, 1975, pp. 17-21. (C.D.I.U.P.A., n° 79816).

62. LIND H. Anlagen zur Getreideentwesung mit Hochfrequenz. *Mühle Mischfuttertech.*, 112, n° 20, 1975, pp. 263-265. (C.D.I.U.P.A., n° 83712).

63. LOWE J. Disinfesting process Hertz only insects. *Design News*, n° 8, 1975, pp. 62-63.

64. MEISEL N. Nouvelles applications des micro-ondes à l'industrie alimentaire. *Ind. Aliment. Agric.*, n° 9-10, 1970, pp. 1059-1067, Désinsectisation, pp. 1064-1065. (C.D.I.U.P.A., n° 25542).

65. MENGES R., WAYLAND J. UHF electromagnetic energy for weed control in vegetables. *Weed Sci.*, 22, n° 6, 1974, pp. 584-590.

66. MONARD A., VIDAL J. Communication personnelle. 17 juin 1987.

67. MUNZEL F. Schadlingsbekampfung in Zeralienlagern mittels physikalischen Methoden. *Alimenta*, 14, n° 1, 1975, pp. 7-9. (C.D.I.U.P.A., n° 79814).

68. NELSON S. Radiation processing in agriculture. *Trans. Am. Soc. Agric. Eng.*, 5, 1962, pp. 20-30.

69. NELSON S. Electromagnetic and sonic energy for insect control. *Trans. Am. Soc. Agric. Eng.*, 9, 1966, pp. 398-405.

70. NELSON S. Electromagnetic energy. In : KILGORE W., DOUTT R. Pest control. Biological, physical and selected chemical methods. New York, Academic Press, 1967, pp. 89-145.

71. NELSON S. Effects of radiofrequency electrical treatment on germination of vegetable seeds. *J. Am. Soc. Hortic. Sci.*, 95, 1970, pp. 359-366.

72. NELSON S. Frequency dependence of the dielectric properties of wheat and the rice weevil. Thesis, Ph. D., Ann Arbor (Michigan), Iowa State Univ., 1972. (Dissertation Abstracts International n° 72-19997).

73. NELSON S. Insect-control possibilities of electromagnetic energy. *Cereal Sci. Today*, 17, n° 12, 1972, pp. 377-387. (C.D.I.U.P.A., n° 56033).

74. NELSON S. Possibilities for controlling stored-grain insects with RF-energy. *J. Microw. Power*, 7, n° 3, 1972, pp. 231-239.

75. NELSON S. Electrical properties of agricultural products (a critical review). *Trans. ASAE*, 16, n° 2, 1973, pp. 1-21.

76. NELSON S. Microwave dielectric properties of grain and seed. *Trans. Am. Soc. Agric. Eng.*, 16, n° 5, 1973, pp. 902-905.

77. NELSON S. Possibilities for controlling insects with microwaves and lower-frequency RF energy. *IEEE Trans. Microw. Theory Tech.*, MTT-22, n° 12, 1974, pp. 1303-1305.

78. NELSON S. Electromagnetic radiation effects on seeds. In : Proceedings of the Conference on Electromagnetic radiation in agriculture, Roanoke (Virginia Illuminating Engineering Society) American Society of Agricultural Engineers, 1975, pp. 60-63.

79. NELSON S. Microwave dielectric properties of insects and grain kernels. *J. Microw. Power*, 11, n° 4, 1976, pp. 299-303.

80. NELSON S. Use of microwave and lower frequency RF energy for improving alfalfa seed germination. *J. Microw. Power*, 11, n° 3, 1976, pp. 271-277.

81. NELSON S. Frequency and moisture dependence of the dielectric properties of high-moisture corn. *J. Microw. Power*, 13, n° 2, 1978, pp. 213-218.

82. NELSON S. RF-and microwave dielectric properties of shelled field corn. In : Proceedings of the IMPI Symposium, Ottawa, 1978, pp. 41-43. (Paper 5.1.).

83. NELSON S. Review of factors influencing the dielectric properties of cereal grains. *Cereal Chem.*, 58, n° 6, 1981, pp. 487-492. (C.D.I.U.P.A., n° 163567).

84. NELSON S. Factors affecting the dielectric properties of grain. *Trans. ASAE*, 25, n° 2, 1982, pp. 1045-1049, 1056.

85. NELSON S. A mathematical model for estimating the dielectric constant of hard red winter wheat. ASAE paper n° 84-3546, Am. Soc. Agric. Eng., 1984.

86. NELSON S. Density dependence of the dielectric properties of wheat and whole-wheat flour. *J. Microw. Power*, 19, n° 1, 1984, pp. 55-64. (C.D.I.U.P.A., n° 198100).

87. NELSON S. Moisture, frequency, and density dependence of the dielectric constant of shelled, yellow-dent field corn. *Trans. ASAE*, 27, n° 5, 1984, pp. 1573-1578.

88. NELSON S. RF and microwave energy for potential agricultural applications. *J. Microw. Power*, 20, n° 2, 1985, pp. 65-70.

89. NELSON S., BALLARD L., STETSON L., BUCHWALD T. Increasing legume seed germination by VHF and microwave dielectric heating. *Trans. Am. Soc. Agric. Eng.*, 19, n° 2, 1976, pp. 369-371.

90. NELSON S., BOVEY R., STETSON L. Germination response of some woody plant seeds to electrical treatment. *Weed Sci.*, 26, 1978, pp. 286-291.

91. NELSON S., CHARITY L. Frequency dependence of energy absorption by insects and grain in electric fields. *Trans. Am. Soc. Agric. Eng.*, 15, n° 72-346, 1972, pp. 1099-1102.

92. NELSON S., KANTACK B. Stored grain insect control studies with radiofrequency energy. *J. Econ. Entomol.*, 59, 1966, pp. 588-594.

93. NELSON S., NUTILE G., STETSON L. Effects of radio-frequency treatment on germination of vegetable seeds. *J. Am. Soc. Hortic. Sci.*, 95, n° 3, 1970, pp. 359-366.

94. NELSON S., PAYNE J. Pecan weevil control by dielectric heating. *J. Microw. Power*, 17, n° 1, 1982, pp. 51-55.

95. NELSON S., POUR-EL A., STETSON L., PECK E. Effect of 42- and 2450 MHz dielectric heating on nutrition-related properties of soybeans. *J. Microw. Power*, 16, n° 3-4, 1981, pp. 313-318.

96. NELSON S., STETSON L. Comparative effectiveness of 39- and 2450 MHz electric fields for control of rice weevils in wheat. *J. Econ. Entomol.*, 67, n° 5, 1974, pp. 9-12.

97. NELSON S., STETSON L. Possibilities for controlling insects with microwaves and lower-frequency RF energy. *IEEE Trans. Microw. Theory Tech.*, MTT-22, n° 12, 1974, pp. 1303-1305.

98. NELSON S., STETSON L. Use of microwave and lower frequency RF energy for improving alfalfa seed germination. In : Proceedings of the IMPI Symposium, Milwaukee, 1974, PS 1-1/ 2-4.

99. NELSON S., STETSON L. 250-Hz to 12-GHz dielectric properties of grain and seed. *Trans. Am. Soc. Agric. Eng.*, 18, 1975, pp. 714-715, 718.

100. NELSON S., STETSON L. Frequency and moisture dependence of the dielectric properties of hard red winter wheat. *J. Agric. Res. Eng.*, 21, n° 2, 1976, pp. 181-192.

101. NELSON S., STETSON L. Germination responses of selected plant species to RF electrical seed treatment. ASAE paper n° 84-3058, Am. Soc. Agric. Eng., 1984.

102. NELSON S., STETSON L., RHINE J. Factors influencing effectiveness of radio-frequency electric fields for stored-grain insect control. *Trans. Am. Soc. Agric. Eng.*, 1966, pp. 809-815, 817.

103. NELSON S., WHITNEY W. Radio-frequency electric fields for stored grain insect control. *Trans. Am. Soc. Agric. Eng.*, 3, 1960, pp. 133-137,144.

104. OLSEN R. A theoretical investigation of microwave irradiation of seeds in soil. *J. Microw. Power*, 10, n° 3, 1975, pp. 281-296.

105. PATAY L. Communication personnelle, 8 fév. 1986.

106. PATAY L., LONNE J. Elément localisateur de micro-ondes traitant et chauffant sur toute sa largeur des matériaux en place. Brevet français n° 2 481 561 (80 08357), 11 avril 1980.

107. PATAY L., LONNE J. Désherbage et désinfection des sols par localisation d'énergie micro-ondes. *Econ. Progrès Electr.*, n° 8/9, 1983, pp. 7-8.

108. PELLISSIER J. Communication personnelle. 30 janv. 1987.

109. PEREDEL'SKIJ A. Le problème des mesures électrotechniques permettant de combattre les insectes nuisibles. *Uspehi Sovrem. Biol.*, 41, 1956, pp. 228-245.

110. PERSON N. Microwave radiation as a source of energy in the field of agriculture. *J. Microw. Power*, 7, 1972, pp. 252-272.

111. POUR-EL A., NELSON S., PECK E., TJHIO B., STETSON L. Biological properties of VHF- and microwave-heated soybeans. *J. Food Sci.*, 46, n° 3, 1981, pp. 880-895. (C.D.I.U.P.A., n° 156695).

112. RAI P. An investigation of the morphological and reproductive changes induced by exposure of the yellow mealworm, Tenebrio molitor L., to radiofrequency energy. Thesis doct. Ph. D., Univ. Nebraska, 1970.

113. RAI P., BALL H., NELSON S., STETSON L. Morphological changes in adult Tenebrio molitor (Coleoptera : Tenebrionidae) resulting from radio-frequency or heat treatment of larvae or pupae. *Ann. Entomol. Soc. Am.*, 64, 1971, pp. 1116-1121.

114. RAI P., BALL H., NELSON S., STETSON L. Lethal effects of radio-frequency energy on eggs of Tenebrio molitor (Coleoptera : Tenebrionidae). *Ann. Entomol. Soc. Am.*, 65, 1972.

115. RZEPECKA M., STUCHLY S., HAMID M. Applications for microwave treatment of granular materials in agriculture. IMPI Technical Report Series, TR-72-1, 1972, pp. 1-35.

116. SCHRADER D., Mac NELIS D. Microwave irradiation of plant roots in soil. *J. Microw. Power*, 10, n° 1, 1975, pp. 77-91.

117. SENTER S., FORBUS W., NELSON S., WILSON R., HORVAT R. Effects of dielectric and steam heating treatments on the storage stability of pecan kernels. *J. Food Sci.*, 49, n° 3, 1984, pp. 893-895. (C.D.I.U.P.A., n° 190639).

118. SHACKELFORD R. Studies to determine the electromagnetic spectrum effects on arthropods. Engineering Experiment Station, Georgia Institute of Technology, Final Report, Project A-985, 1970.

119. SMITH C. How to prevent insect contamination. *Intern. Conf.*, 54, n° 9, 1944, pp. 30, 48, 52.

120. STETSON L., NELSON S. Effectiveness of hot air, 39 MHz dielectric, and 2450 MHz microwave heating for hard-seed reduction in alfalfa. *Trans. Am. Soc. Agric. Eng.*, 15, n° 3, 1972, pp. 530-535.

121. STUCHLY M., STUCHLY S. Dielectric properties of biological substances, tabulated. *J. Microw. Power*, 15, n° 1, 1980, pp. 19-26.

122. TERRIEN V. L'huître et les inventeurs. *Sud-Ouest*, 24 nov. 1986.

123. THAYER D. Application of radiant energy in pest management. *Cereal Food World*, 30, n° 10, 1985, pp. 714-721. (C.D.I.U.P.A., n° 204736).

124. THOMAS A. Pest control by high-frequency electric field - critical résumé. Leatherhead, British Electrical and Allied Industries Research Association, Technical Report n° W/T23, 1952.

125. THOMAS A., WHITE M. The sterilization of insect-infested wood by high-frequency heating. *Wood*, 24, 1959, pp. 407-410 and in Technical Report n° W/T37 of the British Electrical and Allied Industries Research Association, Leatherhead, 1959.

126. THOUREL L. Les hyperfréquences et leurs applications industrielles. In : Comptes-rendus des Journées d'Informations Electroindustrielles du Centre et du Sud-Ouest, 15-16 oct. 1975. Publication du Comité Français d'Electrothermie, 21 p. ; Désinsectisation, pp. 13-14.

127. THOUREL L. Désherbage. In : Utilisation des ondes électromagnétiques dans l'industrie agro-alimentaire. Publication du Centre d'Etudes et de Recherches de Toulouse, CERT-ONERA, 1979, pp. 13-14.

128. THOUREL L. Désinsectisation. In : Utilisation des ondes électromagnétiques dans l'industrie agro-alimentaire. Publication du Centre d'Etudes et de Recnerches de Toulouse, CERT-ONERA, 1979, pp. 15-16.

129. THOUREL L., PATAY L. Désherbage micro-ondes. Publication du Centre d'Etudes et de Recherches de Toulouse, 1979, 4 p.

130. TILTON E., BROWER J., BROWN G., KIRKPATRICK R. Combination of gamma and microwave radiation for control of the Angoumis grain moth in wheat. *J. Econ. Entomol.*, 65, 1972, p. 531.

131. TILTON E., VARDELL H. An evaluation of a pilot-plant microwave vacuum-drying unit for stored-product insect control. *J. Georgia Entomol. Soc.*, 17, 1982, p. 133.

132. TRAN V. Optimising the microwave treatment of acacia seeds. *J. Microw. Power*, 16, n° 3-4, 1981, pp. 277-281.

133. TRAN V., CAVANAGH A. Effects of microwave energy on Acacia Longifolia; *J. Microw. Power*, 14, n° 1, 1979, pp. 191-246.

134. TRAN V., CAVANAGH A. Structural aspects of dormancy. *Seed Physiol.*, 2, chap. 1, 1984, pp. 1-43.

135. TRAN V., DI SCIASCIO M. Microwave energy as an effective means of inducing the germination of some Australian hard seeds. In : Comptes-rendus du Symposium international sur les applications énergétiques des micro-ondes, IMPI-CFE, 14, Monaco, 1979, Abstract, pp. 70-73.

136. VAN DEN BRUEL W., BOLLAERTS D., PIETERMAAT F., VAN DIJCK W. Etude des facteurs déterminant les possibilités d'utilisation du chauffage diélectrique à haute fréquence pour la destruction des insectes et des acariens dissimulés en prcfondeur dans les denrées alimentaires empaquetées. *Parasitica*, 16, 1960, pp. 29-61.

137. VELA G., WU J., SMITH D. Effect of 2450 MHz microwave radiation on some soil microorganisms in situ. *Soil Sci.*, 121, 1976, pp. 44-51.

138. WANG J., SCHMUGGE T., WILLIAMS D. Dielectric constants of soils at microwave frequencies. II. NASA technical paper 1238, NASA Scientific and Technical Information Office, May 1978, 35 p.

139. WATTERS F. Control of insects in foodstuffs by high-frequency electric fields. In : Proceedings of the Entomological Society of Ontario, 92, 1962, pp. 26-32.

140. WATTERS F. Microwave radiation for control of Tribolium confusum in wheat and flour. *J. Stored Prod. Res.*, 12, n° 1, 1976, pp. 19-25. (C.D.I.U.P.A., n° 92797).

141. WAYLAND J., DAVIS F., MERKLE M. Toxicity of an UHF device to plant seeds in soil. *Weed Sci.*, 21, 1973, pp. 161-162.

142. WAYLAND J., DAVIS F., YOUNG L.,
MERKLE M. Effects of UHF fields on
plants and seeds of mesquite and
beans. *J. Microw. Power*, $\underline{7}$, n° 4,
1972, pp. 385-388.

143. WAYLAND J., MERKLE M., DAVIS F.,
MENGES R., ROBINSON R. Control of
weeds with electromagnetic fields.
Weed Res., $\underline{15}$, 1975, pp. 1-5.

144. WEBER H., WAGNER R., PEARSON A.
High-frequency electric fields as
lethal agents for insects. *J. Econ.
Entomol.*, $\underline{39}$, n° 4, 1946, pp..487-498.

145. WHATLEY T., WAYLAND J., DAVIS F.,
MERKLE M. Effects of soil moisture
on phytotoxicity of microwave fields.
In : Proceedings of the Weed Science
Society Symposium, $\underline{26}$, 1973, p. 389.

146. WHATLEY T., WAYLAND J., MERKLE M.,
Factors affecting the phytotoxicity
of a microwave field. In : Proceedings of the Weed Science Society,
$\underline{27}$, 1974, p. 344.

147. WHITNEY W., NELSON S., WALKDEN H.
Effects of high-frequency electric
fields on certain species stored-
grain insects. USDA Publication, AMS,
Market Quality Research Division,
Marketing Research Report n° 455,
1961.

148. A la coopé, on teste le four à macération. *Le Narbonnais*, 17 oct. 1985.

149. Electromagnetic and sonic energy for
pest control. In : Scientific Aspects
of Pest Control. Publication n° 1402,
National Academy of Sciences, National Research Council, 1966, pp. 135-
166.

150. Electronic zapper may hold the key
to control of weeds, nematode, and
fungi. *Sugarbeet*, 1972, p. 21.

151. Jusqu'au coeur du raisin. *Dépêche
Midi*, 12 oct. 1986, *Pages Rég. Aude*,
p. 7.

152. Les micro-ondes, au service de la
macération carbonique. *Le Narbonnais*,
17 oct. 1985, F5.

153. Les micro-ondes au service du vin.
Le Narbonnais. 28 oct. 1985, p. 2.

154. Macération carbonique au microfour,
première mondiale. *Dépêche Midi*,
17 oct. 1985, *Pages Rég. Aude*, p. 18.

155. Monographie du four micro-ondes
pour macération carbonique. Publication Sté SARRE SARL, 1987, 2 p.

156. Pince micro-ondes pour désinsectisation des poutres de charpentes.
Publication Sté Rayonnements Systèmes, sans réf., 1986, 2 p.

157. Procédé de vinification par macération carbonique et installation de
vinification. Abrégé descriptif
(demande de brevet), sans réf.,
s.d., 5 p.

158. UHF energy destroys weed seeds.
Sans réf., dec. 1976, 1 p.

Part IV

Biological effects and medical applications

Chapter 1

Interactions with the organism

Interactions with the living organism

The microwaves that we are likely to encounter interacting with the living organism are those emitted by a medical applicator or stray radiation inadvertently leaking from a domestic or industrial applicator. In the former case, the emitting antenna could be an open-ended waveguide, a rectangular horn, a corrugated circular horn, converging lenses, a dipole protruding from a coaxial cable, a resonant slot, or a phased array. In the latter case, the source is usually a radiating aperture, e.g., the access to a tunnel applicator or a slot (defective soldering, bad door closure, bad connections between waveguide flanges). The characteristics of such sources were discussed in Part I, Section 2.12.

In this context, the human or animal organism is a hetero-geneous accumulation of biological tissues with very dissimilar dielectric properties, in which localized heating is opposed by passive and active mechanisms of temperature regulation. The response to microwaves must be evaluated by complementary techniques involving experiments and numerical modeling. In addition, more specific nonthermal interactions may be oper-ating at the cellular or molecular level.

1.1 Dielectric behavior of biological material

1.1.1 Biomolecules

At the molecular level, we can describe quite precisely the dielectric behavior of the constituents of the cell, i.e., amino acids, protides, and nucleic acids (Part I, Section 3.5.5). Con-versely, studies of dielectric behavior provide an insight into the structure and hydration of these molecules.

Protides can be subdivided into peptides and proteids. The latter are further subdivided into heteroproteids and holo-proteids or proteins. Peptides and proteins are formed by the bonding of amino acids that incorporate carboxyl groups ($-COOH$) and amino groups ($-NH_2$). Most of these are α amino acids in which the group NH_2 is bonded to the carbon atom immediately adjacent to the group $-COOH$. Their as-sociation into protides is made by the formation of an amino peptide bond $-CO - NH-$ between the groups $-NH_2$ and $-COOH$ with the loss of one water molecule.

The protein is a very large macromolecule with molecular mass between 6000 and 10^6, so that its relaxation frequency is low and lies between 100 kHz and 50 MHz (β relaxation). Its mobility is totally hindered at microwave frequencies. How-ever, between 200 MHz and 2 GHz (i.e., at the start of the γ relaxation of free water) a narrow zone known as δ relaxation can be observed and may be explained in terms of the free ro-tation of certain parts of the molecule. In hemoglobin at 25°C,

δ relaxation occurs at 850 MHz which corresponds to the frequency range in which its constituent peptides and amino acids exhibit absorption [196], [323].

Peptides and amino acids, much smaller than proteins, resonate in the microwave region. The relaxation frequencies for glycine and three glycylpeptides (di-, tri-, and alanyl-glycine) at 25°C and pH6 are 3.2 GHz, 1.2 GHz, 770 MHz, and 960 MHz, respectively [1]. These are inversely proportional to the size of the molecules. Measurements of the viscocity of these four solutions in the range 0–50°C confirm the validity of the Debye equation [Part I, (3.43)]:

$$\tau = \frac{1}{2\pi f_r} = \frac{4\pi a^3 \xi}{kT}$$

where ξ is the viscocity in $Pa\,s[L^{-1}MT^{-1}], \tau$ is the relaxation time, k is Boltzmann's constant, and a is the radius of the molecule.

The nucleic acids DNA and RNA play a fundamental role in cellular processes. The simple helix of RNA is involved in the synthesis of proteins, especially enzymes, at the level of the cytoplasm. DNA is the carrier and transmitter of genetic information. Although its double helix is symmetric and nonpolar, it exhibits a low frequency relaxation because of the induced displacement of ions at the surface of the molecule under the influence of the applied electric field. The relaxation time is thus proportional to the square of the length of the helix [196], [392], [377], [408], [119].

A higher frequency relaxation of a chemical nature is produced when the biopolymer undergoes an abrupt transition from one configuration to another under the influence of an applied field on the order of 1 kV cm^{-1}. For example, the α helix of the polypeptides forming fibrous proteins consists of CHR-C=ON-H segments. The peptide bonds and the helix assembly are stabilized by hydrogen bonds attaching each carboxyl group C=O to the NH group of the third preceeding segment (Fig. 1.1). When these bonds are broken, the structure collapses. This reversible mechanism relies on protein denaturation and on the switching of biopolymers and membranes under the influence of nerve impulses (Section 1.3.1) [357], [358], [196], [323].

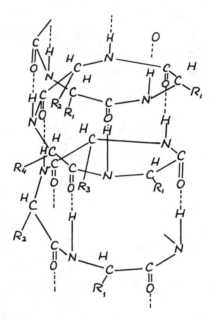

Figure 1.1 α Helix chain of polypeptides making up a fibrous protein (keratine).

The dielectric properties of biomolecules depend on the quantity and type of bound water molecules associated with them. The relaxation frequency lies between 100 MHz and 1 GHz, depending on the orientation of these water molecules, their bonding to the macromolecules, and the solvent viscocity [320]. Further information may be found in [14], [281], [165], [409], [410], [217], [218], [238], [322], [8], [58], [243], [261], [372], [167], [23] and [239].

1.1.2 Cells and membranes

In general, cellular suspensions may be described as conducting particles in a dielectric medium. Blood is an example of such a mixture and has been the subject of several studies since 1951 [97], [99], [312], [348], [424]; It has been found that $\epsilon'_r = 56$ and $\epsilon''_r = 15.9$ at 3 GHz for human blood at 35°C. These values vary little with frequency in the ultrahigh frequency band. This was also independently confirmed in [68] (Table 1.1).

Table 1.1

Permittivity at 21°C of mouse blood [68, 384]

f	100 MHz	200 MHz	500 MHz	1 GHZ	2.45 HGz	3 GHz	5 GHz	10 GHz
ϵ'_r	86	70	63	62	62	62	61	48
ϵ''_r	180	90	35	28	18	19	22.5	33.5

The cell dimensions are on the order of 10–100 μm, and the cells themselves are basically made from four types of molecule, namely, proteins, nucleic acids, lipids, and glucides. Of all its basic components (cytoplasm, membrane, nucleus, mitochondria, and so on), it is the membrane of the animal cell that seems to be the site of the most interesting phenomena in the electromagnetic field. It is a double layer of semifluid phospholipids, approximately 10 nm thick, with protein molecules imbedded in it [196]. Its function is to protect the cell and actively to control the ionic and molecular exchanges between the cell and its ambient medium. The difference between the dielectric properties of the cytoplasm and the ambient medium is responsible for the accumulation of electric charges on the interface. Indeed, the membrane behaves like a leaky capacitor with a surface capacitance on the order of 1 μF cm^{-2} and parasitic current due to the ion flows [133], [134], [72], [196], [356], [398], [397], [320], [103].

The dielectric behavior of the membrane is described by Hasted [196]. The double layer model predicts the relaxation time

$$t = (\epsilon_i' + 2\epsilon_c') \frac{r}{2d\sigma} \qquad (1.1)$$

where r is the average radius of the cell, d is the thickness of the double layer, σ is its conductivity, and ϵ_i' and ϵ_c'' are the internal and external permittivities, respectively. The membrane is the site of very extensive β relaxation at frequencies between a few tens of kilohertz and a few hundred megahertz. At frequencies in excess of this limit, its capacitor component is essentially short circuited and, theoretically, no potential can be induced. For a fuller treatment of this subject, the following references can be consulted: [333], [344], [128].

1.1.3 Tissues

Biological tissues are structured assemblies of cells whose dielectric properties in the microwave region are well known. The earliest work on this subject was concerned with excised tissues [123], [98], [351], [333], [334], [336], [311], [225], [164], [269], [348]. These *in vitro* measurements produced results that are difficult to use because of the absence of blood circulation and the natural degradation of cellular structure. More recently, implanted probes and open-ended coaxial lines have made possible *in vivo* measurements [215], [68], [383], [380], [24], [249], [386].

The dielectric properties of tissues depend greatly on their water content: high water content tissues (muscle, skin) have permittivities and loss factors that are much higher than those of low water content tissues (bone, fat). In general, β relaxation appears in the region of a few hundred megahertz and is followed by a region in which the permittivity is relatively independent of frequency. Above 3 GHz, the γ relaxation of free water is the dominant phenomenon and reaches its maximum between 20 and 24 GHz [237].

Table 1.2 [384] summarizes the large number of measurements made between 1952 and 1979. The loss factor of muscle tissue (75% of water) at 2.45 GHz and 3 GHz is greater by a factor of 17 than that of fat or bone marrow, and greater by a factor of about 12 than that of bone tissue.

The references given in Part I, Chapter 3 are relevant to this subject. The following may also be consulted: [123], [99], [351], [333], [325], [276], [337], [371], [401], [355], [402], [68], [349], [62], [381], [387], [67], [248], [382], [376], [385], [378], [237], [131], [69], [130], [246], [238], [166] and [109].

1.2 Quantum aspects

Electromagnetic radiation of frequency ν has associated with it a photon of energy

$$W = h\nu \tag{1.2}$$

where $h = 6.626 \times 10^{-34}\,\text{J}\,\text{s}$ is Planck's constant. Thus, at the limits of the microwave range ($300\,\text{MHz}$–$300\,GHz$), the photon energies are $1.99 \times 10^{-25}\,\text{J}\,(1.24 \times 10^{-6}\,\text{eV})$ and $1.99 \times 10^{-22}\,\text{J}(1.24 \times 10^{-3}\,\text{eV})$, respectively.

The energy that has to be expended to ionize an atom (i.e., to remove its least strongly bound electron) is

$$W = e\varphi \tag{1.3}$$

where e is the electron charge ($e = 1.602 \times 10^{-19}\,\text{C}$) and φ is the first ionization potential. For the C, H, O, and N atoms, $e\varphi$ is $11.26\,\text{eV}, 13.59\,\text{eV}, 13.62\,\text{eV}$, and $14.53\,\text{eV}$, respectively, which sets the threshold for ionization of biological material at $13.6\,\text{eV}$ [404]. The lowest-energy ionizing radiation for biological material corresponds to

$$h\nu \geq e\varphi = 13.6\,\text{eV} = 2.18 \times 10^{-18}\,\text{J} \tag{1.4}$$

and

$$\nu \geq 3.3 \times 10^{15}\,\text{Hz}$$

or

$$\lambda \leq 91.2\,\text{nm}$$

Table 1.2

Measured permittivity of various organic tissues [384]

Frequency:	500 MHz	1 GHz	2.45 GHz	3 GHz	5 GHz	10 GHz	Reference
Human lung	35	35					[344, 363]
in vitro, 37°C	26						
Canine muscle	48	47	45				[70, 420]
in vitro, 37°C	35	20	11				
Human muscle	52-54	49-52	47.5	45-48	44	40-42 (1)	[344, 345]
in vitro, 37°C	41	23-24	13.5	13-14	14	15 (1)	[363, 416, 419]
Human muscle				50.0			[334]
pectoralis major, 37°C				17.1			
Human muscle				41.0			[334]
soleus, 37°C				18.0			
Mouse muscle	63	61	58	56.0	53	41	[70]
in vivo, 31°C	40	23.5	17.5	17.3	19.2	23	
Human skin	46.5	43-46	43	40-45	40.7	36 (1)	[344, 416]
in vitro, 37°C	26.5	16.4-20	14	12-16	14	15 (1)	
Cunine fat	14.5	14.3	12				[70]
in vivo, 37°C	8.6	6.7	5.1				
Human fat	4.0-7.0	5.3-7.5	5.75	3.9-7.2	4.7		[344, 345, 348]
in vitro, 37°C	2.2	1.5-2.7	0.8	0.67-1.36	0.7		[363, 367]
Human abdominal				4.92			[334]
wall fat, 37°C				1.46			
Human chest				3.94			[334]
fat, 37°C				0.87			
Canine kidney	55	49.5	47.5	47	43	34	[70]
in vitro	25	14.4	13.2	12	14	18	
Canine kidney	57	53	50	49			[70]
in vivo, 37°C	30	21.5	20	18			
Human liver	43-51	46-47		42-43		34-38 (1)	[344, 348]
in vitro, 37°C		17-18.4		12-12.2		10.6-12 (1)	[363]
Human tibia				8.35			[334]
37°C				1.32			
Bone marrow		4.3-7.3		4.2-5.8		4.4-5.4	[344]
in vitro, 37°C		0.8-1.8		0.7-1.35		0.35-1.0	
Human brain			32	33	33 (2)		[275]
in vitro, 37°C			15.5	18	15.8 (2)		
Mouse brain	57.5	55	52.5	52.5	53	42	[70]
in vivo, 32°C	35	21.5	14	13.7	16	18	

(1) at 8.5 GHz, (2) at 4 GHz, (3) at 2 GHz

Table 1.2
continued

Frequency	500 MHz	1GHz	2.45 GHz	3 GHz	5 GHz	10 GHz	Reference
Canine brain			50 (3)			40	[134]
grey matter, 25°C			11 (3)			20	
Canine brain			37 (3)			28	[134]
white matter, 25°C			8.5 (8)			12.4	
Eye	36	32	30	30	30	28	[345]
cornea	15	9.5	8	9	10	18	
Eye	70	70	70	70	68	62	[345]
vitreous body	65	36	20.5	18	20	32	

(1) at 8.5 GHz, (2) at 4 GHz, (3) at 2 GHz

which is in the ultraviolet range.

Microwave energies are also much lower than the activation energy associated with chemical bonds: 5 eV for the covalent bonds O-H; 2–10 eV for the hydrogen bond; and just less than 2 eV for Van der Waals bonds. For Brownian motion at 37°C, $kT = 0.027$ eV. Direct effects of microwaves at the molecular level can therefore be excluded [320], [53], unless it were to be accepted that dielectric saturation reduces the energy of certain bonds between the principal chain and a lateral branch of the macromolecule.

1.3 Basic interaction with cell membranes

1.3.1 Continuous wave

Cell membranes are extremely sensitive to potential differences and, consequently, to electrostatic and electromagnetic fields. This sensitivity manifests itself by the change in membrane currents that results from the ion current between the interior and exterior of the cell.

At rest, the ionic gradient across the membrane produces the electrochemical energy necessary for cell activities. It regulates the secretion of substances that control the enzymatic activity of the cell. An applied potential difference has a complex effect on the conductance of the membrane: the latter first increases for the current of sodium ions entering the cell and then for the current of potassium ions leaving it, until an equilibrium is established [103].

This charge propagation mechanism plays a fundamental part in the transmission of nerve impulses in axons. It is described by the Hodgkin-Huxley formula that gives the membrane current I (in $\mu A\ cm^{-2}$) as a function of the applied voltage V (in mV):

$$I = C_M \left(\frac{dV}{dt}\right) + g_K n^4 (V - V_K) + g_{Na} m^3\, h(V - V_{Na}) + g_1(V - V_1)$$

(1.5)

where $g_K n^4$ and $g_{Na} m^3\, h$ are the conductances of the membrane for potassium and sodium, and C_M is the surface capacitance of the membrane $(1\,\mu F\ cm^{-2})$. The potentials V_K, V_{Na}, and V_1 are $+12\,mV$, $-115\,mV$, and $-10.6\,mV$, respectively, and the quantities g_K, g_{Na}, and g_1 are 36, 120, and $0.3\ mS\ cm^{-2}$, respectively. The dimensionless parameters n, m, and h are solutions of the differential equations

$$\frac{dn}{dt} = \alpha_n(1 - n) - \beta_n n \qquad (1.6)$$

$$\frac{dm}{dt} = \alpha_m(1 - m) - \beta_m m \qquad (1.7)$$

$$\frac{dh}{dt} = \alpha_h(1 - h) - \beta_h h \qquad (1.8)$$

where, at 6.3°C,

$$\alpha_n = 0.01 \frac{V + 10}{\exp\left(\dfrac{V + 10}{10}\right) - 1} \qquad \text{and} \qquad \beta_n = 0.125\exp\left(\frac{V}{80}\right)$$

(1.9)

$$\alpha_m = 0.1 \frac{V + 25}{\exp\left(\dfrac{V + 25}{10}\right) - 1} \qquad \text{and} \qquad \beta_m = 4\exp\left(\frac{V}{18}\right)$$

(1.10)

$$\alpha_h = 0.07 \exp\left(\frac{V}{20}\right) \quad \text{and} \quad \beta_h = \frac{1}{\exp\left(\dfrac{V+30}{10}\right)+1}$$

(1.11)

The model remains valid for alternating potentials

$$V(t) = V_0 + V_m \cos \omega t \tag{1.12}$$

so that, by replacing V with $V(t)$ in (1.9), (1.10), and (1.11), we obtain

$$\beta_m(\omega t) = \sum_{k=-\infty}^{+\infty} \beta_{m_k} e^{jk\omega t} \tag{1.13}$$

which is a Fourier series with coefficients

$$\beta_{m_k} = \frac{1}{2\pi} \int_0^{2\pi} \beta_m(\omega t) e^{-jk\omega t} d(\omega t) \tag{1.14}$$

The average value of $\beta_m(\omega t)$ is

$$\beta_{m_0} = \frac{1}{2\pi} \int_0^{2\pi} \beta_m(\omega t)\, d(\omega t) \tag{1.15}$$

and is a monotonically increasing function of the peak voltage V_m, so that the application of an alternating current produces a continuous variation in the parameters α and β and, hence, new values of n, m, and h, thus modifying the conductances and the current I. This means that an applied alternating current can induce a continuous change in the ion current [70], [71].

Physically, ion transfer takes place along microchannels or microtubes that are specific for different types of ion. The potentials are monitored by certain molecules with high dipole moments, which open and close the tubules by switching the protein molecules away from a stable configuration when the threshold potential is reached (Section 1.1.1). The thermodynamic probability that one of the blocking dipoles is in a given state is a nonlinear function of the applied potential, and the application of the alternating field produces a constant shift in this probability. The changes in permeability due to this nonlinear mechanism are thought to be consistent with the Hodgkin-Huxley model, modified for alternating fields [70].

Early work on the mathematical modeling of this interaction was dominated by the work of Hodgkin and Huxley and their collaborators, and was followed by a number of other studies [212], [211], [216], [63], [93], [125], [94], [114], [157], [199], [7], [214], [235], [400], [312], [197], [198].

The evaluation of local fields requires the use of the model described in [35]. For example, a power flux of $10 \, \mathrm{mW \, cm^{-2}}$ induces a field of the order of $1 \, \mathrm{kV \, m^{-1}}$ in a lipid double layer with permittivity $\epsilon_r' = 2.1$, which, for a thickness of $10 \, \mathrm{nm}$, gives a potential difference of $10 \, \mu\mathrm{V}$ [70]. The effect of microwaves on the ion current was analyzed in [422]. It was shown that a 90-min exposure at 9.2-GHz at the rate of $20 \, \mathrm{mW \, g^{-1}}$ could change the permeability of membranes for the cations $K^+, Na^+,$ and Rb^+, and could facilitate the motion of Na^+ and Rb^+.

Another study [282] has shown that the intercellular saline fluid may play a part in the interaction mechanism. An intracranial field of $200 \, \mathrm{V \, m^{-1}}$, induced by a power flux of $10 \, \mathrm{mW \, cm^{-1}}$ at $100 \, \mathrm{MHz}$ in the intercellular fluid ($\sigma = 2S \, \mathrm{m^{-1}}$) gives rise to a current density of $36 \, \mathrm{mA \, cm^{-2}}$. If we suppose that one-half of this current enters the nerve cell, a relatively high membrane potential can result (on the order of 10–100 mV).

It is difficult at present to be more specific about this type of interaction, the study of which is closely linked to advances in microbiology in general. Research papers are for the most part American or Soviet, the latter being in all probability the more advanced, but virtually inaccessible. The literature on related subjects includes [27], [344], [359], [56], [168], [296], [315], [411], [39], [40], [73], [295], [6], [35], [42], [57], [271], [310], [345], [15], [203], [314], [46], [282], [45], [70], [266], [305], [13],

[21], [22], [71], [309], [370], [346], [139], [347], [88] and 422].

As for plant cells whose inert cellulosic surface is insensitive to radiation, certain interactions are more likely to occur at the vacuole level. The vacuole, a body of water pushing the cytoplasm toward the surface, is separated from it by a specialized membrane. It allows the cell to survive without exploding under the influence of osmotic pressure in aqueous environments with very low concentrations of dissolved substances [124]. The resistivity of the membrane and the electric potential of the vacuole constitute the critical parameters related to the vitality of the cell. An attempt to demonstrate the possible influence of the microwave field on these parameters in the case of the giant cells of green algae (*Chara braunii, Nitella flexilis*) was reported in [3]6, [275], [158] and [159]. The last two of these references describe distinct variations in vacuole potential at frequencies below 1 GHz for power densities of only 1 mW m^{-2}, and the absence of any detectable effect for 10 mW cm^{-2} at 2.45 GHz.

1.3.2 The modulated wave

Although this will take us outside the scope of our book, it is interesting to mention the well-attested effects of low-frequency fields, or UHF fields amplitude-modulated by a square wave of 1–100 Hz. Even power levels that are thermally insignificant may induce large variations in calcium flows in cerebral tissue. The possibility of nonlinear effects with resonance between the applied ELF wave and the slow encephalic waves whose amplitude is only on the order of 10 mV cm^{-1} [266] is discussed in [168].

Another study [66] reported that microwave irradiation was found to alter the birefringence signal accompanying the propagation of the action potential. This is probably due to the Kerr effect that produces an anisotropy in a dielectric that is normally optically isotropic in a steady electric field. The change in the birefringence signal is directly correlated with the membrane potential and the protein configuration in the axoplasm of the neuron. In contrast to other parameters such as the amplitude of the action potential or the speed of the nerve impulse, the amplitude of this signal is independent of temperature and, hence, of any possible thermal effects. It was shown that an absorbed dose (Section 1.4.2) of $123\,\mathrm{mW\,g^{-1}}$ had little or no effect on birefringence, whereas the same average power modulated at $1\,\mathrm{kHz}$ by a square signal of $1\,\mu\mathrm{s}$ duration ($5\,\mathrm{kW}$ peak power) produces a significant effect. This suggests that the wave exerts some influence on the membrane potential.

This interaction is probably nonthermal because membrane cannot store the heat generated during the impulse. For example, it was reported in [135] that the temperature difference induced across a 10-nm-thick, blood-cell membrane by a modulated wave of $1.75\,\mathrm{kW\,g^{-1}}$ peak power was only $10^{-4}\,\mathrm{K}$.

Further information on this subject can be found in [41], [40], [12], [360], [70], [71], [320] and [160].

1.3.3 Pearl chain formation

This very curious phenomenon was first described by Herrick in 1958 [200] and was later observed by many others [330], [331], [138], [137], [320], [178]. A magnetic or electrostatic field, or a continuous or modulated radiofrequency wave, tends to assemble the cells in a colloidal suspension in chains. This so-called pearl-chain formation seems to be irreversible and may be explained by a drop in the membrane potential that is responsible for the repulsive force between cells.

1.4 Thermal interaction with the living organism

The thermal interaction between an electromagnetic wave and a living being is an exceedingly complex phenomenon. In fact, the very word *interaction* implies that the wave modifies the properties of the medium by which it is transmitted and, at the same time, is itself modified. In addition, biological mechanisms of passive and active thermal regulation are part of this interaction.

It is, however, important to have some quantitative appreciation of the characteristics of this interaction before one can examine the physiological effects of microwaves and develop useful applicators for therapeutic hyperthermia. This is achieved by experimental and mathematical modeling techniques. These techniques are briefly described below.

1.4.1 Reflection and transmission

The living organism is a stratified medium consisting of N superposed layers of tissue with different dielectric properties and, in general, linear dimensions that are less than or comparable to the wavelength. This means that the classical laws of reflection and transmission at the interface of two semi-infinite media are no longer sufficient, and a more complex theory must be invoked [412]. This takes into account the surface impedances Z_n of the successive interfaces (Fig. 1.2) and allows us to compute the reflection and transmission coefficients of the corresponding interfaces. For a plane wave at normal incidence

$$\Gamma = \frac{H_r}{H_i} = -\frac{E_r}{E_i} = \frac{\eta_0 - Z_1}{\eta_0 + Z_1} \tag{1.16}$$

$$T_E = \frac{E_t}{E_i} = 1 - \Gamma \tag{1.17}$$

$$T_H = \frac{H_t}{H_i} = 1 + \Gamma \tag{1.18}$$

where

$$Z_1 = \eta_1 \frac{Z_2 + \eta_1 \tanh\gamma_1 d_1}{\eta_1 + Z_2 \tanh\gamma_1 d_1} \tag{1.19}$$

$$Z_2 = \eta_2 \frac{Z_3 + \eta_2 \tanh\gamma_2 d_2}{\eta_2 + Z_3 \tanh\gamma_2 d_2} \tag{1.20}$$

$$\vdots$$

$$Z_n = \eta_n \frac{Z_{n+1} + \eta_n \tanh\gamma_n d_n}{\eta_n + Z_{n+1} \tanh\gamma_n d_n} \tag{1.21}$$

$$\vdots$$

$$Z_{N-1} = \eta_{N-1} \frac{\eta_N + \eta_{N-1} \tanh \gamma_{N-1} d_{N-1}}{\eta_{N-1} + \eta_N \tanh \gamma_{N-1} d_{N-1}} \tag{1.22}$$

in which η_n is the characteristic impedance of medium n [Part I, (2.54)] and γ_n is the propagation constant of the medium [Part I, (2.28)]. The model becomes more complicated for a plane wave at oblique incidence; or for a cylindrical or spherical wave; or when the interaction takes place in the near-field zone of the antenna. The other relevant parameters are the polarization of the wave, the shape and dimensions of the target, its orientation in the field, and its environment (nature of the ground, presence of reflecting surfaces or other living organisms, and so on).

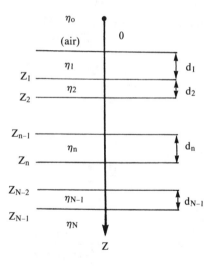

Figure 1.2 A stratified medium [412].

Depending upon the characteristics of the wave, the environment, and the target, the reflection coefficient Γ lies between 0 and 1, which corresponds to perfect matching and total mismatching, respectively. Figure 1.3 shows Γ as a function of frequency for the air-skin interface and two internal interfaces.

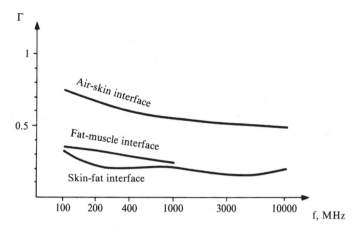

Figure 1.3 Reflection coefficients of interfaces [404].

At 2.45 MHz the reflection coefficient of the air-skin interface is on the order of 0.55, which means that the fraction of power transmitted is approximately 70%. Stratified media are also discussed in [64].

1.4.2 Absorption and dosimetry

The depth of penetration of microwaves in low water content tissue (fatty tissue), as defined in Part I, (2.83), is about 30 cm at 100 MHz and 3 cm at 10 GHz. The corresponding figures for high water content tissue (muscular tissue) at 100 MHz and 10 GHz are 4 cm and 2 mm, respectively [53]. The transmitted wave creates local fields that are very difficult to determine analytically, but may be measured *in situ* (Section 1.4.3) or estimated by numerical modeling (Section 1.4.4). The power absorbed per unit mass, or the *specific absorption rate* (SAR), is a useful concept in the interpretation of the measured power deposition in tissue [175], [208], [228], [234], [391]. It is a function of only the average local field and the dielectric loss factor:

$$P_M = SAR \stackrel{\mathrm{d}}{=} \frac{\sigma}{\rho}\bar{E}_{\mathrm{loc}}^2 = \frac{1}{\rho}\omega\epsilon_0\epsilon_r''\bar{E}_{\mathrm{loc}}^2 \qquad (\text{in W kg}^{-1}) \quad (1.23)$$

which may be compared with (3.56) in Part I.

The above definition of SAR in terms of the dose of radiation, accumulated thermally in a finite biological medium, mirrors the nomenclature used for ionizing radiation. It encompasses, however, a more complex physical reality that is difficult to grasp because of the extreme complexity of the interaction. The evaluation of actual local heating rates is very laborious, and the simple formula for the rise in temperature, that is,

$$\Delta T = \frac{1}{C} P_M \Delta t \qquad (1.24)$$

where C is the specific heat (in $J\,kg^{-1}\,K^{-1}$), is at best an approximation [118], [300], [61], [374], [375], [53], [104].

1.4.3 Experimental aspects

1.4.3.1 Biological materials

Experiments with excised human or animal tissue lack the effects of thermal regulation, metabolism, respiration, and circulation. Except for measurements of permittivity (Section 1.1.3) and the study of certain isolated organs (e.g., the eye [25]), such experiments are of limited value and were conducted only in older studies [100], [334], [263], [181], [284], [262].

Laboratory animals are widely used in studies of physiological effects and, to a lesser extent, as models of the interaction with electromagnetic radiation.

Determinations of internal fields and absorbed power employ standing-wave waveguides for small specimens [136], [280], [279], anechoic chambers, and TEM cells in which small animals can be exposed to plane waves [180], [177], [182], [286], [233], [361]. In the case of *in situ* measurements, temperature, power, or field sensors can be implanted [92], [366], [321], [201]. Noninvasive techniques rely on microwave radiometry or thermography [55], [413], [272], [320].

There is an extensive literature on the subject. Early work was concerned with details of the temperature rise [120], [174], [240]. Dosimetric aspects were considered later [327], [183], [185], [278], [148], [149]. The specific dependence on frequency and polarization was discussed in [141], [143], [144] and [332], and, finally, the effects of the ambient temperature were examined in [298].

Such experiments raise obvious ethical questions and the corresponding results must be treated with caution for the following reasons. First, they cannot be extended to the human body at the same frequency, but the Gandi extrapolation rule can be used at lower frequency [142]. For example, a 2.45-GHz wave induces geometric resonances in small animals; similar resonances appear in humans, but only at much lower frequencies (on the order 300 MHz). Second, the thermal regulation mechanism of fur-covered animals such as mice and rabbits is less sophisticated than in humans [307]. Third, the surface to volume ratio of small animals is necessarily larger than that of the human body, and this necessarily affects heat-loss data. The results obtained in animal experiments are therefore only very approximately valid for humans [2], [3].

1.4.3.2 Biological phantoms

Phantom materials are useful in the development of microwave applicators, measuring probes, and the validation of numerical models. Phantoms are chemical preparations with dielectric and thermal properties close to those of biological tissue. Some very elaborate phantoms made of several layers are sometimes used to simulate the human body or parts of it, e.g., the head or limbs [295], [172], [95], [245], [96], [10], [229], [184], [205], [87], [186], [206], [207], [140], [34], [33], [223], [85], [89], [417]. For example, muscle tissue can readily be simulated by polystyrene foam containing a saline solution [417], or a mixture of water, gelatine, and salt [406]; blood can be simulated by a mixture of water and glucose, or water and isopropyl alcohol [406]. Some organs are more difficult to simulate. For example, the lungs seem to obey the mixture formula of Polder and Van Santen for spherical bubbles [246]

$$\epsilon' = \epsilon'_c \left(1 - \frac{3p}{2}\right) \qquad (1.25)$$

where ϵ'_c is the permittivity of the tissue and p is the volume filling factor that varies between 0.45 (full lungs) and 0.25 (empty lungs). In the frequency range 100 MHz–2 GHz, this organ can be satisfactorily simulated by hollow silica globules suspended in a semi-liquid mixture of water, sugar, salt, and gelatine.

1.4.4 Modeling

Numerical modeling of the interaction of electromagnetic radiation with the living organism is currently based on very sophisticated algorithms that produce increasingly precise descriptions of the phenomena.

Early simulations employed simplistic models of organs and limbs. They were concerned with simple shapes such as spheres, layered spheres, ellipsoids, cylinders, and so on [320], [25], [252], [363], [252], [253], [96], [141], [232], [277], [283], [419], [115], [205], [228], [254], [350], [395], [415], [28], [47], [80], [81], [84], [121], [122], [146], [147], [156], [169], [170], [206], [255], [273], [292], [301], [302], [29], [30], [37], [82], [83], [155], [183], [185], [190], [194], [207], [289], [290], [299], [420], [421], [31], [129], [328], [32], [117], [153], [188], [189], [291], [287], [306], [319], [326], [418], [79], [116], [329], [250].

The adoption of SAR as a dosimetric unit and the advances in numerical techniques in the early 1980s led to much more satisfactory models. These "dosimetric" models do not rely on analytic methods, but use discrete numerical techniques specially adapted for inhomogeneous media and irregular shapes. They are the method of moments and the method of finite elements.

In the method of moments, the starting point is a vectorial integral equation derived from Maxwell's equations [Part I, (2.10)–(2.13)] and the relevant boundary conditions [Part I, (2.91)] with the induced electric field as the unknown. The body is subdivided into N cells in the form of parallelepipeds, each with its own values of ϵ', μ, and σ, and the initial equation is written in matrix form:

$$[G_{mn}][E_m] = -[E_m^i] \tag{1.26}$$

in which the matrices describe the electromagnetic coupling between the cells, the induced field, and the incident field, respectively. Since we are dealing with a vector field, $[G_{mn}]$ is a $3N \times 3N$ matrix, which means that, in practice, the number of cells is restricted to a few hundred. The matrix equation is then transformed into a set of linear equations with the aid of basis functions or trial distributions, usually pulses or Dirac δ distributions [367], [60], [369], [390]. Iterative methods (e.g., the method of conjugated gradients) are particularly effective. Matrix storage and inversion are no longer necessary, and the number of iterations for convergence to the exact solution is at most equal to N for any basis function [390]. The technique can be improved still further by using linear basis functions and a net of tetrahedral cells that provide a finer model of the body [288], [367].

The second approach is based on the method of finite differences [111]. In principle, this takes less time and requires less computer memory than the other technique. The body is still represented by a lattice of cells and Maxwell's equations are discretized by finite-difference approximations for both time and space derivatives. The field components are positioned along the directions defined by the Yee lattice (Fig. 1.4), and the finite-difference expressions corresponding to one of the scalar components of Maxwell's equations are written in the form

$$
H_x \left(x, y + \frac{\Delta y}{2}, z + \frac{\Delta z}{2}, t + \frac{\Delta t}{2} \right)
$$

$$
= H_x \left(x, y + \frac{\Delta y}{2}, z + \frac{\Delta z}{2}, t - \frac{\Delta t}{2} \right) +
$$

$$
+ \frac{\Delta t}{\mu_0 \Delta z} \left[E_y \left(x, y + \frac{\Delta y}{2}, z + \Delta z, t \right) - E_y \left(x, y + \frac{\Delta y}{2}, z, t \right) \right]
$$

$$
+ \frac{\Delta t}{\mu_0 \Delta y} \left[E_z \left(x, y, z + \frac{\Delta z}{2}, t \right) - E_z \left(x, y + \Delta y, z + \frac{\Delta z}{2}, t \right) \right]
$$

etc $\hspace{4cm}$ (1.27)

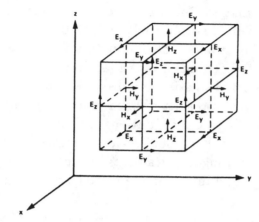

Figure 1.4 The Yee lattice [367].

The new value of each of the six E and H field components depends only on its previous value and the previous values of adjacent components. A solution is thus achieved by stepping through the entire lattice.

The main weakness of the method is that the space around the body must be included in the model, or at least a large enough volume must be used so that the waves that are reradiated by the organism can be regarded as plane waves. This is enough to counterbalance the advantage of smaller computer memory. One way to overcome this difficulty is to use expansion techniques. The calculation is first performed with a very coarse, low-resolution model. A part (subvolume) of the body is then isolated and subdivided into a much finer grid (Fig. 1.5), and the model is rerun, using as boundary conditions the tangential fields at the boundary of the subvolume calculated in the first run.

The finite-difference technique yields transient solutions of Maxwell's equations, but it is possible to obtain steady-state solutions by iterating the calculation to convergence [367], [369].

The above dosimetric models may be combined with thermoregulatory models involving not only the classical heat-

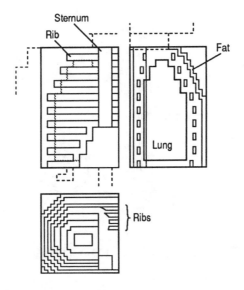

Figure 1.5 Model of the rib cage [367].

transfer equations (conduction, convection, and radiation) but also the biological phenomena of vasoconstriction, vasodilation, and perspiration. For example, the modified Stolwijk model (Fig. 1.6) gives a good simulation of axial and radial heat transfer, starting with the standard one-dimensional heat transfer equation [Part I, (3.63)]

$$\rho(z)C(z)\frac{\partial T(z,t)}{\partial t} = \frac{\partial}{\partial z}\left[\Lambda(z)\frac{\partial T(z,t)}{\partial z}\right] + q(T,z,t) \quad (1.28)$$

where $q(T,z,t)$ is the algebraic sum of the heat generated by the metabolism and by the microwave radiation, and the heat released by the skin and by respiration.

The solution involves discretization into sublayers and the evaluation of second-order finite differences. This gives the following expression for the m-th layer:

$$\Delta^2 \rho_m C_m \frac{\partial T_m}{\partial t} = \Lambda_m\left[T_{m+1}(t) - 2T_m(t) + T_{m-1}(t)\right]$$

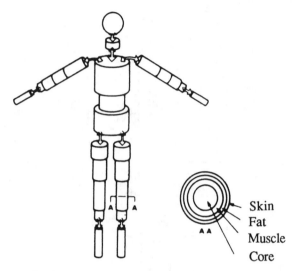

Figure 1.6 Stolwijk thermal model of man [367].

$$+ \Delta^2 q_m(T,t) \qquad (1.29)$$

subject to the boundary condition that the heat flux across the surface is proportional to the temperature difference between the surface and the ambient medium, namely

$$\Lambda \frac{\partial T}{\partial n} + \lambda_s(T - T_0) = 0 \qquad (1.30)$$

where $\partial/\partial n$ is the normal derivative on the surface and λ_s is the surface thermal conductivity in $\mathrm{W\,m^{-2}\,K^{-1}}[\mathrm{MT^3\Theta^{-1}}]$ [113], [367].

The metabolism and the blood flow can be modeled by the equations

$$P_m = \rho_t P_0 (1.1)^{\Delta T} \qquad (1.31)$$

$$P_b = \varphi_s C_b \rho_b \rho_t \Delta T \; (\mathrm{in W\,m^{-3}}) \qquad (1.32)$$

where P_m and P_b are, respectively, the power generated by the metabolism and the power removed by blood; P_0 the metabolic base rate at 37°C; ΔT is the rise in temperature; φ_b is the blood flow rate in $m^3 \, kg^{-1} (tissue) s^{-1}$; C_b is the specific heat of blood in $J \, kg^{-1} \, K^{-1}$, and ρ_b and ρ_t are the densities of blood and tissue, respectively [407].

All these models yield the local values of the SAR, calculated from the local electric fields and whole-body SAR obtained by averaging the local values over the whole body [258]. The following references can be consulted on this subject: [190], [191], [74], [227], [368], [393], [187], [151], [221], [226], [110], [222], [259], [59], [178], [389], [86], [193], [293], [369], [396], [416].

1.4.5 Near-field interaction

Plane-wave irradiation was implicitly assumed in all the above models [Part I, (2.12.5)]. When the exposed body is in the near field of the radiating source, the interaction becomes much more difficult to model because of the very rapid field variation.

This situation is rarely encountered in industrial or domestic microwave applications. By contrast, it is the normal state of affairs in microwave hyperthermia and high-frequency industrial applications. The problem was examined in a number of studies [76], [77], [75], [78] which showed that the induced fields were considerably weaker than in the far-field zone. The SAR is given by the following empirical expression in this case:

$$(SAR)_n = \frac{(SAR)_f}{\left[1 + \left(\dfrac{A_V}{\Delta_V}\right)^2\right]\left[1 + \left(\dfrac{A_H}{\Delta_H}\right)^2\right]} \qquad (1.31)$$

where SAR_f and SAR_n are the specific absorption rates in the far and near fields, Δ_V and Δ_H are the spatial half-periods of the vertical and horizontal variation of the sinusoid that best approximates the electric field in air in the position occupied by the exposed body, and A_V and A_H are dimensional constants (in meters) evaluated at between 10 and 915 MHz (Fig. 1.7). It is clear that SAR_n is always less than SAR_f, which is the asymptotic value obtained when Δ_H and Δ_V become inifnite [77]. The following references may be consulted on this subject: [173], [44], [304], [204], [140], [76], [219], [236], [223], [260], [16], [2], [22], [257], [259], [220], [285], [405], [423], [379], [224].

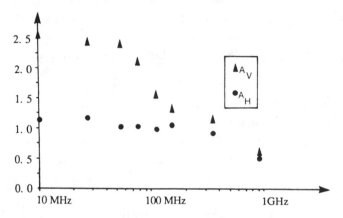

Figure 1.7 The dimensional constants A_V and A_H [77].

1.4.6 Main results

A decade after the development of the first numerical simulation programs, it may be considered that the fundamentals of the interaction of electromagnetic radiation with the human body are now well established. The interaction is found to be far more complex than was suggested by the simple models developed in the seventies. The analytic solution of Maxwell's equations for simple geometric models is therefore no longer acceptable. The numerical models now used, for which matrix inversion is no longer necessary, disclose fine details because the body is subdivided into hundreds of thousands of cells. In the future, these models will have to take into account more closely the process of thermal regulation and the role of biological mechanisms.

The extrapolation to the human body of experimental results obtained with laboratory animals is unsatisfactory [2],

[3], contrary to the opinions expressed in [161], [162]. The difficulty lies in the difference between the metabolic and thermal-regulation systems, as well as in the frequency shift of resonance effects and the extreme difficulty in defining and detecting thresholds.

The different techniques used to study the interaction must be regarded as complementary. They must always be compared and mutually validated. For example, a numerical model of a simple shape must be compared for calibration with an analytic study of the same shape; measurements performed on tissue or phantom material must be compared with results obtained on experimental animals, which, in turn, must be compared with numerical modeling data for the animal and its environment [60].

The specific absorption rate constitutes a valuable dosimetric unit that facilitates the interpretation of correlations between results obtained under different experimental conditions. To this day, the most significant result is the confirmation of resonant absorption phenomena that are functions of frequency and polarization. The specific absorption rate at resonance t is greater by a factor on the order of 3.3 than the average value [320]. Figure 1.8 shows the SAR as a function of frequency for the human body and for mice. It is clear that resonances occur at 70 and 700 MHz, respectively. In general, resonance occurs when the largest linear dimension of the body is parallel to the electric field and is on the order of 0.36λ–0.4λ. The effect is less pronounced in the transverse direction, and occurs when the linear dimension is on the order of λ. Energy absorption can never be a maximum for the human body in the microwave region [320], [151]. The resonance frequency is given by the empirical Gandhi equation

$$f_r \simeq 1.14 \times 10^8 \frac{1}{L} \qquad (1.34)$$

where L is the height of the body. For a man 180 cm tall, $f_r = 63$ MHz. For smaller parts of the body, resonances occur at higher frequencies. For example, for the head,

$$f_r \simeq 0.75 \times 10^8 \frac{1}{D} \tag{1.35}$$

where D is the mean diameter. This gives a resonance frequency of approximately 400–500 MHz for man.

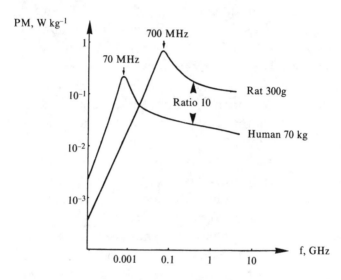

Figure 1.8 Average SAR for man and mouse in a power flux of $1\,\text{mW}\,\text{cm}^{-2}$ [53].

The specific absorption rate can be calculated approximately from the following empirical formulas for an incident flux of $1\,\text{mW}\,\text{cm}^{-2}$:

$$0.5\,f_r < f < f_r: \qquad \text{SAR} \simeq 5.2 \frac{L^2}{M} \left(\frac{f}{f_r}\right)^{2.75} \tag{1.36}$$

$$f_r < f < 1.6\,\delta_r f_r: \qquad \text{SAR} \simeq 5.95 \times 10^5 \frac{L}{Mf} \quad (\text{in W kg}^{-1}) \tag{1.37}$$

in which M is the mass of the body and

$$\delta_r \overset{\text{d}}{=} 4.8 \sqrt{\frac{10L^3}{M}} \tag{1.38}$$

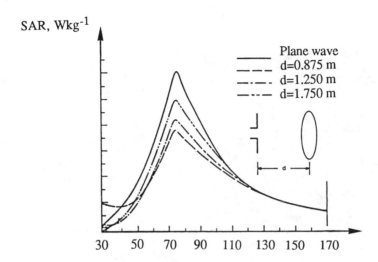

Figure 1.9 Average SAR for an ellipsoidal model ($a = 0.875\,\text{m}$, $a/b = 6.34$) exposed to the near field of an electric dipole parallel to its long axis [224].

The presence of metal pieces or other bodies in the surroundings may increase the SAR by a factor of up to 5–20 [320].

The interaction in the near-field zone is more difficult to model. The internal fields and the SAR are then much lower than in the far field. Figure 1.9 shows the power absorbed by an ellipsoid as a function of frequency after normalization relative to the far zone.

Internal hotspots may correspond to SARs five to ten times greater than the average [320]. For example, in a bird, the hotspots are located in the head, the neck, and at the tip of the wings (i.e., in parts with minimum cross sectional area) [91]. The situation is no different for the human body in which the hotspots appear in the vicinity of the wrists, the ankles, and the neck [179], [367]. For an average SAR of $50\,\text{m W kg}^{-1}$, obtained for an incident flux of $1\,\text{mW cm}^{-2}$ at $450\,\text{MHz}$, the maximum absorption at the wrists may reach $650\,\text{mW kg}^{-1}$ [247]. Hotspots were also demonstrated in the hypothalamus [367] and in the vitreous body of the eye, as shown in Figure 1.10 at $1\,\text{GHz}$ [60].

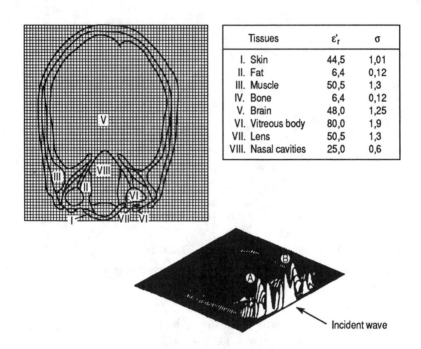

Tissues	ε'_r	σ
I. Skin	44,5	1,01
II. Fat	6,4	0,12
III. Muscle	50,5	1,3
IV. Bone	6,4	0,12
V. Brain	48,0	1,25
VI. Vitreous body	80,0	1,9
VII. Lens	50,5	1,3
VIII. Nasal cavities	25,0	0,6

Figure 1.10 SAR in a model of the head [60].

References

1. AARON M., GRANT E., YOUNG S. The dielectric properties of some amino-
acids, peptides, and proteins at
decimetre wavelengths. In : Molecular
relaxation processes. Proceedings of
the Chemical Society Symposium.
Aberystwyth, 7-9 July 1965. London,
Academic Press, 1966, pp. 77-82.

2. ADAIR E., SPIERS D., STOLWIJK J.,
WENGER C. Technical note : on changes
in evaporative heat loss that result
from exposure to nonionizing electromagnetic radiation. *J. Microw. Power*,
18, n° 2, 1983, pp. 209-211.

3. ADAIR E., WENGER C., SPIERS D. Technical note : beyond allometry. *J.
Microw. Power*, 19, n° 2, 1984, pp.
145-148.

4. ADEY W. Frequency and power windowing
in tissue interactions with weak
electromagnetic fields. *Proc. IEEE*,
68, 1980, pp. 119-125.

5. ADEY W. Tissue interaction with non-
ionizing electromagnetic fields.
Physiol. Rev., 61, 1981, pp. 453-514.

6. ADEY W., BAWIN S. Brain interactions
with weak electric and magnetic
fields. *NRP Bull.*, 15, 1977, pp. 1-29.

7. AGIN D. Some comments on the Hodgkin-
Huxley equations. *J. Theor. Biol.*,
n° 5, 1963, pp. 161-170.

8. AHMED N., SMITH C., CALDERWOOD J.,
FROLICH H. Dielectric properties and
measurement techniques for biomolecules. In : Dielectric materials,
measurements and applications. Cambridge, 1975, pp. 44-47.

9. ALLEN S. Comparison of theoretical
and experimental absorption of radio-
frequency power. Report SAM-75-52.
Salt Lake City, Univ. Utah, 1975.

10. ALLEN S. Measurement of power absorption by human phantoms immersed in
radiofrequency fields. In : TYLER P.
Biologic effects of non-ionizing
radiation. Proceedings of a conference. *Ann. New York Acad. Sci.*, 247,
1975, pp. 494-498.

11. ALLIS J., BLACKMAN C., FROMME M.,
BENANE S. Measurement of microwave
radiation absorbed by biological systems, analysis of heating and cooling
data. *Radio Sci.*, 12, n° 6 (S), 1977,
pp. 1-8.

12. ALLIS J., FROMME M. Activity of mem-
brane-bound enzymes exposed to amplitude-modulated 2450 MHz radiation.
In : Biological effects of electromagnetic waves. International symposium, Airlie, 30 Oct.-4 Nov. 1977,
Abstracts of scientific papers.

13. ALLIS J., SINHA B. Fluorescence depo-
larisation studies of red cell membrane fluidity : the effect of exposure to 1 GHz microwave radiation.
Bioelectromagn., 2, 1981, pp. 13-22.

14. ALMASSY G., MISIK M. Determination
of the properties of the bound water
in biological samples by microwave
methods. In : Proceedings of the IMPI
Symposium, Loughborough, 1973, pp.
1-2. (Paper 1 B3).

15. ALMERS W. Gating currents and charge
movements in excitable membranes.
Rev. Physiol. Biochem. Pharmacol.,
82, 1978, pp. 96-190.

16. AMEMIYA Y., UEBAYASHI S. The distribution of absorbed power inside a
sphere simulating human head in the
near field of a λ/2 dipole antenna.
*Trans. Inst. Electron. Commun. Eng.
Jpn.*, J66/B, Sept. 1983, pp. 1115-
1122.

17. ANDERSEN J., BALLING P. Admittance
and radiation efficiency of the
human body in the resonance region.
Proc. IEEE, 60, n° 7, 1972, pp. 900-
901.

18. ANNE A. Scattering and absorption of
microwaves of dissipative dielectric
objects : the biological significance and hazard to mankind. Thesis,
Ph. D., Philadelphia, Univ. Pennsylvania, doc. 55105, ASTIA n° 408997,
1963.

19. ANNE A., SAITO M., SALATI O., SCHWAN H. Relative microwave absorption cross sections of biological significance. In : PEYTON M. Proceedings of the 4th annual Tri-Service conference on biological effects of microwave-radiating equipments, biological effects of microwave radiations, 1961. New York, Plenum Press, 1961, RADC-TR-60-180, pp. 153-176.

20. ANNE A., SAITO M., SALATI O., SCHWAN H. Penetration and thermal dissipation of microwaves in tissues. Technical report RADC-TDR-62-244. Philadelphia, Univ. Pennsylvania, Contract AF3-(602)-2344, ASTIA n° 284981, 1962.

21. ARBER S. The effect of microwave radiation on passive membrane properties of snail neurons. J. Microw. Power, 16, n° 1, 1981, pp. 15-20.

22. ARBER S. The effect of microwaves on dimensions and nucleic-acid concentration in snail neurons. J. Microw. Power, 16, n° 1, 1981, pp. 21-23.

23. ARUNA R., BEHARI J. Dielectric loss in biogenic steroids at microwave frequencies. IEEE Trans. Microw. Theory Tech., MTT-29, n° 11, 1981, pp. 1209-1213.

24. ATHEY T., STUCHLY M., STUCHLY S. Measurement of radio-frequency permittivity of biological tissues with an open-ended coaxial line : part I. IEEE Trans. Microw. Theory Tech., MTT-30, n° 1, 1982, pp. 82-86.

25. BAILLIE H., HEATON A., PAL D. The dissipation of microwaves as heat in the eye. In : CLEARY S. Biological effects and health implications of microwave radiation. Symposium proceeding, Commonwealth Univ., Richmond, 17-19 Sept. 1969, US Dep. Health, Educ. Welfare, BRH, DBE 70-2, PB 193 898, Library of Congress card n° 76-607340, June 1970, pp. 85-89.

26. BAKER V., WIANT D., TABOADA O. Electromagnetic energy on biological material. In : Symposium on physiological and pathological effects of microwaves. Rochester, Mayo Clinic, Sept. 1955, pp. 23-29.

27. BARANSKI S., SZMIGIELSKI S., MONETA J. Effects of microwave radiation in vitro on cell-membrane permeability. In : CZERSKI P., OSTROWSKI K., SHORE M., SILVERMAN C., SÜSS M., WALDESKOG B. Biological effects and health hazards of microwave radiation. Proceedings of an international symposium, Warszawa, 15-18 Oct. 1973. Warszawa, Polish Medical Publishers, 1974.

28. BARBER P. Numerical study of electromagnetic power deposition in biological-tissue bodies. In : JOHNSON C., SHORE M. Biological effects of electromagnetic waves. Selected papers of the USNC/URSI annual meeting, Boulder, 20-23 Oct. 1975, US Dep. Health, Educ., Welfare, FDA-77-8011. Washington, US Government Printing Office, Vol. 2, 1976, pp. 119-134.

29. BARBER P. Electromagnetic power absorption in prolate spheroidal models of man and animals at resonance. IEEE Trans. Biomed. Eng., BME-24, 1977, pp. 513-521.

30. BARBER P. Resonance electromagnetic absorption by nonspherical dielectric objects. IEEE Trans. Microw. Theory Tech., MTT-25, n° 5, 1977, pp. 373-381.

31. BARBER P. Scattering and absorption efficiencies for nonspherical dielectrical objects - Biological models. IEEE Trans. Biomed. Eng., BME-25, 1978, pp. 155-159.

32. BARBER P., GANDHI O., HAGMANN M., CHATTERJEE I. Electromagnetic absorption in a multilayered model of man. IEEE Trans. Biomed. Eng., BME-26, 1979, pp. 400-405.

33. BARDATI F. Time-dependent microwave heating and surface cooling of simulated living tissues. IEEE Trans. Microw. Theory Tech., MTT-29, n° 8, 1981, pp. 825-828.

34. BARDATI F., GEROSA G., LAMPARIELLO P. Temperature distribution in simulated living tissues irradiated electromagnetically. Alta Frequenza, XLIX, marzo-aprile 1980, pp. 61-67.

35. BARNES F., HU C. Model for some nonthermal effects of radio and microwave fields on biological membranes. IEEE Trans. Microw. Theory Tech., MTT-25, n° 9, 1977, pp. 742-746.

36. BARSOUM Y., PICKARD W. The vacuolar potential of characean cells subjected to electromagnetic radiation in the range 200-8200 MHz. Bioelectromagn., 3, 1982, pp. 393-400.

37. BASSEN H., CHEUNG A. Comments on "Experimental and theoretical studies on electromagnetic fields induced inside finite biological bodies" by GURU B. and CHEN K. IEEE Trans. Microw. Theory Tech., MTT-25, n° 7, 1977, pp. 623-624.

38. BASSETT H., ECKER H., JOHNSON R., SHEPPARD A. New techniques for implementing microwave biological-exposure systems. *IEEE Trans. Microw. Theory Tech.*, MTT-19, n° 2, 1971, pp. 197-205.

39. BAWIN S., ADEY W. Interactions between nervous tissues and weak environmental electric fields. In : JOHNSON C., SHORE M. Biological effects of electromagnetic waves. Selected papers of the USNC/URSI annual meeting, Boulder, 20-23 Oct. 1975, US Dep. Health, Educ., Welfare, FDA-77-8010. Washington, US Government Printing Office, Vol. 1, 1976, pp. 323-330.

40. BAWIN S., ADEY W. Sensitivity of calcium binding in cerebral tissue to weak environmental electric fields oscillating at low frequency. In : Proceedings of the National Academy of Science, 73, 1976, pp. 1999-2003.

41. BAWIN S., KACZMAREK L., ADEY W. Effects of modulated VHF fields on the central nervous system. In : TYLER P. Biologic effects of nonionizing radiation. Proceedings of a conference. *Ann. New York Acad. Sci.*, 247, 1975, pp. 74-81.

42. BAWIN S., SHEPPARD A., ADEY W. Models of long-range order in cerebral macromolecules : effects of ELF, VHF and UHF fields in calcium binding. In : Biological effects of electromagnetic waves. International symposium, Airlie, 30 Oct.-4 Nov. 1977, Abstracts of Scientific papers.

43. BEISCHER D. Microwave reflection and diffraction by man. In : CZERSKI P., OSTROWSKI K., SHORE M., SILVERMAN C., SÜSS M., WALDESKOG B. Biological effects and health hazards of microwave radiation. Proceedings of an international symposium, Warszawa, 15-18 Oct. 1973. Warszawa, Polish Medical Publishers, 1974.

44. BEISCHER D., RENO V. Microwave energy distribution measurements in proximity to man and their practical application. In : TYLER P. Biologic effects of non-ionizing radiation. Proceedings of a conference. *Ann. New York Acad. Sci.*, 247, 1975, pp. 473-480.

45. BELTRAME F., CHIABRERA A., GRATTAROLA M. Electromagnetic control of cell function. *Alta Frequenza*, 49, 1980, pp. 101-114.

46. BERKOWITZ G., BARNES F. The effects of nonlinear membrane capacity on the interaction of microwave and radio frequencies with biological materials. *IEEE Trans. Microw. Theory Tech.*, MTT-27, n° 2, 1979, pp. 204-207.

47. BERNARDI P., GIANNINI F., SORRENTINO R. Effects of the surroundings on electromagnetic-power absorption in layered-tissue media. *IEEE Trans. Microw. Theory Tech.*, MTT-24, n° 9, 1976, pp. 621-625.

48. BERTEAUD A. The effect of microwaves on biological media. *Electrón. Fís. Apl.*, 16, n° 3, 1973, pp. 517-522.

49. BERTEAUD A. Les effets biologiques des micro-ondes : une confirmation. *La Recherche*, 9, n° 85, 1978, pp. 65-66.

50. BERTEAUD A. Electro-magnetic waves and living systems. In : Comptes-rendus du Symposium international sur les Applications énergétiques des micro-ondes, IMPI-CFE. 14, Monaco, 1979, pp. 29-34.

51. BERTEAUD A. Interaction des micro-ondes avec les milieux vivants. *Radioprot.*, n° 1, suppl., 1984, pp. 7-13.

52. BERTEAUD A., DARDALHON M. Biological effects of microwaves. In : 7th European Microwave Conference, København, 1977.

53. BERTEAUD A., DARDALHON M. Effets biophysiques et biologiques associés à l'utilisation des micro-ondes. 1. Interaction des micro-ondes avec les milieux vivants. Aspects biophysiques. *Econ. Progrès Electr.*, n° 13, jan.-fév. 1985, pp. 21-23.

54. BERTEAUD A., JULLIEN H., DAVID R., HER C. Organic system hydration studied by microwave dielectric absorption. *J. Microw. Power*, 12, n° 3, 1977, pp. 231-239.

55. BIGU DEL BLANCO J., ROMERO-SIERRA C., TANNER J. Colour thermography : a powerful technique in the evaluation of microwave field radiation patterns in biological systems. It use as a microwave energy density monitor. In : Proceedings of the IMPI Symposium, Milwaukee, 1974, pp. 1-4. (Paper A3-3).

56. BLACKMAN C., BENANE S., WEIL C., ALI J. Effects of nonionizing electromagnetic radiation on single-cell biologic systems. In : TYLER P. Biologic effects of nonionizing radiation. Proceedings of a conference. *Ann. New York Acad. Sci.*, 247, 1975, pp. 352-366.

57. BLACKMAN C., ELDER J., WEIL C., BENANE S., EICHENGER D. Two parameters affecting radiation induced calcium efflux from brain tissue. In : Biological effects of electromagnetic waves. International symposium, Airlie, 30 Oct.-4 Nov. 1977, Abstracts of scientific papers.

58. BONE S., GASCOYNE P., LEWIS J., PETHIG R. Dielectric measurements on hydrated biomacromolecules. In : Dielectric Materials, Measurements and Applications. Cambridge, 1975, pp. 48-51.

59. BORUP D., GANDHI O. Fast-Fourier-transform method for calculation of SAR distributions in finely discretized inhomogeneous models of biological bodies. *IEEE Trans. Microw. Theory Tech.*, MTT-32, n° 4, 1984, pp. 355-360.

60. BORUP D., GANDHI O. Calculation of high-resolution SAR distributions in biological bodies using FFT algorithm and conjugate-gradient method. *IEEE Trans. Microw. Theory Tech.*, MTT-33, n° 5, 1985, pp. 417-419.

61. BOWMAN H. Heat transfer and thermal dosimetry. *J. Microw. Power*, 16, n° 2, 1981, pp. 121-133.

62. BRADY M., SYMONS S., STUCHLY S. Dielectric behavior of selected animal tissues in vitro at frequencies from 2 to 4 GHz. *IEEE Trans. Biomed. Eng.*, BME-28, n° 3, 1981, pp. 305-307.

63. BRADY A., WOODBURY J. The sodium-potassium hypothesis as the basis of electrical activity in frog ventricle. *J. Physiol.*, n° 154, 1960, pp. 385-407.

64. BREKHOVSKIKH L. Waves in layered media. New York, Academic Press, 1960, pp. 44-61.

65. BRODWIN M., TAFLOVE A., MATZ J. A passive electrodeless method for determining the interior field of biological materials. *IEEE Trans. Microw. Theory Tech.*, MTT-24, n° 8, 1976, pp. 514-521.

66. BROWN P., LARSEN L. Differing effects of pulsed and CW microwave energy upon nerve function as detected by birefringence measurement. *IEEE Trans. Microw. Theory Tech.*, MTT-28, n° 10, 1980, pp. 1126-1133.

67. BURDETTE E. Electromagnetic and acoustic properties of tissues. In : NUSSBAUM G. Physical aspects of hyperthermia. American Association of Physicists in Medicine, 3-7 Aug. 1982, pp. 105-150.

68. BURDETTE E., CAIN F., SEALS J. In vivo probe measurement technique for determining dielectric properties at VHF through microwave frequencies. *IEEE Trans. Microw. Theory Tech.*, MTT-28, n° 4, 1980, pp. 414-427.

69. BURDETTE E., FRIEDERICH P., SEAMAN R., LARSEN L. In situ permittivity of canine brain : regional variations and postmortem changes. *IEEE Trans. Microw. Theory Tech.*, MTT-34, n° 1, 1986, pp. 38-50.

70. CAIN C. A theoretical basis for microwave and RF field effects on excitable cellular membranes. *IEEE Trans. Theory Tech.*, MTT-28, n° 2, 1980, pp. 142-147.

71. CAIN C. Correction to "A theoretical basis for microwave and RF field effects on excitable cellular membranes". *IEEE Trans. Microw. Theory Tech.*, MTT-29, n° 1, 1981, p. 74.

72. CARSTENSEN E. Passive electrical properties of micro-organisms. II. Resistance of the bacterial membrane. *Biophys. J.*, 7, 1967, pp. 493-503.

73. CHAMNESS A., SCHOLES H., SEXAUER S., FRAZER J. Metal ion content of specific areas of the rat brain after 1600 MHz radiofrequency irradiation. *J. Microw. Power*, 11, n° 4, 1976, pp. 333-338.

74. CHAN A., SIGELMANN R., GUY A., LEHMANN J. Calculation by the method of finite differences of the temperature distribution in layered tissues. *IEEE Trans. Biomed. Eng.*, BME-26, n° 4, 1979, pp. 244-250.

75. CHATTERJEE I., GANDHI O., HAGMANN M. Numerical and experimental results for near-field electromagnetic absorption in man. *IEEE Trans. Microw. Theory Tech.*, MTT-30, n° 11, 1982, pp. 2000-2005

76. CHATTERJEE I., HAGMANN M., GANDHI O. Electromagnetic-energy deposition in an inhomogeneous block model of man for near-field irradiation conditions. *IEEE Trans. Microw. Theory Tech.*, MTT-28, n° 12, 1980, pp. 1452-1459.

77. CHATTERJEE I., HAGMANN M., GANDHI O. An empirical relationship for electromagnetic energy absorption in man for near-field exposure conditions. *IEEE Trans. Microw. Theory Tech.*, MTT-29, n° 11, 1981, pp. 1235-1238.

78. CHATTERJEE I., HAGMANN M., GANDHI O. Corrections to "An empirical relationship for electromagnetic energy absorption in man for near-field exposure conditions". *IEEE Trans. Microw. Theory Tech.*, MTT-30, n° 5, 1982, p. 838.

79. CHEN K. Comments on "Numerical calculation of electromagnetic energy deposition for a realistic model of man". *IEEE Trans. Microw. Theory Tech.*, MTT-28, n° 9, 1980, p. 1034.

80. CHEN K., GURU B. Induced EM field and absorbed power density inside a human torso. *Proc. IEEE*, 64, n° 9, 1976, pp. 1450-1453.

81. CHEN K., GURU B. Internal EM field induced by VHF and UHF EM waves in a human torso (realistic model). In : Annual meeting of the International Union of Radio Science, URSI. Amherst, Univ. Massachusetts, 11-15 Oct. 1976.

82. CHEN K., GURU B. Induced EM fields inside human bodies irradiated by EM waves of up to 500 MHz. *J. Microw. Power*, 12, n° 2, 1977, pp. 173-183.

83. CHEN K., GURU B. Internal EM field and absorbed power density in human torsos induced by 1-500 MHz EM waves. *IEEE Trans. Microw. Theory Tech.*, MTT-25, n° 9, 1977, pp. 746-756.

84. CHEN K., LIVESAY D., GURU B. Induced current in and scattered field from a finite cylinder with arbitrary conductivity and permittivity. *IEEE Trans. Antennas Propag.*, AP-24, n° 5, 1976, pp. 330-336.

85. CHEN K., RUKSPOLLMUANG S., NYQUIST D. Measurement of induced electric fields in a phantom model of man. *Radio Sci.*, 17, n° 5 (S), 1982, pp. 49-59.

86. CHEN K., WANG H., LIANG Y. New methods for quantification of induced EM fields in finite biological bodies. In : Proceedings of the IMPI Symposium, Chicago, 1985, pp. 108-111.

87. CHEUNG A., KOOPMAN D. Experimental development of simulated biomaterials for dosimetry studies of hazardous microwave radiation. *IEEE Trans. Microw. Theory Tech.*, MTT-24, n° 10, 1976, pp. 669-673.

88. CHIABRERA A., GRATTAROLA M., VIVIANI R. Interaction between electromagnetic fields and cells : microelectrophoretic effect on ligands and surface receptors. *Bioelectromagn.*, 5, 1984, pp. 173-191.

89. CHOU C., CHEN G., GUY A., LUK K. Formulas for preparing phantom muscle tissue at various radiofrequencies. *Bioelectromagn.*, 5, 1984, pp. 435-441.

90. CHOU C., GUY A. Effects of electromagnetic fields on isolated nerve and muscle preparations. *IEEE Trans. Microw. Theory Tech.*, MTT-26, n° 3, 1978, pp. 141-147.

91. CHOU C., GUY A. Absorption of microwave energy by muscle models and by birds of differing mass and geometry. *J. Microw. Power*, 20, n° 2, 1985, pp. 75-84.

92. CHRISTMAN C., HO H., YARROW S. A microwave dosimetry system for measuring sampled integral-dose rate. *IEEE Trans. Microw. Theory Tech.*, MTT-22, n° 12, 1974, pp. 1267-1272.

93. COLE K. Ionic current measurement in the squid giant axon membrane. *J. Gen. Physiol.*, 44, 1960, pp. 123-167.

94. COLE K. An analysis of the membrane potential along a clamped squid axon. *Biophys. J.*, 1, 1961, pp. 401-417.

95. CONOVER D. Temperature distributions induced by 2450 MHz microwave radiation in a trilayered spherical phantom. *Health Phys.*, 27, n° 6, 1973, pp. 632-633.

96. CONOVER D., VETTER R., WEEKS W., ZIEMER P., LANDOLT R. Temperature distributions induced by 2450 MHz microwave radiation in a trilayered spherical phantom. *J. Microw. Power*, 9, n° 2, 1974, pp. 69-78.

97. COOK H. The dielectric behavior of human blood at microwave frequencies. *Nature*, 168, 1951, pp. 247-248.

98. COOK H. The dielectric behavior of some types of human tissues at microwave frequencies. *Br. J. Appl. Phys.*, 2, 1951, p. 295.

99. COOK H. A comparison of the dielectric behavior of pure water and human blood at microwave frequencies. *Br. J. Appl. Phys.*, 3, 1952, pp. 249-255.

100. COOK H. A physical investigation of heat production in human tissues when exposed to microwaves. *Br. J. Appl. Phys.*, 3, 1952, pp. 1-6.

101. COPE F. Superconductivity - a possible mechanism for non-thermal biological effects of microwaves. *J. Microw. Power*, 11, n° 3, 1976, pp. 267-270.

102. COPSON D. Athermic microwave power effects. In : Proceedings of the IMPI Symposium, Stanford, 1967, abstracts p. 42.

103. COPSON D. Informational bioelectromagnetics. Beaverton, Matrix Publishers, 1982, 746 p.

104. CZERSKI P. The development of biomedical approaches and concepts in radiofrequency radiation protection. *J. Microw. Power*, 21, n° 1, 1986, pp. 9-23.

105. DARDALHON M., AVERBECK D., BERTEAUD A. Determination of a thermal equivalent of millimeter microwaves in living cells. *J. Microw. Power*, 14, n° 4, 1979, pp. 307-312.

106. DARDANONI L., TORREGROSSA V., TAMBURELLO C., ZANFORLIN L., SPALLA M. Biological effects of millimeter waves at spectral singularities. In : Proceedings of the 3rd symposium on Electromagnetic compatibility. Wrocław, Wrocławskiej Wydawnictwo Politechniki, 1976, pp. 308-313.

107. DAVYDOV B. Rayonnement électromagnétique radiofréquence (microonde) ; règles et critères de définition de normes et de seuils de niveau de dose. (En russe). *Kosmič. Biol. Aviakosmič. Med.*, 19, 1985, pp. 8-21.

108. DAVYDOV B., TIHONČUK V., ANTIPOV V. Effets biologiques : normalisation et protection contre le rayonnement électromagnétique. (En russe). *Medicina*, Moskva, 1984.

109. DAWINKS A., NIGHTINGALE H., SOUTH G., SHEPPARD R., GRANT E. The role of water in microwave absorption by biological material with particular reference to microwave hazards. *Phys. Med. Biol.*, 24, 1979, pp. 1168-1176.

110. DEFORD J., GANDHI O., HAGMANN M. Moment-method solutions and SAR-calculations for inhomogeneous models of man with large number of cells. *IEEE Trans. Microw. Theory Tech.*, MTT-31, 1983, pp. 848-851.

111. DEMIDOVIČ B., MARON I. Eléments de calcul numérique. (Edition française). Izdatel'stvo Mir, Moskva, 1973, 677 p.

112. DEVJATKOV N. Influence of millimeter-band electromagnetic radiation on biological objects. (En russe). Rapports des sessions scientifiques de la division de physique générale et d'astronomie de l'Académie des Sciences de l'URSS, 17-18 Jan. 1973 et in *Usp. Fiz. Nauk*, 110, n° 3, 1973, pp. 453-454, traduction anglaise in *Sov. Phys. Usp.*, 16, n° 4, Jan.-Feb. 1974, pp. 568-569.

113. DE WAGTER C. Computer simulation predicting temperature distributions generated by microwave absorption in multilayered media. *J. Microw. Power*, 19, n° 2, 1984, pp. 97-105.

114. DODGE F. Ionic permeability changes underlying nerve excitation. American Association for the Advancement of Sciences. *Biophys. Physiol. Pharmacol. Actions*, 1961, pp. 113-143.

115. DURNEY C. Long-wavelength analysis of plane wave irradiation of a prolate spheroid model of man. *IEEE Trans. Microw. Theory Tech.*, MTT-23, n° 2, 1975, pp. 246-253.

116. DURNEY C. Electromagnetic dosimetry for models of humans and animals : a review of theoretical and numerical techniques. *Proc. IEEE*, 68, n° 1, 1980, pp. 33-40.

117. DURNEY C., ISKANDER M., MASSOUDI H., JOHNSON C. An empirical formula for broad-band SAR calculations of prolate spheroidal models of humans and animals. *IEEE Trans. Microw. Theory Tech.*, MTT-27, n° 8, 1979, pp. 758-763.

118. DURNEY C., JOHNSON C., BARBER P., MASSOUDI H., ISKANDER M., LORDS J., RYSER D., ALLEN S., MITCHELL J. Radiofrequency radiation dosimetry handbook. Salt Lake City, Univ. Utah, Dep. Electr. Eng. Bioeng., 2nd ed., 1978, USAF School Aerosp. Med., Brooks Air Force Base, report SAM-TR-78-22, 1978.

119. EDWARDS G., DAVIS C., SAFER J., SWICORD M. Microwave-field-driven acoustic modes in DNA. *Biophys. J.*, 47, 1985, pp. 799-807.

120. ELY T., GOLDMAN D., HEARON J., WILLIAMS R., CARPENTER H. Heating characteristics of laboratory animals exposed to ten-centimeter microwaves. Bethesda, Nat. Nav. Med. Cent., Nav. Med. Res. Inst., report NM001-056-13-02, 21 March 1947, Nav. Med. Res. Inst. Reports, 15, 1957, pp. 77-138. *IEEE Trans. Biomed. Eng.*, BME-11, n° 10, 1964, pp. 123-137

121. EMERY A. The numerical thermal simulation of the human body when absorbing non-ionizing microwave irradiation with emphasis on the effect of different sweat models. In : JOHNSON C., SHORE M. Biological effects of electromagnetic waves. Selected papers of the USNC/URSI annual meeting, Boulder, 20-23 Oct. 1975, US Dep. Health, Educ., Welfare, FDA-77-8011, Washington, US Government Printing Office, Vol. 2, 1976, pp. 96-118.

122. EMERY A., SHORT R., GUY A., KRANING K., LIN J. The numerical thermal simulation of the human body when undergoing exercise or nonionizing electromagnetic radiation. *J. Heat Transfer*, 48, 1976, pp. 284-291.

123. ENGLAND T., SHARPLES N. Dielectric properties of the human body in the microwave region of the spectrum. *Nature*, 163, 1949, pp. 487-488.

124. FIRKET H. La cellule vivante. Paris, P.U.F., n° 989, 1978, 127 p. (Coll. Que sais-je ?).

125. FITZHUGH R. Thresholds and plateaus in the Hodgkin-Huxley nerve equations. *J. Gen. Physiol.*, 43, 1960, pp. 867-896.

126. FLEMING J. Microwave radiation in relation to biological systems and neural activity. In : PEYTON M. Proceedings of the 4th annual Tri-Service conference on biological effects of microwave-radiating equipments, biological effects of microwave radiations, 1961. New York, Plenum Press, 1961, RADC-TR-60-180, p. 239.

127. FOSTER K., BARNES F., WACHTEL H., BOWMAN R., FRAZER J., CHALKER R. An exposure system for variable electromagnetic-field orientation electrophysiological studies. *IEEE Trans. Microw. Theory Tech.*, MTT-33, n° 8, 1985, pp. 674-680.

128. FOSTER K., BIDINGER J., CARPENTER D. The electrical resistivity of cytoplasm. *Biophys. J.*, 16, 1976, pp. 991-1001.

129. FOSTER K., KRITIKOS H., SCHWAN H. Effect of surface cooling and blood flow on the microwave heating of tissue. *IEEE Trans. Biomed. Eng.*, BME-25, n° 5, 1978, pp. 313-316.

130. FOSTER K., SCHEPPS J., SCHWAN H. Microwave dielectric relaxation in muscle : a second look. *Biophys. J.*, 29, 1980, pp. 271-282.

131. FOSTER K., SCHEPPS J., STOY R., SCHWAN H. Dielectric properties of brain tissue between 0.01 and 10 GHz. *Phys. Med. Biol.*, 24, n° 6, 1979, pp. 1177-1187.

132. FRANK-KAMENECKIJ D. Les effets de plasma dans les semi-conducteurs et l'effet biologique des ondes radio. (En russe). *Doklady Akad. Nauk*, 136, 1961, pp. 476-478.

133. FRICKE H. A mathematical treatment of the electric conductivity and capacity of disperse systems. *Physiol. Rev.*, 40, 1924, pp. 575-587.

134. FRICKE H., MORSE S. An experimental study of the electrical conductivity of disperse systems. *Physiol. Rev.*, 41, 1925, pp. 361-367.

135. FRIEND A., GARTNER S., FOSTER K., HOWE H. The effects of high-power microwave pulses on red blood cells and the relationship to transmembrane thermal gradients. *IEEE Trans. Microw. Theory Tech.*, MTT-29, n° 12, 1981, pp. 1271-1277.

136. FUJIWARA O., AMEMIYA Y. Microwave power absorption in a biological specimen inside a standing-wave irradiation waveguide. *IEEE Trans. Microw. Theory Tech.*, MTT-30, n° 11, 1982, pp. 2008-2012.

137. FUREDI A., OHAD I. Effects of high-frequency electric fields on the living cell. *Biochim. Biophys. Acta*, 79, 1964, p. 1.

138. FUREDI A., VALENTINE R. Factors involved in the orientation of microscopic particles in suspensions influenced by radio-frequency fields. *Biochim. Biophys. Acta*, 56, 1962, p. 33.

139. FURMANIAK A. Quantitative changes in potassium, sodium and calcium in the submaxillary salivary gland and blood serum of rats exposed to 2880 MHz microwave radiation. *Bioelectromagn.*, 4, 1983, pp. 55-62.

140. GAJDA G., STUCHLY M., STUCHLY S. Mapping of the near-field pattern in simulated biological tissues. *Electron. Lett.*, 15, n° 4, 1979, pp. 120-121.

141. GANDHI O. A model of measuring RF absorption of whole animals and bodies of prolate spheroidal shapes. In : Proceedings of the IMPI Symposium, Milwaukee, 1974, pp. 1-4. (Paper B2.2).

142. GANDHI O. Polarization and frequency effects on whole animal absorption of RF energy. *Proc. IEEE*, 62, 1974, pp. 1171-1175.

143. GANDHI O. Conditions of strongest electromagnetic power deposition in man and animals. *IEEE Trans. Microw. Theory Tech.*, MTT-23, n° 12, 1975, pp. 1021-1029.

144. GANDHI O. Frequency and orientation effect on whole animal absorption of electromagnetic waves. *IEEE Trans. Biomed. Eng.*, BME-22, 1975, p. 536.

145. GANDHI O. Strong dependence of whole animal absorption on polarization and frequency of radio-frequency energy. In : TYLER P. Biologic effects of non-ionizing radiation. Proceedings of a conference. *Ann. New York Acad. Sci.*, 247, 1975, pp. 532-538.

146. GANDHI O. Deposition of electromagnetic energy in animals and models of man. In : USNC/URSI annual meeting, Amherst, Oct. 1976.

147. GANDHI O. Electromagnetic power deposition in man and animals with and whithout ground and reflector effects. In : USNC/URSI annual meeting, Amherst, Oct. 1976.

148. GANDHI O. Dosimetry - The absorption properties of man and experimental animals. *Bull. New York Acad. Med.*, 55, 1979, pp. 999-1020.

149. GANDHI O. State of the knowledge for electromagnetic absorbed dose in man and animals. *Proc. IEEE*, 68, n° 1, 1980, pp. 24-32.

150. GANDHI O. Biological effects and medical applications of RF electromagnetic fields. *IEEE Trans. Microw. Theory Tech.*, MTT-30, n° 11, 1982, pp. 1831-1847.

151. GANDHI O. Electromagnetic absorption in an inhomogeneous model of man for realistic exposure conditions. *Bioelectromagn.*, 3, 1982, pp. 81-90.

152. GANDHI O. Some basic properties of biological tissues for potential biomedical applications of millimeter waves. *J. Microw. Power*, 18, n° 3, 1983, pp. 295-304.

153. GANDHI O., HAGMANN M., D'ANDREA J. Part-body and multibody effects on absorption of radiofrequency electromagnetic energy by animals and by models of man. *Radio Sci.*, 14, n° 6 (S), 1979, pp. 15-22.

154. GANDHI O., HAGMANN M., HILL D., PARTLOW L., BUSH L. Millimeter-wave absorption spectra of biological samples. *Bioelectromagn.*, 1, 1980, pp. 285-298.

155. GANDHI O., HUNT E., D'ANDREA J. Deposition of electromagnetic energy in animals and in models of man with and without ground and reflector effects. *Radio Sci.*, 12, n° 6 (S), 1977, pp. 39-47.

156. GANDHI O., SEDIGH K., BECK G., HUNT E. Distribution of electromagnetic energy deposition in models of man with frequencies near resonance. In : JOHNSON C., SHORE M. Biological effects of electromagnetic waves. Selected papers of the USNC/URSI annual meeting, Boulder, 20-23 Oct. 1975, US Dep. Health, Educ. Welfare, FDA-77-8011, Washington, US Government Printing Office, Vol. 2, 1976, pp. 44-67.

157. GEORGE E., JOHNSON E. Solutions of the Hodgkin-Huxley equations for squid axon treated with tetraethylammonium and in potassium-rich media. *Aust. J. Exp. Biol. Med. Sci.*, 39, 1961, pp. 275-293.

158. GOKHALE A., BRUNKARD K., PICKARD W. Vacuolar hyperpolarizing offsets in characean cells exposed to mono- and bichromatic CW and to squarewave-modulated electromagnetic radiation in the band 200-1000 MHz. *Bioelectromagn.*, 5, 1984, pp. 357-360.

159. GOKHALE A., PICKARD W., BRUNKARD K. Low-power 2.45-GHz microwave radiation affects neither the vacuolar potential nor the low frequency excess noise in single cells of Characean algae. *J. Microw. Power*, 20, n° 1, 1985, pp. 43-46.

160. GOODMAN R., BASSET A., HENDERSON A. Pulsing electromagnetic fields induce cellular transcription. *Science*, 220, 1983, pp. 1283-1285.

161. GORDON C. Effects of ambient temperature and exposure to 2450-MHz microwave radiation on evaporative heat loss in the mouse. *J. Microw. Power*, 17, n° 2, 1982, pp. 145-150.

162. GORDON C. Note : further evidence of an inverse relation between mammalian body mass and sensitivity to radio-frequency electromagnetic radiation. *J. Microw. Power*, 18, n° 4, 1983, pp. 377-383.

163. GORDON C., SCHWAN H. Statement on microwave intensities. In : CZERSKI P., OSTROWSKI K., SHORE M., SILVERMAN C., SÜSS M., WALDESKOG B. Biological effects and health hazards of microwave radiation. Proceedings of an international symposium, Warszawa, 15-18 Oct. 1973. Warszawa, Polish Medical Publishers, 1974, pp. 318-319.

164. GRANT E. Dielectric studies on biological materials near 1 GHz : biophysical and medical implications. In : Proceedings of the IMPI Symposium, Loughborough, 1973, p. 1. (Paper 1.B2).

165. GRANT E., KEEFE S., TAKASHIMA S. The dielectric behaviour of aqueous solutions of bovine serum albumin from radio wave to microwave frequencies. *J. Phys. Chem.*, 72, n° 13, 1968, pp. 4373-4380.

166. GRANT E., SHEPPARD R., SOUTH G. Importance of bound water studies in the determination of energy absorption by biological tissue. In : 5th European Microwave Conference, Ed. Sevenoaks, 1975, pp. 366-370.

167. GRANT E., SHEPPARD R., SOUTH G. Dielectric behavior of biological molecules in solution. Monographs on Physical Biochemistry, Oxford University Press, 1978, pp. 1-237.

168. GRODSKY I. Possible physical substrates for the interaction of electromagnetid fields with biologic membranes. In : TYLER P. Biologic effects of non-ionizing radiation. Proceedings of a conference. *Ann. New York Acad. Sci.*, 247, 1975, pp. 117-124.

169. GURU B. An experimental and theoretical study of the interaction of electromagnetic fields with arbitrary shaped biological bodies. Thesis, Ph. D., Michigan State Univ., 1976.

170. GURU B., CHEN K. Experimental and theoretical studies on electromagnetic fields induced inside finite biological bodies. *IEEE Trans. Microw. Energy Tech.*, MET-24, n° 7, 1976, pp. 433-440.

171. GUY A. Analyses of electromagnetic fields induced in biological tissues by thermographic studies on equivalent phantom models. *IEEE Trans. Microw. Theory Tech.*, MTT-19, n° 2, 1971, pp. 205-214.

172. GUY A. Electromagnetic fields and relative heating patterns due to a rectangular aperture source in direct contact with bilayered biological tissue. *IEEE Trans. Microw. Theory Tech.*, MTT-19, n° 2, 1971, pp. 214-223.

173. GUY A. Measurement of absorbed power patterns in the head and eyes of rabbits exposed to typical microwave sources. In : C.P.E.M. 74 Digest : Conference on Precision Electromagnetic Measurements, 1974, pp. 255-257.

174. GUY A. Quantitation of induced electromagnetic fields in tissue and associated biological effects. In : CZERSKI P., OSTROWSKI K., SHORE M., SILVERMAN C., SÜSS M., WALDESKOG B. Biological effects and health hazards of microwave radiation. Proceedings of an international symposium. Warszawa, 15-18 Oct. 1973. Warszawa, Polish Medical Publishers, 1974, pp. 203-216.

175. GUY A. Correspondence on Dr. Justesen's prescriptive grammar for the radiobiology of non-ionizing radiation. *J. Microw. Power*, 10, n° 4, 1975, pp. 358-359.

176. GUY A. RF cell culture irradiation system with controlled temperature and field strength. Research report, US Dep. Health, Educ., Welfare. Publication n° 77-182, 1977, 46 p.

177. GUY A. Miniature anechoic chamber for chronic exposure of small animals to plane-wave microwave fields. *J. Microw. Power*, 14, n° 4, 1979, pp. 327-338.

178. GUY A. History of biological effects and medical applications of microwave energy. *IEEE Trans. Microw. Theory Tech.*, MTT-32, n° 9, 1984, pp. 1182-1200.

179. GUY A., CHOU C., NEUHAUS B. Average SAR and SAR distributions in man exposed to 450-MHz radiofrequency radiation. *IEEE Trans. Microw. Theory Tech.*, MTT-32, n° 8, 1984, pp. 752-763.

180. GUY A., KORBEL S. Dosimetry studies on a UHF cavity exposure chamber for rodents. In : Proceedings of the IMPI Symposium, Ottawa, 1972, pp. 180-193. (Paper 10.4.).

181. GUY A, LEHMANN J. Quantitative methods for analyzing electromagnetic fields and the associated heating patterns in human tissues. In : Proceedings of the IMPI Symposium, Edmonton, 1969, pp. 45-53. (Paper C.3).

182. GUY A., WALLACE J., Mac DOUGALL J. Circularly-polarized 2450-MHz waveguide system for chronic exposure of small animals to microwaves. *Radio Sci.*, 14, n° 6 (S), 1979, pp. 63-74.

183. GUY A., WEBB M., EMERY A., CHOU C. Measurement of power distribution at resonant and non-resonant frequencies in experimental animals and models. US Air Force School Aerosp. Med., Aerosp. Med. Div., Brooks Air Force Base, Contract F41609-76-C-0032, report, 1977.

184. GUY A., WEBB M., Mac DOUGALL J. A new technique for measuring power deposition patterns in phantoms exposed to EM fields of arbitrary polarization : example, the microwave oven. In : Proceedings of the IMPI Symposium, Waterloo (Can.), 1975, pp. 36-47. (Paper 3.3.).

185. GUY A., WEBB M., Mac DOUGALL J. RF radiation absorption pattern : human and animal modeling data. US Dep. Health, Educ., Welfare, NIOSH, 77-183, Sept. 1977.

186. GUY A., WEBB M., SORENSEN C. Determination of power absorption in man exposed to high-frequency electromagnetic fields by thermographic measurements on scale models. *IEEE Trans. Biomed. Eng.*, BME-23, 1976, pp. 361-371.

187. HAGMANN M., CHATTERJEE I., GANDHI O. Dependence of electromagnetic deposition upon angle of incidence for an inhomogeneous block model of man under plane-wave irradiation. *IEEE Trans. Microw. Theory Tech.*, MTT-29, n° 3, 1981, pp. 252-255.

188. HAGMANN M., GANDHI O. Numerical calculation of electromagnetic energy deposition in man with ground and reflector effects. *Radio Sci.*, 14, n° 6 (S), 1979, pp. 23-29.

189. HAGMANN M., GANDHI O., D'ANDREA J., CHATTERJEE I. Head resonance : numerical solutions and experimental results. *IEEE Trans. Microw. Theory Tech.*, MTT-27, n° 9, 1979, pp. 809-813.

190. HAGMANN M., GANDHI O., DURNEY C. Upper bound on cell size for moment-method solutions. *IEEE Trans. Microw. Theory Tech.*, MTT-25, n° 10, 1977, pp. 831-832.

191. HAGMANN M., GANDHI O., DURNEY C. Improvement of convergence in moment-method solutions by the use of interpolants. *IEEE Trans. Microw. Theory Tech.*, MTT-26, n° 11, 1978, pp. 904-908.

192. HAGMANN M., GANDHI O., DURNEY C. Numerical calculation of electromagnetic energy deposition for a realistic model of man. *IEEE Trans. Microw. Theory Tech.*, MTT-27, n° 9, 1979, pp. 804-809.

193. HAGMANN M., LEVIN R. Criteria for accurate usage of block models. In : Proceedings of the IMPI Symposium, Chicago, 1985, pp. 117-119.

194. HAND J. Microwave heating patterns in simple tissue models. *Phys. Med. Biol.*, 22, n° 5, 1977, p. 981.

195. HARRISON G. Mac CULLOCH D., BALCER-KUBICZEK E., ROBINSON J. Far-field 2.45 GHz irradiation system for cellular monolayers in vitro. *J. Microw. Power*, 20, n° 3, 1985, pp. 145-151.

196. HASTED J. Biomolecules and tissue. In : HASTED J. Aqueous dielectrics. London, Chapman and Hall, 1973, pp. 204-233.

197. HELLER J. Chemical aspects of certain biological effects. In : Proceedings of the IMPI Symposium, Boston, 1968, abstracts, p. 2. (Paper A.2.).

198. HELLER J. Cellular effects of microwave radiation. In : CLEARY S. Biological effects and health implications of microwave radiation. Symposium proceedings, Richmond, Commonwealth Univ., 17-19 Sept. 1969. US Dep. Health, Educ., Welfare, BRH, DBE 70-2, PB 193898, Library of Congress card n° 76-607340, June 1970, pp. 116-121.

199. HELLER J., MICKEY G. Non thermal effects of radio-frequency in biological systems. *Dig. Int. Conf. Electron.*, 21, 1961, p. 2.

200. HERRICK J. Pearl-chain formation. In : Proceedings of the 2nd annual Tri-Service conference on biological effects of microwave energy. Charlottesville, Univ. Virginia, 1958, RADC-1-TR-58-54, ASTIA 131477, pp. 83-93.

201. HILL D. Waveguide technique for the calibration of miniature implantable electric-field probes for use in microwave-bioeffects studies. *IEEE Trans. Microw. Theory Tech.*, MTT-30, n° 1, 1982, pp. 92-99.

202. HILL D. Human whole-body radiofrequency absorption studies using a TEM-cell exposure system. *IEEE Trans. Microw. Theory Tech.*, MTT-30, n° 11, 1982, pp. 1847-1854.

203. HILLE B. Ionic channels in excitable membranes. *Biophys. J.*, _22_, 1978, pp. 283-294.

204. HIZAL A., BAYKAL Y. Heat potential distribution in an inhomogeneous spherical model of a cranial structure exposed to microwaves due to loop or dipole antennas. *IEEE Trans. Microw. Theory Tech.*, MTT-_26_, n° 8, 1978, pp. 607-612.

205. HO H. Dose rate distribution in triple-layered dielectric cylinder with irregular cross section irradiated by plane wave source. *J. Microw. Power*, _10_, n° 4, 1975, pp. 421-432.

206. HO H. Energy absorption patterns in circular triple-layered tissue cylinders exposed to plane wave sources. *Health Phys.*, _31_, 1976, pp. 97-108.

207. HO H., FADEN J. Experimental and theoretical determination of absorbed microwave dose rate distributions in phantom heads irradiated by an aperture source. *Health Phys.*, _33_, n° 1, 1977, p. 13.

208. HO H., GUY A. Development of dosimetry for RF and microwave radiation. *Health Phys.*, _29_, 1975, pp. 317-324.

209. HO H., GUY A., SIGELMANN R., LEHMANN J. Microwave heating of simulated human limbs by aperture sources. *IEEE Trans. Microw. Theory Tech.*, MTT-_19_, n° 2, 1971, pp. 224-231.

210. HO H., Mac MANAWAY M. Heat-dissipation rate of mice after microwave irradiation. *J. Microw. Power*, _12_, n° 1, 1977, pp. 93-100.

211. HODGKIN A., HUXLEY A. A quantitative description of membrane current and its application to conduction and excitation in nerve. *J. Physiol.*, _117_, 1952, pp. 500-544.

212. HODGKIN A., HUXLEY A., BATZ B. Ionic currents underlying activity in the giant axon of the squid. *Arch. Sci. Physiol.*, _3_, 1949, pp. 129-163.

213. HOLLIS M., BLACKMAN C., WEIL C., ALLIS J., SCHAEFER D. A swept-frequency magnitude method for the dielectric characterization of chemical and biological systems. *IEEE Trans. Microw. Theory Tech.*, MTT-_28_, n° 7, 1980, pp. 791-801.

214. HOLODOV J. Effect of a UHF electromagnetic field on the electrical activity of a neuronally isolated region of the cerebral cortex. *Bjul.*

Eksp. 'noj Biol. Med., _57_, 1964, pp. 98-104, trad. Library of Congress, ATF-P-65-68, Washington.

215. HRYCAK P. Microwave dielectric constant measurement system for in-vivo tissue. Thesis, Ph. D., Univ. Maryland, 1979.

216. HUXLEY A. Ion movements during nerve activity. *Ann. New York Acad. Sci.*, _81_, 1959, pp. 221-246.

217. ILLINGER K. Molecular mechanisms for microwave absorption in biological systems. In : CLEARY S. Biological effects and health implications of microwave radiation. Symposium proceedings. Richmond, Commonwealth Univ., 17-19 Sept. 1969, US Dep. Health, Educ., Welfare, BRH, DBE 70-2, PB 193898, Library of Congress card n° 76-607340, June 1970, pp. 112-116.

218. ILLINGER K. Interaction between microwave and millimeter wave electromagnetic fields and biological systems, molecular mechanisms. In : CZERSKI P., OSTROWSKI K., SHORE M., SILVERMAN C., SÜSS M., WALDESKOG B. Biological effects and health hazards of microwave radiation. Proceedings of an international symposium, Warszawa, 15-18 Oct. 1973. Warszawa, Polish Medical Publishers, 1974, pp. 160-172.

219. ISKANDER M., BARBER P., DURNEY C., MASSOUDI H. Irradiation of prolate spheroidal models of humans and animals in the near field of a short electric dipole. *IEEE Trans. Microw. Theory Tech.*, MTT-_28_, n° 7, 1980, pp. 801-807.

220. ISKANDER M., LAKHTAKIA A. Extension of the iterative EBCM to calculate scattering by low-loss or lossless elongated dielectric objects. *Appl. Opt.*, _23_, 1984, pp. 948-952.

221. ISKANDER M., LAKHTAKIA A., DURNEY C. A new iterative procedure to solve for scattering and absorption by lossy dielectric objects. *Proc. IEEE*, _70_, 1982, pp. 1361-1362.

222. ISKANDER M., LAKHTAKIA A., DURNEY C. A new procedure for improving the solution stability and extending the frequency range of the EBCM. *IEEE Trans. Antennas Propag.*, AP-_31_, 1983, pp. 317-324.

223. ISKANDER M., MASSOUDI H., DURNEY C., ALLEN S. Measurements of the RF power absorption in spheroidal human and animal phantoms exposed to the near field of a dipole source. *IEEE Trans. Biomed. Eng.*, BME-_28_, n° 3, 1981, pp. 258-263.

224. ISKANDER M., OLSON S., Mac CALMONT J. Near-field absorption characteristics of biological models in the resonance frequency range. *IEEE Trans. Microw. Theory Tech.*, MTT-35, n° 8, 1987, pp. 776-779.

225. ISKANDER M., STUCHLY S. A time-domain technique for measurement of the dielectric properties of biological substances. *IEEE Trans. Instrum. Meas.*, IM-21, 1972, pp. 425-429.

226. ISKANDER M., TURNER P., DUBOW J., KAO J. Two-dimensional technique to calculate the EM power deposition pattern in the human body. *J. Microw. Power*, 17, n° 3, 1982, pp. 175-185.

227. JANNA W., RUSSO E., Mac AFEE R., DAVOUDI R. A computer model of temperature distribution inside a lossy sphere after microwave radiation. *Bioelectromagn.*, 1, 1980, pp. 337-343.

228. JOHNSON C. Recommendations for specifying EM wave irradiation conditions in bioeffects research. *J. Microw. Power*, 10, 1975, pp. 249-250.

229. JOHNSON C., DURNEY C., MASSOUDI H. Electromagnetic power absorption in anisotropic tissue media. *IEEE Trans. Microw. Theory Tech.*, MTT-23, n° 6, 1975, pp. 529-532.

230. JOHNSON C., DURNEY C., MASSOUDI H. Long-wavelength electromagnetic power absorption in prolate spheroidal models of man and animals. *IEEE Trans. Microw. Theory Tech.*, MTT-23, n° 9, 1975, pp. 739-747.

231. JOHNSON C., GUY A. Nonionizing electromagnetic wave effects in biological materials and systems. *Proc. IEEE*, 60, n° 6, 1972, pp. 692-718.

232. JOINES W., SPIEGEL R. Resonance absorption of microwaves by human skull. *IEEE Trans. Biomed. Eng.*, BME-21, n° 1, 1974, pp. 46-48.

233. JUROSHEK J., HOER C. A high-power automatic network analyzer for measuring the RF power absorbed by biological samples in a TEM cell. *IEEE Trans. Microw. Theory Tech.*, MTT-32, n° 8, 1984, pp. 818-824.

234. JUSTESEN D. Towards a prescriptive grammar for the radiobiology of non-ionizing radiations : quantities, definitions, and units of absorbed electromagnetic energy. *J. Microw. Power*, 10, n° 4, 1975, pp. 343-356.

235. KAMEN'SKIJ J. L'influence des micro-ondes sur les conditions fonctionnelles du nerf. (En russe). *Biofiz.*, 9, n° 6, 1964, pp. 695-700.

236. KARIMULLAH K., CHEN K., NYQUIST D. Electromagnetic coupling between a thin-wire antenna and a neighboring biological body : Theory and experiment. *IEEE Trans. Microw. Theory Tech.*, MTT-28, n° 11, 1980, pp. 1218-1225.

237. KAROLKAR B., BEHARI J., PRIM A. Biological tissues characterization at microwave frequencies. *IEEE Trans. Microw. Theory Tech.*, MTT-33, n° 1, 1985, pp. 64-66.

238. KENT M. Complex permittivity of protein powders at 9,4 GHz as a function of temperature and hydration. *J. Phys.*, D : Appl. Phys., 5, 1972, pp. 394-409.

239. KENT M., MEYER W. Complex permittivity spectra of protein powders as a function of temperature and hydration. *J. Phys.*, D : Appl. Phys., 17, 1984, pp. 1687-1698.

240. KING N., HUNT E., PHILLIPS R. Biological dosimetry of 2450 MHz irradiation with mice. In : Proceedings of the IMPI Symposium, Milwaukee, 1974, pp. 1-4. (Paper B2.5).

241. KING H., WONG J. Effects of a human body on a dipole antenna for the 240- to 400-MHz frequency range. *IEEE Trans. Antennas Propag.*, AP-25, n° 3, 1977, pp. 376-379.

242. KISELEV R., ZALJUBOV N. Effet des ondes électromagnétiques de longueur d'onde millimétrique sur la cellule et ses éléments structurels. (En russe). *Usp. Fiz. Nauk*, 110, n° 3, 1973, pp. 464-466.

243. KLAINER S., FRAZER J. Raman spectroscopy of molecular species during exposure to 100 MHz radio-frequency fields. In : TYLER P. Biologic effects of non-ionizing radiation. Proceedings of a conference. *Ann. New York Acad. Sci.*, 247, 1975, pp. 323-326.

244. KLEINSTEIN B. Biological effects of nonionizing electromagnetic radiation. *Dig. Curr. Lit.*, 8, n° 4, and 9, n° 1, Sept.-Dec. 1984, 150 p.

245. KOVAČ R. Distribution de température en chauffage par micro-ondes dans un modèle à deux couches d'objet biologique. (En russe). *Med. Teh.*, 7, n° 1, 1973, pp. 18-21.

246. KRASZEWSKI A., HARTSGROVE G. A macroscopic model of lungs and material simulating their dielectric properties at radio and microwave frequencies. *J. Microw. Power*, 21, n° 2, 1986, pp. 96-98.

247. KRASZEWSKI A., STUCHLY M., STUCHLY S., HARTSGROVE G., ADAMSKI D. Specific absorption rate distribution in a full-scale model of man at 350 MHz. *IEEE Trans. Microw. Theory Tech.*, MTT-32, n° 8, 1984, pp. 779-783.

248. KRASZEWSKI A., STUCHLY M., STUCHLY S., SMITH A. In-vivo and in-vitro dielectric properties of car tissues at radio-frequencies. *Bioelectromagn.*, 3, 1982, pp. 421-432.

249. KRASZEWSKI A., STUCHLY S., STUCHLY M., SYMONS S., On the measurement accuracy of the tissue permittivity in vivo. *IEEE Trans. Instrum. Meas.*, IM-32, n° 1, 1983, pp. 37-42.

250. KRITIKOS H., FOSTER K., SCHWAN H. Temperature profiles in spheres due to electromagnetic heating. *J. Microw. Power*, 16, n° 3-4, 1981, pp. 327-344.

251. KRITIKOS H., SCHWAN H. The possibility of non uniform temperature rise resulting from microwave exposure. In : Proceedings of the IMPI Symposium, Monterey, 1971, (Paper 7.3.).

252. KRITIKOS H., SCHWAN H. Hot spots generated in conducting spheres by electromagnetic waves and biological implications. *IEEE Trans. Biomed. Eng.*, BME-19, n° 1, 1972, pp. 53-58.

253. KRITIKOS H., SCHWAN H. The differential temperature rise at hot spots generated in lossy spheres by electromagnetic waves. In : Proceedings of the IMPI Symposium, Ottawa, 1972, pp. 194-195. (Paper 10.5.).

254. KRITIKOS H., SCHWAN H. The distribution of heating potential inside lossy spheres. *IEEE Trans. Biomed. Eng.*, BME-22, 1975, pp. 457-463.

255. KRITIKOS H., SCHWAN H. Formation of hot spots in multilayered spheres. *IEEE Trans. Biomed. Eng.*, BME-22, 1976, pp. 168-172.

256. KRITIKOS H., SCHWAN H. Potential temperature rise induced by electromagnetic field in brain tissues. *IEEE Trans. Biomed. Eng.*, BME-26, n° 1, 1979, pp. 29-34.

257. LAKHTAKIA A., ISKANDER M. Scattering and absorption characteristics of lossy dielectric objects exposed to the near fields of aperture sources. *IEEE Trans. Antennas Propag.*, AP-31, 1983, pp. 111-120.

258. LAKHTAKIA A., ISKANDER M. Theoretical and experimental evaluation of power absorption in elongated biological objects at and beyond resonance. *IEEE Trans. Electromagn. Compat.*, EMC-25, n° 4, 1983, pp. 448-453.

259. LAKHTAKIA A., ISKANDER M., DURNEY C. An iterative extended boundary condition method for solving the absorption characteristics of lossy dielectric objects of large aspect ratios. *IEEE Trans. Microw. Theory Tech.*, MTT-31, n° 8, 1983, pp. 640-647.

260. LAKHTAKIA A., ISKANDER M., DURNEY C., MASSOUDI H. Near-field absorption in prolate spheroidal models of humans exposed to a small loop antenna of arbitrary orientation. *IEEE Trans. Microw. Theory Tech.*, MTT-29, n° 6, 1981, pp. 588-594.

261. LAWINSKI C., SHEPHERD J., GRANT E. Measurement of permittivity of solution of small biological molecules at radiowave and microwave frequencies. *J. Microw. Power*, 10, n° 2, 1975, pp. 147-162.

262. LAWRENCE J. Effect of microwaves on isolated skin. In : Proceedings of the IMPI Symposium, Loughborough, 1973. (Paper 3.B.3).

263. LEHMANN J., GUY A., JOHNSTON V., BRUNNER G., BELL J. Comparison of relative heating patterns produced in tissues by exposure to microwave energy at frequencies of 2450 and 900 Mc. *Arch. Phys. Med. Rehabil.*, 43, 1962, pp. 69-76.

264. LEHMANN J., Mac MILLAN J., BRUNNER G., GUY A. A comparative evaluation of temperature distributions produced by microwaves at 2456 and 900 megacycles in geometrically complex specimens. *Arch. Phys. Med. Rehabil.*, 43, 1962, pp. 502-507.

265. LEHMANN J., SILVERMAN D., BAUM B., KIRK N., JOHNSTON V. Temperature distribution in human thigh, produced by infrared, hot pack and microwave applicators. *Arch. Phys. Med. Rehabil.*, 47, n° 5, 1966, pp. 291-299.

266. LERNER E. RF radiation : biological effects. *IEEE Spectrum*, n° 12, 1980, pp. 51-59.

267. LIN J. A cavity-backed slot radiator for microwave biological effect research. *J. Microw. Power*, 9, n° 2, 1974, pp. 63-67.

268. LIN J. Interaction of electromagnetic transient radiation with biological materials. *IEEE Trans. Electromagn. Compat.*, EMC-17, n° 1, 1975, pp. 93-97.

269. LIN J. Microwave properties of fresh mammalian brain tissues at body temperature. *IEEE Trans. Biomed. Eng.*, BME-22, n° 1, 1975, pp. 74-76.

270. LIN J. Microwave biophysics. In : Transactions of the IMPI, 8, Microwave bioeffects and radiation safety. 1978, pp. 15-54.

271. LIN J., CLEARY S. Effect of microwave radiation on erythrocyte membranes. In : Biological effects of electromagnetic waves. International symposium, Airlie, 30 Oct.-4 Nov. 1977, Abstracts of scientific papers.

272. LIN J., GUY A., CALDWELL L. Thermographic and behavioral studies of rats in the near field of 918-MHz radiations. *IEEE Trans. Microw. Theory Tech.*, MTT-25, n° 10, 1977, pp. 833-836.

273. LIN J., LAM C. Coupling of gaussian electromagnetic pulse into a muscle-bone model of biological structure. *J. Microw. Power*, 11, n° 1, 1976, pp. 67-75.

274. LINDAUER G., LIU L., SKEWES G., ROSENBAUM F. Further experiments seeking evidence of nonthermal biological effects of microwave radiation. *IEEE Trans. Microw. Theory Tech.*, MTT-22, n° 8, 1974, pp. 790-793.

275. LIU L., GARBER F., CLEARY S. Investigation of the effects of continuous-wave, pulse- and amplitude-modulated microwaves on single excitable cells of Chara corallina. *Bioelectromagn.*, 3, 1982, pp. 303-312.

276. LIVENSON A. Electrical parameters of biological tissue in the microwave range. (En russe). *Med. Prom.*, 18, 1964, pp. 14-20, traduction anglaise, US Joint Publ. Res. Serv. Rep., JPRS/26429-TT64-41450, ATD-P-65-68.

277. LIVESAY D., CHEN K. Electromagnetic fields induced inside arbitrarily shaped biological bodies. *IEEE Trans. Microw. Theory Tech.*, MTT-22, n° 12, 1974, pp. 1273-1280.

278. LOVELY R., MYERS D., GUY A. Irradiation of rats by 918-MHz microwaves at 2.5 mW/cm^2 : delineating the dose response relationship. In : USNC/URSI annual meeting, Amherst, 1976. *Radio Sci.*, 12, n° 6 (S), 1977, pp. 139-146.

279. LOVISOLO G., ADAMI M., ARCANGELI G., BORRANI A., CALAMAI G., CIVIDALLI A., MAURO F. A multifrequency water-filled waveguide applicator : thermal dosimetry in vivo. *IEEE Trans. Microw. Theory Tech.*, MTT-32, n° 8 1984, pp. 893-896.

280. LU S., ROBERTS N., MICHAELSON S. A dual vial waveguide exposure facility for examining microwave effects in vitro. *J. Microw. Power*, 18, n° 2, 1983, pp. 121-131.

281. Mac CLEAN V., SHEPPARD R., GRANT E. A generalized model for the interaction of microwave radiation with bound water in biological material. *J. Microw. Power*, 16, n° 1, 1981, pp. 1-7.

282. Mac GREGOR R. A possible mechanism for the influence of electromagnetic radiation on neuroelectric potentials. *IEEE Trans. Microw. Theory Tech.*, MTT-27, n° 11, 1979, pp. 914-921.

283. Mac REE D. Determination of energy absorption of microwave radiation using the cooling curve technique. In : Proceedings of the IMPI Symposium, Loughborough, 1973, pp. 1-3, (Paper 4B.5). and in *J. Microw. Power*, 9, n° 3, 1974, pp. 263-270.

284. Mac REE D., WALSH P. Microwave exposure system for biological specimens. *Rev. Sci. Instrum.*, 42, 1971, pp. 1860-1864.

285. MALABIAU R. Proximal region power density assessment in the scope of non-ionizing radiation hazards prediction. In : Comptes-rendus des journées d'études de la SEE, Champs électromagnétiques dans le proche environnement des équipements industriels microonde et des antennes, Univ. Toulon Var, 19-20 jan. 1984, réf. 842520, 4 p.

286. MARSHALL S., BROWN R., HUGHES C., MARSHALL P. Environmentally controlled exposure system for irradiation of mice at frequencies below 500 MHz. In : IEEE Electromagnetic Compatibility Society Symposium, 1981, pp. 99-104.

287. MASSOUDI H., DURNEY C., BARBER P., ISKANDER M. Electromagnetic absorption in multilayered cylindrical models of man. *IEEE Trans. Microw. Theory Tech.*, MTT-27, n° 10, 1979, pp. 825-830.

288. MASSOUDI H., DURNEY C., ISKANDER M. Limitation of the cubical block model of man in calculating SAR distributions. *IEEE Trans. Microw. Theory Tech.*, MTT-32, n° 8, 1984, pp. 746-752.

289. MASSOUDI H., DURNEY C., JOHNSON C. Long-wavelength analysis of plane-wave irradiation of an ellipsoidal model of man. *IEEE Trans. Microw. Theory Tech.*, MTT-25, n° 1, 1977, pp. 41-46.

290. MASSOUDI H., DURNEY C., JOHNSON C. Long-wavelength electromagnetic power absorption in ellipsoidal models of man and animals. *IEEE Trans. Microw. Theory Tech.*, MTT-25, n° 1, 1977, pp. 47-52.

291. MASSOUDI H., DURNEY C., JOHNSON C. Geometrical-optics and exact solutions for internal fields and SAR's in a cylindrical model of man irradiated by an electromagnetic plane wave. *Radio Sci.*, 14, n° 6 (S), 1979, pp. 35-42.

292. MASSOUDI H., DURNEY C., JOHNSON C., ALLEN S. Theoretical calculations of power absorbed by monkey and human prolate spheroidal phantoms in an irradiation chamber. In : JOHNSON C., SHORE M., Biological effects of electromagnetic waves. Selected papers of the USNC/URSI annual meeting, Boulder, 20-23 Oct. 1975. US Dep. Health, Educ., Welfare, FDA-77-8011. Washington, US Government Printing Office, Vol. 2, 1976, pp. 135-157.

293. MASSOUDI H., DURNEY C., TSAI C. A comparison of two different integral equations and basis functions for evaluation of SAR distributions in biological models. In : Proceedings of the IMPI Symposium, Chicago, 1985, pp. 115-116.

294. MERMAGEN H. Phantom experiments with microwaves at the University of Rochester. In : PEYTON M. Proceedings of the 4th annual Tri-Service conference on biological effects of microwave-radiating equipments, biological effects of microwave radiations, 1961. New York, Plenum Press, 1961, RADC-TR-60-180, pp. 143-152.

295. MERRITT J., HARTZELL R., FRAZER J. The effect of 1.6 GHz radiation on neurotransmitters in discrete areas of the rat brain, US Air Force report SAM-TR-76-3, USAF Sch. Aerosp. Med., Aerosp. Med. Div. (AFSC), Brooks Air Force Base, 1976.

296. MIRUTENKO V., BOGACH P. Changements du potentiel de la membrane de cellules nerveuses de ganglions isolés du mollusque "Planorbis corneus" sous l'effet d'un champ électromagnétique. (En russe). *Fiziol. Ž.*, 21, 1975, pp. 528-531.

297. MISRA D., CHEN K. Responses of electric-field probes near a cylindrical model of the human body. *IEEE Trans. Microw. Theory Tech.*, MTT-33, n° 6, 1985, pp. 447-452.

298. MONAHAN J., HO H. The effect of ambient temperature on the reduction of microwave energy absorption by mice. *Radio Sci.*, 12, n° 6 (S), 1977, pp. 257-262.

299. NEUDER S. A finite element technique for calculating induced internal fields and power deposition in biological media of complex irregular geometry. In : HAZZARD D. Biological effects and measurement of RF/microwaves. Symposium Proceedings, 16-18 Feb. 1977, US Dep. Health, Educ. Welfare, BRH, FDA-77-8026, July 1977, pp. 170-190.

300. NEUDER S. Electromagnetic fields in biological media. Part 1 : dosimetry - a primer on bioelectromagnetics. US Dep. Health, Educ., Welfare, Bureau Radiol. Health. Publication (FDA) n° 78-8068, 1978, 21 p.

301. NEUDER S., KELLOGG R., HILL D. Microwave power density absorption in a spherical multilayered model of the head. In : JOHNSON C., SHORE M. Biological effects of electromagnetic waves. Selected papers of the USNC/URSI annual meeting, Boulder, 20-23 Oct. 1975, US Dep. Health, Educ., Welfare, FDA-77-8011. Washington, US Government Printing Office, Vol. 2, 1976, p. 199.

302. NEUDER S., MEIJER P. Finite element-variational calculus approach to the determination of electromagnetic fields in irregular geometry. In : JOHNSON C., SHORE M. Biological

effects of electromagnetic waves. Selected papers of the USNC/URSI annual meeting, Boulder, 20-23 Oct. 1975, US Dep. Health, Educ., Welfare, FDA-77-8011. Washington, US Government Printing Office, Vol. 2, 1976, pp. 193-198

303. NEUKOMM P. Wechselwirkung zwischen Mensch und elektromagnetischen Feldern im Frequenzbereich 10 bis 1000 MHz. *Bull. SEV/VSE*, 17, 1980, pp. 924-932.

304. NYQUIST D., CHEN K., GURU B. Coupling between small thin-wire antennas and a biological body. *IEEE Trans. Antennas Propag.*, AP-25, n° 6, 1977, pp. 863-865.

305. OLCERST R., BELMAN S., EISENBUD M., MUMFORD W., RABINOWITZ J. The increased passive efflux of sodium and rubidium from rabbit erythrocytes by microwave radiation. *Radiation Res.*, 82, 1980, pp. 244-256.

306. OLSEN R. Preliminary studies : far-field microwave dosimetric measurements of a full-scale model of man. *J. Miarow. Power*, 14, n° 4, 1979, pp. 383-388.

307. ORLOWSKI A. Effets biologiques des ondes électromagnétiques courtes et ultra-courtes. *Usine Nouvelle*, n° de Printemps, 1969, pp. 579-606.

308. OSEPCHUK J. Biological effects of electromagnetic radiation. Publication Institute of Electrical and Electronics Engineers, inc. Réimpression de 61 articles publiés par l'IEEE, 1983, 608 p.

309. PARTLOW L., BUSH L., STENSAAS L., HILL D., RIAZI A., GANDHI O. Effects of multimeter-wave radiation on monolayer cell cultures : I. Design and validation of a novel exposure system. *Bioelectromagn.*, 2, 1981, pp. 123-140.

310. PAULSSON L., HAMNERIUS Y., Mac LEAN W. The effects of microwave radiation on microtubules and axonal transport. *Radiation Res.*, 70, 1977, pp. 212-223.

311. PAULY H., SCHWAN H. The dielectric properties of the bovine eye lens. *IEEE Trans. Biomed. Eng.*, BME-11, n° 3, 1964, pp. 103-109.

312. PAULY H., SCHWAN H. Dielectric properties and mobility in erythrocytes. *Biophys. J.*, 6, 1966, pp. 621-639.

313. PETHIG R. Electronic conduction in biological systems. *Electron. Power*, 19, n° 18, 1973, pp. 445-449.

314. PICKARD W., ROSENBAUM F. Biological effects of microwaves at the membrane level. 2 : possible athermal mechanisms and a proposed experimental test. *Math. Biosci.*, 39, 1978, pp. 235-254.

315. PORTELA A., VACCARI J., MICHAELSON S., LLOBERA O., BRENNAN M. Transient effects of low-level microwave irradiation on muscle-cell bioelectric properties, water permeability, and water distribution. *Studia Biophys.*, n° 53, 1975, pp. 197-224.

316. POZAS R., RICHARDSON A., KAPLAN H. Non-uniform biophysical heating with microwaves. In : CLEARY S. Biological effects and health implications of microwave radiation. Symposium proceedings. Richmond, Commonwealth Univ., 17-19 Sept. 1969, US Dep. Health, Educ., Welfare, BRH, DBE 70-2, PB 193898, Library of Congress card n° 76-607340, June 1970, pp. 70-75.

317. PRESMAN A. The physical basis for the biological action of centimeter waves. (En russe). *Arch. Biol. Mol.*, 11, 1956, pp. 40-54, traduction anglaise, Washington, Library of Congress, ATD-P-65-17.

318. PRESMAN A. The action of microwaves on living organisms and biological structures. *Soviet Phys. Usp.*, 8, n° 3, 1965, pp. 463-488. (Traduction anglaise).

319. PRIEBE L., WELCH A. A dimensionless model for the calculation of temperature increase in biologic tissues exposed to nonionizing radiation. *IEEE Trans. Biomed. Eng.*, BME-26, n° 4, 1979, pp. 244-250.

320. PRIOU A. Interactions avec la matière vivante, applications thérapeutiques et au diagnostic, approche des mécanismes des effets biologiques des ondes électromagnétiques. In : PRIOU A. Ensemble des travaux, Thèse de Doctorat d'Etat, Toulouse, Univ. Paul Sabatier, 1981, Chap. III, pp. 489-506.

321. PRIOU A., DEFICIS A. Dosimétrie sous rayonnement électromagnétique non ionisant : mesure des puissances de champs et des températures. *Rev. Gen. Therm.*, n° 226, oct. 1980, pp. 799-808.

322. RABINOWITZ J. Possible mechanisms for the biomolecular absorption of microwave radiation with functional implications. *IEEE Trans. Microw. Theory Tech.*, MTT-21, n° 12, 1973, pp. 850-851.

323. RELYVELD E., CHERMANN J., HERVE G. Les protéines. Paris, P.U.F., 2ème éd., n° 679, 1980, 127 p. (Coll. Que sais-je ?).

324. RICHMOND J. Scattering by a dielectric cylinder of arbitrary cross section and shape. *IEEE Trans. Antennas Propag.*, AP-13, 1965, pp. 334-341.

325. ROBERTS J., COOK H. Microwaves in medical and biological research. *Br. J. Appl. Sci.*, 3, 1952, pp. 33-40.

326. ROWLANDSON G., BARBER P. Absorption of higher-frequency RF energy by biological models : calculations based on geometrical optics. *Radio Sci.*, 14, n° 6 (S), 1979, pp. 43-50.

327. RUGHT R., HO H., Mac MANAWAY M. The relation of dose rate of microwave radiation to the time of death and total absorbed dose in the mouse. *J. Microw. Power*, 11, n° 3, 1976, pp. 280-281.

328. RUPPIN R. Calculation of electromagnetic energy absorption in prolate spheroids by the point matching method. *IEEE Trans. Microw. Theory Tech.*, MTT-26, n° 2, 1978, pp. 87-90.

329. RYAN C., CAIN F., WANG J., COWN B., COOKE W. Electromagnetic models for antenna performance, EMC, and biological effects. *IEEE Trans. Electromagn. Compat.*, EMC-22, n° 4, 1980, pp. 244-255.

330. SAITO M. RF field induced forces on microscopic particles. *Dig. Int. Conf. Med. Electron.*, 21, 1961, p. 3.

331. SAITO M., SCHWAN H. The time constants of pearl-chain formation. In : PEYTON M. Proceedings of the 4th annual Tri-Service conference on biological effects of microwave-radiating equipments, biological effects of microwave radiations, 1961. New York, Plenum Press, 1961, Vol. 1, RADC-TR-60-180, pp. 85-97.

332. SCHROT J., HAWKINS T. Interaction of microwave frequency and polarization with animal size. In : JOHNSON C., SHORE M. Biological effects of electromagnetic waves. Selected papers of the USNC/URSI annual meeting, Boulder, 20-23 Oct.

1975, US Dep. Health, Educ., Welfare, FDA-77-8011. Washington, US Government Printing Office, Vol. 2, 1976, pp. 184-192.

333. SCHWAN H. Electrical properties of tissues and cell suspensions. In : LAWRENCE J., TOBIAS C. Advances in biological and medical physics. Vol. 5, New York, Academic Press, 1957, pp. 147-209.

334. SCHWAN H. Survey of microwave absorption characteristics of body tissues. In : PATTISHALL E., BANGHART F. Proceedings of the 2nd annual Tri-Service conference on biological effects of microwave energy, RADC-TR-58-54, AD 131 477, 1958, pp. 126-145.

335. SCHWAN H. Absorption and energy transfer of microwaves and ultrasonics in tissue : characteristics. In : GLASSER O. Medical physics. Vol. 3, Chicago, Yearbook Publication, 1960, pp. 1-7.

336. SCHWAN H. Determination of biological impedances. In : Physical techniques in biological research, 6, chap. 6, New York, Academic Press, 1963, pp. 323-407.

337. SCHWAN H. Electrical characteristics of tissues, a survey. *Biophysik*, 1, 1964, pp. 198-208.

338. SCHWAN H. Microwave biophysics. In : OKRESS E. Microwave power engineering, New York, Academic Press, 1968, pp. 213-244.

339. SCHWAN H. Radiation biology, medical applications and radiation hazards. In : OKRESS E. Microwave power engineering. Vol. 2, New York, Academic press, 1968, pp. 215-532.

340. SCHWAN H. Effects of microwave radiation on tissue ; a survey of basic mechanisms. *Non-ioniz. Radiat.*, 1, n° 1, 1969, pp. 23-31. (Traduction INRS 172 A 70).

341. SCHWAN H. Interaction of microwave and radiofrequency radiation with biological systems. In : CLEARY S. Biological effects and health implications of microwave radiation. Symposium proceedings, Richmond, Commonwealth Univ., 17-19 Sept. 1969, US Dep. Health, Educ., Welfare, BRH, DBE 70-2, PB 193898, Library of Congress card n° 76-607340, June 1970, pp. 13-20.

342. SCHWAN H. Interaction of microwave and radio frequency radiation with biological systems. *IEEE Trans. Microw. Theory Tech.*, MTT-19, n° 2, 1971, pp. 146-152.

343. SCHWAN H. Microwave radiation : biophysical considerations and standard criteria. *IEEE Trans. Biomed. Eng.*, BME-19, 1972, pp. 302-312.

344. SCHWAN H. Principles of interaction of microwave fields at the cellular and molecular level. In : CZERSKI P., OSTROWSKI K., SHORE M., SILVERMAN C., SÜSS M., WALDESKOG B. Biological effects and health hazards of microwave radiation. Proceedings of an international symposium, Warszawa, 15-18 Oct. 1973. Warszawa, Polish Medical Publishers, 1974, pp. 152-159.

345. SCHWAN H. Field interaction with biological matter. In : TAKASHIMA S., FISHMAN H. Electrical properties of biological polymers, water and membranes. *Ann. New York Acad. Sci.*, 303, 1977, pp. 198-213.

346. SCHWAN H. Nonthermal cellular effects of electromagnetic fields : AC-field induced ponderomotoric forces. *Br. J. Cancer*, 45, suppl. V, 1982, pp. 220-224.

347. SCHWAN H. Biophysics of the interaction of electromagnetic energy with cells and membranes. In : GRANDOLFO M., MICHAELSON S., RINDI A. Biological effects and dosimetry of nonionizing radiation. New York, Plenum Press, 1983, pp. 213-232.

348. SCHWAN H., FOSTER K. Microwave dielectric properties of tissue. Some comments on rotational mobility of tissue water. *Biophys. J.*, 17, n° 2, 1977, pp. 193-197.

349. SCHWAN H., FOSTER K. RF-field interactions with biological systems : electrical properties and biophysical mechanisms. *Proc. IEEE*, 68, n° 1, 1980, pp. 104-113.

350. SCHWAN H., KRITIKOS H. Energy deposition in homogeneous and multilayer tissue spheres and effect of circulation. In : Proceedings of the IMPI Symposium, Waterloo (Can.), 1975, pp. 5-9. (Paper 1.2.).

351. SCHWAN H., LI K. Capacity and conductivity of body tissues at ultrahigh frequencies. In : Proceedings of the IRE, 41, Dec. 1953, pp. 1735-1740.

352. SCHWAN H., LI K. Mechanism of absorption of ultrahigh frequency electromagnetic energy in tissues, as related to the problem of tolerance dosage. Institution of Radio Engineers and Medical Electronics, P.G.M.E., 4, 1956, pp. 45-49.

353. SCHWAN H., PIERSOL G. The absorption of electromagnetic energy in body tissues. Part I : Biophysical aspects. *Am. J. Phys. Med.*, 33, n° 12, 1954, pp. 370-404.

354. SCHWAN H., PIERSOL G. The absorption of electromagnetic energy in body tissues. Part II : Physiological and clinical aspects. *Am. J. Phys. Med.*, 34, 1955, pp. 425-448.

355. SCHWAN H., SHEPPARD R., GRANT E. Complex permittivity of water at 25°C. *J. Chem. Phys.*, 64, mars 1976, pp. 2257-2258.

356. SCHWARTZ M. The theory of impedance in biological systems. *J. Biol. Phys.*, 1, 1973, pp. 123-142.

357. SCHWARZ G. Ultrasonic and dielectric relaxation due to conformational changes in biopolymers. In : Molecular relaxation processes. Proceedings of the Chemical Society Symposium. Aberystwyth, 7-9 July 1965. London, Academic Press, 1966, pp. 191-197.

358. SCHWARZ G. On dielectric relaxation due to chemical rate processes. *J. Phys. Chem.*, 71, 1967, pp. 4021-4030.

359. SEAMAN R. Neuronal effects of low-level microwaves. Thesis, Ph. D., Durham, Duke Univ., 1974.

360. SEAMAN R., WACHTEL H. Slow and rapid responses to Cw and pulsed microwave radiation by individual aplysia pacemakers. *J. Microw. Power*, 13, n° 1, 1978, pp. 77-86.

361. SEGAL A., MAGIN R., CAIN C. Electromagnetic dosimetry in a Crawford cell : 225 to 500 MHz. *J. Microw. Power*, 19, n° 1, 1984, pp. 15-24.

362. SHAPIRO S. Josephson currents in superconductive tunneling : the effect of microwaves and other observations. *Phys. Rev. Lett.*, 11, 1963, pp. 80-82.

363. SHAPIRO A., LUTOMIRSKI R., YURA H. Induced fields and heating within a cranial structure irradiated by an electromagnetic plane wave. *IEEE Trans. Microw. Theory Tech.*, MTT-19, n° 2, 1971, pp. 187-196.

364. SHER L. Interaction of microwave and RF energy on biological material. In : Electronic product radiation and the health physicist, US Dep. Health, Educ., Welfare, BRH, DEP 70-26, 1970, pp. 431-461.

365. SHER L., KRESCH E., SCHWAN H. On the possibility of non-thermal biological effects of pulsed electromagnetic radiation. *Biophys. J.*, 10, n° 10, 1970, pp. 970-979.

366. SMITH G. The electric-field probe near a material interface with application to the probing of fields in biological bodies. *IEEE Trans. Microw. Theory Tech.*, MTT-27, n° 3, 1979, pp. 270-278.

367. SPIEGEL R. A review of numerical models for predicting the energy deposition and resultant thermal response of humans exposed to electromagnetic fields. *IEEE Trans. Microw. Theory Tech.*, MTT-32, n° 8, 1984, pp. 730-746.

368. SPIEGEL R., DEFFENBAUGH D., MANN J. A thermal model of the human body exposed to an electromagnetic field. *Bioelectromagn.*, 1, n° 3, 1980, pp. 253-270.

369. SPIEGEL R., FATMI M., KUNZ-LAWRENCE K. Application of a finite-difference technique to the human microwave dosimetry problem. *J. Microw. Power*, 20, n° 4, 1985, pp. 241-254.

370. STENSAAS L., PARTLOW L., BUSH L., IVERSEN P., HILL D., HAGMANN M., GANDHI O. Effects of millimeter-wave radiation on monolayer cell cultures : II. Scanning and transmission electron microscopy. *Bioelectromagn.*, 2, 1981, pp. 141-150.

371. STOGRYN A. Equations for calculating the dielectric constant of saline water. *IEEE Trans. Microw. Theory Tech.*, MTT-19, n° 8, 1971, pp. 733-736.

372. STRAUB K. Molecular absorption of non-ionizing radiations in biological systems. *Bull. Am. Phys. Soc.*, 22, n° 1, 1977, p. 45.

373. STUCHLY M. Interaction of radiofrequency and microwave radiation with living systems : a review of mechanisms. *Radiat. Environ. Biophys.*, 16, 1979, pp. 1-14.

374. STUCHLY M. Les radiofréquences et la santé. *Rev. Gén. Electr.*, 11, nov. 1981, pp. 849-851.

375. STUCHLY M. Dosimetry of radiofrequency and microwave radiation : theoretical analysis. In : GRANDOLFO M., MICHAELSON S., RINDI A. Biological effects and dosimetry of nonionizing radiation. New York, Plenum Press, 1983, pp. 163-177.

376. STUCHLY M. Fundamentals of the interaction of radio-frequency and microwave energies with matter. In : GRANDOLFO M., MICHAELSON S., RINDI A. Biological effects and dosimetry of nonionizing radiation. New York, Plenum Press, 1983, pp. 75-93.

377. STUCHLY M. Microwave dielectric absorption of DNA in aqueous solution. *Biopolym. J.*, 23, 1984, 593-599.

378. STUCHLY M. Permittivity of mammalian tissues in vivo and in vitro : advances in experimental techniques and recent results. *Int. J. Electron.*, 56, 1984, pp. 443-456.

379. STUCHLY M. Exposure of human models in the near and far field - a comparison. *IEEE Trans. Biomed. Eng.*, BME-32, n° 8, 1985, pp. 609-616.

380. STUCHLY M., ATHEY W., SAMARAS G., TAYLOR G. Measurement of radio-frequency permittivity of biological tissues with an open-ended coaxial line : part II. Experimental results. *IEEE Trans. Microw. Theory Tech.*, MTT-30, n° 1, 1982, pp. 87-92.

381. STUCHLY M., ATHEY T., STUCHLY S., SAMARAS G., TAYLOR G. Dielectric properties of animal tissues in vivo at frequencies 10 MHz-1 GHz. *Bioelectromagn.*, 2, 1981, pp. 93-103.

382. STUCHLY M., KRASZEWSKI A., STUCHLY S., SMITH A. Dielectric properties of animal tissue in vivo at radio and microwave frequencies - Comparison between species. *Phys. Med. Biol.*, 27, 1982, pp. 927-936.

383. STUCHLY M., STUCHLY S. Coaxial-line reflection methods for measuring dielectric properties of biological substances at radio and microwave frequencies - A review. *IEEE Trans. Instrum. Meas.*, IM-29, n° 3, 1980, pp. 176-183.

384. STUCHLY M., STUCHLY S. Dielectric properties of biological substances - tabulated. *J. Microw. Power*, 15, n° 1, 1980, pp. 19-26.

385. STUCHLY M., STUCHLY S. Industrial, scientific, medical and domestic applications of microwaves. *IEE Proc.*, 130, part A, n° 8, 1983, pp. 467-503.

386. STUCHLY M., STUCHLY S. A six-octave experimental system for in vivo and in vitro dielectric measurements. In : CHIABRERA A., NICOLINI C., SCHWAN H. Interactions between electromagnetic fields and cells. New York, Plenum Publishing Corp., 1985, pp. 117-129.

387. STUCHLY M., STUCHLY S., ATHEY T. In vivo electrical properties of animal tissues at radio and microwave frequencies. In : Proceedings of the V° ICEBI, Tokyo, August 1981, pp. 105-108.

388. STUCHLY S., KRASZEWSKI A., STUCHLY M. Energy deposition in a model of man in the near field. *Bioelectromagn.*, 6, 1985, pp. 115-130.

389. SULTAN M., MITTRA R. Iterative methods for analyzing electromagnetic scattering from dielectric bodies. Urbana, Univ. Illinois, Electromagn. Lab., report n° 84-4, 1984.

390. SULTAN M., MITTRA R. An iterative moment method for analyzing the electromagnetic field distribution inside inhomogeneous lossy dielectric objects. *IEEE Trans. Microw. Theory Tech.*, MTT-33, n° 2, 1985, pp. 163-168.

391. SÜSSKIND C. Correspondence on Dr. Justesen's "Prescriptive grammar for the radiobiology of non-ionizing radiation". *J. Microw. Power*, 10, n° 4, 1975, p. 357.

392. SWICORD M., EDWARDS G., SAGRIPANTI J., DAVIS C. Chain-length dependent microwave absorption in DNA. *Biopolym. J.*, 22, 1983, pp. 2513-2516.

393. TAFLOVE A. Application of the finite-difference time-domain method to sinusoidal steady state electromagnetic penetration problems. *IEEE Trans. Electromagn. Compat.*, EMC-22, n° 3, 1980, pp. 191-202.

394. TAFLOVE A., BRODWIN M. Numerical solution of steady-state electromagnetic scattering problems using the time-dependent Maxwell's equations. *IEEE Trans. Microw. Theory Tech.*, MTT-23, n° 8, 1975, pp. 623-630.

395. TAFLOVE A., BRODWIN M. Computation of the electromagnetic fields and induced temperatures within a model of the microwave-irradiated human eye. *IEEE Trans. Microw. Theory Tech.*, MTT-23, n° 11, 1975, pp. 888-896.

396. TAFLOVE A., UMASHANKAR K. Comment on "Fast-Fourier-transform method for calculation of SAR distributions in finely discretized inhomogeneous models of biological bodies". *IEEE Trans. Microw. Theory Tech.*, MTT-33, n° 4, 1985, pp. 345-346.

397. TAKASHIMA S., FISHMAN H. Electrical properties of biological polymers, water and membranes. *Ann. New York Acad. Sci.*, 303, 1977, 446 p.

398. TAKASHIMA S., SCHWAN H. Passive electrical properties of squid axon membrane. *J. Membr. Biol.*, 17, 1974, pp. 51-68.

399. TEODORIDIS V., SPHICOPOULOS T., GARDIOL F. The reflection from an open-ended rectangular waveguide terminated by a layered dielectric medium. *IEEE Trans. Microw. Theory Tech.*, MTT-33, n° 5, 1985, pp. 359-366.

400. TERZUOLO C., BULLOCK T. Measurement of imposed voltage gradient adequate to modulate neuronal firing. In : Proceedings of the National Academy of Science, Washington, 42, 1965, pp. 687-694.

401. TINGA W., NELSON S. Dielectric properties of materials for microwave processing-tabulated. *J. Microw. Power*, 8, n° 1, 1973, pp. 23-65.

402. TOLER J., SEALS J. RF dielectric properties measurement system : human and animal data. US Dep. Health, Educ. Welfare, NIOSH, 77-176, 1977.

403. TOMBERG V. Biochemical effects of microwave fields. In : Proceedings of the IMPI Symposium, Boston, 1969, Abstract p. 28. (Paper E.3.).

404. TREMOLIERES J. Effets biologiques des champs électromagnétiques non ionisants. *Electron. Appl.*, n° 7, automne 1978, pp. 71-77.

405. UEBAYASHI S., AMEMIYA Y. Power deposition in a block model of man exposed to the near field of a dipole antenna. *Trans. Inst. Electron. Commun. Eng. Jpn. (IECEJ)*, J 67-B, August 1984, pp. 877-883.

406. VAN LOOCK W. Useful phantom recipes for the 2.45-3 GHz range. *J. Microw. Power*, 21, n° 2, 1986, pp. 94-96.

407. VAN SLIEDREGT M. Computer calculations of a one-dimensional model, useful in the application of hyperthermia. *Microw. J.*, 26, n° 6, 1983, pp. 113-126.

408. VARMA M., TRABOULAY E. Comparison of native and microwave-irradiated DNA. *Experientia*, 33, 1977, pp. 1649-1650.

409. VOGELHUT P. Microwave techniques in biophysical measurements. *J. Microw. Power*, 3, n° 3, 1968, pp. 143-147.

410. VOGELHUT P. Interaction of microwave and radio frequency radiation with molecular systems. In : CLEARY S. Biological effects and health implications of microwave radiation. Symposium proceedings, Richmond, Commonwealth Univ., 17-19 Sept. 1969, US Dep. Health, Educ., Welfare, BRH, DBE 70-2, PB 193898, Library of Congress card n° 76-607340, June 1970, pp. 98-100.

411. WACHTEL H., SEAMAN R., JOINES W. Effects of low-density microwaves on isolated neurons. In : TYLER P. Biologic effects of non-ionizing radiation. Proceedings of a conference, *Ann. New York Acad. Sci.*, 247, 1975, pp. 46-62.

412. WAIT J. Electromagnetic waves in stratified media. National Series of monographs on Electromagnetic Waves. Vol. 3. London, Pergamon Press, 1962, 372 p.

413. WEBB M., GUY A., Mac DOUGALL J. Assessment of the EM field coupling of 915 MHz oven leakage to human subjects by thermographic studies on phantom models. In : Proceedings of the IMPI Symposium, Leuven, 1976. (Paper 5.A.1.).

414. WEBB M., GUY A., Mac DOUGALL J. Assessment of the EM field coupling of 915 MHz oven leakage to human subjects by thermographic studies on phantom models. *J. Microw. Power*, 11, 1976, pp. 162-164.

415. WEIL C. Absorption characteristics of multilayered sphere models exposed to VHF/microwave radiation. *IEEE Trans. Biomed. Eng.*, BME-22, 1975, pp. 468-476.

416. WILSON D., LANGAN D., SCHAUBERT D., GLISSON A. Computation of electromagnetic absorption in biological material via triangular and tetrahedral cell modelling. In : Proceedings of the IMPI Symposium, Chicago, 1985, pp. 112-114.

417. WONG G., STUCHLY S., KRASZEWSKI A., STUCHLY M. Probing electromagnetic fields in lossy spheres and cylinders. *IEEE Trans. Microw. Theory Tech.*, MTT-32, n° 8, 1984, pp. 824-828.

418. WU T. Electromagnetic fields and power deposition in body-of-revolution models of man. *IEEE Trans. Microw. Theory Tech.*, MTT-27, n° 3, 1979, pp. 279-283.

419. WU T., TSAI L. Numerical analysis of electromagnetic fields in biological tissues. *Proc. IEEE*, 62, n° 8, 1974, pp. 1167-1168.

420. WU T., TSAI L. Electromagnetic fields induced inside arbitrary cylinders of biological tissue. *IEEE Trans. Microw. Theory Tech.*, MTT-25, n° 1, 1977, pp. 61-65.

421. YAMAURA I. Measurements of 1.8-2.7 GHz microwave attenuation in the human torso. *IEEE Trans. Microw. Theory Tech.*, MTT-25, n° 8, 1977, pp. 707-710.

422. YOVA D., LOUKAS S., BOUDOURIS G. Changes in liposomes permeability induced by Gramicidin D after microwave irradiation. *IEEE Trans. Microw. Theory Tech.*, MTT-32, n° 8, 1984, pp. 891-893.

423. ZHU S., CHEN K., CHUANG H. Interaction of the near-zone fields of a slot on a conducting sphere with a spherical model of man. *IEEE Trans. Microw. Theory Tech.*, MTT-32, n° 8, 1984, pp. 784-795.

424. ZORE V., KIMEL'FELD' O., SUDALEVA V., KOBYZEVA L., GENKINA E. Complex dielectric permeability in the frequency range 100 Mcps. to 500 Mcps. of human blood serum in normal conditions and certain diseases. *Biophys.*, 12, n° 12, 1967, pp. 142-144.

425. Les radiofréquences et la santé. Health aspects of radio-frequency and microwave exposure. Ministère de la Santé Nationale et du Bien-être Social du Canada, 1ère partie, nov. 1977, 77-EHD-13, 80 p. ; 2ème partie, mars 1978, 78-EHD-22, 107 p.

426. Ondes électromagnétiques et biologie Actes du Symposium International de l'Union Radio-Scientifique Internationale et du Comité National Français de Radioélectricité Scientifique, 1980, 330 p.

Chapter 2

Biological effects

Until the turn of the century, our knowledge and applications of electromagnetic radiation were mostly confined to the infrared and the visible ranges. By now, the whole of the electromagnetic spectrum has been mastered and is used routinely for communication, production, therapy, diagnostics, protection, and destruction.

Human and animal physiology is the product of many millions of years of survival in the midst of natural electromagnetic emissions, mostly of solar and cosmic origin. It is not adapted to radiation that is not normally a significant part of the natural background in the biosphere. Any modification of this additional radiation may produce harmful effects whose nature, evaluation, and regulation are the preoccupation of scientists and public bodies alike.

The first experimental work in this area was carried out about sixty years ago by E. Pflomm (1931) who studied bradycardia induced by high-frequency radiation. As early as 1932, J. Audiat put forward the hypothesis of nonthermal interaction, which was disputed by L. Hill and H. J. Taylor (1936) for whom the observed effects could be explained by a pure thermal interaction. Indeed, this was the start of a continuing controversy between the supporters of exclusively thermal interactions and those who believed that electromagnetic radiation had a thermal effect for high doses and an athermal (also called *specific*) effect for very low doses.

The advent of radar and its applications during World War II gave rise to some serious ethical dilemmas related to possible health risks to radar operators. However, the United States Naval Research Laboratory set up an inquiry into this subject and concluded that there were no risks. Hyperthermia machines and domestic microwave ovens were therefore developed very rapidly in the United States in the few years just after the war.

Research into the physiological effects of electromagnetic radiations had, however, continued to be active throughout the mid-forties [191], [483], [268]. The risk of cataract was demonstrated in 1948 [192] on a patient treated with diathermia and, in 1952 on an operator of a UHF power supply [381]. These results were subsequently confirmed by a series of experiments with animals [715], [193], [194]. By the end of the 1940s, there were reports of testicular degeneration in irradiated rats [397] and of necrosing coagulations in the brain of rabbits irradiated at 2.45 GHz with behavorial consequences [650].

The United States Navy held a conference in 1953 to evaluate the risks associated with electromagnetic irradiation. The conference was held in Bethesda (Maryland) and was attended by specialists on microwaves, bioelectricity, and thermal regulation (H. Schwan, T. Ely, J. Hardy, and D. Goldman). The outcome of this conference was inconclusive because of lack of data. Equally inconclusive was another meeting held in 1955 at the Mayo Clinic of Rochester, Minnesota, the pioneer in microwave hyperthermia.

In 1957, the accidental death of a young member of the military forces, which might have been caused by overexposure to radar, provided disturbing evidence for the danger of excessive microwave irradiation [545], [660].

At the same time, researchers in the Soviet Union, in particular, Z. Gordon, were accumulating evidence for neurological symptoms in personnel exposed to extremely low doses. Following certain American reports carrying similar evidence, the United States Navy formed a study group led by Colonel G. M. Knauf. This was the Tri-Service Program, which concluded that the interaction of electromagnetic radiation with the body was purely thermal, and that the nonirrigated tissues were particularly at risk [672], [836], [680].

In 1965, the American Secret Service revealed a strange episode in the Cold War. The American embassy in Moscow was, since 1953, submitted by the Soviets to a very low level of radiofrequency irradiation, modulated in amplitude and frequency by an ELF signal. The power flux was $5\,\mu\mathrm{W\,cm^{-2}}$, $9\,\mathrm{hr\,d^{-1}}$ between 1953 and May 3, 1975. It rose to $15\,\mu\mathrm{W\,cm^{-2}}$, 18 hours per day between June 1975 and February 7, 1976, and was then reduced to $1\,\mu\mathrm{W\,cm^{-2}}$. There was serious concern about the health of embassy personnel who became involuntary guinea pigs. However, a long-term clinical study, carried out under the umbrella of the secret Project Pandora, compared their state of health with that of others in selected embassies throughout Eastern Europe and concluded that there were no significant differences [484]. In the United States, all this fueled anew the debate on possible nonthermal effects, whose principal supporters were A. H. Frey and W. R. Adey.

The controversy, fed by prejudices and by ignorance of East European work, was formulated in terms of an East *v.* West, thermal *v.* nonthermal altercation, and lasted until the early eighties despite tentative Soviet-American rapprochement [184]. A semantic confusion was then added to the original misunderstanding: for physicists, a nonthermal or athermal effect can only be caused by a wave that integrally retains its electromagnetic form without any thermal degradation of tissue. For biologists, an athermal effect is a phenomenon that involves no measurable rise in temperature. A very strict definition has finally been adopted whereby a distinction is made between true athermal or nonthermal effects produced by the direct action of the electric field and the microthermal interactions in which there is possibly a nonmeasurable rise in temperature. The existence of the latter had long been postulated in the Soviet Union [667].

There is currently a consensus of international opinion on the possibility of certain nonthermal effects (modulation of bacterial growth as a function of frequency, modulation of transmembrane calcium fluxes, and arrhythmia in the isolated heart), but a complete understanding of these phenomena has continued to elude us. The widest divergence of opinion relates to effects on hematopoiesis and lymphopoiesis, the reproduction and modulation of the cardiac rhythm in animals, the autonomous nervous systems, and to possible psychological effects. Western scientists are beginning to show some interest in long-term, low-dose irradiation effects that have until now been the exclusive preserve of East-European researchers.

The more promising research topics for the future include the effects of modulation, the power and modulating-frequency windows reported by Adey, interactions with biorhythms, analyses of membrane interactions, the effect of hotspot, and the central role of the hypothalamus [342], [955].

2.1 Cells and micro-organisms

At the lowest level of life, the effect of electromagnetic radiation on the living cell is to cause its crude destruction or to produce changes in its morphology, metabolism, lifespan, and reproduction, which may in certain cases indicate some effect on the genoma.

Research into the bacteria *E. coli* and *B. subtilis*, reported in [317], revealed only a lethal thermal effect, whereas [905] reported that power levels as low as $7\,\mu$W at 136 GHz produced a perturbation of cellular division and an inhibition of metabolic processes that could not be explained in thermal terms. The same group [904] reported a higher bacterial growth rate at 68 GHz, and a lower one at 60, 71, and 73 GHz under the same irradiation conditions. This variability in frequency response has also been reported by other workers [226], [82]: a three-hour exposure of *E. coli* to $10\,\text{mW cm}^{-2}$ at 73 GHz produced a fall in the growth rate of between 6% and 30%, with a peak probability of between 20% and 30% (Fig. 2.1).

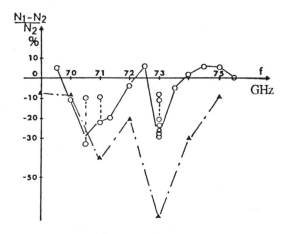

Figure 2.1 Relative number of *E. coli* colonies as a function of frequency [82]. Triangles - results from [904], open circles - results from [82], N_1 - number of colonies in irradiated samples, N_2 - number of colonies in control samples.

In another study of *E. coli* and the yeast *S.cerevesiae* [201], the growth rate was measured in the frequency range 69.5–75.5 GHz in steps of 500 MHz. A slowing in the growth rate occurred at 70.5 and 73.0 GHz (3 hr, $10\,\text{mW cm}^{-2}$).

A study of the yeast *S.cerevesiae* [872], [871] examined its growth rate as a function of frequency in steps of 100 MHz between 2 and 12.4 GHz, and showed considerable changes in the number of colonies between and after irradiations (Fig. 2.2).

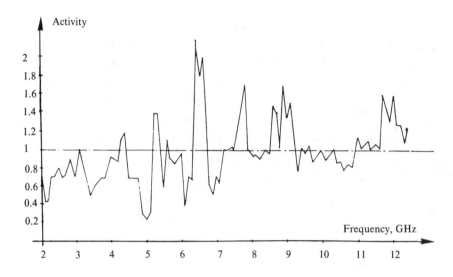

Figure 2.2 Variation of the activity of beer yeast as a function of frequency [872].

At frequencies corresponding to the greatest growth (2.1 and 5 GHz), the yeast activity showed a clear minimum (without cellular destruction) after about 90 s of exposure (Fig. 2.3).

The same phenomenon was observed in *S.cerevesiae* [334], [336]: there was a decrease or increase in growth rate as a function of frequency in a 10-MHz window in the vicinity of 42 GHz [80].

In contrast, other studies [92], [245] chose as a test of vitality of *E. coli* and *S. Typhimurium* their ability to form colonies, and found no particular effects other than thermal at 1.7, 2.45, 68.0, 74.0, and 136.0 GHz CW $(20\,\mathrm{mW\,cm^{-2}}, 40\,\mathrm{W\,kg^{-1}}, 90\,\mathrm{min})$, and at 8.5 and 9.6 GHz PW $(1, 5, \mathrm{and}\ 45\,\mathrm{m\,W\,cm^{-2}}$ average flux density). Similar conclusions were reached in a study of *E. coli* B at 2.6-4.0 GHz $(10\,\mathrm{mW\,cm^{-2}}, 20\,\mathrm{W\,kg^{-1}}, 10\,\mathrm{hr})$, but the small number of frequencies tested reduced the overall significance of these measurements. All in all, the existence of specific effects as functions of frequency is relatively well established.

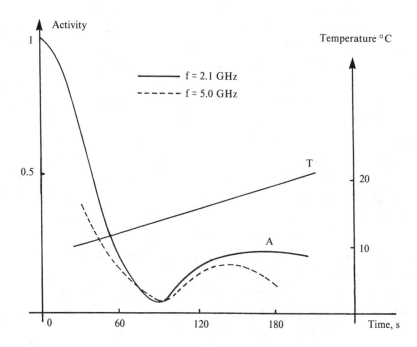

Figure 2.3 Activity of beer yeast as a function of exposure time at constant frequency and power flux density[872].

With regard to genetic effects, the consensus rather negative and only a few researchers believe in their existence. It is reported in [902] and in several Soviet periodicals [812], [813], [66], [227] that there is evidence for a mutagenic effect in *E. coli* and *Aspergillus oryzae*. Some adverse effects have been found [138] on the somatic cells of hamster lungs irradiated at very high levels ($500\,\mathrm{mW\,cm^{-2}}$ at $2.45\,\mathrm{GHz}$ in a waveguide, $1059\,\mathrm{W\,kg^{-1}}$ for 20 min). The cell growth rate fell by 30%, the division rate became faster, and starlike giant cells were found to appear. Finally, after 12 days of incubation at 37°C, the irradiated cells assumed an elongated form and they began to proliferate parallel to each other. These anomalies appeared to be irreversible and were undoubtedly due to a thermal effect.

Most of the data on micro-organisms and on somatic cells

are in remarkable agreement. For instance, no particular mutagenic activity was observed in bacteria exposed to 15 or $70 \, \mathrm{W \, kg^{-1}}$ at $2.45 \, \mathrm{GHz}$ and to $3 \, \mathrm{W \, kg^{-1}}$ at $1.7 \, \mathrm{GHz}$ [94]; comparable negative results were reported for the yeast *S.cerevesiae* and the bacteria *S.typhimurium* irradiated with continuous waves ($2.45 \, \mathrm{GHz}, 20 \, \mathrm{mW.cm^{-2}}, 40 \, \mathrm{W \, kg^{-1}}, 90 \, \mathrm{min}$) and modulated waves ($8.5$–$9.6 \, \mathrm{GHz}$, 1, 5, and $45 \, \mathrm{mW \, cm^{-2}}$ average flux) [245]; no effect was observed for the fungus *Aspergillus nidulans* ($2.45 \, \mathrm{GHz}$, 1 hr in $10 \, \mathrm{mW \, cm^{-2}}$ CW or PW, and $1\text{-}\mu s$ pulses at $600 \, \mathrm{Hz}$) [572]. In the fungus *Physarium polycephalum*, the accelerated synthesis of DNA was indicated [572] by the absorption of thymidine-^{3}H, a tritium-labeled precursor of DNA. This acceleration, twice as high as that encountered in classical hyperthermia, suggests that radiation had an effect on cellular metabolism.

Several studies [82], [38], [201] have demonstrated the absence of a mutagenic effect even in strains of *E. coli* having a deficient enzymatic mechanism for the reparation of DNA damage ($28 \, \mathrm{W \, kg^{-1}}$ at 9.4, 17, 70 and $75 \, \mathrm{GHz}$). A study of the fungus *Aspergillus amstelodami* [227] confirms the absence of morphologic mutations and of mutations relative to the resistance to 8-azaguanine, a growth inhibitor similar to guanine that is itself a precursor of DNA ($8717.5 \, \mathrm{MHz}, 2.3$ and $2.9 \, \mathrm{mW \, cm^{-2}}$ for 3 and 6 hr).

The persistence of these negative experimental reports led a number of workers [586], [306], [80] to conclude that a genetic effect at the cellular level had not been demonstrated, at least for power densities on the order of $10 \, \mathrm{mW \, cm^{-2}}$. Further information may be found in [51], [405], [929], [930], [369], [605], [50], [257], [370], [395], [812], [43], [896], [253], [848], [53], [136], [137], [492], [887], [889], [335], [341], [390], [501], [845], [201], [830], [202], [206], [500], [507], [134], [203], [217], [302], [311], [365], [110], [333], [723], [714], [753], [766], [564], [870].

2.2 Blood and hematopoiesis

The liquid component of blood (i.e., plasma) contains a suspension of proteins, lipids, sugars, vitamins, pigments, coagulation factors, and so on. Three types of cell evolve in this liquid, namely, (1) erythrocytes, i.e., nonnucleated red blood cells that transport oxygen and carbon dioxide, (2) thrombocytes, i.e., nonnucleated platelets, that are responsible for coagulation and hemostasis, and (3) leucocytes, i.e., white blood cells that can be further subdivided into granulocytes, lymphocytes, and monocytes. The granulocytes are polynuclear and can be subdivided into basophils, eosinophils, and neutrophils, the last being in charge of the phagocytosis of bacteria. The function of lymphocytes is to identify antigens and produce antibodies; the monocytes or macrophages play their part in the immune system by phagocyting antigenes [171].

There has been a number studies of the action of microwaves on plasma components, including [337], [643], [28], [841], [67], [897], [743], [210], [211], [244], [343], [305], [344], [296], [932] and [81].

Soviet studies have shown that microwaves, whether CW or PW, could influence the composition of blood even at very low levels. This observation was confirmed by R. Lovely of the University of Washington, who was able to repeat one of the Soviet experiments on rats exposed for 50 hours per week for three months, using $500\,\mu\mathrm{W\,cm^{-2}}$ at 2375 MHz, with identical results. In particular, the activity of cholinesterase present in the serum fell, the sodium level rose, and the potassium level fell [643], [548].

A change in the concentration of Na^+, and K^+, and Ca^{++} ions has also been demonstrated with much higher power densities [296], [81]. Another result has been a reduction in the activity of cholinesterase due to a fall in its concentration after 42 days exposure to a modulated 3-GHz wave $(10\,\mathrm{mW\,cm^{-2}}, 1\,\mathrm{hr\,per\,day}, 1\,\mathrm{hr\,per\,day})$ [820], [548]. An effect on the level of triglycerides has also been reported [210],

[211], [244], as well as on the glycemia of rabbits subjected to microwave hyperthermia at 43°C [932]: The level of glucose rose from 0.97 to 2.05 g dm^{-3} between the third and eighth hour following the end of treatment, and then fell rapidly to 1.10 g dm^{-3}, which was followed by another small maximum of 1.3 g dm^{-3} between the fourteenth and the seventeenth hours. The mechanism implicated by these observations is obviously thermal because hot air produces a similar phenomenon. The thermal nature of all the microwave-induced modifications of composition was demonstrated in a study involving exposure to 5, 10, and 25 mW cm^{-2} at 2.45 GHz [897].

The absence of any athermal effect was strongly suggested by a series of analyses, repreated 11 times in 15 months, of the blood of rats submitted, during this period, to 2.45-GHz electromagnetic radiation modulated by 10-μs pulses (average power density 500 μW cm^{-2}, 0.4 W kg^{-1}) [343], [344]. The results of these experiments, reported by Gandhi [305], showed that there was no change in the protein level or the level of twenty-eight other parameters of plasma and structure elements. Further studies [199], [198], [305] revealed no hematologic effects in rats irradiated at 915 and 2450 MHz (5 mW cm^{-2}, 2.46 and 1.23 W kg^{-1}, respectively). Given our present state of knowledge, it is still difficult to be sure that there are no athermal effects.

Hematopoiesis, or the production of blood cells, relies on the bone marrow as the source of erythrocytes, thrombocytes, granulocytes, and lymphocytes, whereas the spleen is the source of monocytes. Lymphoid tissue (spleen, thymus, Peyer's patches, and the bursa of Fabricius in birds) contribute to the maturation of lymphocytes (Fig. 2.4). The effect of electromagnetic radiation on hematopoiesis, if it were to be shown to exist, would be reflected in the blood count. There is no unanimity in the literature on this point, even for very similar exposures. For example, one study [435] showed that an exposure at 3 GHz (10 mW cm^{-2}, 1 hr per day for 20 days and 5 min per day for six days) produced an increase in the granulocyte level and a reduction in lymphocytes and monocytes, without recovery to normal values for several months. Another study [213] reported that rats exposed to 24-GHz radiation ($20\,\text{mW cm}^{-2}$, one exposure of 10 min or 7 hr) showed an increase in neutrophils and a reduction in leucocytes and lymphocytes, with a recovery to normal values in two days and one week, respectively. The erythrocytic count and the hemoglobin level were found to rise for some strains of rat and fall for others. Another study [735] reported leucocytic proliferation in mice four days after a single 5-min exposure to $100\,\text{mW cm}^{-2}$ at 2.45-GHz. Soviet studies [677], reviewed in [548], revealed reduced hemoglobin, leucocyte, and lymphocyte counts in rats during the second half of a seven-week exposure to 2736-MHz radiation modulated at 395 Hz by 2.6-μs pulses (20 hr per week at $24.4\,\text{mW cm}^{-2}$). Recovery to normal occurred in approximately 10 weeks after the end of the experiment. After the first week of irradiation, the neutrophils exhibited increased activity of the alkaline phosphatase enzyme.

On the negative side, the authors of [229], [349] reported that the blood count of mice was unaffected by exposure to 2.45-GHz radiation ($10\,\text{mW cm}^{-2}$, $17\,\text{W kg}^{-1}$, 180 days of exposure). Negative results were also reported in [199] and [198]: In spite of certain variations with time in the leucocyte and erythrocyte counts in mice, the conclusion was that there were no statistically significant effects at 2.45 GHz ($5\,\text{mW cm}^{-2}$, $1.23\,\text{W kg}^{-1}$, 40 hr per week for 16 weeks) and at 915 MHz ($5\,\text{mW cm}^{-2}$, $2.46\,\text{W kg}^{-1}$, same duration).

These negative results are evidently difficult to interpret in the light of East European studies that reported significant effects for power densities below $1\,\text{mW cm}^{-2}$. For instance, Gončar [320], [548], whose results were reviewed by MacRee [548], observed an acceleration in the metabolism of neutrophils (glycogene and alkaline phosphatase) for three weeks, followed by a return to normal for rats in an electromagnetic flux of 10 or $50\,\mu\text{W cm}^{-2}$ at 2375 MHz for thirty days, seven hours per day. Glycolyse activation was observed in the fourth

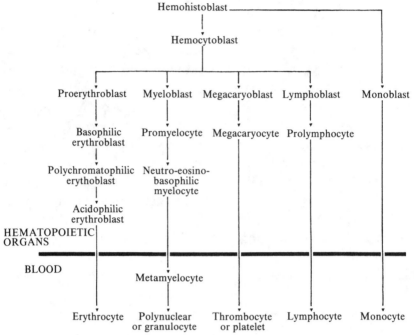

Figure 2.4 Hematopoiesis.

week of exposure to $500\,\mu\mathrm{W}\,\mathrm{cm}^{-2}$.

There are very few results that relate directly to hematopoietic tissues. A study of mice [691] revealed the induction of lymphoid leukemia in a year of chronic exposure to X-band radiation. Anomalies in nuclear structure and in the mitosis of bone marrow erythroblasts were observed in [44], [45], [46]. ($3\,\mathrm{GHz}, 3.5\,\mathrm{mW}\,\mathrm{cm}^{-2}$, 3 hr per day for 3 months). The proliferation of hematopoietic cells in spleen and bone marrow was observed in [735] ($2.45\,\mathrm{GHz}, 100\,\mathrm{mW}\,\mathrm{cm}^{-2}$, single 5-min exposure).

It is rather difficult to conclude this section: some effects have been reported at very low irradiation levels by East European scientists. At medium and high levels, Western studies on the whole tend to suggest that irradiation has no particular effect, whilst the corresponding East European studies report definite effects in hematopoietic tissue and certain changes in blood count. These effects, if indeed they exist, remain difficult to identify and quantify.

There is a wealth of literature on this subject, including [483], [268], [337], [28], [596], [869], [402], [679], [96], [183], [188], [734], [185], [186], [228], [744], [186], [187], [925], [182], [743], [676], [551], [189], [678], [684], [304], [640], [815] and [725].

2.3 Immune system

Immune response is the collective phrase used to describe the defensive reaction of an organism to invasion by a foreign substance or antigen. In the face of such aggression, a mammalian organism mobilizes first the nonspecific immune resources of natural resistance that act through vascular and cellular mechanisms. An inflammation appears at the vascular level that is chemically mediated and is due particularly to pro-inflammatory amines (histamine, serotonin) released by tissue cells known as mastocytes, whose excitation is mediated by certain antibodies (IgE) and by the C3a and C5a components of complement. The latter is a complex system of about 20 plasma proteins that play an important part in natural immunity due to its components C5 through C9 that can attack and lyse cellular membranes by causing an increase in enzyme concentration (Fig. 2.5). In addition to these defense mechanisms, there is also the limited bacteriocidal action of certain plasma proteins such as lysozyme, secreted by macrophages, the transformation of fibrinogen into fibrin, and the fixation of blood iron that deprives bacteria of their growth factor. At the cellular level, natural immunity relies on phagocytosis, which is provided by the polynuclear neutrophils and by monocytes or macrophages. Monocytes are large cells that are able to divide, have long lifespans, and display high phagocytic activity. Polynuclear neutrophils originate from cellular strains of bone marrow that differentiate by segmentation of the nucleus, by enrichment with enzymes, and by acquisition of receptors

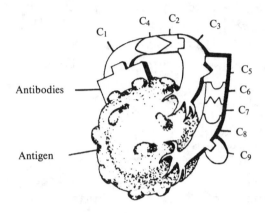

Figure 2.5 Lysis of an antigen by complement [624].

destined for the C3b component of complement and for antibodies.

When natural resistance is insufficient, the body mobilizes its acquired immunity and targets selectively a given antigen by a mechanism that recognises specific molecular structures, the so-called antigenic determinants. This immunity is essentially determined by circulating antibodies, or immunoglobulins (Ig), consisting of four polypeptide chains of which one extremity, the variable region, presents a sequence of amino acids, which is a function of the antibody under consideration. Immunoglobulins can be cut into two fragments, Fab, that include the variable region and are responsible for fixation on the antigen, and a fragment Fc that is able to fix the complement or bond to the membranes of phagocytes that carry suitable receptors. In human antibodies are classified as IgG, IgA, IgM, IgD, and IgE, depending on their structure. They attach themselves to antigens, and sensitize them to the joint lytic action of lysozyme and complement. They also activate the complement, which enhances inflammatory action through its components C3a and C5a. These act as chemotactic agents that attract phagocytes and stimulate mastocytes. These processes constitute the humoral immunity system. A second type of process is recruited when the first fails: this is cell-mediated immunity that activates two lines of specialized cells (i.e., B and T lymphocytes that originate in bone marrow). It is probably there that B lymphocytes acquire their immunologic competence, whereas T lymphocytes acquire theirs in the thymus. Immunologic competence manifests itself by the appearance of receptors, i.e., molecular groups capable of identifying and fixating specifically complement derivatives (CR), an antigen, and an Fc antibody fragment (FcR). The receptors of B lymphocytes are themselves membrane immunoglobulins, essentially IgD and IgM; those of T lymphocytes, recently identified [562], are also found to consist of polypeptide chains but with a different structure (Fig. 2.5). The lymphocytes migrate toward secondary lymphoid organs (spleen, ganglions, etc.) where they are sensitized to a particular antigen and then circulate within the body along blood and lymphatic vessels (Fig. 2.6).

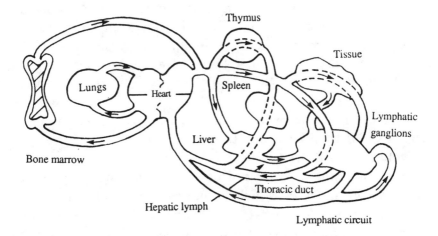

Figure 2.6 Lymphatic system [624].

A complex process of cellular cooperation is initiated when an infection appears. The antigens, already partially degraded by macrophages, associate themselves to a protein of the major histocompatibility complex (MHC). A T lymphocyte possessing a specific receptor for the antigen-protein complex attaches itself to this complex and provokes the secretion by the macrophage of the immuno-hormone interleukine 1 (IL 1) that simulates the multiplication and differentiation of T lymphocytes. These latter produce several subpopulations, namely, auxilliary T lymphocytes, known as suppressors and activators. The activator lymphocytes act as killer or cytotoxic lymphocytes (CAL); they attach themselves to an infected cell and destroy it if the antigen is present at the same time as a histocompatability protein that authenticates the infected cell as endogenous. Certain activator T lymphocytes that are not subject to MHC restriction are capable of spontaneously killing certain cellular varieties *in vitro*, without prior immunization. These natural killer lymphocytes (NK) seem to autoregulate by secretion of interferon.

The auxillary and suppressor T lymphocytes (Ta and Ts) control antagonistically the stimulation of B lymphocytes. The latter identify antigens by their receptors, fixate them, and proceed to degrade them. This results in an antigen-MHC protein complex to which the auxillary T lymphocyte binds. It becomes activated and releases interleukins, including the proliferation factor BCGF and the factor BCDF which is responsible for differentiation in plasmocytes. The transmission of the proliferation order to the lymphocyte nucleus seems to be ensured by a cascade of intracellular enzymatic reactions involving the protein kinase C enzyme. The B lymphocyte transforms into a young cell or lymphoblast and then into a clone of plasmocytes, which are large nondividing cells that secrete antibodies at the rate of 2,000 molecules per second, with the same antigen fixation properties as the initial B lymphocyte. These antibodies mark the antigens in preparation for their destruction. The process continues until it is inhibited by suppressor T lymphocytes. It must be added that certain B and T lymphocytes differentiate into circulating memory cells capable of reacting to antigens of the same species [624], [258], [272], [261], [273], [562], [612], [736], [878].

2.3.1 Natural resistance

Soviet and Polish researchers have noticed a depression in phagocytic activity of neutrophils in small mammals exposed to a power flux density of 1–3 mW cm^{-2} at 46 GHz or lower. The exposures were 20 days at 46 GHz (1 mW cm^{-2}) [935] and six to twelve weeks at 3 GHz (3 mW cm^{-2}) for six hours per day [846]. The latter study showed that the rabbit infected with *S. aureus* had lower resistance. Other studies involved 1–3 mW cm^{-2} exposures of only 20–60 min at several wavelengths in the meter and centimeter bands [402]. After exposures that were two or three orders of magnitude lower (1, 5, 10, and 50 μW cm^{-2} at 2375 MHz, 30 days), the activity of neutrophils was found to be higher [749] and, unexpectedly, the effect was inversely proportional to the incident power. A stimulation of the phagocytic activity of macrophages was observed for intermediate power levels of around 10 mW cm^{-2}, with better resistance to viral infections [46], [77], [709], [108]. The subject is also discussed in [688] and [565].

The inflammatory reaction seems to be inhibited by microwaves [479], which could be an indication of an effect on histamine-releasing mastocytes [754], [661], [662], [303]. There are reports [882] of degenerative forms such as the giant mastocytes found in the peritoneal cavity of young rats (2.45 GHz, 5 min, rectal temperature at the end of the treatment 42°C). The facts in this area are not well established: for example, an increase in the brain histamine level was reported in [389] after local irradiation.

As for the titres of complement and of lysozyme, Soviet workers quoted in [548] found an irreversible rise in the former in guinea pigs exposed for thirty days to 1, 5, 10, and 50 mW cm^{-2} (2375 MHz) with maximum effect at the lowest level [749] and a reduction by 50% in both in mice exposed for twenty days to 1 mW cm^{-2} (46 GHz).

The effects on acquired immunity are better documented.

They relate to lymphopoiesis and to the characteristics and activity of lymphocytes.

2.3.2 Lymphopoiesis

Morphologic modifications of lymphoid organs are now believed to be very improbable despite certain published observations [479], [935]. By contrast, accelerated production of lymphocytes by bone marrow was reported in several studies [183], [185], [186], [187], [182], [189]: mice immunized to sheep erythrocytes (thymus dependent antigen) and exposed to low levels at 2.95 GHz radiation ($500 \, \mu W \, cm^{-2}, 2 \, hr \, d^{-1}$, 6 to 12 weeks) showed a very significant (by a factor of up to 2) increase in the quantity of lymphoblastoid cells and lymphocytes. The effect was more pronounced after six weeks than after twelve. Analogous results were obtained with a rabbit submitted to $5 \, mW \, cm^{-2}$, two hours per day for six months. The maximum effect was reached after one or two months and was followed by a return to normal, suggesting that an adaptation process was taking place.

2.3.3 Multiplication of lymphocytes FcR^+ and CR^+

A series of publications [914], [915], [916], [918], [757], [919] reported on the peculiar effect of 2.45-GHz radiation, either CW- or amplitude-modulated at 12 Hz, specifically concerning positive lymphocytes with Fc receptors (FcR^+) or complement receptors (CR^+). Six-day-old mice exposed to $14 \, W \, kg^{-1}$ for thirty minutes in a waveguide were found to show an increase of approximately 30% in the normal cell count after three days, and this continued for five to six days before the count returned to normal. The threshold for the observation of this phenomenon is an absorbed dose of $10 \, J \, g^{-1}$ (or $5.6 \, W \, kg^{-1}$ for 30 min) with the possibility of a cumulative effect [831]. It was suggested that a number of exposures lower than the threshold were equivalent to a single high exposure of the same duration. There was apparently no effect on the proliferation of the cells themselves because there was no modification of the synthesis of DNA, RNA, or proteins by the bone-marrow and spleen lymphocytes. The mechanism responsible for this phenomenon could be athermal, membrane-related, or chromosonic interaction with lymphopoietic cells of the B lymphocyte family at the stage of their maturation corresponding to the appearance of Fc or complement receptors [917]. Endocrine regulation by corticosteroids does not intervene [755] and, strangely, this effect that may be conditioned by the presence of a peculiar gene in the MHC. The use of a particular strain of mice (CBA/J) with this genetic capacity could explain [108] the failure of other researchers [801] to reproduce the phenomenon on another strain of mice (BALB/c; 2.45 GHz, 15, 20, 30, and 40 mW cm^{-2}, 30 min, 0.7 W kg^{-1} per mW cm^{-2}). The latter studies showed that a significant increase was observed only in 16-week-old mice under 40 mW cm^{-2} (i.e., a level high enough to induce other types of interaction) (e.g., an increase in endocrine secretion of corticosteroids at the suprarenal level).

The interaction of low-frequency fields or of amplitude-modulated RF fields with the membrane receptors of lymphocytes was studied in [142], [143], [329] and [504]. Modulated microwaves (2.45 GHz, 60 Hz) had no effect on the fixation of antigen-antibody complexes on the surface of LcB *in vitro* or *in vivo* in mice (2.25–45 W kg^{-1}, 30 min) [832], [833], [834].

2.3.4 Stimulation of the response of lymphocytes to mitogens

Immunologists use specific lymphocyte mitogens that provoke proliferation and differentiation into lymphoblasts: phytohemagglutinin-P (PHA) and concanavalin A for T lymphocytes, lipopolysaccharose for *E.coli*, and the derivatives of tuberculin for B lymphocytes. The lymphoblastoid transformation is manifested by enhanced cellular DNA synthesis that may be detected by means of the radioactive tracer methylthymidine ^{3}H.

A lymphocyte preparation was exposed in [823] and [824] to 2.95 and 3 GHz radiation modulated by 1-μs, 1.2 kHz pulses (7 mW cm^{-2} at 4 hr per day and 20 mW cm^{-2} at 15 min per day for 3 to 5 days). An increase was found in the mitotic index after stimulation by PHA. The number of lymphoblasts increased by a factor of 2–6, depending on the duration of exposure. The phenomenon was confirmed by other studies [21], [252], [182]. In the latest of these, the experiments were performed *in vivo* on a rabbit (2.95 GHz, modulated wave, $5 \, mW \, cm^{-2}$, 2 hours per day for 6 months) and on a mouse ($500 \, \mu W \, cm^{-2}$, 2 hr per day for 6 to 12 weeks), in which there was a doubling of the number of lymphoblasts derived from circulating lymphocytes after six weeks of irradiation, followed by a return to normal, despite the continuation of treatment, which seemed equally to indicate an adaptation to radiation. Comparable results were reported in [803] on rat lymphocytes after exposure *in utero* to CW radiation (425 or $2450 \, MHz, 5 \, mW \, cm^{-2}$, 1–$5 \, W \, kg^{-1}$ 4hr per day). In a later study, the same authors [810] demonstrated a strong response to ganglion lymphocytes, but no change in circulating lymphocytes in rat exposed *in utero* from day 12 of gestation until day 20 or day 40 of life ($425 \, MHz, 3.1$–$6.7 \, W \, kg^{-1}$ 4hr per day) or in a 42-day old rat irradiated *in utero* from day 6 to day 19, for 16 hr per day. In contrast, other work by the same authors reported the absence of the effect under $2.45 \, GHz, 10 \, mW \, cm^{-2}, in \, vitro, 19 \, W \, kg^{-1}$, one to four hours of treatment [797]; there was also no effect in mice exposed to $2.45 \, GHz, 5$–$35 \, mW \, cm^{-2}, 4$-$25 \, W \, kg^{-1}$, $15 - 30$ min per day for 1–22 days [806]; in rats exposed to $100 \, MHz$ but below basic metabolic rate [800]; and in rats exposed *in utero* to $2.45 \, GHz, 28 \, mW \, cm^{-2}, 16.5 \, W \, kg^{-1}$ [807].

Other work reporting no effect includes [362] (rats exposed to $2.45 \, GHz$ at $5, 10,$ and $20 \, mW \, cm^{-2}$, (i.e., $0.7, 1.4$) and $2.8 \, W \, kg^{-1}$ for 4, 24, or 44 hr) and [720] which points out the difficulty of *in vitro* dosimetry and the importance of hotspots. An experiment with hamsters ($2.45 \, GHz, 15 \, mW \, cm^{-2}$) showed increased response to mitogenic stimulation in the absence of macrophages, the latter acting as regulators of mitosis [391]. On the other hand, [549] reports on reduced response in rabbits, 42 days after the last irradiation session

($2.45\,\mathrm{GHz}, 7\,\mathrm{mW\,cm^{-2}}, 1.5\,\mathrm{W\,kg^{-1}}$ average, 23 hours per day for 6 months).

It seems probable, in the light of these results, that certain irradiation conditions may have a stimulating effect on the lymphoblastoid transformation. If confirmed, such an effect could have therapeutic implications [586].

2.3.5 Modulation of the activity of activator T lymphocytes

Microwaves are thought to have have a positive or negative effect on CAL and NK activity, depending on the circumstances. Despite some negative conclusions (hamsters at $2.45\,\mathrm{GHz}, 15$ and $30\,\mathrm{mW\,cm^{-2}}$) [391], the effect was confirmed by a number of other reports albeit contradictory in some cases. For instance, a temporary lowering of NK activity was reported for hamsters under $25\,\mathrm{mW\,cm^{-2}}$ at $2.45\,\mathrm{GHz}$ [927]; the effect was absent under $15\,\mathrm{mW\,cm^{-2}}$, which was in agreement with experiments on rats irradiated *in utero* ($2.45\,\mathrm{GHz}, 28\,\mathrm{mW\,cm^{-2}}, 16.5\,\mathrm{W\,kg^{-1}}$) [807] and adult mice treated for two to nine days with $30\,\mathrm{mW\,cm^{-2}}$ at $2.45\,\mathrm{GHz}$ and without any apparent effect at 5 and $15\,\mathrm{mW\,cm^{-2}}$ [809]. Similar observations were reported for amplitude modulated waves [532].

In contrast to all the above conclusions, the experiments described in [756] revealed apparent NK hyperactivity in mice exposed to $2.45\,\mathrm{GHz}$. The cytoxicity of CAL and NK in two-month old BALB/c mice was tested against target $3T_3SV_{40}$ and B16 melanoma cells, respectively. The mice were irradiated at $2.45\,\mathrm{GHz}$ (5, 10, 15, and $20\,\mathrm{mW\,cm^{-2}}$, or 5, 7.5, 11, and $14\,\mathrm{W\,kg^{-1}}$, 4 hr per day for 4 days). A net reduction occured in CAL activity after exposure to $20\,\mathrm{mW\,cm^{-2}}$ with a simultaneous increase in NK activity after exposure to $15\,\mathrm{mW\,cm^{-2}}$.

It seems that the results are too few, and the mechanisms involved are too complex for a precise description of the phenomenon at present. The selective stimulation of the lymphopoiesis of NK was postulated in [219] and, if confirmed, could be of very great interest in the field of cancer therapy.

Apart from the effects mentioned above, a lowering of the

enzymatic acitivity of proteinkinase in B lymphocytes was reported in [111] and [112] together with possible action on the production of interferon [77]. On the other hand, a microwave-induced increase in the amount of corticosterone could be an indication of the stimulation of the pituitary-suprarenal axis that indirectly affects the immune system [516], [480].

It is not easy to draw a conclusion from the above data. If microwaves do produce some effects, which seems incontestable, the nature of these effects, their amplitude, the sign of their variation with the applied field, and the impact they may have on health, do not form a coherent picture. Experimental results are too few in number, they are difficult to exploit dosimetrically, and they cannot readily be extrapolated to man. There are, however, certain common characteristics:

- The observed effects can be reversible, and the return to normal, which often occurs in the course of a long exposure, suggests some form of adaptation by the organism.

- There is a discontinuous transition near $1\,\mathrm{mW\,cm^{-2}}$ between the extremely low-level effects reported by East European researchers and effects observed at moderate and higher levels.

- It may be considered that there are two types of effect that rely on fundamentally different interactions. A number of researchers [516], [580], [724], [725], [799], [108] have postulated a response to thermal stress, whatever its origin, through an increase in the corticosterone level. The primary interaction would then involve the pituitary-suprarenal axis, and its thermal character would sometimes be masked by thermal regulation in the animal. Other types of interaction could be occuring in parallel and could be microthermal, acting on lymphopoietic and immunocompetent cells, as well as on phagocytes and mastocytes. Some nonthermal interactions that are functions of frequency and modulation are also possible (e.g., at the level of the cellular membrane). Finally, certain interactions may depend on a genetic predisposition.

- The case for adverse effects in humans has not been proved. The current trend in this field of research is to try to explore the therapeutic potential of the observed phenomena whilst remembering that they may be harmful.

The following further references may also be consulted: [180], [239], [230], [45], [237], [238], [43], [241], [746], [253], [843], [846], [848], [137], [842], [390], [459], [804], [845], [189], [401], [798], [220], [234], [481], [482], [802], [849], [917], [113], [301], [731], [805], [528], [807], [808], [844], [321], [705], [719], [721].

2.4 Nervous system

The effects of microwaves on the nervous system form a heterogeneous ensemble of facts that are not readily classified in terms of thermal *v.* nonthermal interactions.

The first category includes interactions with the peripheral nervous system and certain neurovegetative functions, and alterations in EEG, in animal behavior, and, possibly, in the permeability of the blood-brain barrier. The second category could include membrane interactions that affect the calcium fluxes, the modulation of neuronal impulse activity, and, possibly, induced arrhythmias in isolated hearts. Also included in this category could be the somewhat confused collection of dystonias and behavioral effects that are often referred to in Eastern Europe as the *microwave syndrome*.

It is also possible that there are some subtle very low level effects of microthermal or nonthermal origin that are masked by the more apparent thermal effects, when the power absorbed is of the same order of magnitude as the basic metabolic rate.

2.4.1 Fluxes of calcium ions

Ca^{2+} ions relay electrochemical messages to the cell surface and its biochemical mechanisms. Evolution seems to have favored calcium in preference to other neighboring ions $(Mg^{2+}, Na^+, K^+, Cl^+)$ because they are able to bond without causing deformation to membrane proteins and to soluble proteins of cytoplasm or organelles. These proteins themselves act as intermediaries: they have several ion bonding sites and, by changing their configuration as a function of the occupied site, they excite a specific target enzyme.

Ca^{2+} ions use special channels to cross the membrane (Section 1.3) along the concentration gradient. These voltage-gated channels are normally closed in excitable cells, but they open in response to the action potential (i.e., the transmembrane voltage pulse induced by the arrival of a messenger on the cellular surface). The membrane resting potential is -90 mV, but it may reach $+40$ mV; the channels open from -30 mV. The phenomenon lasts for a millisecond and allows the passage of 3000 Ca^{2+} ions, after which the outward migration of K^+ ions returns the potential to its equilibrium value. The calcium-ion flux excites the endoplasmic reticulum, which itself liberates Ca^{2+} ions. The evacuation of the ions must take place against the concentration gradient. It is supported by the enzyme ATPase, or calcium pump, which acquires the necesssary energy by dissociating *adenosine triphosphate* (ATP) molecules [116].

Section 1.3, which discusses possible interactions between RF waves and calcium transport, is complemented here with a summary of experimental results now available.

The nonlinear effect of modulated waves on chicken cerebral tissue was demonstrated in [64]. The samples were impregnated with radioactive $^4 5Ca^{2+}$ and were exposed to 0.8 mW cm^{-2} at 147 MHz, amplitude modulated by a sinusoidal signal of 0.5–35 Hz. A statistically significant increase in net $^{45}Ca^{2+}$ transport was observed for modulating frequencies of 6–16 Hz, followed by a fall in the range 20–35 Hz (Fig. 2.7). Similar observations were later reported by the same group [62] and also in [93] which confirmed the existence of *frequency windows*. For an incident power flux of 1 mW cm^{-2}, there was a positive response when the modulation lay between 6 and 12 Hz and there was very little response at 0.5 and 20 Hz. A power window was also shown to exist at constant frequency

(Fig. 2.8): When the chicken cerebral tissue was submitted to a 450-MHz carrier wave modulated at 16 Hz there was a significant increase in $^{45}Ca^{2+}$ transport for power levels of 0.1 and $1\,mW\,cm^{-2}$, while no effect was observed for power levels of 0.05 and $5\,mW\,cm^{-2}$ [65]. The limits of frequency and power windows seem to be 6–20 Hz and 0.1–$1\,mW\,cm^{-2}$, respectively [305]. The carrier frequency must be less than 1 GHz, but has itself little effect; however, the optimum appears to lie between 150 and 450 MHz [7]. For example, for a 450-MHz carrier modulated by a 16-MHz sine wave with an average power density of $0.5\,mW\,cm^{-2}$, the amplitude of the phenomenon reaches 38% of its value at rest in 10 min [503], [725].

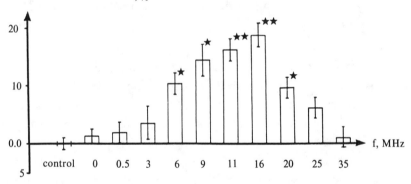

Figure 2.7 Variation in calcium $^{45}Ca^{2+}$ transport in chicken cerebral tissue as a function of modulation frequency for a power flux density of $0.8\,mW\,cm^{-2}$ at 147 MHz. $^{*}p < 0.05$; $^{**}p < 0.1$ [7].

This low-level effect, possibly the best reported, is thought to be athermal in character and due to the direct interaction between neuron membranes and the local electric field. In the vicinity of a membrane, the local field is not very different from its value in free space [543], [7]:

$$E = \sqrt{120\pi\,\phi} \qquad (2.1)$$

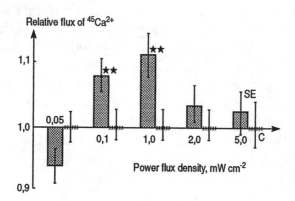

Figure 2.8 Variation of $^{45}Ca^{++}$ in chicken cerebral tissue as a function of power flux density for a 450-MHz wave amplitude-modulated at 16 Hz. $^{**}p < 0.05$ [7].

which gives 61 and 194 $V\,m^{-1}$ for 0.1 and $1\,mW\,cm^{-2}$, respectively. This is negligible in comparison with the figure of $2 \times 10^7\,V\,m^{-1}$ characterizing the static transmembrane potential (90 mV across 4 nm), but is nevertheless large in comparison with slow brain waves ($1\,V\,m^{-1}$ [474]) or with terrestrial ELF fields (10^{-3}–$10^{-6}\,V\,m^{-1}$) picked up by fish, birds, and mammals and used by them for navigation, the detection of prey, and the regulation of the circadian rhythm. A model developed in [543] examines the possiblity of induced transmembrane potential on the order of 10–100 mV. Other studies [644] have also shown that, because of the spike or edge effect, the interface field may locally exceed the macroscopic field by up to two orders of magnitude, which is sufficient to reduce by a factor of 10^4 the threshold of sensitivity to the electromagnetic flux.

This phenomenon is confined to a very narrow ELF modulation band and is therefore unlikely to cause any damage. It should also be noted that the ANSI subcommittee in charge of the revision of the American safety standards did not judge it necessary to take this effect into account [890].

The following references may be consulted for further information on this topic: [6], [91], [630], [32], [639], [724], [785], [9].

2.4.2 Neurons and synapses

The electric activity of neurons manifests itself by the presence of ion currents and electric fields which can interact with an applied radio-frequency wave. The first study of these phenomena [926] was concerned with the nervous ganglions of crustacean animals. It has since become common practice to use the giant neurons of the abdominal ganglions of gastropods. The observed effects relate not only to the specific activity of the cell, but also to its structure, passive electric properties, and metabolism.

In *Aplysia californica*, some of these neurons act as metronomes generating action potentials with a regular rhythm. Other neurons produce voltage bursts triggered by the endogenous depolarizing potential or by signals produced by other ganglion neurons. For neurons with an endogeneous rhythm, a CW or PW radiation modulated by 0.5–10 μs pulses of 1–at 5 kHz perturbs the rhythm and establishes a slower regime in about one minute, followed by a return to normal one to two minutes after the end of the irradiation. The phenomenon is not correlated with frequency or modulation. Although it requires an absorbed power of 50 W kg^{-1}, it is more likely to be an athermal interaction because of the very short time lag in its establishment, which is negligible in comparison with the time constant of the thermal gradient. The response of externally stimulated neurons is slow, gradual, and nonuniform; it is not always accompanied by a return to normal: depending on each particular case, there can be an increase or decrease in the time intervals between the voltage bursts. However, the effect is also incompatible with a response to heat [771].

Attempts have been made to confirm these results [770] by measuring the voltage spectrum of cardiac cell membranes extracted from chicken embryos and exposed to 2.45–GHz radiation (122 and 137 W kg^{-1}). The results of this study were inconclusive because of the very large spread of the measured values, but they did suggest an athermal interaction with an increase in potentials.

Two studies [26], [27] have demonstrated a number of simultaneous modifications of the parietal neuronal ganglions of *Helix lucorum*. After an hour's exposure to 2.45-GHz radiation (15.5 W kg^{-1}), a 33% drop was observed in the concentration of nucleic acids in the nucleus, as well as a 16% drop in the level of cytoplasmic RNA. There was also an apparent doubling of the nuclear to cytoplasmic surface ratio. Moreover, membrane conductivity increased nonlinearly by an average of 22% in 15–20 min without recovering to its previous value, and its time constant fell by 26%. These results, and also those in [771] and in Soviet publications, suggest that there is a high probability that Ca^{2+} ions act as mediators. In fact, it is well known that the injection of calcium ions leads to a reduction in cell-membrane resistivity and hyperpolarization (increase in the resting potential). This leads to an inhibition of RNA synthesis which, in turn, is responsible for a reduction in neuron impulse activity. It was also postulated in [26], [27] that there was a localized interaction in the endoplasmic reticulum, with the calcium ions undergoing a bound-free transition. The reason may be a direct effect on macromolecules that would switch to a configuration with less affinity to these ions. Further information may be found in [865] and [895].

Some of the Soviet [331] and American [571], [570] work has been concerned with the effect of 2375- and 1600-MHz radiation on the metabolism of chemical neurotransmitters in rats. For instance, a 25% increase in the adrenalin level was observed after 20 days of exposure to 50 μW cm^{-2}, at 7 hr per day [331]. A slightly smaller increase in the dopamine and norepinephrine levels was also observed. The level of catecholamine increased on day 5 of exposure to 500 μW cm^{-2} and then fell steadily until day 30. American researchers [570] reported a lowering in the concentration of norepinephrine (16% after 10 min, 29% after 1 hr) and dopamine at the hypothalamus in a power flux density of $20\,mW\,cm^{-2}$. In contrast, no changes were found on exposure to $10\,mW\,cm^{-2}$. The phenomenon was explained by an increase in the temperature of the brain, but this explanation is clearly unconvincing for the very low power levels and longer exposures reported by Soviet researchers.

2.4.3 Blood-brain barrier

A very special interface separates blood capillaries from cerebral parenchymea. This interface or hematoencephalic barrier, also known *as blood-brain barrier* (BBB), actively regulates the selective transfer of chemical substances. Its properties are directly related to the nature of the endothelial cellular membrane forming brain capillaries whose double lipid layer consists of unbonded molecules. This double layer forms a two-dimensional liquid that the liposoluble molecules can cross by simple diffusion at a rate proportional to their lipid affinity and inversely proportional to their size.

However the brain consumes substances that are nonliposoluble (e.g., glucose). Such substances cross the barrier with the help of stereospecific transporters, probably membrane proteins protruding on both sides. The driving mechanism for the transfer is the concentration gradient. The process can saturate when a certain concentration is reached on the luminal side of the membrane, and the rate of transport slows down. The evacuation by the brain of exogenous substances that have crossed the barrier, and of its own secretions, requires an input of energy. The evacuation of glycine, for example, results from the reverse transport of calcium ions, forced by their own concentration gradient. In contrast to the phenomena involved in the transmembrane transport of potassium ions, the ATPase enzyme plays no part here.

The protection afforded by the blood-brain barrier has as a side effect the prevention of a large number of chemotherapeutic agents from reaching the parenchyma, which is a special disadvantage in the case of brain cancer. In contrast, it seems that certain viruses, including HIV, have a great affinity for the brain and can cross the barrier without hindrance. The indentification of a mechanism controlling the permeability of this barrier is thus of enormous therapeutic importance [14], [419], [318], [759].

The permeability of the interface can be evaluated by three methods: intraveneous injection of fluorescein followed by visual or fluorometric detection in brain sections, injection of tracers, histologic examination in an electron microscope, and the use of radioisotopes [14].

The first method has been used to demonstrate [291] a significant increase in the level of fluorescein, compared with control samples, in lateral and third brain ventricles of anesthetized mice placed in a 1.2-kHz CW field ($2.4 \, \mathrm{mW \, cm^{-2}}$ for 30 min) or a PW field ($0.5 \, \mathrm{ms}, 1 \, \mathrm{kHz}, 2.1 \, \mathrm{mW \, cm^{-2}}$ peak and $0.2 \, \mathrm{mW \, cm^{-2}}$ average). The phenomenon was more pronounced in the pulsed field despite the significantly lower average power level. Attempts by another research group to record this phenomenon [569] at lower power levels were unsuccessful. Its negative results, as reinterpreted in [285], showed good correlation between the power flux density (1–$37.5 \, \mathrm{mW \, cm^{-2}}$) and rectal temperature: the increase in the permeability of the BBB may thus be exclusively due to thermal stress and should occur in experimental animals placed in an enclosure at $43°C$. Later studies [818], [819] confirmed that no effects could be observed at low power levels and reported that at moderate levels ($918 \, \mathrm{MHz}, 10 \, \mathrm{mW \, cm^{-2}}$) there was an apparent increase in the permeability of the BBB to certain amino acids. In contrast, other researchers [495], [496], [497] found alterations in the cerebral microvessels of mice only at power densities as high as $3 \, \mathrm{W \, cm^{-2}}$ at $2.45 \, \mathrm{GHz}$, resulting in absorption of $240 \, \mathrm{W \, kg^{-1}}$ with final intracerebral temperature above $43°C$. No changes were found at power densities of 0.5–$1000 \, \mathrm{mW \, cm^{-2}}$ (0.04 to $80 \, \mathrm{W \, kg^{-1}}$) with final intracerebral temperature of $41°C$ at most. Similar results were reported in [568], showing that no effects occurred on the BBB for power densities of $10\,\mathrm{mW\,cm^{-2}}$ or less.

The second method is more sensitive than the first and has proved capable of revealing cellular lesions by means of localized tracer deposition, usually horseradish peroxydases. Some investigators [17] have suggested that these lesions correlate with the intensity and number of exposures (from 30 min per day to 14 hr per day, 1–22 times; 2.45 GHz; 25–50 mW cm^{-2}). These lesions could be evidence for a higher permeability that is also observed after a two-hour exposure to $10\,\mathrm{mW\,cm^{-2}}$, only at the frequencies of 2.45 and 2.8 GHz [12, 13, 14]. This involves the thalamus, hypothalamus, the cerebellum, and some of the cortex and hippocampus. The integrity of the BBB is apparently unaffected, and the lesions usually disappear about two hours after the end of the experiments. Other investigators [839] confirmed the thermal nature of the interaction: The absorption of peroxydase is strongly correlated with the product of exposure time and irradiation intensity. The threshold is around 40°C for 60 min or 45°C for 10 min, but the latency time increases if the animal is placed in a hypothermic state prior to the experiement.

In the third method, the BBB permeability is determined by measuring nonmetabolizable radoisotopes: D-mannitol, inulin, dextran (molecular mass 182, 5,000, and 60,000), and sucrose-^{14}C. This technique has shown [665] that the permeability for mannitol and inulin is raised by very low levels of irradiation at 1.3 GHz CW (0.3 and 3 mW cm^{-2} for 20 min) or PW ($10\,\mu s, 50\,\mathrm{Hz}, 0.3\,\mathrm{mW\,cm^{-2}}$ average and $100\,\mathrm{Hz}, 2\,\mathrm{mW\,cm^{-2}}$ average; $0.5\,\mu s, 1\,\mathrm{kHz}, 0.3\,\mathrm{mW\,cm^{-2}}$ average for 20 min). The pulses were found to be more effective at 50 Hz than at 100 Hz despite the much lower power density, and the maximum effect was obtained for the 5-Hz modulation ($10\,\mu s, 30\,\mathrm{mW\,cm^{-2}}$ average for 20 min). The experiments were repeated and similar results obtained for an unanesthetized rat. The response curves displayed a window effect. Other investigators failed to reproduce these results [663], perhaps because they used a higher frequency (2.8 GHz), a different strain of mice, and the measurement of ratios rather than *Brain Uptake Index* (BUI) of Oldendorf. The permeability of the BBB to sucrose-^{14}C was also investigated in [710] (2.45 and 2.8 GHz, 1–40 mW cm^{-2}) and found to be unaffected by this level of irradiation. Other investigators reported similarly negative results [700], [332], [699].

These experimental data are contradictory and seem to depend strongly on the particular experimental conditions, including the following:

- the characteristics of experimental animals used;
- the use of anesthetics (pentobarbital) that can cause hypothermia and thus modify the cerebral fluxes;
- environmental control, including temperature, manipulation, psychological state of the animals, induction of acoustic sensation by modulated waves, and so on;
- the techniques used to visualize and quantify the phenomenona.

The existence of windowing with an optimum temperature is not confined to microwave-induced phenomena. Indeed, it is well-established for other biological processes that include variations in the BBB permeability.

The very low-level effects obtained under very different conditions in [291], [665] are difficult to reproduce and cannot be used to draw any definite conclusions. A possible interaction through enzymatic inactivation or via the membrane calcium fluxes has been suggested but not verified.

The conclusion is that the permeability variations that appear under moderate and high levels of irradiation could constitute a simple adaptation to thermal stress imposed by microwaves above a certain threshold of temperature and duration of exposure [419], [899], [923], [81].

Further information on this subject may be found in [65], [77], [135], [93], [95], [284], [701], [711], [664], [900] and [286].

2.4.4 Central nervous system

Perturbations of cerebral function can be clearly seen on the electroencephalogram. The quiescent-state regular α waves

(8–12 Hz) are replaced by β waves (15 to 30 Hz) under sensory or psychic excitation. Disfunction is detected by studying the response to stimuli such as *intermittent light stimulation* (ILS). Pathologic response is characterized by fast peaks or slow waves of large amplitude, or by a combination of the two.

Under microwave irradiation, metal electrodes may in some cases increase the local field by several orders of magnitude in their immediate vicinity, thus provoking important electrophysiologic perturbation of neurons. The electrodes also act as antennas, and their susceptibility to radiation may give rise to gross errors in the interpretation of the EEG. These problems are avoided by alternating irradiation and measurement. Alternatively, very low conductivity electrodes may be used, e.g., PTFE electrodes [152] loaded with carbon particles (the conductivity of this material is only $1–4\,S\,m^{-1}$).

With such precautions, , the effect of microwaves can be seen on the EEG in the form of parasitic periodic components. For instance, the authors of [778] detected a signal at the repetition frequency of the modulated field (1-μs impulses at 500 or 600 Hz, modulating a 3-GHz wave; average power density $5\,mW\,cm^{-2}$). This signal persisted intermittently after the end of a ten-day treatment. Another study by the same group [779] revealed paroxysmic surges with fast peaks and trains of peaks, indicating the presence of important perturbations (modulated 3 GHz, $5\,mW\,cm^{-2}$ or CW 2.45 GHz, $5\,mW\,cm^{-2}$ or modulated 9 GHz, $0.7\,mW\,cm^{-2}$).

Soviet researchers [548] have reported an effect on the central structures of the brain and a particular sensitivity of the hippocampus and hypothalamus. Rabbits and rats [410] irradiated at 2375 MHz for $8\,hr\,d^{-1}$ for three or four months (1, 5, and $10\,mW\,cm^{-2}$), presented changes on the EEG and in conditioned reflexes at the higher power levels, and so did rabbits [259, 260] irradiated with 460 MHz at 2 and $5\,mW\,cm^{-2}$ for one week. The review presented in [548] records claims of modified spontaneous bioelectric activity of the brain, accompanied by stimulation of activity (2375 MHz, 10 and

$50\,\mathrm{mW\,cm^{-2}}$, 7 hr per day for one month) and also, by contrast, an inhibition of this activity in higher flux densities $(500\,\mathrm{mW\,cm^{-2}})$.

Older Western research reported microwave effects at 300 and 920 MHz in long low-level exposures in response to ILS [368]. More recently, an amplified response to ILS has been observed under amplitude-modulated radiation at a frequency close to that of evoked potentials $(147\,\mathrm{MHz},\ 0.8\,\mathrm{mW\,cm^{-2}})$ [63]. The reflex intensity declines in 45–60 d if the irradiation is continuous and in 5–7 d if it is interrupted.

Further information can be found in [650], [703], [322], [508], [509], [510], [57], [41], [745], [773], [386], [387], [669], [106], [324], [159], [54], [308], [83], [346], [31], [47], [461], [512], [840], [48], [319], [388], [5], [64], [149], [291], [857], [947], [133], [553], [563], [265], [16], [354], [437], [750].

2.4.5 Peripheral nervous system and sensory perception

A decerebrated cat was found in [533] to exhibit a strong perception reflex accompanied by an accelerated respiratory rhythm and limb movements during exposure to 10 GHz modulated at 1 kHz with an average power flux of $200\,\mathrm{mW\,cm^{-2}}$. At these high power levels, the thermal effect is due to the heating of nerves in the peripheral system to a temperature of 42–47°C. An anesthetized animal with brain intact, submitted to the same power flux at the level of its posterial members, presents a violent analeptic effect with immediate recovery of consciousness and, sometimes, spasm, asphyxia, convulsions, and death. The same result may be obtained with continuous 2.45-GHz radiation or even a traditional heat source [536], [280], [575]. The same type of analeptic response accompanied by nociceptive and escape reflexes was reported in [535]. In other experiments (10 GHz at rest, subcutaneous temperature 45°C) human subjects experienced a sensation which, although not painful, made them unwilling to repeat the ordeal. The thermal character of the interaction was confirmed in [733].

The cutaneous perception of radiation is the result of interaction between the electric field and the heat-sensitive nerve endings. These endings are distributed in the skin at depths between 0.1 and 1 mm, and perception is a function of not only the duration and intensity of exposure, but also of the penetration depth (i.e., of frequency): at higher frequencies, the sensation appears faster and for a lower power density. In the far infrared, for example, the threshold of perception is 0.67 mW cm^{-2} on the chest and 0.84 mW cm^{-2} on the face, and perception begins three seconds after the onset of irradiation [307]. For centimeter waves, the threshold at 10 GHz is 4–6 mW cm^{-2} in 5 s or 10 mW cm^{-2} in 0.5 s, over the entire face [577]. Table 2.1 compares the thresholds at 3 GHz, 10 GHz, and in the far infrared over 37 cm^2 of frontal surface. References reporting on this subject include [891], [535], [373], [472], [372], [422]. In general, other things being equal, a longer interval of time is required to perceive microwave heat than heat from a traditional source. Moreover, the sensation persists for several minutes after the interruption of microwave irradiation [107]. At the lowest ISM frequencies (434 and 915 MHz), there is practically no heat sensation because energy is dissipated in deeper layers and does excite the nerve endings. Except for very high frequencies, the perception of heat is not therefore an adequate protection against electromagnetic irradiation.

The threshold of pain at 3 GHz was measured in 1952 [174] over 9.5 cm^2 of the forearm. The corresponding exposure times are listed in Table 2.2 for different incident power densities. On a 53-cm^2 surface, the sensation of pain appears in 3 min in a flux density of 560 mW cm^{-2}.

The time required for the perception of pain can be expressed [610] simply as a function of the specific absorption rate:

$$\Delta t = 42 \left(\frac{420}{\text{SAR}} \right)^m \tag{2.2}$$

where $m = 2.09 \pm 0.28$, SAR is W kg^{-1}, and Δt is in seconds. Effects on the peripheral nervous system are also discussed in [534], [726], [146], []148].

Table 2.1

Skin sensitivity threshold to RF radiation applied to
37 cm^2 of frontal surface [373]

Exposure time (sec.)		1	2	4
3GHz				
Power flux	(mW cm^2)	58.6	46.0	33.5
10 GHz				
Power flux	(mW cm^2)	21.0	16.7	12.6
ΔT skin	(°C)	0.025	0.040	0.060
Far-field IR				
Power flux	(mW cm^2)	4.2-8.4	4.2	4.2
ΔT skin	(°C)	0.035	0.025	

Table 2.2

Pain threshold at 3 GHz over 9.5 cm^2 of skin surface [174]

Power density, W cm^2	Exposure times, s
3.1	20
2.5	30
1.8	60
1.0	120
0.83	>180

2.4.6 Auditory perception

Pulse-modulated microwaves produce perceptible acoustic sensations in man and other mammals, even in subjects with impaired hearing. The phenomenon was discovered by Frey in 1961 [275], [276], [277] and was reported for power densities on the order of $100\,\mu\mathrm{W\,cm}^{-2}$ (Fig. 2.9). It depended significantly on peak power, carrier frequency, and modulation [280]. It was subsequently verified in [907], [910], [911], [816], [328], [279], [280], [668], [782], [289], [292], [293], [282], [351], [353], [467], [468], [271], [283], [856], [783]. The last of these draws an interesting parallel with the appearance of sound in a carbon polyurethane microwave absorber exposed to 1.2–2.45 GHz waves modulated by 1-μs pulses, with an audibility threshold corresponding to a peak power density of 46–92 mW cm^{-2}. It also envisages the possibility that the signal was transmitted to the inner ear by bone conduction. The levels of perception in the cat were reported in [345] to be 18 μW cm^{-2} for 0.5μs pulses and 47 μW cm^{-2} for 42 μs at 2.45 GHz (similar results were obtained at 915 MHz). The temperature rises associated with this phenomenon did not exceed 10^{-5}–10^{-6} deg [342], [97]. Finally, cochlear microphonics in the cat were reported in [145] and [153].

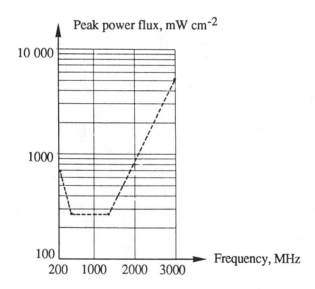

Figure 2.9 Auditory perception [280].

There was a continuous stream of publications on this subject in the late seventies (e.g., [485], [717], [251], [416], [486], [487], [156], [469], [470], [488], [489], [114]). One study [771] suggested a direct interaction between the neuron and the electric field, whereas another [288] analyzed holographically the vibrations induced in the internal ear of the cat or guinea pig, and suggested a process based on the ion flux at the membrane level. A mechanical explanation of the phenomenon was, however, gaining ground. It was noted [98] that three possible mechanisms, namely, electrostriction, radiation pressure, and thermal expansion, could be modeled by a solution consisting of a steady component and a propagation term. The medium was assumed homogeneous and isotropic in its dielectric and elastic properties. Of these three mechanisms, only thermal expansion is consistent with the conversion of microwave energy into acoustic energy. Calculations show, for example, that a 3-GHz wave modulated by $10 - \mu s$ pulses, with peak power density of $27\,W\,cm^{-2}$, induces mechanical vibrations with a peak pressure of $6.5\,Pa$ and a spectral content concentrated below $20\,kHz$. This vibration can reach the cochlea by bone conduction [145], [345], [493]. Several studies [488], [499], [490] confirmed the validity of this model and showed that signals generated in this way were dependent on only the size of the head and on the acoustic properties of tissue.

In addition, they defined the optimum pulse size for energy conversion. There could be a dilation of cervical matter for power flux densities below $100\,mW\,cm^{-2}$, inducing pressure waves at acoustic frequencies, detected by the cochlea [702], [827]. The EEG of the cat demonstrates [498], [499] considerable similarities between evoked potentials produced by simple acoustic stimulation and by 2.45-GHz microwaves modulated at 10–100 Hz by 0.5- and 25-μs pulses. The responses reveal three identical successive wave trains with latent times of 8 ms, 8–50 ms, and 50–300 ms, corresponding respectively to the successive activation of the cochlea and the acoustic centers of the brain stem, cerebral tissue, and, finally, the front of the cortex. The transduction of the RF pulses into acoustic pulses is actually due to the thermoelastic expansion and to bone conduction. The key role of bone conduction and of cochlean activity was confirmed in [44] by showing that guinea pigs continue to perceive a 918-MHz wave modulated at 30 Hz by 10-μs pulses even after the removal of the middle ear.

The process was modeled with phantom muscular and cerebral tissue [656], [657]. The phantom brain was in the form of a sphere consisting of 62.1% water, 0.58% sodium chloride, 29.8% polyethylene powder, and 7.01% gelling agent, with a characteristic frequency of 16 kHz (i.e., close to that of the brain). The sphere was exposed to a 4-kW peak, 1.1-GHz radiation modulated by 10-μs pulses, which produced an average absorption of 824 W kg^{-1}. The phantoms acted as amplifiers for the pressure waves induced by incident radiation. The amplitude depended on the duration and frequency of the pulses and also on peak power, with a linear relation of 1 Pa per kW peak. These results are consistent with [499], which reports that the amplitude of evoked potentials is directly related to pulse duration and amplitude, and is inversely proportional to pulse frequency.

Thermodynamic studies of this phenomenon [340] have shown that acoustic waves can appear if there is a temporal or spatial discontinuity in the irradiation power density, the attentuation coefficient α [Part I, (2.40)], or thermodynamic variables such as the density ρ, specific heat at constant pressure or constant volume C_p and C_v, compressibility x, and expansion coefficient at constant pressure α_p. The phenomenon may also occur in small dielectric objects in which energy deposition is uniform. The coupling of acoustic waves between a medium with a high dielectric loss factor (brain) and a lossless medium (cochlea) was also clarified.

At frequencies corresponding to millimeter waves, bone conduction of auditory sensations requires much higher power densities than for decimeter or centimeter waves. Table 2.3 [307] compares the perception threshold W/S (in J m^{-2}) and the depth of penetration d [Part I, (2.83)] at different frequencies, and shows that the increase in the perception threshold with frequency is slower than the reduction in the depth of penetration.

2.4.7 Autonomic nervous system

The nuclei of the ANS are located in the cerebral stem and operate in relation to the central nervous system (hypothalamus and rhinencephalon). The system consists of two an-

Table 2.3

Depth of penetration (d) and auditory perception threshold (W/S) at various frequencies [307]

f, GHz	d, cm	W/S, μJcm^{-2}	$d_{0.915}/d_f$	$(W/S)/(W/S)_{0.915}$
0.915	3.03	17.3	1.0	1.0
2.450	2.05	20.9	1.48	1.21
3.0	1.97	20.6	1.54	1.19
30.0	0.078	143.3	31.4	8.28
100.0	0.032	376.4	76.6	21.76
300.0	0.023	579.1	106.5	33.47

tagonistic entities, namely, the sympathetic and parasympathetic subsystems, each involving two neurons that originate from a synapse situated in a vegetative ganglion. The postganglionic sympathetic nerves use adrenalin as the neurotransmitter. They accelerate the cardiac rhythm and slow down the digestive transit. Parasympathetic nerves have the opposite effect, the corresponding signals being transmitted through the intermediary of acetylcholine [232].

A number of studies [179], [148], [151] of isolated nerve tissue have shown that continuous waves (2.45 GHz, 0.3, 3, 30, 300, and 1500 W kg^{-1}) and modulated waves (2.45 GHz, 100 Hz, and 1 kHz in 1- and 10-μs pulses; 0.3, 3, 30 and 220 W kg^{-1}) have no effect on the action potential or the propagation time. The tissues used were vagus nerves or saphena and the rabbit superior cervical ganglion, held at constant temperature during irradiation.

However, at the functional level, microwaves have been shown [752] to have an effect on the parasympathetic system in rats: there was an acceleration of the intestinal transit that began eight hours after irradiation (2.45 GHz, 3–4 mW cm^{-2}), reached 34.8%, continued for 24 hours after the end of treatment, and gradually disappeared after two days. The frequency of slow duodenal waves increased by 9% in 70% of all cases under exposure to 5 mW cm^{-2}, and by 19% in 90% of all cases under exposure to 7.5 mW cm^{-2}, whilst the action potential exhibited a degree of inhibition [751]. This last result, attributed by the author to an athermal effect, is in agreement with the relaxation of peristaltic movement reported in [49]. By contrast, classical hyperthermia seems to induce an acceleration of the action potential, parallel to that of slow waves. It is interesting to compare these results with the use of microwaves in the stimulation of uterine contractions during childbirth [190].

A number of ANS symptoms are often reported by people exposed to microwaves in their professional lives. These symptoms are digestive (nausea, vomiting, loss of appetite), peripheral (brittle nails, loss of hair), and cardiovascular (hypo- or hypertension, brady- or tachycardia). They constitute part of the microwave syndrome postulated by East European researchers [660], [794]. Its evaluation is a very delicate matter, except for the cardiovascular symptoms, which are readily quantifiable and have therefore been the subject of numerous clinical and experimental research projects.

The Soviets were the first to envision the possibility of a low-level, nonthermal mechanism affecting the cardiac rhythm of experimental animals exposed to continuous or pulsed radiation [697], [698], [477], [478], [235]. A number of American researchers have attempted to reproduce these experiments on isolated hearts and on whole animals. Some of them [278], [294] reported a faster rhythm in isolated turtle hearts after exposure to a field of 1425 MHz modulated and synchronized with certain phases of the electrocardiogram. Turtle hearts were also used in a number of other studies [511], [874], [658] with exposures to 960-MHz radiation. The experiments revealed a dual effect: a net slowing down when the power absorbed was on the order of 3 mW and an acceleration when it reached 10 mW. The phenomenon was confirmed by other studies [712] of isolated rat hearts (960 MHz, 1–10 mW g^{-1}). Since the sympathetic and parasympathetic systems control tachycardia and bradycardia, respectively, it was suggested that some nerve remnants were stimulated by the radiation, and that the natural predominance of the parasympathetic system imposed a slowdown of the rhythm. At higher power levels, this mechanism could be masked by tachycardia induced by the higher temperature. This hypothesis is supported by the combined effect of microwaves and chemical inhibitors: β blockers for the sympathetic and atropine for the parasympathetic systems. The observed effects correspond to the noninhibited system, with microwaves playing a part equivalent to the release of neurotransmitters (adrenaline or acetylcholine). As for the temperature-induced tachycardia, this is imperceptible at low levels, but becomes predominant at higher levels ($15\,\mathrm{mW\,cm^{-3}}$).

Soviet workers have managed to induce very marked effects in experimental animals, using very long exposures and very low doses. They included tachycardia followed by a slowing down after exposure to $2.4\text{--}10\,\mu\mathrm{W\,cm^{-2}}$ ($2375\,\mathrm{MHz}$) for 60 d and bradycardia at between 66 and $265\,\mathrm{nW\,cm^{-2}}$ [774], [548]. s of 20). The chronotropic coefficient was introduced in [697], [698] and is defined by

$$K = \frac{100 + m_a}{100 + m_d} \tag{2.3}$$

where m_a and m_d are, respectively, the percentage of cases in which the rhythm increased and decreased in comparison with controls. This coefficient was 1.3 during irradiation ($7\text{--}12\,\mathrm{mW\,cm^{-2}}$ continuous or $3\text{--}5\,\mathrm{mW\,cm^2}$ pulsed) and 1.42 immediately after exposure, which indicated significant tachycardia. In American studies [434] the corresponding values were only 0.95 and 0.94.

The same experiment was taken up by another American group [157] which used rabbits placed ventrally and dorsally at $25\,\mathrm{cm}$ from a $12 \times 12\,\mathrm{cm^2}$ horn. Ventral irradiation produced no effect, but dorsal treatment with a continuous wave ($2.45\,\mathrm{GHz}, 80\,\mathrm{mW\,cm^{-2}}$) provoked transient tachycardia due to the thermal effect. In a pulsed field ($2.45\,\mathrm{GHz}, 10\,\mu\mathrm{s}, 13.7\,\mathrm{W\,cm^{-2}}$ peak) some random variations were observed when the pulses were synchronized with the maximum of the R-wave of the ECG, and a chronotropic effect giving a 3% increase was observed for only one animal (which is evidently not significant) when the pulses were at 100 or 200 ms after this maximum.

By contrast, a French study [775] reported a slowing down in the cardiac rhythm of the white rat after a 5–15 day exposure to $5\,\mathrm{mW\,cm^{-2}}$ at 2.45 GHz, 3 GHz, and 9.4 GHz.

Further information on this subject may be found in [471], [101], [534], [230], [670], [683], [89], [160], [433], [521], [655], [681], [506], [939], [908], [302], [269], [300] and [877].

2.4.8 Psychophysiology

Research on the psychologic effects of microwaves displays a strong objectivist trend: behaviorist in the West and Pavlovian in the East. In both cases it has relied on systematic animal experimentation.

The differences between East and West go beyond experimental methodology. They reside in the definition of the field of study itself: long exposures at very low levels or short exposures at high-power densities. East European researchers have favored the former approach, which aims at testing the presence of athermal effects and in so doing tends to parallel the tendency of epidemiologic studies of populations professionally exposed to low-level fields.

The change in animal behavior after exposure to a power density capable of producing a significant thermal effect is often associated with a reduction in activity. In rats trained to accomplish a given task, the slowing down in activity occurs after one hour of intermittent exposure (5 min on, 5 min off) to 2.45-GHz radiation, with an absorbed power density of $6.2\,\mathrm{W\,kg^{-1}}$ [425]. The same phenomenon can be obtained after 15 min at 918 MHz and $40\,\mathrm{mW\,cm^{-2}}$ with $8.4\,\mathrm{W\,kg^{-1}}$ absorbed [494]. Another study [199] has shown a significant slowing down in activity, measured by stabilimetry, after each irradiation of rats ($5\,\mathrm{mW\,cm^{-2}}$ at 2.45 GHz, $1.23\,\mathrm{W\,kg^{-1}}$ for 80 eight-hour days). In contrast, judging by the number of revolutions of the Wahmann freewheel, the nocturnal motor activity went up by 30% for animals irradiated during the day. Exposure to $5\,\mathrm{mW\,cm^{-2}}$ at $915\,\mathrm{MHz}$ ($2.46\,\mathrm{W\,kg^{-1}}$, same duration) produced an increase in stabilimetric activity by only 25% despite a specific absorption rate twice as large, and the freewheel test indicated a higher average nocturnal activity, but with such scatter among individual performances that the results were judged unreliable [198]. The stabilimetric deficit at the end of the session may indicate adaptation to the thermal load, similar to the reduced activity reported in another study [623] after exposure to $10\,\mathrm{mW\,cm^{-2}}$ at $918\,\mathrm{MHz}$. Exposure to $2.5\,\mathrm{mW\,cm^{-2}}$ at $918\,\mathrm{MHz}$ in [520] produced no observable effect.

Another American study [494] determined the threshold for activity slowdown in rats trained to reach their food target after a series of head movements. The threshold was found to be $32\,\mathrm{mW\,cm^{-2}}$ at $918\,\mathrm{MHz}$ ($29\,\mathrm{W\,kg^{-1}}$) after a delay of 20 min. When the power density was $40\,\mathrm{mW\,cm^{-2}}$, the activity fell abruptly after 5 min. The animals foamed at the mouth, were short of breath, and showed signs of great tiredness.

It seems that an absolute relationship between absorbed power and the resulting perturbation will be impossible to establish because of the complexity of experimental factors. On the other hand, the mediating role of induced heat is very well established. This has been confirmed by observations of the synergetic reinforcement when microwave irradiation is combined with high ambient temperature. For example, the response of rats to the random supply of food after exposure to 5, 10, and $15\,\mathrm{mW\,cm^{-2}}$ at $2.45\,\mathrm{GHz}$ was studied in [298] at ambient temperatures of 22–28°C. Nothing happened at 22°C, but at 28°C the response time increased with the power density and the number of responses tended to fall in good statistical correlation with the power density applied. The basic metabolic rate of the rat is only $7\,\mathrm{W\,kg^{-1}}$ [589], so that its behavior can be expected to be affected for absorbed powers of the same order. Thus, these effects may well be due to a thermal load that overwhelms the thermal regulation capacity. The latter can be aggravated by resonant absorption and hotspots, or by high ambient temperature.

Another type of synergetic effect was reported in [866], affecting the performance of rats when modulated radiation ($2.45\,\mathrm{GHz}, 2\,\mu\mathrm{s}, 500\,\mathrm{Hz}, 1\,\mathrm{W\,cm^{-2}}$ peak) was accompanied by the uptake of psychoactive drugs [305], [460], [81]. However, other researchers have been unable to confirm these results [531], [438].

Higher power densities give rise to a perception of pain and to a tendency to escape [850], [853], [854], [289], [563], [627], [626]. In the case of a modulated wave, these reactions might be due to the auditory effect. The effect of a continuous wave may be explained by the thermal excitation of peripheral nerves. For very high power densities, short wavelengths, and intermittent exposures (9.31-GHz modulated at $1.05\,\mathrm{kHz}$ by 500-ns pulses, $300\,\mathrm{W\,cm^{-2}}$ peak, $150\,\mathrm{mW\,cm^{-2}}$ average), Macaca mulatta monkeys show no signs of perturbation in their social, sexual, and maternal behavior [541].

For humans exposed to very low power densities (between a few $\mu\mathrm{W\,cm^{-2}}$ and a few $\mathrm{mW\,cm^{-2}}$), East European epidemiologic studies [696], [561], [241], [746], [49] have revealed a variety of reversible asthenic problems that constitute the hypothetic microwave syndrome (headache, transpiration, emotional instability, irritability, tiredness, somnolence, sexual problems, loss of memory, concentration and decision difficulties, insomnia, and depressive hypocondriac tendencies. The evaluation of these very subjective complaints is difficult, particularly in the absence of a control group and well-established dosimetric data. These problems may well be due to environmental factors unrelated to microwaves, but a possible nonthermal mechanism cannot be completely ruled out [909], [585], [794].

Further information on this subject may be found in [703], [322], [508], [509], [510], [57], [428], [669], [447], [324], [445], [380], [425], [444], [426], [436], [446], [642], [289], [394], [537], [293], [417], [792], [141], [181], [290], [291], [299], [393], [518],

[718], [867], [215], [623], [776], [898], [197], [494], [520], [620], [519], [199], [866], [868], [132], [760], [420], [448], [531], [780], [894], [269], [305], [625].

2.5 Endocrine system

Endocrine glands produce hormones that pass directly into the blood stream. Their function is controlled by other hormones produced by the pituitary gland, a very small gland at the base of the skull in the *sella turcica*. The pituitary gland is itself controlled by the nervous system via the hypothalamus, which transmits chemical signals in the form of releasing hormones. Endocrine glands constitute a feedback system (Fig. 2.10): the quantity of hormones released by a terminal gland controls the pituitary in a feedback loop (Fig. 2.11). For example, the pituitary produces the *thyrotrophic hormone* (TSH) which stimulates the secretion of thyroxine by the thyroid. When the amount of thyroxine in the blood exceeds the normal level, the production of TSH decreases automatically and the system reaches an equilibrium level either from below or from above.

2.5.1 Pituitary-thyroid axis

Hormones secreted by the thyroid, i.e., thyroxine (T_4) and triiodothyronine (T_3) activate the metabolism and stimulate growth. Their levels are regulated through the intermediary of TSH.

There is a great variety of published experimental data. Some studies [56], [48] report increased thyroid activity in rabbits exposed to microwave radiation (3 GHz, $5\,mW\,cm^{-2}$, 3 hr per day for four months), whilst other studies [555], [591] reveal that localized exposure of dog thyroid (2.45 GHz, 13–$100\,mW\,cm^{-2}$ for 30 s to 12 min) is accompanied by a temperature increase of 20°C and a significant increase in secretions, including the release of thyroxine. Another study [524] reports a transient increase in the thyroxine level in rats (2.45 GHz, $1\,mW\,cm^{-2}$, single 4-hr exposure) without change

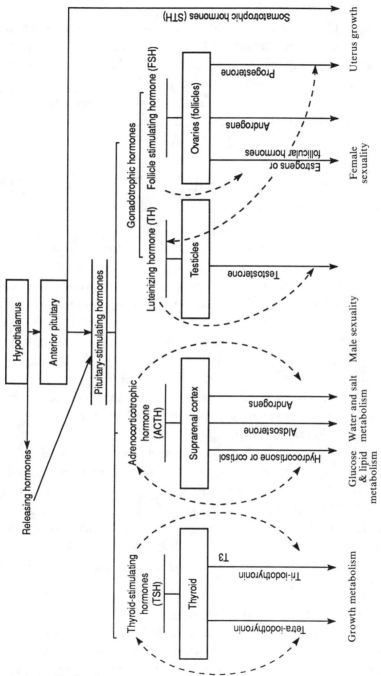

Figure 2.10 Endocrine system.

in the TSH level, which may indicate that microwaves have a direct effect on the thyroid. A stimulating effect can still be found in rats in which TSH production has been hampered by T_3 supply. This could be seen in relation to certain cases of hyperthyroidism reported in the Soviet Union among professionals working with microwave generators [811].

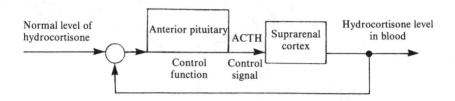

Figure 2.11 Block diagram of the feedback control system involving the anterior lobe of the pituitary gland and the suprarenal cortex.

In contrast to the above results, rats have been observed to have low levels of T3 and T4 (2.45 GHz, 20 and $25\,\mathrm{mW\,cm^{-2}}$ for 16 hr, or $15\,\mathrm{mW\,cm^{-2}}$ for 60 hr) and a lower capacity for the fixation of iodine, which indicate a reduction in thyroid activity [671]. Other researchers [522] used a different radiation protocol (2.45 GHz, 10 mW cm^{-2} for 1 or 2 hr, or $20\,\mathrm{mW\,cm^{-2}}$ for 2 and 8 hr) and concluded that the reduction in thyroid activity was a manifestation of inhibited TSH secretion; another group [876] reported good correlation between the fall in TSH and T4 levels in rat (2.45 GHz, 8 mW cm^{-2}, 8 hr d^{-1}, for 21 days). When pulsed radiation was employed (2.45 GHz, 10 μs, 0.48 mW cm^{-2} on average, SAR = 0.4 W kg^{-1} in rat), there was no observable effect on T_4 levels [343].

2.5.2 Pituitary-suprarenal axis

The cortical part of suprarenal glands secretes three types of hormone or corticoid, namely, the mineralocorticoids (aldosterone and desoxycorticosterone) that act on the metabolism of water and salts; the glycocorticoids (cortisone, hydrocortisone, and cortisol), which act on the metabolism of glucides, lipids, and protides; and the sexual corticoids or suprarenal androgens [232]. This secretion is controlled by the pituitary via the *adrenocorticotrophic hormone* (ACTH).

Microwaves can stimulate, at sufficiently high power densities, the hypothalamus-pituitary-corticosuprarenal axis, thus initiating an increase in the corticosterone level in plasma [596]. Research indicates [338] that there is a latent time of 15–30 min for both the onset of the effect and for its dissipation. An increase in the corticosterone level has been found to be significant [524] when compared to the circadian variations in rats ($2.45\,\text{GHz}, 20\,\text{mW cm}^{-2}$ for eight hours). It disappears completely after hypophysectomy of the rat or after the administration of dexamethasone, a synthetic corticoid, at the rate of $3.2\,\mu\text{g kg}^{-1}$ ($2.45\,\text{GHz}, 50\text{–}60\,\text{mW cm}^{-2}$, 1 hr) [515], showing the absence of direct stimulation of the suprarenal gland. The interaction is more likely to be with the pituitary gland, and there could even be an intervention of the hypothalamus with a parallel increase in the levels of releasing hormones [646], [526]. In all cases, the effect is thermal, as shown by the correlation between rectal temperature and the level of corticosterone. The threshold for this effect at 2.45 GHz is on the order of $20\,\text{mW cm}^{-2}$ ($3.2\,\text{W kg}^{-1}$) for 120 min or $50\,\text{mW cm}^{-2}$($8\,\text{W kg}^{-1}$) for 30–60 min in rats [516]. At 918 MHz nothing happens in rats irradiated for $10\,\text{hr d}^{-1}$ at $10\,\text{mW cm}^{-2}$($3.6\,\text{W kg}^{-1}$) for 3 weeks or at $2.5\,\text{mW cm}^{-2}$($0.9\,\text{W kg}^{-1}$) for 13 weeks. In pulsed fields ($1.29\,\text{GHz}, 2\,\mu\text{s}, 500\,\text{Hz}$), the effects appear under $15\,\text{mW cm}^{-2}$ after 30 min in rats, but not in the rhesus monkey exposed to an average of $38\,\text{mW cm}^{-2}$ for 8 hr, with an increase in rectal temperature of 1.5°C [216].

Electromagnetic radiation may produce certain effects at much lower levels, but these effects are not well-defined. For example, chickens exposed for 23 days to $550\,\mu\text{W cm}^{-2}$ (880 MHz) were found [309] to lose 22.4% of the weight of their suprarenal gland. An attempt to reproduce one of the

Soviet experiments [548] reveals some hypertrophies and local vacuolation of the rat suprarenal glands after 50 hr per week under $500\,\mu W\,cm^{-2}$ (2375 MHz). Soviet experiments with rats [241], [646] have detected maximum corticosuprarenal stimulation at $1\,mW\,cm^{-2}$ (0.5 and 2.5 GHz, 2.6 GHz, respectively) with very different durations of exposure (i.e., $8\,hr\,d^{-1}$ for 120 days [241] and 30 min only) [646]. In both cases, there could have been a cumulative effect for the absorbed doses. According to [646] there was also a parallel increase in the level of releasing hormones in plasma, namely, ACTH and 11-oxycorticosteroid, which tends to confirm an effect on the hypothalamus. This increase appears for an exposure to $100\,mW\,cm^{-2}$, passes a maximum at $1\,\mu W\,cm^{-2}$, and continues to decline until $75\,mW\,cm^{-2}$. In contrast, radiation has an inhibiting effect at low levels (0.1–$1\,mW\,cm^{-2}$ for 4 hr in rats) with a rectification effect on the natural circadian maxima [527]. This led to the suggestion that there are two modes of interaction, namely, one at low and moderate levels (less than $10\,mW\,cm^{-2}$) and another at higher levels (more than $25\,mW\,cm^{-2}$).

2.5.3 Pituitary-ovarian and pituitary-testicular axes

Ovaries and testicles are controlled by two pituitary hormones, namely, the *follicle stimulating hormone* (FSH) and the *luteinizing hormone* (LH). The ovary secretes oestrogen and, after ovulation, progesterone. The testicle has an exocrine function (i.e., spermatogenesis) and an endocrine function (i.e., the secretion of testosterone). The two functions are controlled by FSH and LH, respectively.

Extremely low levels of microwave power (0.02–$400\,pW\,cm^{-2}$ at 6 GHz) could have an effect on the pituitary-ovarian axis of chicken, which is seen as an increase in ovulation frequency [443]. This result must be treated with caution because it has been reported only once and also because of the extremely low fields involved ($0.3 - 39\,mV\,m^{-1}$).

One of the oldest recorded microwave effects is testicular

degeneration of thermal origin, the risk of which is enhanced, as in the case of the cornea, by the absence of blood supply. In 1957, an accident with a 10-MW peak power radar of the US Air Force is thought to have caused the loss of spermatogenesis in eight military personnel [660]. In 1968, testicular biopsy on a metereologic radar technician, who was accidently exposed to very high field levels, revealed atrophies and necroses in seminal tubes, and a permanent decline of spermatogenesis [730]. A clinical inquiry [462] conducted on 31 employees of radar stations did not exclude the possibility of changes in exocrine function after long-term irradiation.

It has been shown experimentally [397] that an intratesticular temperature of 35°C produced by microwaves is sufficient to cause rat tissue changes that occur only at a temperature of around 40°C in the case of the infrared. Moderate microwave levels have an effect on dog and rat testicles that is fundamentally no different from the effect of immersion in a hot bath, and all the symptoms, including sterility, seem to be reversible [255]. Morphologic variations in mouse testicles were reported in [889] (1.7 and 3 GHz, $10 \, \text{mW cm}^{-2}$ for 100 min or $50 \, \text{mW cm}^{-2}$ for 30–40 min). Another study [221] showed no histologic modifications in mouse testicles and no changes in spermatogenesis (2.45 GHz, $10 \, \text{mW cm}^{-2}$, 1, 2, 3, and 4 hr, five consecutive days). The level of sorbitol dehydrogenase, a characteristic enzyme in testicular metabolism, was not affected, but there was an increase in LH and testosterone levels with a return to normal five days after the last irradiation. Similarly, stimulation of LH secretion was reported in [608] (2.87 GHz, $10 \, \text{mW cm}^{-2}$, 36 hr per week for six weeks) and also in [223] (2.45 GHz, $10 \, \text{mW cm}^{-2}$, 24 hr). These effects again suggest a direct interaction with the pituitary gland and even with the hypothalamus.

2.5.4 Growth hormones

The secretion by the anterior lobe of the pituitary gland of *somatotrophin* (or growth) *hormone* (STH) is controlled by the pituitary gland by means of the corresponding releasing hormone and an inhibiting hormone (i.e., somatostatin). The response to radiation is variable and depends on the species. In general, rodents show a decrease in response, whereas dogs, primates, and humans show a rise [526]. In young rats there is a rise in STH levels for exposures below $10\,\mathrm{mW\,cm^{-2}}$ ($2.45\,\mathrm{GHz}$) and a sharp decline for exposures to $36\,\mathrm{mW\,cm^{-2}}$ after 60 min. Inhibition occurs after 30–60 min of exposure to $50\,\mathrm{mW\,cm^{-2}}$ ($8\,\mathrm{W\,kg^{-1}}$). STH levels in mice were observed to decline in [587] after 180 days of irradiation at 2–$5\,\mathrm{mW\,cm^{-2}}$ ($2.45\,\mathrm{GHz}$).

On average, the observed effects are transient. They produce an increase in the secretion of corticosuprarenals and a reduction in thyrotropic and somatotrophic secretions. This is compatible with the hypothesis of a response to a thermal overload. Discrepancies between the results (e.g., on the pituitary-thyroid axis) may be due to different experimental conditions, the random formation of hotspots in the glands and the hypothalamus (Section 1.4.6), and to a variety of other factors, such as the circadian rhythm and differences between species. With the exeption of the high power effects on testicles, which do not belong to the endocrine ensemble, the interaction seems to involve the pituitary gland or even the CNS rather than the terminal glands. [862], [526], [221].

Further information on this subject may be found in [689], [619], [601], [68], [679], [784], [442], [606], [615], [356], [671], [377], [457], [458], [607], [521], [554], [592], [681], [583], [608], [637], [842], [168], [555], [584], [223], [513], [516], [558], [559], [514], [586], [222] and [523].

2.6 Thermal regulation and metabolism

The wave-body interaction and the corresponding temperature rise were treated in Chapter 1, which highlighted the complexity of this interaction, the existence of resonance phenomena, and the random appearance of hotspots.

The body defends itself against thermal overload by means of its thermal regulation system, which is controlled by a dedicated nervous center in the diencephalon, and by the endocrinal metabolism control exerted by the hypothalamus-thyroid axis.

Intense transpiration is observed first. Weight loss may reach 2% per hour for a dog exposed to $165\,\mathrm{mW\,cm^{-2}}$ at $2.8\,\mathrm{GHz}$, and the rectal temperature rises by 1–2°C. In the second phase, the body activates a range of compensation mechanisms, including dyspnea, tachycardia, and increased blood supply. Rectal temperature stabilizes at about 40°C and thermal equilibrium is maintained until the breakdown point, at which the thermal regulation system suddenly collapses. Temperature then reaches a critical value, with the animal experiencing extrasystole and low blood pressure, followed by convulsions [660], [576]. At room temperature, deaths occur at the end of a period of total exposure given by (2.2) with $m = 1.29 \pm 0.79$, SAR in $\mathrm{W\,kg^{-1}}$ and Δt in seconds [610].

Further information on acute hyperthermia may be found in [161], [36], [545], [214] and [722].

This fatal chain of events is initiated when the heat generated by radiation is higher than the basic metabolic rate by a factor of about two. So long as it remains of the same order as the metabolic rate, the regulating mechanisms are sufficient to maintain effective compensation [862]. Environmental conditions (temperature, relative humidity, and air circulation) play an important part in all this. The survival times found for rats exposed to $250\,\mathrm{mW\,cm^{-2}}$ at $24\,\mathrm{GHz}$ range from 17 to 34 min, depending on whether the ambient temperature is 35 or 15°C, and amount to 24 hr in the presence of a stream of air at 15°C [576]. Similarly, rats exposed to $100\,\mathrm{mW\,cm^{-2}}$ at $2.45\,\mathrm{GHz}$ die in 17 min with a rectal temperature of 45°C. When their cage is cooled by circulating liquid nitrogen, the rectal temperature remains normal; no symptoms are observed after several hours of irradiation, and autopsy does not show any pathologic signs [748].

The possible hotspots generated in the hypothalamus may be such as to modify normal response to hyperthermia: it is well known that temperature gradients produced experimentally in the hypothalamus cause important variations in perspiration and vasodilation rates [765]. It has also been reported [570] that there has been an accelerated turnover of norepinephrine, an adrenergic neurotransmitter present in the hypothalamus of rat submitted to $20\,\mathrm{mW\,cm^{-2}}$ at $1.6\,\mathrm{GHz}$. This phenomenon could be related to the homeothermal function.

Further information on this may be found in [212], [633], [634], [635], [636], [683], [875], [494], [383], [821], [1], [2], [901], [3] and [4].

Alterations of metabolism have been noted by a number of East European researchers and are reviewed in [548]. They involve vitamins (lower levels of B_2 and B_6 under $570\,\mathrm{muW\,cm^{-2}}$ at $2\,\mathrm{GHz}$ [617]), metals (variation in the levels of Cu, Mn, Ni, and Mo in rats after three months at the rate of eight hours per day under 10, 100, and $1000\,\mu\mathrm{W\,cm^{-2}}$ at $2374\,\mathrm{MHz}$ [297]), enzymatic activity ($2375\,\mathrm{MHz}$, 100 and $1000\,\mu\mathrm{W\,cm^{-2}}$ at two hours per day for one month [242]), and respiration and phosphorylation of the mitochondria of the liver ($2375\,\mathrm{MHz}, 50\,\mu\mathrm{W\,cm^{-2}}$, 40 min, three times per day for four months [240]). The possible effects of very low radiation levels on digestion and on the oxygen and CO_2 cycle remain unclear [382], [383], [305].

Further information on this subject may be found in [683], [945], [681], [902], [243] and [586].

2.7 Effects on growth

Although no effect has ever been observed on human development, experimental research on growth effects is very active and is concentrated on insects, eggs, and embryos of small mammals irradiated *in utero*.

2.7.1 Insects

Insects are particularly sensitive to teratogenic factors. This makes them convenient objects of study although, of course, there is no direct connection with any effects on humans.

The first experiments in this field were performed in 1930 [573] using silkworms and a 107-MHz RF oscillator. They showed very significant length and weight differences between irradiated insects and a control group. The former were submitted to 30–60 min of irradiation, which also delayed hatching of the butterfly. The weevil *Tenebrio molito* shows [430] abnormal morphologic development of the larva after exposure to RF treatment, whereas 51% of irradiated pupae displayed [129] incomplete metamorphosis of their posterior part, as against 90% normal development of a control group. These larvae were exposed in a waveguide to 80 mW at 10 GHz for 20–30 min or 20 mW for 120 min. Once adult, the insects showed evidence of a cytopathologic effect on the reproductive tissue [707], and a teratogenic effect was also reported at low power levels in [502], [505]. Experiments on *T. molitor* referred to earlier in [129] were repeated in [502] (9 GHz continuous or modulated at 16 Hz and 1.6 kHz, $0.25\,\mu s, 20$ mW average in the WR90 waveguide, $40\,W\,kg^{-1}$). The increase in abdominal temperature did not exceed 2°C. After hatching, the level of morphologic anomaly in the adult insect was above the expected level by an average factor of three in the case of exposure to 72 J. These results have been questioned [686] by researchers who hold the view that teratogenic effects are clearly correlated with a rise in the temperature of larvae by at least 10°C, and also by researchers who have found no evidence for any effects in the viability of eggs and larvae or in the morphology of adults (10 GHz, 13 and $130\,J\,cm^{-2}$) [912]. No effects were recorded in another study [768] of the morphology of tobacco hornworm (*Manduca sexta*) from irradiated larvae (2695 MHz, modulated at 500 Hz, $2\,\mu s, 23\,W\,kg^{-1}$, 12 hr per day).

The provisional conclusion is that there *may* be certain terategenic effects on insects, but they are not well defined at these levels of exposure. The variety of experimental conditions used has not helped in this respect. Further information can be found from [602], [429], [675], [933], [621], [622], [562], [330], [361], [687], [195], [653] and [654].

2.7.2 Birds

The effect of heat on eggs was the subject of some very old (1928) studies (e.g., those reported in [858] and recently reviewed in [163]). They showed that thermal gradients produced no deformity, but had some effect on growth and differentiation of the vascular zone. More studies [884], [885] have reported clear morphologic effects, i.e., retarded development of the whole or parts of the body in 65% of chicks issued from irradiated eggs (2.45 GHz, 400 mW cm^{-2}, 270 s or 280 mW cm^{-2}, 510 s, or 200 mW cm^{-2}, 8 min on day two of embryonic growth). Subsequent studies [309] indicated a reduced growth rate for chicks *in vivo* exposed for one week to about 550 μW cm^{-2} at 880 MHz. However, rapid recovery was observed when irradiation occurred at intervals. Yet another study [552] showed no effect on the eggs of Japanese quail treated with 30 mW cm^{-2} at 2.45 GHz for four hours every day of the first five days of incubation. The absence of nonthermal teratogenic effects was also reported in [264] and reviewed in [163]. It relates to eggs on days four and five of embryonic development, brought by microwaves to 32-36°C as compared with 39°C for normal incubation. Tests on 60 unfertilized hen eggs (2.45 GHz, 80–100 W kg^{-1}, 300 s) revealed [163] very persistent and reproducible temperature differences of 4°C and 2°C, respectively, 5 and 60 min after the end of exposure. They also demonstrated the possiblity of hotspots at the egg center (Section 1.4.6), but did not draw any conclusions with regard to thermally or nonthermally induced teratogenic effects. However, nonthermal effects of very low level continuous radiation cannot be disregarded completely in view of the experiments reported in [852] in which hens were irradiated for 248 days at 7.06 GHz with very low power densities in the range 0.19–360 muW cm^{-2}. Egg production rose by 13.7% for the irradiated birds, but at the cost of twice the mortality rate of chicks due to leukemias and brain tumors and undesirable effects on the health of the surviving chicks. As before [727], [851], it was considered that the onset of nonthermal effects at very low levels may have been inhibited by the experimental power levels commonly used, which are higher by three or four orders of magnitude. This hypothesis, if confirmed, would imply the risk of hazardous athermal effects, constituting a significant potential danger.

Further information may be found in [359], [360], [363], [552], [207], [162], [263].

2.7.3 Mammals

A number of studies have attempted to find evidence for effects on growth *in utero* in the case of mammals [739], [740], [738], [741], [742]. They found a greater number of foetal resorptions and the appearance of exencephaly on mice born of mothers exposed to high levels of radiation (2.45 GHz, 100 mW cm^{-2}, 100 W kg^{-1}, 4–5 min sessions). The effects of irradiation are at their worst on day 8 of gestation, (i.e., at the onset of organogenesis): 68% of the embryos were resorbed, dead, or abnormal as compared with 28% and 20%, respectively, in the case of irradiation on days 10 and 14. The probability of anomalies increases linearly with absorbed dose [647]. Another study [141] found no increase in neonatal mortality, no malformation and no apparent retardation in young mice exposed *in utero* (2.45 GHz, 38 mW cm^{-2}, 10 min per day on days 10 and 14). When combined with cortisone (a teratogenic product), microwaves improve the survival rate in comparison with cortisone alone. The worst case of malformation was encountered when the combined treatment was given on day 14. In another series of experiments, performed under similar conditions with only one 20-min exposure between days 10 and 16 [140], it was found that there was an increase in the number of foetal resorptions without teratogenic effects. The brain of the newly born had a lower norepinephrine content. The reduction in average weight at birth was identical to that produced by exposure to infrared.

A study of mice [407], [408], [409] with lower levels of irradiation [i.e., $8\,\mathrm{hr\,d^{-1}}$ during the whole period of gestation ($10\,\mathrm{mW\,cm^{-2}}$ at $915\,\mathrm{MHz}$ or $20\,\mathrm{mW\,cm^{-2}}$ at $2.45\,\mathrm{GHz}$)] revealed no effect on maternal, foetal, or placental masses, and no effect on the frequency of resorption, foetal death rate, size of litter, sex of the newly born, and their ability to perform. Other studies [590] reported a faster developement of rat foetuses exposed daily to 36–$144\,\mathrm{J\,cm^{-2}}$ at $2.45\,\mathrm{GHz}$. This agrees with another report [415] that noted an increase in the weight of newly born rats and a premature opening of the eyes after prenatal irradiation ($918\,\mathrm{MHz}, 5\,\mathrm{mW\,cm^{-2}}, 380\,\mathrm{hr}$), as well as an impaired ability to learn. In contrast, another study [73] found lower average weight at birth and a higher frequency of encephalopathies, depending on the absorbed dose *in utero* ($2.45\,\mathrm{GHz}$, $100\,\mathrm{min\,d^{-1}}$ from day 1 to day 18 of gestation at 3.4, 13.6, and $28\,\mathrm{mW\,cm^{-2}}$ with the absorption of 2, 8.1, and $22.2\,\mathrm{W\,kg^{-1}}$). However, the same authors subsequently reported [70] the absence of differences between 70 rats irradiated *in utero* ($2.45\,\mathrm{GHz}$, $28\,\mathrm{mW\,cm^{-2}}, 4.2\,\mathrm{W\,kg^{-1}}$, $100\,\mathrm{min\,d^{-1}}$ from day 6 to day 15) and 68 rats in a control group: the survival levels were the same, and there was a total absence of anomalies. The same researchers [71] drew identical conclusions from a series of experiments on hamsters ($2.45\,\mathrm{GHz}, 20\,\mathrm{mW\,cm^{-2}}, 6\,\mathrm{W\,kg^{-1}}$), but, in contrast, found a larger number of foetal resorptions, 10% lower weight, and some bone immaturity after exposure to slightly higher doses ($2.45\,\mathrm{GHz}, 30\,\mathrm{mW\,cm^{-2}}, 9\,\mathrm{W\,kg^{-1}}, 100\,\mathrm{min\,d^{-1}}$ on days 6 and 7). The rectal temperature of irradiated females rose by $0.4°\mathrm{C}$ and $1.6°\mathrm{C}$ for 6 and $9\,\mathrm{W\,kg^{-1}}$ absorbed power, respectively. In a study [399] reviewed in [81] it was reported that embryos of preimplanted mice were unaffected by irradiation ($2.45\,\mathrm{GHz}, 19\,\mathrm{mW\,cm^{-2}}$, 3 hr/d on days 2 and 3).

At very low levels ($2375\,\mathrm{MHz}$, 10, 20, and $50\,\mu\mathrm{W\,cm^{-2}}$), the reproductive capacity of mice was found by Soviet researchers to be somewhat impaired, with smaller litter size and a rise in neonatal mortality, which is a direct function of the power flux density [396], [548]. Another study [309] showed for the rat (as it had been shown for the hen), that growth was reduced after exposure for one week to $0.55\,\mu\text{W cm}^{-2}$ at880 MHz.

These heterogeneous results seem to indicate a hypothetic threshold and lead us to conclude tentatively that

- there is a general depressive effect at extremely low power levels $(5–500\,\mu\text{W cm}^{-2})$, which is a direct function of these levels;
- there are no effects at moderate power levels on the order of $10\,\text{mW cm}^{-2}$;
- there is a depressive effect on development after exposure to power levels higher than $30\,\text{mW cm}^{-2}$ for short intervals of time; and
- there is an acceleration of development with subsequent adverse effects after long or very long *in utero* exposures.

The true state of affairs is probably far more complex, but the available data are not sufficient to allow us to outline it more clearly. Further information may be found in [441], [172], [339], [689], [618], [247], [248], [465], [102], [538], [540], [786], [928], [139], [69] and [406].

All attempts to extrapolate these results to humans lead to very high power densities, partly because geometric resonance effects are very significant in small animals. The few available clinical data relate to pregnant women treated with hyperthermia for chronic pelvic inflammation, after which all full-term pregnancies (six cases out of eight) produced normal children [737], [398], [647].

2.8 Lesions and cataracts

RF burns are characterized by the depth of tissue affected, the latent time, which may be up to several days before the appearance of symptoms, and the onset of hotspots due to multiple reflections at internal interfaces, which can cause very localized lesions [107]. Experiments with animals demonstrated the presence of deep burns with nerve destruction and irreversible blood stases leading to tissue necrosis [598], [107]. Human accidents are, fortunately, very rare. One case of hand and forearm burns due to a defective domestic oven was reported in [266]. The subject suffered nerve lesions accompanied by painful burning sensations and a hypersensitivity that persisted for several years. The lower the frequency, the deeper the lesions: A particular case of accidental exposure to an RF field (27 MHz) necessitated the amputation of the hand because of necrosis by coagulation and arterial occlusion [158], [107]. Other data are presented in [545], [103] and [940].

Cataracts derive from two properties of the crystalline lens, namely, high water content (60–70%) and poor blood supply that is not sufficient to remove the deposited heat. The crystalline lens is biconvex, deforms easily, and grows continually by the regeneration of the epithelial layer of the anterior surface. The cells stretch into hexagonal fibers between the anterior and posterior poles, and lose their nuclei. In the course of this process they push older fibers toward the center. If this mechanism is disturbed (by trauma, diabetes, toxic products, X-rays, ultraviolet, infrared, or microwaves, or simply by age), the epithelial cells cease to differentiate and the entire region loses its transparency. Multiplication of these opaque regions may lead to blindness [384].

The first observations of microwave-induced cataracts were reported in the late 1940s [192], [715] and in the 1950s [193], [194], [381]. Experimental research on laboratory animals can be divided into two classes: far-field interactions involving approximately plane wave, the animal being wholly exposed; and near-field interactions in which the eye is in direct contact with the source or separated from it by a thin layer of air.

The initial conclusion from this research is that opacification is indeed a thermal process [450], [352], [451], [453], [452]. For a given frequency, the relationship between absorbed power and time is nonlinear, and there is a threshold below which even permanent exposure has no apparent cataractogenic effect. The precise determination of this threshold presents certain difficulties because of anatomic differences between primates and rabbits, which are the most commonly used experimental animals. Continuous whole-body exposure in the far-field zone produces no observable lesions in rabbits submitted to $10 \, \text{mW cm}^{-2}$ at $2.45 \, \text{GHz}$ ($17 \, \text{W kg}^{-1}$) in 23 hr of daily exposure for 180 hr [349] or 40 hr weekly for 8 to 17 weeks [262]. No effect was observed after a 30-min exposure to $500 \, \text{mW cm}^{-2}$ at $3 \, \text{GHz}$, despite the fact that the animal exhibited acute thermal lesions [23]. In the near-field zone, the cataractogenic threshold for rabbits is reported in [352] as $150 \, \text{mW cm}^{-2}$ at $2.45 \, \text{GHz}$ ($138 \, \text{W kg}^{-1}$ absorbed), and no lesions appeared so long as the intraocular temperature did not exceed $42°\text{C}$. The same power density at $9.31 \, \text{GHz}$, modulated with $0.5\text{-}\mu s$ pulses at $1.05 \, \text{kHz}$, produced no effect on the eye in monkeys [541], and unmodulated near-field radiation of $500 \, \text{mW cm}^{-2}$ at $2.45 \, \text{GHz}$ had no effect on the eye either, despite some severe facial burns [452]. The cataractogenic thresholds that were established quite precisely for rabbits (Fig. 2.12) are not well defined for monkeys, but are certainly much higher in all cases. The time necessary for the opacification of the rabbit eye lens is given by (2.2) with $m = 4.11 \pm 1.79$ [610].

Experimental studies suggest the absence of any discernible differences between the effects of continuous and modulated waves with the same average power. Moreover, there is no evidence for a cumulative effect, as least so long as a sufficient interval is introduced between irradiation sessions. The cataractogenic thresholds are higher at lower frequencies [88], and there are practically no risks below $500 \, \text{MHz}$ [173], [262].

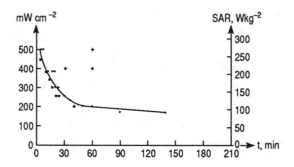

Figure 2.12 Threshold for cataract formation at $2.45\,\text{GHz}$, $5\,\text{cm}$ from a radiating slot [452]. Black circles - onset of opacification in the rabbit eye; open circles - absence of opacification on the rabbit eye; crosses - absence of opacification in the eye of monkeys.

The picture is very different for millimeter waves which produce epithelial and stromal lesions at relatively low power levels: 5–$60\,\text{mW}\,\text{cm}^{-2}$ at $35\,\text{GHz}$ and $107\,\text{GHz}$ after 15–60 min of exposure [90], [729]. Spontaneous regression in 24–48 hr is not enough to rule out all risks of long-term exposure to EHF waves [307]. Further information on this subject can be found in [117], [796], [123], [124], [566], [118], [42], [119], [130], [131], [87], [886], [120], [547], [126], [127], [567], [921], [128], [256], [594], [922], [355], [729], [539], [673], [542] and [822].

The first systematic clinical studies on the subject were reported in [942] and [941], [170] and [943]. It was concluded that there was real risk of cataract in professionals working with microwaves. A direct effect was not envisaged in [170], but it was suggested that repeated exposure to sub-threshold power densities could induce microlesions and then premature ageing of the eye. Others [34], [863], [33], [35] have reported changes in the eye lens and even in the retina of 68 employees of a Swedish factory producing microwave materials. In contrast, several much more extensive inquiries showed total absence of microwave-induced cataracts (3,000 [946], 697 [649], 841 [789], [790], [799], and 817 persons [781]). This was also the conclusion of another study [25] performed for the American Army at Fort Monmouth. The methodology of this study, the composition of the selected group, and the validity of the statistical calculations were refuted in [287], which in turn was strongly criticized in [913] where a statistical re-examination of the original study confirmed a correlation with the subject's age alone.

Finally, the hypothesis of extremely low level effects (below $10 \,\mu\mathrm{W}\,\mathrm{cm}^{-2}$) was rejected in [30] as reviewed in [548]. This involved clinical examinations of 269 young women exposed to such power fluxes over 5 to 23 years of professional life. Actually, all the cataract cases mentioned in [943], [579] and [49] correspond to accidental exposures in excess of $100\,\mathrm{mW}\,\mathrm{cm}^{-2}$ at 2.45 GHz. The Canadian Ministry of Health [950] emphasizes that the therapeutic microwave treatment of the eye, which is generally performed at 2.45 GHz with 90–$240\,\mathrm{mW}\,\mathrm{cm}^{-2}$ for $10-15\,\mathrm{min}$ per day for several weeks, and sometimes several months, does not appear to cause any lesions in the eye lens. Further information on this subject may be found in [249], [39], [378], [463], [578], [614], [22], [99], [121], [233], [595], [937], [938], [24], [270], [906], [348], [122], [169] and [474].

References

1. ADAIR E., ADAMS B. Microwaves induce peripheral vasodilatation in squirrel monkey. *Science,* 207, 1980, pp. 1381-1383.

2. ADAIR E., ADAMS B. Microwaves modify thermoregulatory behavior in squirrel monkey. *Bioelectromagn.,* 1, 1980, pp. 1-20.

3. ADAIR E., ADAMS B. Adjustments in metabolic heat production by squirrel monkeys exposed to microwaves. *J. Appl. Physiol.,* 52, n° 4, 1982, pp. 1049-1058.

4. ADAIR E., SPIERS D., STOLWIJK J., WENGER C. Technical note : on changes in evaporative heat loss that result from exposure to nonionizing electromagnetic radiation, *J. Microw. Power,* 18, n° 2, 1983, pp. 209-211.

5. ADEY W. Introduction : effects of electromagnetic radiation on the nervous system. In : TYLER P. Biologic effects of nonionizing radiation, Proceedings of a conference. *Ann. New York Acad. Sci.,* 247, 1975, pp. 15-20.

6. ADEY W. Experiment and theory in long range interactions from electromagnetic fields at brain cell surfaces. In : Proceedings of Biomagnetic effects workshop, Lawrence Berkeley lab., ed. Tenforde, publ. LBL-7452, 1978, pp. 53-78.

7. ADEY W. Frequency and power windowing in tissue. Interactions with weak electromagnetic fields. *Proc. IEEE,* 68, n° 1, 1980, pp. 119-125.

8. ADEY W., BAWIN S. Brain interactions with weak electric and magnetic fields. *NRP Bull.,* 15, 1977, pp. 1-29.

9. ADEY W., BAWIN S., LAWRENCE A. Effects of weak amplitude-modulated microwave fields on calcium efflux from awake cat cerebral cortex. *Bioelectromagn.,* 3, 1982, pp. 295-307.

10. AICHMANN H., NIMTZ G., DENNHOFER L., FRUCHT A. On the nonthermal microwave response of Drosophila melanogaster. *IEEE Trans. Microw. Theory Tech.,* MTT-32, n° 8, 1984, pp. 888-891.

11. ALBERT E. Light and electron microscopic observations on hamsters. In: EART F., FORD J. Proceedings of the IMPI Symposium, Waterloo (Can.), 1975, pp. 125-126.

12. ALBERT E. Light and electron microscopic observations on the blood-brain barrier after microwave irradiation. In : HAZZARD D.- Biological effects and measurement of RF/microwaves. Symposium proceedings, 16-18 Feb. 1977, US Dep. Health, Educat., Welfare, BRH, FDA-77-8026, July 1977, pp. 294-304.

13. ALBERT E. Ultrastructural pathology associated with microwave-induced alterations in blood-brain barrier permeability. In : Biological effects of electromagnetic radiation, URSI international symposium, Helsinki, 1978, p. 58.

14. ALBERT E. Current status of microwave effects on the blood-brain barrier. *J. Microw. Power,* 14, n° 3, Sept. 1979, pp. 281-285.

15. ALBERT E. Reversibility of microwave-induced blood-brain barrier permeability. *Radio Sci.,* 14, n° 6 (S), 1979, pp. 323-327.

16. ALBERT E., BRAINARD D., RANDALL J., JANATTA F. Neuropathological observations of microwave-irradiated hamsters. In : Biological effects of electromagnetic radiation, URSI international symposium, Helsinki, 1978, p. 59.

17. ALBERT E., DE SANTIS M. Do microwaves alter nervous system structure ? In : TYLER P. Biologic effects of non-ionizing radiation. Proceedings of a conference. *Ann. New York Acad. Sci.,* 247, 1975, pp. 87-108.

18. ALBERT E., KERNS J. Reversible microwave effects on the blood-brain barrier. *Brain Res.,* n° 230, 1981, pp. 153-164.

19. ALBERT E., Mac CULLARS G., SHORE M. The effect of 2450 MHz microwave radiation on liver adenosine triphosphate. *J. Microw. Power,* 9, n° 3, 1974, pp. 205-211.

20. ALLIS J. Irradiation of bovine serum albumin with a crossed-beam exposure detection system. In : TYLER P. Biologic effects of non-ionizing radiation. Proceedings of a conference. *Ann. New York Acad. Sci.*, 247, 1975, pp. 312-322.

21. ANDERSON J., MOLLER G., SJORBERG O. Selective induction of DNA Synthesis in T and B lymphocytes. *Cellular Immunol.*, 4, 1972, pp. 381-393.

22. APPLETON B. Experimental microwave ocular effects. In : CZERSKI P., OSTROWSKI K., SHORE M., SILVERMAN C., SÜSS M., WALDESKOG B. Biological effects and health hazards of microwave radiation. Proceedings of an international symposium, Warszawa, 15-18 Oct. 1973. Warszawa, Polish Medical Publishers, 1974, pp.186-188.

23. APPLETON B., HIRSCH S., BROWN P. Investigation of single exposure microwave ocular effects at 3000 MHz. In : TYLER P. Biologic effects of non-ionizing radiation. Proceedings of a conference. *Ann. New York Acad. Sci.*, 247, 1975, pp. 125-134.

24. APPLETON B., HIRSCH S., KINION R., SOLES M., Mac CROSSAN G., NEIDLINGER R. Microwave lens effects in humans. II. Results of five-year survey. *Arch. Ophtalmol.*, 93, April 1975, pp. 257-258.

25. APPLETON B., Mac CROSSAN G. Microwave lens effects in humans. *Arch. Ophtalmol.*, 88, Sept. 1972, pp. 259-262.

26. ARBER S. The effect of microwave radiation on passive membrane properties of snail neurons. *J. Microw. Power*, 16, n° 1, 1981, pp. 15-20.

27. ARBER S. The effect of microwaves on dimensions and nucleic-acid concentration in snail neurons. *J. Microw. Power*, 16, n° 1, 1981, pp. 21-23.

28. ARONOVA S. K voprosu mehanizma dejstvija električeskogo impulsnogo polja na davlenije krovi. *Vopr. Kurortol. Fizioterap. Lečh. Kul't.*, 3, 1961, pp. 243-246.

29. ARSONVAL J. d'. Influence de la fréquence sur les effets physiologiques des courants alternatifs. *C.R. Acad. Sci.*, tome 116, 1893, pp. 630-633.

30. ARTAMONOVA V., SAMORODOVA L., SAJKINA A. Ophtalmologic changes in persons exposed to microwaves. (trad. du russe) US joint Publicat. Res. Serv., Rep. JPRS L/6791, 12 janv. 1977, pp. 47-50.

31. ASABAJEV C. Seuil de sensibilité du cerveau de lapin aux champs électromagnétique UHF. (En russe). *Bjull. Eksperimental'noj Biolog. Med.*, 74, n° 7, 1972, p. 56.

32. ATHEY T. Comparison of radio-frequency induced calcium efflux from chick brain tissue at different frequencies. *Bioelectromagn.*, 2, 1981, pp. 407-410.

33. AURELL E. Effect of microwave radiation on eye. *Acta Ophtalmol. Kbh. Suppl.*, 51, n° 5, 1973, p. 742.

34. AURELL E., KADEFORS R., MAGNUSSON R., PETERSEN I. Retinal lesions as a result of microwave exposure. In : Digest of the 3rd international conference on medical physics, including medical engineering, Göteborg, 1972, abstract 32.2.

35. AURELL E., TENGROTH B. Lenticular and retinal changes secondary to microwave exposure. *Acta Ophtalmol., Kbh. Suppl.*, 51, 1973, pp. 764-771.

36. AUSTIN G., HORVATH S. Production of convulsions in rats by high frequency electrical currents. *Am. J. Phys. Med.*, 33, 1954, pp. 141-149.

37. AVERBECK D., DARDALHON M. Les effets biologiques des micro-ondes. Comptes rendus du Colloque sur les applications industrielles des hautes fréquences et des micro-ondes, Lyon, 26-27 janv. 1977, 8 p.

38. AVERBECK D., DARDALHON M., BERTEAUD A. Microwave action in procaryotic and eurocaryotic cells and possible interaction with X rays. In : Proceedings of the IMPI Symposium, Leuven, 1976. (Paper 2A.6.).

39. BAILLIE H. Thermal and non-thermal cataractogenesis by microwaves. In : CLEARY S. Biological effects and health implications of microwave radiation, Symposium proceedings, Commonwealth Univ., Richmond, 17-19 Sept. 1969, US Dep. Health, Educat., Welfare, BRH, DBE 70-2, PB 193 898, Library of Congress card n° 76-607 340, June 1970, pp. 59-65.

40. BAILLIE H. A limited review of the recent biological effects of microwave radiation. In : Proceedings of the IMPI Symposium, Loughborough, 1973, p. 1. (Paper 1B1).

41. BALDWIN M., BACH S., LEWIS S. Effects of radio-frequency energy on primate cerebral activity. *Neurol.*, 10, n° 2, 1960, pp. 178-187.

42. BALUTINA A. Lésion expérimentale de l'oeil au moyen d'un champ électromagnétique UHF. (En russe). *Bjull. Eksperimental'noj Biol. Med.*, 60, n° 12, 1965, pp. 41-43.

43. BARANSKA W. The effects of microwaves on human lymphocyte cultures. In : CZERSKI P., OSTROWSKI K., SHORE M., SILVERMAN C., SÜSS M., WALDESKOG B. Biological effects and health hazards of microwave radiation, Proceedings of an international symposium, Warszawa, 15-18 Oct. 1973. Warszawa, Polish Medical Publishers, 1974, pp. 186-195.

44. BARANSKI S. Effect of chronic microwave irradiation on the blood forming system of guinea pigs and rabbits. *Aerosp. Med.*, 42, 1971, pp. 1196-1199.

45. BARANSKI S. Effect of microwaves on reaction of white corpuscule system. *Acta Physiol. Pol.*, 22, n° 6, 1971, p. 898.

46. BARANSKI S. Effects of microwaves on the reactions of the white blood cells system. *Acta Physiol. Pol.*, 23, 1972, pp. 619-629 et 685-695.

47. BARANSKI S. Histological and histochemical effects of microwave irradiation on the central nervous system of rabbits and guinea pigs. *Am. J. Phys. Med.*, 51, n° 4, 1972, pp. 182-192.

48. BARANSKI S. Pharmacologic analysis of microwave effects on the nervous system of experimental animals. In : CZERSKI P., OSTROWSKI K., SHORE M., SILVERMAN C., SÜSS M., WALDESKOG B. Biological effects and health hazards of microwave radiation, Proceedings of an international symposium, Warszawa, 15-18 Oct. 1973, Warszawa, Polish Medical Publishers, 1974.

49. BARANSKI S., CZERSKI P. Biological effects of microwaves. In : CZERSKI P., OSTROWSKI K., SHORE M., SILVERMAN C., SÜSS M., WALDESKOG B. Biological effects and health hazards of microwave radiation, Proceedings of an international symposium, Warszawa, 15-18 Oct. 1973..Warszawa, Polish Medical Publishers, 1974 and Stroudsburg, Dawden-Hutchinson and Ross inc. publ., 1976, pp. 136-137.

50. BARANSKI S., CZERSKI P., JANIAK U., SZMIGIELSKI S. Wpływ naświetlania mikrofalami na rytm dobowy podziału komórek szpiku. *Med. Lotnicza*, 39, 1972, p. 67.

51. BARANSKI S., CZERSKI P., SZMIGIELSKI S. Wpływ mikrofal na mitoze in vitro i in vivo. *Postępy Fiz. Med.*, 6, 1971, p. 93 and *Genet. Pol.*, 10, n° 3-4, 1969.

52. BARANSKI S., DEBIEC H., KWARECKI K., MEZYKOWSKI T. Influence of microwaves on genetical processes of Aspergillus nidulans. In : Proceedings of the IMPI Symposium, Leuven, 1976. (Paper 3 A.2).

53. BARANSKI S., DEBIEC H., KWARECKI K., MEZYKOWSKI T. Influence of microwaves on genetical processes of Aspergillus nidulans. *J. Microw. Power*, 11, n° 2, 1976, pp. 146-147.

54. BARANSKI S., EDELWEJN Z. Studies on the combined effect of microwaves and some drugs on bioelectric activity of the rabbit CNS. *Acta Physiol. Pol.*, 19, 1968, pp. 37-50.

55. BARANSKI S., EDELWEJN Z. Pharmacologic analysis of microwave effects on the central nervous system in experimental animals. In : CZERSKI P., OSTROWSKI K., SHORE M., SILVERMAN C., SÜSS M., WALDESKOG B. Biological effects and health hazards of microwave radiation. Proceedings of an international symposium, Warszawa, 15-18 Oct. 1973. Warszawa, Polish Medical Publishers, 1974, pp. 119-127.

56. BARANSKI S., OSTROWSKI K., STODOLNIK-BARANSKA W. Functional and morphological studies of the thyroid gland in animals exposed to microwave irradiation. *Acta Physiol. Pol.*, 23, 1972, pp. 1029-1039.

57. BARONENKO V., TIMOFEJEVA K. Vlijanije električeskih polej VČ i UVČ na uslovnoreflektornuju dejatel'nosti njekotoryje bezuslovnyje funkcii životnych i čeloveka. *Fiziol. Ž. SSSR im. I.M. Sehonova*, 45, n° 2, 1959, pp. 203-207.

58. BARRON C., BARAFF A. Medical considerations on exposure to microwaves (radar). *J. Am. Med. Ass.*, 168, n° 9, 1958, pp. 1194-1199.

59. BARRON C., LOVE A., BARAFF A. Physical evaluation of personnel exposed to microwave emanations. *J. Aviat. Med.*, 26, 1955, pp. 442-452.

60. BASSEN H. RF/microwave radiation measurement, risk assessment and control activities at the Bureau of Radiological Health. *IEEE Antennas Propag. Soc. Newsl.*, April 1981, pp. 5-10.

61. BAUM S., EKSTROM M., SKIDMORE W., WYANT D., ATKINSON J. Biological measurements in rodents exposed continuously throughout their adult life to pulsed electromagnetic radiation. *Health Phys.*, 30, n° 2, 1970, pp. 161-166.

62. BAWIN S., ADEY W., SABROT I. Ionic factors in release of $^{45}Ca^2$ from chicken cerebral tissues by electromagnetic fields. *Proc. Natl. Acad. Sci. U.S.A.*, 75, n° 12, 1978, pp. 6314-6318.

63. BAWIN S., GAVALAS-MEDICI R., ADEY W. Effects of modulated very high frequency fields on specific brain rhythms in cats. *Brain Res.*, n° 58, 1973, pp. 365-384.

64. BAWIN S., KACZMAREK L., ADEY W. Effects of modulated VHF fields on the central nervous system. In : TYLER P. Biologic effects of non-ionizing radiation, Proceedings of a conference. *Ann. New York Acad. Sci.*, 247, 1975, pp. 74-81.

65. BAWIN S., SHEPPARD A., ADEY W. Possible mechanisms of weak electromagnetic field coupling in brain tissue. *Bioelectrochem. Bioenerg.*, 5, 1978, pp. 67-76.

66. BAZANOVA E., BRJUHOVA A., VILENSKAJA R., GELVIČ E., GOLANT M., LANDAU N., MELNIKOVA V., MIKAELJAN N., OHOHONINA G., SEVASTJANOVA L., SMOLJANSKAJA A., SYČEVA N. Certain methodological problems and results of experimental investigation of the effects of microwaves on micro-organisms and animals. *Soviet Phys. Uspekhi*, 16, n° 4, 1974, pp. 569-570.

67. BELKHODE M., MUC A., JOHNSON D. Thermal and athermal effects of 2.8 GHz microwaves on three human serum enzymes. *J. Microw. Power*, 9, n° 1, 1974, pp. 23-29.

68. BEREZINSKAJA A. Indices de fertilité de souris femelles irradiées par des micro-ondes de 10 cm. (En russe). *Gig. Tr. Prof.'nije Zabol.*, n° 9, 1968 p. 33.

69. BERMAN E., CARTER H., HOUSE D. Tests of mutagenesis and reproduction in male rats exposed to 2450 MHz (Cw) microwaves. *Bioelectromagn.*, 1, 1980, pp. 65-76.

70. BERMAN E., CARTER H., HOUSE D. Observations of rat fetuses after irradiation with 2450-MHz (CW) microwaves. *J. Microw. Power*, 16, n° 1, 1981, pp. 9-13.

71. BERMAN E., CARTER H., HOUSE D. Observations on Syrian hamster fetuses after exposure to 2450-MHz microwaves. *J. Microw. Power*, 17, n° 2, 1982, pp. 107-112.

72. BERMAN E., CARTER H., HOUSE D. Reduced weight in mice offsprings after in-utero exposure to 2450 MHz (CW) microwaves. *Bioelectromagn.*, 3, 1982, pp. 285-291.

73. BERMAN E., KINN J., CARTER H. Observations of mouse fetuses after irradiation with 2.45-GHz microwaves. *Health Phys.*, 35, 1978, pp. 791-801.

74. BERNHARDT J. Zur Gefährdung des Menschen durch nichtionisierende Strahlung. *PTB Mitt.*, 90, 1980, pp. 416-433.

75. BERNHARDT J. Gefährdung von Personen durch Elektromagnetische Felder. *STH Ber.*, 2, 1983, pp. 1-38.

76. BERNHARDT J. Elektromagnetische Felder, potentielle Gefahrdung. *Bundesarbeitsblatt.*, 10, 1984, pp. 17-25.

77. BERTEAUD A. Les effets biologiques des micro-ondes : une confirmation. *Rech.*, 9, n° 85, 1978, pp. 65-66.

78. BERTEAUD A. Caractéristiques physiques des rayonnements électromagnétiques non ionisants. *Ann. Ass. belge Radioprot.*, 6, n° 1, 1981, pp. 17-27.

79. BERTEAUD A., DARDALHON M. Effets biophysiques et biologiques associés à l'utilisation des micro-ondes, 1. Interaction des micro-ondes avec les milieux vivants, aspect biophysique. Rapport établi pour le Club Micro-Ondes EDF. *Econ. Progrès Electr.*, n° 13 janv.-fév. 1985, pp. 21-23.

80. BERTEAUD A., DARDALHON M. Effets biophysiques et biologiques associés à l'utilisation des micro-ondes, 2. Principaux effets biologiques des micro-ondes et hautes fréquences. Rapport établi pour le Club Micro-Ondes EDF. *Econ. Progrès Electr.*, n° 14, mars-avril 1985, pp. 16-19.

81. BERTEAUD A., DARDALHON M. Effets biophysiques et biologiques associés à l'utilisation des micro-ondes, 3. Etude de l'action des micro-ondes au niveau pluricellulaire. Rapport établi pour le Club Micro-Ondes EDF. *Econ. Progrès Electr.*, n° 15, mai-juin 1985, pp. 18-21.

82. BERTEAUD A., DARDALHON M., REBEYROTTE N., AVERBECK D. Action d'un rayonnement électromagnétique à longueur d'onde millimétrique sur la croissance bactérienne. *C.R. Hebd. Séances Acad. Sci.*, 281, 22 sept. 1975, série D, pp. 843-846.

83. BERTHARION G., SERVANTIE B., JOLY R. Electro-cortico-graphic modifications after exposure to microwave fields, on the white rat. *J. Microw. Power*, 6, n° 1, 1971, pp. 62-63.

84. BIGÚ DEL BLANCO J., ROMERO-SIERRA C. Effect of 3 cm. wavelength microwave radiation on the nastic response of the plant Mimosa Pudica. In : 5th Canadian Medical and Biological Engineering Conference, 1974. p. 12, (abstracts), (Paper 6.A-6.B.).

85. BIGÚ DEL BLANCO J., ROMERO-SIERRA C. Les propriétés des plumes d'oiseaux en tant que transducteurs piézo-électriques et récepteurs de micro-ondes. *Onde Electr.*, 57, n° 8-9, 1977, pp. 532-536.

86. BIGÚ DEL BLANCO J., ROMERO-SIERRA C., TANNER J. Effect of microwave radiation on the nastic response of some sensitive plants. In : Proceedings of the IMPI Symposium, Loughborough, 1973, 2 p. (Paper 3.B.6.).

87. BIRENBAUM L., GROSOF G., ROSENTHAL S., ZARET M. Effect of microwaves on the eye. *IEEE Trans. Biomed. Eng.*, BME-16, n° 1, 1969, pp. 7-14.

88. BIRENBAUM L., KAPLAN I., METLAY W., ROSENTHAL S., SCHMIDT H., ZARET M. Effect of microwaves on the rabbit eye. *J. Microw. Power*, 4, n° 4, 1969, pp. 232-243.

89. BIRENBAUM L., KAPLAN I., METLAY W., ROSENTHAL S., ZARET M. Microwave and infrared effects on heart rate, respiration rate and subcutaneous temperature of the rabbit. *J. Microw. Power*, 10, n° 1, 1975, pp. 3-9.

90. BIRENBAUM L., ROSENTHAL S., KAPLAN I., METLAY W., ZARET M. Ocular effects of 35 and 107 GHz Cw microwaves. In : Proceedings of the IMPI Symposium, Waterloo (Can.), 1975, pp. 105-110. (Paper 7.1.).

91. BLACKMAN C., BENANE S., ELDER J., HOUSE D., LAMPE J., FAULK J. Induction of calcium-ion efflux from brain tissue by radio-frequency radiation : effect of sample number and modulation frequency on the power-density window. *Bioelectromagn.*, 1, 1980, pp. 35-43.

92. BLACKMAN C., BENANE S., WEIL C., ALI J. Effects of nonionizing electromagnetic radiation on single-cell biologic systems. In : TYLER P. Biologic effects of non-ionizing radiation. Proceedings of a conference. *Ann. New York Acad. Sci.*, 247, 1975, pp. 352-366.

93. BLACKMAN C., ELDER J., WEIL C., BENANE S., EICHINGER D., HOUSE D. Modulation-frequency and field-strength dependent induction of calcium-ion efflux from brain tissue by radio-frequency radiation. *Radio Sci.*, 14, n° 6 (S), 1979, pp. 93-98.

94. BLACKMAN C., SURLES M., BENANE S. The effects of microwave exposure on bacteria : mutation induction. In : JOHNSON C., SHORE M. Biological effects of electromagnetic waves, Selected papers of the USNC/URSI annual meeting, Boulder, 20-23 Oct. 1975, US Dep. health, Educ., Welfare, FDA-77-8010. Washington, US Government Printing Office, Vol. 1, 1976, pp. 406-413.

95. BLASBERG R. Problems of quantifying effects of microwave irradiation on the blood-brain barrier. *Radio Sci.*, 14, n° 6 (S), 1979, pp. 335-344.

96. BOGGS R. Effects of 2450 MHz microwave radiation on human blood coagulation processes. *Health Phys.*, 22, n° 3, 1972, pp. 217-224.

97. BORTH D., CAIN C. The generation of acoustic signals in materials irradiated with microwave pulses - a theoretical analysis. In : Proceedings of the IMPI Symposium, Waterloo (Can.), 1975, pp. 95-98. (Paper 5.4.).

98. BORTH D., CAIN C. Theoretical analysis of acoustic signal generation in materials irradiated with microwave energy. *IEEE Trans. Microw. Theory Tech.*, MTT-25, n° 11, 1977, pp. 944-954.

99. BOUCHAT J. Cataracts following use of microwave oven. Comments. *New York State J. Med.*, 74, n° 11, 1974, pp. 2038-2039.

100. BOURGEOIS A. The effect of microwave exposure upon the auditory threshold of humans. Thesis, Ph. D., Waco, Baylor Univ., 1967 (Dissertation Abstract International n° 67-2927).

101. BRADY A., WOODBURY J. The sodium-potassium hypothesis as the basis of electrical activity in frog ventricle. *J. Physiol.*, n° 154, 1960, pp. 385-407.

102. BRENT F., FRANKLIN J., WALLACE J.
The interruption of pregnancy using
microwave radiation. *Teratol.*, 4,
1971, Abstract, p. 48A.

103. BRODKIN R., BLEIBURG J. Cutaneous
microwave injury. *Acta Dermatovenerea*,
53, 1973, pp. 50-52.

104. BROWN P., LARSEN L. Differing effects
of pulsed and CW microwave energy
upon nerve function as detected by
birefringence measurement. *IEEE
Trans. Microw. Theory Tech.*, MTT-28,
n° 10, 1980, pp. 1126-1133.

105. BROWN-WESTERDAHL G., GARY N. Chronic
exposure of a honey bee colony to
2.45 GHz continuous-wave microwaves.
Space Solar Power Rev., 2, 1981, pp.
283-295.

106. BRYAN R. Retrograde amnesia : effects
of handling and microwave radiation.
Sci., 153, 1966, pp. 897-899.

107. BUDD R. Can microwave/radiofrequency
radiation (RFR) burns be distingui-
shed from conventional burns ? *J.
Microw. Power*, 20, n° 1, 1985, pp.
9-11.

108. BUDD R., CZERSKI P. Modulation of
mammalian immunity by electromagnetic radiation. *J. Microw. Power*, 20,
n° 4, 1985, pp. 217-231.

109. BURNER A. Biologic effects of radio
and microwaves : present knowledge ;
future directions. In : Proceedings
of the IMPI Symposium, Edmonton,
1969, pp. 57-58. (Paper DA1).

110. BUSH L., HILL D., RIAZI A., STENSAAS
L., PARTLOW L., GANDHI O. Effects of
millimeter-wave radiation on monolayer cell cultures : III. A search
for frequency-specific athermal biological effects on protein synthesis.
Bioelectromagn., 2, 1981, pp. 151-
159.

111. BYUS C., LUNDAK R., FLETCHER R.,
ADEY W. Exposure to a modulated
microwave field results in a decrease
in protein kinase activity in human
lymphocytes. *Abstr. Bioelectromagn.
Soc.*, 2, 5th Annual conference,
Boulder, 1983.

112. BYUS C., LUNDAK R., FLETCHER R.,
ADEY W., Alterations in protein
kinase activity following exposure
of cultured human lymphocytes to
modulated microwave fields. *Bioelectromagn.*, 5, 1984, pp. 341-351.

113. CAIN C., RAMA-RAO G., TOMPKINS W.
Enhancement of antibody-complement
cytotoxicity against virus-trans-
formed hamster PARA-7 cells treated
with heat and microwave radiation.
Radiat. Res., 88, 1981, p. 96.

114. CAIN C., RISSMAN W. Mammalian auditory response to 3.0 GHz microwave
pulses. *IEEE Trans. Biomed. Eng.*,
BME-25, 1978, pp. 288-293.

115. CAIN C., YANG H., RAMA-RAO G., FAN
S., MAGIN R., TOMPKINS W. Effect
of in-vivo microwave exposure on
the hamster immune system. *Bioelectromagn.*, n° 1, 1980, p. 246.

116. CARAFOLI E., PENNISTON J. Le signal
calcium. *Pour Sci.*, n° 99, janv.
1986, pp. 78-88.

117. CARPENTER R. Studies of the effects
of 2450 Mc. radiation on the eye
of the rabbit. In : SUSSKIND C. Proceedings of the 3rd annual Tri-Service Conference on biological hazards of microwave radiating equipments. Univ. California, Berkeley,
25-27 August 1959, RADC-TR-59-140,
p. 279.

118. CARPENTER R. An experimental study
of the biological effects of microwave radiation in relation to the
eye. In : RADC-TRD-62-131, Tufts
Univ., 1962, (AD 275840).

119. CARPENTER R. The action of micro-
wave radiation on the eye. In :
Proceedings of the IMPI Symposium,
Stanford, 1967, abstracts pp. 35-36.

120. CARPENTER R. Experimental microwave
cataract : a review. In : CLEARY S.
Biological effects and health impli-
cations of microwave radiation, Sym-
posium proceedings, Commonwealth Univ.,
Richmond, 17-19 Sept. 1969, US Dep.
Health, Educ., Welfare, BRH, DBE 70-2,
PB 193 898, June 1970, pp. 76-81.

121. CARPENTER R. Cataracts following
use of microwave oven. Comments.
New York State J. Med., 74, n° 11,
1974, pp. 2035-2036.

122. CARPENTER R. Ocular effects of microwave radiation. *Bull. New York Acad.
Med.*, 55, 1979, pp. 1048-1057.

123. CARPENTER R., BIDDLE D., VAN UMMER-
SEN C. Biological effects of microwave radiation with particular reference to the eye. In : Proceedings
of the 3rd International Conference
on Medicine Electronics, London,
Vol. 3, 1960, pp. 401-408.

124. CARPENTER R., BIDDLE D., VAN UMMER-SEN C. Opacities in the lens of the eye experimentally induced by exposure to microwave radiation. Institute of Radio Engineers Transactions, P.G.M.E. 7, 1960, pp. 152-157.

125. CARPENTER R., CLARK V. Responses to radio-frequency radiation. In : ALTMAN P., DITTMER D. Environmental biology, 1966, pp. 131-138.

126. CARPENTER R., FERRI E., HAGAN G. Lens opacities in eyes of rabbits following repeated daily irradiation at 2.45 GHz. In : Proceedings of the IMPI Symposium, Ottawa, 1972, pp. 143-144. (Paper 8.6.).

127. CARPENTER R., FERRI E., HAGAN G. Assessing microwaves as a hazard to the eye : progress and problems : CZERSKI P., OSTROWSKI K., SHORE M., SILVERMAN C., SÜSS M., WALDESKOG B. Biological effects and health hazards of microwave radiation, Proceedings of an international symposium, Warszawa, 15-18 Oct. 1973. Warszawa, Polish Medical Publishers, 1974, pp. 178-185.

128. CARPENTER R., HAGAN G., FERRI E. Use of a dielectric lens for experimental microwave irradiation for the eye. In : TYLER P. Biologic effects of non-ionizing radiation, Proceedings of a conference. *Ann. New York Acad. Sci.*, 247, 1975, pp. 142-154.

129. CARPENTER R., LIVSTONE E. Evidence for nonthermal effects of microwave radiation : abnormal development of irradiated insect pupae. *IEEE Trans. Microw. Theory Tech.*, MTT-19, n° 2, 1971, pp. 173-178.

130. CARPENTER R., VAN UMMERSEN A. The action of microwave radiation on the eye. In : Radiation Control for Health and Safety Act of 1967, Hearings before the Committee of Commerce, Senate of the United States of America, 90th Congress, 2nd session, S.2067, S.3211, HR 10790, serial n° 90-49. Washington, US Government Printing office, 1968, part 2.

131. CARPENTER R., VAN UMMERSEN A. The action of microwave radiation on the eye. *J. Microw. Power*, 3, n° 1, 1968, pp. 3-19.

132. CARROLL D., LEVINSON D., JUSTESEN D., CLARKE R. Failure of rats to escape from a potentially lethal microwave field. *Bioelectromagn.*, 1, 1980, pp. 101-115.

133. CATRAVAS G., KATZ J., TAKENAGA J., ABBOTT J. Biochemical changes in the brain of rats exposed to microwaves of low power density. In : Proceedings of the IMPI Symposium, Leuven, 1976. (Paper 3A.3.).

134. CHANG B., HUANG A., JOINES W. Inhibition of DNA synthesis and enhancement of the uptake and action of methotrexate by low-power density microwave radiation in L 1210 leukemia cells. *Cancer Res.*, 40, 1980, pp. 1002-1005.

135. CHANG B., HUANG A., JOINES W., KRAMER R. The effect of microwave radiation (1.0 GHz) on the blood-brain barrier in dogs. In : Blood-Brain Barrier Workshop, Dep. Navy, Annapolis, 30-31 Oct. 1978.

136. CHEN K., LIN C. Cytotoxic effects of electromagnetic radiation on Chinese hamster cells in culture. In : Proceedings of the IMPI Symposium, Leuven, 1976. (Paper 2A.3.).

137. CHEN K., LIN C. Cytotoxic effects of electromagnetic radiation on Chinese hamster cells in culture. *J. Microw. Power*, 11, 1976, pp. 140-141.

138. CHEN K., LIN C. A system for studying effects of microwaves on cells in culture. *J. Microw. Power*, 13, n° 3, 1978, pp. 251-256.

139. CHERNOVETZ M., JUSTESEN D., LEVINSON D. Acceleration and deceleration of fetal growth of rats by 2450-MHz microwave irradiation. In : STUCHLY S. Electromagnetic fields in biological systems. Edmonton, IMPI, 1979, pp. 175-193.

140. CHERNOVETZ M., JUSTESEN D., OKE A. A teratological study of the rat : microwave and infrared radiations compared. *Radio Sci.*, 12, n° 6 (S), 1977, pp. 191-197.

141. CHERNOVETZ M., JUSTESEN D., WAGNER J., KING N. Teratology, survival, and reversal learning after fetal irradiation of mice by 2450 MHz microwave energy. *J. Microw. Power*, 10, n° 4, 1975, pp. 391-409.

142. CHIABRERA A., GRATTAROLA M., VIVIANI R. Modeling of the perturbation induced by low-frequency electromagnetic fields on membrane receptors of stimulated human lymphocytes. *Studia Biophys.*, 91, 1982, pp. 125-131.

143. CHIABERA A., GRATTAROLA M., VIVIANI R. Interaction between electromagnetic fields and cells : microelectrophoretic effect on ligands and surface receptors. *Bioelectromagn.*, 5, 1984, pp. 173-191.

144. CHOU C., GALAMBOS R. Middle-ear structures contribute little to auditory perception of microwaves. *J. Microw. Power*, 14, n° 4, 1979, pp. 321-326.

145. CHOU C., GALAMBOS R., GUY A., LOVELY R. Cochlear microphonics generated by microwave pulses. *J. Microw. Power*, 10, n° 4, 1975, pp. 361-367.

146. CHOU C., GUY A. Effect of 2450 MHz microwave fields on peripheral nerves. In : IEEE International Symposium on the Biological Effects of Microwaves, Boulder, 1973, pp. 318-320.

147. CHOU C., GUY A. Effect of microwave field on muscle contraction. In : Proceedings of the IMPI Symposium, Waterloo (Can.), 1975, pp. 79-86. (Paper 5.2.).

148. CHOU C., GUY A. Effect of microwave fields on rabbit vagus nerves and superior cervical ganglia. In : IEEE International Microwave Symposium Digest, Palo Alto, 1975, pp. 292-294.

149. CHOU C., GUY A. The effects of electromagnetic fields on the nervous system. Technical and scientific report n° 6. Seattle, Bioelectromagn. Res. Lab., Dep. Rehabil. Med., Washington Univ., Sch. Med., August 1975.

150. CHOU C., GUY A. Quantitation of microwave biological effects. In : HAZZARD D. Biological effects and measurement of RF/microwaves. Symposium proceedings, 16-18 Feb. 1977, US Dep. Health, Educ., Welfare, BRH, FDA-77-8026, July 1977, pp. 81-103.

151. CHOU C., GUY A. Effects of electromagnetic fields on isolated nerve and muscle preparations.*IEEE Trans. Microw. Theory Tech.*, MTT-26, n° 3, 1978, pp. 141-147.

152. CHOU C., GUY A. Carbon-loaded teflon electrodes for chronic EEG recordings in microwave research. *J. Microw. Power*, 14, n° 4, 1979, pp. 399-404.

153. CHOU C., GUY A., GALAMBOS R. Microwave-induced auditory response : cochlear microphonics. In : JOHNSON C., SHORE M. Biological effects of electromagnetic waves, selected papers of the USNC/URSI annual meeting, Boulder, 20-23 Oct. 1975, US Dep. Health, Educ., Welfare, FDA-77-8010. Washington, US Government Printing Office, Vol. 1, 1976, pp. 89-103

154. CHOU C., GUY A., GALAMBOS R. Microwave-induced cochlear microphonics in cats. In : Proceedings of the IMPI Symposium, Leuven, 1976. (Paper 6A.1.).

155. CHOU C., GUY A., GALAMBOS R. Microwave-induced cochlear microphonics in cats. *J. Microw. Power*, 11, 1976, pp. 171-173.

156. CHOU C., GUY A., GALAMBOS R. Characteristics of microwave-induced cochlear microphonics. *Radio Sci.*, 12, n° 6 (S), 1977, pp. 221-228.

157. CHOU C., HAN L., GUY A. Microwave radiation and heart-beat rate of rabbits. *J. Microw. Power*, 15, n° 2, 1980, pp. 87-93.

158. CIANO M., BURLIN J., PARDOE R., MILLS R., HENTZ V. High-frequency electromagnetic radiation injury to the upper extremity : local and systemic effects. *Ann. Plastic Surg.*, 7, n° 2, 1981, pp. 128-135.

159. ČIŽENKOVA R. Rôle des diverses formations du cerveau dans les réponses électroencéphalographiques du lapin à un champ magnétique constant et à des champs électromagnétiques UHF et SHF. (En russe). *Ž. Vyššej Njervnoj Djejatjel'nosti*, 17, n° 2, 1967, pp. 313-321.

160. CLAPMAN R., CAIN C. Absence of heart-rate effects in isolated frog heart irradiated with pulse modulated microwave energy. *J. Microw. Power*, 10, n° 4, 1975, pp. 411-419.

161. CLARK J. Effects of intense microwave radiation on living organisms. *Proc. IRE*, 38, n° 9, 1950, pp. 1028-1032.

162. CLARKE R. Behavioral, ethological, and teratological effects of electromagnetic radiation on an avian species..Thesis, Ph..D., Dep. Psychol., Univ. Kansas, 1977.

163. CLARKE R., JUSTESEN D. Temperature gradients in the microwave-irradiated egg : implications for avian teratogenesis. *J. Microw. Power*, 18, n° 2, 1983, pp. 169-180.

514 Part IV - Biological effects and medical applications

164. CLEARY S. Biological effects and
health implications of microwave
radiation. Symposium proceedings,
Commonwealth Univ. Richmond, 17-19
Sept. 1969, US Dep. Health, Educ.,
Welfare, BRH, DBE-70-2, PB 193898,
Library of Congress card n° 76-607
340, June 1970.

165. CLEARY S. Biological effects of
microwave and radiofrequency radiation. *CRC Crit. Rev. Environ. Control*, 1, n° 2, 1970, pp. 257-306.

166. CLEARY S. Considerations in the
evaluation of the biological effects
of exposure to microwave radiation.
Am. Ind. Hyg. Ass. J., 31, 1970, pp.
52-59.

167. CLEARY S. Uncertainties in the evaluation of the biological effects
of microwave and radiofrequency
radiation. *Health Phys.*, 25, n° 4,
1973, pp. 387-404.

168. CLEARY S. Biological effects of
microwave and radio-frequency radiation. *CRC Crit. Rev. Environ.
Control*, 2, 1977, pp. 121-166.

169. CLEARY S. Microwave cataractogenesis.
Proc. IEEE, 68, 1980, pp. 49-55.

170. CLEARY S., PASTERNAK B. Lenticular
changes in microwave workers. *Arch.
Environ. Health*, 12, 1966, pp. 23-29.

171. CLOAREC M. Sang. In : Alpha-Encyclo-
pédie. Paris, Ed. Grange Batelière,
Vol.,13, 1972, pp. 5297-5299.

172. COCOZZA G., BLASIO A., NUNZIATA B.
Rilievi sulle embriopatie da onde
corte. *Pediatr., Riv. Ig. Med. Chir.
Infanz.*, 68, n° 1, 1960, pp. 7-23.

173. COGAN D., FRICKER S., LUBIN M.,
DONALDSON D., HARDY H. Cataracts
and ultra-high frequency radiation.
Arch. Ind. Health, 18, 1958, pp.
299-302.

174. COOK H. The pain threshold for microwave and infrared radiation. *J. Physiol.*, 118, 1952, pp. 1-11.

175. COOPER T., PINAKATT T., JELLINEK M.,
RICHARDSON M. Effect of adrenalec-
tomy, vagotomy, and ganglionic bloc-
kade on the circulatory response to
microwave hyperthermia. *Aerosp. Med.*,
33, 1962, p. 794.

176. CORELLI J., GUTMANN R., KOHAZI S.,
LEVY J. Effects of 2.6-4.0 GHz microwave radiation on E. coli B. *J.
Microw. Power*, 12, n° 2, 1977, pp.
141-144.

177. CORY W., FREDERICK C. Effects of
electromagnetic energy on the environment, a summary report. *IEEE Trans.
Aerosp. Electron. Syst.*, AES-10, n°
5, 1974, pp. 738-742.

178. COURTADE. Oreille. In : GALTIER-
BOISSIERE. Larousse Médical illustré.
Paris, Larousse, 1912, pp. 817-828.

179. COURTNEY K., LIN J., GUY A., CHOU C.
Microwave effect on rabbit superior
cervical ganglion. *IEEE Trans. Microw. Theory Tech.*, MTT-23, n° 10,
1975, pp. 809-813.

180. ČUHLOVIN B. Formation d'anticorps
spécifiques dans l'organisme sous
certaines conditions d'exposition
aux émissions micro-ondes. (En russe)
In : Gigijena truda i biologičeskoje
djejstvije elektromagnitnych voln
radiočastot SV. materialov, III.
Vsesimpoz., Moskva, 1968, pp. 24-28.

181. CUNITZ R., GALLOWAY W., BERMAN C.
Behavioral suppression by 383-MHz
radiation. *IEEE Trans. Microw.
Theory Tech.*, MTT-23, n° 3, 1975,
pp. 313-316.

182. CZERSKI P. Microwave effects on the
blood-forming system with particular reference to the lymphocyte.
In : TYLER P. Biologic effects of
non-ionizing radiation, Proceedings
of a conference. *Ann. New York Acad.
Sci.*, 247, 1975, pp. 232-241.

183. CZERSKI P., BARANSKI S., SIEKIER-
ZYNSKI M. Microwave irradiation and
bone-marrow function. In : Digest
of the 3rd International conference
on medical physics, including medical engineering, Göteborg, 1972,
Abstract 39.9.

184. CZERSKI P., OSTROWSKI K., SHORE M.,
SILVERMAN C., SÜSS M., WALDESKOG B.
Biological effects and health hazards of microwave radiation. Proceedings of an international symposium, Warszawa, 15-18 Oct. 1973.
Warszawa, Polish Medical Publishers,
1974, 350 p.

185. CZERSKI P., PAPROCKA-SŁONKA E.
Microwave irradiation and the circadian rythm of bone marrow cell
mitosis. In : Proceedings of the
IMPI Symposium, Loughborough, 1973,
pp. 1-2. (Paper 2B1).

186. CZERSKI P., PAPROCKA-SŁONKA E.,
SIEKIERZYNSKI M., STOLARSKA A. In-
fluence of microwave radiation on
the hematopoietic system. In : CZER-
SKI P., OSTROWSKI K., SHORE M.,
SILVERMAN C., SÜSS M., WALDESKOG B.
Biological effects and health hazards

of microwave radiation, Proceedings of an international symposium, Warszawa, 15-18 Oct. 1973. Warszawa, Polish Medical Publishers, 1974, pp. 67-74.

187. CZERSKI P., PAPROCKA-SŁONKA E., STOLARSKA A. Microwave irradiation and the circadian rythm of bone marrow cell mitosis. *J. Microw. Power*, 9, n° 1, 1974, pp. 31-37.

188. CZERSKI P., PIENKOWSKI M. Etude sur les effets des micro-ondes sur le système hématopoíétique. (En russe). In : GORDON Z. Biologičeskoje Djejstvije Elektromagnitnych Voln Radiočastot, Gigijena Truda, Medicina, Moskva, 56, 1972.

189. CZERSKI P., WIKTOR-JEDRZEJCZAK W. Effet de l'exposition aux micro-ondes d'un système cellulaire immunocompétent. In : Effets biologiques des rayonnements non ionisants ; utilisation et risques associés. 9ème congrès international de la Société Française de Radioprotection, Nainville-les-Roches, 22-26 mai 1978, p. 255.

190. DAELS J. Microwave heating of the uterine wall during parturition. In : Proceedings of the IMPI Symposium, Leuven, 1976. (Paper 5A.4.).

191. DAILY L. A clinical study of the results of exposure of laboratory personnel to radar and high-frequency radio. *US Nav. M. Bull.*, 41, 1943, pp. 1052-1056.

192. DAILY L., WAKIM K., HERRICK J., PARKHILL E. Effects of microwave diathermy on the eye. *Am. J. Physiol.*, n° 155, 1948, p. 482.

193. DAILY L., WAKIM K., HERRICK J., PARKHILL E., BENEDICT W. The effects of microwave diathermy on the eye. *Am. J. Ophtalmol.*, 33, 1950, pp. 1241-1254.

194. DAILY L., WAKIM K., HERRICK J., PARKHILL E., BENEDICT W. The effects of microwave diathermy on the eye of the rabbit. *Am. J. Ophtalmol.*, 35, 1952, pp. 1001-1017.

195. D'AMBROSIO G., DI MEGLIO F., FERRARA G., TRANFAGLIA A. Entomological experiments on the teratogenic effects of electromagnetic fields. *Alta Frequenza*, 49, n° 2, 1980, pp. 115-119.

196. D'ANDREA J., BAILEY C., THIMAKIS T., DEWITT J., GANDHI O. Behavioral and physiological effects of prolonged exposure to 500-μW/cm^2 2450-MHz microwaves. In : 3rd annual Bioelectromagnetics Society meeting, Washington, 9-12 August 1981.

197. D'ANDREA J., GANDHI O., LORDS J. Behavioral and thermal effects of microwave radiation at resonant and non-resonant wavelengths. *Radio Sci.*, 12, n° 6 (S), 1977, pp. 251-256.

198. D'ANDREA J., GANDHI O., LORDS J., DURNEY C., ASTLE L., STENSAAS L., SCHOENBERG A. Physiological and behavioral effects of prolonged exposure to 915 MHz microwaves. *J. Microw. Power*, 15, n° 2, 1980, pp. 123-135.

199. D'ANDREA J., GANDHI O., LORDS J., DURNEY C., JOHNSON C., ASTLE L. Physiological and behavioral effects of chronic exposure to 2450 MHz microwaves. *J. Microw. Power*, 14, n° 4, 1979, pp. 351-362.

200. DARDALHON M. Les effets biologiques des micro-ondes. *Radioprot.*, n° 1, suppl., 1984, pp. 15-30.

201. DARDALHON M., AVERBECK D. Quelques actions des micro-ondes chez les bactéries, les levures et les drosophiles. In : Effets biologiques des rayonnements non ionisants ; utilisation et risques associés, 9ème congrès international de la Société Française de Radioprotection, Nainville-les-Roches, 22-26 mai 1978, pp. 279-299.

202. DARDALHON M., AVERBECK D., BERTEAUD A. Determination of a thermal equivalent of millimeter microwaves in living cells. *J. Microw. Power*, 14, n° 4, 1979, pp. 307-312.

203. DARDALHON M., AVERBECK D., BERTEAUD A. Action des ondes centimétriques seules ou combinées avec les rayons ultraviolets sur les cellules eucaryotiques. In : BERTEAUD A., SERVANTIE B. Ondes électromagnétiques et biologie, Symposium international URSI, Jouy-en-Josas, juil. 1980, pp. 17-24.

204. DARDALHON M., AVERBECK D., BERTEAUD A. Studies on possible genetic effects of microwaves on procaryotic and eucaryotic cells. *Radiat. Environ. Biophys.*, 20, 1981, pp. 37-51.

205. DARDALHON M., BERTEAUD A., AVERBECK D. Microwave effects in Drosophila melanogaster. *Radioprot.*, 14, n° 3, 1979, pp. 145-159.

206. DARDANONI L., TORREGROSSA M., TAMBURELLO Z., ZANFORLIN L. Sensitivity of C. albicans cells to frequency of modulation in the 72-74 GHz band. In : Bioelectromagnetics symposium, USNC/URSI, Seattle, 18-22 June 1979.

207. DAVIDSON J., KONDRA P., HAMID M. Effects of microwave radiation on eggs, embryos and chickens. *Can. J. Anim. Sci.*, 56, n° 4, 1976, pp. 709-713.

208. DEB A. Biological effects of microwave irradiation : a reexamination. *Electro-Technol.*, 19, n° 1, 1975, pp. 1-11.

209. DEFICIS A., DUMAS J., LAURENS S. Biological alterations observed under microwave irradiation. In : Proceedings of the IMPI Symposium, Waterloo (Can.), 1975, pp. 48-58. (Paper 3.4.).

210. DEFICIS A., DUMAS J., LAURENS S., PLURIEN G. Variation of serum triglycerides rate under the action of electromagnetic waves : power level influence. In : Proceedings of the IMPI Symposium, Leuven, 1976. (Paper 1A. 3a.).

211. DEFICIS A., DUMAS J., LAURENS S., PLURIEN G. Effect of electromagnetic energy on the formation of triglycerides : frequency influence. In : Proceedings of the IMPI Symposium, Leuven, 1976. (Paper 1A.3b.).

212. DEICHMANN W., BERNAL E., KEPLINGER M. Effects of environmental temperature and air volume change on survival of rats exposed to microwave radiation of 24000 Mc/s. *Ind. Med. Surg.*, 28, 1959, p. 535.

213. DEICHMANN N., MIALE J., LANDEEN K. Effect of microwave radiation on the hemopoietic system of the rat. *Toxicol. Appl. Pharmacol.*, 6, 1964, pp. 71-77.

214. DEICHMANN W., STEPHENS F., KEPLINGER M., LAMPE K. Acute effects of microwave radiation on experimental animals (24,000 megacycles). *J. Occup. Med.*, 1, n° 7, 1959, pp. 369-381.

215. DE LORGE J. The effects of microwave radiation on behavior and temperature in Rhesus monkeys. In : JOHNSON C., SHORE M. Biological effects of electromagnetic waves. Selected papers of the USNC/URSI annual meeting, Boulder, 20-23 Oct. 1975, US Dep. Health, Educ., Welfare, FDA-77-8010. Washington, US Government Printing Office, Vol. 1, 1976, pp. 158-174.

216. DE LORGE J. Disruption of behavior in mammals of three different sizes exposed to microwave : extrapolation to larger mammals. In : STUCHLY S. Electromagnetics fields in biological systems. Proceedings of a symposium, Ottawa, 27-30 June 1978, Abstracts of scientific papers, p. 37.

217. DE LORGE J., EZELL C. Observing responses of rats exposed to 1.28 and 5.62 GHz microwaves. *Bioelectromagn.*, 1, 1980, pp. 183-198.

218. DEROCHE. Etude des perturbations biologiques chez les techniciens ORTF dans certains champs électromagnétiques de hautes fréquences. *Arch. Mal. Prof.*, 32, n° 10-11, pp. 679-683.

219. DESCHAUX P., DOUSS T., SANTINI R., BINDER P., FONTANGES R. Effect of microwave irradiation (2450 MHz) on murine cytotoxic lymphocyte and natural killer (NK) cells. *J. Microw. Power*, 19, n° 2, 1984, pp. 107-110.

220. DESCHAUX P., DUMONT J., PELLISSIER J., FONTANGES R. Etude de la réponse immunitaire de la souris irradiée à 2450 MHz en chambre anéchoïque : influence de la puissance appliquée. In : BERTEAUD A., SERVANTIE B. Ondes électromagnétiques et biologie. Symposium international URSI, Jouy-en-Josas, Juil. 1980, pp. 139-143.

221. DESCHAUX P., JIMENEZ C., SANTINI R., PELLISSIER J. Effet d'un rayonnement micro-onde (2450 MHz) sur la reproduction de la souris mâle. *Econ. Progrès Electr.*, n° 8-9, mars-juin 1983, pp. 15-17.

222. DESCHAUX P., JOUVE de GUIBERT C., SANTINI R., PELLISSIER J. Effet d'un rayonnement microonde (2450 MHz) sur certaines fonctions endocrines de la souris. *Rev. Roum. Biol. Anim.*, 26, 1981, pp. 159-167.

223. DESCHAUX P., PELLISSIER J., SANTINI R., FONTANGES R. Effets d'un rayonnement dû aux micro-ondes sur l'axe hypophysotesticulaire chez le rat blanc. In : Effets biologiques des rayonnements non ionisants ; utilisation et risques associés. 9ème congrès international de la Société Française de Radioprotection, Nainville-les-Roches, 22-26 mai 1978, pp. 265-270.

224. DEVANEY J., BEERWINKLE K. Effects of microwave and various combinations of ambient temperature and humidity exposures on off-host survival of northern fowl mites. *Poult. Sci.*, 59, n° 10, 1980, pp. 2198-2201.

225. DEVJATKOV N. Effet des rayonnements
électromagnétiques de longueur
d'onde millimétrique sur les objets
biologiques. (En russe). *Usp. Fiz.
Nauk*, 110, n° 3, 1973, pp. 453-454.

226. DEVJATKOV N. Influence of millimeter-
band electromagnetic radiation on
biological objects. *Soviet Phys.
Uspekhi*, 16, n° 4, 1974, pp. 568-
569.

227. DHAHI S., HABASH R., AL-HAFID H.
Lack of mutagenic effects on conidia
of Aspergillus amstelodami irradiated
by 8.7175-GHz CW microwaves. *J. Microw. Power*, 17, n° 4, 1982, pp.
345-351.

228. DJORDJEVIĆ Z., KOLAK A. Changes in
the peripheral blood of the rat
exposed to microwave radiation (2400
MHz) in conditions of chronic exposure. *Aerosp. med.*, 44, 1973, pp.
1051-1054.

229. DJORDJEVIĆ Z., LAZAREVIĆ N., DJOKVIĆ
V. Studies on the hematologic effects
of long-term, low-dose microwave
exposure. *Aviat., Space Environ.
Med.*, 48, 1977, pp. 516-518.

230. DODGE C. Clinical and hygienic as-
pects of exposure to electromagnetic
fields : a review of Soviet and East-
European literature. In : CLEARY S.
Biological effects and health im-
plications of microwave radiation.
Symposium proceedings, Commonwealth
Univ., Richmond, 17-19 Sept. 1969,
US Dep. Health, Educ., Welfare, BRH,
DBE 70-2, PB 193 898, Library of
Congress card n° 76-607340, June
1970, pp. 140-149.

231. DODGE C., GLASER Z. Trends in non-
ionizing electromagnetic radiation
bio-effects research and related
occupational health aspects. *J. Microw. Power*, 12, n° 4, 1977, pp.
319-334.

232. DOMART A., BOURNEUF J. Petit Larousse
de la médecine. Paris, Larousse,
1976, 842 p.

233. DONALDSON D. Cataracts following
use of microwave oven. Comments.
New York State J. Med., 74, n° 11,
1974, p. 2037.

234. DRAUSSIN M., CHEVALIER J., LAGOUTTE
J., SERRE A., GOUJET P., SERVANTIE
B., MIRO L., BUREAU J. Effets des
ondes centimétriques à 3 GHz pulsés
sur la réponse immunitaire in vivo
chez la souris. In : BERTEAUD A.,
SERVANTIE B. Ondes électromagnétiques
et biologie. Symposium international
URSI, Jouy-en-Josas, juil. 1980, pp.
133-137.

235. DROGIČINA E., KONČALOVSKAJA N.,
GLOTOVA K., SADČIKOVA M., SNEGOVA G.
Autonomic and cardiovascular disorders
during chronic exposure to super-
high-frequency electromagnetic
fields. (En russe). *Gig. Tr. Prof.
Zabol.*, n° 10, 1966, p. 13.
Traduct. angl. LC-ATD-66-124, Li-
brary of Congress, Aerosp. Technol.
Div., Washington, 1966.

236. DROGIČINA E., SADČIKOVA M. Clinical
syndromes arising under the effect
of various radio-frequency bands.
(En russe). *Gig. Tr. Prof. Zabol.*,
9, 1965, p. 17. (JPRS 29694,
Joint publications Res. Serv., Washington).

237. DRONOV I., KIRICEVA A. Changements
immunologiques chez les animaux
immunisés après exposition de longue durée aux SHF. (En russe). *Gig.
Sanit.*, 36, 1971, pp. 63-65.

238. DRONOV I., KIRICEVA A. Réactivité
immunologique des animaux soumis
à une irradiation prolongée par UHF.
(En russe). *Gig. Tr. Prof. Zabol.*,
9, 1972, pp. 15-18.

239. DRONOV I., KIRICEVA A., SERADSKAJA
L. Effets des micro-ondes sur la
formation d'anticorps chez le lapin.
(En russe). In : Gigiena truda i
biologičeskoje djejstvije elektromagnitnyh voln radiočastot SV. materialov. III. Vsesimpoz, Moskva,
1968, pp. 46-48.

240. DUMANSKIJ J., RUDIŠENKO V. Dependence of functional activity of
liver mitochondria on microwave
radiation. (Trad. du russe). US
joint Pub. Res. Serv. Rep., JPRS
72606, 12 Jan. 1979, pp. 27-32.

241. DUMANSKIJ J., ŠANDALA M. The biological action and hygienic significance of electromagnetic fields of
super-high and ultra-high frequencies in densely populated areas.
In : CZERSKI P., OSTROWSKI K., SHORE
M., SILVERMAN C., SÜSS M., WALDESKOG
B. Biological effects and health
hazards of microwave radiation.
Proceedings of an international
symposium, Warszawa, 15-18 Oct. 1973.
Warszawa, Polish Medical Publishers,
1974, pp. 289-293.

242. DUMANSKIJ J., TOMAŠEVSKAJA. Investigation of the activity of some enzymatic systems in response to a
superhighfrequency electromagnetic
field. (Trad. du russe). US joint
Pub. Res. Serv. Rep. JPRS 72606,
12 jan. 1979, pp. 1-7.

243. DUMAS J., LAURENS S., PLURIEN G. Comparaison de l'effet des micro-ondes sur le métabolisme lipidique de plusieurs souches de souris. In Effets biologiques des rayonnements non ionisants ; utilisation et risques associés. 9ème congrès international de la Société Française de Radioprotection, Nainville-les-Roches, 22-26 mai 1978, pp. 271-276.

244. DUMAS J., NOUGAROLIS S., STOLL M., PLURIEN G. Influence de la durée d'exposition aux microondes sur la triglycéridémie de la souris. In : BERTEAUD A., SERVANTIE B. Ondes électromagnétiques et biologie. Symposium international URSI, Jouy-en-Josas, juil. 1980, pp. 75-77.

245. DUTTA S., NELSON W., BLACKMAN C., BRUSICK D. Lack of microbial genetic response to 2.45 GHz CW and 8.5-to 9.6-GHz pulsed microwaves. J. Microw. Power, 14, n° 3, 1979, pp. 275-280.

246. DWIVEDI R., OGUNWUYI S., Mac LEOD W. Some biological and cytological observations on effect of non-ionizing (microwave) radiation. J. Cell Biol., 63, n° 2, 1974, p. 90.

247. EDWARDS M. Congenital malformation in the rat following induced hyperthermia during gestation. Teratol., 1, 1968, pp. 173-178.

248. EDWARDS M. Congenital defects in guinea pigs : fetal resorptions, abortions, and malformations following induced hyperthermia during early gestation. Teratol., 2, 1969, pp. 313-328.

249. EGAN W. Eye protection in radar fields. Electr. Eng., 76, n° 2, 1957, pp. 126-127.

250. EGYPTIEN H. Effects of electrical and electromagnetic fields on organisms. Electrotech. Z., Ausgabe B, 27, n° 26, 1975, pp. 729-730.

251. EICHERT E., FREY A. Human auditory system response to lower power density, pulse modulated, electromagnetic energy : a search for mechanisms. In : Proceedings of the IMPI Symposium, Leuven, 1976. (Paper 2A.4).

252. ELDER J. Biological effects of non-ionizing radiation. In : 2nd joint US/USSR symposium on the comprehensive analysis of the environment. Honolulu, 22-25 Oct. 1975

253. ELDER J., ALI J. The effect of microwaves (2450 MHz) on isolated rat liver mitochondria. In : TYLER P.

Biologic effects on non-ionizing radiation. Proceedings of a conference. Ann. New York Acad. Sci., 247, 1975, pp. 251-262.

254. ELDER J., CAHILL D. Biological effects of radiofrequency radiation. US Environ. Prot. Agency, Rapport EPA-600/8-83-026F, Cincinnati, 1984.

255. ELY T., GOLDMAN D., HEARON J., WILLIAMS R., CARPENTER H. Heating characteristics of laboratory animals exposed to ten-centimeter microwaves. IEEE Trans. Biomed. Eng., 11, 1964, pp. 123-137.

256. EMERY A., KRAMAR P., GUY A., LIN J. Microwave-induced temperature rises in rabbit eyes in cataract research. J. Heat Transf., 97, 1975, pp. 123-128.

257. EVERTS J., HERMAN W., COLVIN M., PORTER C., PHILLIPS C. Cytogenetic effect of microwave radiation on Chinese hamsters. In : Proceedings of the IMPI Symposium, Ottawa, 1972. pp. 139-142. (Paper 8.5.).

258. FABIANI G. Les défenses de l'organisme. Paris, P.U.F., n° 5, 1984, 128 p. (Coll. Que sais-je ?).

259. FAJTEL'BERG-BLANK V., PEREVALOV G. Selective action of decimeter waves on central structures of the brain. (Trad. du russe). US joint Pub. Res. Serv. Rep., JPRS L/7467, 15 Nov. 1977, pp. 16-23.

260. FAJTEL'BERG-BLANK V., PEREVALOV G. Effects of superhigh-frequency electromagnetic fields on central structures of the brain. (Trad. du russe). US joint Pub. Res. Serv. Rep., JPRS L/7899, 19 July 1978, pp. 5-7.

261. FELDMANN M. Les cellules qui suppriment l'immunité. La Recherche, n° 177, mai 1986, pp. 692-699.

262. FERRI E., HAGAN G. Chronic low-level exposure of rabbits to microwaves. In : JOHNSON C., SHORE M. Biological effects of electromagnetic waves, Selected papers of the USNC/URSI annual meeting, Boulder, 20-23 Oct. 1975, US Dep. Health, Educ., Welfare, FDA-77-8010. Washington, US Government Printing Office, Vol. 1, 1976, pp. 129-142.

263. FISHER P., LAUBER J., VOSS W. Effects of chronic exposure to microwaves at low power densities on the mass and cranial size of developing chicken embryos. In : Biological effects of electromagnetic waves, international symposium, Airlie, 30 Oct.-4 Nov. 1977, Abstracts of sci. papers.

264. FISHER P., LAUBER J., VOSS W. The effect of low-level 2450-MHz CW microwave radiation and body temperature on early embryonic development in chickens. *Radio Sci.*, 14, n° 6 (S), 1979, pp. 159-163.

265. FLANIGAN W., BOWMAN R., LOWELL W. Non-metallic electrode system for recording EEG and ECG in electromagnetic fields. *Physiol. Behav.*, 18, 1977, pp. 531-533.

266. FLECK M. Microwave oven burn. *Bull. New York Acad. Med.*, 59, n° 3, 1983, pp. 313-317.

267. FLEURAT-LESSARD F., LESBATS M., LAVENSEAU L., CANGARDEL H., MOREAU R., LAMY M., ANGLADE P. Effets biologiques des micro-ondes sur deux insectes : Tenebrio molitor L.(Col.: Tenebrionidae) et Pieris brassicae L. (Lep. : Pieridae). *Ann. Zool. Ecol. Anim.*, 11, n° 3, 1979, pp. 457-478.

268. FOLLIS R. Studies on the biological effect of high-frequency radio waves (radar). *Am. J. Physiol.*, n° 147, 1946, pp. 281-283.

269. FORMAN S., HOLMES C., Mac MANANON T., WEDDING W. Psychological symptoms and intermittent hypertension following acute microwave exposure. *J. Occupat. Med.*, 24, 1982, pp. 932-934.

270. FOSTER M. A model for thermal cataractogenesis. In : Proceedings of the IMPI Symposium, Waterloo (Can.), 1975, pp. 115-116. (Paper 7.4.).

271. FOSTER K., FINCH E. Microwave hearing : evidence for thermoacoustical auditory stimulation by pulsed microwaves. *Science*, 185, 1974, pp. 256-258.

272. FOUGEREAU M. L'immunologie. Paris, P.U.F., n° 1358, 2ème ed., 1985, 128 p. (Coll. Que sais-je ?).

273. FRADELEZI D. Les messagers de l'immunité. *La Recherche*, n° 177, mai 1986, pp. 668-678.

274. FRAZER J., MITCHELL J., GRASS A., HURT W. Exposure of biological specimens to high power HF band fields. In : Proceedings of the IMPI Symposium, Ottawa, 1972, pp. 167-168. (Paper 10.1.).

275. FREY A. Auditory system response to radio-frequency energy. *Aerosp. Med.*, 32, 1961, pp. 1140-1142.

276. FREY A. Human auditory system response to modulated electromagnetic energy. *J. Appl. Physiol.*, 17, 1962, pp. 689-692.

277. FREY A. Some effects on human subjects of ultra-high-frequency radiation. *Am. J. Med. Electron.*, 2, 1963, pp. 28-31.

278. FREY A. Behavioral biophysics. *Psychol. Bull.*, 63, 1965, pp. 322-337.

279. FREY A. Brain stem evoked responses associated with low-intensity pulsed UHF energy. *J. Appl. Physiol.*, 23, n° 6, Dec. 1967, pp. 984-988.

280. FREY A. Biological function as influenced by low power modulated RF energy. *IEEE Trans. Microw. Theory Tech.*, MTT-19, n° 2, 1971, pp. 153-164.

281. FREY A. Recent findings on the biological effects of low power microwaves. In : Proceedings of the IMPI Symposium, Monterey, 1971. (Paper 7.1.).

282. FREY A. Human perception of illumination with pulsed UHF electromagnetic energy. *Science*, 181, 1973, pp. 356-358.

283. FREY A. Differential biological effects of pulsed and continuous electromagnetic fields and mechanisms of effect. *Ann. New York Acad. Sci.*, 238, 1974, pp. 273-279.

284. FREY A. Studies of the blood-brain barrier : preliminary findings and discussion. *Radio Sci.*, 14, n° 6 (S), 1979, pp. 349-350.

285. FREY A. Laboratory notes : on microwave effects at the blood-brain barrier. *Bioelectromagn.*, 1, 1980, p. 4.

286. FREY A. Comments on "Microwaves and the blood-brain barrier". *J. Bioelectr.*, 2, n° 1, 1983, pp. 83-88.

287. FREY A. Science and standards. Data analysis reveals significant microwave-induced eye damage in humans. *J. Microw. Power*, 20, n° 1, 1985, pp. 53-56.

288. FREY A., COREN E. Holographic assessment of a hypothesized microwave hearing mechanics. *Science*, 206, 12 Oct. 1979, p. 232.

289. FREY A., FELD S. Perception and avoidance of illumination with low-power pulsed UHF electromagnetic energy. In : Proceedings of the IMPI Symposium, Ottawa, 1972, pp. 130-138.

290. FREY A., FELD S. Avoidance by rats of illumination with low power non-ionizing electromagnetic energy. *J. Comp. Physiol. Psychol.*, **89**, 1975, pp. 183-188.

291. FREY A., FELD S., FREY B. Neural function and behavior : defining the relationship. In : TYLER P. Biologic effects of non-ionizing radiation. Proceedings of a conference. *Ann. New York Acad. Sci.*, **247**, 1975, pp. 433-439.

292. FREY A., MESSENGER R. Psychophysical data on the RF hearing effect. In : Proceedings of the IMPI symposium, Ottawa, 1972, pp. 169-177. (Paper 10.2.).

293. FREY A., MESSENGER R. Human perception of illumination with pulsed ultra-high-frequency electromagnetic energy. *Science*, **181**, 1973, pp. 356-358.

294. FREY A., SEIFERT E. Pulse modulated UHF energy illumination of the heart associated with change in heart rate. *Life Sci.*, **7**, 1968, part II, pp. 505-512.

295. FUREDI A., OHAD I. Effects of high frequency electric fields on the living cell. I. Behaviour of human erythrocytes in high frequency electric fields and its relation to their age. *Biochim. Biophys. Acta*, **79**, n° 1, 1964, pp. 1-8.

296. FURMANIAK A. Quantitative changes in potassium, sodium and calcium in the submaxillary salivary gland and blood serum of rats exposed to 2880 MHz microwave radiation. *Bioelectromagn.*, **4**, 1983, pp. 55-62.

297. GABOVIČ R. Effects of super high frequency fields of different intensity on the balance and metabolism of copper, manganese, molybdenum and nickel in the organs of experimental animals. (Trad. du russe). US joint Pub Res. Serv. Rep., JPRS 66512, 7 Jan. 1976, pp. 33-42.

298. GAGE M. Microwave irradiation and ambient temperature interact to alter rat behavior following over-night exposure. *J. Microw. Power*, **14**, n° 4, 1979, pp. 389-398.

299. GALLOWAY W. Microwave dose-response relationships on two behavioral tasks. In : TYLER P. Biologic effects of non-ionizing radiation. Proceedings of a conference. *Ann. N.Y. Acad. Sci.*, **247**, 1975, pp. 410-416.

300. GALVIN M., DUTTON M., Mac REE D. Influence of 2.45 GHz CW microwave radiation on spontaneously beating rat atria. *Bioelectromagn.*, **3**, 1982, pp. 219-226.

301. GALVIN M., Mac REE D., HALL C., THAXTON J., PARKHURST C. Humoral and cell-mediated immune function in adult Japanese quail following exposure to 2.45-GHz microwave irradiation during embryogeny. *Bioelectromagn.*, **2**, 1981, pp. 269-278.

302. GALVIN M. Mac REE D., LIEBERMAN M. Effects of 2.45 GHz microwave radiation on embryonic quail hearts. *Bioelectromagn.*, **1**, 1980, pp. 389-396.

303. GALVIN M., ORTNER M. Effect of 2450 MHz microwave radiation on concanavalin A or ionophore-induced histamine release from rat peritoneal mast cells. *Int. J. Radiat. Biol.*, **39**, n° 6, 1981, pp. 671-675.

304. GALVIN M., ORTNER M., Mac REE D. Studies on acute in-vivo exposure of rats to 2450-MHz microwave radiation : biochemical and hematologic effects. *Radiat. Res.*, **90**, 1982, pp. 558-563.

305. GANDHI O. Biological effects and medical applications of RF electromagnetic fields. *IEEE Trans. Microw. Theory tech.*, MTT-**30**, n° 11, 1982, pp. 1831-1847.

306. GANDHI O. Some basic properties of biological tissues for potential biomedical applications of millimeter waves, *J. Microw. Power*, **18**, n° 3, 1983, pp. 295-304.

307. GANDHI O., RIAZI A. Absorption of millimeter waves by human beings and its biological implications. *IEEE Trans. Microw. Theory Tech.*, MTT-**34**, n° 2, 1986, pp. 228-235.

308. GAVALAS R., WALTERS D., HAMER J., ADEY W. Effect of low-level, low-frequency electric fields on EEG and behavior in Macaca nemestrina. *Brain Res.*, n° 18, 1970, pp. 491-501.

309. GIAROLA A., KRUEGER W. Continuous exposure of chicks and rats to electromagnetic fields. *IEEE Trans. Microw. Theory Tech.*, MTT-**22**, n° 4, 1974, pp. 432-437.

310. GIAROLA A., KRUEGER W., FANGUY R., SCHREKENHAMER A. Exposure of Gallus domesticus to various electromatic fields. In : Proceedings of the IMPI Symposium, Milwaukee, 1974, pp. 1-7. (Paper B 2.3.).

311. GILLOIS M., AUGE C., CHEVALET C.
Effet des ondes électromagnétiques
non ionisantes sur la viabilité et
l'hérédité des cellules de mammi-
fères en culture établie. In : BER-
TEAUD A., SERVANTIE B. Ondes électromagnétiques et biologie. Symposium international URSI, Jouy-en-
Josas, juil. 1980, pp. 9-15.

312. GINNS E., RUGH R., HO H., LEACH W.,
GILLESPIE L., BUDD R., HAZZARD D.
Microwave biological effects under
reproducible dosimetric and environmental conditions. In : Proceedings of the IMPI Symposium, Loughborough, 1973, pp. 1-2. (Paper 2.B 2).

313. GLASER Z. Studies on the biomedical
effects of microwave radiation :
past, present, and future. In :
Proceedings of the IMPI Symposium,
Milwaukee, 1974, abstract in *IMPI
Newsl.*, 2, n° 3, 1974, pp. 14-15.

314. GLASER Z., BROWN P., BROWN M. Bibliography of reported biological phenomena ("effects") and clinical manifestations attributed to microwave
and radio-frequency radiation : com-
pilation and integration of report
and seven supplements. Naval Med.
Res. Inst. Report, A.D. n° A029-430,
Sept. 1976, 180 p. and eighth Suppl.,
A.D. n° A029-430, Aug. 1976, 19 p.

315. GLASER Z., DODGE C. Biomedical
aspects of radio-frequency and microwave radiation : a review of selected
Soviet, East European, and Western
references. In : JOHNSON C., SHORE
M. Biological effects of electromagnetic waves, Selected papers of the
USNC/URSI annual meeting, Boulder,
20-23 Oct. 1975, US Dep. Health,
Educ., Welfare, FDA-77-8010. Washington, US Government Printing Office,
1976, Vol. 1, pp. 2-34.

316. GOKHALE A.,.BRUNKARD.K., PICKARD W.
The absence of significant short-
term electromagnetic bioeffects in
giant algal cells exposed to CW and
pulse-modulated X-band bursts. *IEEE
Trans. Microw. Theory Tech.*, MTT-32,
n° 8, 1984, pp. 795-797.

317. GOLDBLITH S., WANG D. Effect of
microwaves on Escherichia coli and
Bacillus subtilis. *Appl. Microbiol.*,
15, n° 6, 1967, pp. 1371-1375.
(C.D.I.U.P.A., n° 21961)

318. GOLDSTEIN G., BETZ L. La barrière
qui protège le cerveau. *Pour la
Science*, n° 109, Nov. 1986, pp.
84-94.

319. GOLDSTEIN L., SISKO Z. A quantitative electroencephalographic study
of the acute effects of X-band microwaves in rabbits. In : CZERSKI P.,
OSTROWSKI K., SHORE M., SILVERMAN C.,
SÜSS M., WALDESKOG B. Biological
effects and health hazards of microwave radiation. Proceedings of an
international symposium, Warszawa,
15-18 Oct. 1973. Warszawa, Polish
Medical Publishers, 1974, pp. 128-
133.

320. GONČAR N. Distinction referable to
the effects of electromagnetic energy in the superhigh-frequency range
on cytochemical blood indices. (Trad.
du russe), US joint Pub. Res. Serv.
Rep., JPRS, L/7957, 15 July 1978,
pp. 12-16.

321. GOODMAN R., BASSET A., HENDERSON A.
Pulsing electromagnetic fields induce cellular transcription. *Science*,
220, 1983, pp. 1283-1285.

322. GORDON Z. Voprosy gigijeny truda pri
rabote s generatorami santimetrovyh
voln. *Ž. Gig. Epidemiol. immunol.*,
1, 1957, pp. 399-404.

323. GORDON Z. The biological effects of
microwaves and protective measures
against them. Medical Electronics
Conference 3, 1960, pp. 410-411.
(abstracts 1126 of 1962).

324. GORDON Z. Voprosy gigijeny truda i
biologičeskogo djejstvija elektro-
magnitnyh poljej sverhvysokih častot. *Medicina*, 1966, pp. 25-125.

325. GORDON Z. Pathological effects of
radio waves. New York, Plenum Press,
1973.

326. GORDON Z., SCHWAN H. Statement on
microwave intensities. In : CZERSKI
P., OSTROWSKI K., SHORE M., SILVER-
MAN C., SÜSS M., WALDESKOG B. Biological effects and health hazards
of microwave radiation. Proceedings
of an international symposium, Warszawa, 15-18 Oct. 1973. Warszawa,
Polish Medical Publishers, 1974, pp.
318-319.

327. GORODECKAJA S. K harakteristike
biologičeskogo djejstvija trëhsan-
timetrovyh radiovoln na životnyj
organizn. In : Voprosy Biofiziki i
Mehanizma Djejstivija Ionizirujuščej
Radiacii, Kiev, 1964, pp. 70-74.

328. GOURNAY L. Conversion of electromagnetic to acoustic energy by surface heating. *J. Acoust. Soc. Am.*,
40, 1966, pp. 1322-1330.

329. GRATTAROLA M., VIVIANI R.,
CHIABRERA A. Modelling of the perturbation induced by low-frequency
electromagnetic fields on the membrane receptors of stimulated human
lymphocytes. 1. Influence of the
fields on the system's free energy.
Studia Biophys., 91, 1982, p. 117.

330. GREEN D., ROSENBAUM F., PICKARD W.
Intensity of microwave irradiation
and the teratogenic response of
Tenebrio molitor. *Radio Sci.*, 14,
n° 6 (S), 1979, pp. 165-171.

331. GRIN' A. Effects of microwaves on
catecholamine metabolism in brain.
(Trad. du russe). US joint Pub. Res.
Serv. Rep., JPRS, 72606, 12 Jan.
1976, pp. 14-16.

332. GRUENAU S., OSCAR K., FOLKER M.,
RAPOPORT S. Absence of microwave
effect of blood-brain barrier permeability to ^{14}C sucrose in the con-
scious rat. *Exp. Neurol.*, 75, 1982,
pp. 299-307.

333. GRUNDLER W. Recent results of experiments on nonthermal effects of
millimeter microwaves on yeast
growth. *Coll. Phenom.*, 3, 1981, pp.
181-186.

334. GRUNDLER W., KEILMANN F. Non thermal effects of millimeter microwaves
on yeast growth. *Ztg. Nat.forsch.*.
33C, 1978, pp. 15-22.

335. GRUNDLER W., KEILMANN F., FRÖHLICH H.
Resonant growth rate response of
yeast cells irradiated by weak microwaves. *Phys. Lett.*, 62 A, 1977,
pp. 463-466.

336. GRUNDLER W., KEILMANN F., PUTTERLIK V.,
SANTO L., STRUBE D., ZIMMERMANN I.
Non thermal resonant effects of 42
GHz microwaves on the growth of
yeast cultures. In : FROHLICH H.,
KREMER F. Coherent excitations in
biological systems. Heidelberg,
Springer-Verlag, 1983, pp. 21-37.

337. GRZESIK J., KUMASZA F. Wlyw pole
elektromagnetycznego średniej często-
totliwosci na organy miesowe i
białka krwi bieżych myszy. *Med.
Pracy*, 11, n° 5, 1960, pp. 323-330.

338. GUILLET R., LOTZ W., MICHAELSON S.
Time course of adrenal response in
microwave-exposed rats. In : JOHNSON
C., SHORE M. Biological effects of
electromagnetic waves, Selected
papers in the USNC/URSI annual meeting, Boulder, 20-23 Oct. 1975, US
Dep. Health, Educ., Welfare, FDA
77-8011. Washington, US Government
Printing Office, 1976, p. 316.

339. GUNN S., GOULD T., ANDERSON W. The
effects of microwave radiation on
morphology and function of the rat
testes. *Lab. Invest.*, 10, 1961, pp.
301-314.

340. GUO T., GUO W., LARSEN L. Microwave-
induced thermoacoustic effect in
dielectrics and its coupling to external medium. A thermodynamical
approach. *IEEE Trans. Microw. Theory
Tech.*, MTT-32, n° 8, 1984, pp. 835-
843.

341. GUY A. A method for exposing cell
cultures to electromagnetic fields
under controlled conditions of temperature and field strength. *Radio
Sci.*, 12, n° 6 (S), 1977, pp. 87-96.

342. GUY A. History of biological effects
and medical applications of microwave energy. *IEEE Trans. Microw.
Theory Tech.*, MTT-32, n° 9, 1984,
pp. 1182-1200.

343. GUY A., CHOU C., JOHNSON R., KUNZ L.
Study of effects of long-term low-
level RF exposure on rats : a plan.
Proc. IEEE, 68, 1980, pp. 92-97.

344. GUY A., CHOU C., JOHNSON R., KUNZ L.,
CROWLEY J., RILEY V., SPACKMAN D.,
HELLSTROM K. Effects of long-term
RFR exposure on rats. Washington
Univ., Bioelectromagn. Res. Lab.,
Technical Report, Contract F33 615-
80-C-0612, for the USAF Sch. Aerosp.
Med., Aerosp. Med. Div. (AFSC),
Brooks Air Force Base, March 1982.

345. GUY A., CHOU C., LIN J., CHRISTEN-
SEN D. Microwave-induced acoustic
effects in mammalian auditory systems and physical materials. In :
TYLER P. Biologic effects of non-
ionizing radiation. Proceedings of
a conference. *Ann. New York Acad.
Sci.*, 247, 1975, pp. 194-218.

346. GUY A., HARRIS F., HO H. Quantitation of the effects of microwave
radiation on central nervous system
function. In : Proceedings of the
IMPI Symposium, Monterey, 1971.
(Paper 7.6.).

347. GUY A., HARRIS C., KRAMAR P.,
EMERY A. Study of the effects of
chronic low level microwave radiation on rabbits. In : Proceedings
of the IMPI Symposium, Leuven, 1976.
(Paper 1A.1.).

348. GUY A., KRAMAR P., EMERY A., LIN J.,
HARRIS C. Quantitation of microwave
radiation effects on the eyes of
rabbits and primates at 2450 MHz and
918 MHz. Sci. report n° 8, Contract
N 00014-75-C-0414, Office Naval Res.,
March 1976.

349. GUY A., KRAMAR P., HARRIS C., CHOU C. Long-term 2450-MHz CW microwave irradiation of rabbits : methodology and evaluation of ocular and physiologic effects. *J. Microw. Power*, 15, n° 1, 1980, pp. 37-44.

350. GUY A., LIN J. CHOU C. Electrophysiological effects of electromagnetic fields on animals. In : GUY A. Fundamental and applied aspects of nonionizing radiation. New York, Plenum Press, 1975, pp. 167-211.

351. GUY A., LIN J., HARRIS F. The effect of microwave radiation on evoked tactile and auditory CNS response in cats. In : Proceedings of the IMPI Symposium, Ottawa, 1972, pp. 120-129. (Paper 8.3.).

352. GUY A., LIN J., KRAMAR P., EMERY A. Effect of 2450-MHz radiation on the rabbit eye. *IEEE Trans. Microw. Theory Tech.*, MTT-23, n° 6, 1975, pp. 492-498.

353. GUY A., TAYLOR E., ASHLEMAN B., LIN J. Microwave interaction with the auditory systems of human and cats. In : IEEE Symposium on Microwave Theory and Techniques, Digest of technical paper, June 1973, pp. 321-323.

354. GWÓŹDŹ B., DYDUCH A., GRZYBEK H., PANZ D. Structural changes in brain mitochondria of mice subjected to hyperthermia. *Exp. Pathol. Bd.*, 15, 1978, pp. 124-126.

355. HAGAN G., CARPENTER R. Relative cataractogenic potencies of two microwave frequencies (2.45 and 10 GHz. In : JOHNSON C., SHORE M. Biological effects of electromagnetic waves. Selected papers of the USNC/URSI annual meeting, Boulder, 20-23 Oct. 1975, US Dep. Health, Educ., Welfare, FDA-77-8010, Washington, US Government Printing Office, 1976, Vol. 1, pp. 143-155.

356. HAIDT S., Mac TIGHE A. The effect of chronic, low-level microwave radiation on the testicles of mice. In : IEEE Symposium on Microwave Theory and Techniques, Digest of technical papers, 1973, pp. 324-326.

357. HALL C. Mac REE D., GALVIN M., WHITE N., THAXTON J., CHRISTENSEN V. Influence of in-vitro microwave radiation on the fertilizing capacity of turkey sperm. *Bioelectromagn.*, 4, 1983, pp. 43-54.

358. HAMID M., BADOUR S. The effects of microwaves on green algae. *J. Microw. Power*, 8, 1973, pp. 267-271.

359. HAMID M., BOULANGER R., HODGSON G., KONDRA P., SMITH K., BRAGG D. The effect of microwave radiation on the growth and reproduction of chickens. *J. Microw. Power*, 4, n° 4, 1969, pp. 253-256.

360. HAMID M., KONDRA P., SMITH K., HODGSON G., BRAGG D., GOVORA J., BOULANGER R. Growth and reproduction of chickens subjected to microwave radiation. *Can. J. Anim. Sci.*, 50, n° 3, Dec. 1970, pp. 639-644.

361. HAMNERIUS Y., OLOFSSON H., RASMUSON A., RASMUSON B. A negative test for mutagenic action of microwave radiation in Drosophila melanogaster. *Mutat. Res.*, 68, 1979, pp. 217-223.

362. HAMRICK P., FOX S. Rat lymphocytes in cell culture exposed to 2450 MHz (CW) microwave radiation. *J. Microw. Power*, 12, n° 2, 1977, pp. 125-132.

363. HAMRICK P., Mac REE D. Exposure of the Japanese quail embryo to 2.45 GHz microwave radiation during the second day of development. *J. Microw. Power*, 10, n° 2, 1975, pp. 211-221.

364. HANZON K. Biological effects of electromagnetic fields. *Elektro*, 88, n° 21, 1975, pp. 24-25

365. HARRISON G., ROBINSON J., Mac CULLOCH D., CHEUNG A. Comparison of hyperthermal cellular survival in the presence or absence of 2.45 GHz microwave radiation. In : BERTEAUD A., SERVANTIE B. Ondes électromagnétiques et biologie. Symposium international URSI, Jouy-en-Josas, juil. 1980, pp. 41-45.

366. HAUF R. Latest studies on effect of electromagnetic fields on human subjects. *Elektrotech. Z.*, Ausgabe B, 28, n° 6-7, 1976, pp. 181-183.

367. HAZZARD D. Biological effects and measurement of RF/microwaves. Symposium proceedings, 16-18 Feb. 1977, US Dep. Health, Educ., Welfare, BRH, FDA-77-8026, July 1977.

368. HEARN G. Effects of UHF radio fields on visual acuity and critical flicker fusion in the albino rat. Thesis, Ph. D., Baylor Univ., Waco, 1965. (Dissertation Abstract International n° 65-10 002).

369. HELLER J. Cellular effects of microwave radiation. In : CLEARY S. Biological effects and health implications of microwave radiation. Symposium proceedings, Commonwealth Univ., Richmond, 17-19 Sept. 1969, US Dep. Health, Educ., Welfare, BRH, DBE 70-2, BP 193 898, June 1970, pp. 116-121.

370. HELLER J. Experiments on the effect of electromagnetic waves on the reactivity of cells and tissues. III : Communication : effect of irradiation with red light and microwave on pinocytosis in FL cell cultures. *Zent.bl. Bakteriol.*, 1. Abteil, A, 221, 1972, pp. 386-397.

371. HELLER J., TEIXEIRA-PINTO A. A new physical method of creating chromosomal aberrations. *Nature*, 183, 1959, pp. 905-906.

372. HENDLER E. Cutaneous receptor response to microwave irradiation. In : HARDY J. Thermal problems in aerospace medicine. Hemel Hempstead, Unwin Ltd., 1968, pp. 149-161.

373. HENDLER E., HARDY J., MURGATROYD D. Skin heating and temperature sensation produced by infra-red and microwave irradiation. In : HARDY J. Temperature measurement and control in science and industry. 3rd part : Biology and medicine. New York, Rheinhold Publisher, 1963, p. 191.

374. HERRICK J., KRUSEN F. Certain physiologic and pathologic effects of microwaves. *Elect. Eng.*, 72, n° 3, 1953, pp. 239-244.

375. HEYNICK L., POLSON P. Bioeffects of radio-frequency radiation : a review pertinent to air-force operations. US Air Force report SAM-TR-83-1, USAF Sch. Aerosp. Med., Aerosp. Med. Div. (AFSC), Brooks Air Force Base, 1983.

376. HEYNICK L., POLSON P. USAFSAM review and analysis of radiofrequency radiation bioeffects literature. US Air Force report SAM-TR-84-17, USAF Sch. Aerosp. Med., Aerosp. Med. Div. (AFSC), Brooks Air Force Base, 1984.

377. HILLS G., HAMID M., KONDRA P. Effects of microwave radiation on hatchability and growth in chickens and turkeys. *Can. J. Anim. Sci.*, 54, n° 4, 1974, pp. 573-578.

378. HIRSCH F. Microwave cataracts - a case report reevaluated. In : CLEARY S. Biological effects and health implications of microwave radiation. Symposium proceeding, Commonwealth Univ., Richmond, 17-19 Sept. 1969, US Dep. Health, Educ., Welfare, BRH, DBE 70-2, PB 193 898, Library of Congress card n° 76-607340, June 1970, pp. 111-140.

379. HIRSCH F., BRUNER A. Absence of electromagnetic pulse effects on monkeys and dogs. *J. Occup. Med.*, 14, n° 5, 1971, pp. 380-386.

380. HIRSCH F., Mac GIBONEY D., HARHISH T. The psychological consequences of exposure to high density pulsed electromagnetic energy. *Int. J. Biom.*, 12, 1968, pp. 263-270.

381. HIRSCH F., PARKER J. Bilateral lenticular opacities occurring in a technician operating a microwave generator. *Arch. Ind. Hyg.*, 6, 1952, pp. 512-517.

382. HO H., EDWARDS W. Oxygen-consumption rate of mice under differing dose rates of microwave radiation. *Radio Sci.*, 12, n° 6 (S), 1977, pp. 131-138.

383. HO H., EDWARDS W. The effect of environmental temperature and average dose rate of microwave radiation on the oxygen-consumption rate of mice. *Radiat. Environ. Biophys.*, 16, 1979, pp. 323-338.

384. HOCKWIN O. La cataracte. *La Recherche*, n° 173, janv. 1986, pp. 30-38.

385. HODGKIN A., HUXLEY A. A quantitative description of membrane current and its application to conduction and excitation in nerve. *J. Physiol.*, n° 117, 1952, pp. 500-544.

386. HOLODOV J. The effect of an electromagnetic field on the central nervous system. (En russe). *Priroda*, 4, 1962, pp. 104-105. Traduction anglaise ASTIA n° 284 123, Library of Congress, ATD-P-65-68, FTD-TT 62-1107.

387. HOLODOV J. Effet d'un champ électromagnétique UHF sur l'activité électrique d'une région neuronalement isolée du cortex cérébral. *Bjul. Eksp. Biol. Med.*, 57, 1964, pp. 98-104.

388. HOLODOV J. Electromagnetic fields and brain. *Impact Sci. Soc.*, 24, n° 4, 1974, pp. 291-297.

389. HOUGH H., DOMINO E. Elevation in rat brain histamine content after focused microwave irradiation. *J. Neurochem.*, 29, n° 2, 1977, pp. 199-205.

390. HUANG A., ENGLE M., ELDER J., KINN J., WARD T. The effect of microwave radiation (2450MHz) on the morphology and chromosomes of lymphocytes. *Radio Sci.*, 12, n° 6 (S), 1977, pp. 173-177.

391. HUANG A., MOLD N. Immunologic and hematopoietic alterations by 2450 MHz electromagnetic radiations. *Bioelectromagn.*, 1, 1980, pp. 77-87.

392. HÜBNER R. Neue Erkenntnisse über biologische Wirkungen durch Hochfrequenz. *Elektron. Rundsch.*, 14, n° 6, 1960, pp. 229-230.

393. HUNT E., KING N., PHILLIPS R. Behavioral effects of pulsed microwave radiation. In : TYLER P. Biologic effects of non-ionizing radiation. Proceedings of a conference. *Ann. New York Acad. Sci.*, 247, 1975, pp. 440-453.

394. HUNT E., PHILLIPS R., CASTRO R., KING N. General activity of rats immediately following exposure to 2450 MHz microwaves. In : Proceedings of the IMPI Symposium, Ottawa, 1972, p. 119. (Paper 8.2.).

395. IGNATOV V., ŠENDERO B., PANASENK V., PIDENKO A., MAGAGINA A. Elimination de déterminants génétiques de résistance chez Staphylococcus aureus sous l'effet d'un champ électromagnétique intense. (En russe). *Genetika*, 9, n° 4, 1973, pp. 57-61.

396. IL'ČEVIČ N., GORODECKAJA S. Effects of the chronic application of electromagnetic microwave fields on the function and morphology of the reproductive organs of animals. (Trad. du russe). US joint Pub. Res. Serv. Rep., JPRS, L/5615, 10 Feb. 1976, pp. 5-7.

397. IMIG C., THOMPSON J., HINES H. Testicular degeneration as a result of microwave irradiation. In : Proceedings of the Society of Experimental Biology and Medicine, 69, n° 357, 1948, pp. 382-386.

398. IMRIE A. Pelvic short wave diathermy given inadvertently in early pregnancy. *J. Obstet. Gynecol.*, 78, 1971, pp. 91-92.

399. INOUYE M., MATSUMOTO N., GALVIN M., Mac REE D. Lack of effect of 2.45 GHz microwave radiation on the development of preimplanted embryos of mice. *Bioelectromagn.*, 3, 1982, pp. 275-283.

400. ISMAILOV E., ZOBKOVA S. Physiochemichal mechanisms of the biological activity of microwaves. (Trad. du russe). US joint Pub. Res. Serv. Rep., JPRS 7010, 7 Nov. 1977, pp. 33-50.

401. IVANOFF B., ROBERT D., DESCHAUX P., PELLISSIER J., FONTANGES R. Effet des micro-ondes sur la réponse immunitaire cellulaire de la souris Swiss. *C.R. Soc. Biol.*, 173, 1979, pp. 932-936.

402. IVANOV A., ČUHLOVIN B. Etat fonctionnel des leucocytes d'un organisme exposé aux micro-ondes. In : Gigiena truda i biologičeskoje djejstvije elektromagnitnyh voln radiočastot SV. materialov, III. Vsesimpoz., Moskva, 1968, pp. 62-63.

403. JACOB. Cerveau. In : GALTIER-BOISSIERE. Larousse médical illustré. Paris, Larousse, 1912, pp. 186-193.

404. JAKIMENKO D. Lečenije njekotoryh njejrotrofičeskih kožnyh zabolevanij ultrafioletovym oblučenijem i tokami ultravysokoj častoty v malyh dozirovkah. *Vjestn. Dermatov Venerol.*, 3, n° 5, 1961, pp. 33-36,

405. JANES D., LEACH W., MILLS W., MOORE R., SHORE M. Effects of 2450 MHz microwaves on protein synthesis and on chromosomes in Chinese hamsters. *Non-ioniz. Radiat.*, 1, 1969, pp. 125-130.

406. JENSH R. Studies of the teratogenic potential of exposure of rats to 6000 MHz microwave radiation. 1. Morphologic analysis at term. *Radiat. Res.*, 97, n° 2, 1984, pp. 272-281.

407. JENSH R., LUDLOW J., WEINBERG I., VOGEL W., RUDDER T., BRENT R. Teratogenic effects on rat offspring of non-thermal chronic prenatal microwave irradiation. *Teratol.*, 15, n° 2, 1977, abstract, p. 14 A.

408. JENSH R., LUDLOW J., WEINBERG I., VOGEL W., RUDDER T., BRENT R. Studies concerning the postnatal effects of protracted low dose prenatal 915 MHz microwave radiation. *Teratol.*, 17, n° 2, 1978, Abstract, p. 21 A.

409. JENSH R., LUDLOW J., WEINBERG I., VOGEL W., RUDDER T., BRENT R. Studies concerning the effects of protracted prenatal exposure to a non-thermal level of 2450 MHz microwave radiation in the pregnant rat. *Teratol.*, 17, n° 2, 1978, Abstract, p. 48 A.

410. JERŠOVA L., DUMANSKIJ J. Physiological changes in the central nervous system of animals under the chronic effect of continuous microwave fields. (Trad. du russe). US joint Pub. Res. Serv. Rep., JPRS L/5615, 10 Feb. 1976, pp. 1-4.

411. JOHNSON C. Recommendations for specifying EM wave irradiation conditions in bioeffects research. *J. Microw. Power*, 10, n° 3, 1975, pp. 249-250.

412. JOHNSON C., DURNEY C., BARBER P., MASSOUDI H., ALLEN S., MITCH J. Descriptive summary : radio-frequency radiation dosimetry handbook. *Radio Sci.*, 12, n° 6 (S), 1977, pp. 57-59.

413. JOHNSON C., GUY A. Nonionizing electromagnetic wave effects in biological materials and systems. *Proc. IEEE*, 60, n° 6, 1972, pp. 692-718.

414. JOHNSON C., SHORE M. Biological effects of electromagnetic waves. Selected papers of the USNC/URSI annual meeting, Boulder, 20-23 Oct. 1975, US Dep. Health, Educ., Welfare, FDA-77-8010 (Vol. 1), FDA-77-8011 (Vol. 2). Washington, US Government Printing Office, 1976, AD A045-044.

415. JOHNSON R., MIZUMARI S., MYERS D., GUY A., LOVELY R. Effects of pre- and post-natal exposures to 918 MHz microwave radiation on development and behavior in rats. In : International Symposium on Biological effects of electromagnetic waves, Airlie, 1977.

416. JOHNSON R., MYERS D., GUY A., LOVELY R., GALAMBOS R. Discriminative control of appetitive behavior by pulse microwave radiation in rats. In : JOHNSON C., SHORE M. Biological effects of electromagnetic waves. Selected papers of the USNC/URSI annual meeting, Boulder, 20-23 Oct. 1975, US Dep. Health, Educ., Welfare, FDA-77-8010. Washington, US Government Printing Office, Vol. 1, 1976, pp. 238-247.

417. JUSTESEN D. Microwaves and behavior. *Am. Psychol.*, 30, 1973, pp. 391-401.

418. JUSTESEN D. Behavioral and psychological effects of microwave radiation. *Bull. New York Acad. Med.*, 55, 1979, pp. 1058-1078.

419. JUSTESEN D. Microwave irradiation and the blood-brain barrier. *Proc. IEEE*, 68, n° 1, 1980, pp. 60-67.

420. JUSTESEN D. Behavioral effects of radio-frequency electromagnetic fields. In : Proceedings of the IMPI Symposium, Toronto, 1981, Digest pp. 136-139.

421. JUSTESEN D. Scientific and hygienic issues in biological research on microwaves : toward rapprochement between East and West. *Radio Sci.*, 17, 1982, pp. 1-12.

422. JUSTESEN D., ADAIR E., STEVENS J., BRUCE-WOLF V. A comparative study of human sensory thresholds : 2450 MHz microwave vs. far-infrared radiation. *Bioelectromagn.*, 3, 1982, pp. 117-125

423. JUSTESEN D., BAIRD R. Biological effects of electromagnetic waves. Special issue. *Radio Sci.*, 14, n° 6 (S), Nov.-Dec. 1979, 350 p.

424. JUSTESEN D., GUY A. Biological effects of electromagnetic radiation. Special issue. *Radio Sci.*, 12, n° 6 (S), Nov.-Dec. 1977, 293 p.

425. JUSTESEN D., KING N. Behavioral effects of low-level microwave irradiation in the closed-space situation. In : CLEARY S. Biological effects and health implications of microwave radiation. Symposium proceedings, Commonwealth Univ., Richmond, 17-19 Sept. 1969, US Dep. Health, Educ., Welfare, BRH, DBE 70-2, PB 193 898, June 1970, pp. 154-179.

426. JUSTESEN D., LEVINSON D., CLARKE R., KING N. The microwave oven in behavioral and biological research : electrical and structural modifications, dosimetry, and functional evaluation. *J. Microw. Power*, 6, n° 3, 1971, pp. 237-258.

427. JUSTESEN D., LEVINSON D., JUSTESEN L. Psychologic stressors are potent mediators of the thermal response to microwave irradiation. In : CSERSKI P., OSTROWSKI K., SHORE M., SILVERMAN C. SÜSS M., WALDESKOG B. Biological effects and health hazards of microwave radiation. Proceedings of a symposium, Warszawa, 15-18 Oct. 1973. Warszawa, Polish Medical Publishers, 1974, pp. 134-140.

428. JUSTESEN D., PENDLETON R., PORTER P. Effects of hyperthermia on activity and learning. *Psychol. Reports*, 9, 1961, pp. 99-102.

429. KADOUM A. Electrophoretic studies of serum proteins obtained from radio-frequency-treated yellow mealworm larvae. *J. Econ. Entomol.*, 62, 1969, pp. 745-746.

430. KADOUM A., BALL H., NELSON S. Morphological abnormalities resulting from radiofrequency treatment of larvae of Tenebrio molitor. *Ann. Entomol. Soc. Am.*, 60, 1967, pp. 889-892.

431. KALANT H. Physiological hazards of microwave radiation : a survey of published literature. *Can. Med. Ass. J.*, 81, 1959, p. 575.

432. KAMEN'SKIJ J. Effet des micro-ondes sur l'état fonctionnel du nerf. (En russe). *Biofiz.*, 9, n° 6, 1964, pp. 758-764.

433. KAPLAN I., METLAY W., ROSENTHAL S., ZARET M. Microwave and infrared effects on heart-rate, respiration rate and subcutaneous temperature of the rabbit. *J. Microw. Power*, 10, n° 1, 1975, pp. 3-18.

434. KAPLAN I., METLAY W., ZARET M., BIRENBAUM L., ROSENTHAL S. Absence of heart-rate effects in rabbits during low-level microwave irradiation. *IEEE Trans. Microw. Theory Tech.*, MTT-19, n° 2, 1971, pp. 168-173.

435. KIČOVSKAJA I. L'effet des ondes centimétriques à divers niveaux de puissance sur le sang et les organes hémopoïétiques de la souris blanche. (En russe). *Gig. Tr. Prof. Zabol.*, 8, n° 14, 1964.

436. KING N., JUSTESEN D., CLARKE R. Behavioral sensitivity to microwave irradiation. *Science*, 172, 1971, pp. 398-401.

437. KLEIN M., MILHAUD C. Résultats préliminaires des effets des micro-ondes sur le système nerveux central d'un primate (Macaca mulatta). In : Effets biologiques des rayonnements non ionisants ; utilisation et risques associés. 9ème Congrès international de la Société Française de Radioprotection, Nainville-les-Roches, 22-26 mai 1978, pp. 315-336.

438. KLEIN M., STERN L., MILHAUD C., ROSOLEN S. Application des techniques pharmacologiques à l'étude des effets biologiques des micro-ondes. *Radioprot.*, 17, n° 4, 1982, pp. 225-241.

439. KLEINSTEIN B. Biological effects of nonionizing radiation. Philadelphia, Franklin Res. Cent., V, 1980, pp. 1-4.

440. KLEINSTEIN B. Biological effects of nonionizing electromagnetic radiation. *Dig. Curr. Lit.*, 8, n° 4 and 9, n° 1, Sept.-Dec. 1984, 150 p.

441. KNUDSON A., SCHAIBLE P. The effect of exposure to an ultra-high frequency field on growth and reproduction in the white rat. *Arch. Pathol.*, 11, 1931, p. 728.

442. KONDRA P., HAMID M., HODGSON G. Effects of microwave radiation on growth and reproduction of two stocks of chickens. *Can. J. Anim. Sci.*, 52, n° 2, 1972, pp. 317-320.

443. KONDRA P., SMITH W., HODGSON G., BRAGG D., GAVORA J., HAMID M., BOULANGER R. Growth and reproduction of chickens subjected to microwave radiation. *Can. J. Anim. Sci.*, 50, 1970, pp. 639-644.

444. KORBEL S. Behavioral effects of low-intensity UHF radiation. In : CLEARY S. Biological effects and health implications of microwave radiation. Symposium proceeding, Commonwealth Univ., Richmond, 17-19 Sept. 1969, US Dep. Health, Educ., Welfare, BRH, DBE 70-2, PB 193 898, June 1970, pp. 180-184.

445. KORBEL S., FINE H. Effects of low intensity UHF radio fields as a function of frequency. *Psychon. Sci.*, 9, 1967, pp. 527-528.

446. KORBEL S., JOHNSON K., ROWLAND P. After-effects of low intensity UHF radiation. *Psychon. Sci.*, 23, n° 1A, 1971, pp. 50-51.

447. KORBEL S., THOMPSON W. Behavior effects of stimulation by UHF radio fields. *Psychol. Rep.*, 17, 1965, pp. 595-602.

448. KÖSSLER F., LANGE F., KUPFER J., ROTHE R. Wirkung einer Mikrowellenstrahlung niedriger Intensität auf Körpergewicht, Spontanaktivität und Blutwerte von Mäusen. *Z. Ges. Hyg. Grenzgeb.*, 27, 1981, pp. 44-47.

449. KÖTWITSCH G. Funkstörungsgrenzwerte hier und drüben. *Fernmeldepraxis*, 38, n° 6, 1961, pp. 211-221.

450. KRAMAR P., EMERY A., GUY A., LIN J. The ocular effects of microwaves on hypothermic rabbits : a study of microwave cataractogenic mechanisms. In : TYLER P. Biologic effects of non-ionizing radiation. Proceedings of a conference. *Ann. New York Acad. Sci.*, 247, 1975, pp. 155-165.

451. KRAMAR P., GUY A., EMERY A., HARRIS P. Acute microwave irradiation and cataract formation in rabbits and monkeys. In : Proceedings of the IMPI Symposium, Leuven, 1976. (Paper 1A.2.).

452. KRAMAR P., HARRIS C., EMERY A., GUY A. Acute microwave irradiation and cataract formation in rabbits and monkeys. *J. Microw. Power*, 13, n° 3, 1978, pp. 239-249.

453. KRAMAR P., HARRIS C., GUY A., EMERY A. Mechanisms of microwave cataractogenesis in rabbits. In : JOHNSON C., SHORE M. Biological effects of electromagnetic waves.

Selected papers of the USNC/URSI annual meeting, Boulder, 20-23 Oct. 1975, US Dep. Health, Educ., Welfare, FDA-77-8010. Washington, US Government Printing Office, Vol. 1, 1976, pp. 40-60.

454. KRAUSE N. Sicherheitsmassnahmen bei der Anwendung hochfrequenter elektrischer Felder. *Moderne Unf.verhüt.,* 26, 1982, pp. 101-109.

455. KREMER F. Biologische Effekte elektromagnetischer Felder. *Umschau,* 1984, pp. 402-403.

456. KRITIKOS H. Effects of RF fields on nervous activities. In : Proceedings of the IMPI Symposium, Waterloo (Can.), 1975, pp. 64-65. (Paper 3.7.).

457. KRUEGER W., GIAROLA A., BRADLEY J., SCHREKENHAMER A. Effect of various electromagnetic fields on reproductive performance of chicken. *Poult. Sci.,* 53, n° 4, 1974, p. 1640.

458. KRUEGER W., GIAROLA A., BRADLEY J., SCHREKENHAMER A. Effects of electromagnetic fields of fecundity in chicken. In : TYLER P. Biologic effects of non-ionizing radiation. Proceedings of a conference. *Ann. New York Acad. Sci.,* 247, 1975, pp. 391-400.

459. KRUPP J. The relationship of thermal stress to immune system response in mice exposed to 2.6 GHz radiofrequency radiation. In : Biological effects of electromagnetic waves. International symposium, Airlie, 30 Oct.-4 Nov. 1977, Abstracts of scientific papers.

460. LAI H., HORITA A., CHOU C., GUY A. Psychoactive drug response is affected by acute low level microwave irradiation. *Bioelectromagn.,* 4, 1983, pp. 205-214.

461. LAMBERT P., NEALEIGH R., WILSON M. Effects of microwave exposures on the central nervous system of beagles. *J. Microw. Power,* 7, 1972, pp. 367-380.

462. LANCRACJAN I. Gonadic function in workman with long-term exposure to microwaves. *Health Phys.,* 29, 1975, pp. 381-385.

463. LA ROCHE L., ZARET M., BRAUN A. An operational safety program for ophtalmic hazards of microwaves. *Arch. Environ. Health,* 20, 1970, pp. 350-355.

464. LARSEN L., MOORE R., ACEVEDO J. A microwave decoupled electrode for the encephalogram. *IEEE Trans.*

Microw. Theory Tech., MTT-22, n° 10, 1974, pp. 884-887.

465. LASKEY J., DAWES D., HOWES M. Progress report on 2450 MHz irradiation of pregnant rats and the effect on the fetus. In : Radiation bio-effects summary report, PHS, US Dep. Health, Educ., Welfare, BRH/DBE-70, Rockville, 1970, pp. 167-173.

466. LEACH W. Genetic, growth and reproductive effects of microwave radiation. *Bull. New York Acad. Med.,* 56, 1980, pp. 249-257.

467. LEBOVITZ R. Caloric vestibular stimulation via UHF-microwave irradiation. *IEEE Trans. Biomed. Eng.,* BME-20, 1973, pp. 117-126.

468. LEBOVITZ R. Significance of microthermal effects derived from low level UHF-microwave irradiation of head ; indirect caloric vestibular stimulation. *J. Theor. Biol.,* 41, n° 2, 1973, pp. 209-221.

469. LEBOVITZ R., SEAMAN R. Microwave hearing : the response of single auditory neurons in the cat to pulsed microwave radiation. *Radio Sci.,* 12, n° 6 (S), 1977, pp. 229-236.

470. LEBOVITZ R., SEAMAN R. Single auditory unit responses to weak, pulsed microwave radiation. *Brain Res.,* n° 126, 1977, pp. 370-375.

471. LEDEN U., HERRICK J., WAKIM K., KRUSEN F. Preliminary studies on the heating and circulating effects of microwaves (radar). *Br. J. Phys. Med.,* 10, 1947, pp. 177-184.

472. LEHMANN J., SILVERMAN D., BAUM B., KIRK N., JOHNSTON V. Temperature distribution in human thigh produced by infrared, hot pack and microwave applicators. *Arch. Phys. Med. Rehabil.,* 47, n° 5, 1966, pp. 291-299.

473. LENKO J., DOLATOWSKI A., GRUSZECKI L., KLAJMAN S., JANUSZKIEWICZ L. Effect of 10-cm radarwaves on the level of 17-ketosteroids and 17-hydroxy-corticosteroids in the urine of rabbits. *Prz. Lek.,* 22, 1966, pp. 296-299.

474. LERNER E. RF radiation : biological effects. *IEEE Spectrum,* 17, n° 12, 1980, pp. 51-59.

475. LERNER E. Biological effects of electromagnetic fields : the drive towards regulation. *IEEE Spectrum,* 21, n° 3, 1984, pp. 63-70.

476. LERNER E. Biological effects of EM fields : the evidence mounts. *IEEE Spectrum*, 21, n° 5, 1984, pp. 57-69.

477. LEVITINA N. Action des micro-ondes sur le rythme cardiaque du lapin en irradiation locale. (En russe). *Bjul. Eksp. Biol. Med.*, 58, 1964, pp. 820-822.

478. LEVITINA N. Action non-thermique des micro-ondes sur le rythme cardiaque de la grenouille. (En russe) *Bjul. Eksp. Biol. Med.*, 62, 1966, pp. 1386-1387.

479. LIBURDY R. Effects of radio-frequency radiation on inflammation. *Radio Sci.*, 12, n° 6 (S), 1977, pp. 179-183.

480. LIBURDY R. Radiofrequency radiation alters the immune system. I. Modulation of T- and B-lymphocyte levels and cell-mediated immunocompetence by hyperthermic radiation. *Radiat. Res.*, 77, 1979, pp. 34-46.

481. LIBURDY R. Radiofrequency radiation alters the immune system. II. Modulation of in-vivo lymphocyte circulation. *Radiat. Res.*, 83, 1980, pp. 66-73.

482. LIDDLE C., PUTNAM J., ALI J., LEWIS J., BELL B., WEST M., LEWTER O. Alteration of circulating antibody response of mice exposed to 9-GHz pulsed microwaves. *Bioelectromagn.*, 1, 1980, pp. 397-404.

483. LIDMAN B., COHN C. Effect of radar emanations on the hematopoietic system. *Air Surgeon's Bull.*, 2, 1945, pp. 448-449.

484. LILIENFELD A., TONASCIA J., TONASCIA S. Foreign service health status study. Evaluation of health status of foreign service and other employees from selected Eastern European posts. Final report, Contract n° 6025-619073, US Dep. State, 31 July 1978.

485. LIN J. Biomedical effects of microwave radiation - a review. In : Proceedings of the National Electr. Conf., 30, 1975, pp. 224-232.

486. LIN J. Microwave auditory effect - a comparison of some possible transduction mechanisms. *J. Microw. Power*, 11, n° 1, 1976, pp. 77-81.

487. LIN J. Microwave-induced hearing : some preliminary theoretical observations. *J. Microw. Power*, 11, n° 3, 1976, pp. 295-298.

488. LIN J. On microwave-induced hearing sensations. *IEEE Trans. Microw. Theory Tech.*, MTT-25, n° 7, 1977, pp. 605-613.

489. LIN J. Further studies on the microwave auditory effect. *IEEE Trans. Microw. Theory Tech.*, MTT-25, n° 11, 1977, pp. 938-943.

490. LIN J. Theoretical calculations of frequencies and thresholds of microwave-induced auditory signals. *Radio Sci.*, 12, n° 6 (S), 1977, pp. 237-242.

491. LIN J. The microwave auditory phenomenon. *Proc. IEEE*, 68, n° 1, 1980, pp. 67-73.

492. LIN J., CHEN K. Effect of microwave radiation on mammalian cells in vitro. In : JOHNSON C., SHORE M. Biological effects of electromagnetic waves. Selected papers of the USNC/URSI annual meeting, Boulder, 20-23 Oct. 1975, US Dep. Health, Educ., Welfare, FDA-77-8010/8011. Washington, US Government Printing Office, 1976, p. 132.

493. LIN J., CHUAN-LIN W., LAM C. Transmission of electromagnetic pulse into the head. *Proc. IEEE*, 63, n° 12, 1975, pp. 1726-1727.

494. LIN J., GUY A., CALDWELL L. Thermographic and behavioral studies of rats in the near field of 918-MHz radiations. *IEEE Trans. Microw. Theory Tech.*, MTT-25, n° 10, 1977, pp. 833-836.

495. LIN J., LIN M. Power-time relations of microwave-induced blood-brain barrier permeation. *Bioelectromagn.*, 1, 1980, p. 207.

496. LIN J., LIN M. Studies on microwave and blood-brain barrier interaction. *Bioelectromagn.*, 1, 1980, pp. 313-323.

497. LIN J., LIN M. Microwave hyperthermia-induced blood-brain barrier alterations. *Radiat. Res.*, 89, 1982, pp. 77-87.

498. LIN J., MELTZER R., REDDING F. Microwave-evoked brainstem auditory responses. In : Proceedings of San Diego Biomedical Symposium, 17, 1978, pp. 461-465.

499. LIN J., MELTZER R., REDDING F. Microwave-evoked brainstem potentials in cats. *J. Microw. Power*, 14, n° 3, 1979, pp. 291-296.

530 Part IV - Biological effects and medical applications

500. LIN J., OTTENBREIT M., WANG S., INOUE S., BOLLINGER R., FRACASSA M. Microwave effects on granulocyte and macrophage precursor cells of mice in vitro. *Radiat. Res.*, 80, 1979, pp. 292-302.

501. LIN J., PETERSON W. Cytological effects of 2450 MHz CW microwave radiation. *J. Bioeng.*, 1, 1977, pp. 471-478.

502. LINDAUER G., LIU L., SKEWES G., ROSENBAUM F. Further experiments seeking evidence of nonthermal biological effects of microwave radiation. *IEEE Trans. Microw. Theory Tech.*, MTT-22, n° 8, 1974, pp. 790-793.

503. LIN-LIU S., ADEY W. Low-frequency amplitude-modulated microwave fields change calcium efflux rates from synaptosomes. *Bioelectromagn.*, 3, 1982, pp. 309-322.

504. LIN-LIU S., ADEY W. Migration of cell surface concanavalin A receptors in pulsed electric fields. *Biophys. J.*, 45, 1984, pp. 1211-1217.

505. LIU L., ROSENBAUM F., PICKARD W. The relation of teratogenesis in Tenebrio molitor to the incidence of low-level microwaves. *IEEE Trans. Microw. Theory Tech.*, 23, n° 11, 1975, pp. 929-931.

506. LIU L., ROSENBAUM F., PICKARD W. The insensitivity of frog heart rate to pulse modulated microwave energy. *J. Microw. Power*, 11, n° 3, 1976, pp. 225-232.

507. LIVINGSTON G., JOHNSON C., DETHLEFSEN L. Comparative effects of waterbath and microwave-induced hyperthermia on survival of Chinese hamster ovary (CHO) cells. *Radio Sci.*, 14, n° 6 (S), 1979, pp. 117-123.

508. LIVŠIC N. Uslovnoreflektornaja dejatel'nost sobak pri lokal'nyh vozdjejstvijah polem UVČ na njekotoryje zony kory bolših polušarij. *Biofiz.*, 2, n° 2, 1957, pp. 197-208.

509. LIVŠIC N. Rol njervnoj sistemy v reakcijah organizma na djejstvije elektromagnitnogo polja ultravysokoj častoty. *Biofiz.*, 2, n° 3, 1957, pp. 372-389.

510. LIVŠIC N. Djejstvije polja ultravysokoj častoty na funkcii njervnoj sistemy. *Biofiz.*, 3, n° 4, 1958, pp. 426-437.

511. LORDS J., DURNEY C., BORG A., TINNEY C. Rate effects in isolated hearts induced by microwave irradiation. *IEEE Trans. Microw. Theory Tech.*, MTT-21, n° 12, 1973, pp. 834-836.

512. LOTT J., Mac CAIN H. Some effects of continuous and pulsating electric fields on brain-wave activity in rats. *J. Biometereol.*, 17, 1973, p. 221.

513. LOTZ W. Neuroendocrine function in Rhesus monkeys exposed to pulsed microwave radiation. In : STUCHLY S. Electromagnetic fields in biological systems. Proceedings of the IMPI Symposium, Ottawa, 1978. Edmonton, IMPI, Abstracts of scientific papers.

514. LOTZ W. Endocrine function in Rhesus monkeys and rats exposed to 1.29 GHz microwave radiation. Bioelectromagnetics symposium, USNC/URSI, Seattle, 18-22 June 1979.

515. LOTZ W., MICHAELSON S. Effects of hypophysectomy and dexamethasone on the rat's adrenal response to microwave irradiation. In : Biological effects of electromagnetic waves. International symposium, Airlie, 30 Oct.-4 Nov. 1977, Abstracts of scientific papers, p. 38.

516. LOTZ W., MICHAELSON S. Temperature and corticosterone relationships in microwave-exposed rats. *J. Appl. Physiol.*, 44, n° 3, 1978, pp. 438-445.

517. LOTZ W., MICHAELSON S., LEBDA N. Growth hormone levels of rats exposed to 2450 MHz (CW) microwaves. In : Biological effects of electromagnetic waves. International symposium, Airlie, 30 Oct.-4 Nov. 1977, Abstracts of scientific papers, p. 39.

518. LOVELY R., GUY A. Conditioned taste aversions in the rat induced by a single exposure to microwaves. In : Proceedings of the IMPI Symposium, Waterloo (Can.), 1975, pp. 87-94. (Paper 5.3.)

519. LOVELY R., GUY A., JOHNSON R., MATHEWS M. Alteration of behavioral and biochemical parameters during and consequent to 500-µW/cm² chronic 2450-MHz microwave exposure. In : STUCHLY S. Electromagnetic fields in biological systems. Proceedings of the IMPI Symposium, Ottawa, 1978. Edmonton, IMPI, p. 34.

520. LOVELY R., MYERS D., GUY A. Irradiation of rats by 918 MHz microwaves at 2.5 mW/cm² : delineating the dose-response relationship. *Radio Sci.*, 12, n° 6 (S), 1977, pp. 139-146.

521. LU S., JONES J., PETTIT S., LEBDA N., MICHAELSON S. Neuroendocrine and cardiodynamic response of the dog subjected to cranial exposure to 2450 MHz microwaves. In : Proceedings of the IMPI Symposium, Waterloo (Can.), 1975, p. 63. (Paper 3.6.).

522. LU S., LEBDA N., MICHAELSON S. Effects of microwave radiation on the rat's pituitary-thyroid axis. In : Biological effects of electromagnetic waves. International symposium, Airlie, 30 Oct.-4 Nov. 1977, Abstracts of scientific papers, p. 37.

523. LU S., LEBDA N., MICHAELSON S. Microwave- induced temperature, corticosterone, and thyrotropin interrelationships. *J. Appl. Physiol.*, 50, 1981, pp. 399-405.

524. LU S., LEBDA N., MICHAELSON S., PETTIT S., RIVERA D. Thermal and endocrinological effects of protracted irradiation of rats by 2450 MHz microwaves. *Radio Sci.*, 12, n° 6 (S), 1977, pp. 147-156.

525. LU S., LEBDA N., PETTIT S., MICHAELSON S. Modification of microwave biological end-points by increased resting metabolic heat load in rats. In : Bioelectromagnetics Symposium, Seattle, 18-22 June 1979.

526. LU S., LOTZ G., MICHAELSON S. Advances in microwave-induced neuroendocrine effects : the concept of stress. *Proc. IEEE*, 68, n° 1, 1980, pp. 73-77.

527. LU S., PETTIT S., MICHAELSON S. Dual actions of microwaves on serum corticosterone in rats. In : Bioelectromagnetics Symposium, Seattle, 18-22 June 1979.

528. LUBEN R., CAIN C., CHEN M., ROSEN D., ADEY W. Effects of electromagnetic stimuli on bone and bone cells in vitro : inhibition of responses to parathyroid hormone by low-energy low-frequency fields. In : Proceedings of the National Academy of Sciences of the USA, 79, 1982, p. 4180.

529. LUBIN M. Effects of ultra-high-frequency radiation on animals. *Arch. Ind. Health.*, 21, n° 6, 1960, pp. 555-558.

530. LUCZAK M., SZMIGIELSKI S., JANIAK M., KOBUS M., DB CLERQ E. Effect of microwaves on virus multiplication in mammalian cells. In : Proceedings of the IMPI Symposium, Leuven, 1976. (Paper 6A.2.).

531. LUNDSTROM D., LOVELY R., PHILLIPS R. Failure to find synergistic effects of 2.8-GHz pulsed microwaves and chlordiazepoxide on fixed-interval behavior in the rat. In : 3rd Annual Bioelectromagnetics Society meeting, Washington, 9-12 August 1981.

532. LYLE D., SCHECHTER P., ADEY W., LUNDAK R. Suppression of T-lymphocyte cytotoxicity following exposure to sinusoidally amplitude-modulated fields. *Bioelectromagn.*, 4, 1983, pp. 281-292.

533. Mac AFEE R. Neurophysiological effects of microwave irradiation. In : SÜSSKIND C. Proceedings of the 3rd annual Tri-Service conference on biological hazards of microwave-radiating equipments. Univ. California, Berkeley, 25-27 August 1959, RADC-TR-59-140, P. 314.

534. Mac AFEE R. Neurophysiological effect of 3 cm-microwave radiation. *Am. J. Physiol.*, n° 200, 1961, pp. 192-194.

535. Mac AFEE R. Physiological effects of thermode and microwave stimulation of peripheral nerves. *Am. J. Physiol.*, 203, n° 2, 1962, pp. 347-378.

536. Mac AFEE R. Analeptic effect of microwave irradiation on experimental animals. *IEEE Trans. Microw. Theory Tech.*, MTT-19, n° 2, 1971, pp. 251-253.

537. Mac AFEE R. Low power density behavior effects of microwave irradiation of experimental animals : real or artifact ? *J. Microw. Power*, 7, n° 2, 1972, pp. 83-85.

538. Mac AFEE R. The effect of 2450-MHz microwaves on the growth of mice. *J. Microw. Power*, 8, 1973, pp. 111-116.

539. Mac AFEE R. Microwave irradiation, behavioral and ocular effects on Rhesus monkey. *Biophys. J.*, 17, n° 2, 1977, p. A 145.

540. Mac AFEE R., BRAUS R., FLEMING J. The effect of 2450 MHz microwave irradiation on the growth of mice. *J. Microw. Power*, 8, n° 1, 1973, pp. 111-116.

541. Mac AFEE R., LONGACRE A., BISHOP R., ELDER S., MAY J., HOLLAND M., GORDON R. Absence of ocular pathology after repeated exposure of unanesthetized monkeys to 9.3-GHz microwaves. *J. Microw. Power*, 14, n° 1, 1979, pp. 41-44.

542. Mac AFEE R., ORTIZ-LUGO R., BISHOP R., GORDON R. Absence of deleterious effects of chronic microwave radiation on the eyes of Rhesus monkeys. *Ophtalmol.*, 90, 1983, pp. 1243-1245.

543. Mac GREGOR R. A possible mechanism for the influence of electromagnetic radiation on neuroelectric potentials. *IEEE Trans. Microw. Theory Tech.*, MTT-27, n° 11, 1979, pp. 914-921.

544. Mac KINLEY G. Short electric wave length radiation in biology. In : DUGGAR B. Biological effects of radiation. Vol. 1. New York City, McGraw-Hill Book Co., 1936, pp. 541-558.

545. Mac LAUGHLIN J. Tissue destruction and death from microwave radiation (radar). *Calif. Med.*, 86, 1957, pp. 336-339.

546. Mac LEES B., FINCH E. Analysis of reported physiologic effects of microwave radiation. In : LAWRENCE J., GOFMAN J. Advances in biological and Medical Physics. New York, Academic Press, Inc., 1973, pp. 163-223.

547. Mac REE D. Thresholds for lenticular damage in the rabbit eye due to single exposure to CW microwave radiation : An analysis of the experimental information at a frequency of 2.45 GHz. *Health Phys.*, 21, 1971, pp. 763-769.

548. Mac REE D. Soviet and Eastern European research on biological effects of microwave radiation. *Proc. IEEE*, 68, n° 1, 1980, pp. 84-91.

549. Mac REE D., FAITH R., Mac CONNELL E., GUY A. Long-term 2450-MHz CW microwave irradiation of rabbits : evaluation of hematological and immunological effects. *J. Microw. Power*, 15, n° 1, 1980, pp. 45-52.

550. Mac REE D., GALVIN M., HALL C., LIEBERMAN M. Microwave effects on embryonic cardiac tissue of Japanese quail. In : BERTEAUD A., SERVANTIE B. Ondes électromagnétiques et biologie, Symposium international URSI, Jouy-en-Josas, Juil. 1980, pp. 79-84.

551. Mac REE D., HAMRICK P. Exposure of Japanese quail embryos to 2.45-GHz microwave radiation during development. *Radiat. Res.*, 71, 1977, pp. 355-366.

552. Mac REE D., HAMRICK P., TINKL J. Some effects of exposure of the Japanese quail embryo to 2.45 GHz microwave radiation. In : TYLER P. Biologic effects of non-ionizing radiation. Proceedings of a conference. *Ann. New York Acad. Sci.*, 247, 1975, p. 377.

553. Mac REE D., WYATT R., HASEMAN J., SOMJEN G. The transmission of reflexes in the spinal cord of cats during direct irradiation with microwaves. *J. Microw. Power*, 11, n° 1, 1976, pp. 49-60.

554. MAGIN R., LU S., MICHAELSON S. Biological effects of locally applied microwaves on the thyroid gland of dogs. In : Proceedings of the IMPI Symposium, Waterloo (Can.), 1975, pp. 117-124. (Paper 7.5.).

555. MAGIN R., LU S., MICHAELSON S. Microwave heating effect on the dog thyroid gland. *IEEE Trans. Biomed. Eng.*, BME-24, n° 6, 1977, pp. 522-530.

556. MAGIN R., LU S., MICHAELSON S. Stimulation of the dog thyroid by the local application of high intensity microwave. *Am. J. Physiol.*, n° 233, Nov. 1977, pp. E 363-E 368.

557. MALBOYSSON E. Medical control of men working within electromagnetic fields. *Rev. Gén. Electr.*, Ed. spéciale, juil. 1976, pp. 75-80.

558. MANIEZ J., LE RUZ P., PLURIEN G. Effet de l'exposition du rat nouveau-né aux micro-ondes sur le développement ultérieur des gonades. In : Effets biologiques des rayonnements non ionisants ; utilisation et risques associés. 9ème congrès international de la Société Française de Radioprotection, Nainville-les-Roches, 22-26 mai 1978, pp. 257-264.

559. MANIEZ J., LE RUZ P., PLURIEN G. Effets de l'exposition néonatale aux micro-ondes sur l'évolution ultérieure des fonctions corticotrope et gonadotrope chez le rat. In : BERTEAUD A., SERVANTIE B. Ondes électromagnétiques et biologie. Symposium international URSI, Jouy-en-Josas, juil. 1980, pp. 85-88.

560. MARHA K. Biologické účinky elektromagnetických vln o vysoké frekvenci. *Prac. Lék.*, XV, n° 9, 1963, pp. 238-241 et 387.

561. MARHA K., MUSIL J., TUHA H. Electromagnetic fields and the living environment. San Francisco, San Francisco Press Inc., 1971, 138 p.

562. MARRACK P., KAPPLER J. Le lymphocyte T et ses récepteurs. *Pour la Science*, n° 102, avril 1986, pp. 18-29.

563. MATTSON J. Effect of electromagnetic pulse on avoidance behaviour and electroencephalogram of a Rhesus-monkey. *Aviat. Sp. Environ. Med.*, 47, n° 6, 1976, pp. 644-648.

564. MAUPAS P., VIALARD - GOUDOU A., MOTHIRON J. A study of the biochemical and ultrastructural changes of Escherichia coli K.12 under microwave action at 2.45 GHz. Publication Lab. Microbiol., Fac. Med. Pharm. Tours, s.d., 6 p.

565. MAYERS C., HABESHAW J. Marked depression of phagocytosis : a non thermal effect of microwave radiation as a potential hazard to health. *Int. J. Radiat. Biol.*, 24, n° 5, 1973, pp. 444-461.

566. MEROLA L., KINOSHITA J. Changes in the ascorbic acid content in lenses of rabbit eyes exposed to microwave radiation. In : PEYTON M. Proceedings of the 4th annual Tri-Service conference on biological effects of microwave-radiating equipments, biological effects of microwave radiations. New York, Plenum Press, 1, 1961, RADC-TR-60-180, p. 285.

567. MERRIAM G. Cataracts following use of microwave oven : Comments. *New York State J. Med.*, 74, n° 11, 1974, pp. 2036-2037.

568. MERRITT J. Science and standards - another viewpoint. *J. Microw. Power*, 20, n° 1, 1985, pp. 55-56.

569. MERRITT J., CHAMNESS A., ALLEN S. Studies on blood-brain barrier permeability after microwave irradiation. *Radiat. Environ. Biophys.*, 15, 1978, pp. 367-377.

570. MERRITT J., CHAMNESS A., HARTZELL R., ALLEN S. Orientation effects on microwave-induced hyperthermia and neurochemical correlates. *J. Microw. Power*, 12, n° 2, 1977, pp. 167-172.

571. MERRITT J., HARTZELL R., FRAZER J. The effect of 1.6 GHz radiation on neurotransmitters in discrete areas of the rat brain. US Air Force report SAM-TR-76-3, USAF Sch. Aerosp. Med., Aerosp. Med. Div. (AFSC), Brooks Air Force Base, 1976.

572. MEZYKOWSKI T., BAL J., DEBIEC H., KWARECKI K. Response of Aspergillus nidulans and Physarium polycephalum to microwave irradiation. *J. Microw. Power*, 15, n° 2, 1980, pp. 75-80.

573. MEZZADROLI G., VARETON E., LONGO B. Action des ondes électromagnétiques ultra-courtes (λ=2 à 3 m) sur les vers à soie. In : Comptes-rendus de la Reale Academia dei Lincei, Roma, XII, 6a, 2ème sem., fasc. 5-6, sept. 1930, fasc. 9, nov. 1930.

574. MICHAELSON S. Biologic effects of microwave energy. In : Proceedings of the Consumer Microwave Oven Systems Conference. Cornell Univ., Ithaca, 1970, pp. 17-21.

575. MICHAELSON S. Biological effects of microwave exposure - an overview. *J. Microw. Power*, 6, n° 3, 1971, pp. 259-267.

576. MICHAELSON S. The Tri-Service Program - a tribute to George M. Knauf, USAF (MC). *IEEE Trans. Microw. Theory Tech.*, MTT-19, n° 2, 1971, pp. 131-146.

577. MICHAELSON S. Cutaneous perception of microwaves. *J. Microw. Power*, 7, n° 2, 1972, pp. 67-73.

578. MICHAELSON S. The relevancy of experimental studies of microwave-induced cataracts to man. Contract of the US Atomic Energy Commission, Univ. Rochester Atomic Energy Project, Report n° UR-3490-103, In : Proceedings of the IMPI Symposium, Ottawa, 1972, pp. 145-147. (Paper 8.7.).

579. MICHAELSON S. Effects of exposure to microwaves and perspectives. *Environ. Health Perspect.*, 8, 1974, pp. 133-156.

580. MICHAELSON S. Review of a program to assess the effects on man from exposure to microwaves. *J. Microw. Power*, 9, n° 2, 1974, pp. 147-161.

581. MICHAELSON S. Thermal effects of single and repeated exposures to microwaves. In : CZERSKI P., OSTROWSKI K., SHORE M., SILVERMAN C., SÜSS M., WALDESKOG. B. Biological effects and health hazards of microwave radiation. Proceedings of an international symposium, Warszawa, 15-18 Oct. 1973. Warszawa, Polish Medical Publishers, 1974, pp. 1-4.

582. MICHAELSON S. Protection guides and standards for microwave exposure. In : AGARD Lecture Series, n° 78, Radiation hazards, Sept. 1975, pp. 12.1-12.6.

583. MICHAELSON S. Influence of microwave exposure on neuroendocrine function. In : Proceedings of the Federation of American Societies for Experimental Biology, 35, n° 3, 1976, p. 845.

584. MICHAELSON S. Endocrine and biochemical effects. In : Microwave and
radiofrequency radiation. København,
OMS/WHO, Regional Office for Europe,
1977, sect. 7, pp. 18-23.

585. MICHAELSON S. Biologic and patho-
physiologic effects of exposure to
microwaves. In : Microwave bioeffects
and radiation safety. *Trans. IMPI*,
1978, pp. 54-94.

586. MICHAELSON S. Microwave biological
effects : an overview. *Proc. IEEE*,
68, n° 1, 1980, pp. 40-49.

587. MICHAELSON S. Neuroendocrine effects
of exposure to microwave/radiofre-
quency energies. In : Proceedings
of the IMPI Symposium, Toronto, 1981,
Digest pp. 101-103.

588. MICHAELSON S. Human response to
radiofrequency and microwave exposure.
G. Ital. Med. Lavoro, 4, 1982, pp.
7-12.

589. MICHAELSON S., DODGE C. Soviet views
on the biological effects of microwaves - An analysis. *Health Phys.*,
21, n° 7, 1971, pp. 108-111.

590. MICHAELSON S., GUILLET R., CATALLO M.,
SMALL J., INAMINE G., HEGGENESS F.
Influence of 2450 MHz (CW) microwaves
on rats exposed in utero. In : Pro-
ceedings of the IMPI Symposium,
Leuven, 1976. (Paper 5A.3.).

591. MICHAELSON S., GUILLET R., LOTZ W.,
LU S., MAGIN R. Neuroendocrine response in the rat and dog exposed to
2450 MHz (CW) microwaves. In :
HAZZARD D. Biological effects and
measurement of RF/microwaves. Symposium proceedings, 16-18 Feb. 1977,
US Dep. Health, Educ., Welfare, BRH,
FDA-77-8026, July 1977, pp. 263-279.

592. MICHAELSON S., HOUK W., LEBDA N.,
LU S., MAGIN R. Biochemical and
neuroendocrine aspects of exposure
to microwaves. In : TYLER P. Biologic effects of non-ionizing radiation. Proceedings of a conference.
Ann. New York. Acad. Sci., 247,
1975, pp. 21-45.

593. MICHAELSON S., HOWLAND J., DEICHMANN
W. Response of the dog to 24,000 and
1285 MHz microwave exposure. *Ind.
Med. Surg.*, 40, 1971, pp. 18-23.

594. MICHAELSON S., MAGIN S. The ocular
lens and cataract. In : Proceedings
of the IMPI Symposium, Waterloo (Can.),
1975, pp. 111-112. (Paper 7.2.).

595. MICHAELSON S., OSEPCHUK J. Cataracts
following use of microwave oven -
Comments. *New York State J. Med.*,
74, n° 11, 1974, pp. 2034-2035.

596. MICHAELSON S., THOMSON R., EL-TAMAMI
M., SETH H., HOWLAND J. The hematologic effects of microwave exposure.
Aerosp. Med., 35, Sept. 1964, pp.
824-829.

597. MICHAELSON S., THOMSON R., HOWLAND J.
Physiologic aspects of microwave
irradiation of mammals. *Am. J. Physiol.*, n° 201, 1961, pp. 351-356.

598. MICHAELSON S., THOMSON R., HOWLAND J.
Biological effects of microwave
exposure. Final report RADC-TR-67-
461, contract AF 30(602)-2921 cove-
ring the period 1958-1965, ASTIA
doc. n° AD-842-242, 1967, 138 p.

599. MICHAELSON S., THOMSON R., HOWLAND J.
Biological effects of microwave
exposure. In : Radiation Control for
Health and Safety Act of 1967, Hearings before the Committee of Commerce, Senate of the USA, 90th Congress, 2nd Session, S.2067, S.3111,
HR 10790, serial n° 90-49, Washington,
US Government Printing Office, 1968,
pp. 1443-1570.

600. MICHAELSON S., THOMSON R., QUINLAN W.
Effects of electromagnetic radia-
tions on physiological responses.
Aerosp. Med., 37, 1966, pp. 292-302.

601. MICHAELSON S., THOMSON R., QUINLAN W.
Effects of electromagnetic radia-
tions on physiological responses.
Aerosp. Med., 38, 1967, pp. 293-298.

602. MICKEY G. Crossing-over in males
of Drosophila induced by radio-frequency treatment. In : Proceedings
of the 11th international Congress
of Genetics, Den Haag, 1963, pp.
1-72.

603. MICKEY G. Electromagnetism and its
effect on the organism. *New York
State J. Med.*, 63, 1963, pp. 1935-
1942.

604. MIHAILO V., SKRIPAL A., MELNIKOV V.,
KOROTKIJ V., NAIMITEN L. Effets de
petites touches d'énergie électromagnétique sur l'organisme humain.
(En russe). *Dokl. Akad. Nauk SSSR*,
16, n° 12, 1972, p. 1147.

605. MIKOŁAJCZYK H. Podziały mitotyczne
komórek nabłonkowych rogówski oka
u zwierzsat dświadcznalnych podda-
nych działaniu mikrofal. *Med. Pr.*,
21, 1970, p. 15.

606. MIKOŁAJCZYK H. Hormonal responses and changes in endocrine glands induced by microwaves. *Med. Lotnicza,* 39, 1972, p. 39.

607. MIKOŁAJCZYK H. Microwave irradiation and endocrine functions. In : CZERSKI P., OSTROWSKI K., SHORE M., SILVERMAN C., SÜSS M., WALDESKOG B. Biological effects and health hazards of microwave radiation. Proceedings of an international symposium, Warszawa, 15-18 Oct. 1973. Warszawa, Polish Medical Publishers, 1974, pp. 46-51.

608. MIKOŁAJCZYK H. Microwave-induced shifts of gonadotropic activity in the anterior pituitary gland of rats. In : JOHNSON C., SHORE M. Biological effects of electromagnetic waves. Selected papers of the USNC/URSI annual meeting, Boulder, 20-23 Oct. 1975, US Dep. Health, Educ., Welfare, FDA-77-8010. Washington, US Government Printing Office, Vol. 1, 1976, pp. 377-383.

609. MILD K. Occupational exposure to radio-frequency electromagnetic fields. *Proc. IEEE,* 68, 1980, pp. 12-17.

610. MILLER T., BRODWIN M., CEMBER H. An empirical time-intensity relationship for thermal bioelectromagnetic effects. *J. Microw. Power,* 17, n° 3, 1982, pp. 195-202.

611. MILLS W. A program to study the effects of microwave radiation on various biological systems. *J. Microw. Power,* 6, n° 2, 1971, pp. 141-150.

612. MILON G., MARCHAL G. La défense antibactérienne. *La Recherche,* n° 177, mai 1986, pp. 652-661.

613. MILROY W., MICHAELSON S. Biological effects of microwave radiation. *J. Health Phys. Soc.,* 20, 1971, p. 567.

614. MILROY W., MICHAELSON S. Microwave cataractogenesis : a critical review of the literature. *Aerosp. Med.,* 43, 1972, pp. 67-75.

615. MILROY W., MICHAELSON S. Thyroid pathophysiology of microwave radiation. *Aerosp. Med.,* 43, 1972, pp. 1126-1131.

616. MILROY W., O'GRADY T., PRINCE E. Electromagnetic pulse radiation : a potential biological hazard. *J. Microw. Power,* 9, n° 3, 1974, pp. 213-218.

617. MINAJEV V., ŽDANVIČ N., UDALOV J. Effects of SHF fields on enzymatic activities and pyridoxine levels in the organs of white rats. (Trad. du russe). US joint Pub. Res. Serv. Rep., JPRS 66512, 7 Jan. 1976, pp. 77-82.

618. MINECKI L., Działanie pól elektromagnetycznych wielkiej częstotliwości na rozwoj embrionalny. *Med. Pr.,* 15, n° 6, 1964, pp. 391-396.

619. MIRO L., LOUBIERE R., PFISTER A. Recherche des lésions viscérales observées chez des souris et des rats exposés aux ondes ultra-courtes ; étude particulière des effets de ces ondes sur la reproduction. *Rev. Méd. Aéronaut.,* 4, 1965, p. 37.

620. MITCHELL D., SWITZER W., BRONAUGH E. Hyperactivity and disruption of operant behavior in rats after multiple exposures to microwave radiation. *Radio Sci.,* 12, n° 6 (S), 1977.

621. MITTLER S. Failure of 2-and 10-meter radio waves to induce genetic damage in Drosophila melanogaster. *Environ. Res.,* 11, n° 3, 1976, pp. 326-330.

622. MITTLER S. Failure of chronic exposure to non-thermal FM radio waves to mutate Drosophila. *J. Heredity,* 68, n° 4, 1977, pp. 257-265.

623. MOE K., LOVELY R., MYERS D., GUY A. Physiological and behavioral effects of chronic low-level microwave radiation in rats. In : JOHNSON C., SHORE M. Biological effects of electromagnetic waves. Selected papers of the USNC/URSI annual meeting, Boulder, 20-23 Oct. 1975, US Dep. Health, Educ., Welfare, FDA-77-8011. Washington, US Government Printing Office, Vol. 2, 1976, pp. 248-257.

624. MOHLER R., BRUNI C., GANDOLFI A. A system approach to immunology. *Proc. IEEE,* 68, n° 8, 1980, pp. 964-990.

625. MONAHAN J., D'ANDREA J. Behavioral effects of microwave radiation absorption. Proceedings of a workshop, Midway, March 1982, US Dep. Health, Educ., Welfare. Rockville, FDA CDRH, FDA 85-8238.

626. MONAHAN J., HENTON W. Free-operant avoidance and escape from microwave radiation. In : HAZZARD D. Biological effects and measurement of RF/microwaves. Symposium proceedings, 16-18 Feb. 1977, US Dep. Health, Educ., Welfare, BRH, FDA-77-8026, July 1977, pp. 23-33.

627. MONAHAN J., HO H. Microwave-induced avoidance behavior in the mouse. In : JOHNSON C., SHORE M. Biological effects of electromagnetic waves. Selected papers of the USNC/URSI annual meeting. Boulder, 20-23 Oct. 1975, US Dep. Health, Educ., Welfare, FDA-77-8010. Washington, US Government Printing Office, Vol. 1, 1976, pp. 274-283.

628. MONAHAN J., HO H. The effect of ambient temperature on the reduction of microwave energy absorption by mice. *Radio Sci.*, 12, n° 6 (S), 1977, pp. 257-262.

629. MORGAN W. Microwave radiation hazards. *Arch. Ind. Health*, 21, n° 6, 1960, pp. 570-573.

630. MOTZKIN S., MELNICK R., RUBENSTEIN C., ROSENTHAL S., BIRENBAUM L. Effects of CW millimeter wave irradiation on mitochondrial oxidative phosphorylation and Ca++ transport. In : BERTEAUD A., SERVANTIE B. Ondes électromagnétiques et biologie. Symposium international URSI, Jouy-en-Josas, juil. 1980, pp. 109-115.

631. MUC A. Standards, regulations and health effects - infrared, microwave and radiofrequency. In : Radiant Wave Electroheat Worshop, Toronto, 2 Dec. 1985, pp. 63-71.

632. MUMFORD W. Some technical aspects of microwave radiation hazards. Bell Telephone Systems, Technical publication 3865. *Proc. IRE*, 49, Feb. 1961, pp. 427-447.

633. MUMFORD W. Heat stress due to RF radiation. In : Proceedings of the IMPI Symposium, Edmonton, 1969, pp. 59-62. (Paper DA.2.).

634. MUMFORD W. Heat stress due to RF radiation. *Proc. IEEE*, 57, 1969, pp. 171-178.

635. MUMFORD W. Heat stress due to RF radiation. *J. Microw. Power*, 4, n° 4, 1969, pp. 244-253.

636. MUMFORD W. Heat stress due to RF radiation. In : CLEARY S. Biological effects and health implications of microwave radiation. Symposium proceedings, Commonwealth Univ., Richmond, 17-19 Sept. 1969, US Dep. Health, Educ., Welfare, BRH, DBE 70-2, BP 193898, June 1970, pp. 21-31.

637. MURACA G., FERRI E., BUCHTA F. A study of the effects of microwave irradiation of the rat testis. In : JOHNSON C., SHORE M. Biological effects of electromagnetic waves. Selected papers of the USNC/URSI annual meeting, Boulder, 20-23 Oct. 1975, US Dep. Health, Educ., Welfare, FDA-77-8010. Washington, US Government Printing Office, Vol. 1, 1976, pp. 484-494.

638. MURRAY K. Hazard of microwave ovens to transdermal delivery system. *New-Engl. J. Med.*, 310, n° 11, 1984, p. 721.

639. MYERS R., ROSS D. Radiation and brain calcium : a review and critique. *Neurosci. Biobehav. Rev.*, 5, 1981, pp. 503-543.

640. NAGESWARI K., TANDON H., VARMA S., BHATNAGAR V. Effect of chronic microwave radiation on rabbit erythrocytes. *Indian J. Exp. Biol.*, 20, 1982, pp. 13-15.

641. NATARAJAN K., JAGANNATHAN N. Impact of radar irradiation on human systems. *Inst. Electron. Telecommun. Eng. J.*, 22, 1976, pp. 326-329.

642. NEALEIGH R., GARNER R., MORGAN R., CROSS H., LAMBERT P. The effect of microwave on Y-maze learning of the white rat. *J. Microw. Power*, 6, n° 1, 1971, pp. 49-54.

643. NIKOGASJAN S. Vlijanije SVČ na aktivnost holinesterazy v syvorotke krovi i organah u životnyh. In : O Biol. Djejstvii Sverhvysokih Častot, Moskva, 1960, pp. 81-85.

644. NILSSON B., PETTERSSON L. A mechanism for high-frequency electromagnetical field - induced biological damage ? *IEEE Trans. Microw. Theory Tech.*, MTT-27, n° 6, 1979, pp. 616-618.

645. NOVÁK J., ČERNY V. Vliv impulsního elektromagnetického pole na lidský organismus. *ČLČ*, CIIn° 18, 1963, pp. 496-497.

646. NOVICKIJ A., MURAŠOV B., KRASNOBAJEV P., MARKOZOVA N. Utilisation de l'état fonctionnel du système hypothalamus-hypophyse-cortex surrénal comme critère d'établissement des niveaux tolérables d'émission électromagnétique SHF. (En russe). *Vojenn. Med. Ž.*, 8, août 1977, pp. 53-56.

647. O'CONNOR M. Mammalian teratogenesis and radio-frequency fields. *Proc. IEEE*, 68, n° 1, 1980, pp. 56-60.

648. O'CONNOR M. A review of teratogenic studies of radio-frequency electromagnetic fields. In : Proceedings of the IMPI Symposium, Toronto, 1981, Digest pp. 104-105.

649. ODLAND L. Radio-frequency energy : a hazard to workers ? *Ind. Med. Surg.*, 42, 1973, pp. 23-26.

650. OLDENDORF W. Focal neurological lesions produced by microwave irradiations. In : Proceedings of the Society of Experimental Biology and Medicine, 72, 1949, pp. 432-434.

651. OLSEN C. Biological effects of microwaves - future research directions. A special panel discussion. *Microw. Energy Appl. Newsl.*, 1, n° 5, 1968, pp. 9-10.

652. OLSEN R. Insect teratogenesis in a standing-wave irradiation system. *Radio Sci.*, 12, n° 6 (S), 1977, pp. 199-207.

653. OLSEN R. Microwave - induced developmental defects in the common mealworm (Tenebrio molitor). A decade of research. Pensacola, Naval Aerosp. Med. Res. Lab., Naval Air Station, n° 32508, 1981, pp. 1-10.

654. OLSEN R. Constant-dose microwave irradiation of insect pupae. *Radio Sci.*, 17, n° 5 (S), 1982, pp. 145-148.

655. OLSEN R., DURNEY C., LORDS J., JOHNSON C. Low-level microwave interaction with isolated mammalian hearts. In : Proceedings of the IMPI Symposium, Waterloo (Can.), 1975, pp. 76-78. (Paper 5.1.).

656. OLSEN R., HAMMER W. Evidence for microwave - induced acoustical resonances in biological material. *J. Microw. Power*, 16, n° 3-4, 1981, pp. 263-269.

657. OLSEN R., LIN J. Microwave pulse - induced acoustic resonances in spherical head models. *IEEE Trans. Microw. Theory Tech.*, MTT-29, n° 10, 1981, pp. 1114-1117.

658. OLSEN R., LORDS J., DURNEY C. Microwave - induced chronatropic effects in the isolated rat heart. *Ann. Biomed. Eng.*, 5, 1977, pp. 395-409.

659. ONDRAČEK J., RUPES V., ZDAREK J., LINEVA V. Lethal effects of microwaves on strains of Musca domestica, having different resistance to insecticides. *Acta Entomol. Bohemoslov.*, 73, n° 4, 1976, pp. 283-285.

660. ORLOWSKI A. Effets biologiques des ondes électromagnétiques courtes et ultra-courtes. *Usine Nouv.*, n° de printemps, 1969, pp. 579-606.

661. ORTNER M., GALVIN M. The effect of 2450 MHz microwave radiation on histamine secretion by rat peritoneal mast cells. *Cell Biophys.*, 2, 1980, pp. 127-138.

662. ORTNER M., GALVIN M., Mac REE D. Some studies on acute in-vivo exposure of rats to 2450 MHz microwave radiation : mast cells and basophils. *Radiat. Res.*, 86, 1981, pp. 580-588.

663. OSCAR K., GRUENAU S. Regional cerebral blood flow and blood-barrier permeability after low-level microwave energy exposure. Blood-Brain Barrier Workshop, Annapolis, Dep. Navy, 30-31 Oct. 1978.

664. OSCAR K., GRUENAU S., FOLKER M., RAPOPORT S. Local cerebral blood flow after microwave exposure. *Brain Res.*, n° 204, 1981, pp. 220-225.

665. OSCAR K., HAWKINS T. Microwave alterations of the blood-brain barrier system of rats. *Brain Res.*, n° 126, 1977, pp. 281-293.

666. OSEPCHUK J. Biological effects of electromagnetic radiation. New York, IEEE Press, 1983, 593 p.

667. OSIPOV J. La santé des travailleurs exposés au rayonnement radio-fréquence en hygiène du travail et les effets des champs électromagnétiques radio-fréquence sur les travailleurs. (En russe). Leningrad, Izdatel'stvo Medicina, 1965, pp. 104-144.

668. OSWALD R., Mac LEAN F., SCHALLHORN D., BUXTON L. One-dimensional thermoelastic response of solids to pulsed energy deposition. *J. Appl. Phys.*, 42, 1971, pp. 3463-3473.

669. PALLADIN A., SPASKAJA F. K voprosu vlijanija UVF poljej na specifičeskije funkcii ženščin rabotajuščih s generatory UVF. *Akust. Ginekol.*, 4, 1962, pp. 69-74.

670. PARKER L. Changes in sympathetic nervous system activity in rats following whole-body exposure to low-intensity microwave radiation at 2450 MHz. In : Proceedings of the IMPI Symposium, Monterey, 1971. (Paper 7.4.).

671. PARKER L. Thyroid suppression and adrenomedullary activation by low intensity microwave radiation. *Am. J. Physiol.*, n° 224, 1973, pp. 1388-1390.

672. PATTISHALL E., BANGHART F. Proceedings of the 2nd annual Tri-Service conference on biological effects of microwave energy, ASTIA n° AD-131-477, 1958.

673. PAULSSON L., HAMNERIUS Y., HANSSON H., SJOSTRAND J. Retinal damage experimentally induced by microwave radiation. *Acta Ophtalmol. Kbh. and Suppl.*, 57, 1979, pp. 183-197.

674. PAY T., ANDERSEN F., JESSUP G. A comparative study of the effects of microwave radiation and conventional heating on the reproductive capacity of Drosophila melanogaster. *Radiat. Res.*, 76, 1978, pp. 271-282.

675. PAY T., BEYER E., REICHELDERFER C. Microwave effects on reproductive capacity and genetic transmission in Drosophila melanogaster. *J. Microw. Power*, 7, n° 2, 1972, pp. 75-82.

676. PAZDEROVA-VEJLUPKOVA J., FRANK Z. Influence of pulsed microwaves on haematopoiesis of adolescent rats. In : Proceedings of the IMPI Symposium, Leuven, 1976. (Paper 2A.1.).

677. PAZDEROVA-VEJLUPKOVA J., JOSIFKO M. Changes in the blood count of growing rats irradiated with a microwave pulse field. *Arch. Environ. Health*, Jan.-Feb. 1979, pp. 44-50.

678. PETERSON D., PARTLOW L., GANDHI O. An investigation of thermal and athermal effects of microwave radiation on erythrocytes. *IEEE Trans. Biomed. Eng.*, BME-26, n° 7, 1979, pp. 428-436.

679. PETROV I., SYNGAJEVSKAJA V. Endocrine glands. In : PETROV I. Influence of microwave radiation on the organism of man and animals. (En russe). Izdatel'stvo Medicina, Leningrad, 1970. Traduction anglaise NASA-TT-F-708, National Technical Information Service, Springfield, 1970, ch. 2, pp. 31-41.

680. PEYTON M. Proceedings of the 4th annual Tri-Service Conference on biological effects of microwave-radiating equipments, biological effects of microwave radiations. New York, Plenum Press, 2 vol., RADC-TR-60-180.

681. PHILLIPS R., HUNT E., CASTRO R., KING N. Thermoregulatory, metabolic, and cardiovascular response of rats to microwaves. *J. Appl. Physiol.*, 38, n° 4, 1975, pp. 630-635.

682. PHILLIPS R., HUNT E., KING N. Vital functions of rats after exposure to 2450 MHz microwaves. In : Proceedings of the IMPI Symposium, Ottawa, 1972, p. 118. (Paper 8.1.).

683. PHILLIPS R., KING N., HUNT E. Thermoregulatory, cardiovascular and metabolic responses of rats to single and repeated exposure to 2450 MHz microwaves. In : Proceedings of the IMPI Symposium, Loughborough, 1973, pp. 1-4. (Paper 3B.5.).

684. PIANA M., HELLUMS S., WILSON W. Effects of microwave irradiation on human-blood platelets. *IEEE Trans. Biomed. Eng.*, BME-28, n° 9, 1981, pp. 661-664.

685. PICKARD W., OLSEN R. Developmental effects of microwaves in Tenebrio molitor : experiments to detect possible influences of radiation frequency and of culturing protocols. In : JOHNSON C., SHORE M. Biological effects of electromagnetic waves. Selected papers of the USNC/URSI annual meeting, Boulder, 20-23 Oct. 1975, US Dep. Health, Educ., Welfare, FDA-77-8010. Washington, US Government Printing Office, 1976, p. 66.

686. PICKARD W., OLSEN R. Development effects of microwaves in Tenebrio molitor : experiments to detect possible influences of radiation frequency and of culturing protocols. In : Biological effects of electromagnetics waves, international symposium, Airlie, 30 Oct.-4 Nov. 1977, Abstracts of scientific papers.

687. PICKARD W., OLSEN R. Developmental effects of microwaves on Tenebrio : influence of culturing protocol and of carrier frequency. *Radio Sci.*, 14, n° 6 (S), 1979, pp. 181-185.

688. PLURIEN G., SENTENAC-ROUMANOU H., JOLY R., DOUET J. Influence du rayonnement électromagnétique d'un émetteur radar sur la fonction phagocytaire des cellules du système réticulo-endothélial au contact du sang chez la souris. *C.R. Séance Soc. Biol.*, 160, 1966, pp. 597-602.

689. POBZITKOV V., TJAGIN N., GREBESEČMIKOVA A. Influence des champs SHF en impulsions sur la conception et le cours de la gestation chez la souris blanche. (En russe). *Bjul. Eksp. Biol. Med.*, 51, 1961, p. 105.

690. PORTELA A. Transient effects of low level microwave radiation on bioelectric muscle cell properties and on water permeability and its destruction. In : GUY A. Fundamental and applied aspects of non-ionizing radiation. New York, Plenum Press, 1975, pp. 215-231.

691. PRAUSNITZ S., SÜSSKIND C. Effects of chronic microwave radiation on mice. *IRE Trans. Biomed. Eng.*, BME-4, 1962, pp. 104-108.

692. PREDMERSKY T., BALLAY L., BÖLÖNI E., SZÁBO L., VAMOS L. Investigations on microwave radiation exposure. *Acta Physiol. Acad. Sci. Hung.*, 52, 1982, pp. 479-486.

693. PRESMAN A. Biologičeskoje djejstvije mikrovoln. *Usp. Sovrem. Biol.*, 51, n° 1, 1961, pp. 82-103.

694. PRESMAN A. Le rôle des champs électromagnétiques dans les processus vitaux. (En russe). *Biofiz.*, 9, 1964, pp. 131-134.

695. PRESMAN A. The action of microwaves on living organisms and biological structures. (En russe). *Usp. Fiz. Nauk*, 86, 1965, pp. 262-302. Traduction anglaise in *Soviet Phys. Uspekhi*, 8, 1966, pp. 463-488.

696. PRESMAN A. Electromagnetic fields and life. Moskva, Izdatel'stvo Nauka, 1970. Traduction anglaise, New York, Plenum Press, 1970, 336 p.

697. PRESMAN A., LEVITINA N. Action non thermique des micro-ondes sur le rythme des contractions cardiaques chez l'animal. 1ère partie : étude des effets des micro-ondes entretenues. (En russe). *Bjul. Eksp. Biol. Med.*, 53, 1962, pp. 36-39.

698. PRESMAN A., LEVITINA N. Action non thermique des micro-ondes sur le rythme des contractions cardiaques chez l'animal. 2ème partie : recherche de l'action des micro-ondes en impulsions. (En russe). *Bjul. Eksp. Biol. Med.*, 53, 1962, pp. 154-157.

699. PRESTON E. Failure of hyperthermia to open rat blood-brain barrier : reduced permeation of sucrose. *Acta Neuropathol.*, 57, 1982, pp. 255-262.

700. PRESTON E., PREFONTAINE G. Cerebrovascular permeability to sucrose in the rat exposed to 2450-MHz microwaves. *Appl. Physiol. Respir. Environ. Exercise Physiol.*, 49, 1980, pp. 218-223.

701. PRESTON E., VAVASOUR E., ASSENHEIM H. Permeability of the blood-brain barrier to mannitol in the rat following 2450-MHz microwave irradiation. *Brain Res.*, n° 174, 1979, pp. 109-118.

702. PRIOU A. Ensemble des travaux. Thèse de Doctorat d'Etat. Toulouse, Univ. Paul Sabatier, 1981, 699 p., chap. III, § IV : Effets biologiques et approches des mécanismes. pp. 609-664.

703. PROMTOVA T. Vlijanije njepreryvnogo električeskogo polja UVČ navysšuju nervnuju djejatjel'nost sobak v norme i patologii. *Ž. Vysš. Nerv. Djejatel'.*, 6, n° 6, 1956, pp. 846-854.

704. PYLE S., NICHOLS D., BARNES S., GAMOW E. Threshold effects of microwave radiation on embryo cell systems. In : TYLER P. Biologic effects of non-ionizing radiation. Proceedings of a conference. *Ann. New York Acad. Sci.*, 247, 1975, pp. 401-407.

705. RAGEN H., PHILLIPS R., BUSCHBOM R., BUSCH R., MORRIS J. Hematologic and immunologic effect of pulsed microwaves in mice. *Bioelectromagn.*, 4, 1983, pp. 383-396.

706. RAI P., BALL H., NELSON S., STETSON L. Morphological changes in adult Tenebrio molitor (Coleoptera : Tenebrionidae) resulting from radiofrequency or heat treatment of larvae or pupae. *Ann. Entomol. Soc. Am.*, 64, n° 5, 1971, pp. 1116-1121.

707. RAI P., BALL H., NELSON S., STETSON L. Cytopathological effects of radiofrequency electric fields on reproductive tissue of adult Tenebrio molitor (Coleoptera : Tenebrionidae). *Ann. Entomol. Soc. Am.*, 67, 1974, pp. 687-690.

708. RAI P., BALL H., NELSON S., STETSON L. Effects of radiofrequency electrical treatment on fecundity of Tenebrio molitor L. (Coleoptera : Tenebrionidae). *Ann. Entomol. Soc. Am.*, 68, 1975, pp. 542-544.

709. RAMA-RAO G., CAIN C., LOCKWOOD J., TOMPKINS W. Effects of microwave exposure on the hamster immune system. III : Peritoneal macrophage function. *Bioelectromagn.*, 4, 1983, pp. 141-145.

710. RAPOPORT S., OHNO K., FREDERICKS W., PETTIGREW K. Regional cerebrovascular permeability to ^{14}C sucrose after osmotic opening of the blood-brain barrier. *Brain Res.*, n° 150, 1978, pp. 653-657.

540 Part IV - Biological effects and medical applications

711. RAPOPORT S., OHNO K., FREDERICKS W., PETTIGREW K. A quantitative method for measuring altered cerebrovascular permeability. *Radio Sci.*, 14, n° 6 (S), 1979, pp. 345-349.

712. REED J., LORDS J., DURNEY C. Microwave irradiation of the isolated rat heart after treatment with ANS blocking agents. *Radio Sci.*, 12, n° 6 (S), 1977, pp. 161-165.

713. REPACHOLI M. Differentiation between biological effects and health hazards - scaling from animals. Advances in biological effects and dosimetry of low-energy electromagnetic fields. NATO courses, E. Majorana School of Advanced Study, Sicilia April 1981.

714. RIAZI A., HILL D., HAGMANN M., GANDHI O., D'ANDREA J. A broadband temperature-controlled system for the study of cellular bioeffects of microwaves. *IEEE Trans. Microw. Theory Tech.*, MTT-30, 1982, pp. 1996-1998.

715. RICHARDSON A., DUANE T., HINES H. Experimental lenticular opacities produced by microwave irradiation. *Arch. Phys. Med.*, 29, Dec. 1948, pp. 765-769.

716. RICHARDSON A., LOMAX D., NICHOLS J., GREEN H. The role of energy, pupillary diameter and alloxan diabetes in the production of ocular damage by microwave radiation. *Am. J. Ophtamol.*, 35, 1952, p. 993.

717. RISSMAN W., CAIN C. Microwave hearing in mammals. In : Proceedings of the National Electrical Conference, 30, 1975, pp. 239-244.

718. ROBERTI B., HEEBELS G., HENDRICX J., DE GREEF A., WOLTHUIS O. Preliminary investigations of the effect of low-level microwave radiation on spontaneous motor activity in rats. In : TYLER P. Biologic effects of non-ionizing radiation. Proceedings of a conference. *Ann. New York Acad. Sci.*, 247, 1975, p. 417.

719. ROBERTS N. Radiofrequency and microwave effects on immunological and hematopoietic systems. In : Biological effects and dosimetry of non-ionizing radiation. New York, Plenum Press, 1983, pp. 429-460.

720. ROBERTS N., LU S., MICHAELSON S. Human leukocyte function and the U.S.. safety standard for exposure to radio-frequency radiation. *Science*, 220-4594, 1983, pp. 318-320.

721. ROBERTS N., MICHAELSON S., LU S. Exposure of human mononuclear leukocytes or microwave energy pulse-modulated at 16 or 60 Hz. *IEEE Trans. Microw. Theory Tech.*, MTT-32, n° 8, 1984, pp. 803-807.

722. ROBINETTE C., SILVERMAN C. Causes of death following occupational exposure to microwave radiation (radar), 1954-1974. In : HAZZARD D. Biological effects and measurement of RF/microwaves. Symposium proceedings, 16-18 Feb. 1977, US Dep. Health, Educ., Welfare, BRH, FDA-77-8026, July 1977, pp. 338-344.

723. ROBINSON J. HARRISON G., Mac CULLOCH D., Mac CREADY W., CHEUNG A. The effects of microwaves on cell survival at elevated temperature. *Radiat. Res.*, 88, 1981, pp. 542-551.

724. ROMERO-SIERRA C. Bioeffects of electromagnetic waves. In : BOWHILL S. Review of radio science, 1978-1980. Bruxelles, URSI, 1981, ch. 2, pp. A.B.1-A.B.17.

725. ROMERO-SIERRA C. Bioeffects of electromagnetic waves. In : BOWHILL S. Review of radio science, 1981-1983. Bruxelles, URSI, 1984, ch. 9, pp. K.1-K.17.

726. ROMERO-SIERRA C., HALTER S., TANNER J. Effect of an electromagnetic field on the sciatic nerve of the rat. In : WULFSOHN N., SANCES A. Nervous system and electrical currents. Vol. 2. New York, Plenum Press, 1971.

727. ROMERO-SIERRA C., TANNER J. Biological effects of non-ionizing radiation : outline of fundamental laws. In : TYLER P. Biologic effects of non-ionizing radiation. Proceedings of a conference. *Ann. New York Acad. Sci.*, 247, 1975, pp. 263-272.

728. ROSENTHAL S..More research on biological effects of microwaves needed. *J. Microw. Power*, 6, n° 1, 1971, p. 55.

729. ROSENTHAL S. Effects of 35 and 107 GHz CW microwaves on the rabbit eye. In : JOHNSON C., SHORE M. Biological effects of electromagnetic waves. Selected papers of the USNC/URSI annual meeting, Boulder, 20-23 Oct. 1975, US Dep. Health, Educ., Welfare, FDA-77-8010, Washington, US Government Printing Office, Vol. 1, 1976, pp. 110-128.

730. ROSENTHAL D., BEERING S. Hypogonadism after microwave radiation. *J. Am. Med. Assoc.*, 205, n° 4, 1968, pp. 105-108.

731. ROSZKOWSKI W., ROSZKOWSKI K., WREMBEL K., SZMIGIELSKI S., PULVERER G., JELJASZEWICZ J. Effect of whole-body hyperthermia on delayed cutaneous hypersensitivity to oxazolone in mice. *J. Clin. Lab. Immunol.*, 6, 1981, pp. 257-261.

732. ROTH E. Microwave radiation. Compendium of human responses to the Aerospace Environment. Vol. I. NASA CR-1205, n° 1, 1968, Washington, 1968, 26 p.

733. ROTHMEIER J. Effect of microwave radiation on the frog sciatic nerve, In : Nervous system and electric currents. New York, Plenum Press, Vol. 1, 1970, pp. 57-69.

734. ROTKOVSKÁ D., VACEK A. Effect of high frequency electromagnetic field upon hemopoietic stem cells in mice. *Folia Biol.*, 18, 1972, pp. 292-297.

735. ROTKOVSKÁ D., VACEK A. The effect of electromagnetic radiation on the hematopoietic stem cells of mice. In : TYLER P. Biologic effects of non-ionizing radiation. Proceedings of a conference. *Ann. New York. Acad. Sci.*, 247, 1975, pp. 243-250.

736. ROUGEON F. La diversité des anticorps. *La Recherche*, n° 177, mai 1986, pp . 680-689.

737. RUBIN A., ERDMAN W. Microwave exposure of the human female pelvis during early pregnancy and prior to conception. *Am. J. Phys. Med.*, 38, 1959, pp. 219-220.

738. RUGH R. The relation of sex, age, and weight of mice to microwave radiation sensitivity. *J. Microw. Power*, 11, n° 2, 1976, pp. 127-132.

739. RUGH R., GINNS E., HO H., LEACH W. Are microwaves teratogenic ? In : CZERSKI P., OSTROWSKI K., SHORE M., SILVERMAN C., SÜSS M., WALDESKOG B. Biological effects and health hazards of microwave radiation. Proceedings of an international symposium, Warszawa, 15-18 Oct. 1973. Warszawa, Polish Medical Publishers, 1974, pp. 98-107.

740. RUGH R., GINNS E., HO H., LEACH W. Responses of the mouse to microwave radiation during estrous cycle and pregnancy. *Radiat. Res.*, 62, 1975, pp. 225-241.

741. RUGH R., HO H., Mac MANAWAY M. The relation of dose rate of microwave radiation to the time of death and total absorbed dose in the mouse. *J. Microw. Power*, 11, n° 3, 1976, pp. 279-281.

742. RUGH R., Mac MANAWAY M. Are mouse fetuses uniformly sensitive to microwave radiation ? *Teratol.*, 13, 1977, Abstract, pp. 34A-35A.

743. RUPP R., MONTET J., FRAZER J. A comparison of thermal and radiofrequency exposure effects on trace metal content of blood plasma and liver cell fractions of rodents. In : TYLER P. Biologic effects of non-ionizing radiation. Proceedings of a conference. *Ann. New York Acad. Sci.*, 247, 1975, pp. 282-291.

744. RUSJAJEV V., KUKSINSK V. Etude de l'effet d'un champ électromagnétique sur les propriétés coagulatives et fibrinolytiques du sang. (En russe). *Biofiz.*, 18, n° 1, 1973, pp. 160-163.

745. SADČIKOVA M. Sostojanije njervnoj sistemy pri vozdjejstvii poljej vysokih i sverhvysokih častot. In : Fiz. Faktory vnešnjej Sredy, Moskva, 1960, pp. 177-183.

746. SADČIKOVA M. Clinical manifestations of reactions to microwave irradiation in various occupational groups. In : CZERSKI P., OSTROWSKI K., SHORE M., SILVERMAN C., SÜSS M., WALDESKOG B. Biological effects and health hazards of microwave radiation. Proceedings of an international symposium, Warszawa, 15-18 Oct. 1973. Warszawa, Polish Medical Publishers, 1974, pp. 261-267.

747. SALISBURY W., CLARK J., HINES H. Exposure to microwave. *Electron.*, 22, n° 5, 1949, pp. 66-67.

748. SAMARAS G., MUROFF L., ANDERSON G. Prolongation of life during high-intensity microwave exposures. *IEEE Trans. Microw. Theory Tech.*, MTT-19, n° 2, 1971, pp. 245-247.

749. ŠANDALA M., VINOGRADOV G. Immunological effects of microwave action. (Trad. du russe), US joint Pub. Res. Serv. Rep., JPRS 72956, 9 March 1979, pp. 16-21.

750. SANDERS A., SCHAEFFER D., JOINES W. Microwave effects on energy metabolism of rat brain. *Bioelectromagn.*, 1, 1980, pp. 171-181.

751. SANTINI R. Effect of low-level microwave irradiation on the duodenal electrical activity of the unanesthetized rat. *J. Microw. Power*, 17, n° 4, 1982, pp. 329-334.

752. SANTINI R., DESCHAUX P., JACOMINO J., PELLISSIER J. Effects of low-level microwave radiation on the digestive transit of the rat. *J. Microw. Power*, 14, n° 3, 1979, pp. 287-289.

753. SAPARETO S., LI G., WHITER K., HAHN G., VAGUINE V., GIEBELER R., TANABE E. Microwave cytotoxicity : lack of in-vitro evidence for non-thermal effects at high power levels. *Radiat. Res.*, 89, 1982, pp. 124-133.

754. SAWICKI W., OSTROWSKI K. Non-thermal effect of microwave radiation of cells in vitro on peritoneal mast cells of the rats. *Am. J. Phys. Med.*, 47, 1968, p. 255.

755. SCHLAGEL C., AHMED A. Evidence for genetic control of microwave-induced augmentation of complement receptor-bearing B lymphocytes. *J. Immunol.*, 129, n° 4, 1982, pp. 1530-1533.

756. SCHLAGEL C., FOLKS T., WOODY J., LEACH W., SULEK K. Enhancement of natural killer cell activity in mice after exposure to microwave radiation. Meeting abstract. *Fed. Proc.*, 38, n° 2, 1979, part 2, p. 915.

757. SCHLAGEL C., SULEK K., AHMED A., HO H., LEACH W. Kinetics and mechanisms of the induction of an increase in complement receptor positive (CR+) mouse spleen cells following a single exposure to 2450-MHz microwaves. In : Bioelectromagnetics Symposium, Seattle, 18-22 June 1979, abstracts.

758. SCHLAGEL C., SULEK K., HO H., LEACH S., AHMED A., WOODY J. Biologic effects of microwave exposure. II : studies on the mechanisms controlling susceptibility to microwave-induced increases in complement receptor-positive spleen cells. *Bioelectromagn.*, 1, 1980, pp. 405-414.

759. SCHMECK H. Seeking chinks in the blood-brain barrier. *Int. Herald Tribune*, 8 Jan. 1987, p. 7.

760. SCHROT J., THOMAS J., BANVARD R. Modification of the repeated acquisition of response sequences in rats by low-level microwave exposure. *Bioelectromagn.*, 1, 1980, pp. 89-99.

761. SCHUY S., LEITGEB N. Gutachten über Schädliche Wirkungen von Mikrowellen. *Ernähr.*, 9, n° 10, 1985, pp. 710-714.

762. SCHWAN H. Microwave biophysics. In : OKRESS E. Microwave power engineering. New York, Academic Press, Vol. 2, 1968, pp. 213-244.

763. SCHWAN H. Interaction of microwave and radiofrequency radiation with biological systems. *IEEE Trans. Microw. Theory Tech.*, MTT-19, n° 2, 1971, pp. 146-152.

764. SCHWAN H. Keynote address and C.C. Johnson Memorial lecture. Microwave bioeffects research : historical perspectives on productive approaches. *J. Microw. Power*, 14, n° 1, 1979, pp. 1-5.

765. SCHWAN H. Microwave and RF hazard standard considerations. *J. Microw. Power*, 17, n° 1, 1982, pp. 1-9.

766. SCHWAN H. Nonthermal cellular effects of electromagnetic fields : AC-field induced ponderomotoric forces. *Br. J. Cancer*, 45, Suppl. V, 1982, pp. 220-224.

767. SCHWAN H., LI K. Hazards due to total body irradiation by radar. *Proc. IRE*, 44, Nov. 1956, pp. 1572-1581.

768. SCHWARTZ J., PHILOGENE B., STEWART J., MEALING G., DUVAL F. Chronic exposure of the tobacco hornworm to pulsed microwaves. Effects on development. *J. Microw. Power*, 20, n° 2, 1985, pp. 85-93.

769. SCHWARTZ J., STEWART J., PHILOGENE B., Microwave effects on insects - a review and some new experimental results. In : Proceedings of the IMPI Symposium, Toronto, 1981, pp. 106-108.

770. SEAMAN R., AYER R., DE HAAN R. Changes in cardiac-cell membrane noise during microwave exposure. In : IEEE Symposium on Microwave Theory and Techniques, Dallas, 1982, Digest, V-1, pp. 436-437.

771. SEAMAN R., WACHTEL H. Slow and rapid responses to CW and pulsed microwave radiation by individual Aplysia pacemakers. *J. Microw. Power*, 13, n° 1, 1978, pp. 77-86.

772. SEARLE G., DAHLEN R., IMIG C., WUNDER C., THOMSON J., THOMAS J., MORESSI W. Effect of 2450 Mc microwaves in dogs, rats and larvae of the common fruit fly. In : PEYTON M. Proceedings of the 4th annual Tri-Service conference on biological effects of microwave-radiating equipments, biological effects of microwave radiations. New York, Plenum Press, 1961, RADC-TR-60-180, pp. 187-199.

773. ŠERCL M. Zur Wirkung der elektromagnetischen Zentimeterwellen auf
das Nervensystem des Menschen (radar). *Z. Gesamte Hyg. Grenzgeb.*, 7,
n° 12, 1961, pp. 897-907.

774. SERKJUK A. State of the cardiovascular system under the chronic
effect of low-intensity electromagnetic fields. (Trad. du russe).
US joint Pub. Res. Serv. Rep., JPRS
L/5615. 10 Feb. 1976, pp. 8-12.

775. SERVANTIE B., CRETON B., BRUSCHERA D.,
ROUSSEL J. Ralentissement de la fréquence cardiaque après exposition
à un champ de micro-ondes chez le rat
blanc. In : BERTEAUD A., SERVANTIE
B. Ondes électromagnétiques et biologie. Symposium international URSI,
Jouy-en-Josas, juil. 1980, pp. 67-69.

776. SERVANTIE B., GILLARD J., SERVANTIE A.,
OBRENOVITCH J., BERTHARION G.,
PERRIN J., CRETON B. Comparative
study of the action of three types
of microwave fields upon the behavior of the white rat. In : Proceed-
dings of the IMPI Symposium, Leuven,
1976. (Paper 3A.1.).

777. SERVANTIE B., JOLY R., BERTHARION G.
Experimental study of the biological effects of microwave radiation
on the white rat and mouse. *J.
Microw. Power*, 6, n° 1, 1971, pp.
59-61.

778. SERVANTIE B., SERVANTIE A., ETIENNE J.
Synchronization of cortical neurons
by a pulsed microwave field as evi-
denced by spectral analysis of elec-
trocorticograms from the white rat.
In : TYLER P. Biologic effects of
non-ionizing radiation. Proceedings
of a conference. *Ann. New York Acad.
Sci.*, 247, 1975, pp. 82-86.

779. SERVANTIE B., TYLER P. Les champs
électromagnétiques non ionisants
facteurs d'environnement en milieu
militaire. Comptes-rendus du stage
ADERA, Univ. Bordeaux II, 15-17 Nov.
1975.

780. SESSIONS G. Effects of chlordiazepoxide, HCl and 2450-MHz pulse-modulated microwave radiation in waveguide
on fixed- and variable-interval bar-
press responding in rats. In : 3rd
annual Bioelectromagnetics Society
meeting, Washington, 9-12 August
1981.

781. SHACKLETT D., TREDICI T., EPSTEIN D.
Evaluation of possible microwave
induced lens changes in the U.S. Air
Force. *Aviat. Sp. Environ. Med.*, 46,
1975, pp. 1403-1406.

782. SHAPIRO A., LUTOMIRSKI R., YURA H.
Induced fields and heating within
a cranial structure irradiated by
an electromagnetic plane wave. *IEEE
Trans. Microw. Theory Tech.*, MTT-19,
n° 2, 1971, pp. 187-196.

783. SHARP J., GROVE H., GANDHI O. Generation of acoustic signals by pulsed
microwave energy. *IEEE Trans. Microw.
Theory Tech.*, MTT-22, n° 5, 1974,
pp. 583-584.

784. SHARP J., PAPARIELLO C. The effects
of microwave exposure on thymidine-
3H uptake in albino rats. *Radiat.
Res.*, 45, n° 2, 1971, pp. 434-439.

785. SHELTON W., MERRITT J. In-vitro
study of microwave effects on calcium
efflux in rat brain tissue. *Bioelectromagn.*, 2, 1981, pp. 161-167.

786. SHORE M., FELTON R., LAMANNA A. The
effect of repetitive prenatal low-
level microwave exposure on development in the rat. In : HAZZARD D.
Biological effects and measurement
of RF/microwaves. Symposium proceedings, 16-18 Feb. 1977, US Dep.
Health, Educ., Welfare, BRH, FDA-
77-8026, July 1977, pp. 280-289.

787. SIEKIERZYNSKI M. A study of the
health status of microwave workers.
In : CZERSKI P., OSTROWSKI K., SHORE
M., SILVERMAN C., SÜSS M., WALDESKOG
B. Biological effects and health
hazards of microwave radiation. Proceedings of an international symposium, Warszawa, 15-18 Oct. 1973.
Warszawa, Polish Medical Publishers,
1974, pp. 273-280.

788. SIEKIERZYNSKI M., CZARNECKI C.,
DZIUK E., JEDRZEJCZAK W., SZADY J.
Microwave radiation and other
harmful factors of working environment in radiolocation: method of
determination of microwave effects.
In : Proceedings of the IMPI Symposium, Leuven, 1976. (Paper 2A.7.).

789. SIEKIERZYNSKI M., CZERSKI P.,
GIDYNSKI A. Health surveillance of
personnel occupationally exposed to
microwaves. I. Theoretical considerations and practical aspects.
Aerosp. Med., 45, n° 10, 1974, pp.
1137-1142.

790. SIEKIERZYNSKI M., CZERSKI P.,
GIDYNSKI A., ZYDECKI S., CZARNEK C.
Health surveillance of personnel
occupationally exposed to microwaves.
III. Lens tranlucency. *Aerosp. Med.*,
45, n° 10, 1974, pp. 1146-1148

791. SIEKIERZYNSKI M., CZERSKI P.,
MILCZARE H., GIDYNSKI A., CZARNEK C.
Health surveillance of personnel
occupationally exposed to microwaves. II. Functional disturbances.
Aerosp. Med., 45, n° 10, 1974, pp.
1143-1145.

792. SILVERMAN C. Nervous and behavioral
effects of microwave radiation in
humans. *J. Epidemiol.*, 97, 1973,
pp. 219-224.

793. SILVERMAN C. Epidemiologic approach
to the study of microwave effects.
In : Symposium on health aspects
of nonionizing radiation, New York
Academy of Science, 9-10 April 1979,
pp. 1166-1181.

794. SILVERMAN C. Epidemiologic studies
of microwave effects. *Proc. IEEE*,
68, n° 1, 1980, pp. 78-84.

795. SILVERMAN C. Epidemiology of microwave radiation effects in humans.
In : CASTELLANI A. Epidemiology and
quantitation of environmental risks
in humans from radiation and other
agents. New York, Plenum Press,
1985, pp. 433-458.

796. SIMKOVIČ I., ŠILJAJEV V. Cataractes
des deux yeux survenues à la suite
d'expositions courtes et répétées
à un champ électromagnétique de
forte densité. (En russe). *Vjest.
Oftal'mol.*, 72, 1959, pp. 12-16.

797. SMIALOWICZ R. The effects of microwaves (2450 MHz) on lymphocyte blast
transformation in vitro. In :
JOHNSON C., SHORE M. Biological
effects of electromagnetic waves,
Selected papers of the USNC/URSI
annual meeting, Boulder, 20-23 Oct.
1975, US Dep. Health, Educ., Welfare,
FDA-77-8010. Washington, US Government Printing Office, Vol. 1, 1976,
pp. 472-483.

798. SMIALOWICZ R. Hematologic and immunologic effects of non-ionizing
electromagnetic radiation. *Bull.
New York Acad. Med.*, 55, n° 11, 1979,
pp. 1094-1118.

799. SMIALOWICZ R. The effects of radio-
frequency radiation on the immune
system : a review. In : IEEE Symposium on Microwave Theory and Techniques, Dallas, 1982, Digest pp.
140-143.

800. SMIALOWICZ R., ALI J., BERMAN E.,
BURSIAN S., KINN J., LIDDLE C.,
REITER L., WEIL C. Chronic exposure
of rats to 100-MHz (CW) radiofrequency radiation : assessment of biological effects. *Radiat. Res.*, 86,
1981, pp. 488-505.

801. SMIALOWICZ R., BRUGNOLOTTI P.,
RIDDLE M. Complement receptor positive spleen cells in microwave
(2450 MHz)-irradiated mice. *J.
Microw. Power*, 16, n° 1, 1981, pp.
73-77.

802. SMIALOWICZ R., COMPTON K., RIDDLE M.,
ROGERS R., BRUGNOLOTTI P. Microwave
radiation (2450 MHz) alters the en-
dotoxin-induced hypothermic response
of rats. *Bioelectromagn.*, n° 1, 1980,
pp. 353-361.

803. SMIALOWICZ R., KINN J., ELDER J.
Perinatal exposure of rats to 2450
MHz CW microwave radiation : effects
on lymphocytes. *Radio Sci.*, 14,
n° 6 (S), 1979, pp. 147-153.

804. SMIALOWICZ R., KINN J., WEIL C.,
WARD T. Chronic exposure of rats at
425 or 2450 MHz : effects on lymphocytes. In : Biological effects
of electromagnetic waves. International symposium, Airlie, 30 Oct.-
4 Nov. 1977, Abstracts of scientific papers.

805. SMIALOWICZ R., RIDDLE M., BRUGNOLOT-
TI P., ROGERS R., COMPTON K. Detection of microwave heating in 5-hy-
droxytryptamine-induced hypothermic
mice. *Radiat. Res.*, 88, 1981, pp.
108-117.

806. SMIALOWICZ R., RIDDLE M., BRUGNOLOT-
TI P., SPERRAZZA J., KINN J. Evaluation of lymphocyte function in
mice exposed to 2450-MHz (CW) microwaves. In : STUCHLY S. Electromagnetic fields in biological systems.
Proceedings of a symposium, Ottawa,
1978, Abstracts of scientific papers,
pp. 122-152.

807. SMIALOWICZ R., RIDDLE M., ROGERS R.,
STOTT G. Assessment of immune function development in mice irradiated
in utero with 2450-MHz microwaves.
J. Microw. Power, 17, n° 2, 1982,
pp. 121-126.

808. SMIALOWICZ R., RIDDLE M., WEIL C.,
BRUGNOLOTTI P., KINN J. Assessment
of the immune responsiveness of
mice irradiated with continuous-
wave or pulse-modulated 425 MHz
radiofrequency radiation. *Bioelectromagn.*, 3, 1982, pp. 467-470.

809. SMIALOWICZ R., ROGERS R., GARNER M.,
RIDDLE M., LUEBKE R., ROWE D. Microwaves (2450 MHz) suppress murine
natural killer cell activity. *Bioelectromagn.*, 4, 1983, pp. 371-381.

810. SMIALOWICZ R., WEIL C., KINN J., ELDER J. Exposure of rats to 425-MHz (CW) radiofrequency radiation : effects on lymphocytes. *J. Microw. Power*, 17, n° 3, 1982, pp. 211-221.

811. SMIRNOVA M., SADČIKOVA M. Determination of the functional activity of the thyroid by means of radioactive iodine in workers with UHF generators. In LETAVET A., GORDON Z. The biological action of ultrahigh frequencies. Moskva, Akad. Med. Nauk. (Traduit du russe). in US joint Pub. Res. Serv., JPRS 12471, 1960, pp. 47-49.

812. SMOLJANSKAJA A., VILENSKAJA R. Effets du rayonnement électromagnétique de longueur d'onde millimétrique sur l'activité fonctionnelle des éléments génétiques des cellules bactériennes. (En russe). *Usp. Fiz. Nauk*, 110, n° 3, 1973, pp. 458-460.

813. SMOLJANSKAJA A., VILENSKAJA R. Effects of millimetric band electromagnetic radiation on the functional activity of certain genetic elements of bacterial cells. *Soviet Phys. Uspekhi*, 16, n° 4, 1974, pp. 571-572.

814. SNYDER S. The effect of microwave irradiation on the turnover rate of serotonin and norepinephrine and the effect on monoamine metabolizing enzymes. Final Report, Johns Hopkins Univ., Baltimore, 1971 (AD 729161).

815. SOKOLOV V., IVANOVA L., GORIZONTOVA M., NIKONOVA K., SADČIKOVA M. Modifications cytochimiques et cytogénétiques dans le sang des travailleurs exposés aux micro-ondes. (En russe). *Gig. Tr. Prof. Zabol.*, n° 10, 1983, pp. 5-9.

816. SOMMER H., Von GIERKE H. Hearing sensations in electric fields. *Aerosp. Med.*, 35, 1964, pp. 834-839.

817. SOSNICKY A. Sources and biological effects of non-ionizing electromagnetic radiation. M.S. Thesis, Naval Postgraduate School, Monterey, 1976, 75 p.

818. SPACKMAN D., RILEY V. Studies of RF radiation effects on blood-brain barrier permeability using fluoresceine and amino-acids. In : Biological effects of electromagnetic radiation, URSI international symposium, Helsinki, 1978, p. 75.

819. SPACKMAN D., RILEY V., GUY A., CHOU C. The elevation of natural amino acids in brain following exposure of mice to low-level microwave

(RFR) radiation in blood-brain barrier studies. *Fed. Proc.*, 39, 1980, abstracts, p. 1903.

820. STEMLER V. Effects of microwaves on blood serum butyryl cholinesterase activity in vivo. (Trad. du russe). US joint Pub. Res. Serv., JPRS 66512, 7 Jan. 1976, pp. 60-66.

821. STERN S., MARGOLIN L., WEISS B., LU S., MICHAELSON S. Microwaves : effect on thermoregulatory behavior in rats. *Science*, 206, 1979, pp. 1198-1201.

822. STEWART-DE HAAN P., CREIGHTON M., LARSEN L., JACOBI J., ROSS W., SANWAL M., GUO T., GUY W., TREVITHICK J. In vitro studies of microwave-induced cataract : separation of field and heating effect. *Exp. Eye Res.*, 36, 1983, pp. 75-90.

823. STODOLNIK-BARANSKA W. Lymphoblastoid transformation of lymphocytes in vitro after microwave irradiation. *Nature*, 214, 1967, pp. 102-103.

824. STODOLNIK-BARANSKA W. The effects of microwaves on human lymphocyte cultures. In : CZERSKI P., OSTROWSKI K., SHORE M., SILVERMAN C., SÜSS M., WALDESKOG B. Biological effects and health hazards of microwave radiation. Proceedings of an international symposium, Warszawa, 15-18 Oct. 1973. Warszawa, Polish Medical Publishers, 1974, pp. 189-195.

825. STROBEL G., SCHERENZEL H. Die günstigste Funkentstörungsschaltung. *Frequenz*, 7, 269-275, 1953, pp. 295-298.

826. STUCHLY M. Radio waves and microwaves - basic definitions and concepts. *Trans. IMPI*, 8, Microwave bioeffects and radiation safety, 1978, pp. 1-4.

827. STUCHLY M. Interaction of radiofrequency and microwave radiation with living systems : a review of mechanisms. *Radiat. Environ. Biophys.*, 16, 1979, pp. 1-14.

828. STUCHLY M. Les radiofréquences et la santé. *Rev. Gén. Electr.*, 11, nov. 1981, pp. 849-851.

829. STUCHLY M., STUCHLY S. Industrial, scientific, medical and domestic applications of microwaves. *IEEE Proc.*, 130, part A, n° 8, nov. 1983, pp. 467-503.

830. SUBJECK J., HETZEL F., SANDHU T., JOHNSON R., KOWAL H. Nonthermal effect of microwave heating on the cell periphery as revealed by the binding of colloidal iron hydroxide. *Radiat. Res.*, 74, 1978, pp. 584-585.

831. SULEK K., SCHLAGEL C., WIKTOR-JEDR-ZEJCZAK W., HO H., LEACH W., AHMED A., WOODY J. Biologic effects of microwave exposure. 1. Threshold conditions for the induction of the increase in complement receptor positive (CR⁺) mouse spleen cells following exposure to 2450-MHz microwave. *Radiat. Res.*, 83, 1980, pp. 127-137.

832. SULTAN M., CAIN C., TOMPKINS W. Effects of microwaves and hyperthermia on capping of antigen-antibody complexes on the surface of normal mouse B lymphocytes. *Bioelectromagn.*, 4, 1983, pp. 115-122.

833. SULTAN M., CAIN C., TOMPKINS W. Immunological effects of amplitude-modulated radio-frequency radiation : B lymphocyte capping. *Bioelectromagn.*, 4, 1983, p. 157.

834. SULTAN M., TOMPKINS W., CAIN C. Hyperthermic enhancement of antibody-complement cytotoxicity against normal mouse B lymphocytes and its relation to capping. *Radiat. Res.*, 96, 1983, pp. 251-260.

835. SÜSS M., BENWELL D. Institutions and legislation concerned with non-ionizing radiation, health related research and protection, a directory. OMS/WHO, København, Regional Office for Europe and RPB, Ottawa, Health and Welfare Canada, 1984, 45 p.

836. SÜSSKIND C. Proceedings of the 3rd annual Tri-Service conference on biological hazards of microwave-radiating equipments. Berkeley, Univ. California, 25-27 August 1959, RADC-TR-59-140.

837. SÜSSKIND C. The "story" of nonionizing radiation research. *Bull. New York Acad. Med.*, 55, n° 11, 1979, pp. 1152-1163.

838. SÜSSKIND C., PRAUSNITZ S. Non-thermal effects of microwave radiation. Final report 1957-62, Contract AF 41(657)114, Berkeley, Univ. California, ASTIA 269 385, 1957.

839. SUTTON C., CARROLL F. Effects of microwave-induced hyperthermia on the blood-brain barrier of the rat. *Radio Sci.*, 14, n° 6 (S), 1979, pp. 329-334.

840. SUTTON C., NUNNALLY R. Protection of the microwave-irradiated brain with body-core hypothermia. In : 10th annual Meeting on Cryobiology, 10, 1973, pp. 513-514.

841. SWIECICKI W. Effet de l'irradiation micro-onde aux fréquences de 3 et 10 cm sur les protéines du sang de lapins. (En polonais). *Med. Lotnicza*, 11, 1963, pp. 54-59.

842. SZADY J., SIEKIERZYNSKI M., DZIUK E., WIKTOR-JEDRZEJCZAK W., CZARNECKI C. Effect of microwaves on the 24-hour rythm and 24-hour urinary excretion of 17 hydroxycorticoids and 17 keto-steroids. In : Proceedings of the IMPI Symposium, Leuven, 1976. (Paper 2A.2.).

843. SZMIGIELSKI S. Effect of 3 GHz electromagnetic radiation (microwave) on granulocytes in vitro. In : TYLER P. Biologic effects of non-ionizing radiation. Proceedings of a conference. *Ann. New York Acad. Sci.*, 247, 1975, pp. 275-281.

844. SZMIGIELSKI S. Immunological responses of mammals to microwaves. In : ALBERTS E., GAUTHERIE M. Biomedical thermology. New York, Alan R. Liss Inc., 1982, pp. 227-246.

845. SZMIGIELSKI S., JANIAK M. Injury of cell membranes in normal and SV40-virus transformed fibroblasts exposed in vitro to microwave (2450 MHz) or water-bath hyperthermia (43°C). In : Biological effects of electromagnetic waves. International symposium, Airlie, 30 Oct.-4 Nov. 1977, Abstracts of scientific papers.

846. SZMIGIELSKI S., JELJASZEWICA J., WIRANOWSKA M. Acute staphylococcal infections in rabbits irradiated with GHz microwaves. In : TYLER P. Biologic effects of non-ionizing radiation. Proceedings of a conference. *Ann. New York Acad. Sci.*, 247, 1975, pp. 305-311.

847. SZMIGIELSKI S., LUCZAK M., JANIAK M., KOBUS M., LASKOWSKA B., DE CLERCQ E., DE SOMER P. In-vitro and in-vivo inhibition of virus multiplication by microwave hyperthermia. *Arch. Virol.*, 53, 1977, pp. 71-77.

848. SZMIGIELSKI S., LUCZAK M., WIRANOW-SKA M. Effect of microwaves on cell function and virus replication in cell cultures irradiated in vitro. In : TYLER P. Biologic effects of non-ionizing radiation. Proceedings of a conference. *Ann. New York Acad. Sci.*, 247, 1975, p. 263.

849. TA-FU H., MOLD N. Immunologic and hematopoietic alteration by 2450 MHz electromagnetic radiation. *Bioelectromagn.*, 1, 1980, pp. 77-87.

850. TANNER J. Effect of microwave radiation on birds. *Nature*, 210, n° 5037, 1966, p. 636.

851. TANNER J., ROMERO-SIERRA C. Beneficial and harmful accelerated growth induced by the action of non-ionizing radiation. In : TYLER P. Biological effects of non-ionizing radiation. Proceedings of a conference. *Ann. New York Acad. Sci.*, 247, 1975, pp. 171-175.

852. TANNER J., ROMERO-SIERRA C. The effects of chronic exposure to very low intensity microwave radiation on domestic fowl. *J. Bioelectr.*, 1, n° 2, 1982, pp. 195-205

853. TANNER J., ROMERO-SIERRA C., DAVIE S. Non-thermal effects of microwave radiation on birds. *Nature*, 216, n° 5120, 1967, p. 1139.

854. TANNER J., ROMERO-SIERRA C., DAVIE S. The effects of microwaves on birds, preliminary experiments. In : Proceedings of the IMPI Symposium, Edmonton, 1969, pp. 63-67. (Paper DA.3.).

855. TANNER J., ROMERO-SIERRA C., DAVIE S. The effects of microwaves on birds : preliminary experiments. *J. Microw. Power*, 4, n° 2, 1969, pp. 122-128.

856. TAYLOR E., ASHLEMAN B. Analysis of the central nervous system involvement in the microwave auditory effect. *Brain Res.*, n° 74, 1974, pp. 201-208.

857. TAYLOR E., ASHLEMAN B. Some effects of electromagnetic radiation on the brain and spinal cord of cats. In : TYLER P. Biologic effects of non-ionizing radiation. Proceedings of a conference. *Ann. New York Acad. Sci.*, 247, 1975, pp. 63-73.

858. TAZELAAR M. The effect of a temperature gradient on the early development of the chick. *Q. J. Microsc. Sci.*, 72, 1928, pp. 419-446.

859. TCHAO Y., HUET C., LENOIR-ROUSSEAU J. Action de l'hyperfréquence sur la métamorphose du coléoptère Tenebrio molitor. *Bull. Soc. Zool.*, 102, n° 3, 1977, p. 330.

860. TCHAO Y., HUET C., LENOIR-ROUSSEAU J. Effet de l'irradiation hyperfréquence sur le mécanisme de tannage chez Tenebrio molitor coléoptère.

C.R. Acad. Sci. Paris, 284, 1977, pp. 1589-1592.

861. TCHAO Y., RADZISZEWSKI E., SAUZIN-MONNOT M., BRIANÇON C., WIKGREN M. Effet spécifique de l'impact biologique des micro-ondes. In : Effets biologiques des rayonnements non ionisants ; utilisation et risques associés. 9ème congrès international de la Société Française de Radioprotection, Nainville-les-Roches, 22-26 mai 1978, pp. 301-314.

862. TELL R., HARLEN F. A review of selected biological effects and dosimetric data useful for development of radiofrequency safety standards for human exposure. *J. Microw. Power*, 14, n° 4, 1979, pp. 405-424.

863. TENGROTH B. Retinal lesions as result of microwave exposure. *Phys. Med. Biol.*, 17, n° 5, 1972, p. 690.

864. TENGROTH B., AURELL E. Retinal changes in microwave workers. In : CZERSKI.P., OSTROWSKI K., SHORE M., SILVERMAN C., SÜSS M., WALDESKOG B. Biological effects and health hazards of microwave radiation. Proceedings of an international symposium, Warszawa, 15-18 Oct. 1973, Warszawa, Polish Medical Publishers, 1974, p. 302.

865. TERZUOLO C., BULLOCK T. Measurement of imposed voltage gradient adequate to modulate neuronal firing. In : Proceedings of the National Academy of Sciences of the USA, 42, 1965, pp. 687-694.

866. THOMAS R., BURCH L., YEANDLE S. Microwave radiation and chlordiazepoxide : synergistic effects on fixed-interval behavior. *Science*, 203, 1979, pp. 1357-1358.

867. THOMAS J., FINCH E., FULK D., BURCH L. Effects of low-level microwave radiation on behavioral baselines. In : TYLER P. Biologic effects of non-ionizing radiation. Proceedings of a conference. *Ann. New York Acad. Sci.*, 247, 1975, pp. 425-432.

868. THOMAS J., MAITLAND G. Microwave radiation and dextroamphetamine : evidence of combined effects on behavior of rats. *Radio Sci.*, 14, n° 6 (S), 1979, pp. 253-258.

869. THOMSON R., MICHAELSON S., HOWLAND J. Leukocyte response following simultaneous ionizing and microwave (radar) irradiation. *Blood*, 28, 1966, pp. 157-162.

870. THOUREL L., Action des micro-ondes sur les enzymes et les micro-organismes. Publication de l'ONERA, Centre d'Etudes et de Recherches de Toulouse, DERMO, s.d., 7 p.

871. THOUREL L., PAREILLEUX A., PRIOU A., THOUREL B., AUGE C. Microwave specific effects on beer yeast. In : Proceedings of the IMPI Symposium, Waterloo (Can.), 1975, pp. 127-128. (Paper 7.7.), et Publication du CERT-ONERA. Toulouse 12 p.

872. THOUREL L., PAREILLEUX A., THOUREL B., AUGE C. Microwave specific effects on beer yeast. In : Proceedings of the IMPI Symposium, Milwaukee, 1974, pp. 1-2. (Paper A3.4.).

873. TIBBS C. Putting the microwave oven radiation hazard into proper perspective. *Times*, 19 July 1973.

874. TINNEY C., LORDS J., DURNEY C. Rate effects in isolated turtle hearts induced by microwave irradiation. *IEEE Trans. Microw. Theory Tech.*, MTT-24, n° 1, 1976, pp. 18-24.

875. TORRE-BUENO J. Temperature regulation and heat dissipation during flight in birds. *J. Exp. Biol.*, 65, 1976, pp. 471-482.

876. TRAVERS W., VETTER R. Low intensity microwave effects on the synthesis of thyroid hormones and serum proteins. 22nd annual meeting, Health Physics Society, Atlanta, 3-8 July 1977.

877. TREMOLIERES J. Rayonnement électromagnétique et santé. *RGS*, n° 30, Janv. 1984, pp. 16-23.

878. TRUFFA-BACHI P., LECLERC C. Comment les cellules coopèrent pour défendre l'organisme. *La Recherche*, n° 177, mai 1986, pp. 702-717.

879. TYLER P. Overview of the biological effects of electromagnetic radiation. *IEEE Trans. Aerosp. Electron. Syst.*, AES-9, n° 2, 1973, pp. 225-228.

880. TYLER P. Biologic effects of non-ionizing radiation. Proceedings of a conference. *Ann. New York Acad. Sci.*, 247, 1975, 545 p.

881. TYLER P. Overview of electromagnetic radiation research, past, present, and future. In : TYLER P. Biologic effects of non-ionizing radiation, Proceedings of a conference. *Ann. New York Acad. Sci.*, 247, 1975, pp. 6-14.

882. VALTONEN E. Giant mast cells - a special degenerative form produced by microwave radiation. *Exp. Cell Res.*, 43, 1966, pp. 221-224.

883. VALUDE . Oeil. In : GALTIER-BOISSIERE . Larousse médical illustré. Paris, Larousse, 1912, pp. 775-802.

884. VAN UMMERSEN C. The effect of 2450 Mc. radiation on the development of the chick embryo. In : PEYTON M. Proceedings of the 4th annual Tri-Service conference on biological effects of microwave-radiating equipments, biological effects of microwave radiation. New York, Plenum Press, 1961, RADC-TR-60-180, Vol. 1, pp. 201-219.

885. VAN UMMERSEN C. An experimental study of development abnormalities induced in the chick embryo by exposure to radio-frequency waves. Ph. D. Dissertation, Tufts Univ., Dep. Biol., Medford, 1963.

886. VAN UMMERSEN C., KOGAN F. Effects of microwave radiation on the lens epithelium in the rabbit eye. In : Proceedings of the IMPI Symposium, Boston, 1969, abstracts, pp. 26-27. (Paper E 2.).

887. VARMA M., DAGE E., JOSHI S. Mutagenicity induced by nonionizing radiation in Swiss male mice. In : JOHNSON C., SHORE M., Biological effects of electromagnetic waves. Selected papers of the USNC/URSI annual meeting, Boulder, 20-23 Oct. 1975, US Dep. Health, Educ., Welfare, FDA-77-8010. Washington, US Government Printing Office, Vol. 1, 1976, pp. 397-405.

888. VARMA M., TRABOULAY E. Biological effects of microwave radiation on the testes of mice. *Experientia*, 31, 1975, p. 301.

889. VARMA M., TRABOULAY E. Evaluation of dominant lethal test and DNA studies in measuring mutagenicity caused by non-ionizing radiation. In : JOHNSON.C., SHORE M. Biological effects of electromagnetic waves. Selected Papers of the USNC/URSI annual meeting, Boulder, 20-23 Oct. 1975, US Dep. Health, Educ., Welfare, FDA-77-8010. Washington, US Government Printing Office, Vol. 1, 1976, pp. 386-396.

890. VAUTRIN J. Risques liés aux rayonnements électromagnétiques non ionisants. *Econ. Progrès Electr.*, n° 7, janv.-fév. 1983, pp. 15-19.

891. VENDRIK A., VOS J. Comparison of the stimulation of warmth sense organ by microwave and infrared. *J. Appl. Physiol.*, 13, 1958, p. 435.

892. VETTER R. Neuroendocrine response to microwave irradiation. In : Proceedings of the Nat. Electron. Conf., 30, 1975, pp. 237-238.

893. VOGELMAN J. Microwave instrumentation for the measurement of biological effects. In : PEYTON M. Proceedings of the 4th annual Tri-Service conference on biological effects of microwave-radiating equipments, biological effects of microwave radiations. New York, Plenum Press, August 1961, RADC-TR-60-180, pp. 23-31.

894. VOSS W., PARANJAPE R., TURNER R. Can EM-induced behavioral effects in rodents, be detected by monitoring ultrasound emissions ? In : Proceedings of the IMPI Symposium, Toronto, 1981, Digest p. 135.

895. WACHTEL H., SEAMAN R., JOINES W. Effects of low-density microwaves on isolated neurons. In : TYLER P. Biologic effects of non-ionizing radiation. Proceedings of a conference. *Ann. New York. Acad. Sci.*, 247, 1975, pp. 46-62.

896. WALKER C., Mac WHIRTER K., VOSS W. Use of bacteriophage system for investigating the biological effects of low intensity pulsed microwave radiation. *J. Microw. Power*, 9, n° 3, 1974, pp. 221-229.

897. WANGEMANN R., CLEARY S. The in-vivo effects of 2.45 GHz microwave radiation on rabbit serum components and sleeping time. *Health Phys.*, 27, n° 6, 1974, pp. 633-634.

898. WANGEMANN R., CLEARY S. The in-vivo effects of 2.45 GHz microwave radiation on rabbit serum components and sleeping times. *Radiat. Environ. Biophys.*, 13, 1976, pp. 89-103.

899. WARD T., ELDER J., LONG M. A comparative study of microwave and high ambient temperature exposures on the blood-brain barrier. *Bioelectromagn.*, 1, n° 2, 1980, p. 207.

900. WARD D., ELDER J., LONG M., SVENDS-GAARD D. Measurement of blood-brain barrier permeation in rats during exposure to 2450-MHz microwaves. *Bioelectromagn.*, 3, 1982, pp. 371-383.

901. WAY W., KRITIKOS H., SCHWAN H. Thermoregulatory physiologic responses in the human body exposed to microwave radiation. *Bioelectromagn.*, 2, 1981, pp. 341-356.

902. WEBB S. Genetic continuity and metabolic regulation as seen by the effects of various microwave and black light frequencies on the phenomena. In : TYLER P. Biologic effects of non-ionizing radiation. Proceedings of a conference. *Ann. New York. Acad. Sci.*, 247, 1975, pp. 327-351.

903. WEBB S. Factors affecting the induction of lambda prophages by millimeter microwaves. *Phys. Letters*, 73 A, 1979, pp. 145-148.

904. WEBB S., BOOTH A. Absorption of microwaves by micro-organisms. *Nature*, 222, n° 5199, 1969, pp. 1199-1200.

905. WEBB S., DODDS D. Inhibition of bacterial cell growth by 136 GHz microwaves. *Nature*, 218, 1968, pp. 374-375.

906. WEITER J., FINCH E., SCHULTZ W., FRATTALI V. Ascorbic acid changes in cultured rabbit lenses after microwave irradiation. In : TYLER P. Biologic effects of non-ionizing radiation. Proceedings of a conference. *Ann. New York Acad. Sci.*, 247, 1975, pp. 175-181.

907. WEVER E., VERNON J., PETERSON E. The high-frequency sensitivity of the guinea pig ear. In : Proceedings of the National Academy of Sciences of the USA, 49, 1963, pp. 319-322.

908. WHITCOMB L., BLACKMAN C., WEIL C. Contraction of smooth muscle in a microwave field. *Radio Sci.*, 14, n° 6 (S), 1979, pp. 155-158.

909. WHITE D. Natural receptors of EMI. In : WHITE D. EMC handbook, Electromagnetic interference and compatibility, Vol. 3 : EMI control methods and techniques. Germantown, Don White Consultants, Inc., 1973, pp. 3.3.-3.11.

910. WHITE D. Elastic wave generation by electron bombardment or electromagnetic wave absorption. *J. Appl. Phys.*, 34, 1963, pp. 2123-2124.

911. WHITE R. Generation of elastic waves by transient surface heating. *J. Appl. Phys.*, 34, 1963, pp. 3559-3567.

912. WHITNEY H., KHARADLY M. Some results on low-level microwave treatment of the mountain pine beetle and the darkling beetle. *IEEE Trans. Microw. Theory Tech.*, MTT-32, n° 8, 1984, pp. 798-803.

913. WIKE E., MARTIN E. Science and standards - Comments on Frey's "Data analysis reveals significant microwave-induced eye damage in humans". *J. Microw. Power*, 20, n° 3, 1985, pp. 181-184.

914. WIKTOR-JEDRZEJCZAK W., AHMED A., CZERSKI P., LEACH W., SELL K. Effects of microwaves (2450 MHz) on immune systems in mice ; increase in complement receptor bearing lymphocytes (LCR). *Exp. Hematol.*, 4, n° 5, 1976, p. 73.

915. WIKTOR-JEDRZEJCZAK W., AHMED A., CZERSKI P., LEACH W., SELL K. Immune response of mice to 2450-MHz microwave radiation : overview of immunology and empirical studies of lymphoid splenic cells. *Radio Sci.*, 12, n° 6 (S), 1977, pp. 209-219.

916. WIKTOR-JEDRZEJCZAK W., AHMED A., CZERSKI P., LEACH W., SELL K. Increase in the frequency of Fc receptor (FcR) bearing cells in the mouse spleen following a single exposure of mice to 2450 MHz microwaves. *Biomed.*, 27, 1977, pp. 250-252.

917. WIKTOR-JEDRZEJCZAK W., AHMED A., LEACH W., SELL K. Effect of microwaves (2450 MHz) on the immune system in mice : studies of nucleic acid and protein synthesis. *Bioelectromagn.*, 1, 1980, pp. 161-170.

918. WIKTOR-JEDRZEJCZAK W., AHMED A., SELL K., CZERSKI P., LEACH W. Microwaves induce an increase in the frequency of complement receptor-bearing lymphoid spleen cells in mice. *J. Immunol.*, 118, n° 4, 1977, pp. 1499-1502.

919. WIKTOR-JEDRZEJCZAK W., SCHLAGEL C., AHMED A., LEACH W., WOODY J. Possible humoral mechanism of 2450-MHz microwave-induced increase in complement receptor positive cells. *Bioelectromagn.*, 2, 1981, pp. 81-84.

920. WILLIAMS D., MONAHAN J., NICHOLSON W., ALDRICH J. Biologic effects studies on microwave radiation : time and power thresholds for the production of lens opacities by 12.3 cm microwaves. IRE Trans. on Biomedical Electronics, P.G.M.E., 4, 1956, pp. 17-22.

921. WILLIAMS R., FINCH E. Examination of the cornea following exposure to microwave radiation. *Aerosp. Med.*, 45, 1974, pp. 393-396.

922. WILLIAMS R., Mac KEE A., FINCH E., FULK D. Electron microscopic evaluation of the lenses of rabbits exposed to long-term 2450 MHz continuous microwave energy at 10mW/cm². In : Proceedings of the IMPI Symposium, Waterloo (Can.), 1975, pp. 113-114. (Paper 7.3.).

923. WILLIAMS W., LU S., del CERRO M., HOSS W., MICHAELSON S. Effect of 2450-MHz microwave energy on the blood-brain barrier : an overview and critique of past and present research. *IEEE Trans. Microw. Theory Tech.*, MTT-32, n° 8, 1984, pp. 808-818.

924. WORDEN R. New biological effects of RF energy. *Electron.*, 32, n° 49, 1959, pp. 38-39.

925. YAGI K., KEYAMA R., KUROHANE S., HIRAMINE N., ITOH H., UMEHARA S. Hazardous effects of microwave radiation on the bone marrow. In : CZERSKI P., OSTROWSKI K., SHORE M., SILVERMAN C., SÜSS M., WALDESKOG B. Biological effects and health hazards of microwave radiation. Proceedings of an international symposium, Warszawa, 15-18 Oct. 1973. Warszawa, Polish Medical Publishers, 1974.

926. YAMAURA I., CHICHIBU S. Super-high frequency electric field and crustacean ganglionic discharges. *Tohoku J. Exp. Med.*, 93, 1967, pp. 249-259.

927. YANG H., CAIN C., LOCKWOOD J., TOMPKINS W. Effects of microwave exposure on the hamster immune system. 1. Natural killer cell activity. *Bioelectromagn.*, 4, 1983, pp. 123-139.

928. YAO K. Microwave-radiation-induced chromosomal aberrations in corneal epithelium of Chinese hamsters. *J. Heredity*, 69, 1978, pp. 409-412.

929. YAO K., JILES M. Effects of 2450-MHz microwave radiation on cultivated rat kangaroo cells. In : CLEARY S. Biological effects and health implications of microwave radiation. Symposium proceedings. Commonwealth Univ., Richmond, 17-19 Sept. 1969, US Dep. Health, Educ., Welfare, BRH, DBE 70-2, PB 193 898, June 1970, pp. 123-133.

930. YAO K., JILES M. Mortality of 2450 MHz microwave irradiated rat kangaroo cells in vitro. In : Proceedings of the IMPI Symposium, Monterey, 1971. (Paper 7.5.).

931. YEAGERS E., LANGLEY J., SHEPPARD A., HUDDLESTONE G. Effects of microwave radiation on enzymes. In : TYLER P. Biologic effects of non-ionizing radiation. Proceedings of a conference. *Ann. New York Acad. Sci.*, 247, 1975, pp. 301-304.

932. YERUSHALMI A., KATZAP I., GOTTESFELD F., BASS D. Systemic alteration in blood glucose levels following localized deep microwave hyperthermia. *J. Microw. Power*, 19, n° 1, 1984, pp. 73-76.

933. ZALJUBOVSKAJA N. Reactions of living organisms to exposure to millimeter-band electromagnetic waves. *Soviet Phys. Uspekhi*, 16, n° 4, 1974, pp. 574-576.

934. ZALJUBOVSKAJA N., KISELEV R. Biological oxidation in cells under the influence of radiowaves in the millimeter range. (Trad. du russe). US joint Pub. Res. Serv. Rep., JPRS, L/7957, 15 August 1978, pp. 6-11.

935. ZALJUBOVSKAJA N., KISELEV R. Effect of radio waves of a millimeter frequency range on the body of man and animals. (Trad. du russe). US joint Pub. Res. Serv. Rep., JPRS, 72956, 9 March 1979, pp. 9-15.

936. ZARET M. Comments on papers delivered at the 3rd Tri-Service conference on biological effects of microwave radiation. In : SÜSSKIND C. Proceeding of the 3rd annual Tri-Service conference on biological hazards of microwave-radiating equipments. Berkeley, Univ. California, 25-27 August 1959, RADC-TR-59-140, p. 334.

937. ZARET M. Cataracts following use of microwave ovens. *New York State J. Med.*, 74, n° 11, 1974, pp. 2032-2034.

938. ZARET M. Selected cases of microwave cataract in man associated with concomitant annotated pathologies. In : CZERSKI P., OSTROWSKI K., SHORE M., SILVERMAN C., SÜSS M., WALDESKOG B. Biological effects and health hazards of microwave radiation. Proceedings of an international symposium, Warszawa, 15-18 Oct. 1973. Warszawa, Polish Medical Publishers, 1974, p. 294.

939. ZARET M. Electronic smog as a potentiating factor in cardiovascular diseases : a hypothesis of microwaves as an etiology for sudden death from heart attack in North Karelia. *Med. Res. Eng.*, 12, 1976, pp. 13-16.

940. ZARET M. Lésions provoquées sur des êtres humains par des rayonnements non ionisants. In : Effets biologiques des rayonnements non ionisants ; utilisations et risques associés. 9ème Congrès international de la Société Française de Radioprotection, Nainville-les-Roches, 22-26 mai 1978, pp. 337-356.

941. ZARET M., CLEARY S., PASTERNAK B., EISENBUD M., SCHMIDT H. A study of lenticular imperfections in the eyes of a sample of microwave workers and a control population. Final Report, Contract AF 30(602)2215, RADC-TDR-63-125, ASTIA doc. AD 413 294. New York, New York Univ., 1963.

942. ZARET M., EISENBUD M. Preliminary results of studies of the lenticular effects of microwaves among exposed personnel. In : PEYTON M. Proceedings of the 4th annual Tri-Service conference on biological effects of microwave-radiating equipments, biological effects of microwave radiations. New York, Plenum Press, 1961, RADC-TR-60-180, pp. 293-308.

943. ZARET M., KAPLAN I., KAY A. Clinical microwave cataracts. In : CLEARY S. Biological effects and health implications of microwave radiation. Symposium proceedings, Commonwealth Univ., Richmond, 17-19 Sept. 1969, US Dep. Health, Educ., Welfare, BRH, DBE 70-2, PB 193 898, Library of Congress card n° 76-607340, June 1970, p. 82.

944. ZDAREK J., ONDRAČEK J., DATLOV J. Differential sensitivity to microwaves in the fleshfly larva tissues. *Entomol. Exp. Appl.*, 20, 1976, pp. 270-274.

945. ZEMAN G., CHAPUT R., GLAZER Z., GERSHMAN L. Gamma aminobutyric acid metabolism in rats following microwave exposure. *J. Microw. Power*, 8, n° 3-4, 1973, pp. 211-216.

946. ZYDECKI S. Assessment of lens translucency in juveniles, microwave workers, and age-matched groups. In : CZERSKI P., OSTROWSKI K., SHORE M., SILVERMAN C., SÜSS M., WALDESKOG B. Biological effects and health hazards of microwave radiation. Proceedings of an international symposium, Warszawa, 15-18 Oct. 1973. Warszawa, Polish Medical Publishers, 1974, pp. 306-308.

947. A summary of the ERMAC work session on nervous system and behavioral effects of non-ionizing electromagnetic radiations. *J. Microw. Power*, 10, n° 2, 1975, pp. 127-140.

948. Bureau study shows cataracts non linked to low-level microwave exposure. US Dep. Health, Educ., Welfare, Bureau of Radiological Health, 12, n° 13, 1978, et tiré à part de l'International Microwave Power Institute, New Release, 1 p.

949. Health aspects of radio-frequency and microwave radiation exposure, Ministère de la Santé et du Bien-être social du Canada, Section des rayonnements non-ionisants, Bureau de Protection contre les Rayonnements, 77-EHD-13, 1977, 78-EHD-22, 1978.

950. Les radiofréquences et la santé. Health aspects of radio-frequency and microwave exposure. Ministère de la Santé Nationale et du Bien-être Social du Canada, 1ère partie, novembre 1977, 77-EHD-13, 80 p. ; 2ème partie, mars 1978, 78-EHD-22, 107 p.

951. New biological effects of RF energy. *Electronics*, 32, n° 49, 1959, pp. 38-39.

952. Occupational exposure to microwave radiation (radar). Final report, Contract FDA 223-76-6003, National Research Council, 27 Dec. 1978.

953. Possible effects of electromagnetic radiations. Naval Research Reserve. *Naval Res. Rev.*, 27, n° 7, 1974, pp. 20-22.

954. Questions and answers about biological effects and potential hazards of radio-frequency radiation, Federal Communications Commission, Office of Science and Technology, Technical Analysis Division, *OST Bull.*, n° 56, July 1982.

955. Soviet bioelectromagnetics research blossoming. *Microw. News*, V, n° 8, Nov.-Dec. 1985, pp. 1 and 9.

Chapter 3

Safety standards

The physiological effects that result from exposure to microwaves create the need for rules and regulations that will protect the general public and professionals routinely exposed to electromagnetic radiation.

The safety of a source in relation to a human being must be examined in two complementary ways. First, it can be expressed in terms of the emission of radiation and requires the regulation of radiation emitted by the source, so that it does not present any danger to the user or to the public. Second, it can be expressed in terms of susceptibility to radiation, which requires complete exclusion or limited access of people to regions in which electromagnetic field exceeds a certain threshold, so that protective clothing may have to be used. The corresponding safety regulations incorporate emission and exposure standards, respectively.

The question becomes very complicated when we have to

establish the criteria for proper operation (i.e., to evaluate the field levels that can be tolerated by the human being). The setting up of safety standards would be greatly simplified if we were able to establish a clear relationship between exposure levels and pathophysiological effects [114].

For example, if we test a potentially toxic agent on 100,000 laboratory animals, taking as our criterion the increase in death rate over a certain period, a figure of 10^{-5} will not be significant. We then have to increase the administered dose by several orders of magnitude and deduce the tolerance threshold by linear interpolation [181]. As previously noted, this protocol can be criticized on several counts:

- effects observed on laboratory animals cannot be reliably extrapolated to humans or, sometimes, even to other animals belonging to a different strain; in particular, differences in weight play an important part in the capacity to absorb radiation (Fig. 3.1).
- linear interpolation is not valid because biological processes are strongly nonlinear.

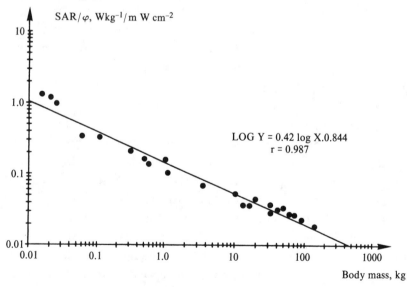

Figure 3.1 Ratio of SAR to power density as a function of body mass [194].

- in general, the basic criterion is not death, but some subtle biological phenomenon that is more difficult to observe and measure.
- the number of animals tested often lacks statistical significance.

There are also deep differences of opinion (Chapter 2) over the existence of nonthermal effects, and this has led American and Soviet workers to adopt, over the past 50 years, incompatible positions that have resulted in a difference of three orders of magnitude between their original exposure standards. With few exceptions, countries belonging to the two power blocks have been content to adopt the respective standards. A certain convergence has occurred when the United States adopted a very strict emission standard for domestic microwave ovens, which is significantly more stringent than their general exposure standard. The East-West dialog began soon after the 1972 Brezhnev-Nixon talks on health and the environment, with an international symposium held in Warsaw by the *World Health Organization* (WHO), the Polish Ministry of Health, and the US Department of Health, Education, and Welfare. This dialog was subsequently continued through a number of international bodies such as the *International Microwave Power Institute* (IMPI), the *International Union of Radio Science* (URSI), the *Bioelectromagnetic Society* (BEMS), founded in 1978, the WHO, and the *International Radiation Protection Association* (IRPA). A certain number of fundamental points are accepted in both East and West (i.e., the validity of the dosimetric approach, the concept of specific absorption rate, the importance of resonance and focusing phenomena, the still unexplained role of synergetic effects, and the existence of nonthermal or microthermal effects at the membrane level, and the nonlinearity of the biological response to pulsed waves, expressed in terms of power and frequency windows). These common points, and the recent recommendations of IRPA, may finally lead to a convergence toward internationally recognized standards.

3.1 Soviet Union

Research in the Soviet Union over the past 50 years has suggested the possibility of athermal interactions and even of cumulative effects. This has been based on a considerable amount of data obtained in long-term experiments, at extremely low power levels, that were concerned mainly with behavior and with the physiology of the central nervous system. The interaction is essentially a dual phenomenon: it is manifested by the stimulation or inhibition of biological effects, which depend on the power level and become more differentiated as the frequency increases. Particular attention has been devoted by Soviet researchers to the synergetic effects of microwaves that are associated with certain drugs or environmental factors [106].

In 1953, the Soviet government created in Moscow a special laboratory for the study of biological effects, next to the Institute for Safety at Work and for Occupational Diseases, which is run under the aegis of the Academy of Medical Sciences and is currently directed by Professor V. M. Savin [35]. This laboratory has developed the first Soviet professional standard which, because of the disquiet generated by experimental and epidemiologic studies, had to be established at a very low level. The *Provisional Rules for the Safety of Personel Working in the Vicinity of Microwave Sources (No. 273–58)* were introduced on 26 November 1958 by the Ministry of Hygiene. They specify a maximum permitted exposure of $10\,\mu\mathrm{W\,cm}^{-2}$ in a working day of eight hours. Working in an environment with a higher power density must be limited to two hours per day between 10 and $100\,\mu\mathrm{W\,cm}^{-2}$, and 15–20 min, with compulsory wearing of protective glasses, between $100\,\mu\mathrm{W\,cm}^{-2}$ and $1\,\mathrm{mW\,cm}^{-2}$ [279], [142], [106], [43]. These regulations were adopted without modification for the military personnel of the Warsaw Pact. The standard for nonprofessional exposure added a safety margin amounting to a factor of 10, and was thus established at $1\,\mu\mathrm{W\,cm}^{-2}$.

The reasoning behind these very stringent exposure standards was as follows:

(*a*) experimental data show that there is a threshold beyond which functional effects can be observed in animals. This threshold is on the order of $1\,\mathrm{mW\,cm^{-2}}$ for 1 hr at 3 GHz.

(*b*) extrapolation to an exposure of 10 hours reduces this threshold to $100\,\mu\mathrm{W\,cm^{-2}}$.

(*c*) the safety margin by a factor of 10 gives $10\,\mu\mathrm{W\,cm^{-2}}$ for the exposure of professionals.

(*d*) a further factor of 10 is introduced for the general public; hence $1\,\mu\mathrm{W\,cm^{-2}}$ [118], [148].

The professional exposure limit was subsequently confirmed by the standards 30.03.848.70 published [290] by the Ministry of Health and GOST 12.1.006.76 issued [297] by the State Committee for Standards of the USSR Council of Ministers. It was recently revised and relaxed slightly by groups headed by V. Savin and M. Šandala, along the lines recommended by INIRC/IRPA [158], [38]. The re-evaluation of experimental data produced a threshold of hazard amounting to $2\,\mathrm{mW\,cm^{-2}}$ per hour of exposure. By including a safety factor of 10, they obtained a limit of $2\,\mathrm{W\,hr\,m^{-2}}$ in a steady field of $25\,\mu\mathrm{W\,cm^{-2}}$ for 8 hours and $100\,\mu\mathrm{W\,cm^{-2}}$ for 2 hours, the incident power density being less than $1\,\mathrm{mW\,cm^{-2}}$. For scanning antennas, $20\,\mathrm{W\,hr\,m^{-2}}$ could be tolerated. A new professional standard (GOST 12.1.006-84) [298] was adopted on 29 November 1984 and became compulsory on 1 January 1986. The standard for nonprofessional exposure was modified in the same direction [259], [257], with the new accepted level fixed at $10\,\mu\mathrm{W\,cm^{-2}}$.

3.2 United States of America

American preoccupation with the possible hazards of microwaves dates from World War II: the negative conclusions reached at that time by the US Naval Research Laboratory were instrumental in bringing about a rapid extension of civil applications.

In 1953, J. MacLaughlin of Hughes Aircraft drew the attention of the Department of Defense to the dangers of radar. Following his intervention, a commission of the Office of Naval Research met in Bethesda (Maryland) in April 1953 and suggested an exposure limit of $100\,\mathrm{mW\,cm^{-2}}$, which was subsequently reduced to $10\,\mathrm{mW\,cm^{-2}}$ at the instigation of Dr. H. Schwan of the University of Pennsylvania [165]. According to Schwan [164] this choice was dictated by a number of factors, including:

(a) the thermal character of interactions

(b) the fact that the human metabolic rate corresponds to an outward heat flux on the order of $5\text{--}10\,\mathrm{mW\,cm^{-2}}$, which rises by a factor of 2–10 during physical excersize

(c) exposure to the sun provides a heat flux of $40\,\mathrm{mW\,cm^{-2}}$

(d) the rise in temperature produced by a flux of $10\text{--}20\,\mathrm{mW\,cm^{-2}}$ is about $1°\mathrm{C}$

(e) the energy absorbed is a small fraction of the incident energy

(f) there is insufficient evidence for harmful effects due to power densities below $10\,\mathrm{mW\,cm^{-2}}$; the best documented phenomenon is the formation of cataracts, which is known to require a power density of at least $100\,\mathrm{mW\,cm^{-2}}$ [148].

Two years later, Schwan and Li proposed, without success, that the limit of exposure should be a function of frequency: $30\,\mathrm{mW\,cm^{-2}}$ below $1\,\mathrm{GHz}$, $10\,\mathrm{mW\,cm^{-2}}$ between 1 and $3\,\mathrm{GHz}$, and $20\,\mathrm{mW\,cm^{-2}}$ above $3\,\mathrm{GHz}$. The controversy on the hazards of microwaves was thus born and the first publications on this subject bear witness to that [79], [24], [166], [7], [167], [222], [101]. The death of a young military person accidentally exposed to a radar beam at $10\,\mathrm{GHz}$ was disclosed in [101] and has not been conclusively interpreted to date [179], [55].

The Tri-Service Program introduced in 1956 by the US Navy had as its remit the replacement of the voluntary empirical limit of $10\,\mathrm{mW\,cm^{-2}}$ by a true standard based on all the then available experimental data. The responsibility for this revision was assigned to Committee C95 of the *American Stan-*

dards Association (ASA), which has since become the *USA Standards Institute* (USASI) and then the *American National Standards Institute* (ANSI) under the control of the US Navy Bureau of Ships and the IRE (now IEEE). Until 1965 the president was H. Schwan and 17 of the 29 members were military personnel or industrialists [179]. In the absence of new data, the Committee could only recommend the official adoption of the old limit of $10\,\mathrm{mW\,cm^{-2}}$ for the armed forces. This became standard C95.1 - 1966 (*Safety Level of Electromagnetic Radiation with Respect to Personnel*). The safety limit of $10\,\mathrm{mW\,cm^{-2}}$ was deemed applicable between 10 MHz and 100 GHz for integrated exposures of 6 min or more. For time intervals less than 6 min, the incident energy must not exceed $1\,\mathrm{mW\,hr\,cm^{-2}}$. In the spirit of these documents, one assumes that this was not a standard of emission, but of professional exposure, although this was not explicitly stated [234].

In 1968, the US Congress adopted a new civil regulation, known as *the Radiation Control for Health and Safety Act, Public Law 90-602* [271] as a consequence of the increasing number of domestic microwave ovens. Apparently close to standard C95.1, this new regulation was in reality much more stringent. It was actually a standard of emission, limiting the level of leakage from microwave equipment to $10\,\mathrm{mW\,cm^{-2}}$ at 5 cm from any point on its surface. If we consider that power is inversely proportional to the square of distance or, at worst, inversely proportional to distance (for a cylindrical wave escaping from a radiating slot), we find that $10\,\mathrm{mW\,cm^{-2}}$ at 5 cm represents only $625\,\mu\mathrm{W\,cm^{-2}}$ ($2.5\,\mathrm{mW\,cm^{-2}}$ for a slot) at 1 m and $6\,\mu\mathrm{W\,cm^{-2}}$ ($250\,\mu\mathrm{W\,cm^{-2}}$ for a slot) at 2 m [197]. Statistical analysis of typical distances between a domestic microwave oven and its user reveals an absorption of $5-20\,\mu\mathrm{W\,cm^{-2}}$ when the oven produces $5\,\mathrm{mW\,cm^{-2}}$ at 5 cm [33]. In terms of energy density (power density times duration of exposure), it is found [130] that the dose absorbed in 30 min of daily use is lower by a factor of at least 20 than the $10\,\mu\mathrm{W\,cm^{-2}}$ cited in the Soviet exposure standard.

The metamorphosis that occurred between the publication of standard C95.1 and the 90–602 law did not escape the attention of those interested: the manufacturers of equipment joined forces under the banner of the *Association of Home Appliance Manufacturers* (AHAM) and made their disapproval known before half-heartedly admitting in the *Microwave Energy Application Newsletter* [301] that the new stringent safety standards were perhaps a necessary evil.

The 90.602 law instituted the *Technical Electronic Product Radiation Safety Standards Committee* (TEPRSSC) consisting of 15 members who had to be consulted by the US Secretary for Health, Education and Welfare before any modification of the law could be made. Despite alarming data collected by the Committee relating to the level of leakage from domestic microwave ovens [217], the *US Department of Health, Education and Welfare* (USDHEW) estimated that a revision of the standards would not be necessary [295]. However, to the great dismay of the AHAM a revision did take place a short time later, and the emission from domestic microwave ovens was reduced from $10\,\text{mW}\,\text{cm}^{-2}$ to 1 and $5\,\text{mW}\,\text{cm}^{-2}$ at delivery and at the end of working life, respectively [280].

In 1969, USDHEW and the Medical College of Virginia organised at Richmond the first civil symposium on the effects of microwaves, but a consensus could not be reached on tolerable power levels. The ANSI C95 committee was meanwhile carrying out work under Professor Saul Rosenthal with the view to quinquennial re-examination of C95.1-1966. Since no conclusion was reached within the prescribed time, and despite numerous criticisms, the standard was brought back into action in 1971 [289] and published in 1974 by the IEEE without any changes.

In the early 1970s, a number of different organizations began to include in their literature communications concerning the effects of microwaves. They included the National Science Foundation, USNC-URSI, and the *Microwave Theory and Technique Society* (MTT) of the *Institute of Electrical and Elec-*

tronic Engineers (IEEE). This period also saw a proliferation of specialized bodies, including the *Electromagnetic Radiation Management Advisory Council* (ERMAC), controlled by the Office of Telecommunications Policy, the Department of Commerce Interagency Task Force on *Biological Effects of Nonionizing Electromagnetic Radiation* (BENER), the *Environmental Protection Agency* (EPA), and the Technical Committee on the Biological Effects of Microwave Radiation of the IEEE/MTT. Following some irresponsible articles in the press and the threat of a boycott by the Consumers Union, the IEEE created the Committee on Man and Radiation (COMAR) under the presidency of Mark Grove assisted by A. Ecker, D. Justesen, and J. Osepchuk [72], [134]. The antimicrowave campaign by the Consumers Union [249], [265] was fueled by the great confusion between standards of exposure and standards of emission, and by disturbing declarations of M. Zaret during the Senate hearings in connection with the re-examination of the 90.602 law [232]. A number of specialists wrote to the Senate [240], and it was argued by the authors of [135] and by Polish scientists that the American emission standard was consistent with the most stringent East European regulations [130].

In 1974, the USDHEW and the *Food and Drug Administration* (FDA) confirmed the above regulations [280] in relation to domestic microwave ovens, catering ovens, restaurants, food stalls, trains, aeroplanes and so on, in the frequency range 890–6000 MHz [282], [201]. Industrial microwave ovens and the medical diathermic equipment were excluded and continued in operation, following the IMPI suggestion that they should be governed by the old standard of $10\,\mathrm{mW\,cm^{-2}}$ on a specified test load [235], [246], [185].

The second quinquennial re-examination of the ANSI C95.1 exposure standards was performed with only three years' delay [216]. In the light of new experimental discoveries, including in particular the role of resonance and hotspots, H. Schwan himself admitted in 1979 that the factor of safety between 3 MHz and 3 GHz could well be less than 10. The C95-4 subcommittee

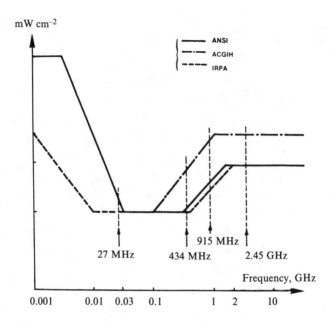

Figure 3.2 Limits of exposure recommended by ANSI, ACGIH, and IRPA [18].

reached agreement on the following points:

(a) there is no evidence that the exposure limits specified by the C95.1 1974 standard can cause any lesions or harmful effects.

(b) the 1974 recommendations were based on a limited amount of data relating exclusively to thermal effects, although the interaction mechanisms were not fully understood and certain data indicated the possible existence of nonthermal effects.

(c) the C95.1-1974 standard was generally considered applicable to professionals only; since members of the public were being exposed to sources of radiation whose numbers were increasing all the time, it was necessary to set a safety limit for the general public.

(d) although the C95.1 1974 standard had served as the basis for numerous regulations both in the United States and abroad, it was generally accepted that its revision was necessary despite the remaining gaps in our knowledge.

The new limit (Fig. 3.2) takes into account the latest scientific data including the following:

(a) Dosimetric studies using the specific absorption rate (SAR) adopted in 1972 by the SC-39 Committee of the *National Council on Radiation Protection and Measurements* (NCRP).

(b) The expanding scientific database on biological effects, despite the shortage of reliable information on the effects of chronic exposure. Reversible behavioral problems seem to appear for absorbed doses of 4–8 W kg^{-1} which, when integrated over time, correspond to 10–50 mW cm^{-2}. Some East European researchers have reported effects for under 4 W kg^{-1}. Their conclusions were not taken into account because their techniques and procedures were not deemed entirely satisfactory, and there was a lack of information on environmental parameters and measurement techniques. Effects relating to field modulation, including changes in the ion flux in brain tissue, occur only for very specific conditions of frequency modulation. They were reversible and believed to be nonhazardous, and were not, therefore, taken into consideration.

(c) A safety factor of 10 applied at the 4 W kg^{-1} threshold gives a limit of 0.4 W kg^{-1}.

(d) Resonance phenomena and hotspots reduce the accepted power density limit in the frequency band in which they occur. Below resonance, absorption increases rapidly with frequency, (roughly, as f^2). Above resonance, it is proportional to $1/f$. In the latter case, some hotspots can appear as a result of resonances at lower frequencies or quasioptical focusing at higher frequencies. In the resonance zone itself, a power density of 2.5 mW cm^{-2} produces a thermal load of the same order as the basic metabolic rate (1.1 W kg^{-1}) (Fig. 3.3).

(e) For modulated waves, field peak values are not taken into consideration.

(f) The standard is based on plane waves, absorption being always lower in the near-field region (Section 1.4.5) [216], [164], [285], [233].

The COMAR responded [233] to the new standards by not-

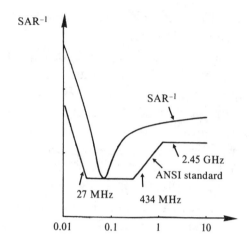

Figure 3.3 Comparison of the ANSI C95.1 - 1982 and of the inverse of the specific absorption rate [17].

ing that the IEEE recognized the disquiet of the general public about the health risks that may arise from the ever- increasing use of radiofrequency and microwave equipment. The safety standards recently proposed by the ANSI C95 Committee seemed to be entirely adequate in the light of the current understanding of biological effects of electromagnetic fields. Because of the number of beneficial applications currently in use, as well as the prospect of many more to be developed on the basis of these technologies, and in view of the remaining gaps in current knowledge, the IEEE recognised the need for research into biological effects in order to ensure proper utilization of this equipment.

Nevertheless, a number of criticisms could be leveled at the existing safety limits that were judged, on the whole, to be too permissive.

The EPA published a federal directive ("Federal Guidance") that was much more stringent in relation to nonprofessional exposure. A number of local jurisdictions have followed its example and produced their own more restrictive measures. For example, the Commonwealth of Massachusetts [242], the

Planning Commission of Portland (Oregon), and the County of Multnomah have adopted a limit of $200\,\mu\mathrm{W\,cm^{-2}}$ for the exposure of the public [72], [266]. The *National Council on Radiation Protection and Measurements* (NCRP) has recommended a safety factor of 50 for the exposure of the public [81]. In contrast, the New Jersey Commission on Radiation Protection accepted without modification the majority of the ANSI recommendations [181]. In relation to professional exposure, the *American Conference of Governmental Hygienists* (ACGIH) published in 1983 a slightly different proposal for the frequency subdivision with a recommended figure above 1 GHz that is twice as high (Fig. 3.2) [302]. The *National Institute of Occupational Safety and Health* (NIOSH) was recently trying, under the guidance of EPA and the USDHEW, to define an exposure standard that would be legally binding [72], [258]. The EPA envisages three possible approaches based on absorbed power limits of 0.04, 0.08, and $0.4\,\mathrm{W\,kg^{-1}}$ [226], [81].

3.3 Eastern Europe

In Czechoslovakia, nonprofessional exposure to modulated waves (radar) was first provisionally controlled by the 1958 Soviet regulations. Professional exposure to continuous waves was limited to $25\,\mu\mathrm{W\,cm^{-2}}$ for eight hours, a value that was later adopted in the Soviet standard GOST 12.1.006-84 [298]. In 1970, Czechoslovakia introduced its own standard [238], [105], [107], [35] which was formulated by the Institute for Industrial Hygiene and Occupational Diseases in Prague. This referred to both professional and nonprofessional exposures, and the limiting values were as listed in Table 3.1. According to [35], the nonprofessional standard is currently under revision, and the new limits for 24-hr exposure are likely to be $(0.01\,\mathrm{f^{-2}})\,\mathrm{mW\,cm^{-2}}$ between 300 and 500 MHz, and $3\,\mu\mathrm{W\,cm^{-2}}$ between 500 MHz and 300 GHz, where the frequency f is in MHz.

Table 3.1

Czechoslovakian limits of exposure in 1970

	Professional exposure		Nonprofessional exposure	
	8 hours of work	**Shorter periods**	**24 hours of exposure**	**Shorter periods**
Continuous wave	25	200/t	2.5	60/t
Modulated wave	10	80/t	1	2/t

Flux in mWcm⁻², t in hours.

In Poland, the Institute for Industrial Hygiene in Łódź has been very active in the investigation of biological effects and the definition of standards. The regulations currently in force in Poland try to reconcile Soviet and American approaches and distinguish between four zones relative to the radiation source, namely, a safety zone and three controlled zones designated as intermediate, at risk, and danger zones. The limits of these zones are indicated in Table 3.2 and their physical limits are established *in situ*. The limits of stay authorized for professional reasons are eight hours in the intermediate zone and $t = 0.32/\Phi^2$ and $t = 8/\Phi^2$ for fixed and scanning sources in the risk zone respectively, where Φ is the power density in $mW\,cm^{-2}$. All exposure is forbidden in the danger zone.

Table 3.2

Definition of exposure zones in Poland (300 MHz–300 GHz)

Zone	Fixed emitters	Scanning emitters
safety	≤10 μW cm⁻²	≤100 μW cm⁻²
intermediate	10-200 μW cm⁻²	0.1-1 mW cm⁻²
risk	0.2-10 mW cm⁻²	1-10 mW cm⁻²
danger	>10 mW cm⁻²	>10 mW cm⁻²

The distinction between fixed and scanning sources involves the parameter $C = t_i/T$, which is respectively higher or lower than 0.1, where T is the scanning period and t_i the duration of irradiation of the subject over that period. For nonprofessional users, permanent occupation of premises is permitted below $25\,\mu\mathrm{W\,cm^{-2}}$, whereas intermittent occupation (less than a day) is allowed below $100\,\mu\mathrm{W\,cm^{-2}}$. The public must not be exposed to power densities in excess of $100\,\mu\mathrm{W\,cm^{-2}}$ [286], [287], [305], [276], [306], [37], [239], [35].

3.4 Canada

The Canadian Standard Association adopted the Z65-1966 regulations under the heading *Radiation Hazards from Electronic Equipment*, essentially the same as the American ANSI C95.1-1966. A series of regulations was issued at the same time by the Ministry of Health and Welfare. They appeared on the 14 February 1972 [247] and 13 July 1974 [248], and aligned the position of Canada with that of the United States, reproducing without modification the 90.602 law. Subsequent directives [253], [260], [277] gave more precise details. The CP1974-2330 law of 27 October 1974 [241] and CP1979-3259 of September 29, 1979 [278] limit leakages to $1\,\mathrm{mW\,cm^{-2}}$ with a test load specified by the manufacturer and to $5\,\mathrm{mW\,cm^{-2}}$ without load. These limits also apply to industrial equipment.

In 1977, the Ministry of Health and Welfare proposed a reduction to $1\,\mathrm{mW\,cm^{-2}}$ of the Z65-1966 exposure limit for the following reasons.

(a) The limit of $10\,\mathrm{mW\,cm^{-2}}$ is inadequate even on the assumption of purely thermal interaction. In the case of total absorption, the power absorbed is 58 W, which is a large percentage of the basic metabolic rate (73–88 W at rest, and 293 W during moderate exercise).

(b) The existing limit did not take into account the increasing evidence for nonthermal effects obtained in East European countries. Although most of these effects have not been confirmed by Western researchers, their possible existence forces the choice of a very conservative standard.

(c) The proposed limit constitutes a compromise between the Soviet and American standards [288], [43], [148].

3.5 Australia

The Standards Association of Australia has recently produced a standard of exposure limiting power densities to $1\,\mathrm{mW\,cm^{-2}}$ between 30 MHz and 300 GHz for professionals, and $200\,\mu\mathrm{W\,cm^{-2}}$ for the public [243], [35].

3.6 Sweden

Swedish regulations limit professional exposure to an average of $1\,\mathrm{mW\,cm^{-2}}$ between 300 MHz and 300 GHz for any 6-min interval in a working day of eight hours. They also give a ceiling of $25\,\mathrm{mW\,cm^{-2}}$ that must never be exceeded [148], [77], [187].

3.7 European Community

Most of the members of the European Community have professional exposure limits: Belgium since 4 June 1979, Denmark since 29 January 1979, France since 2 May 1979, Italy since 4 July 1979, and the United Kingdom since 28 January 1979 [187].

In Germany, the standard of exposure is governed by DIN 57 848 of 1982 [228]. It applies to professionals and to members of the public. For durations in excess of 6 min, the power density must not exceed $2.5\,\mathrm{mW\,cm^{-2}}$ between 30 MHz and 3 GHz, $10\,\mathrm{mW\,cm^{-2}}$ above 12 GHz, and between 3 and 12 GHz the limit is $(2.5f/3000)\,\mathrm{mW\,cm^{-2}}$, where f is in MHz [154], [36].

In France, the radiation emitted by domestic microwave ovens is regulated by NF-C73-601 [292], ratified on 5 December 1980 and taking effect on 5 January 1981. This standard is fairly close to the American 90.602 law. For professional exposures, the Ministry of Health examined safety regulations close to the ANSI C95.1 1982 standard [18]. For the Armed Forces, a military circular authorizes $1\,\mathrm{mW\,cm^{-2}}$ for permanent exposures and $10\,\mathrm{mW\,cm^{-2}}$ for any working interval longer than one hour. When the power density is between 10 and $55\,\mathrm{mW\,cm^{-2}}$, the maximum allowed time in minutes is $t = 60(10/\Phi)^2$ where Φ is in $\mathrm{mW\,cm^{-2}}$ [204, 285]. In the United Kingdom, the average power density of professional exposure in any 6-min interval must not exceed $\Phi = f/100\,\mathrm{mW\,cm^{-2}}$ between 100 MHz and 1 GHz where f is in MHz. The limit between 1 GHz and 300 GHz is set at $10\,\mathrm{mW\,cm^{-2}}$. For the general public, the corresponding 6-min averages are $\Phi = f/300\,\mathrm{mW\,cm^{-2}}$ between 300 MHz and 1.5 GHz, and $5\,\mathrm{mW\,cm^{-2}}$ between 1.5 GHz and 300 GHz [269], [154]. These national regulations are destined to disappear when the new community standard is ready [18].

3.8 International organisations

The *International Electrotechnical Commission* (CEI) has published emission standards for mobile ovens with a mass of less than 18 kg. According to standard 335-25 [291], microwave leakages from such ovens must not exceed $5\,\mathrm{mW\,cm^{-2}}$ at 5 cm from the surface at maximum power for a load of $275 \pm 15\,\mathrm{cm^3}$ of water maintained at $20 \pm 5°\mathrm{C}$ with 1% NaCl. With no load present, the emission must not exceed $10\,\mathrm{mW\,cm^{-2}}$ over 10 min or until the safety mechanism is triggered. The document specifies *inter alia* periodic tests on moving parts, door seals, leakage currents from transformers, and so on.

Industrial equipment is also expected to conform to the voluntary standard IS-1 [235] of the IMPI, which specifies maximum leakage of $10\,\mathrm{mW\,cm^{-2}}$ at 5 cm from the surface for the manufacturers' maximum power ratings and minimum loads. The equipment must also be provided with at least two safety door locks that are electrically and mechanically independent.

In the field of exposure standards, the *International Non-ionizing Radiation Committee of the International Radiation Protection Association* (INIRC-IRPA) plays an increasing part in the definition of safety thresholds. According to its Interim Guidelines [237] approved in July 1983 by the Executive Council of the organisation, and adopted by the WHO as part of the United Nations Environmental Program, the specific absorption rate in professional whole-body exposure must not exceed $0.4 \, \text{W kg}^{-1}$ or $4 \, \text{mW}$ in $1 \, \text{g}$ of tissue over any 6-min interval. For the general public, the exposure must not exceed one-twentieth of these values (i.e., 0.08 and $0.8 \, \text{W kg}^{-1}$). These doses take into consideration environmental conditions, the differences in the size of the subjects, reflection, diffraction, focusing, and synenergetic effects due to certain drugs and environmental factors, possible athermal interactions, the presence of Adey windows in frequency and power [1], and the effect of modulation on the central nervous system. The corresponding power densities can be deduced by taking into account resonance effects (30–300 MHz for the whole body, 400 MHz for the head and other parts of the body), and possible hotspots that may be attributed (depending on frequency) to resonance or focusing (400 MHz to 2 or 3 GHz [154]). These power densities amount to $1 \, \text{mW cm}^{-2}$ between 10 and 400 MHz, $5 \, \text{mW cm}^{-2}$ between 2 and 300 GHz, and $f/400$ between 400 MHz and 2 GHz of professional exposure. The corresponding figures for the general public are 0.2, 1.0, and $(f/2000) \, \text{mW cm}^{-2}$, respectively (Fig. 3.2).

We note that IRPA has a number of Soviet and East European specialists among its members, and this may finally contribute to the development of a world consensus on standards of exposure [72], [35], [36].

References

1. ADEY W. Frequency and power windowino in tissue, interactions with weak electromagnetic fields. *Proc. IEEE*, 68, n° 1, 1980, pp. 119-125.

2. ALLAM D. Radio and microwave radiations, applications and potential hazards. Lecture given at the University of Surrey, 2-3 Jan. 1969. *J. Microw. Power*, 4, n° 2, 1969, pp. 108-114.

3. ALLAM D. Caution required to avoid electromagnetic pollution. *J. Microw. Power*, 6, n° 1, 1971, p. 57.

4. ANDERSON B., PRITCHARD R., ROWSON B. Spurious radiation from microwave ovens. *Nature*, n° 282, 1979, pp. 594-596.

5. BALDWIN B., CONSTANT P., JONES B., RUNGE L., WAIDELICH D. Survey of radio frequency radiation hazards. Kansas City, Midwest Res. Inst., US Navy Bur. Ships, Contract NOBSR-77142, Interim Report 1, Oct. 1960, Report 2, 20 June 1961.

6. BARRON C., BARAFF A. Medical considerations on exposure to microwaves (radar). *J. Med. Assoc.*, 168, n° 9, 1958, pp. 1194-1199.

7. BARRON C., LOVE A., BARAFF A. Physical evaluation of personnel exposed to microwave emanations. *J. Aviat. Med.*, 26, 1955, pp. 442-452.

8. BASSEN H. RF/microwave radiation measurement, risk assessment and control activities at the Bureau of Radiological Health. *IEEE Antennas Propag. Soc. Newsl.*, 23, n° 2, 1981, pp. 5-10.

9. BASSEN H., COAKLEY R. United States radiation safety and regulatory considerations for radiofrequency hyperthermia systems. *J. Microw. Power*, 16, n° 2, 1981, pp. 215-226.

10. BASSEN H., KANTOR G., RUGGERA P., WITTERS D. Leakage in the proximity of microwave diathermy applicators used on humans or phantom models. US Dep. Health, Educ., Welf., Bur. Radiol. Health, Publication (FDA) n° 79-8073, 1978, 9 p.

11. BERNARDI P., GIANNINI F. Analyse de modèles d'évaluation des dangers provenant de l'action des champs électromagnétiques sur les tissus biologiques. *Alta Freq.*, 45, n° 3, 1976, pp. 167-176.

12. BERNARDI P., GIANNINI F., SORRENTINO R. Effects of the surroundings on electromagnetic power absorption in layered-tissue media. *IEEE Trans. Microw. Theory Tech.*, MTT-24, n° 9, 1976, pp. 621-625.

13. BERNHARDT J. Zur Gefährdung des Menschen durch nichtionisierende Strahlung. *PTB Mitt.*, 90, 1980, pp. 416-433.

14. BERNHARDT J. Gefährdung von Personen durch elektromagnetische Felder. *STH Ber.*, 2, 1983, pp. 1-38.

15. BERNHARDT J. Elektromagnetische Felder, potentielle Gefährdung. *Bundesarbeitsblatt*, 10, 1984, pp. 17-25.

16. BERNHARDT J., DAHME M., ROTHE F. Gefährdung von Personen durch elektromagnetische Felder. STH-Berichte n° 2, Inst. Strahlenhyg. BDA. Neuherberg, D. Reimer Verlag, 1983.

17. BERTEAUD A., DARDALHON M. Effets biophysiques et biologiques associés à l'utilisation des micro-ondes. 1. Interaction des micro-ondes avec les milieux vivants. Aspects biophysiques. *Econ. Progrès Electr.*, n° 13, janv.-fév. 1985, pp. 21-23.

18. BERTEAUD A., DARDALHON M. Effets biophysiques et biologiques associés à l'utilisation des micro-ondes. 4. Limites d'exposition recommandées et normes. *Econ. Progrès Electr.*, n° 16, juil.-sept. 1985, pp. 13-16.

19. BOVILL C. Are radar radiations dangerous ? A survey of the possible hazards. *Br. Commun. Electron.*, 7, n° 5, 1960, pp. 363-365.

20. BOWMAN R. Some recent developments in the characterization and measurement of hazardous electromagnetic fields. In : CZERSKI P., OSTROWSKI K., SHORE M., SILVERMAN C., SÜSS M., WALDESKOG

B. Biological effects and health hazards of microwave radiation. Proceedings of an international symposium, Warszawa, 15-18 Oct. 1973. Warszawa, Polish Medical Publishers, 1974, pp. 217-227.

21. BRADY M. The question of microwave radiation hazard. *Tek. Ukebl.*, 118, n° 37, 1971, pp. 17-25.

22. BRODEUR P. A reporter at large : microwaves I and II. *New Yorker Magaz.*, 13-20 Dec. 1976.

23. BRODEUR P. The zapping of America : microwaves, their deadly risks and the cover up. New York, W.W. Norton Co., Inc., 1977, 312 p.

24. BRODY S. The operational hazard of microwave radiation. *J. Aviat. Med.*, 24, 1953, pp. 328-333.

25. BROWN W. IMPI position on leakage standards. *J. Microw. Power*, 5, n° 2, 1970, p. 149.

26. CARLEY W., STURGILL L. Calculations of hazardous zones of electromagnetic radiation. Report US Navy Bur. Ships, from Jansky and Bailey, Div. Atlantic Res., Washington, Alexandria, Contract NOBSR-81162, Feb. 1961.

27. CHOU C., GUY A., Mac DOUGALL J. Shielding effectiveness of improved microwave-protective suits. *IEEE Trans. Microw. Theory Tech.*, MTT-35, n° 11, 1987, pp. 995-1001.

28. CHRISTIANSON C., RUTKOWSKI A. Electromagnetic radiation hazards in the Navy. Brooklyn, Naval Appl. Sci. Lab. Techn. memo. 3, SF 013-15-04, report R-NASL-TM-3, task 2162, project 9400-20, AD n° 645695, 24 jan. 1967.

29. CORY W., FREDERICK C. Effects of electromagnetic energy on the environment - A summary report. *IEEE Trans. Aerosp. Electron. Syst.*, AES-10, n° 5, 1974, pp. 738-742.

30. CRAPUCHETTES P. Microwaves leakage - its nature and means for its measurement. In : Proceedings of the IMPI Symposium, Edmonton, 1969, pp. 85-86. (Paper DA8).

31. CRAPUCHETTES P. Microwave heating. its nature and means for its measurement. *J. Microw. Power*, 4, n° 3, 1969, pp. 138-151.

32. CZERSKI P. Health surveillance of personnel occupationally exposed to microwaves. 1. Theoretical consideration and practical aspects. *Aerosp. Med.*, 45, n° 10, 1974, pp. 1137-1142.

33. CZERSKI P. Microwave radiation protection standards in various countries. In : Nouvelles applications de l'électricité, Colloque international, Journées d'Electronique de Toulouse, 7-11 mars 1977, III-4, 14 p.

34. CZERSKI P. Microwave and radio-frequency protection standards. In : Overviews on non-ionizing radiation. International Radiation Protection Association, printed by US Dep. Health, Educ., Welf., April 1977, pp. 4-29.

35. CZERSKI P. Radiofrequency radiation exposure limits in Eastern Europe. *J. Microw. Power*, 20, n° 4, 1985, pp. 233-239.

36. CZERSKI P. The development of biomedical approaches and concepts in radiofrequency radiation protection. *J. Microw. Power*, 21, n° 1, 1986, pp. 9-23.

37. CZERSKI P., PIOTROWSKI W. Założenia dla ustlenia dopuszczalnych dawek mikrofal dla ludzi. *Med. Lotnicza*, n° 39, 1972, pp. 127-139.

38. DAVYDOV B. Rayonnement électromagnétique radiofréquence (micro-onde) ; recommandations, critères pour l'établissement de normes et niveaux dosimétriques seuils. (En russe). *Kosm. Biol. Aviakosm. Med.*, 19, 1985, pp. 8-21.

39. DECAREAU R. Caution : radiation may be hazardous to your health. *Microw. Energy Appl. Newsl.*, 1, n° 5, 1968, pp. 2, 10.

40. DECAREAU R. Microwave ovens are innocent until proven guilty. *Microw. Energy Appl. Newsl.*, 2, n° 3, 1969, pp. 2-4, 14.

41. DECAREAU R. The microwave oven is the safest appliance in the kitchen. *Microw. Energy Appl. Newsl.*, 6, n° 2, 1973, pp. 3-6.

42. DE MINEO A. Generation and detection of pulsed X rays from microwave sources. In : PEYTON M. Proceedings of the 4th annual Tri-Service conference on biological effects of microwave-radiating equipments, biological effects of microwave radiations, 1961. New York, Plenum Press, 1961, RADC-TR-60-180, pp. 33-46.

43. DODGE C., GLASER Z. Trends in non-ionizing electromagnetic radiation bioeffects research and related occupational health aspects. *J. Microw. Power*, 12, n° 4, 1977, pp. 319-334.

44. DONDERO R. Determination of power density at microwave frequencies. *US Navy Med. News Letter*, 31, n° 2, 1958, p. 22.

45. DURNEY C., JOHNSON C., BARBER P., MASSOUDI H., ISKANDER M., LORDS J., RYSER D., ALLEN S., MITCHELL J. Radio-frequency radiation dosimetry handbook. US Air Force report SAM-TR-78-22, USAF Sch. Aerosp. Med., Aerosp. Med. Div. (AFSC), Brooks Air Force Base, 1978.

46. EBNETH H. Metallized textile fabrics : their properties and technical applications. *J. Ind. Fabr.*, 3, n° 3, 1985, pp. 30-35.

47. EBNETH H., FITZKY H. Metallized textile fabrics for microwave-protective suits. *J. Ind. Fabr.*, 3, n° 3, 1985, pp. 36-42.

48. EDEN W. Microwave oven repair : hazard evaluation. In : Electronic product radiation and the health physicist. US Dep. Health, Educ., Welf., Bur. Radiol. Health, BRH-DEP-70-26, 1970, pp. 159-172.

49. EDMUNDS F. Naval exposure environment. In : PEYTON M. Proceedings of the 4th annual Tri-Service conference on biological effects of microwave-radiating equipments, biological effects of microwave radiation, 1961. New York, Plenum Press, RADC-TR-60-180, Vol. 1, 1961.

50. EFFERT S. The effect of electric and electromagnetic fields on the organism. In : 3rd international colloquium of the international section of the ISSA for the prevention of occupational risks due to electricity, Marbella, 27-29 Oct. 1975, B1240.

51. ELDER R., EURE J., NICOLLS J. Radiation leakage control of industrial microwave power devices. *J. Microw. Power*, 9, n° 2, 1974, pp. 51-61.

52. ELDER R., GUNDAKER W. Microwave ovens and their public health significance. *J. Milk Food Technol.*, 34, n° 9, 1971, pp. 444-446. (C.D.I.U.P.A., n° 40373).

53. ELDER R., KLEIN H. Product testing of microwave ovens for safety assurance. *J. Microw. Power*, 6, n° 4, 1971, pp. 291-295.

54. ELY T. Microwave death. *J. Am. Med. Assoc.*, 217, n° 1394, 1971. Letter to the editor.

55. ELY T. A letter to the editor. *J. Microw. Power*, 20, n° 2, 1985, p. 137.

56. EURE J., NICOLLS J., ELDER R. Radiation exposure from industrial microwave applications. *Am. J. Public Health, and Nations Health*, 62, n° 12, 1972, pp. 1573-1577.

57. FAGO E. Evaluation of radio-frequency protective clothing and measuring instruments. Kansas City, Midwest Res. Inst., Final report to Naval Ships Systems Command, Contract NOBSR-93169, March 1965-Aug. 1966.

58. FREY A. Science and standards - Data analysis reveals significant microwave-induced eye damage in humans. *J. Microw. Power*, 20, n°1, 1985, pp. 53-56.

59. GALLAGHER J. Standards and legislative aspects of microwave radiation hazards. IEEE-IMPI (Europe) Meeting, 26 Feb. 1974.

60. GANDHI O. The ANSI RF safety guideline - its rationale and some of its problems. Proceedings of the Symposium on Biological effects of electropollution. Washington, Howard Univ., 9 Sept. 1985.

61. GEORGE W. The need for voluntary standards. *Microw. Energy Appl. Newsl.*, 11, n° 4, 1978, pp. 3-6.

62. GERLING J. Microwave oven standards. *J. Microw. Power*, 13, n° 1, 1978, pp. 37-41.

63. GHOSH S. Microwave diathermy units - Are they hazardous ? *Am. Ind. Hyg. Assoc. J.*, 33, n° 2, 1972, p. 45.

64. GHOSH S., MUC A. The biological effects of electromagnetic waves and their relationship to pending Canadian regulations. In : Proceedings of the IMPI Symposium, Ottawa, 1972, pp. 25-28. (Paper 2.4.).

65. GHOSH S., MUC A., LETOURNEAU E., BELKHODE M. Safety considerations for the use and operation of microwave blood warmers. In : Proceedings of the IMPI Symposium, Ottawa, 1972, pp. 17-19. (Paper 2.1.).

66. GLASER Z., HEIMER G. Determination and elimination of hazardous microwave fields aboard naval ships. *IEEE Trans. Microw. Theory Tech.*, MTT-19, n° 2, 1971, pp. 232-238.

67. GORDON Z. Hygienic evaluation of the working conditions of workers with UHF generators. In : LETAVET A., GORDON Z. The biological action of

ultrahigh frequencies. *Akad. Med. Nauk.* Traduit du russe. US joint Pub. Res. Serv. Rep., JPRS 12471, 1960, pp. 22-25.

68. GORDON Z. The biological effects of microwaves and protective measures against them. Medical Electronics Conference 3, 1960, pp. 410-411. (Abstracts 1126 of 1962).

69. GORDON Z. Biological effects of microwaves in occupational hygiene. *Izdatel'stvo Med,*, Leningrad, traduction NASA TT 70-50087, TT-F-633, 1970.

70. GUNDAKER W., MOORE T., COPPOLA S. A study to determine the necessity for development of standards for the use of selected non-medical electronic products for commercial and industrial purposes. US Dep. Health, Educ., Welf., Bur. Radiol. Health, DEP 70-30, PB 196 448, 1970.

71. GUY A. Engineering considerations and measurements. In : AGARD Lecture Series n° 78 on Radiation Hazards, 1973, pp. 9:1-9:36.

72. GUY A. History of biological effects and medical applications of microwave energy. *IEEE Trans. Microw. Theory Tech.*, MTT-32, n° 9, 1984, pp. 1182-1200

73. GUY A., CHOU C., Mac DOUGALL J., SORENSEN C. Measurement of shielding effectiveness of microwave-protective suits. *IEEE Trans. Microw. Theory Tech.*, MTT-35, n° 11, 1987, pp. 984-994.

74. HANKIN N. An evaluation of selected satellite communication systems as sources of environmental microwave radiation. Technical Report of the Environmental Protection Agency, EPA-520/2-74-008, 1974.

75. HANKIN N. Radiation characteristics of traffic radar systems. Environmental Protection Agency. Technical note ORP/EAD-76-1, 1976.

76. HANKIN N., TELL R., JANES D. Assessing potential for exposure to hazardous levels of microwave radiation from high power sources. *Health Phys.*, 27, n° 6, 1974, p. 633.

77. HANSSON-MILD K. Occupational exposure to radio-frequency electromagnetic fields. *Proc. IEEE*, 68, n° 1, 1980, pp. 12-17.

78. HEIMER G., HOWARD K. Navy radio frequency radiation hazards program. *Safety Rev.*, 18, n° 4, 1961, p. 11.

79. HINES H., RANDALL J. Possible industrial hazards in the use of microwave radiation. *Electr. Eng.*, 71, n° 10, 1952, pp. 879-881.

80. HOFFMANN R. Microwave ovens : applications, dangers, testing. *Elektrotech. Z.*, B, 24, n° 21, 1972, pp. 537-540.

81. HOOLIHAN D. A technical analysis of the United States Environmental Protection Agency's proposed alternatives for controlling public exposure to radiofrequency radiation. In : IEEE international symposium on electromagnetic compatibility, Atlanta, 25-27 Aug. 1987, session 6C, pp. 501-504.

82. HORN L. UK standard for microwave cooking appliances. *J. Microw. Power*, 7, n° 2, 1972, pp. 139-142.

83. INGUS L. Why the double standard ? A critical review on the Russian work on the hazards of microwave radiation. In : IEEE Electromagnetic Compatibility Symposium, 1970, pp. 168-172.

84. JAMMET H. Recommandations sur les principes généraux de protection contre les rayonnements non-ionisants. 6ème Congrès international de l'IRPA, Berlin, 7-12 mai 1984. *Fachver. Strahlenschutz*, Vol. 1, mai 1984, pp. 31-38.

85. JANES D., TELL R., ATHEY T., HANKIN N. Nonionizing radiation exposure in urban areas of the United States. In : Proceedings of the 4th International Radiation Protection Association Congress, Paris, April 1977, pp. 329-332.

86. JOHNSON C. Research needs for establishing a radiofrequency electromagnetic radiation safety standard. *J. Microw. Power*, 8, n° 3-4, 1973, pp. 367-388.

87. JOHNSON C. Research needs for establishing a radiofrequency electromagnetic radiation safety standard. *J. Microw. Power*, 9, n° 3, 1974, pp. 219-220.

88. JOHNSON C. Radiofrequency radiation dosimetry handbook. US Air Force Sch. Aerosp. Med., Brooks Air Force Base, Report SAM-TR-76-35, 1976.

89. JOHNSON C. The role of radio science in investigating electromagnetic biological hazards. *Radio Sci.*, 12, n° 3, 1977, pp. 349-354.

90. JUSTESEN D. The committee on man and radiation : a commitment to the public interest. *Bull. New York Acad. Med.*, 55, n° 11, Dec. 1979, pp. 1267-1273.

91. JUSTESEN D. Scientific and hygienic issues in biological research on microwaves : toward rapprochement between East and West. *Radio Sci.*, 17, n° 1, 1982, pp. 1-12.

92. KALL A. Study of radiation hazards caused by high power, high frequency fields. Final report of a one-year study. Willow Grove, Ark Electronics for US Informat. Agency, contract IA-11651, 1968.

93. KLASCIUS A. Microwave radiation hazards around large microwave antennas. *Am. Ind. Hyg. Assoc. J.*, 33, n° 2, 1972, p. 46.

94. KNAUF G. The biological effects of microwave radiation on Air Force personnel. *Arch. Ind. Health*, 17, 1958, pp. 48-52.

95. KRYLOV V., SOLOVJEJ A. Safety measures recommended for work on radiofrequency generator installations. Gosudarstvennoje Naučnotehničeskoje Izdatel'stvo, Oborongiz, 1961, pp. 3-17 (FTD-TT 62-339, Wright Patterson Air Force Base, Ohio, 1961).

96. LEARY F. Researching microwave health hazards. *Electronics*, 32, n° 8, 1959, pp. 49-53.

97. LERNER E. RF radiation : biological effects. *IEEE Spectrum*, n° 12, 1980, pp. 51-59.

98. Mac CONNELL D. Microwave ovens - revolution in cooking. Part 2. Radiation and safety. *Electron. World*, 84, n° 3, 1970, pp. 37-39, 72.

99. Mac CONNELL D., FOERSTNER R., BUCKSBAUM A. Microwave oven design for radiation safety. In Proceedings of the 21st Appliance Technical Conference, G.A., 1970, 31 p. (Paper n° 70 CP 335-I).

100. Mac KAY H. Electromagnetic pollution measurement. *EID Electron. Instrum.*, V/I, 7, n°12, 1971, pp. 14-17.

101. Mac LAUGHLIN J. Tissue destruction and death from microwave radiation (radar). *California Med.*, 86, 1957, p. 336.

102. Mac REE D. Environmental aspects of microwave radiation. *Environ. Health Perspect.*, 6, 1972, pp. 41-53.

103. MAFRICI D., BERNSTEIN S., MILLER C. Nassau county microwave oven study, June 1969-March 1970. *Radiol. Health Data Rep.*, 11, n° 2, 1970, pp. 667-670.

104. MALABIAU R. Proximal region power density assessment in the scope of non ionizing radiation hazards prediction. In : Champs électromagnétiques dans le proche environnement des équipements industriels micro-onde et des antennes. Journées d'études de la SEE, Univ. Toulon Var, 19-20 janv. 1984, réf. 84-2520, 4 p.

105. MARHA K. Maximum admissible values of HF and UHF electromagnetic radiation at work places in Czechoslovakia. In : CLEARY S. Biological effects and health implications of microwave radiation. Symposium proceeding. Richmond, Commonwealth Univ., 17-19 Sept. 1969, US Dep. Health, Educ., Welf., BRH, DBE 70-2, PB 193898, Library of Congress card n° 76-607340, June 1970, pp. 188-196.

106. MARHA K. Microwave radiation safety standards in Eastern Europe. *IEEE Trans. Microw. Theory Tech.*, MTT-19, n° 2, 1971, pp. 165-168.

107. MARHA K., MUSIL J., TUHÁ H. Electromagnetic fields and the life environment. (Traduit du Tchèque). San Francisco, San Francisco Press, 1971, 140 p.

108. MEISEL N. Les micro-ondes, savoir.. et..bien faire. *Ind. Alim. Agric.*, 101, n° 4, 1984, pp. 259-264. (C.D.I.U.P.A., n° 221917)

109. MERCKEL C. Microwave and man - Direct and indirect hazards, and precautions. *California Med.*, 117, n° 1, 1972, p. 20.

110. MERRITT J. Science and standards- another viewpoint. *J. Microw. Power*, 20, n° 1, 1985, pp. 55-56.

111. MICHAELSON S. Microwave hazards evaluation - Concepts and criteria. In : Proceedings of the IMPI Symposium, Edmonton, 1969, pp. 83-84. (Paper DA.6.).

112. MICHAELSON S. Microwave hazards evaluation - Concepts and criteria. *J. Microw. Power*, 4, n° 2, 1969, pp. 114-119.

113. MICHAELSON S. Biomedical aspects of microwave exposure. *Am. Ind. Hyg. Assoc. J.*, 32, 1971, pp. 338-345.

114. MICHAELSON S. Human exposure to non-ionizing radiant energy - Potential hazards and safety standards. *Proc. IEEE*, <u>60</u>, n° 4, 1972, pp. 389-421.

115. MICHAELSON S. Microwave exposure safety standards - Physiologic and philosophic aspects. *Am. Ind. Hyg. Assoc. J.*, <u>33</u>, n° 3, 1972, p. 156.

116. MICHAELSON S. Protection guides and standards for microwave exposure. *AGARD Lecture Series*, n° 78, Radiation hazards, Sept. 1975, pp. 12.1-12.6.

117. MICHAELSON S., SÜSS M. An international program for microwave exposure protection. *IEEE Trans. Microw. Theory Tech.*, MTT-<u>22</u>, n° 12, 1974, pp. 1301-1302.

118. MININ B. SVČ i bjezopastnost' čelovjeka. *Sovetskoje Radio*, Moskva. Traduction anglaise, Microwaves and human safety, Arlington, US Joint Pub. Res. Serv. Rep., JPRS 65506-1, 65506-2, 1975.

119. MOBLEY M. F.C.C. Equipment type approval program with reference to microwave oven tests. *J. Microw. Power*, <u>6</u>, n° 4, 1971, pp. 297-303.

120. MOBLEY M. Revision of federal communications-commission type approval tests procedures. In : Proceedings of the IMPI Symposium, Loughborough, 1973, pp. 1-2. (Paper 4B.6.).

121. MOORE W. Literature survey - Biological aspects of microwave radiation. A review of hazards. In : Radiation Control for Health and Safety Act of 1967. Hearings before the Committee on Commerce, US Senate. Washington, US Government Printing Office, Part 2, Serial n° 90-49, pp. 1191-1215.

122. MORGAN W. Microwave radiation hazards. *Arch. Ind. Health*, <u>21</u>, 1960, pp. 570-573.

123. MOSELEY H., DAVIDSON M. Exposure of physiotherapists to microwave radiation during microwave diathermy treatment. *Clinical Phys. Physiol. Meas.*, <u>2</u>, 1981, pp. 217-221.

124. MUC A. Standards, regulations and health effects - infrared microwave and radiofrequency. In : Radiant Wave Electroheat Workshop, Toronto, 2 Dec. 1985, pp. 63-71.

125. MUMFORD W. Some technical aspects of microwave radiation hazards. *Proc. IRE*, <u>49</u>, n° 2, 1961, pp. 427-447.

126. OATES W., SNELLINGS D., WILSON E. Microwave oven survey results in Arkansas during 1970. *Am. J. Public Health*, <u>63</u>, n° 3, 1973, pp. 193-198.

127. ODLAND L. Radio-frequency energy : a hazard to workers ? *Ind. Med. Surg.*, <u>42</u>, 1973, pp. 23-26.

128. OSEPCHUK J. A microwave radiation safety standard for industrial equipment. *J. Microw. Power*, <u>6</u>, n° 3, 1971, pp. 191-192.

129. OSEPCHUK J. Characteristics of principal microwave radiation safety standards exposure. Characteristics of principal microwave oven leakage performance standards. In : Microwave Engineers' Handbook. Vol. 2, Dedham, Artech House Inc., 1971, pp. 200-201.

130. OSEPCHUK J. A review of microwave oven safety. *J. Microw. Power*, <u>13</u>, n° 1, 1978, pp. 13-26.

131. OSEPCHUK J. A review of microwave oven safety. *Microw. J.*, <u>22</u>, n° 5, 1979, pp. 25-37.

132. OSEPCHUK J. Radiofrequency hazards and side effects : towards the ultimate electromagnetic compatibility. In : Proceedings of the IMPI Symposium, Vienna, (USA), 1983, pp. 3-4.

133. OSEPCHUK J. Safety standards. In : OSEPCHUK J. Biological effects of electromagnetic radiation. Piscataway, IEEE Press, IEEE Serv. Center, 1983, 608 p., section VI.

134. OSEPCHUK J. A history of microwave heating applications. *IEEE Trans. Microw. Theory Tech.*, MTT-<u>32</u>, n° 9, 1984, pp. 1200-1224.

135. OSEPCHUK J., FOERSTNER R., Mac CONNELL D. Computation of personnel exposure in microwave leakage fields and comparison with personnel exposure standards. In : Proceedings of the IMPI Symposium, Loughborough, 1973, pp. 1-2. (Paper 4B.4.).

136. OSIPOV J. Gigijena truda i vlijanije na rabotajuščih elektromagnitnyh elektročastot. *Izdatel'stvo Med.*, Leningrad, 1965, pp. 104-144.

137. PHARR M. Safety factors of microwave ovens. Technical report, Bâton Rouge, Gulf Southern Res. Inst., 1970, 3 p.

138. PHILLIPS R. Problems of physical and biological dosimetry of microwave irradiation. In : Proceedings of the IMPI Symposium, Monterey, 1971. (Paper 7.7.).

139. PHILLIPS R., HUNT E., KING N. Field measurement, absorbed dose, and biological dosimetry of microwaves. In : TYLER P. Biologic effects of non-ionizing radiation. Proceedings of a conference, *Ann. New York Acad. Sci.*, 247, 1975, pp. 499-509.

140. PIOTROWSKI M., DZIECIOLOWSKI D. Microwave dosimetry. In : Proceedings of the IMPI Symposium, Waterloo (Can.), 1975, pp. 144-147. (Paper 8.5.).

141. POWELL C., VERNON E. Health surveillance of microwave hazards. *Am. Ind. Hyg. Assoc. J.*, 31, 1970, pp. 358-361.

142. PRESMAN A. Temporary sanitation rules when working on cm-wave generators. *Gigijena Sanit.*, 1, 1958, p. 21. Traduit du russe, US Joint Pub. Res. Serv. Rep., JPRS 12471.

143. PRUCHA R. Human thermal loading by exposure to emissions from a microwave oven. In : Proceedings of the IMPI Symposium, Leuven, 1979. (Paper 4A.7.).

144. QUON K. Hazards of microwave radiation. *Ind. Med. Surg.*, 29, July 1960, pp. 315-318.

145. QUON K. Hazards of microwave radiation. *US Navy Med. News Lett.*, 36, 18 Nov. 1960, pp. 29-34.

146. REINS D., WEISS R. Physiological evaluation of effects of personnel wearing the microwave protective suit and overgarment. Natick, US Navy Clothing and Textile Res. Unit, Work order 523-003-10, July 1969.

147. REPACHOLI M. Control of microwave exposure in Canada. *Trans. IMPI*, 8. Microwave bioeffects and radiation safety, 1978, pp. 102-112.

148. REPACHOLI M. Proposed exposure limits for microwave and radio-frequency radiations in Canada. *J. Microw. Power*, 13, n° 2, 1978, pp. 199-211.

149. REPACHOLI M. Les fours à micro-ondes. *Object. Prév.*, 2, n° 7, 1979, pp. 7-9.

150. REPACHOLI M. Taking the heat out of microwaves. *Emerg. Plan. Dig.*, 6, n° 2, 1979, pp. 7-9.

151. REPACHOLI M., STUCHLY M. Emission and exposure standards for microwave radiation. IEEE Conference and exposition, Toronto, Sept. 1977, pp. 92-93, paper 77114.

152. REXFORD-WELCH S., LINDSAY I. The practice of microwave radiation safety. In : Proceedings of the IMPI Symposium, Leuven, 1976. (Paper 4A.8.).

153. ROMERO-SIERRA C. Bioeffects of electromagnetic waves. In : BOWHILL S. Review of Radio Science, 1978-1980. Bruxelles, URSI, 1981, ch. 2, pp. A.B.1-A.B.17.

154. ROMERO-SIERRA C. Bioeffects of electromagnetic waves. In : BOWHILL S. Review of Radio Science, 1981-1983. Bruxelles, URSI, 1984, ch. 9, pp. K.1-K.17.

155. ROSE V., GELLIN G., POWELL C., BOURNE H. Evaluation and control of exposures in repairing microwave ovens. *Am. Ind. Hyg. Assoc. J.*, n° 3-4, 1969, pp. 137-142.

156. RUTKOWSKI A., CHRISTIANSON C. Development of radiation hazard suit and RF measuring techniques. Brooklyn US Naval Appl. Sci. Lab., NASL-9400-20-1, 1965.

157. SADLACK J. Microwave standards - Consumer's viewpoint. *Microw. Energy Appl. Newsl.*, 9, n° 3, 1978, pp. 3-4.

158. SAVIN V., NIKONOVA K., LOBANOVA E. Nouvelles tendances en matière de normes de rayonnement électromagnétique micro-ondes. (En russe). *Gigijena Truda Prof. Zabolevanija*, 3, 1983, pp. 1-3.

159. SCHUY S., LEITGEB N. Gutachten über schädliche Wirkungen von Mikrowellen. *Ernährung*, 9, n° 10, 1985, pp. 710-714.

160. SCHWAN H. Radiation biology, medical applications and radiation hazards. In : OKRESS E. Microwave Power Engineering, Vol. 2, New Tork, Academic Press, 1968, pp. 215-532.

161. SCHWAN H. Interaction of microwave and radio-frequency radiation with biological systems. *IEEE Trans. Microw. Theory Tech.*, MTT-19, n° 2, 1971, pp. 146-152.

162. SCHWAN H. Non-ionizing radiation hazards. *J. Franklin Inst.*, 296, n° 6, 1973, pp. 495-497.

163. SCHWAN H. Keynote address and C.C. Johnson memorial lecture. Microwave bioeffects research : historical perspectives on productive approaches. *J. Microw. Power*, 14, n° 1, 1979, pp. 1-5.

164. SCHWAN H. Microwave and RF hazard standard consideration. *J. Microw. Power*, 17, n° 1, 1982, pp. 1-9.

165. SCHWAN H. Science and standards - RF-hazards and standards : an historical perspective. *J. Microw. Power,* 19, n° 4, 1984, pp. 225-231.

166. SCHWAN H., LI K. Capacity and conductivity of body tissues at ultra-high frequencies. *Proc. IRE,* 41, 1953, pp. 1735-1740.

167. SCHWAN H., LI K. Hazards due to total body irradiation by radar. *Proc. IRE,* 44, Nov. 1956, pp. 1572-1581.

168. SCOTT G. The development of a microwave oven performance standard. Problems and possibilities. In : Proceedings of the IMPI Symposium, Loughborough, 1973, pp. 1-2. (Paper 5B.1.).

169. SCOTT J. Is today's standard for microwave radiation safe for humans ? *Microw.,* 10, n° 1, 1971, pp. 9-14.

170. SERVANTIE B., TYLER P. Les champs électromagnétiques non ionisants facteurs d'environnement en milieu militaire. Comptes-rendus du stage ADERA, Univ. Bordeaux II. 15-17 nov. 1977.

171. SETTER L. Regulations, standards, and guides for microwaves, ultraviolet, and radiation from lasers and television receivers - an annotated bibliography. Washington, Public Health Serv., 1969, p. 63.

172. SHINN D. Avoidance of radiation hazards from microwave antennas. *Marconi Rev.,* 39, 1976, pp. 61-80.

173. SLINEY D., PALMISANO W. Microwave hazards bibliography. Edgewood Arsenal, Army environmental Hygiene Agency, AD-n° 652-708, 20 April 1967.

174. SLINEY D., WOHLBARSHT M., MUC A. Differing radiofrequency standards in the microwave region - Implications for future research. *Health Phys.,* 49, 1985, Editorial.

175. SMITH S., NICHOLS J., MOORE R., GUNDAKER W. Laboratory testing and evaluation of microwave ovens. US Dep. Health, Educ., Welf., Bur. Radiol. Health, BRH-DEP-70-25, 1970.

176. SOLON L. A local Health Agency approach to a permissible environmental level for microwave and radiofrequency radiation. *Bull. New York Acad. Med.,* 55, 1979, pp. 1251-1266.

177. STANLEY J. Radiation exposure considerations when employing microwave excited spectroscopic sources. *Appl. Spectrosc.,* 27, 1973, pp. 265-267.

178. STENECK N. Values in standards : the case of ANSI C95.1-1982. *Microw. RF,* May 1983, pp. 137-167.

179. STENECK N. Science and standards - The case of ANSI C95.1-1982. *J. Microw. Power,* 19, n° 3, 1984, pp. 153-158.

180. STENECK W., COOK H., VANDER A., KANE G. The origins of US Safety Standards for microwave radiation. *Science,* 208, 1980, pp. 1230-1237.

181. STERZER F. Radiation protection in New Jersey. *Microw. RF,* 23, n° 10, 1984, pp. 40-43.

182. STUCHLY M. Potentially hazardous microwave radiation sources - a review. *J. Microw. Power,* 12, n° 4, 1977, pp. 369-381.

183. STUCHLY M., REPACHOLI M. Microwave and radio-frequency protection standards. *Trans. IMPI,* 8, Microwave bioeffects and radiation safety, 1978, pp. 95-101.

184. STUCHLY M., REPACHOLI M., LECUYER D. The impact of regulations on microwave ovens in Canada. *Health Phys.,* 37, 1979, pp. 137-144.

185. STUCHLY M., STUCHLY S. Industrial, scientific, medical and domestic applications of microwaves. *IEE Proc.,* 130, part A, n° 8, Nov. 1983. pp. 467-503.

186. SUROVIEC H. Microwave oven radiation hazards in food vending establishments. *Arch. Environ. Health,* 14, n° 3, 1967, pp. 469-472.

187. SÜSS M., BENWELL D. Institutions and legislation concerned with nonionizing radiation, health related research and protection - A directory. København, OMS/WHO, Reg. Office Europe, Ottawa, WHO Collab. Agency Radiat. Prot. Bur., Minist. Santé Bien-être Social/Health Welf. Canada, 1985, 60 p.

188. SÜSSKIND C. Review of "Electromagnetic fields and life" (Presman, A.S., 1970), Review of "Electromagnetic field and the life environment" Marka, K et al., 1971). *IEEE Trans. Microw. Theory Tech.,* MTT-19, n° 2, 1971, p. 248.

189. SWANSON J., ROSE V., POWELL C. A review of international microwave exposure guides. In : Electronic product radiation and the health physicist. US Dep. Health, Educ., Welf., Bur. Radiol. Health, DEP-70-26, 1970, pp. 95-110.

190. TAYLOR J. Microwave radiations hazards control in the US Army. In : Proceedings of the IMPI Symposium, Ottawa, 1972, pp. 31-32. (Paper 2.6.).

191. TAYLOR U. Evaluation of hazards from microwave ovens. *Am. Ind. Hyg. Assoc. J.*, 33, n° 2, 1972, p. 45.

192. TELL R. Environmental non-ionizing exposure. A preliminary analysis of the problem and continuing work with EPA. EPA/ORP 73-2, Environmental exposure to non-ionizing radiation, 1973, pp. 47-68.

193. TELL R. An analysis of radio-frequency and microwave absorption data with consideration of thermal safety standards. EPA Technical Note, ORP/EAD 78-2, 1978.

194. TELL R., HARLEN F. A review of selected biological effects and dosimetric data useful for development of radiofrequency standards for human exposure. *J. Microw. Power*, 14, n° 4, 1979, pp. 405-424.

195. TELL R., NELSON J. Microwave hazard measurements near various aircraft radars. *Radiat. Data Rep.*, 15, 1974, pp. 161-179.

196. TELL R., NELSON J., HANKIN N. HF spectral activity in the Washington, D.C., area. *Radiat. Data Rep.*, 15, 1974, pp. 549-558.

197. THOUREL L. Réglementation française sur les utilisations industrielles des ondes électromagnétiques. In : Nouvelles applications de l'électricité. Colloque international, Journees d'Electronique de Toulouse, 7-11 mars 1977, III-3, 9 p.

198. TILTON R. The health hazards of microwaves with special reference to microwave ovens. *Q. Bull. Assoc. Food Drug. Off. US*, 35, n° 4, 1971, pp. 271-275.

199. TURNER J. The effects of radar on the human body ; results of Russian studies on the subject. ASTIA AD 278172, JPRS-12471, Washington, 1962. (Résumé des travaux de Letavet et Gordon).

200. VAN DE GRIEK A., BRITAIN R. Amendments to the U.S. Department of Health, Education, and Welfare microwave oven performance standard. In : Proceedings of the IMPI Symposium, Loughborough, 1973, pp. 1-2. (Paper 4.B.3.).

201. VAN DE GRIEK A., BRITAIN R. Amendments to the US Department of Health, Education, and Welfare microwave oven performance standard. *J. Microw. Power*, 9, n° 1, 1974, pp. 3-11.

202. VAUTRIN J. Les problèmes de sécurité posés par l'emploi des hautes fréquences et des micro-ondes. In : Comptes rendus du Colloque sur les applications industrielles des hautes fréquences et des micro-ondes, Lyon, 1977, 19 p.

203. VAUTRIN J. Risques liés aux rayonnements électromagnétiques non ionisants. *Econ. Progrès Electr.*, n° 7, janv.-fév. 1983, pp. 15-19.

204. VAUTRIN J., CAVELIER C., CLAUZADE B. Le rayonnement électromagnétique non ionisant. Domaine des radiofréquences et hyperfréquences, applications et risques. In : Travail et Sécurité, et tiré à part de l'I.N.R.S., éd. n° 110, Code 8432, 1977, 7 p.

205. VAUTRIN J., CAVELIER C., CLAUZADE B. Le rayonnement électromagnétique "radiofréquences", applications et risques. *Cahier Notes Doc.*, n° 92, 1978, pp. 361-385, et tiré à part de l'I.N.R.S., note n° 1127-92-78 (39-14), CDU 621.37:538.56.

206. VAUTRIN J., CLAUZADE B., CAVELIER C. Radiofréquences et hyperfréquences. Applications et risques. In : Effets biologiques des rayonnements non ionisants ; utilisation et risques associés, 9ème congrès international de la Société Française de Radioprotection, Nainville les Roches, 22-26 mai 1978, pp. 357-364.

207. VILLFORTH J. US Public Health Service concern and activities related to health implications of microwave radiation. In : CZERSKI P., OSTROWSKI K., SHORE M., SILVERMAN C., SÜSS (M.)., WALDESKOG B. Biological effects and health hazards of microwave radiation. Proceedings of an international symposium, Warszawa, 15-18 Oct. 1973. Warszawa, Polish Medical Publishers, 1974.

208. VOGELMAN J. Physical aspects of biological hazards from RF sources. In : Proceedings of the IMPI Symposium, Monterey, 1971. (Paper 7.2.).

209. VOSS W. Microwave safety, education required. *Microw. World*, 2, n° 1, Jan.-Feb. 1981, pp. 12-15.

210. VOSS W. Microwave safety : continuing education is required. *Microw. World*, 2, n° 2, March-April 1985, pp. 10-13.

211. WACKER P., BOWMAN R., Quantifying hazardous electromagnetic fields : scientific basis and practical considerations. *IEEE Trans. Microw. Theory Tech.*, MTT-19, n° 2, 1971, pp. 178-187.

212. WHITE D. A handbook on EMI control methods and techniques. Germantown, Don White Consultants, Inc., 1973, § 3.1. Natural receptors of EMI, pp. 3.3-3.10.

213. ZARET M. Investigation of personnel hazard associated with radio-frequency fields encountered in naval operations. Final Report for ONR, Contract N00014-69-C-0358, 1971.

214. ZUK W., REPACHOLI M. How safe are microwave ovens. *Bull. Am. Phys. Soc.*, 21, n° 5, 1976, p. 836.

215. Amendments to the microwave ovens standard : measurement and test conditions. US Dep. Health, Educ., Welf., Bur. Radiol. Health. *Fed. Regist.*, 45, n° 87, 1980, § 1030, pp. 29307-29308.

216. American national safety levels with respect to human exposure to radio frequency electromagnetic fields, 300 kGz to 100 GHz. ANSI standard C95.1-1982, American National Standards Institute, approved 30 July 1982.

217. 1969 annual report to the Congress on the administration of the Radiation Control for Health and Safety Act of 1968, public law 90-602. Rockville, US Dep. Health, Educ., Welf., Environ. Health Serv., Bur. Radiol. Health, BRH/OBD 70-3, 1 April 1970, 54 p., plus Attachment 1, Report to the Congress on studies conducted pursuant to section 357 of the Radiation Control of Health and Safety Act of 1968, public law 90-602, 1 Jan. 1970, revised 1 July 1970, 42 p., Attachment 2, List of BRH detailed study reports prepared pursuant to section 357, public law 90-602, 1 p.

218. Bureau study shows cataracts not linked to low-level microwave exposure. US Dep. Health, Educ., Welf., Bur. Radiol. Health, 12, n° 13, 1978, et tiré à part de l'IMPI, News Release, 1 p.

219. Canada-wide survey of non-ionizing radiation-emitting medical devices. Part I : Short-wave and microwave devices. Ottawa, Non-ionizing Radiat. Sect., Radiat. Prot. Bur., Ministère Santé Bien-être Social/Health Welfare Canada, 80-EHD-52, 1980

220. Construction safety and health regulations, 2nd part. Dep. Labor, Occup. Safety Health Adm., *Fed. Regist.*, 39, n° 122, 24 June 1974, Subpart D, § 1926-54, 1, p. 22810.

221. Control of hazards to health from microwave radiation. Washington, US Air Force, report n° TB-MED-270/AFM-161-7.

222. Critique of the biological hazards of microwave radiation. Washington, US Air Force, George Washington Univ., report n° 56-21 AF 18 (600) 1180, 1956.

223. Electronic product radiation and the health physicist. In : Annual Midyear Topical Symposium. 4. US Dep. Health, Educ., Welf., Bur. Radiol., Health, BRH, DEP 70-26, PB 195 772, 1970

224. Exposure to electromagnetic pulses. US Dep. Labor, *Fed. Regist.*, 39, 1974, p. 7499.

225. Federal radiation protection guidance for public exposure to radiofrequency radiation. Advance notice of proposed recommendations, Electromagnetic Energy Policy Alliance, *Fed. Regist.*, 47, n° 247, 23 Dec. 1982, pp. 57338-57340.

226. Federal radiation protection guidance ; proposed alternatives for controlling public exposure to radiofrequency radiation. *Fed. Regist.*, 51, n° 146, 30 July 1986.

227. Fundamentals of non-ionizing radiation protection. US Dep. Health, Educ., Welf., Bur. Radiol. Health, BRH, S-16 04, 1971.

228. Gefahrdung durch elektromagnetische Felder, Deutsche Norn DIN-57848. Berlin, VDE Verlag, 1984.

229. Guide for submission of information on microwave heating equipment required pursuant to 21 CFR 1002.10. US Dep. Health, Educ., Welf., Bur. Radiol. Health, OMB n° 57 R 0068, 1971, 6 p.

230. Guide for the filing of annual reports. US Dep. Health, Educ. Welf., Bur. Radiol. Health, OMB n° 57 R 0068, 1977, p. 12. (Page A-2 § 4.2. Microwave ovens).

231. Guidelines for microwave diathermy devices, Lettre d'information n° 585, Ottawa, Non-ionizing Radiat. Sect., Radiat. Prot. Bur., Ministère Santé Bien-être social/Health and Welfare Canada, 19 sept. 1980.

232. Hearings of the Senate of the United States of America on the "Radiation Control for Health and Safety Act" of 1968. 8, 9, 12 March 1973, Washington, US Government Printing Office, ref. 96-087, 1973, pp. 100-109.

233. Human exposure to microwaves and other radiofrequency electromagnetic fields, COMAR position paper. *IEEE Microw. Theory Tech.*, MTT-S, winter 1982, pp. 28-29.

234. IMPI Microwave safety : hazards in perspective. IMPI Publication 15-2, 1975 et *J. Microw. Power*, 10, n° 4, 1975, pp. 333-341.

235. IMPI performance standard on leakage from industrial microwave systems. Edmonton, IMPI, Publication IS-d, Aug. 1973, 6 p.

236. Importations of electronic products subject to radiation emission standards established by the Department of Health, Education, and Welfare. Washington, Dep. Treas., US Customs Serv., RES-2-0-DC, 1974, 6 p.

237. Interim guidelines on limits of exposure to radiofrequency fields in the frequency range from 100 kHz to 300 GHz. Int. Nonionizing Radiat. Comm., Int. Radiat. Prot. Assoc., INIRC/IRPA. *Health Physics*, 46, n° 4, 1984, pp. 975-984.

238. Jednotná metodika stanovení intensity pole a ozáření elektromagnetickými vlnami v pásma vysokých frekvencí a velmi vysokých frekvencí k hygienickým účelům. Výnos hlavního hygienika ČSR čj. HE 344.-5 - 3.2. 70. Příloha Č.3 k Informačním zprávám o oboru hygieny práce a nemocí z povolání, Praha, 1970.

239. L'environnement électromagnétique. *J. Télécomun.*, 48, VII/1981, pp. 414-415.

240. Letters addressed to Sen. TUNNEY by Pr. MICHAELSON, LEHMANN, Dr. APPLETON, GUY et al., in Senate Hearing Record, Public n° 93-24. Washington, US Government Printing Office, 8, 9, 12 March 1973.

241. Loi sur les dispositifs émettant des radiations. Règlement sur les dispositifs émettant des radiations-modification, C.P. 1974-2330, 22 Oct. 1974, partie III, pp. 2822-2825, *Canada Gaz./Gaz. Canada*, II, 108, n° 21, SOR/DORS/74-601, 1974.

242. Massachusetts non-ionized radiation regulations. Special Report. *Microw. J.*, Euro-global edition, 25, n° 5, 1982, pp. 170-171.

243. Maximum exposure levels-radiofrequency radiation, 300 kHz-300GHz. Australian Standard 2772. Sydney, Standards Association of Australia, 1985.

244. Measurements of emission levels during microwave and shortwave diathermy treatments. US Dep. Health, Educ., Welf., Bur. Radiol. Health, BRH-FDA-80-8119, 1980, 16 p.

245. Microwave diathermy devices. Lettre d'information n° 556. Ottawa, Non-ionizing Radiat. Sect., Radiat. Prot. Bur., Ministère Santé Bien-être Social/Health Welfare Canada, 5 Juil. 1979.

246. Microwave diathermy products ; performance standard. *Fed. Regist.*, 45, n° 147, 29 July 1980, § 1030. pp. 50359-50368.

247. Microwave oven. Canadian Ministry National Health Welfare, *Canada Gaz./Gaz. Canada*, II, 106, n° 5, 1972, pp. 266-271.

248. Microwave oven. Canadian Ministry National Health Welfare, Radiation emitting devices regulations, schedule B, part III, *Canada Gaz./ Gaz. Canada*, I, 13 juil. 1974, pp. 2673-2676.

249. Microwave oven is not recommended. *Consum. Rep. Magaz.*, 38, n° 4, 1973, pp. 221-230.

250. Microwave oven leakage : federal regulations soon. *Microwaves*, 9, n° 9, 1970, pp. 17-24.

251. Microwave oven regulations, amendment, part III, schedule II, *Canada Gaz./Gaz. Canada*, I, 12 nov. 1983, p. 1011.

252. Microwave ovens. AMA Committee on occupational toxicology, editorial. *J. Am. Med. Assoc.*, 215, n° 10, 1971, pp. 1661-1662.

253. Microwave ovens. Canadian Standards Association. Standard C 22.2, n° 150. 1974, 19 p.

254. Microwave ovens. US. Dep. Health, Educ., Welf., Bur. Radiol. Health, *Fed. Regist.*, 40, n° 124, 1975, § 1030.10, pp. 27038-27039.

255. Microwave ovens. US Dep. Health, Educ. Welf., Bur. Radiol. Health, *Fed. Regist.*, 40, n° 216. 1975. p. 52007.

256. Microwave safe exposure level criticized. *Microwaves*, 8, n° 9, 1969, pp. 13-14.

257. New Soviet population standard : 10 µW/cm^2 at MW frequencies. *Microw. News*, 5, n° 5, 1985, pp. 1-5.

258. NIOSH draft occupational RF/MW exposure standard. *Bioelectromagn.Soc. Newsl.*, 62, Oct. 1985, pp. 4-5.

259. Normes sanitaires temporaires et règles de protection de la population générale vis-à-vis des effets des champs électromagnétiques engendrés par les équipements de radio-émission. (En russe). Moskva, Ministjerstvo Zdravohranjenija, ref. 2963-84, 1984.

260. Occupational safety and health standards. 2nd part. Dep. Labor, Occupat. Safety Health Adm., *Fed. Regist.*, 39, n° 25, 1974, subpart G, § 1910. 97-1910.100, subpart i, § 1910.137-1910.140.

261. Performance standard for microwave ovens, Code of federal regulation. *Fed. Regist.*, 35, n° 194, 1970, title 42, part 78, subpart C, section 78.212, pp. 15642-15644.

262. Performance standard proposed for microwave diathermy equipment. Dep. Health, Educ., Welf., Bur. Radiol. Health, *BRH Bull.*, XIV, n° 15, 1980, 6 p.

263. Performance standards for microwave and radio-frequency emitting products. US Dep. Health, Educ. Welf., Bur. Radiol. Health, *Fed. Regist.*, 40, n° 124, 26 June 1975, part I, title 21, subpart J, § 1030.

264. Performance standards for microwave and radio-frequency emitting products. US Dep. Health, Educ., Welf., Bur. Radiol. Health, *Fed. Regist.*, 40, n° 216, 7 Nov. 1975, part I, title 21, subpart J, § 1030.

265. Petition by consumers union to amend the performance standard for microwave ovens. *Fed. Regist.*, 38, n° 177, 13 Sept. 1973, pp. 25421-25654.

266. Portland RF/MW proposal changed to 200 µW/cm^2. *Microw. News*, V, n° 8, Nov.-Dec. 1985, p. 7.

267. Program for control of electromagnetic pollution of the environment. Canadian Office of Telecommunication Policy, The assessment of biological hazards of non-ionizing electromagnetic radiation. 1er Rapport de mars 1973, 2ème Rapport de mai 1974, 3ème Rapport d'avril 1975, 4ème Rapport de juin 1976.

268. Progress in radiation protection. US Dep. Health, Educ., Welf., Bur. Radiol. Health, BRH, FDA-75-8023, 1974, 7 p. (Page 2, Microwave ovens).

269. Proposals for the health protection of workers and members of the public against the dangers of extra low frequency, radiofrequency and microwave radiations : a consultative document. National Radiological Protection Board, NRPB-82, 1982.

270. Provisorische Sicherheitsvorschriften für Mikrowellen-Backöfen. Schweizerischer elektrotechnischer Verein, TP-61/1A-D, 1975, 12 p.

271. Radiation control for health and safety act, public law 90-602, 90th Congress, H.R. 10790, 18 Oct. 1968.

272. Radiation emitting devices regulations, microwave ovens, amendments. *Canada Gaz./Gaz. Canada*, II, 2 Déc. 1979.

273. Radiofrequency and microwaves, environmental health criteria n° 16, Genève, OMS/WHO, 1981.

274. Recommended practice for the measurement of hazardous electromagnetic fields - RF and microwave. American National Standards Institute, Standard ANSI C95.5-1981, 1981.

275. Recommended safety procedures for the installation and use of radiofrequency and microwave devices in the frequency range 10 MHz-300 GHz, Safety code 6. Ottawa, Non-ionizing radiat. Sect., Radiat. Prot. Bur., Ministère Santé Bien-être Social/ Health Welfare Canada, 79-EHD-30, 1979.

276. Règlement du 25 mai 1972 concernant
la sécurité et l'hygiène du travail
en utilisation d'équipements émettant
des champs électromagnétiques dans
la gamme des micro-ondes. (En polonais). *Rad Minist.*, *Dz. Ustaw PRL*,
n° 21, 1972, pp. 193-195.

277. Règlement - Fours à micro-ondes.
Lettre de renseignements du Minis-
tère Canadien de la Santé et du
Bien-être Social, Dir. Gen. Prot.
Santé, n° 450, 1978.

278. Règlement - Fours à micro-ondes.
Lettre de renseignements du Minis-
tère Canadien de la Santé et du
Bien-être Social, Dir. Gén. Prot.
Santé, n° 580, 1980 citant le décret
C.P. 1979-3259 du 29 sept. 1979.
Canada Gaz./Gaz. Canada, partie II,
12 déc. 1979.

279. Règlements provisoires de sécurité
concernant le personnel travaillant
en présence de générateurs de micro-
ondes. (En russe). *Minist. Zdravohr.
SSSR*, n° 273-58, 20 nov. 1958.

280. Regulations for the administration
and enforcement of the Radiation
Control for Health and Safety Act of
1968. US Dep. Health, Educ., Welf.,
Bur. Radiol. Health, BRH-OBD 71-1,
1971, 20 p.

281. Regulations for the administration
and enforcement of the Radiation Control for Health and Safety Act of
1968. US Dep. Health, Educ., Welf.,
Bur. Radiol. Health. *Fed. Regist.*, 38,
n° 151, 1973, part I, subpart F, 278.

282. Regulations for the administration
and enforcement of the Radiation
Control for Health and Safety Act
of 1968. US Dep. Health, Educ., Welf.,
Bur. Radiol. Health, BRH-FDA-75-8003,
§ 1030.10, Microwave ovens, July
1974, pp. 36-37.

283. Regulations for the administration
and enforcement of the Radiation
Control for Health and Safety Act
of 1968. US Dep. Health, Educ., Welf.,
Bur. Radiol. Health, BRH-FDA-76-8035,
1976, 58 p., Microwave ovens, p. 40.

284. Regulations for the administration
and enforcement of the Radiation
Control for Health and Safety Act
of 1968. US Dep. Health, Educ., Welf.,
Bur. Radiol. Health, BRH-FDA-80-8035,
1980, 80 p.

285. Risques liés aux rayonnements électromagnétiques non ionisants. Inst. Nat.
Rech. Sécurité (INRS), tiré à part des
Cahiers de Notes Documentaires, n° 107,
2° trim. 1982, pp. 263-268.

286. Rozporządzenie rady ministrów z dnia
20.10.1961 w sprawie bezpieczeństwa
i higieny pracy uzywaniu urządzeń
mikrofalowych. Dziennik Usław PRL.
(*J. Lois RP Pologne*), n° 48, poz.
255, 1961.

287. Rozporządzenie rady ministrów z dnia
25.2.1972 w sprawie bezpieczeństwa
i higieny pracy przy stosowaniu
urządzeń wytwarzających pola elektro-
magnetyczne w zakresie mikrofalowym.
Dziennik Usław PRL. (*J. Lois RP
Pologne*), n° 21, n° 2, poz. 153,
1972.

288. Safety code. Recommended installation and safety procedure for all
closed and open beam microwave
devices. Health Welf. Canada, Feb.
1977, 40 p.

289. Safety level of electromagnetic
radiation with respect to personnel.
American National Standards Institute,
Standard n° ANSI C 95-1, *Fed. Regist.*,
36, n° 105, 1971, pp. 10522-10523,
taking effect 15 Feb. 1972.

290. Sanitarnyje normy i pravila pri
rabote s istočnikami elektromagnitnyh poljejvysokih, ultravysokih i
sverhvysokih častot. *Minict. Zdravohr.*,
SSSR, 30, 03, n° 848-70, 1970.

291. Sécurité des appareils électrodo-
mestiques et analogues. 2ème partie :
règles particulières pour les appareils de cuisson à micro-ondes. (En
français et anglais). Norme de la
CEI, Bur. Cent. Comm. Electrotech.
Int., Publication 335-25, 1ère éd.,
Genève, 1976, 53 p.

292. Sécurité des appareils électrodo-
mestiques et analogues. 2ème partie :
appareils de chauffage des aliments
par micro-ondes. Norme française
homologuée NF-C73-601 adoptée le 21
janv. 1980, homologuée par arrêté du
5 déc. 1980 pour prendre effet à
compter du 5 janv. 1981 (*J.O.* 14
déc. 1980), Paris, Union Technique
de l'Electricité, janv. 1981, 20 p.

293. Sicherheitsbestimmung für Funksender.
Deutsche Norm DIN-57866, Deutsche
Elektronische Kommission im DIN und
VDE, VDE 0866/12, 1978, 50 p.

294. Spécifications pour la sécurité dans
les installations électrothermiques
industrielles à hyperfréquences.
In : Sécurité dans les installations
électrothermiques. 6ème partie. (En
français et anglais). Norme de la CEI,
Bur. Cent. Comm. Electrotech. Int.
Publication 519-6, Genève, 1982.

295. Status of microwave oven standards. *Microw. Energy Appl. Newsl.*, <u>2</u>, n° 6, 1969, pp. 3-5. (C.D.I.U.P.A., n° 30585).

296. Survey of selected industrial applications of microwave energy. US Dep. Health, Educ., Welf., Bur. Radiol. Health, BRH, DEP-70-10, PB 191 394, 1970.

297. Système de normes de sécurité professionnelle, champs électromagnétiques RF, exigences générales de sécurité. (En russe).Comité d'Etat de l'URSS sur les Normes, Gosudar' stvjennij Standard Sojuza GOST 12.1.006-76, Moskva, 1976.

298. Système de normes de sécurité propessionnelles, champs électromagnétiques RF, niveaux autorisés sur les lieux de travail et spécifications pour leur contrôle. (En russe). Comité d'Etat de l'URSS sur les Normes, Gosudar'stvennij Standard Sojuza GOST 12.1.006-84, Moskva, 29 nov. 1984.

299. Technical manual for RF radiation hazards. Dep. Navy, Naval Ships Syst. Commands, NAVSHIPS 09-005-8000, July 1966.

300. Techniques and instrumentation for the measurement of potentially hazardous electromagnetic radiation at microwave frequencies. American National Standards Institute, Standard n° ANSI 95.3-1973.

301. The microwave oven radiation hazards. Fact or fancy ? *Microw. Energy Appl. Newsl.*, <u>1</u>, n° 5, 1968, pp. 3-8.

302. Threshold limit values. Cincinnati, American Conference of Governmental Industrial Hygienists, ACGIH, 1982, pp. 89-93.

303. Variances from performance standards. *Fed. Regist.*, <u>39</u>, n° 76, 1974, parts 1010-1020, § 1010.4, pp. 13879-13880.

304. We want you to know about microwave oven radiation. US Dep. Health, Educ. Welf., FDA-73-8049, 1973.

305. Wyjaśnienia interpretacyjne do rozporządzenia rady ministrów z dnia 20.10.1961. Ministerstwo Zdrowia i Opieki Społecznej. Dep. Sanit.-Epidemiol. n° EP-44647-31/6629, Kwiecień 1966.

306. Zarządzenie ministra zdrowia i opieki społecznej z dnia 9.8.1972 w sprawie pkreslenia pól elektromagnetycznych w zakresie mikrofalowym oraz dopuszczalnego czasu pracy w strefie zagrożeń. *Dziennik Ustaw MZiOS*, n° 17, poz. 78, 1972.

Biomedical applications

Microwaves have important therapeutic applications in tissue heating. They may also activate some specific athermal mechanisms.

Several applications will be described below in Section 4.3, but the most promising is hyperthermia for the treatment of cancer. This technique seems to have no side effects. It continues to be effective because the body does not get accustomed to it, and it produces only a minimum of discomfort for the patient. The results in themselves are very positive and become remarkable when microwaves are combined with radiotherapy or chemotherapy. A synergetic effect appears to ensure that the rate of success of the combined treatment exceeds the sum of the separate treatments. The only limitation is the intrinsically poor penetration of microwaves. However, implantable applicators allow *in situ* treatment of small tumors. Spectacular regressions have already been achieved with malignant brain tumors.

Section 4.2 deals with "specific effects"(for lack of a better name) and is almost completely devoted to the building of immune defences by nonthermal means. It can be looked upon as an extension of Chapter 2, especially Section 2.3. This is still "terra incognita" that is the center of great controversy, an example of which in France is the regrettable loss of the Priore machine.

4.1 Hyperthermia for cancer treatment

4.1.1 Historical development

Although the effect of heat on tumors has been well known and documented since antiquity, its systematic use dates from the last century. Two techniques have been exploited in medical practice: local electric treatment, pioneered in 1830 by B. Fabre-Palaprat, and general hyperthermia or artificial fever. The first documented clinical application of the latter technique was introduced in 1866 [41]: spontaneous regression of multiple facial sarcomas was achieved in a patient suffering from persistent erysipelas with fever [94]. Subsequent therapeutic protocols included a course of injections of bacterial toxins [68], [69] or localized external heating.

The decisive step was taken in 1891, when Arséne d'Arsonval (1851-1940) (Fig. 4.1) passed a high current at 10 kHz through his hand. To his great astonishment, he felt no pain and only a sensation of heat associated with intensive perspiration. This he attributed not to tissue heating, but to a vasodilation effect. Having realised the therapeutic potential of the phenomenon, he devised in 1893 a process known as autoconduction, or induced faradization, without direct contact: the patient was placed inside a giant Ruhmkorff coil (Fig. 4.2) fed by two Leyden jars. According to [79] "there is a very significant effect

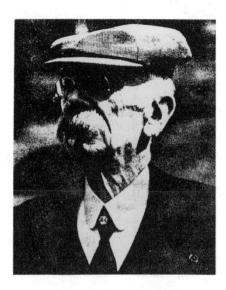

Figure 4.1 Arsène d'Arsonval in 1940 [55].

on the vaso-motor system: the skin sweats and the therapeutic effect is said to be prodigious".

D'Arsonval treated 75 patients in 1894 and 1895 in 15–20-min-per-day appointments and achieved some results with arthritis, gout, and rheumatism. He then developed his condenser bed in which the body of the patient formed one electrode and the metal frame of the bed the other. In between them were cushions that acted as the dielectric mediums. At the time, this arrangement was judged more efficient than the solenoid [100], but its thermal nature was not recognised until 1899 (by von Zynech). It was demonstrated in 1907 at Dresden and in 1908 at Budapest by F. Nagelschmidt, who coined the term *diathermy* [123]. Local applications of high-frequency radiation, or fulguration, relied on a variety of sources. The procedure was used for the first time in 1900 by a Dr. Riviere and then by Drs. Juge and de Keating-Hart at the Marseille hospital. Cauterization by electric spark at 1 MHz gave good results for both superficial cancers and for diverse conditions

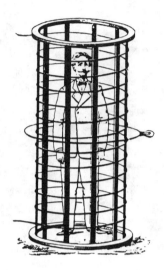

Figure 4.2 Autoconduction by the d'Arsonval system [79].

of the skin such as eczema, psoriasis, lupus, angioma, cysts, and verruca [100].

The switch to meter waves occurred with the construction in 1928 of the first short-wave generator (400–1500 W at 100 MHz) by Esau at Jena. This led to attempts [320] to use these waves to arrest the growth of carcinomas and sarcomas grafted on mice and, in some cases, to eradicate the tumors completely. For example, a 23% remission was reported for 403 mice with sarcomas. A series of experiments was performed on mice in 1928 at the Ecole Supérieure d'Electricitè in Paris, using a number of hot-filament sources rated at 15 W (30–37.5 MHz), 20 W (150–173 MHz), and 100 W (37.5–75 MHz). This eventually lead to human experimentation, in which it was combined with radium therapy of patients deemed inoperable or abandoned. The results were remarkable. There was general improvement, including softening of the tumors and dissipation of acute pain, nausea, headaches, and vomiting, as well as an improvement in the blood picture. After several sessions, the tumor became partially liquified and could be punctured. In some cases it disappeared completely. It was concluded that "short waves have a very special inhibitive and necrotic action on tumors" [191]. Epitheliomas of the tongue that were resistant to radium and to X-rays were treated [179] at the rate of 2 or 3 hr per day for 15–20 days. Healing occurred in several cases between days 12 and 15. Between one- and six-month remissions were obtained [414] in 32 cases of cancer that were deemed fatal. The combined use of X-rays and short waves was reported in the 1930s [3], [14]. The applicators were either metal plates covered with bakelite or soft conductors wound in the form of solenoids around the part of the body to be treated [336]. Despite the excessive enthusiasm with which it was sometimes regarded, high-frequency treatment remained marginal in comparison to radiotherapy and X-ray therapy with which it was frequently combined.

The era of centimeter waves actually began at the end of World War II when high-powered magnetrons developed for radar became available. It became clear that only microwaves were capable of resolving the problem of focusing on tumors [336], [118]. In this field, as in many others, the United States had by then taken the lead from Europe. The first magnetron producing 400 W at 3 GHz and destined for medical research was delivered in June 1946 to the Mayo Clinic in Rochester (Minnesota). In the same year, the *Federal Communication Commission* (FCC) released the frequency of 2.45 GHz for medical applications because of its expected superiority in applications requiring absorption by tissue. In December 1947, the Council on Physical Medicine of the American Medical Association approved the use of the *Microtherm* diathermy equipment manufactured by Raytheon. This delivered 125 W at 2.45 GHz through four types of applicator: two hemispherical reflectors with stubs and a dihedral reflector illuminated by a half-wave dipole [123], [195], [207], [268], [194].

The first reports on the regression of spontaneous mammary carcinoma in mice after exposure to microwaves alone appeared in the late 1940s and early 1950s [268], [194], [105]. By the mid-1950s, a combined microwave plus X-ray treatment was developed [1] and gave good results on carcinomas in rats [123]. Since the early 1960s, the development of applicators has proceeded along three complementary lines, as follows:

- *Animal experimentation.* As early as the early 1960s, several researchers treated rat hepatomas [48], mice sarcomas [251], and Walker 256 carcinomas [72], using a combination of microwaves and radiotherapy, and microwaves and X-rays [272], [203], [204]. The leading groups were at the University of Maryland School of Medicine (Baltimore) [305], [306], [19], [60], [303], [326], the Warsaw Faculty of Medicine, and the Warsaw Center for Radiology and Radio Protection in association with the University of Louvain [384], [386], [383], [312], the interdisciplinary group of the RCA Laboratories, Princeton (New Jersey), the Montefiore Hospital and Medical Center, Bronx (New York) [237], [238], and a Franco American group that included researchers of the Biophysical Laboratory of Nancy and of the Denver Research Institute [91]. The results were fairly positive e.g., there were 100% regressions of implanted adenocarcinomas after two 45-min sessions (43°C) [235] and 50% regressions of sarcomas 180 after 10 daily 2-hr sessions [383].

- *Experiments on tissues and phantom material.* This line was pursued by the group at the University of Washington School of Medicine at Seattle, which compared the efficacy of 915 and 2450 MHz and decided in favor of the lower frequency [208], [209], [210], [125]. This was followed by the development of a number of direct-contact applicators at 915 MHz [81], [212], [121], [122], [141], [127], [214], [128], [211], [213].

- *Characterization and evaluation of applicators.* This was carried out by the Division of Electronic Products, which later became the National Center for Devices and Radiological Health, under the control of the *Bureau of Radiological Health* (BRH), Food and Drug Administration at Rockville (Maryland) [170], [168], [171], [165], [169], [177], [178], [227],

[166], [172], [173], [174], [167].

In France, the *Direction Générale de la Recherche Scientifique et Technique* (DGRST) initiated in 1978 the *Hyperthermia* triannual program to promote the development of irradiation systems [102]. In the United States, the *National Cancer Institute* (NCI) has been funding important research projects since 1980 [107].

4.1.2 Mode of action

It has been known for a long time that the therapeutic effect of heat requires a temperature of 43–45°C in the tumor. Once this temperature is reached, and despite the continuing energy absorption, the temperature of the tissues tends to decrease because of the increase in blood flow [209], [81]. The optimum range is actually between 42–43.5°C [354], [76], [97], [116], [202], [47], [82], [437]. Below 42°C, there is no effect on the tumor and there is even a risk of increasing the proliferation of malignant cells. Above 42.5°C, the time required for the destruction of neoplasms decreases by a factor of two per each additional degree [354], so that only a few minutes would theoretically be sufficient at 45°C or 46°C if there were no risk of damage to healthy cells in the vicinity of the tumor [328]. In practice, a ceiling of 42–43°C is imposed with treatment durations on the order of one hour (Fig. 4.3).

The thermal sensitivity that is peculiar to tumors may be due to their inhomogeneity: The periphery is fed by the vascular system of healthy surrounding tissue; the center is necrosed; and in the intermediate zone the cells are anoxic, badly supplied because of insufficient vascularization consisting of fragile vessels with thin, unsupported walls [97]. Under the influence of heat, the blood flow in normal tissue is greater by a factor of four to six, and accelerates still further by vasodilation, but it diminishes or remains constant in the neoplasm, which therefore becomes even more vulnerable [370], [301], [362], [333], [332], [410], [163].

Hyperthermia is often combined with one of the traditional cancer treatments (i.e., radiotherapy or chemotherapy). This

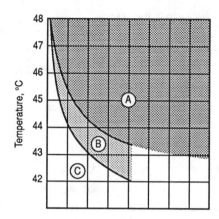

Figure 4.3 Effect of various combinations of temperature and time on healthy tissues and on tumors *in vivo* [362]: *a* - tumors and healthy tissue destroyed, *b* - survival of most healthy tissue and destruction of the tumor without subsequent occurrence, *c* - survival of the tumor.

combination seems to be synergetic in that the combined treatment is often more effective than the two procedures employed individually. The following results have been obtained for human fibroblasts transformed by cytomegalovirus grafted in mice and treated successively with microwaves (20 min) and ionizing radiation (6.4 Gy) [76]:

- complete regression of 50% of tumors as against 8% for hyperthermia alone, 6% for radiation alone, and 7% for controls.
- partial regression of 77% of tumors as against 8% for hyperthermia alone, 13% for radiation alone, and 20% for controls.

Heat and the radiation appear to be complementary: DNA synthethizing cells are particularly resistant to radiation, but are highly sensitive to heat. On the other hand, the anoxic intermediate zone of the tumor, which requires doses of ionizing radiation 2.5 to 3 times higher than the peripheral zone, is very sensitive to heat [362]. Hyperthermia is also thought to block repair mechanisms in cells that have acquired sublethal lesions during irradiation [321]. Other researchers [199], [198] have tried to elucidate the synergetic mechanism in experiments with the insect *Tribolium confusum*. Microwaves (2 hr at 2.45 GHz, 680 or 760 W kg^{-1}) were found to affect the cell repair capacity either by damaging certain enzymes specialized in DNA repairs, or by altering the complex chromatine-DNA. When microwaves were applied before exposure to ^{137}Cs γ rays, their amplifying action was found to persist for eight hours. When they were applied second, they were effective only if the interval of time between the two treatments did not exceed one hour. Medical practice shows that the most marked effect occurs when hyperthermia precedes irradiation rather than the other way around.

When combined with chemotherapy, microwave treatment facilitates the circulation of chemical substances in the blood stream and increases the permeability of cells for these substances. The optimum solution is attained by the simultaneous administration of the two treatments [321]. It has been reported [383] that the level of regression of mice sarcoma 180, which is approximately 50% after microwave treatment, reaches 66% when the microwaves are coupled with the administration of interferon or interferon inductors poly-I and poly-C. The differential absorption of toxic drugs by the tumor can be achieved [428] by placing the carrier animal in a state of hypothermia with rectal temperature of about 12°C, whilst using microwaves to maintain the tumor at 37°C. Experiments at 2.45 GHz in the X band, using titanium dioxide loaded horns, resulted in only one death and 17 instances of regression in the case of spontaneous mammary tumors in mice, compared with three deaths attributed to the toxicity of the product (5-fluorouracil) and two regressions achieved with chemotherapy alone.

The energy necessary for hyperthermia is not excessive: 1.5 W is theoretically sufficient to raise the temperature of a 20-mm diameter tumor by 5°C in one minute [47], if energy is efficiently transmitted to the tumor. Neoplastic tissue has a higher dielectric loss factor than healthy tissue because of its higher water content. For example, skin carcinoma contains 81.6% water as compared with 60.9% in normal epidermis [95]; hepatoma contains 81.9% as compared with 71.4% in liver. These differences are considered in [96], [90], [89] and [302] to be sufficient to enable us to discriminate between different types of carcinoma by diagnostic radio-frequency imaging [379]. In contrast, another study [431] has shown no significant difference between the permittivities of muscular tissue and rhabdomyosarcoma R1H at frequencies in the range 0.2–2.4 GHz. It also showed that ionizing radiation (γ rays, 15 Gy) had no effect on dielectric properties. Preferential heating may occur but, understandably, it may depend on the type of tumor. Precise targeting and focusing of radiation are particularly important in practice.

Macroscopic modeling of wave penetration and of the rise in tissue temperature depends on the position and number of sources (i.e., on the particular applicator used). In contrast to the dosimetric models of Section 1.4.4, the interaction occurs in the near zone. There is an extensive literature on this subject [305], [306], [6], [303], [427], [114], [129], [150], [97], [392], [404], [409], [334]. The role of blood flow and of thermal regulation is discussed in [116], [248], [380]. An example of a numerical calculation of field and temperature distribution can be found in [202] in which finite-element techniques are used to compute isotherms for a section of the arm exposed to 2.45 GHz-radiation. The optimum power density for a temperature in the range 42–43.5°C was thus found to be $200 \, \mathrm{mW} \, \mathrm{cm}^{-2}$.

The risk of burns to healthy tissue arises from a number of factors: (a) hotspots, particularly near bony protrusions such as vertebral apophyses, (b) heating of fatty tissue in which poor vascularization does not contribute significantly to the evacuation of accumulated energy, and (c) higher absorption by internal organs. Other types of hazards are associated with radiation leakage. Applicator design is therefore always subject to two fundamental requirements: The leakage of inci-

dent or reflected power into the immediate environment must
be strictly limited, and the distributions of induced field and
temperature in space and in time must be well controlled.

4.1.3 Applicators

In Section 4.3.2 of Part I we considered the various constraints on the choice of material, impedance matching, and
shielding. They also apply to hyperthermia equipment. The
main difference is the absence of a treatment cavity, the wave
being directly coupled to the subject by an antenna fed by a
waveguide or a coaxial cable. Depending on whether this antenna is simply applied to the skin at the level of the tumor or
inserted in an orifice or into tissue, hyperthermia is said to be
superficial or interstitial. Most antennas belong to one of the
following three main categories.

4.1.3.1 Radiating apertures

The simplest, and one of the earliest to be used, is an open
waveguide for which the propagation coefficient in the tissue,
normalized to the longer side a of the guide, is given by [59]

$$K = ak = a\omega\sqrt{\mu_0 \epsilon} \tag{4.1}$$

where $\epsilon = \epsilon' - j\epsilon''$.

Figure 4.4 shows the depth of penetration [Part I, (2.83)],
normalized to a, as a function of the imaginary (abscissa) and
the real (ordinate) parts of K^2. In the top left-hand corner
of the figure, the penetration is practically independent of the
abscissa. For $\epsilon''/\epsilon' < 0.25$ and $\log(\omega^2 \mu_0 \epsilon' a^2) > 1$, we have

$$d \simeq 0.4 \left[a \frac{\sqrt{\epsilon'_r}}{\lambda_m} \right]^{1.4} \tag{4.2}$$

where λ_m is the wavelength in the medium. On the right
of the diagram, the penetration is almost independent of the
ordinate and tends to the plane-wave limiting value given by

the following expression:

$$d = \frac{\lambda_0}{2\pi a \sqrt{2\epsilon_r'}} \left[\sqrt{1 + \left(\frac{\epsilon''}{\epsilon'}\right)^2} - 1 \right]^{-\frac{1}{2}} \qquad (4.3)$$

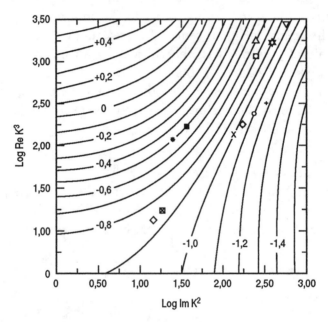

Figure 4.4 Depth of penetration normalized to the width a of a rectangular radiating aperture $a \times b$ [59].

It has been shown [128] that the optimum ratio of deep to superficial heating is achieved for an aperture of width λ and height between λ and 2λ, excited in the TE_{10} mode. Two 915-MHz applicators based on this principle have been developed, namely, (1) a waveguide $12 \times 6 \, cm^2$ enlarged to $12 \times 16 \, cm^2$ by a transition and (2) a rectangular open cavity of $12 \times 15.4 \, cm^2$ fed by a coaxial to waveguide transition of $12 \times 7.75 \, cm^2$. These designs are simple and unencumbered, but are normally under cut-off (Part I, Section 2.9) at 915 MHz. They were filled with a high-permittivity, low-loss dielectric such as aluminum oxide ($\epsilon'_r \simeq 4$), which was equivalent to increasing the dimensions by a factor of $\sqrt{\epsilon'_r}$. The aluminum oxide was subsequently replaced with a similar but lighter material (Eccofoam manufactured by by Emerson and Cuming), whose porosity allows forced air circulation for cooling the skin. The aperture was protected by an acrylic radome with suitable holes. Experiments with phantoms and volunteers have shown that the holed radome and the forced circulation of cold air prevented the appearance of hotspots in the lipid layers and the painful heating of superficial tissue [211]. Maximum temperature was achieved in seven minutes, and the increased blood flow then stabilized the temperature at the therapeutic level. The measured efficiency (i.e., the ratio of specific absorption rate to incident power) showed excellent correlation between the phantom material and real tissues (2.74 and 2.81 kg^{-1}, respectively).

Another type of applicator in this category is a rectangular waveguide filled with a dielectric [7], developed in France by the CNRS and Ecole Supérieure d'Electricité, for which the depth of penetration was shown to be less than that for the plane wave, but could be calculated from the simple formula

$$d \simeq \frac{c}{4\pi f \mathrm{Im}\left(\sqrt{\epsilon_1^* \cos\theta_1}\right)} \tag{4.4}$$

where

$$\cos^2\theta_1 = \frac{\epsilon_1^* - (f_c/f)^2}{\epsilon_1^*} \tag{4.5}$$

where c is the speed of light, ϵ_1^* the conjugate of the complex permittivity of the tissue, f is the working frequency, and f_c is the cut-off frequency of the fundamental mode of the loaded waveguide.

A Transco applicator working at 2.45 GHz and consisting of a circular waveguide terminated by a conical horn with a 152-mm aperture was evaluated for the Food and Drug Administration in [177] and [178]. The aperture was surrounded by an annular wavetrap and featured a series of posts producing circular polarization (Fig. 4.5). In direct contact with the skin, radiation leakage does not exceed $0.8\,\mathrm{mW\,cm^{-2}}$ for $100\,\mathrm{W}$ incident power, and $4\,\mathrm{mW\,cm^{-2}}$ when the applicator is $1\,\mathrm{cm}$ from the subject. Circular polarization eliminates large thermal gradients.

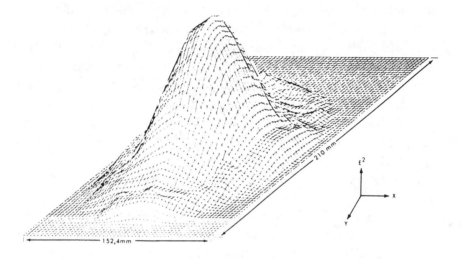

Figure 4.5 Radiation field of the Transco circular applicator operating at $2.45\,\mathrm{GHz}$ [178].

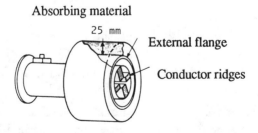

Figure 4.6 The Transco circular applicator operating at $915\,\mathrm{MHz}$ [176].

The 915-MHz device (Fig. 4.6) has two pairs of orthogonal ridges placed in a circular waveguide, 150 or 200 mm in diameter, with two 25-mm concentric layers of absorbing material [176] (Fig. 4.6). Another applicator described in [175] is a WR 430 rectangular waveguide loaded with PTFE blocks and operating at 2.45 GHz. The excitation of the TEM mode in the space between the PTFE blocks gives a remarkably uniform field. The waveguide flanges are replaced with wave traps (Fig. 4.7). Another solution [377], [256] consists of a circular waveguide with dimensions suitable for the transmission of the TE_{11} mode, fitted with an internally corrugated flange

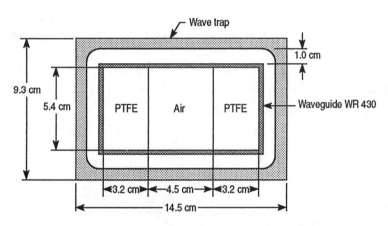

Figure 4.7 Open-ended waveguide applicator loaded with PTFE blocks and operating at 2.45 GHz [173].

(Fig. 4.8). This applicator is light and sturdy, and gives good field uniformity with very little leakage and a VSWR of less than 2.

Researchers at the Universidad Politécnica in Madrid [295], [296] have endeavored to improve open waveguide applicators and their depth of penetration by means of transitions optimized for multilayer loads by variational calculations and modal analysis. Researchers at the Ecole Polytechnique Fédérale in Lausanne have treated the difficult theoretical problem of reflections from multiple layers of lossy dielectrics

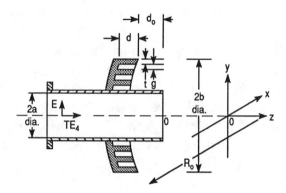

Figure 4.8 Circular applicator with a corrugated flange, operating at 9.47 GHz [256]: $a = 12.5$ mm, $t = 2.0$ mm, $g = 2.0$ mm, $B = 26.5$ mm, $R_0 = 150$ mm, depth of corrugation $d = 8.2$ mm without dielectric and 4.9 mm with dielectric (paraffin wax $\epsilon'_r = 2.73$); d_0 can be adjusted to optimize the VSWR.

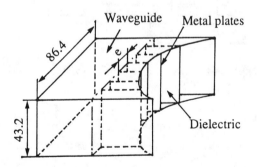

Figure 4.9 Direct contact lens (2.45 GHz) [262]. $\lambda'/2 < e < \lambda/2$ where λ' is the wavelength in the dielectric.

for the wave emitted by an open-ended waveguide [396]. The penetration depth can be further improved by about 40% by means of converging lenses [263], [261], [264], [262] made of parallel metal blades separated by a dielectric (Figs. 4.9 and 4.10).

4.3.2.2 Phased arrays

Phased arrays are ensembles of elementary sources fed through phase shifters. They were originally developed for radar and telecommunications and their use in hyperthermia is relatively recent [233].

The simplest array consists of two identical sources fed by a magic tee and a phase shifter. It was shown in [115] that two sources working in phase provide very superficial heating, but out-of-phase operation generates a deep maximum in fatty tissue (4 cm at 2.45 GHz); heating in muscular tissue remains moderate and superficial because of high absorption. A precise focal point can be produced in fatty tissue only by an array of N sources on the surface of a spheroid pointing toward the focal point F [188]. For arbitrary phases, the *powers* add at F, but, when all the radiation is in phase, the *field* at F is given by

$$E_F = N E_0 e^{-\alpha R} \tag{4.6}$$

where E_0 is the field at the surface of the tissue, α is the attenuation coefficient [Part I, (2.38)], and R is the radial distance. The temperature of the irradiated tissue is proportional to the incident power and is given by

$$T_F = \kappa N^2 E_0^2 e^{-2\alpha R} \tag{4.7}$$

The maximum depth of penetration $R_{\max}$ can be defined as that at which the tissue temperature is equal to the surface temperature T_0. Thus, in the vicinity of a source

$$T_0 = \kappa E_0^2 \tag{4.8}$$

Hence

$$T_F = T_0 = N^2 T_0 e^{-2\alpha R_{\max}} \tag{4.9}$$

so that

$$R_{\max} = \frac{\ln N}{\alpha} = d \ln N \tag{4.10}$$

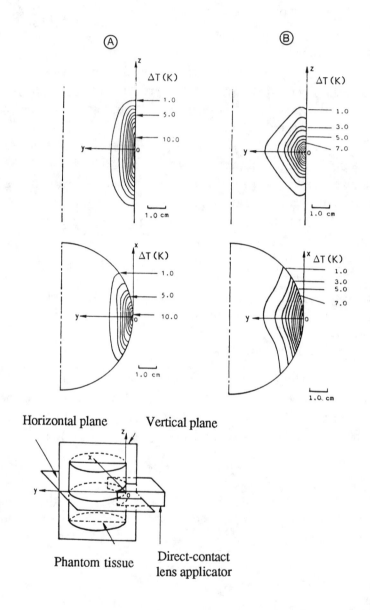

Figure 4.10 Rise in the temperature of a phantom muscle tissue in contact with an open-ended waveguide (A) and with the lens of Fig. 4.9 (B) [262].

For dephased sources, $R_{\max}$ falls to half this value. The expression given by (4.10) is only a rough estimate, valid for a small number of sources. Actually, the depth of penetration is reduced by the divergence of the rays, but this is partially compensated by blood flow and by convective cooling of the surface. The presence of hotpoints in fatty tissue was verified experimentally in [188]. They were found to appear at a depth of $0.7d$ and $1.1d$ for two and three sources, respectively.

The sources themselves are often half-wave dipoles [412], [397]. The array of three dipoles used in [402] had a current distribution of the form 9, -1, 9 or 2, -1, 2 to ensure good penetration in the HF range without excessive overheating of the surface. Open waveguides filled with dielectrics [421] and microstrip antennas [35] have also been used. The essential element of the array is the phase shifter. As the units available for radar and telecommunications are costly and large, it was suggested in [31] and [32] that microstrip phase shifters on dielectric substrates could be used for hyperthermia at 915 MHz with low losses. However, the power rating is poor (0.25–1 W), which means that each element in the array must be provided with an amplifier. Phased arrays are discussed in [11], [103], [119], [155] and [291]. Although they operate in the high-frequency band, the annular phased arrays made by BSD Corporation of Salt Lake City (Utah) can be mentioned here. They are dipole sources mounted on a dielectric cylinder surrounding part or all of the body, and are used for deep-tumor treatment [106], [405], [52], [53], [163].

4.3.2.3 Dipoles and slots

In situ treatment of neoplasms relies on interstitial applicators inserted in tissue or conducted by catheterization to the vicinity of the tumor through natural cavities or channels. One such development [27] is the syringe antenna used to inject chemotherapeutic preparations during hyperthermia. Another device described in [374] uses a quarter-wave antenna connected by coaxial cable to a variable source (3–10 W between 500 and 1300 MHz) and continuously matched by means of a double stub. The 2.45-GHz applicator described in [273] and [420] is a semi-rigid coaxial line whose inner conductor protrudes beyond the end of the line. It is inserted in a Fischer tube, 2.8 mm in diameter. The temperature is measured by a thermocouple on a PTFE substrate, mounted on the outer conductor of the coaxial line. This applicator can remain implanted for several months for periodic hyperthermia sessions.

A simple open coaxial line without a protruding inner conductor, or a coaxial line fixed to a ground plane to form an annular slot, is not suitable for hyperthermia because hotpoints may appear in the vicinity of the aperture [258]. When the central conductor protrudes beyond the open end, it radiates as a dipole, and much of the power emitted into the high-loss tissue is absorbed within an ellipsoid that is not much larger than the external diameter of the coaxial line[381]. This has led to interest in intracavity applicators for tumors localized on the surface of hollow organs (oesophagus, stomach, rectum, larynx, uterus, and bladder). By introducing the antenna in a catheter, it is possible to bring the applicator close to the tumor and to heat it whilst reducing to a minimum the exposure of healthy tissue and the reflections at interfaces between tissue layers. An applicator based on this principle is described in [37]. It operates at 434 MHz and consists of a dipole 17.76 cm long, inserted in a surgical latex tube, 7.6 mm in diameter (Fig. 4.11).

An equivalent solution [283], [337] is based on radiating slots instead of dipoles, e.g., in an applicator [283], [337] working at 434 MHz and used in the treatment of cancer of the prostate (Fig. 4.12). A slotted cylinder is fed by a semirigid coaxial cable and is enclosed in a polythene tube that provides forced cooling-water circulation at 3°C . Contact between the radiating slot and the wall of the rectum touching the prostate is ensured by a balloon filled with water after the insertion of the applicator. Further information can be found in [19], [238], [253], [393], [9], [86], [12], [13], [394], [395], [356] and [138].

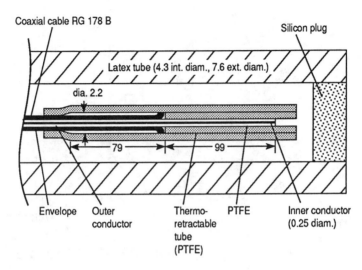

Figure 4.11 Interstitial applicator in the form of a half-wave dipole (434 MHz) [37].

4.1.4 Integrated systems

A practical microwave hyperthermia system consists essentially of a generator and an applicator connected by a waveguide or a coaxial cable, and a number of subsystems for temperature monitoring, power control, and data acquisition and storage.

Temperature measurement in the presence of microwaves is tricky. It must be precise, sensitive, stable, easy to monitor, and above all, it must be protected from the direct effects of microwave power. Noninterfering temperature sensors, including optic fibers, can be employed, the alternatives being thermocouples or low-impedance thermistors, for which emission must be briefly interrupted whilst measurement takes place, and thermographic systems based on radiometry. Infrared radiometry cannot give the temperature of deep tissues, and microwave imaging is often a good choice. It relies on the spontaneous thermal radiation emitted by tissues, which is still of

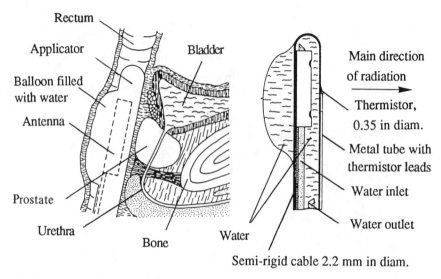

Figure 4.12 Interstitial applicator for the treatment of cancer of the prostate (434 MHz) [337].

sufficient intensity in the microwave range and is enhanced further by hyperthermic treatment [46]. A temperature difference of one degree is produced by a change of only 10^{-14} W in detected power, whereas the power emitted during treatment amounts to several tens of watts. The 150-dB difference means that the hyperthermic and thermographic channels must be carefully decoupled. This decoupling may be achieved in the time or frequency domain. In the former approach [64], [65], the two channels use the same antenna and the emission is automatically turned off for five seconds per minute for temperature homogenization and measurement. In the latter approach, the two channels operate at different frequencies (e.g, 4.6 GHz for emission and 9.6 GHz for reception [260]) or 1.6 GHz and 4.7 GHz [47].

Temperature monitoring is used to modulate the transmitted power according to a simple binary algorithm that turns the generator on or off depending on whether the measured temperature is lower or higher than the preset value. Another method is to use a proportional algorithm, the emitted energy being a function of the difference between the measured and required temperatures [327]. This principle is exploited in [189] where the nominal temperature is held to within ±0.1°C. In cases of accidental runaway, a safety device interrupts or reduces the emission immediately.

The safety and reliability of such equipment rely on the use of proper redundancies, automatic data acquisition (transmitted power, switch status, proximity sensors, and so on), ease of use, and password-protected access.

Existing systems are usually in the form of robotic arms that carry different applicators. The applicator motion in a repetitive sequence over the whole treated surface is often computer controlled [363], [367]. The hyperthermia equipment manufactured by the SAISA/ODAM Company of Wissembourg (Bas-Rhin, France) is fairly typical of the state of the art. It was developed for the treatment of tumors located 2–3 cm below the surface (breast lumps, metastasizing nodules, and so on). There are several versions, designated according to their working frequency HYLCAR 434 (120 W), HYLCAR 915 (60 W), and HYLCAR 2450 (30 W). HYCLAR II (Fig. 4.13) has transmitters (915 and 2450 MHz) and two radiometers (1 and 3 GHz). The hyperthermic and thermographic channels are conveyed by the same applicator in an alternated arrangement, and the precision of temperature control is ±0.2°C. The equipment is mounted on a mobile trolley, measuring $700 \times 380 \times 980$ mm, weighing 55 kg, and consuming 450 VA [298, 433, 434, 435]. In the United States, two manufacturers have each sold about 100 units in ten years. They are the BSD Corporation that has already been mentioned and Clini-Therm Corporation of Dallas (Texas). BSD offers the mobile systems BSD 400 and 500, working at 915 MHz, with solid-state amplifiers delivering respectively 350 W and 8×50 W. The BSD 400 was the first equipment to include a microprocessor and the Luxtron optic-fiber thermometer. Clini-Therm has developed the MARK series, of which the most recent (MARK-IX) is controlled by an IBM PC. It has four 100-W channels for interstitial or superficial hyperthermia; the temperature is measured by implantable GaAs sensors and optic fibers [163].

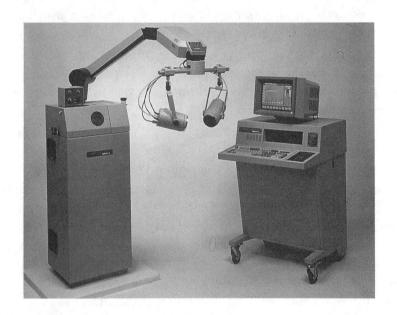

Figure 4.13 HYLCAR II system (915 - 2450 MHz) made by the SAISA ODAM Company (courtesy of V. Ringeisen, 1988).

4.1.5 Clinical results

Pharmacologic regulations require that any new treatment must pass three stages of trials. Stage I is the feasibility stage that includes preliminary clinical trials defining the mode of application, the dose, and any secondary effects. The dose is increased until the required response is obtained, or until secondary effects become intolerable. In the case of cancer, such trials are normally possible only with terminally ill patients.

Stage II is concerned with the efficiency of the new treatment on a statistically significant group of patients. The responses are more readily monitored and evaluated if the available therapies are ineffective and the normal course of the disease is well known.

Stage III is undertaken only if the results of stages I and II suggest that success is likely. It consists of a series of systematic, and therefore long and costly, comparisons between the new treatment and the traditional management of the disease [321], [163].

The first clinical trials were reported in 1977 and 1978 [145], [144]. This was soon followed by trials of interstitial hyperthermia at the University of Maryland Hospital at Baltimore [321], [323], [327]. Patients had intracranial probes implanted for the treatment of brain tumors. The feasibility of the method and the absence of secondary effects (cardiac rhythm, blood pressure, EEG, intracranial pressure, and so on) were demonstrated. The patient suffered no discomfort for intratumor temperatures of 45–47°C. Multiform glioblastoma was treated successfully in one patient [162]. Another group was formed on the initiative of Mark Nowogrodzki as a collaboration between the Microwave Technology Center of the RCA Company at Princeton (New Jersey), the Montefiore Hospital and Medical Center, Bronx (New York), and the Orange Hospital (New Jersey). RCA began funding the development of applicators for superficial tumors (array of dipoles), cervical and uterine tumors (semi-rigid coaxial probes, 2.45 GHz), and choroid melanoma (spiral antenna on a concave substrate, 5.8 GHz) [236], [365], [273], [94]. Several patients, deemed inoperable, were treated with official approval at the rate of one hour per week after radiotherapy. The first results were very positive. On May 11, 1983, the Institutional Review Board of the FDA approved a protocol for the application of microwave hyperthermia [162]. The trials continued with 75% favorable clinical response on 12 patients with very advanced brain tumors [420]. The tumors receded in 3 out of the 12 cases, their growth slowed down in 2 other cases, and in 4 further cases tumor tissue became necrosed. Headaches that were intolerable and very difficult to treat disappeared in 5 cases out of 12. RCA also developed a robotic arm (2.45 GHz, 100 W) for the treatment of breast cancer [367].

The efficacy of microwaves in combination with ionizing radiation was confirmed in 1980 when 14 complete remissions were reported [401] among 17 patients submitted to radiation treatment (5–20 sessions or 18–42 Gy in total) and two or three weekly hyperthermia sessions at 915 or 2450 MHz. As part of stage II, 39 tumors were treated at the Mallinckrodt Institute of Radiology at St. Louis (Missouri) with total doses of 24–40 Gy in steps of 4 Gy every 72 hours, each followed by a 90-min microwave session (915 or 2450 MHz). Complete remission was recorded for eight out of twelve recurrent epidermoid carcinomas of the head and neck, four out of five metastatic melanoma nodules (80% remission in the fifth), and five out of nine recurrent adenocarcinomas of the breast (80% remission in two others).

In France, a group of researchers at the CNRS Laboratory No. 287 and at the Bourgogne Medical Center at Lille treated 45 superficial tumors with a rectangular aperture or a microstrip slot at 434, 915, or 2450 MHz [65]. The protocols were as follows: two weekly sessions of one hour after radiotherapy, or one hour before and after interstitial radiation therapy, and so on. Out of 31 patients who were followed for between 2–24 months, there were 21 complete remissions and ten 50% remissions. Clinical trials also began at the Claudius Regaud Center in Toulouse, using the microstrip antennas reported in [35]. Finally, the HYLCAR system was used in more than 500 hyperthermia sessions on about 30 patients in 18 months [298].

In June 1983, the US National Cancer Institute created the Hyperthermia Physics Center at the Allegheny Hospital of Pittsburgh (Pennsylvania) with the remit of establishing national standards under United States regulations. In 1985, the FDA gave its approval for the use of hyperthermia as an addition to radiotherapy, which also meant that the cost of treatment was thereafter covered by medical insurance. The efficacy of the procedure having been demonstrated, stage III studies of head and neck tumors were begun by a multi-institutional organisation, known as the Radiation Therapy Oncology Group [163].

4.2 Specific effects

4.2.1 Bioelectric vibrations

There is a persistent parascientific belief that the health of the body depends on certain bioelectric vibrations that are susceptible to chemical or physical toxic factors. Although this view is not generally accepted in the world of science, there is some evidence that there might be some truth in it and, perhaps, great therapeutic potential. At any rate, this was the opinion of Frölich [98] who thought that there were coherent electric vibrations in the frequency range 100 GHz to 1 THz, excited in cells by metabolic processes. This idea is based on observations of the inhibition or stimulation of the growth of yeasts and bacteria as functions of the applied frequency, showing very stable and repetitive resonances (Section 2.1).

If such vibrational states are indeed metabolically excited, then they should be manifest in Raman spectroscopy. Actually, their existence was demonstrated during periods of metabolic activity of lysozyme and *E.coli B* (700 GHz to 5 THz). Emissions have also been observed at lower frequencies (i.e., 150 GHz or less). A quantum explanation of all this was given in [98] and a double hypothesis was put forward: (1) these vibrations occur in the tissues of higher organisms and (2) they exercise some control on cellular growth. Cancerization could result from a modification of these vibrations by the invasion of foreign molecules, the presence of free electrons in the conduction bands of proteins, and so on. There is evidence, cited in [98], for the presence of double spectral lines at 1.5 and 6 THz in breast carcinoma, which may be an indication of an interaction between normal cellular vibrations and free electrons.

A verification of these hypotheses would throw open the possibility of a millimeter-wave effect on cancerous tissue, but for one reservation: these waves have a very small penetration depth. The effect should be seen in relation to observations [416] of a periodic variation in the power transmitted by cultured malignant cells between 66 and 76 GHz, the peak separation being 2 GHz; similarly, there is evidence [358] for six absorption peaks with a separation of 1.3 GHz between 76 and 86 GHz. However, it has been suggested [136] that these resonances could be nothing but experimental artefacts.

4.2.2 Antigenicity

A group of researchers from INSA at Villeurbanne (Rhône, France) and the Faculty of Science at Limoges [87], [331] has demonstrated an increase in the antigenicity of malignant cells treated with microwaves. The cells in question were B16 melanoma in suspension, exposed for one hour to 5.4 W at 2.45 GHz ($40 \, \mathrm{mW \, cm^{-2}}$). The final temperature was 44°C and was maintained for 20 minutes. The suspension was then centrifuged and fractioned into liquid and coagulate. The two fractions were inoculated in mice, some of which received after 26 days an injection of viable B16 melanoma. It was found that the injection of treated malignant cells, or of either of their fractions, induced a specific immunization, indicated by an increased level of immunoglobulin relative to controls. The average survival rate was significantly higher for the immunized mice.

It was thought that the microwave treatment could have been responsible for a modification of the nature or chemical configuration of certain molecules, increasing their antigenicity. Another group of researchers [92] has reported that microwaves may induce a thickening of the cell membrane in suspension, possibly due to a change in the configuration of certain lipidic compounds. It is not yet clear whether this is a thermal or a specific microwave effect.

4.2.3 Immune response

It has been reported [389], [388] that chronic exposure to radiation accelerates the development of spontaneous mammary tumors, skin cancer induced by 3,4 benzopyrene, and pulmonary nodules in mice. At 2.45 GHz, the effect is greater for 15 mW cm^{-2} than for 5 mW cm^{-2}. It was suggested that this was a straightforward consequence of thermal stress, particularly on NK cells. Further information may be found in [313], [314], [387], [152] and [15].

On the other hand, it has also been reported [382] that exposure to microwaves at a nonthermal (2.45 GHz, 20 mW cm^{-2}) or thermal (40–60 mW cm^{-2}) level, the latter inducing a rectal temperture of 42°C, inhibited acute viral infection in mice. The mice were submitted to a series of daily exposures either before or after innoculation with Herpes simplex type I (HSV$_1$) or *vaccine virus* (VV). An identical reduction in the mortality due to HSV$_1$ encephalitis was observed in both cases, and proliferation of VV was inhibited. This effect could be explained by a strengthening of immune response. Mice have also been treated *in utero* [287] in four 20-min sessions (2.45 GHz, 35 W kg^{-1}) on days 11, 12, 13, and 14 of gestation. Sarcomas implanted on day 16 *post partum* were found to develop more slowly than on untreated subjects. This was also explained by a stimulation of the immune response, and some pioneers of microwave hyperthermia [313], [314], [362] have suggested that the antineoplastic effect of radiation could be due in part to this stimulation, including in particular the activation of macrophages and T lymphocytes.

According to Trémolières [401], "one has the feeling that electromagnetic radiation has an enormous potential. However, the link between the results obtained and the means employed is still to be clarified. This is the case with the Priore machine which produced very promising results for some cancers, and also in the treatment of the sleep disease, apparently by stimulating the immune response to trypanosoma". Antoine Priore (1912–1983), an Italian engineer who lived in Bordeaux (France), developed from 1957 onward a range of very complex equipments that incorporated several microwave carriers modulated in amplitude with low frequencies, 17 MHz in particular, and also produced a magnetic induction of 22.5 mT and, later, 62 mT (Model P2), 90 mT (Model M235 of 1965), and 500 mT (model M600 of 1972), modulated by ELF impulses of 0.5–2 Hz. The system contained a discharge tube with an annular graphite anode maintained in slow rotational motion by four external magnets and internal polepieces. It emitted neither X-rays nor any particular corpuscular radiation, and the average power was so low that it could not give rise to any thermal effects [24], [110]. The "machine," patented on October 7, 1963 [8], was to become the object of great controversy that that led to the eventual withdrawal of financial support by the government. This lamentable state of affairs was partly due to the secrecy mania and the lack of social graces on the part of the inventor, but this should not overshadow the fact that his "machine" gave very creditable experimental results. An extensive digest of communications to the French Academy of Sciences on this subject is given in [110]. A brief summary is given below.

The first experiments [300] were performed with rats grafted with very malignant uterine tumors (atypical T8 epithelioma), for which there was no treatment. This tumor brought about the death of the animals in 3–5 weeks following the invasion of the ganglion system by metastases: 24 rats grafted in a control group died between days 22 and 30. Groups of 12 rats were exposed to microwave and radio frequency fields of unknown characteristics, and also to magnetic induction of 30 or 62 mT, for 10, 20, 40, or 90 min per day. The treatment began on day 2, 6, 10, or 14 and was continued until day 25 or day 37 after the graft. The remission of tumors and matastases was completed for 20 and 40 min daily in 30 mT and 10, 20, or 40 min daily in 62 mT if the first session had taken place on day 2, 4, or 6, and for 90 min daily in 62 mT if it had taken place on day 14. There were no secondary reactions. A new series of experiments was performed under the control of Professor R. Courrier, secretary of the French Academy of Sciences, and was concerned with lymphoblastic lymphosarcoma 347, a tumor that rapidly invades the ganglions and is accompanied by the leukemic syndrome. Two weeks after the graft, all the controls were dead, and on day 19 all the animals treated for one hour per day were also dead. Rats treated for two hours per day all survived. The sessions started on day 5, as the lymphoblastic cells made their first appearance in the blood. Priore successfully treated another very virulent grafted tumor (lymphosarcoma L52) [73], [299].

The following experiments related to trypanosomiases, which are fatal parasitic diseases. Infested Swiss mice (2×10^4 trypanosoma innoculated on day 4, 10^6 per mm^3 of blood) died in five days. 82.3% of the animals treated 12 hours per day immediately after the infestation survived to day 5, their deaths being probably due to the massive lysis of the trypanosoma. Total recovery was reached between days 10 and 12. Preventive irradiation of one week, prior to infestation, had no effect if it was not carried on afterward, in which case the recovery was total and no trypanosoma could be detected in blood. The treatment thus appeared to produce considerable stimulation of nonspecific immunity [275]. These results were confirmed by further experiments and demonstrated the very high immunity acquired by animals that recovered and were reinfested up to seven times in six months, the last time at the rate of 2×10^6 trypanasoma. Moreover, the blood of these animals had acquired considerable protective power: the injection of 1/50 dilute serum protected three mice out of five against the injection of 2×10^4 trypanosoma [277], [278]. The frequency of 9.4 GHz appeared [24] to play a fundamental part. Positive results were obtained for trypanosomal orchitis in rabbits, 15 to 26 days after innoculation [231]. Other experimental trials [280], [281], [276] produced spectacular results in the case of Priore's system P2 and the short-lived M600. It became clear that there was an impressive strengthening of the immune response. Finally, another study [279] showed that the treatment had a hypocholesterolic effect on rabbits that may possibly be attributable to the activation of lipid catabolism.

However, Priore's unique aim was the treatment of cancer in humans. Trials on hopeless and inoperable cases were begun in 1977 with full agreement of the authorities. Twelve patients were submitted to daily 1-hr sessions, using a small machine designed for the treatment of small mammals. Nevertheless a clinical and biological improvement was observed in all cases, with a few remissions and one final recovery. These encouraging preliminary results were insufficient to counteract hostile criticism. Priore died on May 9, 1983 and, unfortunately, it is doubtful that it will ever be possible reconstruct and analyze "what might have been the most important medical and scientific discovery of this century" [110].

4.3 Miscellaneous applications

4.3.1 Clinical

Microwave hyperthermia is effective in relieving neurologic or arthritic pain [104], [153], [225]. It was reported in [318] and [149] that pregnant women suffering from chronic inflammation of the pelvis could be treated with 2.45-GHz radiation without adverse effects. Microwaves have been found [77], [78] to have an analgesic effect on 10,000 women in labor.

It has been suggested [417], [146], [426] that microwaves could be used for rapid rewarming after accidental hypothermia orheart surgery. Experiments with rhesus monkeys [224] (9.31-GHz modulated with 1050-kHz, 0.5-μs pulses, 150 $mW\,cm^{-2}$ average flux, or 20 $W\,kg^{-1}$) showed no secondary effects.

Microwaves are effective in tissue healing [310], [311], [335], and have also been used (550 W, 2.45 GHz) to coagulate, instead of removing, a lobe from the liver of a rabbit [267], [391]. This new hemostatic technique was found to be effective and accurate. It avoids the usual complications associated with hepatectomy, which include infections, necroses, and biliary effusions.

Another interesting application is the thawing of blood derivatives in vinyl packages kept at temperatures at between $-30°$ and $-60°C$. Normally, these products are immersed in a water bath at 37°C, which creates the risk of bacterial contamination and requires approximately 30 min, which is an unacceptable delay in the case of emergencies. Microwaves offer a practical solution that is rapid and aseptic; there is no contamination of the product by the packaging [56]. The only problem is that domestic ovens produce nonuniform heating. A recent development [180] is a system in which the sachets rotate and are shaken in the vertical plane, and metal guards protect their seams from overheating. In another design [21], a rotating enclosure with a stepping motor is used to produce intermittent irradiation. Nonthermal effects could cause damage to erythrocyte membranes, and this has lead to a study of the threshold for hemolysis by microwaves [57]. This threshold lies between 50°C and 55°C and is purely thermal. It does not depend on the coagulation rate, but only on the temperature reached and on the time for which the temperature is maintained. The use of microwaves to produce up to 37°C can therefore be considered safe. Further information may be found in [353] and [44]. The same method has been proposed for the thawing of organs [130], [40], [120].

4.3.2 Biological

The transmission of nerve impulses between neurons requires the intermediary of chemical vectors (i.e., neurotransmitters) which are emitted by the axon terminal when the neuron is depolarized. The neurotransmitters attach themselves to specific receptors of the postsynaptic membrane in neighboring neurons, giving rise to the depolarization of these neurons and to the propagation of potential. The repolarization of the membrane stops the release of neurotransmitters, whose molecules are subsequently eliminated by enzymatic catabolism. The whole process is so rapid that neurochemical research faces the difficult problem of the dosage of these substances. The conventional process involves the very rapid cooling of the brain of the sacrificed animal. Liquid nitrogen can be used to freeze the cortex of a rat in 35 s and the hypothalamus in 90 s. These delays are acceptable in studies of norepinephrine, dopamine, and serotonine, whose metabolism is relatively slow, but not in the case of *adenosine triphosphate* (ATP), *cyclic 3,5-adenosine monophosphate* (AMP_c), *3,5-guanosine monophosphate* (GMP_c), *γ-aminobutyric acid* (GABA), and *acetylcholine* (Ach) whose concentrations are subject to very rapid *post mortem* variations. The most recent technique, known as freeze blowing, consists of collecting brain matter from a receptacle precooled with liquid nitrogen by means of a compressed-air probe. The disadvantage of this technique is that tissue structures are destroyed, and the method is still not fast enough for some enzymes [241].

Microwaves can be used as a means of fixation [360]. The temperature of the brain tissue must be brought as fast as possible, and as reproducibly and uniformly as possible, to the enzyme inactivation level [252]. The equipment must be easy to use, without risk to nonspecialists, and the death of the animal must be instantaneous and painless. This last point was demonstrated in [361]. It was also realized that there was no significant reduction in the water and protein content of brain tissue [232], and that the temperature rose to 90°C in 300 ms when the radiation was focused on the head of the animal [249]. At this temperature, the dissection of the brain becomes very easy and most enzymes such as adenylcyclase, phosphodiesterase, cholinesterase, and choline acetyltransferase are inactivated. For example, AMP_c is relatively thermostable, whilst the enzymes that are implicated in its metabolism [i.e., adenylate cyclase (AC) and phosphodiesterase (PDE)] are thermolabile and denature between 65°C and 90°C. An irradiation system is described in [218] in which a rat was held in a plexiglass container inserted transversely into a WR 430 waveguide fed by a 3.5-kW, 2.45-GHz generator. A similar system (Fig. 4.14), operating with 800 W or 4.5 kW is presented in [252]. The Hitachi MA-500 "Instat" system, developed in 1975, produces good inactivation in 1.5 s at 3 kW or 300 ms at 5 kW [230]. The only inconvenience with this process is the risk of nonuniform heating, which may result in insufficient inactivation in certain zones and damage to the structural integrity of the brain tissues [244], [245].

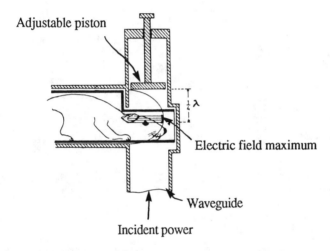

Figure 4.14 Waveguide applicator for the fixation of brain tissue [252].

Microwave inactivation has produced some interesting results on acetylcholine [340], [265], [359], on AMP_c [341], [342], [219], dopamine [243], [164], and fatty acids [17], [18], [50], [297]. Further data may be found in [117], [217], [42], [43], [216], [38], [352], [242] and [339].

References

1. ALLEN F. Biological modification of effects of roentgen rays. Part II. High temperature and related factors. *Am. J. Roentgenol.*, 73, 1955, pp. 836-848.

2. Von ARDENNE M. On a new physical principle for selective local hyperthermia of tumor tissues. In : STREFFER C. Cancer therapy by hyperthermia and radiation. München, Urban und Schwarzenberg, 1978, pp. 96-104.

3. ARONS J., SOKOLOFF B. Combined roentgenotherapy and ultra-shortwave. *Am. J. Surg.*, 36, 1937, pp. 533-543.

4. d'ARSONVAL J. Influence de la fréquence sur les effets physiologiques des courants alternatifs. *C.R. Acad. Sci.*, 116, 1893, pp. 630-633.

5. d'ARSONVAL J. Des courants à haute fréquence. *Arch. Electr. Med.*, 1897, pp. 166-179.

6. ATKINSON E. Microwave-induced hyperthermia dose definition. *IEEE Trans. Microw. Theory Tech.*, MTT-26, n° 8, 1978, pp. 595-598.

7. AUDET J., BOLOMEY J., PICHOT C., N'GUYEN D., ROBILLARD M., CHIVE M., LEROY Y. Electrical characteristics of waveguide applicators for medical applications. *J. Microw. Power*, 15, n° 3, 1980, pp. 177-186.

8. BADER J. Le cas Priore, prix Nobel ou imposture ? Paris, J.C. Lattès, 1984, 224 p.

9. BAHL I., STUCHLY S. The effect of finite size of the ground plane on the impedance of a monopole immersed in a lossy medium. *Electron. Lett.*, 15, 1979, pp. 728-729.

10. BAHL I., STUCHLY S., Microstrip slot radiators for local hyperthermia. In : Proceedings of the IMPI Symposium, Toronto, 1981, Digest pp. 74-76.

11. BAHL I., STUCHLY S., LAGENDIJK J., STUCHLY M. Microstrip loop radiators for medical applications. *IEEE Trans. Microw. Theory Tech.*, MTT-30, n° 7, 1982, pp. 1090-1093.

12. BAHL I., STUCHLY S., STUCHLY M. A new microstrip radiator for medical applications. *IEEE Trans. Microw. Theory Tech.*, MTT-28, n° 12, 1980, pp. 1464-1468.

13. BAHL I., STUCHLY S., STUCHLY M. New microstrip slot radiator for medical applications. *Electron. Lett.*, n° 16, 1980, pp. 731-732.

14. BAKER R., SMITH V., PHILLIPS T., KANE L., KOBE L. A system for developing microwave-induced hyperthermia in small animals. *IEEE Trans. Microw. Theory Tech.*, MTT-26, n° 8, 1978, pp. 541-545.

15. BALCER-KUBICZEK E., HARRISON G. Evidence for microwave carcinogenesis in vitro. *Carcinogenesis 6*, 1985, pp. 859-864.

16. BALCER-KUBICZEK E., ROBINSON J., Mac CULLOCH D. Membrane injury in CHO cells exposed to microwave hyperthermia at 37,5° C and 43° C. In : 4th annual Bioelectromagnetics Society meeting, Los Angeles, 1982, p. 83.

17. BALCOM G., LENOX R., MEYERHOFF J. Regional τ-aminobutyric acid levels in rat determined after microwave fixation. *J. Neurochem.*, 24, 1975, pp. 609-613.

18. BALCOM G., LENOX R., MEYERHOFF J. Regional glutamate levels in mouse brain determined after microwave fixation. *J. Neurochem.*, 26, 1976, pp. 423-425.

19. BASSEN H., HERCHENROEDER P., CHEUNG A., NEUDER S. Evaluation of an implantable electric-field probe within finite simulated tissues. *Radio Sci.*, 12, n° 6 (S), 1977, pp. 15-25.

20. BEER-GABEL M., YERUSHALMI A. Altered glucose metabolism in intact and tumor-bearing rats subjected to local hyperthermia. *Bull. Cancer*, 33, 1975, pp. 91-95.

21. BELLAVOINE R. Procédé et appareillage de chauffage d'un produit par
micro-ondes, notamment à des fins de
décongélation, et application au
plasma sanguin. Brevet français
n° 2 492 210 (80 21619), 9 oct. 1982.

22. BEN-HUR E., ELKIND M., BRONK B.
Thermally enhanced radio response
of cultured Chinese hamster cells :
inhibition of repair of sublethal
damage and enhancement of lethal
damage. Radiat. Res., n° 58, 1974,
pp. 38-51.

23. BERTEAUD A. Les effets biologiques
des micro-ondes : une confirmation.
La Recherche, 9, n° 85, 1978, pp.
65-66.

24. BERTEAUD A., BOTTREAU A., PRIORE A.,
PAUTRIZEL A., BERLUREAU F., PAUTRIZEL
R. Essai de corrélation entre l'évolution d'une affection par Trypanosoma equiperdum et l'action d'une
onde électromagnétique pulsée et
modulée. C.R. Acad. Sci., 272, 15
fév. 1971, pp. 1003-1006.

25. BICHER H., HETZEL F., D'AGOSTINO L.,
JOHNSON R., Ó HARA M. Changes in
tumor tissue oxygenation induced
by microwave hyperthermia. In : 26th
Annual Meeting of the Radiation
Research Society, Toronto, 14-18
May 1978, abstracts p. 74.

26. BICHER H., WOLFSTEIN R. Local hyperthermia for deep tumors. Experience
with three techniques. J. Microw.
Power, 21, n° 2, 1986, (summary),
p. 103.

27. BIGU DEL BLANCO J., ROMERO-SIERRA C.
The design of a monopole radiator
to investigate the effect of microwave radiation in biological systems.
J. Bioeng., n° 1, 1977, pp. 181-184.

28. BIRKNER R., WASCHMANN F. Über die
Kombination von Röntgenstrahlen und
Kurzwellen. Strahlentherapie, 79,
1949, p. 93.

29. BISHOP F., HORTON R., WARREN S. A
clinical study of artificial hyperthermia induced by high-frequency
currents. Am. J. Med. Sci., n° 184,
1932, pp. 512-532.

30. BODEM F., REIDENBACH H., NEHLS M.,
BRAND H., FRUHMORGEN P. Suitability
of 2.45 GHz microwave-heating for
endoscopic coagulation therapy.
Biomedizin. Tech., 20, n° 6, 1975,
pp. 238-239.

31. BOESCH R. Development of a continuous phase shifter for a microwave
hyperthermia system. M.S. Thesis,
Univ. Illinois, Urbana, 1987.

32. BOESCH R., MAGIN R., FRANKE S. Phase
shifter for a 915 MHz phased array
hyperthermia system. IEEE Trans.
Biomed. Eng., BME-34, n° 11, 1987,
pp. 904-907.

33. BOJSZA W. Fred Sterzer : finding new
hope for cancer victims. Microwaves,
24, n° 7, 1981, pp. 15-18.

34. BOLMSJÖ M., HAFSTRÖM L., HUGANDER A.,
JÖNSSON P., PERSSON B. Experimental
set-up for studies of microwave-induced hyperthermia in rats. Phys.
Med. Biol., 27, 1982, pp. 397-406.

35. BONS D., RUTHER V. Réalisation d'un
applicateur à réseau d'antennes micro-
bande à 2,45 GHz. In : Comptes-rendus
du Colloque sur les Applications
énergétiques des rayonnements. Toulouse, Univ. Paul Sabatier, 12 juin
1986. Toulouse, ENSEEIHT, pp. 11-15.

36. BOWMAN H. Heat transfer and thermal
dosimetry. J. Microw. Power, 16, n° 2,
1981, pp. 121-133.

37. BROSCHAT S., CHOU C., LUK K., GUY A.,
ISHIMARU A. An insulated dipole
applicator for intracavitary hyperthermia. IEEE Trans. Biomed. Eng.,
BME-35, n° 3, 1988, pp. 173-178.

38. BROWN P., LENOX R., MEYERHOFF J.
Microwave enzyme inactivation system :
electronic control to reduce dose
variability. IEEE Trans. Biomed. Eng.,
BME-25, 1978, pp. 205-208.

39. BURDETTE E., FRIEDERICH P., SEAMAN R.,
LARSEN L. In situ permittivity of
canine brain : regional variations
and postmortem changes. IEEE Trans.
Microw. Theory Tech., MTT-34, n° 1,
1986, pp. 38-50.

40. BURDETTE E., WIGGINS S., BROWN R.,
KAROW A. Microwave thawing of frozen
kidneys : a theoretically based,
experimentally-effective design.
Cryobiol., 17, 1980, pp. 393-402.

41. BUSCH W. Über den Einfluss welche
heftigere Erysipeln zuweileg auf
organisierte Neubildelungen ausüben.
Verhandl. Naturh. Preuss Rhein.
Westphal., 23, 1866, pp. 28-30.

42. BUTCHER L., BUTCHER S. Brain temperature and enzyme histochemistry
after high-intensity microwave irradiation. Life Sci., 19, n° 7, 1976,
pp. 1079-1088.

43. BUTCHER S., BUTCHER L., HARMS S., JENDEN D. Fast fixation of brain in situ by high-intensity microwave irradiation : application to neuro-chemical studies. *J. Microw. Power*, 11, n° 1, 1976, pp. 61-65.

44. CAMPBELL N., DREWE J. Microwave tha-wing of frozen packed red blood cells. In : IEEE Symposium on Microwave theory and techniques, Los Angeles, 1981, pp. 479-481.

45. CARPENTER C., PAGE A. The production of fever in man by short radio waves. *Science*, 71, 1930, p. 450.

46. CARR K., EL MAHDI A., SHAEFFER J. Dual-mode microwave system to enhance early detection of cancer. *IEEE Trans. Microw. Theory Tech.*, MTT-29, n° 3, 1981, pp. 256-260.

47. CARR K., EL MAHDI A., SHAEFFER J. Passive microwave thermography cou-pled with microwave heating to en-hance early detection of cancer. *Microw. J.*, 25, n° 5, 1982, pp. 125-136.

48. CATER D., SILVER I., WATKINSON D. Combined therapy with 220 kV roentgen and 10-cm MW heating in rat hepatoma. *Acta Radiol.*, 2, 1964, pp. 321-336.

49. CAVALIERE R., CIOCATTO E., GIOVANELLA B. Selective heat sensitivity of cancer cells. Biochemical and clini-cal studies. *Cancer*, n° 20, 1967, pp. 1351-1381.

50. CENEDALLA R., GALLI C., PAOLETTI R. Brain free fatty acid levels in rats sacrificed by decapitation versus focused microwave irradiation. *Lipids*, 10, 1975, pp. 290-293.

51. CHANG B., HUANG A., JOINES W. Inhibi-tion of DNA synthesis and enhancement of the uptake and action of metho-trexate by low-power-density micro-wave radiation in L 1210 leukemia cells. *Cancer Res.*, 40, 1980, pp. 1002-1005.

52. CHARNY C., GUERQUIN-KERN J., HAGMANN M., LEVIN S., LACK E., SINDELAR W., ZABELL A., GLATSTEIN E., LEVIN R. Human leg heating using a mini-annular phased array. *Med. Phys.*, 13, July/Aug. 1986, pp. 449-456.

53. CHARNY C., HAGMANN M., LEVIN R. A whole-body thermal model of man during hyperthermia. *IEEE Trans. Biomed. Eng.*, BME-34, n° 5, 1987, pp. 375-387.

54. CHAUDHARY S., MISHRA R., SWARUP A., THOMAS J. Dielectric properties of normal and malignant human breast tissue at radiowave and microwave frequencies. *Indian J. biochem. bio-phys.*, 21, 1983, pp. 76-79.

55. CHAUVOIS L. D'Arsonval a 19 ans. *La Nature*, n° 3198, oct. 1951, pp. 310-312.

56. CHECCUCCI A., BENELLI G., DUMINUCO M., GAETANI M., PAOLETTI P., VANNINI S., MORFINI M. Reliability of microwave heating for hemoderivative thawing. *J. Microw. Power*, 18, n° 2, 1983, pp. 163-168.

57. CHECCUCCI A., OLMI R., VANNI R. Thermal haemolytic threshold of human erythrocytes. *J. Microw. Power*, 20, n° 3, 1985, pp. 161-163.

58. CHECCUCCI A., PAOLETTI P. Chauffage par micro-ondes de matériaux bio-logiques. In : BERTEAUD A., SERVANTIE B. Ondes électromagnétiques et bio-logie. Symposium international URSI-CNFRS, Jouy-en-Josas, 30 juin-4 juil. 1980, session G.

59. CHEEVER E., LEONARD J., FOSTER K. Depth of penetration of fields from rectangular apertures into lossy media. *IEEE Trans. Microw. Theory Tech.*, MTT-35, n° 9, 1987, pp. 865-867.

60. CHEUNG A., DAO T., ROBINSON J. Dual-beam TEM applicator for direct-con-tact heating of dielectrically encap-sulated malignant mouse tumor. *Radio Sci.*, 12, n° 6 (S), 1977, pp. 81-85.

61. CHEUNG A., GOLDING W., SAMARAS G. Direct-contact applicators for micro-wave hyperthermia. *J. Microw. Power*, 16, n° 2, 1981, pp. 151-159.

62. CHIVÉ M., CONSTANT E., LEROY Y., MAMOUNI A., MOSCHETTO Y., N'GUYEN D., SOZANSKI J. Procédé et dispositif de thermographie-hyperthermie en micro-ondes. Brevet français n° 2 497 947 (81 00682), 9 janv. 1981.

63. CHIVÉ M., LEROY Y., GIAUX G., PRÉVOST B. Microwave thermography for controlled local hyperthermia at 2.5 gigahertz. In : Proceedings of the IMPI Symposium, Toronto, 1981, Digest pp. 179-181.

64. CHIVÉ M., N'GUYEN D., LEROY Y. Une nouvelle application des microondes en génie biologique et médical : l'hyperthermie locale contrôlée par thermographie microonde à 2,45 GHz. *Onde Electr.*, 62, n° 2, 1982, pp. 66-70.

65. CHIVÉ M., PLANCOT M., GIAUX G., PRÉVOST B. Microwave hyperthermia controlled by microwave radiometry : technical aspects and first clinical results. *J. Microw. Power*, 19, n° 4, 1984, pp. 233-241.

66. CHIVÉ M., PLANCOT M., LEROY Y., GIAUX G., PRÉVOST B. Microwave (1 GHz and 2,45 GHz) and radiofrequency (13,56 MHz) hyperthermia monitored by microwave thermography. In : 12th European Microwave Conference, Helsinki, 13-17 Sept. 1982. Microwave Exhibitions and Publishers Ltd., Paper B6.1.

67. CHRISTENSEN D., DURNEY C. Hyperthermia production for cancer therapy : a review of fundamentals and methods. *J. Microw. Power*, 16, n° 2, 1981, pp. 89-105.

68. COLEY W. The treatment of malignant tumors by repeated inoculations of erysipelas, with a report of ten original cases. *Am. J. Med. Sci.*, 105, n° 5, 1893, pp. 487-511.

69. COLEY W. The therapeutic value of the mixed toxins of the Streptococcus of erysipelas and Bacillus prodigious in the treatment of inoperable malignant tumors. *Am. J. Med. Sci.*, 112, n° 3, 1986, pp. 251-281.

70. CONNOR W. Localized current field heating as an adjunct to radiation therapy. *Radiat. Environ. Biophys.*, 17, n° 3, 1980, pp. 219-228.

71. CONWAY J. Assessment of a small microwave (2450 MHz) diathermy applicator as suitable for hyperthermia. *Phys. Med. Biol.*, 28, 1983, pp. 249-256.

72. COPELAND E., MICHAELSON S., Effect of selective tumor heating on the localization of 131I fibrinogen in the Walker carcinoma 256. II : heating with microwaves. *Acta Radiol.*, 9, 1970, pp. 323-336.

73. COURRIER R. Influence de champs électromagnétiques particuliers sur l'évolution de tumeurs expérimentales chez le rat et la souris. Note présentée à l'Académie des Sciences le 1er mars 1965. In : GRAILLE J. Dossier Priore. Paris, Ed. Denoël, 1984, pp. 67-71.

74. CRAVALHO E., FOX L., KAN J. The application of the biological equation to the design of thermal protocols for local hyperthermia. *Ann. New York Acad. Sci.*, 335, 1980, pp. 86-97.

75. CRILE G. Selective destruction of cancers after exposure to heat. *Ann. Surg.*, 156, 1962, pp. 404-407.

76. CUNNINGHAM D., FREY R., VELKLEY D. Microwave hyperthermia potentiates radiation in treatment of radioresistant tumor of human origin in the nude mouse without increasing metastatic frequency. In : Proceedings of the IMPI Symposium, Toronto, 1981, Digest pp. 32-33.

77. DAELS J. Microwave heating of the uterine wall during parturition. *J. Am. Coll. Obstet. Gynecol.*, 42, n° 1, 1973, pp. 76-79.

78. DAELS J. Microwave heating of the uterine wall during parturition. In : Proceedings of the IMPI Symposium, Leuven, 1976. (Paper 5A.4.).

79. DARY G. A travers l'électricité. Paris, Vuibert et Nony éd., 1903, 460 p.

80. DAVIS J. Clinical applications of microwave radiation : hypothermy and diathermy. *Ann. New York Acad. Med.*, 55, 1979, pp. 1187-1192.

81. DELATEUR B., LEHMANN J., STONEBRIDGE J., WARREN C., GUY A. Muscle heating in human subjects with 915-MHz microwave contact applicator. *Arch. Phys. Med. Rehabil.*, 51, n° 3, 1970, pp. 147-151.

82. DE WAGTER C. Computer simulation for local temperature control during microwave-induced hyperthermia. *J. Microw. Power*, 20, n° 1, 1985, pp. 31-42.

83. DEWEY W., FREEMAN M. Rationale for use of hyperthermia in cancer therapy. *Ann. New York Acad. Sci.*, 335, 1980, pp. 372-378.

84. DICKSON J., CALVERDWOOD S. Temperature range and selective sensitivity of tumors to hyperthermia : a critical review. *Ann. New York Acad. Sci.*, 335, 1980, pp. 180-205.

85. DICKSON J., SHAH S. Technology for the hyperthermic treatment of large solid tumors at 50° C. *Clin. Oncol.*, 3, n° 3, 1977, pp. 301-318.

86. DOUPLE E., STROHBEHN J., BOWERS E., WALSH J. Cancer therapy with localized hyperthermia using an invasive microwave system. *J. Microw. Power*, 14, n° 2, 1979, pp. 181-186.

87. DOUSS T., SANTINI R., DESCHAUX P., PACHECO H. Augmentation du pouvoir immunogène de cellules cancéreuses soumises à l'action des micro-ondes. *C.R. Soc. Biol.*, 179, 1985, pp. 299-306.

88. EDDAOUDI B. Dispositifs microondes automatisés pour le traitement par hyperthermie localisée. Thèse de doctorat de 3ème cycle, Inst. Nat. Polytech. Toulouse, ENSEEIHT, EEA n° 177, 1983.

89. ENGLAND T. Dielectric properties of the human body for wavelengths in the 1-10 cm. range. *Nature*, 166, 1950, pp. 480-481.

90. ENGLAND T., SHARPLES N. Dielectric properties of the human body in the microwave region of the spectrum. *Nature*, 163, 1949, pp. 487-488.

91. ESCANYE J., ROBERT J., ITTY C., EDRICH J. Mesure de l'hyperthermie lors d'une irradiation à 432 MHz de souris saines et porteuses de tumeurs par thermographies micro-ondes et infra-rouge et par thermocouples. In : Compte-rendu du Symposium international sur les applications énergétiques des micro-ondes, IMPI-CFE. 14, Monaco, 1979. Résumés pp. 256-258.

92. FARJARDO L., EGBERT B., MARMOR J., HAHN G. Effects of hyperthermia in malignant tumor. *Cancer*, 45, 1980, pp. 613-623.

93. FAYOS J., GOTTLIEB C., BALZANO Q., AHMAD K., KIM Y. Computer-controlled hyperthermia unit for cancer therapy. *Rev. Interam. Radio.*, 6, 1981, pp. 7-10.

94. FINGER P., PACKER S., SVITRA P., PAGLIONE R., CHESS J., ALBERT D. Hyperthermic treatment of intraocular tumors. *Arch. Ophtalmol.*, 102, n° 10, 1984, pp. 1477-1483.

95. FOSTER K., SCHEPPS J. Dielectric properties of tumor and normal tissues at radio through microwave frequencies. *J. Microw. Power*, 16, n° 2, 1981, pp. 107-119.

96. FRICKE H., MORSE S. The electric capacity of tumors of the breast. *J. Cancer Res.*, 16, 1926, pp. 310-376.

97. FRIEDENTHAL E., MENDECKI J., BOTSTEIN C., STERZER F., NOWOGRODZKI M., PAGLIONE R. Some practical considerations for the use of localized hyperthermia in the treatment of cancer. *J. Microw. Power*, 16, n° 2, 1981, pp. 199-204.

98. FRÖHLICH H. Coherent electric vibrations in biological systems and the cancer problem. *IEEE Trans. Microw. Theory Tech.*, MTT-26, n° 8, 1978, pp. 613-617.

99. FUCHS G. Zur Sensibilisierung maligner Tumoren durch Ultra-Kurzwellen. *Strahlentherapie*, 88, 1952, p. 647.

100. GALTIER-BOISSIERE Larousse médical illustré. Paris, Librairie Larousse, 1912, 1294 p.

101. GANDHI O. Biological effects and medical applications of RF electromagnetic fields. *IEEE Trans. Microw. Theory Tech.*, MTT-30, n° 11, 1982, pp. 1831-1847.

102. GAUTHERIE M., GUERQUIN-KERN J. Thermothérapie par radiofréquences et hyperfréquences en cancérologie. *Rev. Gén. Electr.*, 11, nov. 1981, pp. 852-856.

103. GEE W., LEE S., BONG N., CAIN C., MITTRA R., MAGIN R. Focused array hyperthermia applicator : theory and experiment. *IEEE Trans. Biomed. Eng.*, BME-31, n° 1, 1984, pp. 38-46.

104. GERSTEN J., WAKIM K., HERRICK J., KRUSEN F. The effect of microwave diathermy on the peripheral circulation and on tissue temperature in man. *Arch. Phys. Med.*, 30, n° 1, 1949, pp. 7-25.

105. GESSLER A., Mac CARTY K., PARKINSON M. Eradication of spontaneous mouse tumors by high frequency radiation. *Exp. Med. Surg.*, 8, 1950, p. 143.

106. GIBBS F. Clinical evaluation of a microwave/radiofrequency system (BSD Corporation) for induction of local and regional hyperthermia. *J. Microw. Power*, 16, n° 2, 1981, pp. 185-192.

107. GILBERT E., PISTENMA D., MAHONEY F. Hyperthermia. Research supported by the National Cancer Institute. *J. Microw. Power*, 16, n° 2, 1981, pp. 227-231.

108. GOODMAN R., NUSSBAUM G. Studies of applicator-load coupling in microwave induced hyperthermia. In : Proceedings of the IMPI Symposium, Toronto, 1981, Digest p. 41.

109. GOTTLIEB C., FAYOS J., KIM Y., BALZANO Q. Computer-controlled microwave hyperthermia system for cancer therapy. In : Proceedings of the IMPI Symposium, Toronto, 1981, Digest pp. 68-70.

110. GRAILLE J. Dossier Priore, une nou- velle affaire Pasteur ? Paris, Ed. Denoël, 1984, 308 p.

111. GRAILLE J. Communication personnelle. 17 mai 1988.

112. GRANT E. Microwaves. A tool in medi- cal and biologic research. In : CZERSKI P., OSTROWSKI K., SHORE M., SILVERMAN C., SÜSS M., WALDESKOG B. Biological effects and health hazards of microwave radiation. Proceedings of an international symposium, Warszawa, 15-18 Oct. 1973. Warszawa, Polish Medical Pu- blishers, 1974, pp. 309-316.

113. GRUNNER O. Differential effects of electromagnetic fields in treatment of neuroses and depression. Act. Nerv. Super., 17, n° 4, 1975, p. 294.

114. GUERQUIN-KERN J., PALAS L., GAUTHERIE M., FOURNET-FAYAS C., GIMONET E., PRIOU A., SAMSEL M. Etude comparative d'applicateurs hyperfréquences (2450 MHz, 434 MHz) sur fantômes et sur pièces opératoires, en vue d'une utilisation thérapeutique de l'hyper- thermie micro-onde en cancérologie. In : BERTEAUD A., SERVANTIE B. Ondes électromagnétiques et biologie. Symposium international URSI-CNFRS, Jouy-en-Josas, 30 juin-4 juil. 1980, session G, pp. 241-247.

115. GUERQUIN-KERN J., PALAS L., PRIOU A., GAUTHERIE M. Local hyperthermia using microwaves for therapeutic purposes - experimental studies of various applicators. J. Microw. Power, 16, n° 3-4, 1981, pp. 305- 311.

116. GUERQUIN-KERN J., PALAS L., SAMSEL M., GAUTHERIE M. Hyperthermie micro- ondes : influence du flux sanguin et des phénomènes thermorégulateurs. Bull. Cancer, 68, n° 3, 1981, pp. 501-510.

117. GUIDOTTI A., CHENEY D., TRABUCCHI M., DUTEUCHI M., WANG C. Focused micro- wave radiation : a technique to mini- mize postmortem changes of cyclic nucleotides, and choline and to preserve brain morphology. Neuro- pharmacol., 13, 1974, pp. 1115-1122.

118. GUIERRE M. Les ondes et les hommes, histoire de la radio. Paris, Ed. René Julliard, 1951, 278 p.

119. GUO T., GUO W., LARSEN L. A local- field study of a water-immersed microwave antenna array for medical imagery and therapy. IEEE Trans. Theory Tech., MTT-32, n° 8, 1984, pp. 844-854.

120. GUTTMAN F., BOSISIO R., BOLONGO D., SEGAL N., BORZONE J. Microwave illu- mination for thawing frozen canine kidneys : (A) Assessment of two ovens by direct measurement and thermo- graphy, (B) The use of effective dielectric temperature to monitor change during microwave thawing. Cryobiol., 17, 1980, pp. 465-472.

121. GUY A. Analyses of electromagnetic fields induced in biological tissues by thermographic studies on equiva- lent phantom models. IEEE Trans. Microw. Theory Tech., MTT-19, n° 2, 1971, pp. 205-214.

122. GUY A. Electromagnetic fields and relative heating patterns due to a rectangular aperture source in direct contact with bilayered bio- logical tissue. IEEE Trans. Microw. Theory Tech., MTT-19, n° 2, 1971, pp. 214-223.

123. GUY A. History of biological effects and medical applications of micro- wave energy. IEEE Trans. Microw. Theory Tech., MTT-32, n° 9, 1984, pp. 1182-1200

124. GUY A., CHOU C. Physical aspects of localized heating by radiowaves and microwaves. In : STORM F. Hyperther- mia and cancer therapy. Boston, G.K. Hall Medical Publishers, 1983, pp. 270-304.

125. GUY A., LEHMANN J. On the determina- tion of an optimum microwave dia- thermy frequency for a direct-con- tact applicator. IEEE Trans. Biomed. Eng., BME-13, n° 2, 1966, pp. 76-87.

126. GUY A. LEHMANN J., Mac DOUGALL J., SORENSEN C. Studies on therapeutic heating by electromagnetic energy. In : Thermal problems in biotechno- logy. New York, American Society of Mechanical Engineering, 1968, pp. 26-45.

127. GUY A., LEHMANN J., STONEBRIDGE J. Therapeutic applications of electro- magnetic power. Proc. IEEE, 62, n° 1, 1974, pp. 55-75.

128. GUY A., LEHMANN J., STONEBRIDGE J., SORENSEN C. Development of a 915-MHz direct-contact applicator for thera- peutic heating of tissues. IEEE Trans. Microw. Theory Tech., MTT-26, n° 8, 1978, pp. 550-556.

129. HAHN G., KERNAHAN P., MARTINEZ A., POUNDS D., PRIONAS S. Some heat transfer problems associated with heating by ultrasound, microwaves or radiofrequency. Ann. New York Acad. Sci., 335, 1980, pp. 327-345.

130. HAMILTON R., KETTERER F., HOLST H., LEHR H. Rapid thawing of frozen canine kidneys by microwaves. *Cryobiol.*, 4, 1968, pp. 265-266.

131. HAND J., A multichannel system for microwave heating of tissues. *Br. J. Radiol.*, 52, 1979, pp. 984-988.

132. HAND J. Electromagnetic techniques in cancer therapy by hyperthermia. *IEE Proc.*, 128, 1981, pp. 593-601

133. HAND J., HUME S., ROBINSON J., MARIGOLD J., FIELD S, A microwave heating system for improving temperature uniformity in heated tissue. *J. Microw. Power*, 14, n° 2, 1979, pp. 145-149.

134. HAND J., JOHNSON R. Field penetration from electromagnetic applicators for localised hyperthermia. *Recent Results Cancer Res.*, 101, 1986, pp. 7-17.

135. HAND J., TER HAAR G. Heating techniques in hyperthermia. I. Introduction and assessment of techniques. *Br. J. Radiol.*, 54, 1981, pp. 433-436.

136. HERSHBERGER W. Microwave transmission through normal and tumor cells. *IEEE Trans. Microw. Theory Tech.*, MTT-26, n° 8, 1978, pp. 618-619.

137. HILL C. Ultrasound, microwave and radiofrequency radiations : the basis for their potential in cancer therapy. Proceedings of the 10th L.H. Gray Conference. *Br. J. Cancer*, 45, suppl. V, 1982, pp. 1-257.

138. HILL D. Waveguide technique for the calibration of miniature implantable electric-field probes for use in microwave-bioeffects studies. *IEEE Trans. Microw. Theory Tech.*, MTT-30, n° 1, 1982, pp. 92-99.

139. HILL L. Actions of ultra short waves on tumors. *Br. Med. J.*, 2, 1934, pp. 370-371.

140. HIRAOKA M., TAKAHASHI M., ABE M., MATSUDA T., SUGIYAMA A., NAKASA Y., YAMAMOTO I., SUGHARA T. Thermoradiotherapy of refractory malignant tumors : an experience with microwave and RF capacitive hyperthermia. *Med. Instrum.*, 18, n° 3, 1984, pp. 181-186.

141. HO H., GUY A., SIGELMANN R., LEHMANN J. Microwave heating of simulated human limbs by aperture sources. *IEEE Trans. Microw. Theory Tech.*, MTT-19, n° 2, 1971, pp. 224-231.

142. HOCHULI C., KANTOR G. An analysis of minimally perturbing temperature probe and thermographic measurements in microwave diathermy. *IEEE Trans. Theory Tech.*, MTT-29, n° 12, 1981, pp. 1285-1291.

143. HOFER K., CHOPPIN D., HOFER M. Effect of hyperthermia on the radiosensitivity of normal and malignant cells in mice. *Cancer*, n° 38, 1976, pp. 279-287.

144. HORNBACK N., SHUPE R., SHIDNIA H., JOE B., SAYOC E., GEORGE R., MARSHALL C. Radiation and microwave therapy in the treatment of advanced cancer. *Radiol.*, 130, 1978, pp. 459-464.

145. HORNBACK N., SHUPE R., SHIDNIA H., JOE B., SAYOC E., MARSHALL C. Preliminary clinical results of combined 433 megahertz microwave therapy and radiation therapy on patients with advanced cancer. *Cancer*, 40, 1977, pp. 2854-2863.

146. HUANG Z., SUN Q., SHEN M. Rewarming with microwave irradiation in severe cold injury syndrome. *Chin. Med. J.*, 93, 1980, pp. 119-120.

147. HUMBERT La diathermie avec les appareils à ondes amorties et à ondes entretenues. In : Compte-rendu de la 52ème session de l'Association française pour l'Avancement des Sciences, La Rochelle, 1928. Paris, Masson, 1928, résumé pp. 536-537.

148. HUNT J. Applications of microwave, ultrasound, and radiofrequency heating. In : 3rd L.A. Dethlefsen international symposium on cancer therapy by hyperthermia, drugs, and radiation. National Cancer Institute Monographies, 61, 1982, pp. 447-456.

149. IMRIE A. Pelvic short wave diathermy given inadvertently in early pregnancy. *J. Obstet. Gynecol.*, 78, 1971, pp. 91-92.

150. ISHIDA T., KATO H., MIYAKOSHI J., FURUKAWA M., OHSAKI S., KANO E. Physical basis of RF hyperthermia for cancer therapy for distribution in absorbed power from radiofrequency exposure in agar phantom. *J. Radiat. Res.*, 21, 1980, pp. 180-189.

151. ISKANDER M., DURNEY C. An electromagnetic energy coupler for medical applications. *Proc. IEEE*, n° 67, 1979, pp. 1463-1465.

152. JANIAK M., SZMIGIELSKI S. Alteration of the immune reactions by whole-body and local microwave hyperthermia in normal and tumor-bearing animals, review of own 1976-1980 experiments. *Br. J. Cancer*, 45, suppl. V, 1982, pp. 122-126.

153. JOHNSON C., GUY A. Non-ionizing electromagnetic wave effects in biological materials and systems. *Proc. IEEE*, 60, n° 6, 1972, pp. 692-718.

154. JOHNSON H. The action of short radiowaves on tissues. III : A comparaison of the thermal sensitivities of transplantable tumors in vivo and in vitro. *Am. J. Cancer*, 38, 1940, pp. 533-550.

155. JOHNSON R., ANDRASIC G., SMITH D., JAMES J. Field penetration of arrays of compact applicators in localised hyperthermia. *Int. J. Hypertherm.*, 1, 1985, pp. 321-336.

156. JOHNSON R., KRISHNAMSETTY R., YAKAR D., SUBJECK J., KOWAL H., WOJTAS F., CLAY L., DANNELS J. Clinical use of hyperthermia with radiation. In : BERTEAUD A., SERVANTIE B. Ondes électromagnétiques et biologie. Symposium international URSI-CNFRS, Jouy-en-Josas, 30 juin-4 juil. 1980, I-1.

157. JOHNSON R., PREECE A., HAND J., JAMES J. A new type of lightweight low-frequency electromagnetic hyperthermia applicator. *IEEE Trans. Microw. Theory Tech.*, MTT-35, n° 12, 1987, pp. 1317-1321.

158. JOHNSON R., SANDHU T., HETZEL F., SONG A., BICHER H., SUBJECK J., KOWAL H. A pilot study to investigate skin and tumor thermal enhancement ratios of 41,5-42,0°C hyperthermia with radiation. *Int. J. Radiat. Oncol. Biol. Phys.*, 5, 1979, pp. 947-953.

159. JUSTESEN D. Diathermy versus the microwaves and other radio-frequency radiations : a rose by another name is a cabbage. *Radio Sci.*, 12, n° 3, 1977, pp. 355-364.

160. JUSTESEN D., GUY A. Arsène-Jacques d'Arsonval : a brief history. *Bioelectromagn.*, 6, 1985, pp. 111-114.

161. JUSTESEN D., MORANTZ R., CLARK M., REEVES D., MATHEWS M. Effects of handling and surgical treatment on convulsive latencies and mortality of tumor-bearing rats to 2450 MHz microwave radiation. In : Biological effects of electromagnetic waves. International symposium, Airlie, 30 Oct.-4 Nov. 1977, Abstracts of scientific papers.

162. KACHMAR M. Brain tumors succomb to new microwave probes. *Microw. RF*, 22, n° 11, 1983, pp. 30-46.

163. KACHMAR M. The twin faces of microwave hyperthermia. *Microw. RF*, 26, n° 1, 1987, pp. 35-37 et 174.

164. KANT G., LENOX R., MEYERHOFF J. Dopamine diffusion after microwave fixation at 896 MHz. *Neurochem. Res.*, 4, 1979, pp. 529-534.

165. KANTOR G. New types of microwave diathermy applicators. Comparison of performance with conventional types. In : HAZZARD D. Biological effects and measurement of RF/microwaves. Symposium proceedings, 16-18 Feb. 1977, US Dep. Health, Educ. Welf., BRH, FDA-77-8026, July 1977, pp. 230-249.

166. KANTOR G. Direct-contact microwave diathermy applicator. US Patent n° 4 108 147, 22 Aug. 1978.

167. KANTOR G. Evaluation and survey of microwave and radiofrequency applicators. *J. Microw. Power*, 16, n° 2, 1981, pp. 135-150.

168. KANTOR G., BASSEN H., SWICORD M. Mapping of free space and scattered fields in microwave diathermy. In : JOHNSON C., SHORE M. Biological effects of electromagnetic waves. Selected papers of the USNC/URSI annual meeting, Boulder, 20-23 Oct. 1975, US Dep. Health, Educ., Welf., FDA-77-8011. Washington, US Government Printing Office, vol. 2, 1976, pp. 411-422.

169. KANTOR G., CETAS T. A comparative heating pattern study of direct-contact applicators in microwave diathermy. *Radio Sci.*, 12, n° 6 (S), 1977, pp. 111-120.

170. KANTOR G., RUGGERA P. A limited microwave diathermy field survey. US Dep. Health, Educ., Welf., Publication (FDA) n° FRA 75-8018, Bur. Radiol. Health, Dec. 1974, 15 p.

171. KANTOR G., WITTERS D. A comparative study of spaced applicators in microwave diathermy. In : Proceedings of the IMPI Symposium, Leuven, 1976, (Paper 5A.2.).

172. KANTOR G., WITTERS D. Comparative study of 2450 MHz and 915 MHz diathermy applicators with phantoms. In : Bioelectromagnetics Society Symposium, Seattle, 1979. Abstracts p. 411.

173. KANTOR G., WITTERS D. A 2450 MHz slab-loaded direct-contact applicator with choke. *IEEE Trans. Microw. Theory Tech.*, MTT-28, n° 12, 1980, pp. 1418-1422.

174. KANTOR G., WITTERS D. Comparative study of microwave diathermy at 434 MHz, 915 MHz and 2450 MHz. In : Bioelectromagnetics Society Symposium, San Antonio, 1980, abstracts. *Bioelectromagn.*, 1, n°2, 1980, p. 232.

175. KANTOR G., WITTERS D. Comparative study of microwave diathermy at 434 MHz, 915 MHz and 2450 MHz. In : BERTEAUD A., SERVANTIE B. Ondes électromagnétiques et biologie. Symposium international URSI-CNFRS, Jouy-en-Josas, 30 juin-4 juil. 1980, session G.

176. KANTOR G., WITTERS D. The performance of a new 915-MHz direct-contact applicator with reduced leakage. *J. Microw. Power*, 18, n° 2, 1983, pp. 133-142.

177. KANTOR G., WITTERS D., GREISER J. The design and performance of circulary polarized direct contact applicator for microwave diathermy. In : IEEE Symposium on Microwave Theory and Techniques, San Diego, June 1977, pp. 364-367.

178. KANTOR G., WITTERS D., GREISER J. The performance of a new direct-contact applicator for microwave diathermy. *IEEE Trans. Microw. Theory Tech.*, MTT-26, n° 8, 1978, pp. 563-568.

179. KARSIS Contribution au traitement des néoplasies. *Rev. Pathol. Comparée*, 31, n° 413, fév. 1931, pp. 137-144 et In : LAKHOVSKY G. L'oscillation cellulaire. Ensemble des recherches expérimentales. Paris, G. Doin et Cie, 1931, ch. II, pp. 131-136.

180. KASHYAP S. Microwave thawing of blood plasma. *J. Microw. Power*, 21, n° 2, 1986, pp. 99-100.

181. KIM J., ANTICH P., AHMED S., HAHN E. Clinical experiment with radiofrequency hyperthermia. *J. Microw. Power*, 16, n° 2, 1981, pp. 193-197.

182. KIM J., HAHN E., AHMED S. Combination of hyperthermia and radiation therapy for malignant melanoma. *Cancer*, 50, 1982, pp. 478-482.

183. KIM J., HAHN E., TOKITA N. Combination hyperthermia and radiation therapy for cutaneous malignant melanoma. *Cancer*, 41, 1978, pp. 2143-2148.

184. KIM J., HAHN E., TOKITA N., NISCE L. Local hyperthermia in combination with radiation therapy. *Cancer*, 40, 1977, pp. 161-169.

185. KIM S., KIM J., HAHN E. The radiosensitization of hypoxic tumor cells by hyperthermia. *Radiol.*, 114, 1975, pp. 727-728.

186. KISELEV R., ZALJUBOVSKAJA N. Study of inhibiting effect of superhigh frequency millimeter on adenovirus. (Trad. du russe). US Joint Pub. Res. Serv. Rep., JPRS L/5615, 10 Feb. 1976, pp. 71-76.

187. KLINGER H. The minimum inactivated cell ratio : a concept for the estimation of biological effectiveness of hyperthermic treatment of tissue. *J. Microw. Power*, 14, n° 2, 1979, pp. 151-154.

188. KNÖCHEL R. Capabilities of multi-applicator systems for focussed hyperthermia. *IEEE Trans. Microw. Theory Tech.*, MTT-31, n° 1, 1983, pp. 70-76.

189. KNÖCHEL K., MEYER W., ZYWIETZ F. Dynamic "in vivo" performance of temperature-controlled local microwave hyperthermia at 2.45 GHz. In : IEEE Symposium on Microwave Theory and Techniques, 1982, Dallas. 1982, IV.4., Digest pp. 444-447.

190. KOPECKI W., PEREZ C. A microwave hyperthermia treatment and thermometry system. *Int. J. Radiat. Oncol. Biol. Phys.*, 5, 1979, pp. 2113-2115.

191. KOTZAREFF A. Traitement des cancers dits inopérables, incurables et abandonnés, par les ondes hertziennes ultracourtes. Paris, Vigot frères, 1931, résumé In : LAKHOVSKY G. L'oscillation cellulaire. Ensemble des recherches expérimentales. Paris, G. Doin et Cie, 1931, ch. III, pp. 137-139.

192. KOWAL H., KANTOR G., JOHNSON R., SUBJECK J. Microwave hyperthermia delivery systems. In : BERTEAUD A., SERVANTIE B. Ondes électromagnétiques et biologie. Symposium international URSI-CNFRS, Jouy-en-Josas, 30 juin-4 juil. 1980, session G.

193. KRAINIK R. Les ondes courtes et le système nerveux. In : Compte-rendu de la 57ème session de l'Association française pour l'Avancement des Sciences, Chambéry, 1933. Paris, Masson, 1933, résumé p. 384.

194. KRUSEN F. New microwave diathermy director for heating large regions of the human body. *Arch. Phys. Med.*, 32, 1951, pp. 695-698.

195. KRUSEN F., HERRICK J., LEDEN U., WAKIM K. Microkymatotherapy : preliminary report of experimental studies of the heating effect of microwaves (radar) in living tissues. In : Proceedings of Staff Meetings. Mayo, Mayo Clinic, 22, 1947, pp. 209-224.

196. LAGENDIJK J. A microwave heating technique for the hyperthermic treatment of tumors of the eye, especially retinoblastoma. *Phys. Med. Biol.*, 27, 1982, pp. 1313-1324.

197. LAGENDIJK J. A new coaxial TEM radiofrequency/microwave applicator for non-invasive deep-body hyperthermia. *J. Microw. Power*, 18, n° 4, 1983, pp. 367-375.

198. LAI P., CAIN C., DUCOFF H. Interaction between 2450-MHz microwaves and ionizing radiation in Tribolium confusum. *IEEE Trans. Microw. Theory Tech.*, MTT-26, n° 8, 1978, pp. 530-534.

199. LAI P., DUCOFF H. Kinetics of interaction of hyperthermia and ionizing radiation in Tribolium confusum. *Radiat. Res.*, 72, 1977, pp. 296-307.

200. LAI H., HORITA A., CHOU C., GUY A. Microwave-induced post-exposure hyperthermia : involvement of endogenous opioids and serotonin. *IEEE Trans. Microw. Theory Tech.*, MTT-32, n° 9, 1984, pp. 882-887.

201. LAKHOVSKY G. L'oscillation cellulaire. Paris, G. Doin et Cie, 1931, 319 p.

202. LAMBERT N., BRAUER J. Application of microwave diathermy to the human arm. In : Proceedings of the IMPI Symposium, Toronto, 1981, Digest pp. 65-67.

203. LAPPENBUSCH W. Effects of joint microwave and X-rays stress on Chinese hamster. *Health Phys.*, n° 3, 1972, p. 423.

204. LAPPENBUSCH W., GILLESPIE L. Effect of 2450-MHz microwaves on the radiation response of X-irradiated Chinese hamsters. *Radiat. Res.*, 54, 1973, pp. 294-303.

205. LARSEN L. A microwave-decoupled brain-temperature transducer. *IEEE Trans. Microw. Theory Tech.*, MTT-22, n° 4, 1974, pp. 438-444.

206. LAW H., PETTIGREW R. New apparatus for the induction of localized hyperthermia for treatment of neophasia. *IEEE Trans. Biomed. Eng.*, BME-26, 1979, pp. 175-177.

207. LEDEN U., HERRICK J., WAKIM K., KRUSEN F. Preliminary studies on the heating and circulating effects of microwaves (radar). *Br. J. Phys. Med.*, 10, 1947, pp. 177-184.

208. LEHMANN J., BRUNNER G., Mac MILLAN J., GUY A. A comparative evaluation of temperature distributions produced by microwaves at 2450 and 900 megacycles in geometrically complex specimens. *Arch. Phys. Med. Rehabil.*, 43, n° 10, 1962, pp. 502-507.

209. LEHMANN J., BRUNNER G., SILVERMAN D., JOHNSTON V. Modification of heating patterns produced by microwaves at the frequency of 2456 and 900 Mc by physiologic factors in the human. *Arch. Phys. Med. Rehabil.*, 45, n° 11, pp. 555-563.

210. LEHMANN J., GUY A., JOHNSTON V., BRUNNER G., BELL J. Comparison of relative heating patterns produced in tissues by exposure to microwave energy at frequencies of 2450 and 900 megacycles. *Arch. Phys. Med. Rehabil.*, 43, n° 2, 1962, pp. 69-76.

211. LEHMANN J., GUY A., STONEBRIDGE J., DELATEUR B. Evaluation of a therapeutic direct-contact 915 MHz microwave applicator for effective deep-tissue heating in human. *IEEE Trans. Microw. Theory Tech.*, MTT-26, n° 8, 1978, pp. 556-563.

212. LEHMANN J., GUY A., WARREN C., DELATEUR B., STONEBRIDGE J. Evaluation of a microwave contact applicator. *Arch. Phys. Med. Rehabil.*, 51, n° 3, 1970, pp. 143-147.

213. LEHMANN J., STONEBRIDGE J., GUY A. A comparison of patterns of stray radiation from therapeutic microwave applicators measured near tissue-substitute models and human subjects. *Radio Sci.*, 14, n° 6 (S), 1979, pp. 271-283.

214. LEHMANN J., WARREN C. Therapeutic heat and cold. *Clin. Orthop.*, 99, n° 3, 1974, pp. 207-245.

215. LELE P. Induction of deep, local hyperthermia by ultrasound and electromagnetic fields. *Radiat. Environ. Biophys.*, 17, 1980, pp. 205-217.

216. LENOX R. Brain enzymes : in vivo inactivation using microwave heating. *IEEE. Trans. Microw. Theory Tech.*, MTT-24, n° 1, 1976, pp. 58-61.

217. LENOX R., GANDHI O., MEYERHOFF J., BROWN P. Modifications of in vivo rapid microwave inactivation of enzymes in the central nervous system. In : Proceedings of the IMPI Symposium, Milwaukee, 1974, pp. 1-5. (Paper A3.1.).

218. LENOX R., GANDHI J., MEYERHOFF J., GROVE H. A microwave applicator for in vivo rapid inactivation of enzymes in the central nervous system. *IEEE Trans. Microw. Theory Tech.*, MTT-24, n° 1, 1976, pp. 58-61.

219. LENOX R., MEYERHOFF J., GANDHI O., WRAY H. Microwave inactivation : pitfalls in determination of regional levels of cyclic AMP in rat brain. *J. Cyclic Nucleid Res.*, 3, 1977, pp. 367-379.

220. LI G., EVANS R., HAHN G. Modification and inhibition of repair of potentially lethal X-ray damage by hyperthermia. *Radiat. Res.*, n° 67, 1976, pp. 791-801.

221. LIBURDY R. Suppression of murine allograft rejection by whole-body microwave hyperthermia. *Fed. Proc.*, 37, 1978, pp. 1281-1283.

222. LIN J. Temperature-time profile in rats subjected to selective microwave irradiation of the brain. In : BERTEAUD A., SERVANTIE B. Ondes électromagnétiques et biologie. Symposium international URSI-CNFRS, Jouy-en-Josas, 30 juin-4 juil. 1980, H-2.

223. LIN J., GUY A., KRAFT G. Microwave selective brain heating. *J. Microw. Power*, 8, n° 3-4, 1973, pp. 275-286.

224. Mac AFEE R., ORTIZ-LUGO R., BISHOP R., GORDON R. Safety of 9,3-GHz microwave radiant heating for possible caloric supplement and medical treatment. *J. Microw. Power*, 20, n° 1, 1985, pp. 13-16.

225. Mac NIVEN D., WYPER D. Microwave therapy and muscle blood flow in man. In Proceedings of the IMPI Symposium, Leuven, 1976. (Paper 5A.5).

226. MAGIN R. A microwave system for the controlled production of local tumor hyperthermia in animals. *IEEE Trans. Microw. Theory Tech.*, MTT-27, n° 1, 1979, pp. 78-83.

227. MAGIN R., KANTOR G. Comparison of the heating patterns of small microwave (2450 MHz) applicators. *J. Bioeng.*, n° 1, 1977, pp. 493-509.

228. MALLARD J., LAWN D. Dielectric absorption of microwaves in human tissues. *Nature*, 213, 1967, pp. 28-30.

229. MARMOR J., HAHN G. Combined radiation and hyperthermia in superficial human tumors. *Cancer*, 46, 1980, pp. 1986-1991.

230. MARUYAMA Y., IIDA N., HORIKAWA A., HOSOYA E. A new microwave device for rapid thermal fixation of the murine brain. *J. Microw. Power*, 13, n° 1, 1978, pp. 53-57.

231. MAYER G., PRIORE A., MAYER G., PAUTRIZEL R. Action de champs magnétiques associés à des ondes électromagnétiques sur l'orchite trypanosomienne du lapin. *C.R. Acad. Sci.*, 274, 29 mai 1972, pp. 3011-3014.

232. MEDINA M., JONES D., STAVINOHA W., ROSS D. The levels of labile intermediary metabolites in mouse brain following rapid tissue fixation with microwave irradiation. *J. Neurochem.*, 24, 1975, pp. 223-227.

233. MELEK M., ANDERSON A. Theoretical studies of localized tumour heating using focused microwave arrays. *IEE Proc.*, 127, part F, n° 4, 1980, pp. 319-321.

234. MELEK M., ANDERSON P. A thinned cylindrical array for focussed microwave hyperthermia. In : Proceedings of the 11th European microwave Conference, Amsterdam, 1981, pp. 427-432.

235. MENDECKI J., FRIEDENTHAL E., BOTSTEIN C. Effects of microwave-induced local hyperthermia on mammary adenocarcinoma in C3H mice. *Cancer Res.*, 36, June 1976, pp. 2113-2114.

236. MENDECKI J., FRIEDENTHAL E., BOTSTEIN C., STERZER F., PAGLIONE R. Therapeutic potential of conformal applicators for induction of hyperthermia. *J. Microw. Power*, 14, n° 2, 1979, pp. 139-144.

237. MENDECKI J., FRIEDENTHAL E., BOTSTEIN C., STERZER F., PAGLIONE R., NOWOGRODZKI M., BECK E. Microwave applicators for localized hyperthermia treatment of malignant tumors. *J. Bioeng.*, 1, 1977, pp. 511-518.

238. MENDECKI J., FRIEDENTHAL E., BOTSTEIN C., STERZER F., PAGLIONE R., NOWOGRODZKI M., BECK E. Microwave-induced hyperthermia in cancer treatment : apparatus and preliminary results. *Int. J. Radiat. Oncol., Biol., Phys.*, 4, 1978, pp. 1095-1103.

239. MENSI M. Applicateur à 2,45 GHz multisource alimentée séquentiellement : modélisation et mesures. In : Comptes-rendus du Colloque sur les Applications énergétiques des rayonnements. Toulouse, Univ. Paul Sabatier, 12 juin 1986. Toulouse, ENSEEIHT, pp. 122-124.

240. MERRITT J., CHAMNESS A., HARTZELL R., ALLEN S. Orientation effects on microwave-induced hyperthermia and neurochemical correlates. *J. Microw. Power*, 12, n° 2, 1977, pp. 167-172.

241. MERRITT J., FRAZER J. Microwave fixation of brain tissue as a neurochemical technique - a review. *J. Microw. Power*, 12, n° 2, 1977, pp. 133-139.

242. MEYERHOFF J. Brain enzyme inactivation in rats by microwave irradiation : 896 MHz compared with 2450 MHz. *IEEE Trans. Microw. Theory Tech.*, MTT-27, n° 3, 1979, pp. 267-270.

243. MEYERHOFF J., BALCOM G., LENOX R. Increase in dopamine in cerebral cortex and other regions of rat brain after microwave fixation : possible diffusion artifact. *Brain Res.*, n° 152, 1978, pp. 161-169.

244. MEYERHOFF J., GANDHI O., JACOBI J., LENOX R. Comparison of microwave irradiation at 896 versus 2450 MHz for in-vivo inactivation of brain enzymes in rats. *IEEE Trans. Microw. Theory Tech.*, MTT-27, n° 3, 1979, pp. 267-270.

245. MEYERHOFF J., LENOX R., BROWN P., GANDHI O. The inactivation of rodent brain enzymes in vivo using high-intensity microwave irradiation. *Proc. IEEE*, 68, n° 1, 1980, pp. 155-159.

246. MICHAELSON S., THOMSON R., ODLAND L., HOWLAND J. The influence of microwaves on ionizing radiation exposure. *Aerosp. Med.*, 34, 1963, pp. 111-115.

247. MICHEL D., LEFEUVRE S. Simulation d'hyperthermie micro-onde. In : Compte-rendu du Symposium international sur les applications énergétiques des micro-ondes, IMPI-CFE, 14, Monaco, 1979, résumés pp. 263-265.

248. MILLIGAN A., CONRAN P., ROPAR M., Mac CULLOCH H., AHUJA R., DOBEL¬ BOWER R. Predictions of blood flow from thermal clearance during regional hyperthermia. *Int. J. Radiat. Oncol., Biol., Phys.*, 9, 1983, pp. 1337-1343.

249. MODAK A., WEINTRAUB S., Mac COY T., STAVINOHA W. Use of 300-msec. microwave irradiation for enzyme inactivation : a study of effects of sodium pentobarbital on acetylcholine concentration in mouse brain regions. *J. Pharmacol. Exp. Ther.*, 197, 1976, pp. 245-252.

250. MOOR F. Microwave diathermy. In : LICHT S. Therapeutic heat and cold. 2nd ed. New Haven, Elizabeth Licht, 1965, pp. 310-320.

251. MORESSI W. Mortality patterns of mouse sarcoma 180 cells resulting from direct heating and chronic microwave irradiation. *Exp. Cell. Res.*, 33, 1964, pp. 240-253.

252. MOROJI T., TAKAHASHI K., OGURA K., TOISHI T., ARAI S. Rapid microwave fixation of rat brain. *J. Microw. Power*, 12, n° 4, 1977, pp. 273-286.

253. MOUSAVINEZHAD S., CHEN K., NYQUIST D. Response of insulated electric field probes in finite heterogenous biological bodies. *IEEE Trans. Microw. Theory Tech.*, MTT-26, n° 8, 1978, pp. 599-607.

254. MÜLLER C. Die Krebskrankheit und ihre Behandlung mit Röntgenstrahlen und hochfrequenter Elektrizität resp. Diathermie. *Strahlenther.*, 2, 1913, pp. 170-191.

255. MYERS P., BARRETT A. Feasibility of microwave radiometry as a hyperthermia monitor. In : Proceedings of the IMPI Symposium, Toronto, 1981, Digest pp. 171-172.

256. NEELAKANTASWAMY P., RAJARATNAM A. Open-ended circular waveguide with a curved corrugated disk at its aperture as a diathermy applicator. *IEEE Trans. Microw. Theory Tech.*, MTT-30, n° 11, 1982, pp. 2005-2008.

257. NELSON A., HOLT J. Combined microwave therapy. *Med. J. Aust.*, 2, 1978, pp. 88-90.

258. NEVELS R., BUTLER C., YABLON W. The annular slot antenna in a lossy biological medium. *IEEE Trans. Microw. Theory Tech.*, MTT-33, n° 4, 1985, pp. 314-319,

259. N'GUYEN D., CHIVÉ M., LEROY Y. Hyperthermie locale contrôlée par thermographie microondes a 2,5 GHz. In : BERTEAUD A., SERVANTIE B. Ondes électromagnétiques et biologie. Symposium international URSI-CNFRS, Jouy-en-Josas, 30 juin-4 juil. 1980, H-6.

260. N'GUYEN D., MAMOUNI A., LEROY Y., CONSTANT E. Simultaneous microwave local heating and microwave thermography. Possible clinical applications. *J. Microw. Power*, 14, n° 2, 1979, pp. 135-137.

261. NIKAWA Y., KIKUCHI M., IWAMOTO M., MORI S. Waveguide applicator with convergent lens for localized microwave hyperthermia. *Trans. IECEJ*, J 66-B , Aug. 1983, pp. 1035-1042.

262. NIKAWA Y., KIKUCHI M., MORI S. Development and testing of a 2450-MHz lens applicator for localized microwave hyperthermia. *IEEE Trans. Microw. Theory Tech.*, MTT-33, n° 11, 1985, pp. 1212-1222.

263. NIKAWA Y., MIYASHITA T., MORI S., KIKUSHI M., SEKIYA T. Lens applicator for localized microwave hyperthermia. *Trans. IECEJ*, J65-B, Dec. 1982, pp. 1539-1546.

264. NIKAWA Y., MORI S., KIKUCHI M. Direct-contact waveguide applicator with convergent lens for localized microwave hyperthermia. *Strahlenther.*, 159, June 1983, p. 381.

265. NORDBERG A., SUNDWALL A. Biosynthesis of acetylcholine in different brain regions in vivo following alternative methods of sacrifice by microwave irradiation. *Acta Physiol. Scand.*, 98, n° 3, 1976, pp. 307-317.

266. OLIVIER L. Les expériences de M. d'Arsonval sur les propriétés physiques et physiologiques des courants alternatifs. *Rev. Gén. Sci.*, 5, n° 9. 1894, pp. 312-324.

267. OSAKA A. Use of microwave radiation in surgery and cancer therapy. *J. Microw. Power*, 13, n° 2, 1978, pp. 155-161.

268. OSBORNE S., FREDERICK J. Microwave radiations. Heating of human and animal tissues by means of high frequency current with wavelength of twelve centimeters (The Microtherm). *J. Am. Med. Assoc.*, 137, 1948, pp. 1036-1040.

269. OSTEEN R., FINI E., WILSON R. The effect of low-dose hyperthermia and chemiotherapy on systemic micrometastases from B16 melanoma. In : 34th Annual cancer symposium, 11-13 May 1983, Boston, p. 179.

270. OVERGAARD J., NELSON O. The role of tissue environment factors on the kinetics and morphology of tumor cells exposed to hyperthermia. *Ann. N.Y. Acad. Sci.*, 335, 1980, pp. 254-280.

271. OVERGAARD K., OVERGAARD J. Investigations on the possibility of a thermic tumor therapy. I. Short-wave treatment of a transplanted isologous mouse mammary carcinoma. *Europ. J. Cancer*, 8, 1972, pp. 65-78.

272. OVERGAARD K., OVERGAARD J. Investigations on the possibility of a thermic tumor therapy. II. Action of combined heat-roentgen treatment on a transplanted mouse mammary carcinoma. *Europ. J. Cancer*, 8, 1972, pp. 573-576.

273. PAGLIONE R. Medical applications of microwave energy. *RCA Eng.*, 27-5, Sept.-Oct. 1982, pp. 17-21.

274. PALIWAL B., CARDOZO C., JAFARI F., HANSON J., CALDWELL W. Heating patterns produced by 434 MHz Erbotherm UHF 69. *Radiobiol.*, n° 135, 1980, pp. 511-512.

275. PAUTRIZEL R. Influence d'ondes électromagnétiques et de champs magnétiques associés sur l'immunité de la souris infestée par le Trypanosoma equiperdum. *C.R. Acad. Sci.*, 263, 1er août 1966, pp. 579-582.

276. PAUTRIZEL R., MATTERN P., PRIORE A., PAUTRIZEL A., CAPBERN A., BALTZ T. Importance des mécanismes immunitaires dans la guérison de la trypanosomiase expérimentale par stimulation physique. *C.R. Acad. Sci.*, 286, 22 mai 1978, pp. 1487-1492.

277. PAUTRIZEL R., PRIORE A., BERLUREAU F., PAUTRIZEL A. Stimulation, par des moyens physiques, des défenses de la souris et du rat contre la trypanosomose expérimentale. *C.R. Acad. Sci.*, 268, 9 avril 1969, pp. 1889-1892.

278. PAUTRIZEL R., PRIORE A., BERLUREAU F., PAUTRIZEL A. Action des champs magnétiques combinés à des ondes électromagnétiques sur la trypano-somose expérimentale du lapin. *C.R. Acad. Sci.*, 271, 7 sept. 1970, pp. 877-880.

279. PAUTRIZEL R., PRIORE A., DALLOCHIO M., CROCKETT R. Action d'ondes élec-tromagnétiques et de champs magné-tiques sur les modifications lipi-diques provoquées chez le lapin par l'administration d'un régime alimen-taire hypercholestérolé. *C.R. Acad. Sci.*, 274, 17 janv. 1972, pp. 488-491.

280. PAUTRIZEL R., PRIORE A., MATTERN P., PAUTRIZEL A. Stimulation des defenses de la souris trypanosomee par l'action d'un rayonnement associant champ magnétique et ondes électro-magnétiques. *C.R. Acad. Sci.*, 280, 28 avril 1975, pp. 1915-1918.

281. PAUTRIZEL R., PRIORE A., PAUTRIZEL A., CHÂTEAUREYNAUD-DUPRAT P. Influence de l'âge de la souris sur l'efficacité de la stimulation de ses défenses par un rayonnement électromagnétique. *C.R. Acad. Sci.*, 287, 18 sept. 1978, pp. 575-578.

282. PEREZ C. Local hyperthermia and irradiation in cancer therapy. *J. Microw. Power*, 16, n° 2, 1981, pp. 205-214.

283. PETROWICZ O., HEINKELMANN W., ERHARDT W., WRIEDT-LÜBBE I., HEPP W., BLÜMEL G. Experimental studies on the use of microwaves for the loca-lized heat treatment of the prostate. *J. Microw. Power*, 14, n° 2, 1979, pp. 167-171.

284. PIONTEK G., CAIN C., MILNER J. The effects of microwave radiation, hyperthermia, and L-ascorbic acid on Ehrlich ascites carcinoma cell metabolism. *IEEE Trans. Microw. Theory Tech.*, MTT-26, n° 8, 1978, pp. 535-540.

285. PLANCOT M. Contribution à l'étude théorique, expérimentale et clini-que de l'hyperthermie microonde contrôlée par radiométrie microonde. Thèse Doctorat 3ème cycle, Univ. Sci. Tech. Lille, 14 déc. 1983

286. POPOVIĆ V., POPOVIĆ P., WEATHERS D., Effect of localized microwave heat-ing on rat embryos. In : Compte-rendu du Symposium international sur les applications énergétiques des micro-ondes, IMPI-CFE. 14, Monaco, 1979, résumés pp. 260-261.

287. PRESKORN S., EDWARDS W., JUSTESEN D. Retarded tumor growth and greater longivity in mice after fetal irra-diation by 2450-MHz microwaves. *J. Surg. Oncol.*, 10, 1978, pp. 483-492.

288. PRIOU A., FOURNET-FAYAS C., GIMONET E., GUERQUIN-KERN J., SAMSEL M., GAUTHERIE M. Simulation théorique de pénétration des ondes électromagnétiques et applicateur à contact de taille réduite à 434 MHz pour applications au chauffage des tissus. In : BERTEAUD A., SERVANTIE B. Ondes électromagnéti-ques et biologie. Symposium inter-national URSI- CNFRS, Jouy-en-Josas, 30 juin-4 juil. 1980, session G.

289. PRIOU A., GAUTHERIE M. Editorial introduction, Workshop on diagnosis and therapy using microwaves, 8 sept. 1978, París. *J. Microw. Power*, 14, n° 2, 1979, pp. 91-92.

290. RAE J., HERRICK J., WAKIM K., KRUSEN F. A comparative study of the temperatures produced by micro-wave and short wave diathermy. *Arch. Phys. Med.*, 30, 1949, pp. 199-211.

291. RAPPAPORT C., MORGENTHALER F. Localized hyperthermia with EM arrays and the leaky-wave through-guide applicator. *IEEE Trans. Microw. Theory Tech.*, MTT-34, n° 5, 1986, pp. 636-643.

292. RAPPAPORT C., MORGENTHALER F. Optimal source distribution for hyperthermia at the center of a sphere of muscle tissue. *IEEE Trans. Microw. Theory Tech.*, MTT-35, n° 12, 1987, pp. 1322-1327.

293. RAPPAPORT C., MORGENTHALER F., LELE P. Controllable heating pattern of a leaky-wave antenna for local tumor hyperthermia. *J. Microw. Power*, 21, n° 2, 1986, p. 101.

294. RASKMARK P., ANDERSEN B. Focused electromagnetic heating of muscle tissue. *IEEE Trans. Microw. Theory Tech.*, MTT-32, n° 8, 1984, pp. 887-888.

295. REBOLLAR J. Design of waveguide applicators for medical applications considering multilayered configurations of tissues. In : URSI international symposium on electromagnetic theory, Aug. 1983, pp. 669-672.

296. REBOLAR J., ENCINAR J. Design and optimization of multistepped waveguide applicators for medical applications. *J. Microw. Power*, 19, n° 4, 1984, pp. 259-267.

297. RICHARDSON D., SCUDDER C. Microwave irradiation and brain τ-aminobutyric acid levels in mice. *Life Sci.*, 19, 1976, pp. 1431-1440.

298. RINGEISEN V., MATHIEU J., LAMPERT D., CHIVÉ M., LEDÉE R., PLANCOT M., SOZANSKI J., MOSCHETTO Y. The industrial system "Hylcar II" for microwave hyperthermia monitored by microwave radiometry and computer. Wissembourg, Publication Sté ODAM-BRUKER, s.d. 5 p.

299. RIVIÈRE M., GUÉRIN M. Nouvelles recherches effectuées chez des rats porteurs d'un lymphosarcome lymphoblastique soumis à l'action d'ondes électromagnétiques associées à des champs magnétiques. *C.R. Acad. Sci.*, 262, 20 juin 1966, pp. 2669-2672.

300. RIVIÈRE M., PRIORE A., BERLUREAU F., FOURNIER M., GUÉRIN M. Action des champs électromagnétiques sur les greffes de la tumeur T8 chez le rat. *C.R. Acad. Sci.*, 259, 21 déc. 1964, groupe 14, pp. 4895-4897.

301. ROBERT J., ESCANYE J., THOUVENOT P., MARCHAL C. Influence de l'hyperthermie sur le débit vasculaire d'un rhabdomyosarcome chimio-induit de la souris C3H. In : BERTEAUD A., SERVANTIE B. Ondes électromagnétiques et biologie, Symposium international URSI-CNFRS, Jouy-en-Josas, 30 juin-4 juil. 1980, I-2.

302. ROBERTS J., COOK H. Microwaves in medical and biological research. *Br. J. Appl. Physiol.*, 3, 1952, pp. 33-39.

303. ROBINSON J., CHEUNG A., SAMARAS G., Mac CULLOCH D. Techniques for uniform and replicable microwave hyperthermia of a model mouse carcinoma. *IEEE Trans. Microw. Theory Tech.*, MTT-26, n° 8, 1978, pp. 546-549.

304. ROBINSON J., HARRISON G., CHEUNG A., Mac CULLOCH D., Mac CREADY W. A comparison of cellular X-ray sensitivity at elevated temperatures with and without microwave irradia-

tion. In : Proceedings of the IMPI Symposium, Toronto, 1981, Digest pp. 28-31.

305. ROBINSON J., Mac CULLOCH D., CHEUNG A., EDELSACK E. Microwave heating of malignant mouse tumors and tissue-equivalent phantom system. *J. Microw. Power*, 11, n° 2, 1976, pp. 87-98.

306. ROBINSON J., Mac CULLOCH D., CHEUNG A., EDELSACK E. Microwave heating of malignant mouse tumors encapsulated in a dielectric medium. In : Proceedings of the 6th European Microwaves Conference, Roma, 1976, pp. 131-136.

307. ROHDENBURG G., PRIME F. The effect of combined radiation and heat on neoplasms. *Arch. Surg.*, 1921, pp. 116-129.

308. ROMERO-SIERRA C. Bioeffects of electromagnetic waves. In : BOWHILL S. Review of radio science. 1978-1980, Bruxelles, URSI, 1981, chap. 2, pp. A.B.1-A.B.17.

309. ROMERO-SIERRA C. Bioeffects of electromagnetic waves. In : BOWHILL S. Review of radio science, 1981-1983. Bruxelles, URSI, 1984, chap. 9, pp. K.1-K.17.

310. ROMERO-SIERRA C., HALTER S., TANNER J. Effect of an electromagnetic field on the process of wound healing. Nat. Res. Council, DME Control Systems. LTR-CS-83, Ottawa, 1972.

311. ROMERO-SIERRA C., HALTER S., TANNER J., ROOMI M., CRABTREE D. Electromagnetic fields and skin wound repair. *J. Microw. Power*, 10, n° 1, 1975, pp. 59-70.

312. ROSZKOWSKI W., BIELEC M. Treatment of experimental neoplasms by local microwave (2450 MHz) hyperthermia. In : Compte-rendu du Symposium international sur les applications énergétiques des micro-ondes, IMPI-CFE, 14, Monaco, 1979, résumés p. 262.

313. ROSZKOWSKI W., WREMBEL J., ROSZKOWSKI M., JANIAK M., SZMIGIELSKI S. Does whole-body hyperthermia therapy involve participation of the immune system ? *Int. J. Cancer*, 25, 1980, pp. 289-292.

314. ROSZKOWSKI W., WREMBEL J., ROSZKOWSKI M., JANIAK M., SZMIGIELSKI S. The search for an influence of whole-body hyperthermia on antitumor immunity. *Cancer Res. Clin. Oncol.*, 96, 1980, pp. 311-317.

315. ROTKOVSKÁ D., VACEK A. Modification of the repair of radiation damage of haematopoiesis in mice by microwaves. In : Proceedings of the IMPI Symposium, Leuven, 1976. (Paper 2A.5.).

316. ROTKOVSKÁ D., VACEK A. Modification of repair of X-irradiation damage of hemopoietic system of mice by microwaves. *J. Microw. Power*, 12, n°2, 1977, pp. 119-123.

317. ROTKOVSKÁ D., VACEK A., BARTONIC-KOVA A. Therapeutic effect of microwaves on radiation damage to hematopoiesis.*Strahlenther.*, 157, 1981, pp. 677-681.

318. RUBIN A., ERDMAN W. Microwave exposure on the human female pelvis during early pregnancy and prior to conception. *Am. J. Phys. Med.*, 38, 1959, pp. 219-220.

319. RUGGERA P. Measurements of emission levels during microwave and short-wave diathermy treatments. US Dep. Health, Educ., Welf., Bur. Radiol. Health. Rockville, HHS Publication. Technical Information Staff, HFX-28, FDA-80-8119, May 1980.

320. SAIDMAN J., CAHEN R., FORESTIER J. Action des champs électriques de très haute fréquence sur les tissus organiques. *C.R. Acad. Sci.*, 192, 16 fév. 1931, pp. 452-454.

321. SALCMAN M. Clinical hyperthermia trials : design principles and practice. *J. Microw. Power*, 16, n° 2, 1981, pp. 171-177.

322. SALCMAN F. Feasibility of microwave hyperthermia for brain-tumor therapy. *Prog. Exp. Tumor Res.*, 28, 1984, pp. 220-231.

323. SALCMAN M., SAMARAS G. Hyperthermia for brain tumors : biophysical rationale. *Neurosurg.*, 9, 1981, pp. 327-335.

324. SALCMAN M., SAMARAS G., MENA H., MONTEIRO P., GARCIA J. Whole-body hyperthermia potential hazards in its application to glioblastoma. In : PAOLETTI P., WALKER M., BUTTI G., KNERICH R. Multidisciplinary aspects of brain tumor therapy. Amsterdam, Elsevier, 1979, pp. 351-356

325. SAMARAS G., CHEUNG A. Microwave hyperthermia for cancer therapy. *CRC Crit. Rev. Bioeng.*, 5, n° 2, 1981, pp. 123-184.

326. SAMARAS G., CHEUNG A., WEINMANN S. Focussed microwave radiation therapy for deep tumors, in hyperthermia as an antineoplastic treatment modality. NASA Publication 2051, 1978, pp. 67-68.

327. SAMARAS G., SALCMAN M., CHEUNG A. Interstitial microwave hyperthermia in mammalian brain : engineering biophysics, physiology and clinical application. In : Proceedings of the IMPI Symposium, Toronto, 1981, Digest pp. 77-79.

328. SAMARAS G., VAN HORN H., KING V., SLAWSON E., CHEUNG A. Clinical hyperthermia systems engineering. *J. Microw. Power*, 16, n° 2, 1981, pp. 161-169.

329. SANDHU T., BICHER H. Clinical microwave hyperthermia : heating patterns and treatment planning. In : Proceedings of the IMPI Symposium, Toronto, 1981, Digest pp. 37-40.

330. SANDHU T., KOWAL H., JOHNSON R. The development of microwave hyperthermia applicators. *Int. J. Radiat. Oncol., Biol., Phys.*, n° 4, 1978, pp. 515-519.

331. SANTINI R., HOSNI M., DOUSS T., DESCHAUX P., PACHECO H. Increasing antigenicity of B16 melanoma cell fraction with microwave hyperthermia. *J. Microw. Power*, 21, n° 1, 1986, pp. 41-44.

332. SANTINI R., VOULOT C., DESCHAUX P. Hyperthermie micro-ondes et traitement anticancéreux. *Econ. Progrès Electr.*, n° 3, mai-juin 1982, p. 14.

333. SANTINI R., VOULOT C., DESCHAUX P. Incidence de l'hyperthermie micro-ondes sur le mélanome B16 de la souris blanche : études préliminaires. *ITBM*, 3, n° 5, 1982, pp. 542-547.

334. SAPARETO S., DEWEY W. Thermal dose determination in cancer therapy. *Int. J. Radiol.*, 5. 1984, pp. 285-295.

335. ŠAPOŠNIKOV J., JARESKO I., VERNI-GORA J. Etude histomorphologique de la réparation des lésions d'un animal après exposition à long terme aux micro-ondes à bas niveau. (En russe). *Bjull. Eksp. Biol. Med.*, 80, n° 8, 1975, pp. 988-990.

336. SAUVEGRAIN. Les ondes courtes en médecine. *Sci. Vie*, n° H.S. Radio, radar, télévision... 1948, pp. 168-173.

337. SCHEIBLICH J., PETROWICZ O. Radio-frequency-induced hyperthermia in the prostate. *J. Microw. Power*, 17, n° 3, 1982, pp. 203-209.

338. SCHEPPS J., FOSTER K. UHF and microwave dielectric properties of normal and tumor tissues. In : Proceedings of the IMPI Symposium, Toronto, 1981, Digest pp. 34-36.

339. SCHIENDER D., FELT B., GODMAN H. On the use of microwave radiation energy for brain-tissue fixation. *J. Neurochem.*, 38, 1982, pp. 749-752.

340. SCHMIDT D., SPETH R., WELSCH F., SCHMIDT M. The use of microwave radiation in the determination of acetylcholine in the rat brain. *Brain Res.*, 38, 1972, pp. 377-389.

341. SCHMIDT M., SCHMIDT D., ROBINSON G. Cyclic adenosine monophosphate in brain areas : microwave irradiation as a means of tissue fixation. *Sci.*, 173, 1971, pp. 1142-1143.

342. SCHMIDT M., SCHMIDT D., ROBINSON G. Cyclic AMP in the rat brain : microwave irradiation as a means of tissue fixation. *Adv. Cyclic Nucleotide Res.*, 1, 1972, pp. 425-434.

343. SCHNEIDER S., GOTTLIEB C., BALZANO Q., GARAY O. Hyperthermia heating patterns in a tissue-equivalent split phantom produced by an applicator having adjustable aperture and tunable frequency, operated at 482 and 915 MHz. In : Proceedings of the IMPI Symposium, Chicago, 1985, pp. 147-151.

344. SCHWAN H. Biophysics of diathermy. In : LICHT S. Therapeutic heat and cold. 2nd ed. New Haven, Elizabeth Licht, 1965, pp. 63-125.

345. SCHWAN H. Radiation biology, medical applications and radiation hazards. In : OKRESS E. Microwave power engineering. Vol. 2. New York, Academic Press, 1968, pp. 215-532.

346. SCHWAN H. Electromagnetic and ultrasonic induction of hyperthermia in tissue-like substances. *Radiat. Environ. Biophys.*, 17, 1980, pp. 189-203.

347. SCHWAN H., LI K. Variations between measured and biologically effective microwave diathermy dosage. *Arch. Phys. Med.*, 36, 1955, pp. 363-370.

348. SCOTT B. Short-wave diathermy. In : LICHT S. Therapeutic heat and cold. Baltimore, 1958, pp. 255-283.

349. SCOTT R., CHEUNG A., SAMARAS G. Cancer therapy by electromagnetic hyperthermia – special issue : foreword and editorial. *J. Microw. Power*, 16, n° 2, 1981, pp. 85-87.

350. de SEGUIN L. Les propriétés biologiques des micro-ondes. *C.R. Acad. Sci.*, 228, 1949, pp. 235-237.

351. SELKER R. Hyperthermia : possible new tool in cancer therapy. *Surg. Rounds*, 1980, pp. 70-74.

352. SHARPLESS N., BROWN L. Use of microwave irradiation to prevent post-mortem catecholamine metabolism : evidence for tissue disruption artifact in a discrete region of rat brain. *Brain Res.*, n° 140, 1978, pp. 171-176.

353. SHERMAN L., DOERNER I. A new rapid method for thawing fresh frozen plasma. *Transfus.*, 14, n° 6, 1974, pp. 594-597.

354. SHORT J., TURNER P. Physical hyperthermia and cancer therapy. *Proc. IEEE*, 68, n° 1, 1980, pp. 133-142.

355. SHUPE R., HORNBACK N. The friendly fields of RF. *IEEE Spectr.*, 22, n° 6, 1985, pp. 64-69.

356. de SIÉYÈS D., DOUPLE E., STROHBEHN J., TREMBLY B. Some aspects of optimization of an invasive microwave antenna for local hyperthermia treatment of cancer. *Med. Phys.*, 8, 1981, pp. 174- 183.

357. STAMM M., WARREN S., RAND R., WOLFSON W., HARRIS P., IWASAKI R., BLOCK J., SCHOENBERGER J., DOKKEN R., GARRETT J., STOEKEL T., HENDRICKS W., RUTLEGE C. Microwave therapy experiments with B-16 murine melanoma. *IRCS Med. Sci.*, 3, 1975, 392-393.

358. STAMM M., WINTERS W., MORTON D., WARREN S. Microwave characteristics of human tumor cells. *Oncol.*, 29, 1974, pp. 294-301.

359. STAVINOHA W. Microwave fixation for the study of acetylcholine metabolism. In : JENDEN D. Cholinergic mechanisms and psychopharmacology. New York, Plenum Press, 1978, pp. 169-179.

360. STAVINOHA W., PEPELKO B., SMITH P. The use of microwave heating to inactivate cholinesterase in the rat brain prior to analysis for acetylcholine. *Pharmacol.*, 12, 1970, p. 257.

361. STAVINOHA W., WEINTRAUB S., MODAK A. The use of microwave heating to inactivate cholinesterase in the rat brain prior to analysis for acetylcholine. *J. Neurochem.*, 20, 1973, pp. 361-371.

362. STERZER F. Localized hyperthermia treatment of cancer. *RCA Rev.*, 42, Dec. 1981, pp. 727-751.

363. STERZER F., PAGLIONE R., FRIEDEN-THAL E., MENDECKI J. A microwave apparatus for producing uniform hyperthermic temperatures over large surfaces. In : IEEE Symposium on Microwave Theory and Techniques, 1985, C-6, pp. 90-92.

364. STERZER F., PAGLIONE R., MENDECKI J., FRIEDENTHAL E., BOTSTEIN C. RF therapy for malignancy. *IEEE Spectr.*, 17, n° 12, 1980, pp. 32-37.

365. STERZER F., PAGLIONE R., NOWOGROD-ZKI M., MENDECKI J., FRIEDENTHAL E., BOTSTEIN C. Microwave apparatus for the treatment of cancer. *Microw. J.*, n° 1, 1980, pp. 39-44.

366. STERZER F., PAGLIONE R., WOZNIAK F. A self-balancing microwave radiometer for non-invasively measuring the temperature of subcutaneous tissues during localized hyperthermia treatments of cancer. In : IEEE Symposium on Microwave Theory and Techniques,Dallas, 1981, Digest pp. 438-440.

367. STERZER F., PAGLIONE R., WOZNIAK F., FRIEDENTHAL E., MENDECKI J. A robot-operated microwave hyperthermia system for heating large malignant surface lesions. *Microw. J.*, 29, n° 7, 1986, pp. 147-152.

368. STEVENS A., PELUSO F. Temperature changes generated by microwave therapy in the thighs of human subjects. In : Proceedings of the IMPI Symposium, Leuven, 1976. (Paper 5A.6.).

369. STEVENSON H. The effect of heat upon tumor tissue. *J. Cancer Res.*, 4, 1919, p. 54.

370. STORM F., CRIFO C., ROSSI-FANELLI A., MONDOVI B. Biochemical aspect of heat sensitivity of tumor cells : recent results. *Cancer Res.*, 37, 1977, pp. 7-35.

371. STORM F., HARRISON W., ELLIOTT R., HATZITHEOFILOU C., MORTON D. Human hyperthermic therapy : relation between tumor type and capacity to induce hyperthermia by radiofrequency. *Am. J. Surg.*, n° 138, 1979, pp. 170-174.

372. STORM F., HARRISON W., ELLIOTT R., MORTON D. Hyperthermic therapy for human neoplasms. *Cancer*, n° 46. 1980, pp. 1849-1854.

373. STORM F., HARRISON W., MORTON D. Normal tissue and solid tumor effects of hyperthermia in animal models and clinical trials. *Cancer Res.*, 39, 1979, pp. 2245-2251.

374. STROHBEHN J., BOWERS E., WALSH J., DOUPLE E. An invasive microwave antenna for locally-induced hyperthermia for cancer therapy. *J. Microw. Power*, 14, n° 4, 1979, pp. 339-350.

375. STUCHLY M., REPACHOLI M., LECUYER D., MANN R. Electromagnetic fields around short-wave diathermy. In : Proceedings of the IMPI Symposium, Toronto, 1981, Digest pp. 206-208.

376. STUCHLY M., STUCHLY S. Industrial, scientific, medical and domestic applications of microwaves. *IEE Proc.*, 130, part A, n° 8, Nov. 1983, pp. 467-503.

377. STUCHLY M., STUCHLY S., KANTOR G. Diathermy applicators with circular aperture and corrugated flange. *IEEE Trans. Microw. Theory Tech.*, MTT-28, n° 3, 1980, pp. 267-271.

378. STUCHLY S., STUCHLY M. Applications of electromagnetic waves in medicine - A review. In : IREE International Conference, Melbourne, 24-28 Aug. 1981.

379. SUROWIEC A., STUCHLY S., BARR R., SWARUP A. Dielectric properties of breast carcinoma and the surrounding tissues. *IEEE Trans. Biomed. Eng.*, BME-35, n° 4, 1988, pp. 257-263.

380. SUTTON C., CRANE R. Effects of blood flow rate and direction on thermal fields in brain and tumor during microwave hyperthermia. *J. Microw. Power*, 21, n° 2, 1986, p. 104.

381. SWICORD M., DAVIS C. Energy absorption from small radiating coaxial probes in lossy media. *IEEE Trans. Microw. Theory Tech.*, MTT-29, n° 11, 1981, pp. 1202-1209.

382. SZMIGIELSKI S. Inhibition of acute experimental viral infections by whole-body microwave (2450 MHz) hyperthermia. In : Comptes-rendus du Symposium international sur les applications énergetiques des microondes,IMPI-CFE, 14, Monaco, 1979, resumés p. 259.

383. SZMIGIELSKI S., BIELEC M., JANIAK M., KOBUS M., LUCZAK M., DE CLERCQ E. Inhibition of tumor growth in mice by microwave hyperthermia, polyribo-inosinic-polyribocytidylic, and mouse interferon. *IEEE Trans. Microw. Theory Tech.*, MTT-26, n°8, 1978, pp. 520-522.

384. SZMIGIELSKI S., JANIAK M. Injury of cell membranes in normal and SV40-virus-transformed fibroblasts exposed in vitro to microwave (2450 MHz) or waterbath hyperthermia (43° C). In : Biological effects of electromagnetic waves. International symposium, Airlie, 30 Oct.-4 Nov. 1977, Abstracts of scientific papers.

385. SZMIGIELSKI S., LUCZAK M., BIELEC M., JANIAK M., KOBUS M., STEWARD W., DE CLERQ E. Effect of microwaves combined with interferon and/or interferon inducers (poly i-poly c) on development of sarcoma 180 in mice. In : Proceedings of the IMPI Symposium, Leuven, 1976. (Paper 6A.3.).

386. SZMIGIELSKI S., PULVERER G., HRYNIEWICZ W., JANIAK M. Inhibition of tumor growth in mice by microwave hyperthermia, streptolysin S and colcemide. *Radio Sci.*, 12, n° 6,(S), 1977, pp. 185-189.

387. SZMIGIELSKI S., ROSZKOWSKI W., JELJASZEWICZ J. Modification d'infections aiguës de staphylocoques par exposition de longue durée soit aux micro-ondes de faible niveau, soit à une hyperthermie totale par micro-ondes. In : BERTEAUD A., SERVANTIE B. Ondes électromagnétiques et biologie. Symposium international URSI-CNFRS, Jouy-en-Josas, 30 juin-4 juil. 1980, E-1.

388. SZMIGIELSKI S., SZUDZINSKI A., PIETRASZEK A., BIELEC M., JANIAK M., WREMBEL J. Accelerated development of spontaneous and benzopyrene-induced skin cancer in mice exposed to 2450 MHz microwave radiation. *Bioelectromagn.*, 3, 1982, pp. 179-191.

389. SZUDZINSKI A., PIETRASEK A., ROSZKOWSKI W., SZMIGIELSKI S. Exaltation du développement de cancer cutané induit chimiquement par 3,4-benzopyrène chez la souris par exposition de longue durée et à un faible niveau de microondes (2450 MHz). In BERTEAUD A., SERVANTIE B., Ondes électromagnétiques et biologie. Symposium international URSI-CNFRS, Jouy-en-Josas, 30 juin-4 juil. 1980, E-9.

390. SZWARNOWSKI S., SHEPPARD R., GRANT E., BLEEHEN N. A broadband microwave applicator for heating tumours. *Br. J. Radiol.*, 53, 1980, pp. 31-33.

391. TABUSE K., KATSUMI M. Application of a microwave tissue coagulator to hepatic surgery - the hemostatic effects on spontaneous rupture of hepatoma and tumor necrosis. *Arch. Jpn. Chir.*, 50, 1981, pp. 571-579.

392. TANAKA H., KATO H., NISHIDA T., KANO E., SUGAHARA T., ISHIDA T. Physical basis of RF hyperthermia for cancer therapy measurement of distribution of absorbed power from radiofrequency exposure in agar phantom. *J. Radiat. Res.*, 22, 1981, pp. 101-108.

393. TAYLOR L. Electromagnetic syringe. *IEEE Trans. Biomed. Eng.*, BME-25, n° 5, 1978, pp. 308-309.

394. TAYLOR L. Implantable radiators for cancer therapy by microwave hyperthermia. *Proc. IEEE*, 68, n° 1, 1980, pp. 142-149.

395. TAYLOR L. Brain cancer therapy using an implanted microwave radiator. *Microw. J.*, 24, 1981, pp. 66-71.

396. TEODORIDIS V., SPHICOPOULOS T., GARDIOL F. The reflection from an open-ended rectangular waveguide terminated by a layered dielectric medium. *IEEE Trans. Microw. Theory Tech.*, MTT-33, n° 5, 1985, pp. 359-366.

397. TERADA N., AMEMIYA Y. The performance of the dipole array applicator for radio-frequency hyperthermia. *Trans. IECEJ*, J67-B, Feb. 1984, pp. 163-170.

398. THOMPSON J., SIMPSON T., CAULFIELD J. Thermographic tumor detection enhancement using microwave heating. *IEEE Trans. Microw. Theory Tech.*, MTT-26, n° 8, 1978, pp. 573-580.

399. THOMSON R., MICHAELSON S., HOWLAND J. Modification of X-irradiation lethality in mice by microwaves (radar). *Radiat. Res.*, 24, 1965, pp. 631-635.

400. TREMBLY B., STROHBEHN J., de SIÉYÈS D., DOUPLE E. Hyperthermia induced by an array of invasive microwave antennas. In : 3rd International Symposium : Cancer therapy by hyperthermia, drugs, and radiation, Colorado State Univ., Fort Collins, 1980, communication Te 41.

401. .TRÉMOLIÈRES J. Effets biologiques des champs électromagnétiques non ionisants. *Electron. Appl.*, n° 7, automne 1978, pp. 71-77.

402. TSAI C., DURNEY C., CHRISTENSEN D. Calculated power absorption patterns for hyperthermia applicators consisting of electric dipole arrays. *J. Microw. Power*, 19, n° 1, 1984, pp. 1-13.

403. TURNER P. Deep heating of cylindrical or elliptical tissue masses. In : 3rd International Symposium : Cancer therapy by hyperthermia, drugs, and radiation, Colorado State Univ., Fort Collins, 1980, Communication Te 42.

404. TURNER P. Deep heating of cylindrical or elliptical tissue masses. *J. Nat. Cancer Inst; Monogr.*, 61, 1982, pp. 493-495.

405. TURNER P. Hyperthermia and inhomogenous tissue effects using an annular phased array. *IEEE Trans. Microw. Theory Tech.*, MTT-32, n° 8, 1984, pp. 874-882.

406. TURNER P., KUMAR L. Computer solution for applicator heating patterns. *J. Nat. Cancer Inst. Monogr.*, 61, 1982, pp. 521-523.

407. U R., NOELL K., WOODWARD K., WORKE B., FISHBURN R., MILLER L. Microwave-induced local hyperthermia in combination with radiotherapy of human malignant tumors. *Cancer*, 45, 1980, pp. 638-646.

408. VAGUINE V., GIEBELER R., Mac EUEN A. Microwave direct-contact applicator system for hyperthermia therapy research. In : 3rd International Symposium : Cancer therapy by hyperthermia, drugs, and radiation, Colorado State Univ., Fort Collins, 1980, Communication Te 43.

409. VAN DEN BERG P., DE HOOP A., SEGAL A., PRAAGMAN N. A computational model of the electromagnetic heating of biological tissue with application to hyperthermic cancer therapy. *IEEE Trans. Biomed. Eng.*, BME-30, n° 12, 1983, pp. 797-805.

410. VAN SLIEDREGT M. Computer calculations of a one-dimensional model, useful in the application of hyperthermia. *Microw. J.*, 26, n° 6, 1983, pp. 113-126.

411. VOLFOVSKAJA P., OSIPOV J. K voprosu o kombinirovanom vozdjejstvii polja vysokoj častoty i rentgenovskogo izlučenija v proizvodstvjennyh uslovijah. *Gig. Sanit.*, 26, n° 5, 1961, pp. 18-23.

412. WAHID P., HAGMANN M., GANDHI O. Multidipole applicators for regional and whole-body hyperthermia. *Proc. IEEE*, 70, n° 3, 1982, pp. 311-313.

413. WALLEN C., MICHAELSON S. Microwave-induced hyperthermia as an adjuvant to cancer therapy. In : Proceedings of the IMPI Symposium, Leuven, 1976. (Paper 6.A.4.).

414. WARREN S. Preliminary study of the effect of artificial fever upon hopeless tumor cases. *Am. J. Roentgen Radium Ther.*, 33, Jan. 1935, pp. 75-87.

415. WEBB S. Effects of microwaves on normal and tumor cells as seen by laser-raman spectroscopy. In : Proceedings of the IMPI Symposium, Leuven, 1976. (Paper I.A.4.).

416. WEBB S., BOOTH A. Microwave absorption by normal and tumor cells. *Sci.*, 174, 1st Oct. 1971, pp. 72-74.

417. WESTENKOW D., WONG K., JOHNSON C., WILDE C. Physiologic effects of deep hyperthermia and microwave rewarming possible application for neonatal cardiac surgery. *Anesth. Analg.*, 58, 1979, pp. 297-301.

418. WILDERVANG A. Certain experimental observations on a pulsed diathermy machine. *Arch. Phys. Med.*, 40, 1959, p. 45.

419. WILSON D., JAGADEFS P., NEWMAN P., HARRIMAN D. Effects of pulsed electromagnetic energy on peripheral nerve regeneration. *Ann. N.Y. Acad. Sci.*, 238, 1974, pp. 575-585.

420. WINTER A., LAING J., PAGLIONE R., STERZER F. Microwave hyperthermia for brain tumors. *Neurosurg.*, 17, n° 3, 1985, pp. 387-399.

421. WYSLOUZIL W., KASHYAP S. Heating patterns for an array of 915 MHz rectangular waveguide applicators. *J. Microw. Power*, 21, n° 2, 1986, pp. 101-102.

422. YERUSHALMI A. Cure of a solid tumor by simultaneous administration of microwaves and X-ray irradiation. *Radiat. Res.*, 64, 1975, pp. 602-610.

423. YERUSHALMI A., KATZAP I., GOTTES-
FELD B., BASS D. Systemic alteration in blood glucose levels
following localized deep microwave
hyperthermia. *J. Microw. Power*,
19, n° 1, 1984, pp. 73-76.

424. YERUSHALMI A., SHPIRER Z., HOD I.,
GOTTESFELD F., BASS D. Normal
tissue response to localized deep
microwave hyperthermia in the
rabbit's prostate : a preclinical
study. *Int. J. Radiat. Oncol.,
Biol., Phys.*, 9, 1983, pp. 77-82.

425. ZANKER K., JUNG R., STAVROU D.,
BLUMEL G. Influence of microwave
irradiation on cultured glioma cells.
I. An enzymatic and scanning elec-
tron microscopy study. *J. Microw.
Power*, 14, n° 2, 1979, pp. 159-162.

426. ZHONG H., QINYI S., MINGJIANG S.
Rewarming with microwave irradiation
in severe cold injury syndrome. *Chin.
Med. J.*, 93, 1980, pp. 119-120.

427. ZIMMER R. Simulation numérique de
la distribution de l'énergie électromagnétique et de la température
dans un sein cancéreux irradié par
un champ micro-ondes. In : Compte-
rendu du Symposium international
sur les applications énergétiques
des micro-ondes, IMPI-CFE. 14,
Monaco, 1979, résumés pp. 185-187.

428. ZIMMER R., ECKER H., POPOVIĆ V.
Selective electromagnetic heating
of tumors in animals in deep hypothermia. *IEEE Trans. Microw. Theory
Tech.*, MTT-19, n° 2, 1971, pp. 238-
245.

429. ZIMMER R., GROS C. Numerical calculation of electromagnetic energy
and temperature distribution in a
microwave irradiated breast carcinoma : preliminary results. *J. Microw.
Power*, 14, n° 2, 1979, pp. 155-158.

430. ZÝWIETZ F. Effect of microwave heating on the radiation response of
the rhabdomyosarcoma. *Strahlenther.*,
158, 1982, pp. 255-257.

431. ZÝWIETZ F., KNÖCHEL R. The role of
dielectric properties by heating a
rat tumour with microwaves. VIIIth
European Society for Hyperthermic
oncology meeting, Paris, 16-18 Sept.
1985. *Strahlenther.*, 161, n° 9,
1985, pp. 556-557.

432. ZÝWIETZ F., KNÖCHEL R., KORDTS J.
Heating of a rhabdomyosarcoma of
the rat by 2450 MHz microwaves :
technical aspects and temperature
distributions. In Recent results in
cancer research. Heidelberg,
Springer Verlag, 101, 1986, pp.
36-46.

433. Hyperthermic oncology, HYLCAR.
Wissembourg, Publication Sté ODAM,
s.d., 8 p.

434. Hyperthermie localisée - contrôle
atraumatique par radiométrie
(HYLCAR). Wissembourg, Publication
Sté ODAM, s.d., 6 p.

435. Localized microwave hyperthermia
atraumatically controlled through
radiometry, HYLCAR. Wissembourg,
Publication Sté ODAM, s.d., 6 p.

436. Microwave diathermy devices. Ministère de la Santé et du Bien-être
social du Canada, Branche de la
Protection de la Santé. *Lettre Inf.*,
n° 517, 21 déc. 1977.

437. The application of microwave
technology to the detection and
treatment of cancer. *IEEE Microw.
Theory Tech.*, MTT-Newsl., n° 111,
1985, p. 15.

Addresses

The following is a list of some of the industries, organizations, and universities that have made major contributions in the field of microwave technology. It is not meant to be a comprehensive listing.

INDUSTRY

Albatronics Engineering
(Jerome R. White)
44755 Wyandotte Avenue
HEMET, CA92344, USA
Tel (1) 714 927 5098
Fax (1) 714 652 532

Alfastar AB
(Peter Uhnbom)
P.O. Box 500
147 00 Tumba, S
Tel (46) 753 65000
Fax (46) 753 33519
Tx 10260

APV Baker Ltd
(Roger J. Meredith, Chris P. Bass, Gavin B. Kirk)
Subsidiary of APV Corp
Confectionery Cereals and
Snacks
Division (CCS)
Manor Drive
Paston Parkway
Peterborough, PE4 7AP, GB
Tel (44) 733 283 000

Fax (44) 733 283 085

APV France
(Thierry Hoepffner)
6, rue Jacquard
B.P. 684
27006 Evreux La Madeleine, F
Tel (33) 32 28 96 12
Fax (33) 32 28 11 83
Tx 770 880

APV Magnetronics Ltd
(Roger J. Meredith)
9-12 St. Mary's Mills
Evelyn Drive
Leicester, LE3 2BU, GB
Tel (44) 533 891 189
  (44) 533 891 139
Fax (44) 533 823 498
Tx 342 486

APV Parafreeze Ltd
(Roger J. Meredith)
Stephenson Way
Thetford, Norfolk, IP24 3RP, GB
Tel (44) 842 625 11
Fax (44) 842 633 22
Tx 81534

Berstorff Corp
(Wilfried Schlegel, Gerald J. Lynch)
Subsidiary of Hermann Berstorff
Maschineubau GmbH
8200 Arrowbridge Boulevard
P.O. box 240357
Charlotte, NC 28224, USA
Tel (1) 704 523 2614
Fax (1) 704 523 4353

Calorex AB
(Bengt Edin)
OB 2024
Sätraängsvägen 10-12
182 02 Danderyd, S
Tel (46) 875 31601
Fax (46) 875 31602
Tx 12442

Cober Electronics Inc
(Bernard Krieger)
Industrial Microwave Division
102 Hamilton Avenue
Stamford, CT 06902, USA
Tel (1) 203 327 0003
Fax (1) 203 359 6319

David Sarnoff Research Center
(Frank J. Marlowe)
Subsidiary of SRI International
Microwave and Reliability
Laboratory
201 Washington Road
Princeton, NJ 08540-6449, USA
Tel (1) 609 734 3179
Fax (1) 609 734 2050

D. Briggs Associates
(David Briggs)
Manor Farm House
Silchester
Reading, Berks. RG7 2HL, GB
Tel (44) 734 701 186

Digicom Poland
(Juliusz Lukjanik, Krzysztof Piwonski)
Etiudy Rewolucyjnej 9
02-643 Warszawa, PL
Tel (48) 22 48 31 14
    (48) 22 48 55 05
Tx 816 723

Enersyst Development Center Inc
(Donald Paul Smith)
2051 Valley View Lane
Dallas, TX 75 234-8910, USA
Tel (1) 214 247 9624
Fax (1) 214 247 9738

F. Jahn Co Ltd
(Gerard Lupton)
34 York Way
King's Cross
London N1 9AB, GB
Tel (44) 71 837 7792
Fax (44) 71 837 8254
Tx 24265

Gamma Consultants
(Gordon Andrews)
1 Montague Terrace
Durham Road
Bromley, Kent, BR2 0SZ, GB
Tel (44) 81 466 1825
Fax (44) 81 460 7908
Tx 896 827

Gerling Laboratories
(John E Gerling)
1132 Doker Drive
Modesto, CA 95351, USA
Tel (1) 209 521 6549
Fax (1) 209 521 7527
Tx 466 648

**Hermann Berstorff
Maschinenbau GmbH**
(Klaus Koch, Rolf Skubich)
Postfach 629
3000 Hannover 160, D
Tel (49) 511 57 02-0
Fax (49) 511 56 19 16
Tx 921 348

**Hosokawa Micron Europe
BV**
(T. Van Aken)
Vrieco Division
Postbus 10
7020 BL Zelhem, NL
Tel (31) 83 42 10 45
Fax (31) 83 42 34 59
Tx 45 714

**Hosokawa Micron Europe
BV**
*(Antonius Maria Van Aken,
Bert Jan Oosterop)*
P.O. Box 773
2003 RT Haarlem
Nijverheidsweg 25
2031 CN Haarlem, NL
Tel (31) 23 15 74 11
Fax (31) 23 31 83 80
Tx 41 167

**Industries Micro-ondes In-
ternationales (IMI-PREMO)**
(Nicolas Meisel)
Ferme du Mesnil
Route de Pavé
78680 Epône, F
Tel (33) 1 30 95 66 76
Fax (33) 1 30 95 33 87
Tx 695 554

Machines Euraf
(Laurant Skvortzoff)
55-65, rue Émile Deschanel
92400 Courbevoie, F

Tel (33) 1 43 33 41 51
Fax (33) 1 43 33 97 79
Tx 614 122

Microdry Corp
*(Peter Nemeth, Herbie L. Bullis,
Franklin J. Smith, John Wieder-
satz)*
7450 Highway 329
Crestwood, KY 40014, USA
Tel (1) 502 241 8933
Fax (1) 502 241 8648

Microheat Sweden AB
(Larsgöran Gustafsson)
P.B. 7097
172 07 Sundbyberg, S
Tel (46) 8 733 95 10
Fax (46) 8 98 73 51

**Microondes Énergie
Systémes (MES)**
*(André-Jean Berteaud, Patrick
Mahé)*
2 et 4, Avenue de la Cerisaie
Platanes 307
94266 Fresnes cedex, F
Tel (33) 1 46 68 39 39
Fax (33) 1 46 68 45 09
Tx 260 793

Mikrovagsapplikation AB
(Hans Uddborn)
Box 198
67129 Arvika, S
Tel (46) 570 14050
Fax (46) 570 18145
Tx 5758

Microwave Heating Ltd
(Ralph A Shute)
Unit 2
Heron Trading Estate
Whitefield Avenue
Sundon Park
Luton, Beds, LU3 3BB, GB
Tel (44) 582 584 747
Fax (44) 582 584 747
Tx 825 562

Narda Microwave Corp
(Edward Aslan)
435 Moreland Road
Hauppauge, NY 11788, USA
Tel (1) 516 231 1700
Fax (1) 516 213 1711
Twx 510 221 1867

**Office de Distribution
d'Appareils Médicaux
(ODAM/SAISA)**
*(Jean-Pierre Nabire, Victor
Ringeisen)*
34, rue de l'Industrie
67160 Wissembourg, F
Tel (33) 88 94 99 32
Fax (33) 88 54 29 08
Tx 870 639

OMAC SRL
(Carlo Coluccio)
Via Industria 6
42019 Pratissolo di Scandiano
(RE), I
Tel (39) 522 85 73 24
    (39) 522 98 29 87
Tx 531 845

Oshikiri Machinery Ltd
(Michihiko Tanaka)
15-14 4-chome Ohmori-Nishi
Ohta-ku
Tokyo 143, J
Tel (81) 3 3761 9171

Fax (81) 3 3764 1564
Agent in Europe:
Dipack A/S Pakkesystemer
Jyllandsgade 17A
6400 Sönderborg, Denmark
Agent in the USA:
Gemini Bakery Equipment Co.
9990 Gantry Road
Philadelphia, PA 19115, USA
4 Kirihara-cho Fujisawa-shi
Kanagawa 252, J
Tel (81) 4 6644 6011
Tx 386 2234

Pavailler SA
(Jacques Cotte)
B.P. 54
26800 Portes-Lés-Valence, F
Tel (33) 75 57 55 00
Fax (33) 75 57 23 19
Tx 345 738

**P.C. Müller Ingenieurbüro
fur Mikrowellenapplikation
Entwicklung und
Anfertigung von Prototypen**
Schorenstrasse 74
9000 St. Gallen, CH
Tel (44) 71 23 56 31
Tx 71 781

Philips Forschungslaboratorium
*(Reinhard Knöchel, Jürgen
Köhler, Kai-M Lüdeke,
W. Meyer)*
vogt-Kölln Strasse 30
2000 Hamburg 54, D
Tx 213 316

Radio Corporation of America (RCA)
(Mark Nowogrodzki, Robert W. Paglione)
David Sarnoff Research Center
Subsystems and Special Projects
Microwave Technology Center
201 Washington Road
Princeton, NH 08543, USA
Tel (1) 609 734 2000

Rayonnements Systémes
(Jean-Marc Dererian, Madeleine Dubuis)
17 bis, rue Adam
69100 Villeurbanne, F
Tel (33) 78 68 89 02
    (33) 78 68 34 08
Tx 375 263 code R 145

Raytek
(Marc Friaud)
35, rue du Tonkin
69100 Villeurbanne, F
Tel (33) 78 89 77 77
Fax (33) 78 89 27 77
Tx 380 306

Raytheon Co.
(John M. Osepchuk)
Research Division
Electron Beam Devices
131 Spring Street
Lexington, MA 92173, USA
Tel (1) 617 860 3320
Fax (1) 617 860 3345
Tx 948 652
    951 662

Raytheon Co.
(Richard Edgar, Ronald L. Snider, Paul Stevens, Ira Myers)
Industrial Equipment Group
Microwave and Power Tube
Division

190 Willow Street
Waltham, MA 02254, USA
Tel (1) 617 642 3361
Fax (1) 617 642 3718
Tx 923 455

Remy Electronique SA
(Bernard Mangin, Roger Rémy)
17, rue d'Estienne d'Orves
93360 Neuilly-Plaisance, F
Tel (33) 1 43 00 56 69
Fax (33) 1 43 00 63 54
Tx 232 692

R.F. Schiffmann Associates Inc
(Robert F. Schiffmann)
149 West 88th. Street
New York, NY 10024, USa
Tel (1) 212 362 7021
Fax 91) 212 769 4630

Richardson France
165-167, boulevard de Valmy
92706 colombes, F
Tel (33)1 47 60 05 15
Fax (33) 1 47 60 98 04
Tx 615 938

Société d'Applications Electriques (SAE)
(Jacques Lonné)
423, rue de la Ferme du Conte
40000 Mont de Marsan, F
Tel (33) 58 06 31 96
Fax (33) 58 06 88 29

Société d'Application
Industriielle en Recherche
Électronique et Micro-ondes
(SAIREM)
*(J.P. Bernard, Jean-François
Rochas, Jean-Marie Jacomino)*
24 rue Louis Saillant
69120 Vaulx en Velin, F
Tel (33) 72 04 09 25
Fax (33) 72 04 37 65,
Tx 380 157
Agent in the USA:
AJA International
P.O. box 246
890 Country Way
North Scituate, MA 02060
Tel (1) 617 545 7365
Fax (1) 617 545 4105

Société Audoise de
Représentation de
Recherche et d'Eutdes
(SERRE)
(André Monard, Jacques Vidal)
Avenue Ernest Léotard
11150 Bram, F
Tel (33) 68 76 13 27
Tx 500 379

Société d'Energétique,
de Régulation et de
Micro-ondes (SERMO)
(Michel Munoz)
7, rue G. Ramon, Z.I. Nord
B.P. 2098
57052 Metz cedex 2, F
Tel (33) 87 33 17 84
Fax (33) 87 30 52 55
Tx 930 193

Société Française
d'Application des
Micro-ondes (SFAMO)
*(Robert Bellavoine, Guy
Tzifkansky)*

Subsidiary of Cidelcem Industries
Reu de l'Ecossais
B.P. 422, Limas
69653 Villefranche sur Saône, F
Tel (33) 74 62 75 15
Fax (33) 74 62 75 99
Tx 340 601

Sotrimelec
(Georges Batailhou)
Route d'Escalquens
Z.I. de Vic
31320 Castanet-Tolosan, F
Tel (33) 61 27 71 61

Techniques Micro-ondes
(TECHMO)
(Jacques Guillaudeau)
Z.I. de Bois-Pataud
37150 Bléré, F
Tel (33) 47 57 90 53
Fax (33) 47 57 87 83
Tx 751 322

Thomson Tubes
Électroniques
*(Denis Ranque, Pierre Menes,
Charles Kalfon, Héléne Maggiar)*
38, rue Vauthier
B.P. 305
92102 Boulogne-Billancourt
cedex, F
Tel (33) 1 49 09 28 28
Fax (33) 1 46 04 52 09
Tx 633 344

Toshiba Corp
*(Yoshio Amano, Kaichiro Nakai,
Ichiro Namba, Hisao Saito,
Norio Tashiro)*
1-1, Shibaura 1-chome
Minato-ku
Tokyo 104, J
Tel (81) 3 457 33 28
Tx 22587

Voss Associates Engineering
Electromagnetic Systems
(W.A. Geoffrey Voss)
Room 402
1011 Fort Street
Victoria, BC, V8V 3K5, CDN
Tel (1) 604 384 1021
Fax (1) 604 388 9348
Tx 094 7347

Wilmer
(Jerzy Kaliński,
Janusz Rakowski)
Zak/lad Aparatury Mikrofalowej
ZZ
PAN
Ulica Zielna 39
00-108 Warszawa, PL
Tel (48) 22 20 40 51
Tx 814 633

SPECIALIST
ORGANIZATIONS

Agricultural Research
Organization
(U. Merin, I. Rosenthal)
Department of Food Technology
Bet Dagan, IL

Akademia Medyczna
Warszawa
(W. Stodolnik-Baranska)
Instytut Biostruktury
5 ulica Chalubinskiego
02-004 Warszawa, PL
Tx 815 403

Akademija Nauk SSSR
(I.G. Akojev)
Institut Biologičeskoj Fiziki
142292 Moskva, SU
Tx 411 964

Alberta Occupational Health
and Safety

(Dennis Novitcky)
Radiation Health Service
Non-ionizing Radiation
Protection
10709 Jasper Avenue
Edmonton, AB
T5J 3N3, CDN
Tel (1) 403 427 2691
Fax (1) 403 427 5698

Alleghany Singer Research
Institute
(Andrzej J. Surowiec)
Hyperthermia Physics Center
320 E. North Avenue
Pittsburgh, PA 15212-9986, USA
Tel (1) 412 359 3597

Association pour le
Développement
de l'Enseignement et de la
Recherche en Micro-ondes et
Electronique (ADERME)
(Jean-Marc Dererian,
Jean-Pierre Pellissier)
40, rue Colin
69100 Villeurbanne, F
Tel (33) 78 93 27 84
    (33) 78 89 40 87
Banque de données minitel :
3615 - ADERTEL -
Doc'ADERME

Australian Wine Research
Institute
(Paul R. Monk)
Microbiology Group
Private mail bag
Glen Osmond, SA 5064, AUS

**Bundesamt Für
Strahlenschutz**
(Rüdiger Matthes)
Institut für Strahlenhygiene
Ingolstädter Landstrasse 1
8042 Neuherberg Bei Mnchen, G
Tel (49) 89 3187 5243
Fax (49) 89 316 4255

The Royal Infirmary
(Alan W. Preece)
Bristol Radiotherapy Centre
Horfield Road
Bristol, BS2 8ED, GB
Tel (44) 272 230 000 ext. 2469

**Centre de Documentation
Internationale des
Industries Utilisatrices de
Produits Agricoles
(CDIUPA)**
(Gisèle Carra)
1, avenue des Olympiades
91300 Massy, F
Tel (33) 1 69 20 97 38

**Centre d'Information sur les
Utilisations de l'Electricité
dans l'Industrie et
l'Agriculture (CINELI)**
*(Joëlle Aiglin, Francis Julien,
Michel Le Curieux)*
EDF/GDF - Direction de la
Distribution
Tour Atlantique
La Défense 9
92080
Paris La Défense cedex 6, F
Tel (33) 1 49 02 58 35
   (33) 1 49 02 58 34
   (33) 1 49 02 58 40

**Centre National d'Etudes
des Télécommunications
(CNET)**

(Alain Azoulay)
38-40, rue du Général Leclerc
92131 lssy Les Moulineaux, F
Tel (33) 1 45 29 44 44 poste 4442
Tx 631 728

**Centre National de la
Recherche Scientifique
(CNRS - ER 286)**
(Henri Jullien)
Organisation Moléculaire
et Macromoléculaire
Laboratoire de Micro-ondes
2, rue Henri Dunant
94320 Thiais cedex, F
Tel (33) 1 46 87 33 55
Tx 203 855

**Centre de Recherches Tex-
tiles de Mulhouse (CRTM)**
*(Jean Diemunsch, René Freytag,
Pierre Sieger)*
185, rue de l'Illberg
68093 Mulhouse cedex, F
Tel (33) 89 42 74 08
Tx 881 865

**Centrum dla Radiobiologii i
Radioprotekcji**
*(Marian Bielec, Mark Janiak,
Waldemar Roskowski,
Stanisław Szmigielski)*
Laboratorium Biologicznych
Dzia łanij Mikrofal
00-909 Warszawa, PL

Centre Technique des
Industries Aérauliques et
Thermiques (CETIAT)
(Bertrand Meyer)
Plate-forme ERICA
27-29 Boulevard du 11 novembre
1918
B.P. 6084
69604 Villeurbanne cedex, F
Tel (33) 78 93 39 85
Fax (33) 78 89 71 55
Tx 340 310

Club Micro-ondes
(André-Jean Berteaud,
Jean-Claud Gougeuil)
EDF/GDF
Direction Etudes et Recherches
1, avenue du Général De Gaulle
B.P. 408
92141 Clamart cedex, F
Tel (33) 1 47 65 43 21
Tx 204 347

Comité Français de
l'Electricité (CFE
(Monique Henry, Michel Liabot)
Tour Atlantique
92080 Paris La Défense cedex 06,
F
Tel (33) 1 47 75 07 35
Fax (33) 1 47 73 95 53

Commission
Electrotechnique
Internationale (CEI)
(J. Huberdeau)
Bureau Central
3, rue de Varembe
C.P. 131
1211 Geneve 20, CH
Tel (4) 22 34 01 50
Fax (41)22 33 38 43
Tx 28 872

Consiglio Nazionale delle
Ricerche (CNR)
(Alessandro Checcucci, Roberto
Olmi, Riccardo Vanni)
Istituto di Ricerca sulle Onde
Elettromagnetiche
Via Panciatichi 64
50127 Firenze, I
Tel (3) 55 43 78514
Tx 570 231

Electricité de France (EDF)
(Christophe Marchand)
Centre des Renardières
"les Renardières"
Départment Applications
del'Électricité (ADE)
Route de Snes à Écuelles
B.P. 01
77250 Moret sur Loing, F
Tel (33) 60 70 69 04
Fax (33) 60 70 64 40

Electricité de France (EDF)
(Christian Vidal)
Service Etude et de la Promotion
de l'Action Commerciale
2, rue Louis Murat
75009 Paris, F
Tel (33) 1 42 56 94 00 poste 3986

Electricity Council
(Patrick J. Hulls)
British National Committee
on Electroheat
30 Millbank
London SW1P 4RD, GB
Tel (44) 71 834 2333 ext. 6272
Tx 23385

**Food and Drug
Administration**
*(Roger A. Budd, Przemyslaw
Czerski, Zorach Glaser, William
A. Herman, Gideon Kantor,
Charlotte Silverman, Donald
M. Witters)*
National Center for Medical
Devices and Radiological Health
FDA-HFZ/112
12709 Twinbrook Parkway
Rockville, MD 20857, USA
Tel (1) 301 443 4003
      (1) 301 443 7155

**Harlan E. Moore Heart
Research Foundation**
(Dennis E. Leszczynski)
503 South 6th. Street
Champaign, IL 62820, USA
Tel (1) 217 359 0002

Health and Welfare Canada
*(Deirdre A. Benwell, Maria
Stuchly)*
Section de Protection de la Santé
Bureau des Radiations et des
Instruments
Médicaux
Brook Field Road 775
Confederation Heights
Ottawa, ON
K1A 1C1, CDN
Tel (1) 613 954 0306

Institut Curie
(Dietrich Averbeck)
26, rue d'Ulm
75231 Paris cedex 05, F
Tel (33) 1 43 29 12 42
Tx 270471

**Institut National de la
Recherche Agronomique
(INRA)**

(Francis Fleurat-Lessard)
Station de Zoologie
Laboratoire sur les Insectes
des Denrées Alimentaires
B.P. 81
33883 Villenave D'Ornon cedex,
F
Tel (33) 56 84 32 92
Fax (33) 56 84 32 76
Tx 541 521

**Institut National de
Recherche et de Sécurité
(INRS)**
*(Bernard Higel, Jean-Paul
Vautrin)*
Centre de Recherche
Avenue de Bourgogne
B.P. 27 54501 Vandœuvre cedex,
F
Tel (33) 83 51 07 75 postes 253 et
543
Tx 850 778

**Institut Naučnogo
Issljedovanija Obščjej i
Kommunal'noj Gigijeny**
(Y.D. Dumanskij, M.G. Sandala)
252160 Kijev, SU

**Institut Naučnyh
Issljedovanij Truda**
(B.M. Savin)
Laboratorija
Elektromagnetiĕskih Voln
Radiočastot
31 Prospekt Budennogo
105275 Moskva, SU

Institut Textile de France
(ITF)
(Bertrand Meyer, Michel Colrat)
Laboratoire CETRA
Avenue Guy de Collongue
B.P. 60
69132 Ecully cedex, F
Tel (33) 78 33 34 55
Fax (33) 78 43 39 66
Tx 330 316

Institute of Electrical and
Electronics Engineers
(IEEE)
IEEE Service Center
445 Hoes Lane
P.O. box 1331
Piscataway, NJ 08855-1331, USA
Tel (1) 201 981 0060

IEEE Headquarters
345 East 47th. Street
New York, NY 10017, USA
Tel (1) 212 705 7867 (public
information)
Fax (1) 212 752 4929
Tx 236 411

Instituto de Investigaciones
Biofísicas
(A. Portela)
930 Avenida Callao
Buenos Aires Capital, RA

Instytut Medycyny Pracy
(A. Mikolajczyk)
8 ulica Teresy
Łódź, PL

International Electrotechni-
cal Commission (CEI/IEC)
(Philippe Lenoir)
Sub-Committee SC 59H
Microwave Appliances
Committee 61B

Safety of Household Microwave
Ovens
3 rue de Varembé
Boite postale 131
1211 Genève 20, Switzerland
Tel (41) 22 734 01 50
Fax (41) 22 733 38 43
Teletex 228-468 15 102 = CEIEC
CH
Tx 414 121

International Microwave
Power Institute (IMPI)
(Robert Lagasse)
13542 Union Village Circle
Clifton, VA 22024, USA
Tel (1) 703 830 5588
Fax (1) 703 830 0281

Istituto Superiore di Sanità
*(Martino Grandolfo, Eugenio
Tabet)*
Laboratorio delle Radiazioni
Viale Regina Elena 299
00161 Roma, I
Tel (39) 6 49 90
Tx 610 071

John B. Pierce Foundation
Laboratory
(Eleanor R. Adair)
290 Congress Avenue
New Haven, CT 06519, USA
Tel (1) 203 562 9901
Fax (1) 203 624 4950

Journal Français de
l'Electrothermie
PYC Editions
254, rue de Vaugirard
75015 Paris, F
Tel (33) 1 45 32 27 19
Tx 202 639

Laboratoire Central des
Industries Electriques
(LCIE)
(Luc Erard, Jean-Claude Perrin)
33, avenue du Général Leclerc
92260 Fontenay aux Roses, F
Tel (33) 1 46 45 21 84
    (33) 1 40 95 61 05
       (Ligne directe)
Fax (33) 1 40 95 60 95
Tx 250 080

Laboratoire Central des
Ponts et Chaussées
(Xavier Derobert)
Service de Physique
Centre de Nantes
B.P. 19
44340 Bouguenais, F
Tel (33) 40 84 58 00
    (33) 40 84 59 99
Fax (33) 40 84 59 99
Tx 710 805

Laboratoire d'Essais de
Faisabilité sur l'Energie
Electrique sous Forme de
Rayonnements Micro-ondes
(IEFEMO)
(Pierre Goudmand)
Cité Scientifique
59655 Villeneuve d'Ascq cedex, F
Tel (33) 20 43 44 49
Fax (33) 20 43 49 95
Tx 136 339

Laboratoire de Recherche et
de Contrôle du Caoutchouc
et des Plastiques (LRCCP)
(A. Coupard, Pierre Martinon)
60, rue Auber
94408 Vitry sur Seine cedex, F
Tel (33) 1 46 71 91 22
Fax (33) 1 49 60 70 66
Tx 265 560

Medyczna Akademia
Wojskowa
(M. Siekierzynski)
128 ulica Szaserow
Warszawa, PL

Microwave News
(Louis Slesin)
P.O. box 1799
Grand Central Station
New York, NY 10163, USA
Tel (1) 212 517 2800

Ministère de l'Équipement,
du Logement, des Transports
et de l'Espace
*(Jean-Pierre Baron, Richard
Lagabrielle, Xavier Derobert)*
Laboratoire Central des Ponts-et-
Chaussées (LCPC)
Centre de Nantes
Service de Physique
Section de Géophysique
Appliquée
B.P. 19
44340 Bouguenais, F
Tel (33) 40 84 59 05
Fax (33) 40 84 59 99
Tx 710 805

Ministry of Agriculture,
Fisheries and Food
(Michael Kent)
P.O. Box 31
135 Abbey Road
Aberdeen, AB9 8DG, GB
Tel (44) 224 877 071
Fax (44) 224 874 246
Tx 739 719

National Institute of Environmental Health Sciences
(Donald I. Mac Ree)
Division of Environmental Research and Training
P.O. box 12233
Research Triangle Park, NC 27709, USA
Tel (1) 919 541 1442
Fax (1) 919 541 30 26

National Institute for Occupational Safety and Health (NIOSH)
(David L. Conover)
4676 Columbia Parkway
Cincinnati, OH 45226, USa
Tel (1) 513 533 8140
Fax (1) 513 533 8371

National Radiological Protection Board (NRPB)
(Allister McKinlay, J.A. Dennis)
Non-ionising Radiation Department
Chilton, Didcot
Oxon, OX11 0RQ, GB
Tel (44) 235 831 600
Fax (44) 235 833 891
Tx 827 124

National Research Council of Canada
(Geoffrey A.R. Mealing, Jean-Louis Schwartz)
Division of Biological Sciences
Building M-54
Montreal Road
Ottawa, ON
K1A 0R6, CDN
Tel (1) 613 990 0884
Fax (1) 613 952 0583

Naval Aerospace Medical Research Laboratory

(Richard G. Olsen)
U.S. Navy
Research Department
Pensacola, FL 32508-5700, USA
Tel (1) 904 452 8065
Fax (1) 904 452 4479

Office National d'Etudes et de Recherches Aérospatiales (ONERA)
(Florent Christophe, Alain Priou)
Centre d'Etudes et de Recherches de Toulouse (CERT)
EDERMO
Département applications Industrielles
2, avenue Edouard Belin
B.P. 4025
31055 Toulouse cedex, F
Tel (33) 61 55 71 11
Tx 521 596

Ontario Hydro
Michael R. Sanio)
Industrial Product Development Section
620 University Avenue U4-E4
Toronto, ON
M5G 1X6, CDN
Tel (1) 416 592 4765
Tx 06 217 662

Ontario Ministry of Labour
(Maurice E. Bitran)
Occupational health and Safety Branch
Radiation Protection Services
Non-ionizing Radiation Section
81 Resources Road
Weston, ON - M90 3T1, CDN
Tel (1) 416 235 5922
Fax (1) 416 235 5926

Organisation Mondiale de la
Santé (OMS/WHO)
(Michael J. Süss)
Regional Office for Europe
Regional Office for
Environmental Health Hazards
8 Scherfigsvej
2100 København, DK
Tel (45) 1 29 01 11

Société des Electriciens et
des Electroniciens
(Christian Vidal)
Club 14 : Electrothermie
Sous-section Applications
Energétiques
des Micro-ondes
48, rue de la Procession
75724 Paris cedex 15, F
Tel (33) 1 45 67 07 70
Tx 200 565

Svenska Livsmedelsinstitutet
(SIK)
(Thomas Ohlsson)
Institutet för Livsmedelsforskn-
ing
P.B. 5401
Frans Perssons Väg
Delsjöomradet
402 29 Göteborg, S
Tel (46) 31 35 56 00
Fax (46) 31 83 37 82

Union Technique de
l'Electricité (UTE)
(Margverite Miro)
French Member Committee
4 place des Vosges
Covrbevoie-cedex 64
92052 LA Défense, F
Tel (33) 1 46 91 11 11
Fax (33) 1 47 89 47 75

United Bristol Health Trust
(Alan W. Preece, Victor L.
Barley)
Bristol Oncology Centre
Biophysics Department
Horfield Road
Bristol, BS2 8EN, GB
Tel (44) 272 282 469
Fax (44) 272 298 627

US Air Force Occupational
and Environmental Health
Laboratory
(Lieut. Rademacher)
RZC
Radiation Services Branch
Brooks Air Force Base
San Antonio, TX 78 235, USA
Tel (1) 512 536 3486
Fax (1) 512 536 2288

US Bureau of Mines
(Ronald H. Church, William E.
Webb)
Tuscaloosa Research Center
Tuscaloosa, AL 35486-9777, USA
Tel (1) 205 759 9400
Fax (1) 205 759 9447

US Dpt. of
Agriculture
(J.Stan Bailey, Nelson A. Cox,
Stuart O. Nelson)
Agricultural Research Service
Richard B. Russell Agricultural
Research Center
P.O. box 5677
Athens, GA 30613, USA
Tel (1) 404 546 3101
Fax (1) 404 546 3367

US Department of Agriculture
(Robert L. Buchanan, Arthur L. Miller)
Agricultural Research Service
Eastern Regional Research Center
600 East Mermaid Lane
Philadelphia, PA 19118, USA
Tel (1) 215 233 6400
Fax (1) 215 489 6559

US Environmental Protection Agency
(David Janes)
Office of Radiation Program
Analyses and Support Division
ANR-461
401 M Street SW
Washington, DC 20460, USA
Tel 91) 202 475 9626
Fax (1) 202 475 8347

US Environmental Protection Agency
(John Allis, Ezra Berman, Carl Blackman, Joe A. Elder, Christopher J. Gordon, Richard Phillips, Ronald J. Spiegel, Tom Ward)
Experimental Biology Division
Biological Engineering Branch
Physiology Section
Health Effects Research Laboratory
Research Triangle Park, NC 27711, USA
Tel (1) 919 541 2784

US Veterans Administration Medical Center
(Rex L. Clarke, don R. Justesen)
Behavioral Radiology Laboratory
Research Service 151
4801 Linwood Boulevard
Kansas City, MO 64128, USA
Tel (1) 816 861 4700 ext. 466
Fax ext. 257

US Veterans Administration Medical Center
(Rolando Ortiz-lugo)
Engineering
1601 Perdido Street
New Orleans, LA 70146, USA
Tel (1) 504 589 5229
Fax (1) 504 568 0811 ext. 5199

Walter Reed Army Institute of Research
(Col. Edward Elson)
Department of Microwave Research
Washington, DC 20307-5100, USA
Tel (1) 202 427 5125

Wojskowy Instytut Medycyny Lotniczy
(S. Barański)
54 ulica Krasinskiego
01-755 Warszawa 86, PL
Tel (48) 22 33 41 54

Zaret Foundation
(Milton M. Zaret)
1230 Post Road
Scarsdale, NY 10583, USA
Tel (1) 914 723 9129

UNIVERSITIES

Cornell University
(Gertrude Armbruster)
New York State Colleges of Human Ecology and Agricultural and Biological Engineering
Division of Nutritional Sciences
Department of Agricultural and Biological Engineering
366 Martha Van Rensselaer Hall
Ithaca, NY 14853-4401, USA
Tel (1) 607 255 2642
Fax (1) 607 254 1033
Tx WUI 6713054

Deakin University
(Van Nguyen Tran)
School of Sciences
Pidgeons road
Waurn Ponds
Victoria 3217, AUS
Tel (61) 52 471 111
    (61) 52 471 418 (direct line)
Tx 35625

Eastern Virginia Medical School
(James Shaeffer)
Department of Radiation Oncology and Biophysics
600 Gresham Drive
Norfolk, VA 23507-1999, USA
Tel (1) 804 446 5799
    (1) 804 446 5780
Fax (1) 804 446 5172

Ecole Nationale Supérieure de Chimie de Toulouse (EN-SCT)
(Albert Gourdenne)
118, route de Narbonne
31077 Toulouse cedex, F
Tel (33) 61 17 56 75

Fax (33) 61 17 56 00

Ecole Nationale Supérieure d'Electronique, Electrotechnique, Informatique et Hydraulique de Toulouse (ENSEEIHT)
(Serge Lefeuvre, Michèle Audhuy)
Electrothermie Microonde
2, rue Charles Camichel
31071 Toulouse - cedex, F
Tel (33) 61 58 82 03
    (33) 61 58 82 04
Fax (33) 61 63 70 40

Ecole Polytechnique
(Marcel Giroux)
Département de Génie Electrique
Laboratoire d'Aplications Industrielles
des Micro-ondes (LAIMO)
C.P. 6079 suc. A
Montréal, PQ
H3C 3A7, CDN
Tel (1) 514 340 4754
Fax (1) 514 340 3219
Tx 524 146

Ecole Polytechnique Fédérale de Lausanne
(Fred E. Gardiol)
Département Electricité
Laboratoire d'Electromagnétisme et d'Acoustique (LEMA)
1015 Lausanne, Switzerland
Tel (41) 21 693 26 70
Fax (41) 21 693 26 73
Tx 454 062

Eastern Virginia Medical
School
*(Kenneth L. Carr, Anas Morsi
El-Mahdi, James Shaeffer,
Raymond K. Wu)*
Department of Radiation
Oncology and Biophysics
600 Gresham Drive
Norfolk, VA 23 507, USA
Tel (1) 804 628 2075

**Florida International
University**
(Mark J. Hagmann)
Department of Electrical Engineering
VH 140, University Park Campus
Miami, FL 33199, USA
Tel (1) 305 554 2807
Fax (1) 305 554 3582
Tx 522 297

**Institut für Strahlenhygiene
des Bundesgesundheitsamtes**
(Klaus Werner Bögl)
Ingolstädter Landstrasse 1
8042 Neuherberg bei München, D
Tel (49) 89 31871

Institut National des Sciences Appliquées (INSA)
(Roger Santini)
Laboratoire de Physiologie-
Pharmacodynamie
Bâtiment 406
20, avenue Albert Einstein
69621 Villeurbanne cedex, F
Tel (33) 78 94 81 12
Tx 380 856

**Institut des Sciences de la
Matière
et du Rayonnement**
(Bernard Raveau)
Ecole Nationale Supérieure
d'Ingénieurs de Caen
5, avenue
d'Edimbourg/boulevard
du Maréchal Juin
14032 Caen cedex, f
Tel (33) 31 93 37 14

Iowa State University
(Richard Markuszewski)
Ames Laboratory
Institute for Physical Research
and Technology
Fossil Energy Program
158J Metals Development
Building
Ames, IA 50011, USA
Tel (1) 515 294 3758
(1) 515 294 3091

Iowa State University
(Delwyn D. Bluhm)
Ames Laboratory
Institute for Physical Research
and Technology
Engineering Services
158J Metals Development Building
Ames, IA 50011, USA
Tel (1) 515 294 3757
Fax (1) 515 294 0568

Iowa State University
(Glenn A. Norton)
Ames Laboratory
Institute for Physical Research
and Technology
Fossil Energy Program
271Metals Development Building
Ames, IA 50011, USA
Tel (1) 515 294 1035
(1) 515 294 3091

Katholieke Universiteit Leuven
(Peter J. Luypaert, Antoine Van de Capelle)
Departement Elektrotechniek
Aldeling Mikrogolven en Lasers
Kard. Mercierlaan 94
3030 Heverlee, B
Tx 25941

Kyoto University
(Jun-Ichi Azuma, Tetsuo Koshijima, Fumio Tanaka)
Wood Research Institute
Research Section of wood
Chemistry
Gokasho, Uji
Kyoto 611, J
Tx 5422693

Marquette University
(T. Koryu Ishii)
Department of Electrical and computer Engineering
Microwave Power Engineering
Microwave Power Laboratory
1515 West Wisconsin Avenue
Milwaukee WI 53233, USA
Tel (1) 414 244 6998
     (1) 414 244 6820
Fax (1) 414 244 7082

Massachusetts Institute of Technology
*(Padmakar P. Lele, Frederic R. Morgenthaler,
Saul Rappaport)*
Department of Electrical Engineering and Computer Science
Hyperthermia Center
50 Vass Street
Cambridge, MA 02139, USA
Tel (1) 617 253 4600
Tx 921 473
Twx 710 320 0058

Michigan State University
(Robert Y. Ofoli)
Department of Agricultural Engineering
102 Farral Hall
East Lansing, MI 48824-1323, USA
Tel (1) 517 353 3797
Fax (1) 517 353 8982
Twx 810 251 0737

Michigan State University
(Carol A. Sawyer, Mary E. Zabik)
204 Food Science
Department of Food Science and Human Nutrition
East Lansing, MI 48824-1323, USA
Tel (1) 517 355 8474
Fax (1) 517 353 8963

Nagoyo University
(Yoshifumi LAmemiya, Osamu Fujiwara)
Faculty of Engineering
Department of Electrical Engineering
Furo-cho
Chikusa-ku
Nagoya 464, J
Tx 477 73 23-3

Northwestern University
(Morris E. Brodwin)
Department of Electrical Engineering and Computer Science
Room 2736
2145 Sheridan Road
Evanston, IL 60201, USA
Tel (1) 312 491 5410
Fax (1) 312 491 4133

Ohio State University
(Marion L. Cremer)
College of Home Economics
Department of Human Nutrition
and Food Management
1787 Neil LAvenue
Columbus, OH 43210, USA
Tel (1) 614 292 55 88

Oregon State University
(Zoe A. Holmes)
Department of Foods and Nutrition
Corvallis, OR 97331-5103, USA
Tel (1) 503 754 3561

Purdue University
(Karen S. Jamesen,
E. Pratt)
Department of Foods and
Nutrition
West Lafayette, IN 47907, USA
Tel (1) 317 494 8235
    (1) 317 494 8234
Fax (1) 317 949 0327
Tx 272 396

Queen's University
(César Romero-Sierra)
Department of Anatomy
Botterell Hall
Stuart Street
Kingston, ON
K7l 3N6, CDN
Tel (1) 613 545 2600
Fax (1) 613 545 6612

Staatsuniversiteit Gent
(Carlos De Wagter, W. Van
Loock)
Laboratoire
d'Electromagnétisme
et d'Acoustique
St. Pietersnieuwstraat 41
9000 Gent, B

Tufts University
(Karl H. Illinger)
Department of Chemistry
62 Talbot Avenue
Medford, MA 02155, USA
Tel (1) 617 381 3441

Universidad Politécnica de Madrid
(J.A. Encinar, J.M. Rebollar)
Grupo de Electromagnetismo
Aplicado
ETSI Telecommunicación
Ciudad Universitaria
28040 Madrid, E

Universitäts Krankenhaus Eppendorf
(F. Zywietz)
Institut für Biophysik und
Strahlenbiologie
Martinistrasse 52
2000 Hamburg 20, D
Tx 216 45 14

Université d'Assiut
(El-Deek El-Sayed, Adel M.K.
Hashem)
Faculté d'Ingéniérie
Département de Génie Electrique
Assiut, ET
Tx 92 863

Université Bordeaus I
(Claude Marzat, P.A. Bernard)
Laboratoire de Physique
Expérimentale
et de Micro-ondes
33405 Talence cedex, F
Tel (33) 56 84 61 88
    (33) 56 84 61 94

**Université Claude
Bernard-Lyon I**
(Jean-Pierre Pellissier)
Laboratoire HF et Micro-ondes
43, boulevard du 11 Novembre
1918
69621 Villeurbanne, F
Tel (33) 78 89 81 24

Université de Limoges
(Pierre Deschaux)
Faculté des Sciences
Laboratoire
d'Immunophysiologie
123, avenue Albert Thomas
87060 Limoges, F
Tel (33) 55 45 72 00
　　(33) 55 45 74 99
　　　　(ligne directe)
Fax (33) 55 45 72 01

Universiteit Gent
(Daniël de Zutter)
Laboratory of Electromagnetism
and Acoustics (LEA)
Sint-Pietersnieuwstraat 41
9000 lGent, Belgium
Tel 932) 91 64 33 27
　　(32) 91 64 33 16
Fax (32) 91 64 35 93

Université Nancy I
*(Georges Roussy, Jean-Marie
Thiébaut André Zoulalian)*
Laboratoire de Spectroscopie et
des techniques micro-ondes
B.P. 239
54506 Vandœuvre-lés-Nancy
cedex, F
Tel (33) 83 91 20 00
Tx 960 646

**Université de Québec
à Trois-Riviéres**
(Tapan K. Bose, Richard

Chahine)
Département de Physique
Groupe de Recherche en
Diélectriques
C.P. 500
Trois Rivières, PQ
G9A 5H7, CDN
Tel (1) 819 376 5107
　　(1) 819 376 5108
Fax (1) 819 376 5012

**Université des Sciences et
Techniques**
(Yves Leroy)
Centre Hyperfréquences
et Semiconducteurs
LA CNRS n°287
Bâtiment P3 et P4
59655 Villeneuve d'Ascq cedex, F
Tel (33) 20 43 43 43
Fax (33) 20 43 49 94
Tx 13 63 35

Université Louis Pasteur
(Michel Gautherie)
Faculté de Médecine
Laboratoire de Thermologie
Biomédicale
11, rue Human
78085 Strasbourg cedex, F
Tel (33) 88 35 87 00

Université d'Ottawa
*(France M. Duval, Bernard J.R.
Philogène)*
Département de Biologie
30 Avenue Marie Curie
Ottawa, ON,
K1N 6N5, CDN
Tel (1) 613 564 2338
Fax 91) 613 564 5014

Université d'Ottawa
(S. Stuchly)
Département de Génie
Electrique
161 Louis Pasteur
Ottawa, ON
K1N 6N5, CDN
Tel (1) 613 564 2495
Fax (1) 613 564 7681
Tx 0533 338

University of Alberta
(Wayne R. Tinga)
Department of Electrical
Engineering
Edmonton, Alberta
T6J 2G7, CDN
Tel (1) 403 492 2404
   (1) 403 492 4961
Fax (1) 403 492 1811
Tx 037 2979

University of California at Berkeley
(Charles Süsskind)
Department of Electrical
Engineering and Computer
Sciences
Room 269 - Corw
Berkeley, CA 94720, USA
Tel (1) 415 642 3214
Fax (1) 415 643 8426
Twx 910 366 7114

University of Illinois
(Charles A. Cain, Steven J. Franke, Richard L. Magin, Raj Mittra)
Department of Electrical and
Computer Engineering
Bioacoustics Research Labora-
tory
1406 West Green Street
Urbana, IL 61 801, USA
Tel (1) 217 333 7288

Fax (1) 217 244 0105

University of Illinois
(Fred A. Kummerow)
Department of Food Science
Burnsides Research Laboratory
1208 W Pennsylvania
Urbana
IL 61901, USA
Tel (1) 217 333 1806
Tx 206 957

University of Illinois at Chicago Circle
(James C. Lin)
Bioengineering Program
P.O. box 4348
Chicago
IL 60680, USA
Tel (1) 312 996 2331
Fax (1) 312 996 7149

University of Kansas
(Edwin J. Martin)
Department of Psychology
426 Fraser Hall
Lawrence, KS 66045, USA
Tel (1) 913 864 4131

University of Liverpool
Physics Department
Oliver Lodge
P.O. box 147
Liverpool
L69 3BX, GB
Tel (44) 51 709 6022
Tx 627 095

University of London
*(Ronald E. Burge, Brian D.
Carroll, Edward H. Grant,
Rodney J. Sheppard)*
King's College London
Physics Department
Strand
London, WC2R 2LS,GB
Tel (44) 71 873 2836
    (44) 71 873 2154
Fax (44) 71 872 0201

University of Manitoba
(Michael Hamid)
Department of Electrical
Engineering
Winnipeg, MB
R3T 2N2, CDN
Tel (1) 204 474 9641
Fax (1) 204 261 4639
Tx 07587721

University of Maryland
*(Elizabeth K. Balcer-Kubiczek,
Hubert Eddy, George H. Harri-
son, George M. Samaras)*
School of Medicine
Division of Radiation
Research
Department of Radiation
Oncology
650 W Baltimore Street
Baltimore, MD 21201, USA
Tel (1) 301 328 7133
Fax (1) 301 328 8783
Twx 710 234 1610

University of Massachusetts
(Richard E. Mudgett)
Department of Food Science and
Nutrition
Chenowetz Laboratory
Amherst, MA 01003, USA
Tel (1) 413 545 2281
Fax (1) 413 545 1242

Tx 955 355

University of Minnesota
*(Eugenia A. Davis, John
Gordon, Jody Grider)*
Department of Food Science and
Nutrition
1334 Eckles Avenue
St. Paul, MN 55108, USA
Tel (1) 612 624 1290
Fax (1) 612 625 5272
Tx 298 421

University of New Orleans
*(Richard Bishop, Robert D. Mac
Afee)*
College of Engineering
Department of Electrical
Engineering
Lakefront
New Orleans, LA 70148, USA
Tel (1) 504 286 6327

University of Pennsylvania
*(Kenneth R. Foster, Jonathan B.
Leonard, Herman P. Schwan)*
Department of Bioengineering
220 South 33rd. Street
Philadelphia, PA 19104, USA
Tel (1) 215 898 8501
Fax (1) 215 898 1130
Tx 710 670
Twx 710 670 0177
    710 670 0328

University of Queensland
(P.J. Jolly)
Solar Energy Research
Centre
St. Lucia, QLD 4067, AUS
Tx 40315

University of Rochester
(Shin Tsu Lu, Sol M. Michael-son)
School of Medicine and
Dentistry
Department of Radiation
Biology and Biophysics
610 Elmwood Avenue
Rochester, NY 14642, USA
Tel (1) 716 275 3723
Fax (1) 716 275 6207

University of Singapore
(Perambur S. Neelakantaswamy, Arthur Rajaratnam)
Kent Ridge
Singapore 0511, SGP

University of Utah
(Douglas A. Christensen, Carl H. Durney, Om P. Gandhi, Magdy F. Iskander, James L. Lords)
Department of Electrical
Engineering
Department of
Bioengineering
Salt Lake City, UT 84112, USAL
Tel (1) 801 581 6944
Fax (1) 801 581 8692
Twx (1) 925 5283

University of Washington
(Akira Ishimaru)
Department of Electrical and
Computer Engineering
FT/10
Electromagnetics and Remote
Sensing Laboratory
Seattle, WA 98195, USA
Tel (1) 206 543 2169
Fax (1) 206 543 3842
Tx 474 0096

University of Washington
(George M. Pigott)
Institute for Food Science and
Technology college of Fisheries
Mail stop HF/10
Seattle, WA 98195, USA
Tel (1) 206 545 2033

University of Washington
(Arthur W. Guy, Justus F. Lehmann)
Bioelectromagnetics
Research Laboratory
Center for Bioengineering
Mail stop RJ/30
Seattle, WA 98195, USA
Tel (1) 206 543 1060
 (1) 206 543 1071
Fax (1) 206 543 4365
Tx 474 0096

Uniwersytet Rolnictwa
Poznań
(Jan Pikul)
Departament Nauki i
Technologii Spozywczych
Wojska Polskiego 31
60-624 Poznań, PL

Washington University
(F.J. Rosenbaum, Barry E. Spielman)
School of Engineering and
Applied Science Department of
Electrical Engineering
11 Brookings Drive
Campus box 1127
St. Lousi, MO 63130, USA
Tel (1) 314 935 6157
 (1) 314 935 5565
Fax (1) 314 935 4842

Yale University
(Jan A.J. Stolwijk)
School of Medicine
Department of Epidemiology and
Public Health
60 College Street
P.O. box 3333
New Haven, CT 06510, USA
Tel (1) 203 785 2867
Fax (1) 203 785 6917

INDEX

Absorption and dosimetry in thermal
    interaction with organism, 407–408
Acquired immunity, biological effects on, 459–
    460
α-Amylase in wheat, inactivation of, 344–345
Alternating field, dipole alignment polarization
    in, 90–91
Amino resins, polymerization of, 212
Animal products
    microwave cooking of, 235–249
    sterilization of, 350–351
Antigenicity as biomedical application, 612
Aperture, radiation of, 72–75
Applicators, microwave, 117, 128–146
    corona discharges in, 141–146
    design constraints on, 132
    electric arcing in, 141–146
    environmental constraints on, 146
    field homogeneity of, 135–136
    impedance matching of, 132–135
    materials for, 139–141
    shielding of, 136–139
    types of, 128–132
Aqueous dielectrics, 98–100
Arthritic pain, microwave hyperthermia for, 616
Atmospheric pressure, drying food products at,
    307–313
Auditory perception, biological effects on, 478–
    481
Australia, safety standards in, 568
Autonomic nervous system, biological effects
    on, 481–485

Bacon, microwave cooking of, 243–245
Baking, microwave, 255–257
Baldwin nomenclature for enzyme families, 341–
    344
Beans, soya, enzymatic inactivation of, 345–346
Bioelectric vibrations as biomedical application,
    611–612
Biological constituents, dielectric properties of,
    101–102
Biological effects, 443–552
    on blood, 451–455
    on cataracts, 502–505
    on cells, 446–450
    on endocrine system, 488–494
    on growth, 496–501
    on hematopoiesis, 451–455
    on immune system, 455–465
    on lesions, 502–505
    on metabolism, 494–496
    on micro-organisms, 446–450
    on nervous system, 465–488

    on thermal regulation, 494–496
Biological material, dielectric behavior of, 392–
    397
Biomedical applications, 585–630
    antigenicity as, 612
    bioelectric vibrations as, 611–612
    biological, 617–619
    clinical, 616–617
    hyperthermia for cancer treatment as, 586–
        611. See also Hyperthermia for cancer
        treatment
    in immune response, 613–616
Biomolecules, dielectric behavior of, 392–394
Birds, growth of, biological effects on, 498–499
Blanching of fruits and vegetables, 341–344
Blood
    biological effects on, 451–455
    derivatives of, thawing, microwave
        hyperthermia for, 617
Blood-brain barrier (BBB), biological effects on,
    471–474
Boundary conditions, 48
Brain tissue fixation, microwaves for, 618–619
Bread, microwave cooking of, 255–256
Bruggeman equation, 100
Butadiene-styrene rubber (SBR), dielectric
    losses of, 190
Butyl rubber (BR), dielectric losses of, 190

Calcium ion fluxes, biological effects on, 465–
    469
Canada, safety standards in, 567–568
Cancer treatment, hyperthermia for, 586–611.
    See also Hyperthermia for cancer
    treatment
Carbon black, addition of, in microwave
    vulcanization, 194
Carbonic fermentation, wine-making by, 379
Carpets, tufted, drying of, 170–171
Cataracts, biological effects on, 502–505
Catering, microwaves and, 253–255
Cells
    biological effects on, 446–450
    dielectric behavior of, 395–396
    membranes of, interactions of microwaves
        with, 399–404
Cellulosic waste treatment, microwaves in, 224
Central nervous system, biological effects on,
    474–476
Ceramics
    drying of, 172–173
    sintering of, consolidation in, 217–220
Cereals, microwave cooking of, 251–252
Chlorbutadiene rubber (CR), dielectric losses of,
    190

Cigarettes, drying of, 177–178
Clausius-Mosotti formula, 88
Coal, purification of, microwaves in, 221–222
Cole-Cole plot, dielectric relaxation and, 95–96
Concrete
    crushing, microwaves in, 220–221
    drying of, 172–173
    fast-setting, consolidation in, 217
Consolidation, 216–220
    in fast-setting concrete, 217
    in hardening of foundry mouldings, 216–217
    in sintering of ferrites and ceramics, 217–220
Construction, drying in, 172–173
Continuity equation, 35
Continuous wave interaction of microwaves
        with cell membranes, 399–403
Cooking, microwave, 231–274
    of animal products, 235–249
    of bacon, 243–245
    for baking, 255–257
    for catering, 253–255
    of cereals, 251–252
    of dairy products, 248–249
    digestibility of foods after, 257
    of fat, 243–245
    of fish, 246–248
    history of, 17–19
    of meat patties, 245–246
    mechanisms of, 232–235
    of poultry, 239–243
    of red meat, 238–239
    of soya, 251–252
    of vegetable products, 249–253
Corona discharges in microwave applicators,
        141–146
Coulomb's law, 34
Crop protection by microwaves, 378
Crushing of rocks/concrete, microwaves in,
        220–221
Cylindrical waves, 42–44

Debye equation, dielectric relaxation and, 93
Debye plot, dielectric relaxation and, 95–97
Dielectric behavior
    of biological material, 392–397
    of food products, 232
Dielectric polarization, 84–85
Dielectric properties
    of biological constituents, 101–102
    of elastomers, 189–191
    of frozen products, 278–281
    of saline solutions, 101
Dielectric relaxation, 92–97
    Cole-Cole plot and, 95–96
    Debye equation, 93
    Debye plot and, 95–97
    hysteresis and, 92
    intermolecular bonds and, 93–94
    relaxation time and, 94–95
Dielectrics
    aqueous, 98–100
    lowloss, 98
    mixtures and, 100–101
    types of, 97–102

permitivity measurements and, 97
Diethylene glycol (DEG), addition of, in
        microwave vulcanization, 196
Digestibility of foods cooked by microwaves,
        257
Dipole alignment, polarization by, 84
    in alternating field, 90–91
    in static fluid, 85–90
Dipole moment
    induced, 85–88
    permanent, 88–90
Disinfestation, 365–369
Dosimetry and absorption in thermal
        interaction with organism, 407–408
Doughnuts, microwave cooking of, 256–257
Drying, 159–186
    in construction, 172–173
    of food products, 305–338
        at atmospheric pressure, 307–313
        determination of dry content after, 324–
            327
        by expansion in vacuum, 317–324
        freeze, 313–317
        at low pressure, 313–324
        vaporization in, 306–307
    in foundries, 173–174
    humidity and, 159–161
    kinetics of, 161
    in paper industry, 163–164
    in pharmaceutical industry, 176–177
    in plastics industry, 175–176
    in printing industry, 164–165
    in regeneration of zeolites, 178–180
    in rubber industry, 175
    in textile industries, 166–171
    in tobacco industry, 177–178
Dyeing of textiles, drying in, 168–170

Elastomers
    dielectric properties of, 189–191
    treatment of, 187–206
    vulcanization of, 191–201
Electric arcing in microwave applicators, 141–
        146
Electric induction, definition of, 34
Electromagnetic cavities, 58–68
    energy balance in, 61–64
    power loss in walls of, 64–65
    quality factor in, 66–68
    resonant modes of, 58–61
Electromagnetic detection, 14–17
Electromagnetic spectrum, 4–8
Electromagnetism, 8–9
    radiation and, 3–32
Electronic polarization, 84–85
Emulsification, microwaves in, 220
Endocrine system, biological effects on, 488–494
Energy balance in electromagnetic cavities, 61–
        64
Energy limitations in microwave thawing, 283–
        284
Enzymatic inactivation
    in food preservation, 339–346
    in grains and soya beans, 345–346

in wheat, 344–345
Enzymes, families of, 340
Epoxy resins, polymerization of, 211
Ethylene-propylene rubber (EPDM), dielectric losses of, 190
Europe, Eastern, safety standards in, 565–567
European Community, safety standards in, 568–569

Fat(s)
    drying of, 312
    microwave cooking of, 243–245
Fermentation, carbonic, wine-making by, 379
Ferrites, sintering of, consolidation in, 217–219
Field homogeneity of microwave applicators, 135–136
Film, photographic, drying of, 175–176
Fixation of brain tissue, microwaves for, 618–619
Follicle stimulating hormone (FSH), biological effects on, 492–493
Food industry, applications in, 229–388
    for cooking, 231–274. *See also* Cooking, microwave
    for drying, 305–338. *See also* Drying of food products
    for preservation, 339–363
        enzymatic inactivation in, 339–346
        sterilization in, 346–356
    for thawing, 275–304. *See also* Thawing
Foundries, drying of, 173–174
Foundry moulding, hardening of, consolidation in, 216–217
Fraunhofer radiation zone, 77
Freeze drying, 313–317
Frequency bands, 4, 5, 6–7
Fresnel radiation zone, 77, 78
Frozen products, dielectric properties of, 278–281
Fruits, blanching of, 341–344
Fusion, 213–216
    in defrosting ot soil, 216
    dewaxing casting moulds in, 213–214
    in oil/shade oil production, 215
    in road repairs, 215–216
    of viscous materials in metal containers, 214–215

Gauss-Ostrogradskii theorem, 62
Generators, microwave, 117–128
    klystron as, 125–127
    magnetron as, 118–125
    RF energy transmission in, 127–128
    TWT as, 125, 127
Germination, microwaves and, 377–378
Glued products, drying of, 165
Grains, enzymatic inactivation of, 345–346
Growth, biological effects on, 496–501
Growth hormones, biological effects on, 493–494
Guided propagation, 52–56
    discovery of, 16–17

Heart surgery, microwave hyperthermia for, 616
Heat generation, 103–106

Heat transfer, drying and, 161
Hematopoiesis, biological effects on, 451–455
Herbicides, microwaves in place of, 369–376
Horizontal polarization (HP), 50–51
Hormone(s)
    follicle stimulating, biological effects on, 492–493
    growth, biological effects on, 493–494
    luteinizing, biological effects on, 492–493
    somatotrophin, biological effects on, 493–494
    thyroid, biological effects on, 488–490
Horn, radiation from, 75–76
Humidity, drying and, 159–161
Hydrogen bonds, dielectric relaxation and, 93
Hyperthermia
    for accidental hypothermia, 616
    for arthritic pain, 616
    for cancer treatment, 586–611
        applicators for, 595
        clinical results of, 608–611
        dipoles for, 603–605
        historical development of, 586–591
        integrated systems for, 605–608
        mode of action of, 591–595
        phased arrays for, 601–603
        radiating apertures for, 595–601
        slots for, 603–604
    for neurologic pain, 616
Hypothermia, accidental, microwave hyperthermia for, 616
Hysteresis, dielectric relaxation and, 92

Immune response, biomedical applications in, 613–616
Impedance matching of microwave applicators, 132–135
Industrial applications, 157–228. *See also* Elastomers
    cellulosic waste treatment as, 223
    consolidation as, 216–220
    crushing as, 220–221
    drying as, 159–186. *See also* Drying
    emulsification as, 220
    fusion as, 213–216. *See also* Fusion
    nuclear waste treatment as, 222–223
    polymerization as, 207–213. *See also* Polymerization
    purification of coal as, 221–222
    treatment of elastomers as, 187–206
Industrial microwave thawing, advantages of, 296–297
Industry, microwaves in, 19–22
Inflammatory reactions, biological effects on, 459
Inks, printing, drying of, 164–165
Insect growth, biological effects on, 497–498
Intermolecular bonds, dielectric relaxation and, 93–94
International Consultative Committees (CCI), 5
International Electrotechnical Commission, The (IEC), 6
International Electrotechnical Commission (CEl), 569

International Frequency Registration Board (IFRB), 5
International Telecommunication Union, The (ITU), 5–6
International Union of Radio Science, The (URSI), 6
Ionic polarization, 84
ISM bands, 6–7
ISM equipment, classification of, 138

Joule energy losses in walls of electromagnetic cavities, 64

Klystron, 15–16, 125–127
Knockenhauer's spirals, 10, 11
Kotter's equation, 72–73

Leathers, drying of, 166
Lesions, biological effects on, 502–505
Lovyenga's equation, 101
Lowless dielectrics, 98
Luteinizing hormone (LH), biological effects on, 492–493
Lymphocytes
  FcR$^+$ and CR$^+$, multiplication of, biological effects on, 460–461
  response of, to mitogens, stimulation of, biological effects on, 461–463
  T, activator, modulation of activity of, biological effects on, 463–465
Lymphopoiesis, biological effects on, 460

Macromolecules, microwaves and, 188–191
Magnetic tape, drying of, 175–176
Magnetron, 15, 118–125
  current–voltage characteristics of, 125
  efficiency of, 123
  operating chart for, 122–123
  power output of, 124–125
Mammals, growth of, biological effects on, 49–501
Mass, drying and, 161
Matter, microwaves and, 83–116
Maxwell's equations, 36
Meals, prepared, sterilization of, 355–356
Meat, red, microwave cooking of, 235–239
Meat patties, microwave cooking of, 245–246
Media
  polar and dipolar, 85
  propagation, 44–47
Membranes
  cell, interactions of microwaves with, 399–404
  dielectric behavior of, 395–396
Metabolism, biological effects on, 494–496
Micro-organisms, biological effects on, 446–450
Microwave bands, 4, 5
Microwave dessication for determination of dry content, 326–327
Microwave syndrome, 465
Microwaves
  matter and, 83–116
  thermal applications of, 17–19
Modulated wave interaction of microwaves with cell membranes, 403–404

Natural rubber (NR), dielectric losses of, 190
Nervous system, biological effects on, 465–488
Neurologic pain, microwave hyperthermia for, 616
Neurons, biological effects on, 469–471
Nitrile butdiene rubber (NBR), dielectric losses of, 190
Nonpolar media, 85
Nuclear waste treatment, microwaves in, 222–223

Ohm's law, 35
Oil, production of, microwaves in, 215
Oncley's equation, 101
Onion, drying of, 311
Onsager field, 89–90
Organism, living, interactions with, 391–442
  in cell membranes, 399–404
  dielectric behavior of biological material and, 392–397
  quantum aspects of, 397–399
  thermal, 404–420
Oysters, opening of, microwaves in, 379

Paper, manufacture of, drying in, 163–164
Parasites, disinfection of sand in parks for, 376–377
Pasta, drying of, 310–311
Pearl chain formation in interaction of microwaves with cell membranes, 404
Perception
  auditory, biological effects on, 478–481
  sensory, biological effects on, 476–478
Peripheral nervous system, biological effects on, 476–478
Permittivity, relative, definition of, 34
Pharmaceutical industry, drying in, 176–177
Photographic film, drying of, 175–176
Pituitary-ovarian axis, biological effects on, 492–493
Pituitary-suprarenal axis, biological effects on, 490–492
Pituitary-testicular axis, biological effects on, 492–493
Pituitary-thyroid axis, biological effects on, 488–490
Plane wave, 40–42
Plaster, drying of, 172–173
Plastics, drying of, 175–176
Plywood, drying of, 172
Polar activators, additional, in microwave vulcanization, 196
Polar media, 85
Polarization
  dielectric, 84–85
  by dipole alignment, 84
    in alternating field, 90–91
    in static fluid, 85–90
  electronic, 84–85
  ionic, 84
  space charge, 84
Polyesters, polymerization of, 208–209
Polyethylene for microwave applicators, 141
Polymerization, 207–213
  of amino resins, 212

of epoxy resins, 211
of polyesters, 208–209
of polyurethanes, 209–211
of thermoplastic polymers, 207–208, 212–213
of thermosetting polymers, 207–208
Polymers
  drying of, 175
  thermoplastic, polymerization of, 207–208, 212–213
  thermosetting, polymerization of, 207–208
Polynorbornene (PNB), dielectric losses of, 190
Polypropylene for microwave applicators, 141
Polytetrafluorethytene (PTFE) for microwave applicators, 141
Polyurethanes, polymerization of, 209–211
Potato chips, final drying of, 308–310
Poultry, microwave cooking of, 239–243
Poynting vector, 63
Preservation of food, 339–363
  enzymatic inactivation in, 339–346
  sterilization in, 346–356
Printing, drying in, 164–165
Propagation, guided, 52–56
  discovery of, 16–17
Propagation equation, 37–40
Propagation media, 44–47
Psychophysiology, biological effects on, 485–488
Purification of coal, microwaves in, 221–222

Qualitative advantages of microwave thawing, 297–298

Radiation
  of aperture, 72–75
  electromagnetism and, 2–32
  from horn, 75–76
  laws of, 33–82
  pattern, 70
  from slot, 70–72
  sources of, 69–78
    characteristics of, 69–70
  zones of, 76–78
Radiation Control for Health and Safety Act, Public Law 90–602, 559
Radiation Hazards from Electronic Equipment, 567
Radio broadcasting, 10–14
Radio detection finding (RDF) chain, 15
Radio Regulations (RR), 6
Radio wave, thermal applications of, 17–19
Rayleigh radiation zone, 77
Reflection, 49–52
  and transmission in thermal interaction with organism, 405–407
Reheating of rubber, 201
Relative permittivity, definition of, 34
Relaxation mechanisms, 188–189
Relaxation time, dielectric relaxation and, 94–95
Resins
  amino, polymerization of, 212
  epoxy, polymerization of, 211
  thermosetting, microwave reticulation of, 208
Reticulation, microwave, of thermosetting resins, 208

RF energy trans[...]
Road repairs, mic[...]
Roasting, microwa[...]
Rocks, crushing, m[...]
Rubber(s)
  butadiene-styrene,[...]
  butyl, dielectric los[...]
  chlorobutadiene, die[...]
  drying of, 175
  natural, dielectric loss[...]
  nitrile butadiene, dielec[...]
  thawing and reheating o[...]
  vulcanization of, 191–20[...]
Ruhmkroff's coil, 10, 11

Safety standards, 553–584
  in Australia, 568
  in Canada, 567–568
  in Eastern Europe, 565–567
  in European Community, 568–[...]
  international organizations and,[...]
  in Soviet Union, 556–557
  in Sweden, 568
  in United States of America, 557–5[...]
Saline solutions, dielectric properties of,[...]
Schchelkunov's equivalence principle, 72[...]
Sella turcica in hormonal regulation, 488
Sensory perception, biological effects on, 4[...] 478
Shale oil production, microwaves in, 215
Shielding of microwave applicators, 136–139
Silicones for microwave applicators, 141
Slot, radiation from, 70–72
Snell-Descartes law, 50
Soil
  defrosting of, microwaves in, 216
  treatment of, microwaves in, 369–377
Somatotrophin hormone (STH), biological effects on, 493–494
Soviet Union, safety standards in, 556–557
Soya, microwave cooking of, 251–252
Soya beans, enzymatic inactivation of, 345–346
Space charge polarization, 84
Spherical waves, 42–44
Standing wave ratio (SWR), 58
Static fluid, dipole alignment polarization in, 85–90
Stationary wave, 58–59
Stefan's law, 104–105
Sterilization
  of animal products, 350–351
  in food preservation, 346–356
  of prepared meals, 355–356
  of vegetable products, 351–355
Stokes' law, 94
Surface cooling in microwave thawing, 285–290
Sweden, safety standards in, 568
Synapses, biological effects on, 469–471

Teflon for microwave applicators, 141
Textile industries, drying in, 166–171
Thawing
  conventional, 276–277
  microwave, 278–304
    advantages of, 296–298

...ric properties of frozen products
...d, 278–281
...y limitations in, 283–284
...pment available for, 290–295
...290–292, 915 MHz, 290–292
...hanisms of, 278–290
...ubber, 201
...face cooling in, 285–290
...ermal runaway in, 281–282
...applications of radio wave, 17–19
...interaction of microwaves with living
 organisms, 404–420
...orption and dosimetry in, 407–408
...perimental aspects of, 408–410
...in results of, 416–420
...odeling of, 410–415
...ar-field, 415–416
...eflection and transmission in, 405–407
...rmal regulation, biological effects on, 494–
 496
...ermal runaway, 106–108
...in microwave thawing, 281–282
...hermoplastic polymers, polymerization of,
 207–208, 212–213
...hermosetting resins, microwave reticulation of,
 208
Tissues, biological, dielectric behavior of, 396–
 397
Tobacco, drying of, 177–178
Tomato paste, drying of, 312
Transmission, 49–52
 and reflection in thermal interaction with
  organisms, 405–407
Transmitter-receiver, development of, 10–12
Transverse electric (TE) mode of guided
  propagation, 52–56
Transverse magnetic (TEM) wave, 40
Transverse magnetic (TM) mode of guided
  propagation, 52–55
Traveling wave tube (TWT), 125, 127
Triethanol amine (TEA), addition of, in
  microwave vulcanization, 196
Tufts, drying in, 166–168

United States of America, safety standards in,
 557–565

Van der Waals bonds, dielectric relaxation and,
 93–94
Vegetables
 blanching of, 341–344
 microwave cooking of, 249–253
 sterilization of, 351–355
Vertical polarization (VP), 50–51
Voltage standing wave ratio (VSWR), 123
Vulcanization of rubber, 191–201
 microwave, 194–201
  addition of carbon black in, 194
  addition of light-colored fillers in, 194–
   195
  addition of polar activators in, 196
  advantages/disadvantages of, 196–198
  formulation of mixtures in, 194
  materials available for, 199–201
  mixtures of polar and nonpolar

elastomers in, 195

Waste
 celulosic, treatment of, microwaves in, 224
 nuclear, treatment of, microwaves in, 222–
  223
Wheat, α-amylase in, inactivation of, 344–345
Wine-making by carbonic fermentation, 379
Wood, drying of, 172

Yarns, drying in, 166–168

Zeolites, regeneration of, heating in, 178–180

The Artech House Microwave Library

Algorithms for Computer-Aided Design of Linear Microwave Circuits, Stanislaw Rosloniec

Analysis, Design, and Applications of Fin Lines, Bharathi Bhat and Shiban K. Koul

Analysis Methods for Electromagnetic Wave Problems, Eikichi Yamashita, ed.

Automated Smith Chart Software and User's Manual, Leonard M. Schwab

C/NL: Linear and Nonlinear Microwave Circuit Analysis and Optimization Software and User's Manual, Stephen A. Maas

Capacitance, Inductance, and Crosstalk Analysis, Charles S. Walker

Design of Impedance-Matching Networks for RF and Microwave Amplifiers, Pieter L.D. Abrie

Digital Microwave Receivers, James B. Tsui

Electric Filters, Martin Hasler and Jacques Neirynck

E-Plane Integrated Circuits, P. Bhartia and P. Pramanick, eds.

Evanescent Mode Microwave Components, George Craven and Richard Skedd

Feedback Maximization, Boris J. Lurie

Filters with Helical and Folded Helical Resonators, Peter Vizmuller

GaAs FET Principles and Technology, J.V. DiLorenzo and D.D. Khandelwal, eds.

GaAs MESFET Circuit Design, Robert A. Soares, ed.

GASMAP: Gallium Arsenide Model Analysis Program, J. Michael Golio, *et al.*

Handbook of Microwave Integrated Circuits, Reinmut K. Hoffmann

Handbook for the Mechanical Tolerancing of Waveguide Components, W.B.W. Alison

HEMTs and HBTs: Devices, Fabrication, and Circuits, Fazal Ali, Aditya Gupta, and Inder Bahl, eds.

High Power Microwave Sources, Victor Granatstein and Igor Alexeff, eds.

High Power Microwaves, James Benford and John Swegle

Introduction to Microwaves, Fred E. Gardiol

Introduction to Computer Methods for Microwave Circuit Analysis and Design, Janusz A. Dobrowolski

Introduction to the Uniform Geometrical Theory of Diffraction, D.A. McNamara, C.W.I. Pistorius, and J.A.G. Malherbe

LOSLIN: Lossy Line Calculation Software and User's Manual, Fred E. Gardiol

Lossy Transmission Lines, Fred E. Gardiol

Low-Angle Microwave Propagation: Physics and Modeling, Adolf Giger

Low Phase Noise Microwave Oscillator Design, Robert G. Rogers

MATCHNET: Microwave Matching Networks Synthesis, Stephen V. Sussman-Fort

Materials Handbook for Hybrid Microelectronics, J.A. King, ed.

Matrix Parameters for Multiconductor Transmission Lines: Software and User's Manual, A.R. Djordjevic, *et al.*

MIC and MMIC Amplifier and Oscillator Circuit Design, Allen Sweet

Microelectronic Reliability, Volume I: Reliability, Test, and Diagnostics, Edward B. Hakim, ed.

Microelectronic Reliability, Volume II: Integrity Assessment and Assurance, Emiliano Pollino, ed.

Microstrip Lines and Slotlines, K.C. Gupta, R. Garg, and I.J. Bahl

Microwave and RF Component and Subsystem Manufacturing Technology, Heriot-Watt University

Microwave Circulator Design, Douglas K. Linkhart

Microwave Engineers' Handbook: 2 volume set, Theodore Saad, ed.

Microwaves Made Simple: Principles and Applications, Stephen W. Cheung, Frederick H. Levien, *et al.*

Microwave Materials and Fabrication Techniques, Second Edition, Thomas S. Laverghetta

Microwave MESFETs and HEMTs, J. Michael Golio, *et al.*

Microwave and Millimeter Wave Heterostructure Transistors and Applications, F. Ali, ed.

Microwave and Millimeter Wave Phase Shifters, Volume I: Dielectric and Ferrite Phase Shifters, S. Koul and B. Bhat

Microwave Mixers, Stephen A. Maas

Microwave Transmission Design Data, Theodore Moreno

Microwave Transition Design, Jamal S. Izadian and Shahin M. Izadian

Microwave Transmission Line Couplers, J.A.G. Malherbe

Microwave Tubes, A.S. Gilmour, Jr.

MMIC Design: GaAs FETs and HEMTs, Peter H. Ladbrooke

Modern GaAs Processing Techniques, Ralph Williams

Modern Microwave Measurements and Techniques, Thomas S. Laverghetta

Monolithic Microwave Integrated Circuits: Technology and Design, Ravender Goyal, *et al.*

Nonlinear Microwave Circuits, Stephen A. Maas

Optical Control of Microwave Devices, Rainee N. Simons

PLL: Linear Phase-Locked Loop Control System Analysis Software and User's Manual, Eric L. Unruh

Scattering Parameters of Microwave Networks with Multiconductor Transmission Lines: Software and User's Manual, A.R. Djordjevic, *et al.*

Stripline Circuit Design, Harlan Howe, Jr.

Terrestrial Digital Microwave Communications, Ferdo Ivanek, *et al.*

Time-Domain Response of Multiconductor Transmission Lines: Software and User's Manual, A.R. Djordjevic, et al.

Transmission Line Design Handbook, Brian C. Waddell